ANALOG ELECTRONICS-I

By

J.B. Gupta
B.Sc. (Elec. Engg.); C. Eng. (I);
F.I.E. (India); M.I.S.T.E.

Rajeev Manglik
B.E. (Elec.) Pb. University

Rohit Manglik
B.E. (Elec.) Univ. of Roorkee

S.K. KATARIA & SONS®

Publisher of Engineering Books

4885/109, Prakash Mahal, Dr. Subhash Bhargav Lane,
Opposite Delhi Medical Association, Daryaganj, New Delhi–110002 (INDIA)
Phone: +91-11-43551243; *Mobile:* +91-9871775858, 7678425209
e-mail: katariabook@gmail.com, katariabooks@yahoo.com
Website: www.skkatariaandsons.com
(e-books, subscription and bulk buying also available on our website)

Published by:

S.K. Kataria & Sons®

4885/109, Prakash Mahal, Dr. Subhash Bhargav Lane,
Opposite Delhi Medical Association, Daryaganj, New Delhi-110002 (INDIA)
Phone: +91-11-43551243; *Mobile:* +91-9871775858, 7678425209
e-mail: katariabook@gmail.com, katariabooks@yahoo.com
Website: www.skkatariaandsons.com

Edited By : **Shalini Manglik**

First Edition: 2011, Reprint: 2025

Laser Typesetting by :
AISA ENTERPRISES, Delhi-110092

Printed at :
New A.S. Offset Printers, Delhi-110051

PREFACE

The primary aim of presenting this book is to meet the needs of Second Year (Third Semester) students of B.Tech. (E.C.E. and E.E.E.) studying in Engineering Colleges affiliated to G.G.S.I.P. University, Delhi. The book has been written keeping in view the latest revised syllabus. The subject matter has been presented in an easily comprehensible manner, logical sequence using simple language and to provide a detailed exposition to the fundamentals. A large number of workedout examples, theoretical exercises, short answer type questions with answers and numerical problems included in the text are from latest question papers of Universities/Deemed Universities/Technical Universities including G.G.S.I.P. University or are of equivalent standard.

The book consists of 10 chapters divided into 4 Units according to the latest syllabus prescribed by G.G.S.I.P. University, Delhi for the subject of Analog Electronics-I.

The solution to illustrative examples covering almost all topics/subtopics of different chapters makes the entire presentation easy to follow. Selected exercises and numerical problems with answers given at the end of each chapter help the students in evaluating their understanding of the subject. Short answer type questions with answers help the students in preparing for theory and viva voce/ practical examinations.

Every care has been taken to eliminate misprints, omissions and errors but it is too much to expect that no inaccuracy/misprint/error has crept in and the authors would be grateful to the readers for bringing to their notice any such errors/misprints/omissions they may come across while going through the book.

The authors will feel satisfied if the book meets the needs of the students for whom it is meant. Suggestions and criticism for the improvement of the book shall be welcomed.

The authors will like to thank all the Scientists, Engineers, Authors and Publishers whose works and text have been the source of enlightenment, inspiration and guidance to us in the presentation.

Lastly, but not the least, the authors are thankful to their publisher, Sanjeev Kumar Kataria for bringing out the book in such a short time and for pricing it moderately despite heavy cost of paper and printing.

July 14, 2011 **AUTHORS**

To My Loving Grand Son

KUSHAGRA

OTHER USEFUL BOOKS

CONTENTS

UNIT-III
Multistage Amplifiers and Feedback Amplifiers .... 265—337

8. MULTISTAGE AMPLIFIERS .... 265—295

9. FEEDBACK AMPLIFIERS .... 296—337

UNIT I

SEMICONDUCTOR DIODES AND RECTIFIERS

CONTENTS IN THIS UNIT

CHAPTER 1

Basic Semiconductor and P-N Junction Theory

INSIDE THIS CHAPTER

1.1. INTRODUCTION

In an electronic device, movement of electrons is controlled. Basic understanding of the relationship between the electrons and the other components of an atom is essential for study of electronic devices. The movement of electrons within a solid, and the bonding forces between atoms can then be investigated. This leads to a knowledge of the differences between conductors, insulators and semiconductors, and to an understanding of P-type and N-type semiconductor materials.

Junction of P-type and N-type material (P-N junction) are basic to all but a very few semiconductor devices. Forces act upon electrons that are adjacent to a P-N junction, and these forces are altered by the presence of an external voltage.

1.2. STRUCTURE OF AN ATOM

In order to explain many phenomena associated with conduction in gases, metals and semiconductors and the electron emission from a metal surface, it is necessary to assume that the atom has loosely bound electrons which can be torn away from it.

On the basis of experiments on the scattering of α-particles, combined with the discovery of neutron made by Chadwick in 1932, Rutherford suggested a complete model of the atom which is known as *Rutherford's Nuclear Model of Atomic Structure.* According to this model :

(*i*) The centre of an atom consists of a positively charged body which is called *nucleus*. All the protons and neutrons are present in the nucleus. Although the nucleus is small, it is heavy due to the presence of all the protons and neutrons in it. Since the mass of electron is negligible, the mass of an atom is equal to the sum of the masses of protons and neutrons, *i.e., the entire mass of an atom resides in its nucleus.*

(*ii*) The electrons keep revolving in the orbits around the nucleus at extremely high speed and at large distance from the nucleus as the planets revolve around the Sun (Fig. 1.1). Thus Rutherford's model is comparable to solar system and it is for this reason that the electrons revolving around the nucleus are generally referred to as *planetary electrons.*

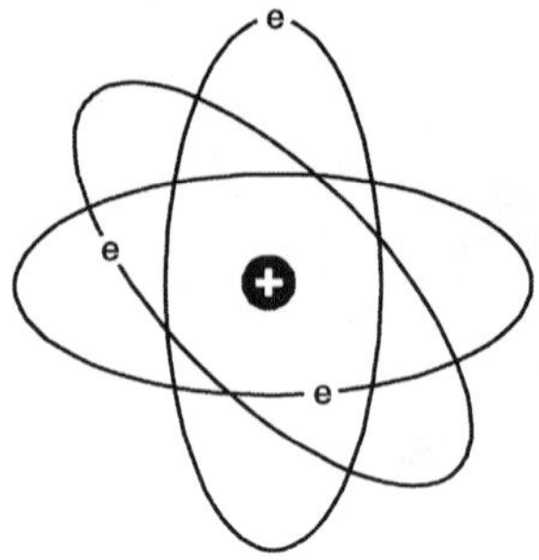

Fig. 1.1. ***Rutherford's Nuclear Model of Atomic Structure***

Since the atom is neutral, the number of electrons which are negatively charged particles is equal to the number of protons in the nucleus or the atomic number of the atom. The electrons revolve around the nucleus with such a speed that the centrifugal force (*i.e.*, repulsive force) acquired by the electrons due to their motion just balances the attractive force between the electrons and the nucleus.

The charge associated with each electron is 1.602×10^{-19} C. Compared to mass of nucleus, electrons are relatively tiny particles of almost negligible mass (each electron weighing 9.107×10^{-31} kg). Protons and neutrons each have mass about 1,800 times the mass of an electron.

The diameter of an atom is approximately 10^{-10} m and that of nucleus is about 10^{-15} m. The ratio of e/m for an electron is equal to $\frac{1.602 \times 10^{-19}}{9.107 \times 10^{-31}} = 1.759 \times 10^{11}$ C/kg. Since it is not very easy to determine either the charge or the mass of an electron, electrons are usually identified by their high ratio of charge to mass. This is so high that a number of electrons making up a mass of 1×10^{-6} kg placed 100 km from a similar mass in air would experience a force of about 3×10^{9} kg. No other particle has so high a value of e/m, so that recognition by this means is easy and certain.

The number of protons (or electrons) in an atom is referred to as the *atomic number* (Z) of the atom. Now the term atomic number is often referred to as the *proton number*. The total number of protons and neutrons in the nucleus of an atom is called the *mass number* A, of the atom. The atomic mass (or atomic weight) is the actual weight of an atom and numerically slightly differs from the mass number of the atom. For example, aluminium has 79 protons and 118 neutrons. Thus its atomic number Z is equal to 79, being equal to the number of protons and the mass number A = 79 + 118 = 197.

The *force of attraction* between the electrons and nucleus follows Coulomb's law. It can be shown from classical mechanics that the resultant closed path will be a circle or an ellipse under the action of such a force.

1.3. ELECTRON ORBITS AND ENERGY LEVELS

Electrons can occupy only certain rings or shells at fixed distances from the nucleus. These shells may be circular or elliptical and are numbered as K, L, M, N, O, P, and Q, K being nearest to the nucleus. Each shell can contain only a particular number of electrons. The formula $2n^2$ (n being the number of shell or level, the first shell or level being K) determines the number of electrons that can be accommodated in any level. For example Kth level can have maximum of 2 electrons. Similarly $L = 2 \times 2^2 = 8$, $M = 2 \times 3^2 = 18$, $N = 2 \times 4^2 = 32$.

The limitation to the formula is that the number of electrons in the outermost shell of an atom cannot exceed 8. The electrons in the outer shell determine the electrical and chemical characteristics of each particular type of atom and are referred to as *valence electrons*. An atom may have its outer or *valence shell* completely filled or only partially filled.

From discussions made above it is clear that electrons can revolve only in the permitted orbits (*i.e.*, orbits of radii r_1, r_2 and r_3) and not in any intermediate orbit. Thus all the radii between r_1 and r_2 and between r_2 and r_3 are forbidden (for explanation refer to quantum physics). Each orbit has fixed amount of energy associated with it and the electrons moving in a particular orbit possess the energy of that orbit. Thus total energy of an electron can be identified with the size of its orbit—the larger the orbit, the greater is its energy *i.e.*, outer orbit electrons possess more energy than the inner orbit electrons.

The level of energy obtainable with different orbits can be conveniently represented by horizontal lines, as illustrated in Fig. 1.2 (*b*). This is known as *energy level diagram*. Electrons in the smallest orbit are on the first energy level; electrons in the second orbit are on the second energy level and so on. Thus energy level is just another way of representing the orbital radius.

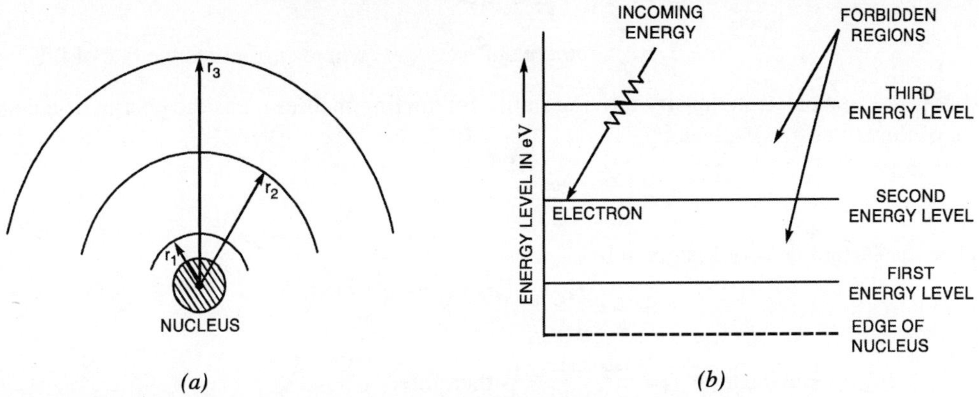

Fig. 1.2. ***Energy Level***

Thus electrons closest to the nucleus need the greatest energy application to extract them from the atom. The electrons in inner orbits, thus, remain bound to the nucleus and are, therefore, called *bound electrons*.

The energy levels considered above are measured in electron volts (*e*V).

Between each energy level a *forbidden region exists*. Electrons cannot orbit in forbidden regions. However, they may quickly pass them. If electrons in a particular orbit acquire enough energy from some outside source (such as an external electric field or thermal energy), they will jump to a high energy level by passing quickly through the forbidden region. *A definite amount of energy is required to be spent in order to remove an electron from its orbit* (*level*).

Excited electrons are unstable, and they tend to fall back to low energy level by giving up their additional energy. The electrons that do leave their parent atom are termed *free electrons.*

1.4. ELECTRON VOLT

The joule (J) is the unit of energy in the SI system. In some engineering power problems this unit is very small, and a factor of 10^3 or 10^6 is introduced to convert from watts (1 W = 1 J/s) to kilowatts or megawatts, respectively. For a discussion of the energies involved in electronic devices, even the joule is too large a unit. This statement is not to be construed to mean that only minute amounts of energy can be obtained from electron devices. It is true that each electron possesses a tiny amount of energy, but an enormous number of electrons is involved even in a small current, so that considerable power may be represented.

A unit of work or energy, called the *electron volt* (*e*V), is given as

$$1\ e\text{V} = 1.602 \times 10^{-19}\ \text{J}$$

The name *electron volt* arises from the fact that, if an electron falls through a potential of 1 V, its kinetic energy will increase by the decrease in potential energy, or by one electron volt.

$$i.e.\quad 1\ e\text{V} = 1.602 \times 10^{-19}\ \text{C} \times 1\ \text{V} = 1.602 \times 10^{-19}\ \text{J}$$

However, the eV unit may be used for any type of energy, and is not restricted to problems involving electrons.

The abbreviations MeV and BeV are used to designate 1 million and billion electron volts respectively.

1.5. CURRENT DENSITY

By definition, the current density, denoted by the symbol J, is the current per unit area of the conducting medium *i.e.*,

$$J = \frac{I}{A} \text{ amperes/m}^2 \qquad \text{where current I is the rate of flow of charge.}$$

If N is the free electron density moving in the conducting medium, e is the charge of electron, then the rate of flow of charge is given by

$$I = \frac{dq}{dt} = \frac{Ne}{T}$$

Thus the current density J is given by

$$J = \frac{I}{A} = \frac{Ne}{TA}$$

$$\text{But the time } T = \frac{\text{Distance}}{\text{Velocity}} = \frac{l}{v}, \text{ therefore,}$$

$$\text{Current density, } J = \frac{Ne}{lA/v} = \frac{Nev}{lA} = \frac{Nev}{\text{Volume}} \qquad \because lA \text{ is the volume of conductor}$$

$$= nev = \rho v \qquad ...(1.1)$$

where $\rho = ne$, the charge density, in coulombs per cubic metre and v is the velocity in metres per second.

The above derivation is independent of the form of conducting medium. Consequently, it is not necessary that the conducting medium is a wire conductor. It may be equally well a portion of gaseous discharge tube or a volume element in the space-charge cloud of a vacuum tube or semiconductor. Furthermore, neither ρ nor v need be constant, but may vary from point to point in space or may vary with time.

1.6. SEMICONDUCTOR MATERIALS

We are familiar with conducting and insulating materials. *Conducting materials* (such as silver, copper, aluminium etc.) are good conductors of electricity and are characterized by a large electrical conductivity and small electrical resistivity. *Insulating materials* (such as porcelain, glass, quartz, rubber, bakelite etc.) are characterized by poor electrical conductivity and are used to block current from flowing where it is not to pass. There is another group of materials (such as germanium, silicon, boron, selenium) which are neither good conductors nor good insulators. At room temperature such materials have conductivities considerably lower than those of conductors and much higher than those of insulators. Such materials are called the *semiconductors*. The resistivity of various semiconductor materials lies in a very wide range from 10^{-4} to about 0.5 Ω-m whereas the resistivities of conductors and insulators are of the order of 10^{-8} Ω-m and 10^{12} Ω-m respectively. An even more important property of semiconductors is that *their resistance depends largely on various factors and therefore, it can be controlled.*

The resistance of conducting materials except that of carbon, increases with the increase in temperature and resistance of insulating materials remains almost constant up to a certain temperature after which it suddenly falls off and insulating materials start conducting. Semiconductors behave in somewhat different way. The resistance of semiconductors decreases with the increase in temperature, as shown in Fig. 1.3, *i.e.* temperature coefficient of resistance of semiconductors is negative. Thus, the semiconductors behave like insulators at very low temperatures but act as conductors at high temperatures.

The resistivity of semiconductors depends on illumination and decreases in brighter surroundings (the "light" resistivity of a semiconductor is smaller than its "dark" resistivity).

The resistivity of semiconductors depends on the magnitude of the electric field in the semiconductor, so that the current in it is not proportional to the voltage, as with conductors that obey Ohm's law, but increases by far more than the voltage, *i.e.*, semiconductors are *nonlinear resistors*.

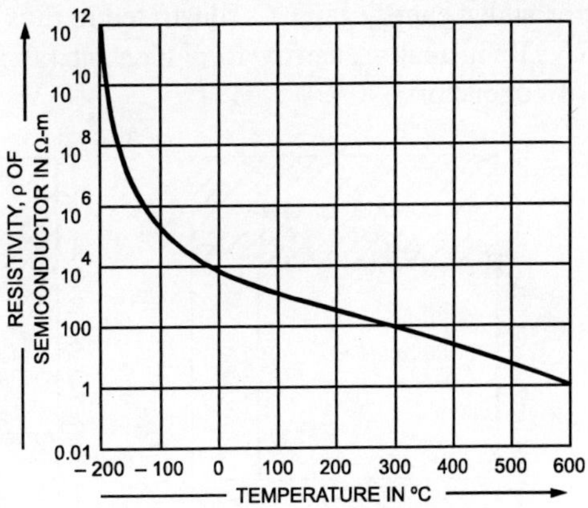

Fig. 1.3

The semiconductors have another very important property—*the resistivity (or conductivity) of semiconductors changes considerably when even minute amounts of certain other substances, called the impurities, are added to them.* Moreover, the introduction of certain impurities (such as arsenic, antimony, gallium, indium etc.) may even change the mechanism of electrical conductivity in the semiconductor, and certain physical phenomena will begin to occur at the junction between sections of the semiconductor having different impurities. Such phenomena make semiconductor devices important as *diodes, transistors, SCRs, UJTs, varistors* and *thermistors.*

The *elemental semiconductors* (*i.e.*, semiconductors comprising a single chemical element) are germanium (Ge), silicon (Si), grey (crystalline) tin, selenium (Se), tellurium (Te), boron (B) etc. *Germanium* and *silicon* are two of the best known and widely used semiconductors.

The devices (such as diodes, transistors etc.) manufactured from semiconductors have many advantages such as compact size, low cost, light weight, rugged construction, more resistive to shocks and vibrations, instantaneous operation (no heating required), low operating voltage, high operating efficiency (no heat loss) and long life with essentially no ageing effect if operated within permissible limits of temperature and frequency.

1.7. ENERGY BANDS IN SOLIDS

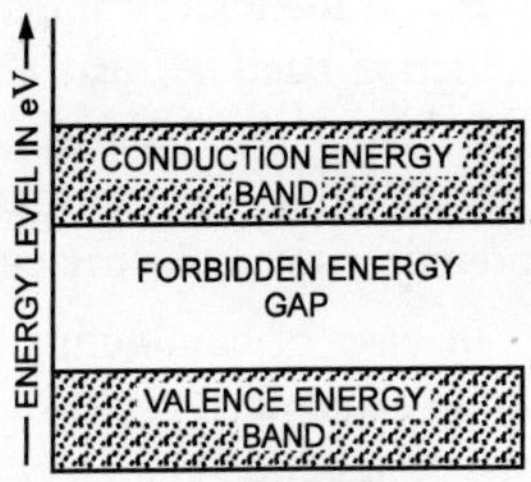

Fig. 1.4

As discussed earlier, the electrons of an isolated atom are acted upon only by the forces within that atom. However, when atoms are brought together, as in a solid, the electrons come under the influence of forces from other atoms. Under these conditions, the energy levels that may be occupied by electrons merge into bands of energy levels. Within any given material there are two *distinct energy* bands in which electrons may exist. These two energy bands are *valence energy band* (or *valence band*) and *conduction energy band* (or *conduction band*) and are separated by an energy gap in which no electron can normally exist, as illustrated in Fig. 1.4. This energy gap is termed as *forbidden energy gap* (or *forbidden gap*).

The energy band of interest is the highest energy band or valence band. If a sufficient amount of energy is given to an electron in the valence band, the electron is freed from the atomic structure. Such an electron is said to possess enough energy to be in the conduction band, where it can take part in electric current flow. Free electrons (electrons in conduction band) can move readily under the influence of an external field. Electrons in the valence band are usually in normal orbit around a nucleus. Normally, electrons do not orbit in the forbidden region, which separates the conduction and valence bands, as illustrated in Fig. 1.4.

For any given material, the forbidden energy gap may be large, small or non-existent. The distinction between conductors, insulators and semiconductors is largely concerned with the relative widths of the forbidden energy gaps. As illustrated in Figs. 1.5 (*a*), (*b*) and (*c*), there is a wide forbidden gap (more than 5 eV) for insulators, narrow forbidden gap (about 1 eV) in case of semiconductors and no forbidden gap in case of conductors.

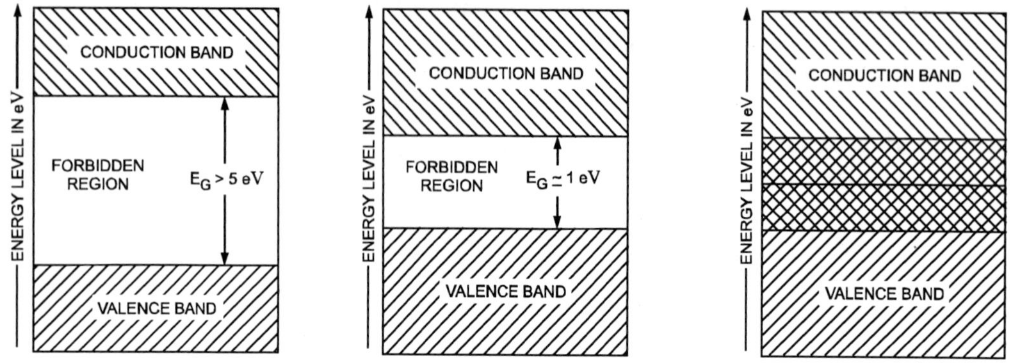

(*a*) *Insulator* (*b*) *Semiconductor* (*c*) *Metal*

Fig. 1.5. *Energy-Band Structure*

In case of insulators, there is practically no electron in the conduction band, and the valence band is filled.* For an insulator, the valence and conduction bands are so apart (more than 5 eV) that at normal temperature (25°C) the energy which can be supplied to an electron from an applied field is too small to carry a particle from the filled band into the vacant band. Since the electron cannot acquire externally applied energy, conduction is impossible. Only at very high temperature or under high external electric field, an electron can jump the gap. This is called the *breakdown of the insulating material.*

For semiconductors at a temperature of absolute zero (–273·15°C) the valence band is usually full and there may be no electron in the conduction band. It means that no current can flow in the semiconductors and, therefore, the semiconductors behave as insulators at absolute zero temperature. However, in semiconductor materials, both bands are so close (about 1 eV apart) that an electron can be lifted from the valence band to the conduction band by imparting some energy to it. This energy must be more than energy gap E_G. If the energy imparted to the electron is less than E_G, it cannot be lifted from valence band to conduction band because no permissible energy level exists between the two bands. The energy imparted by the heat at room temperature is sufficient to lift electrons from the valence band to the conduction band. Some electrons do jump the gap and get into the conduction band. Thus, semiconductors are capable of conducting some electric current, even at normal ambient temperature.

In case of conducting materials, there is no forbidden gap, and the valence and conduction bands overlap each other, as illustrated in Fig. 1.5 (*c*). The orbits in the conduction band are very large. An electron in the conduction band experiences almost negligible nuclear attraction. In fact, an electron in the conduction band does not belong to any particular atom. But it moves randomly throughout the solid. This is the reason that electrons in the conduction band are called the *free electrons*. Due to overlapping of valence and conduction bands, a slight potential difference across the conductor causes the free electrons to constitute electric current.

The noteworthy point in conductors is that, because of absence of forbidden band, there is no structure to create holes and therefore, the total current in conductors is due to flow of electrons alone.

* Filled band means that all the permissible energy levels in the band are occupied with electrons. No electron can move in a filled band because there is no place to move. Thus an electron in a completely filled band cannot contribute to an electric current.

1.8. CONDUCTION IN SOLIDS

Conduction occurs in any given material when an applied electric field causes electrons to move in a desired direction within the material. This may be due to one or both of two processes, *electron motion* and *hole transfer*. In case of former process, *free electrons* in the conduction band move under the influence of the applied electric field. Electrons being negatively charged are repelled from the negative terminal of the applied voltage and attracted towards the positive terminal. Hole transfer involves electrons which are still attached to the atoms *i.e.*, those in valence band.

In case some of the energy levels in the valence band are not occupied by electrons, there are holes where electrons could exist. An electron may jump from one atom to fill the hole in another atom. When it jumps, the electron leaves a hole behind it, and thus the hole has moved in a direction opposite to that of the electron. The current flowing in such a way is said to be due to movement of holes.

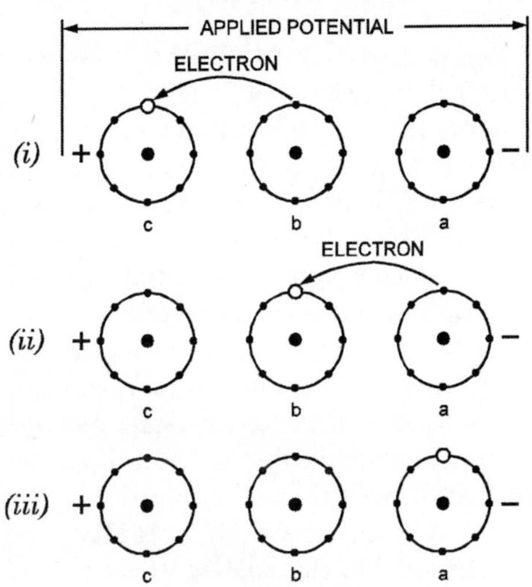

Fig. 1.6. ***Conduction By Transfer of Hole***

In Fig. 1.6 there are no free electrons. However, those electrons in orbit around atoms experience a force of attraction to the positive terminal of the applied potential, and repulsion from the negative terminal. Consequently, an electron can be made to jump from one atom to fill the hole in another atom, so long as it is moving towards the positive terminal. In Fig. 1.6 (*i*), an electron is made to jump from atom *b* to atom *c*. The hole in the valence shell of atom *c* is now filled but a hole is created in the valence shell of atom *b*, as illustrated in Fig. 1.6 (*ii*). If an electron now jumps from atom *a* to fill the hole in atom *b*, a hole is left in the valence shell of atom *a*, as illustrated in Fig. 1.6 (*iii*). Thus, the hole has moved from atom *c* to atom *b* and then to atom *a*.

Holes may be thought of as positive particles, and as such, they move through an electric field in a direction opposite to that of electrons; *i.e.*, positive particles are attracted towards the negative terminal of the applied voltage. In the condition illustrated above (Fig. 1.6) it is more convenient to think in terms of *hole movement* rather than in terms of electrons jumping from atom to atom.

Since the electric current flow is due to movement of electrons in the conduction band and that of holes in the valence band, electrons and holes are referred to as *charge carriers*. Each time a hole moves, an electron must be supplied with sufficient energy to enable it to escape from its atom. Free electrons need less energy to be applied for their movement in comparison to that required by holes. This is true due to the fact that the free electrons are already detached from their atoms. That is why, electrons have *greater mobility* than holes.

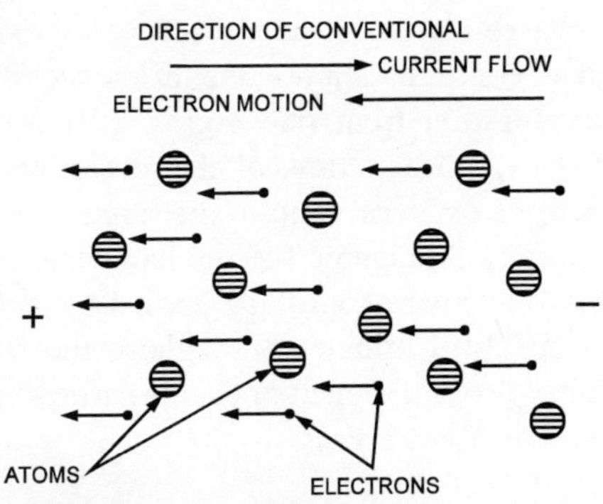

Fig. 1.7

Current flow from positive terminal to negative terminal is referred to as the *conventional current direction* whereas electron flow from negative terminal to positive terminal is referred to as *direction of electron flow*. It is to be noted that every graphic symbol used to represent an electronic device has an arrowhead indicating the direction of flow of conventional current (not the direction of movement of electrons). However, to understand the operation of an electronic device it is necessary to think in terms of movement of electrons.

1.9. DRIFT AND DIFFUSION CURRENTS

The two basic processes which cause electrons and holes to move in a semiconductor are : (*i*) *drift,* which is the movement caused by electric fields, and (*ii*) *diffusion,* which is the flow caused by variations in the concentration, *i.e.*, concentration gradients. Such gradients can be caused by a non-homogeneous doping distribution, or by injection of a quantity of electrons or holes into a region.

1.9.1. Drift Current

When a steady electric field E volts/metre is applied to a metal, the electrons move to the positive terminal of the applied voltage. In their way, they continually collide with the atoms and rebound in a random fashion, as illustrated in Fig. 1.8. Each collision being inelastic *i.e.*, the electrons lose some kinetic energy. After the collision, the electrons are accelerated and gain certain component of velocity in the direction opposite to that of applied electric field (– E) and lose their energy at the next collision. Thus, the applied electric field does not stop collisions and random motion but makes the electrons to drift towards the positive terminal. Consequently, the electrons gain average drift velocity v in the direction opposite to that of the applied electric field. It is observed that the drift velocity v is proportional to the applied electric field E, *i.e.*,

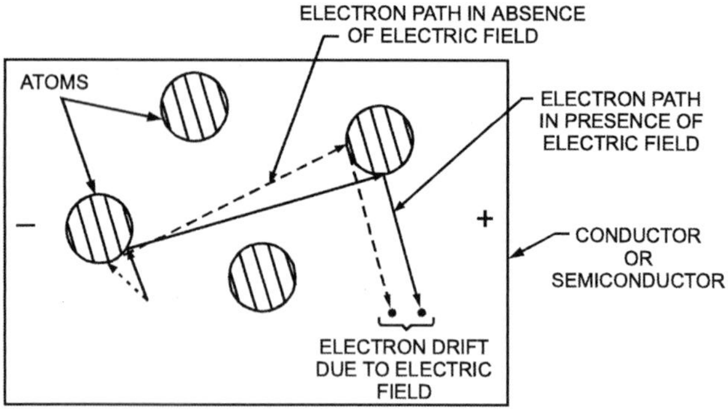

Fig. 1.8

$$v \propto E \quad \text{or} \quad v = \mu_e E \qquad ...(1.2)$$

where μ_e is called the electron *mobility** and is expressed in m^2/V-s.

This steady-state drift velocity is superimposed on the random motion of free electrons caused by thermal agitation. This steady flow of electrons in one direction caused by the applied electric field constitutes an electric current, called the *drift current.*

1.9.2. Diffusion Current

A second type of current, called the *diffusion current,* also exists in a semiconductor. The diffusion of charge carriers is as a result of a gradient of carrier concentration (*i.e.* the difference of carrier concentration from one region to another). In this case concentrations of charge carriers (either electrons or holes) tend to distribute themselves uniformly throughout the semiconductor crystal. This movement continues until all the carriers are evenly distributed throughout the material. This type of movement of charge carriers is called the *diffusion current.*

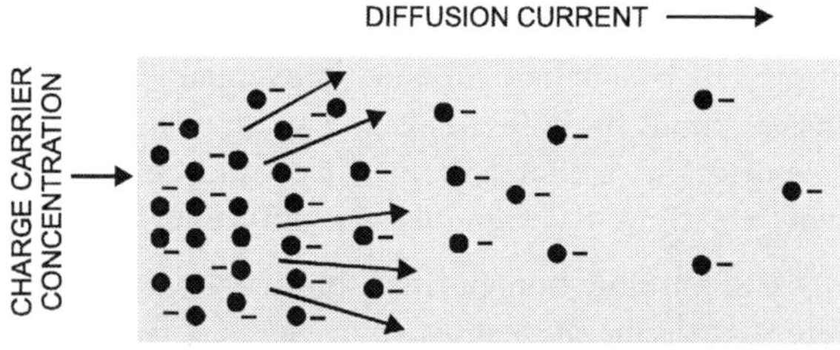

Fig. 1.9. ***Flow of Particles From a Region of High Concentration to a Region of Low Concentration***

* Mobility is defined as the steady-state drift velocity per unit electric field.

Diffusion current occurs when charge carriers diffuse from a point of concentration, to spread uniformly throughout the volume of a piece of material, as illustrated in Fig. 1.9.

Both drift and diffusion occur simultaneously in a semiconductor device.

1.10. ATOMIC BONDS

Whether a material is a conductor, a semiconductor, or an insulator depends largely upon what happens to the outer-shell electrons when the atoms bond themselves together to form a solid.

Atomic bonds are of two types viz. *primary* (or *chemical*) *bonds* and *secondary* (or *molecular*) bonds. Primary bonds are strongest bonds between atoms and can be further subdivided into (*i*) metallic (*ii*) covalent and (*iii*) ionic bonds. Secondary or molecular bonds are weaker than primary bonds. Attraction forces (also called *vander Waal's forces*) exist between atoms or molecules.

In **metallic bonds**, atoms give up their valence electrons to form clouds of electrons which is shared by all atoms. Such a bond exists between elements having loosely held valence electrons so that they can be easily released to the common pool.

For example, in the case of copper, the atoms give up easily detachable electrons and this creates a large mass of free electrons (or electron gas) drifting about through the spaces between the copper atoms. Each atom of copper becomes a positive ion because of losing an electron. The electron gas is, of course, negatively charged; consequently, an electrostatic force of attraction exists between the positive ions and the electron gas. This is the *bonding force* that holds the material together in a solid. In the case of copper and other metals, the bonding force is termed as *metallic bonding* or sometimes *electron gas bonding*. Such a bonding is shown in Fig. 1.10 (*a*).

Covalent bond is formed by sharing of electrons by two or more atoms. It can be formed between atoms of same or different elements. This arrangement is illustrated in Fig. 1.10 (*b*). In covalent bonding every valence shell of every atom appears to be filled, and consequently there are no holes and no free electrons drifting about within the material. The same covalent bonding arrangement exists in silicon and germanium.

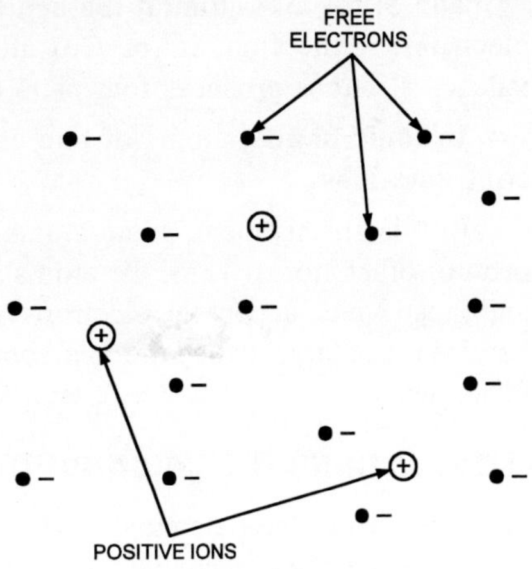

(a) Metallic Bonding

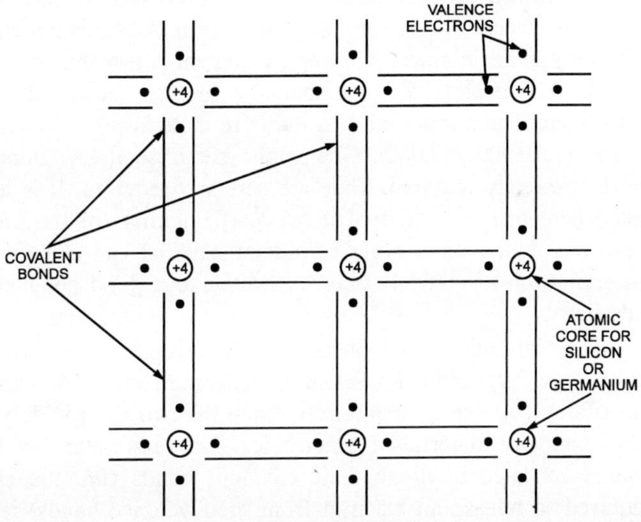

(b) Covalent Bonding

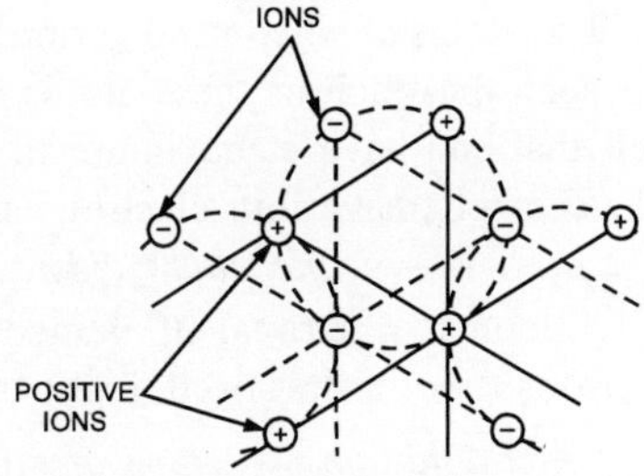

(c) Ionic Bonding

Fig. 1.10. ***Bonding Between Atoms in Solids***

The Si (or Ge) atom at the centre share one of its valence electrons with each of the valence electrons of the four Si (or Ge) atoms surrounding it, as shown in Fig. 1.10 (*b*). This sharing of valence electrons produces four pairs of electrons that are covalently bonded thus completing its octet.

In ionic bonds, atoms of one element transfer electrons to the other so that both have stable configurations.

In this arrangement, some atoms part with outer-shell electrons, but these are accepted into the orbit of other atoms. Thus, the atoms are ionized; those which give up electrons become *positive ions*, and those which accept the electrons become negative ions. This creates an electrostatic bonding force between the atoms, called the *ionic bonding* [Fig. 1.10 (*c*)]. Such bonding occurs in some of the insulating materials.

1.11. INTRINSIC SEMICONDUCTORS

Semiconductor materials may be classified into two classes viz. *intrinsic* (or *pure*) *semiconductors* and *extrinsic* (or *impure*) *semiconductors*.

An *intrinsic semiconductor* is one which is made of the semiconductor material in its extremely pure form. A semiconductor is not truely intrinsic unless its impurity content is less than one part in 100 million parts of semiconductor. In fact, recent advances have reduced impurity levels in the pure material to 1 part in 10 billion (1:10,000,000,000). One might ask if such low impurity levels are really required. They are certainly required. It is to be noted here that the addition of one part impurity (of the proper type) per million in a wafer of silicon material can change that material from a relatively poor conductor to a good conductor of electricity.

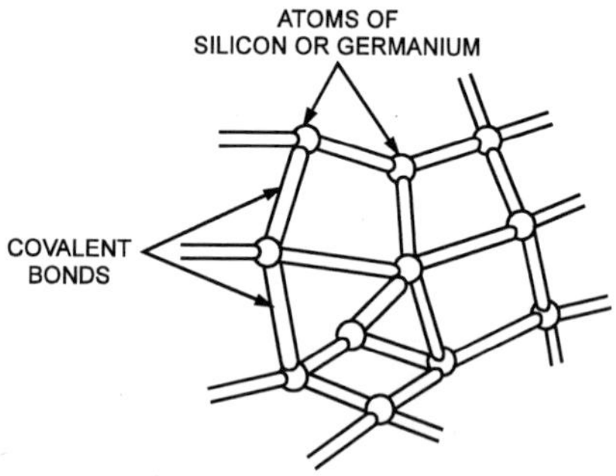

Fig. 1.11. ***Crystal Lattice Structure***

There are many semiconductor materials such as germanium, silicon, grey crystalline tin, selenium, tellurium, boron etc. available but silicon (Si) and germanium (Ge) are the two most widely used semiconductor materials in electronics. The reason for this is that energy required to break their covalent bonds (*i.e.* the energy required to release an electron from their valence bands) is very small (1.12 eV for silicon and 0.72 eV for germanium). Both elements have the same crystal structure (Fig. 1.11) and similar characteristics.

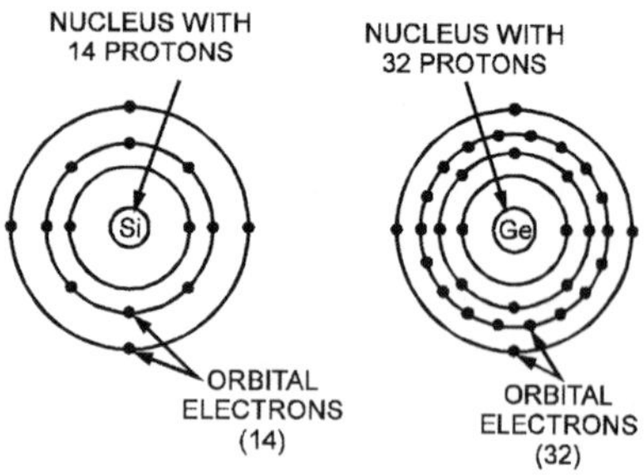

(*a*) *Silicon* (*b*) *Germanium*

Fig. 1.12. ***Bohr's Atomic Model***

The atoms of silicon and germanium are illustrated in Fig. 1.12. It is seen that each of these atoms has four electrons in a valence shell that can have a maximum of eight. Thus, we say that their valence shells have four electrons and four holes. *A hole is defined simply as an absence of an electron in a shell where one could exist.*

Silicon is a tetravalent element and its atomic number is 14. It has, therefore, 14 protons and 14 electrons (two in first orbit, eight in second orbit and four in the third orbit), as shown in Fig. 1.12 (*a*).

Crystal lattice structure of silicon (also applicable to germanium) is shown in Fig. 1.11. Each atom is equidistant from its four neighbouring atoms. In a crystalline structure each atom shares its four valence electrons with the four adjacent atoms. This occurs because each tetravalent semiconductor atom needs a total of eight outer-shell electrons. This sharing of electron pairs is called *covalent bonding*, already explained in Art. 1.10. The covalent bonds are illustrated in Fig. 1.11 by means of rods connecting the adjacent atoms. For simplified discussion, the crystal structure may be reduced to two dimensional, as illustrated in Fig. 1.10 (*b*). The core represents the nucleus and all the orbiting electrons except the valence electrons. Since there are as many protons in the nucleus as there are electrons orbiting around it and there are four valence electrons, the core will have an excess +4 charge. The core of silicon contains 14 protons and 10 electrons (whereas germanium core has 32 protons and 28 electrons).

The energy band diagram for silicon is shown in Fig. 1.13 (*a*).

Germanium is also a tetravalent element *i.e.* it has 4 valence electrons. Its atomic number is 32. It has, therefore, 32, protons and 32 electrons [two in first orbit, eight in second orbit, eighteen in third orbit and four in fourth orbit], as shown in Fig. 1.12 (*b*). Its crystal lattice structure has already been discussed above. Energy band diagram for germanium is shown in Fig. 1.13 (*b*).

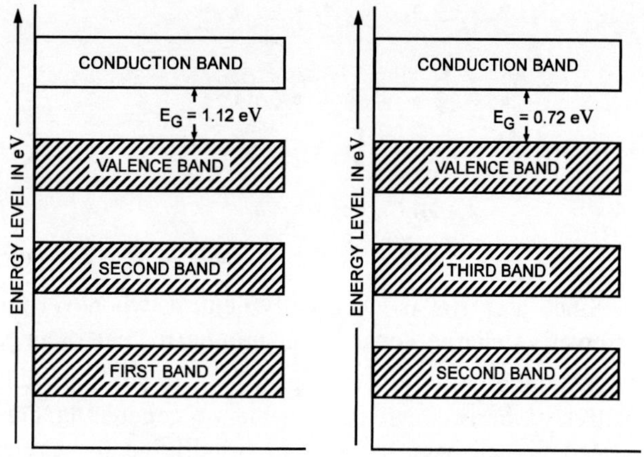

(*a*) *For Silicon* (*b*) *For Germanium*

Fig. 1.13. ***Energy Band Diagrams***

For semiconductors at a temperature of absolute zero (– 273·15°C) the valence band is usually full, and there may be no electrons in the conduction band. At absolute zero temperature E_G for silicon and germanium are 1.21 eV and 0·785 eV respectively and it is difficult to get additional energy of this magnitude from applied electric field. Hence the *conductivity of intrinsic semiconductors at absolute zero temperature is zero*. However, at room temperature (300 K), the energy provided by the heat is sufficient to lift electrons from the valence band to the conduction band. Some electrons do jump the gap and go into the conduction band. So at room temperature, semiconductors are capable of conducting some electric current.

Since energy required in transferring electrons from valence band to conduction band is more in case of silicon than that in case of germanium, the conductivity of silicon will be less than that of germanium at room temperature.

1.11.1. Production of Holes and Free Electrons

Breaking of covalent bond is equivalent to raising an electron from valence band to the conduction band. So the energy required to break a covalent bond is the same as the forbidden energy E_G and equals 0.72 eV in case of germanium and 1.12 eV in case of silicon. This energy may be supplied either (*i*) by raising the temperature and thereby imparting thermal energy (*ii*) by incidence of photons in the visible range or (*iii*) by bombardment of the semiconductor materials by the α-particles, electrons and other particles and X-rays.

When an electron breaks a covalent bond of an intrinsic semiconductor and moves away, a vacancy is created in the broken covalent bond. The vacancy so caused constitutes a hole and is represented by a small circle. Whenever a free electron is generated, a hole is simultaneously created. Thus, free electrons and holes are always generated in pairs. So concentration of free electrons and holes will always be equal in an intrinsic semiconductor. Such generation of free electron-hole pairs* is referred to as *thermal generation*. The formation of electron-hole pairs using crystal structure and energy band diagram is illustrated in Figs. 1.14 (*a*) and (*b*) respectively.

* The liberated electron and the hole left behind is called an electron-hole pair..

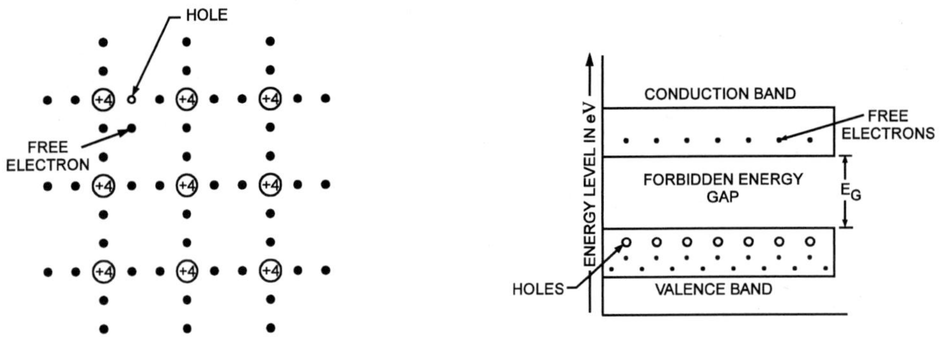

(a) Crystal Structure *(b) Energy Band Diagram*

Fig. 1.14. ***Formation of Electron-Hole Pair in an Intrinsic Semiconductor***

Since a crystal is electrically neutral, whenever an electron moves, a hole is left behind with a net positive charge equal in magnitude to that of negative charge of free electron.

Although, strictly speaking, a hole is not a particle; for all practical purposes it is viewed as a positively charged particle capable of conducting current. This concept of a hole as a positively charged particle merely helps in simplifying the explanation of conduction in semiconductors.

1.11.2. Carrier Generation and Recombination

In an intrinsic semiconductor crystal equal number of free electrons and holes are created by thermal (heat) energy. The reason being that a hole is created only by removing an electron from the covalent bond. Thus the electrons and holes are created in pairs. Since the free electrons and holes move randomly within the crystal lattice, there is always a possibility of an electron meeting a hole. Although this possibility is small, even then when a free electron approaches the hole, it gets attracted and falls into the hole. This merging of a free electron and a hole is called *recombination*. In this process of recombination, both the free electron and hole disappear and energy* is released as a quantum heat or light. The energy so released may in turn be re-absorbed by another electron to break away from its valence band and create a new electron-hole pair.

The recombination rate is proportional to the product of the concentrations or densities of free electrons and holes. Thus the processes of breaking of covalent bonds and recombination of electrons and holes take place simultaneously. The amount of time between the creation and disappearance of a free electron is called the *lifetime*. It varies from a few nanoseconds to several microseconds, depending how perfect the crystal is and other factors.

The rate of production of electron-hole pairs increases with the rise in temperature; the rate of recombination is proportional to the product of concentrations of free electrons and holes and at a given temperature, the concentrations or densities of free electrons and holes remain fixed and equal. So the rate of generation of electron-hole pairs must be equal to the rate of their recombination. With the increase in temperature, the rate of generation of electron-hole pairs increases. The concentration of electrons and holes goes on increasing until rate of recombination equals the increased rate of generation. Thus with the increase in temperature, concentrations of free electrons and holes increase.

* When an electron is freed from valence band, certain amount of energy is absorbed. The same amount of energy is released when an electron falls into a hole

1.11.3. Conduction in Intrinsic Semiconductors

When the electrons are liberated on breaking of covalent bonds, they move randomly through the crystal lattice. These free electrons are neither attracted by the nuclei of the atoms nor repelled by the electrons bound by covalent bonds because their electrical effects are fully engaged in maintaining the covalent bonds. When an external electric field is applied to a pure (intrinsic) semiconductor, the conduction through the semiconductor is by both free electrons and holes. The free electrons in the conduction band move towards the positive terminal of the battery while the holes in the valence band move towards the negative terminal of the battery *i.e.* the electrons and holes move in opposite directions. As the holes reach the negative terminal B, the electrons reaching there combine with the holes which are thus destroyed. At the same time, the loosely held electrons near the positive terminal A are attracted away from their atoms into the positive terminal of the battery. Now new holes are created which again drift towards the negative terminal B. The electric current flows through the intrinsic semiconductor in the same direction as in which the holes are moving (the holes having positive charge). Since electrons are negatively charged, the direction of conventional current is opposite to that of movement of electrons. The total current inside the conductor is thus the sum of currents owing to free electrons and holes (though the two types of charge carriers move in the opposite directions) but current in the external wires is only because of electrons, as illustrated in Fig. 1.15.

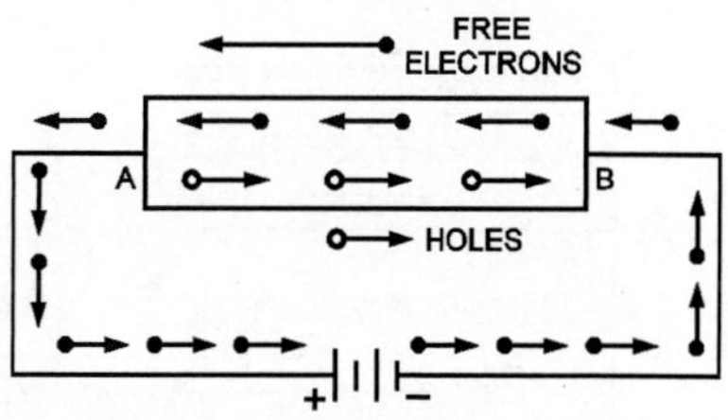

Fig. 1.15. ***Conduction of Current in an Intrinsic Semiconductor***

The mobility of electrons is more than that of holes because the probability of an electron having the energy required to move to an empty state in the conduction band is much greater than the probability of an electron having the energy required to move to the empty state in valence band. Thus the current due to movement of electrons is greater than that due to the hole drift in the semiconductor. *The mobility of hole is about half that of an electron.* The net current is very small.

1.11.4. Effects of Heat and Light on Conductivity of Intrinsic Semiconductors

At absolute zero temperature (–273°C), all the electrons of intrinsic semiconductors (germanium or silicon) are tightly held by their atoms. The inner orbit electrons are bound to nucleus whereas the valence electrons are bound by the forces of covalent bonds. Thus, at absolute zero temperature, no free electron is available in the intrinsic semiconductor and so it behaves like a perfect insulator.

When the material is heated, electrons break away from their atoms and move from the valence band to conduction band. This produces holes in the valence band and free electrons in the conduction band. Conduction can then occur by electron movement and by hole transfer. With the increase in temperature, the rate of generation of electron-hole pairs is increased. This in turn increases the rate of recombination. Thus with the increase in temperature, the concentration of charge carriers (electrons and holes) increases. As more charge carriers are made available, the conductivity of a pure semiconductor increases with the increase in temperature. In other words resistivity of a pure semiconductor decreases with the increase in temperature *i.e.* the semiconductors have *negative temperature coefficient of resistance.*

When the semiconductor is illuminated, its resistance decreases in the same way as in case of increase in temperature.

The forbidden energy gap E_G also depends somewhat on temperature. It has been determined experimentally that E_G for semiconductors decreases with the increase in temperature. For example E_G for Si becomes 1.12 eV and that for Ge becomes 0.72 eV at room temperature while the respective values at absolute zero temperature are 1.21 eV and 0.785 eV.

The resistance of a semiconductor depends largely on (*a*) *temperature*—decreases with the increase in temperature, (*b*) *illumination*—decreases in brighter surroundings and (*c*) *magnitude of the electric field*—the current does not increase in proportion to the applied potential, as in case of conductors obeying Ohm's law, but increases by far more than the voltage *i.e., the semiconductors are nonlinear.*

1.11.5. Fermi Level in an Intrinsic Semiconductor

The *Fermi level* is simply a reference energy level. It is the energy level at which the probability of finding an electron n energy units above it in the conduction band is equal to the probability of finding a hole (electron absence) n energy units below it in the valence band. Very simply, it can be considered to be the average energy level of the electrons, as illustrated in Fig. 1.16 (*a*).

Let at any temperature T K

Number of electrons in the conduction band be n_C

Number of electrons in the valence band be n_V

Total number of electrons in both bands, $n = n_C + n_V$

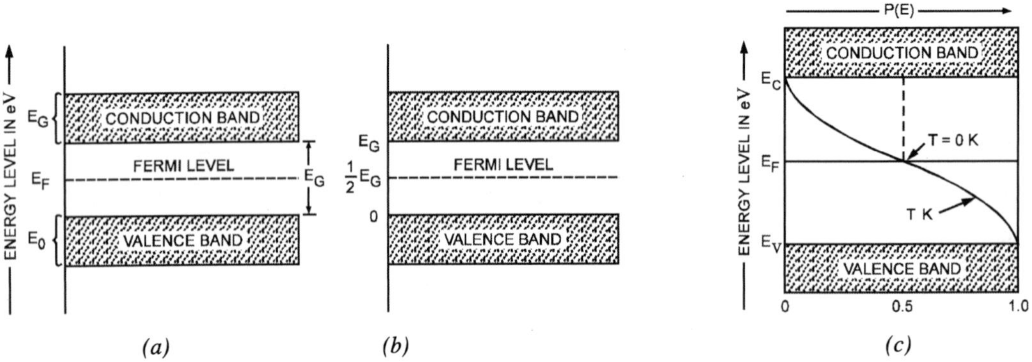

Fig. 1.16

For simplification let us assume that (*i*) widths of energy bands are small in comparison to forbidden energy gap between them (*ii*) all levels in a band have the same energy, bandwidths being assumed to be small (*iii*) energies of all levels in valence band are E_0, as shown in Fig. 1.16 (*a*) and (*iv*) energies of all levels in conduction band are E_G.

Let the zero-energy reference level be taken arbitrarily at the top of the valence band, as shown in Fig. 1.16 (*b*).

Now number of electrons in conduction band,

$$n_C = n.\, P(E_G) \qquad ...(1.3)$$

where P (E_G) represents the probability of an electron having energy E_G. Its value may be determined from Fermi-Dirac probability distribution function given as

$$P(E) = \frac{1}{1 + e^{(E - E_F)/kT}} \qquad ...(1.4)$$

where P (E) is the probability of finding an electron having any particular value of energy E.

$$\text{So } P(E_G) = \frac{1}{1 + e^{(E_G - E_F)/kT}} \qquad ...(1.5)$$ where E_F is the Fermi level.

Substituting the value of P (E_G) from Eq. (1.5) in Eq. (1.3) we have

$$n_C = \frac{n}{1 + e^{(E_G - E_F)/kT}} \qquad ...(1.6)$$

Now number of electrons in valence band,

$$n_V = nP(0) \qquad ...(1.7)$$

The probability P (0) of an electron being found in the valence band with zero energy can be determined by substituting E = 0 in Eq. (1.4)

$$P(0) = \frac{1}{1+e^{(0-E_F)/kT}} = \frac{1}{1+e^{-E_F/kT}} \qquad ...(1.8)$$

$$\text{So } n_V = \frac{n}{1+e^{-E_F/kT}} \qquad ...(1.9)$$

Now total number of electrons in both bands,

$$n = n_C + n_V = \frac{n}{1+e^{(E_G-E_F)/kT}} + \frac{n}{1+e^{-E_F/kT}}$$

$$\text{or } 1 - \frac{1}{1+e^{-E_F/kT}} = \frac{1}{1+e^{(E_G-E_F)/kT}}$$

$$\text{or } E_F = \frac{1}{2}E_G \qquad ...(1.10)$$

i.e. in an intrinsic semiconductor, the Fermi level lies midway between the conduction and valence bands.

Example 1.1. The energy band gap of germanium is 0.72 eV at 300 K. Determine the fraction of the total number of electrons (conduction band as well as valence band) in the conduction band at 300 K. Boltzmann constant is 8.61 × 10⁻⁵ eV/K.

Solution : From Eq. (1.6)

$$\frac{\text{Number of electrons in conduction band, } n_C}{\text{Total number of electrons in both bands, } n} = \frac{1}{1+e^{(E_G-E_F)/kT}}$$

Substituting $E_F = \frac{1}{2}E_G = \frac{1}{2} \times 0.72 = 0.36$ eV, $k = 8.61 \times 10^{-5}$ eV/K and T = 300 K in above equation we have

$$\frac{n_C}{n} = \frac{1}{1+e^{0.36/300\times 8.61\times 10^{-5}}} = 8.85 \times 10^{-7} \textbf{ Ans.}$$

1.12. EXTRINSIC SEMICONDUCTORS

Intrinsic (or pure) semiconductor by itself is of little significance as it has little current conduction capability at ordinary room temperature. However, the electrical conductivity of intrinsic semiconductor can be increased many times by adding very small amount of impurity (of the order of one atom per million atoms of pure semiconductor) to it in the process of crystallization. This process is called *doping* and the doped material is called the *impurity* or *extrinsic* semiconductor.

Germanium and silicon are tetravalent. So the impurity or doping material may be either pentavalent or trivalent. Accordingly the impurity introduced may be of two types viz. (*i*) donor type or N-type and (*ii*) acceptor type or P-type. Depending on the type of impurity added, the extrinsic semiconductors can be divided into two classes namely (*i*) N-type extrinsic semiconductors and (*ii*) P-type extrinsic semiconductors.

1. N-Type Extrinsic Semiconductors. When a small amount of pentavalent impurity, such as *arsenic, antimony, bismuth* or *phosphorus*, is added to a pure semiconductor crystal during the crystal growth, the resulting crystal is called the *N-type extrinsic semiconductor*, where N stands for negative.

When a pentavalent or a donor impurity is added to silicon (or germanium), the impurity atoms form covalent bonds with the silicon (or germanium) atoms; but since intrinsic semiconductor atoms have only four electrons and four holes in their valence shells, one spare valence-shell electron is produced for each impurity atom added. Each spare electron so produced enters the conduction band of pure semiconductor as a free electron. The pentavalent impurity is called the *donor type impurity* as it donates one electron to the conduction band of a pure semiconductor. Though each impurity provides only one free electron yet an extremely small amount of impurity provides enough atoms to supply millions of free electrons.

The effect of adding donor impurity in a semiconductor is illustrated in Fig. 1.17 (*a*). As shown in the figure, the unbound electron remains near its parent atom. This occurs because the impurity atom has a proton in its nucleus which attracts the unbound electron. However, due to the geometry involved, the electron is partially shielded from its parent atom and the attractive force between them is somewhat reduced. The energy required to detach this fifth electron from the atom is of the order of only 0.05 eV for silicon (0.01 eV for germanium). This energy is so small that at room temperature practically all such electrons become free. In other words, at room temperature each impurity atom donates one electron to the conduction band. These donated electrons are called *excess electrons*, because they are in excess to the electrons which are thermally generated by breaking covalent bonds. However, the number of thermally generated electron-hole pairs will be very small in comparison to the number of free electrons due to the impurity atoms. The effect of this doping process on the relative conductivity can best be explained by energy band diagram shown in Fig. 1.17 (*b*). When donor impurities are added to an intrinsic semiconductor, allowable energy levels are introduced at a very small distance below the conduction band, as illustrated in Fig. 1.17 (*b*). These new allowable levels are essentially a discrete level because the added impurity atoms are far apart in the crystal structure and hence their interaction is small. In the case of silicon, the distance of the new discrete allowable energy level is only 0.05 eV (0.01 eV in germanium) below the conduction band and, therefore, at room temperature almost all of the "fifth" electrons of the donor impurity are raised into the conduction band and the conductivity of the material increases considerably. At room temperature in an intrinsic silicon there is about 1 free electron for every 10^{12} atoms (1 to 10^9 for germanium). If the dosage level were 1 in 10 million (10^7), the ratio $10^{12}/10^7$ *i.e.* 10^5 would indicate that the carrier concentration has increased by a ratio 10^5 : 1.

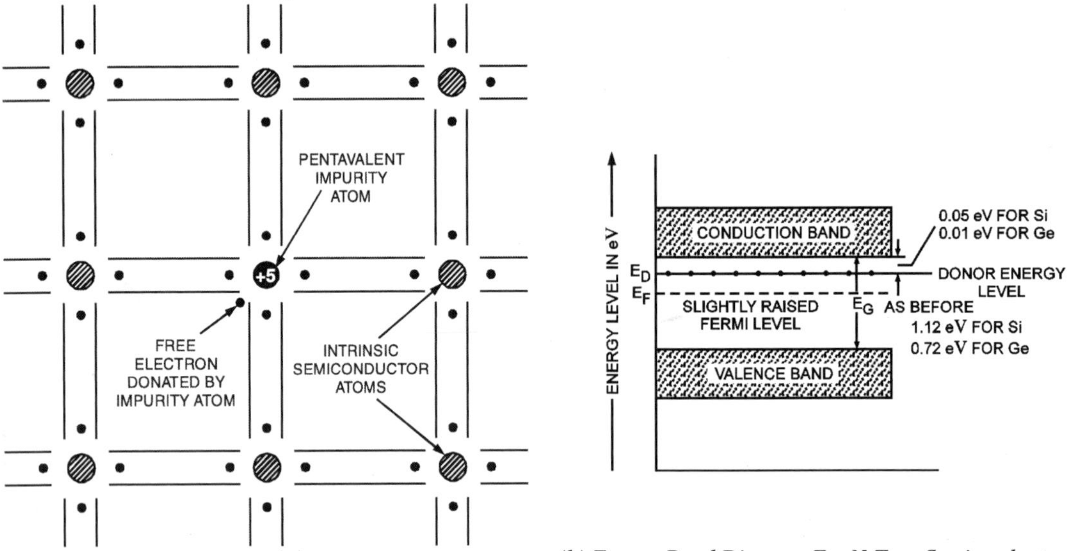

(*a*) (*b*) *Energy Band Diagram For N-Type Semiconductor*

Fig. 1.17

Negative or N-Type Conductivity. When an electric field is applied, the excess electrons donated by impurity atoms move towards the positive terminal of the battery, as shown in Fig. 1.18. This constitutes the electric current. This type of conductivity is called the *negative* or *N-type conductivity* because the current flows through the crystal due to free electrons (negatively charged particles).

Although N-type semiconductor has excess of electrons but it is electrically neutral. This is due to the fact that electrons are created by the addition of neutral pentavalent impurity atoms to the pure semiconductor *i.e.* there is no addition of either negative charges or positive charges.

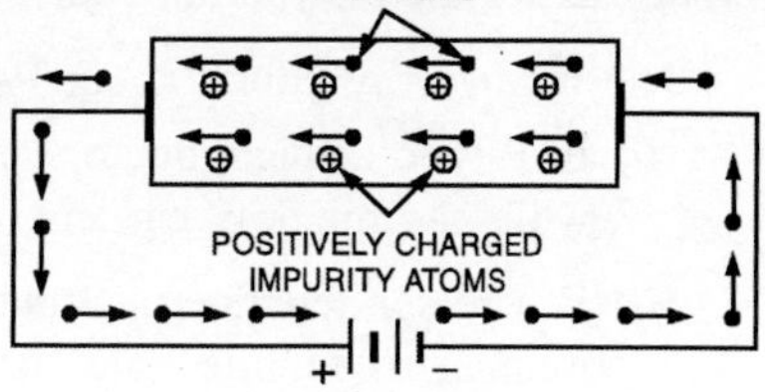

Fig. 1.18

Since pentavalent impurity donates electrons, which have negative charge, to the pure semiconductor, such an impurity is also called the *N-type impurity*.

Since current flowing through the crystal is primarily due to free electrons (negatively charged particles), such a conductor is called the *N-type semiconductor*.

In an N-type semiconductor material, the number of holes is small in comparison to parent intrinsic semiconductor because the larger number of electrons present increase the rate of recombination of electrons with the holes. Thus, the number of free electrons becomes far greater than the number of holes. That is why it is said that an N-type semiconductor has electrons as *majority* carriers, and the holes as *minority* carriers.

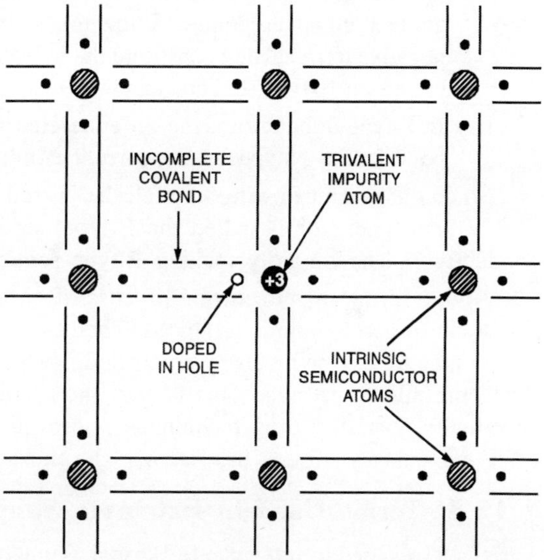

Fig. 1.19

2. P-Type Extrinsic Semiconductors. When a small amount of trivalent impurity, such as *boron, gallium, indium* or *aluminium*, is added to a pure semiconductor crystal during the crystal growth, the resulting crystal is called the *P-type extrinsic semiconductor*, where P stands for positive. The effect of adding one of these impurities in Si or Ge is illustrated in Fig. 1.19.

When a trivalent impurity is added to silicon (or germanium), these impurity atoms form covalent bonds with four surrounding intrinsic semiconductor (silicon or germanium) atoms but one bond is left incomplete and gives rise to a hole, as illustrated in Fig. 1.19. Such impurities make available positive carriers because they create holes which can accept electrons. These impurities are consequently known as *acceptor* or *P-type* impurities. The amount of impurity which is required to be added to have an appreciable effect on the conductivity is very small.

The effect of this doping process on the relative conductivity can best be explained by energy band diagram shown in Fig. 1.20. When acceptor or P-type impurities are added to the intrinsic semiconductor, they produce an allowable discrete energy level which is just above the valence band, as shown in Fig. 1.20. Since a

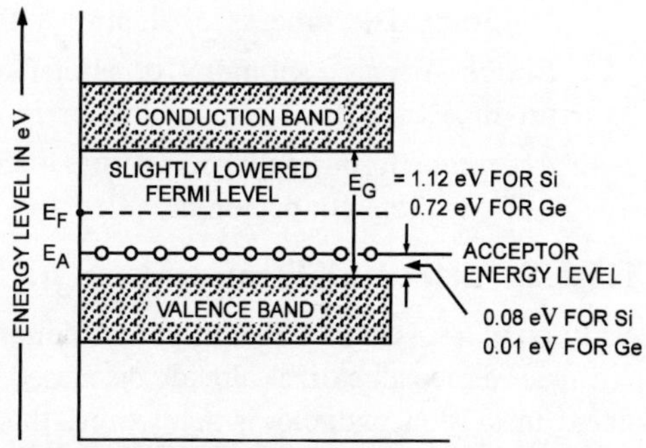

Fig. 1.20. ***Energy Band Diagram For P-Type Semiconductor***

very small amount of energy (0.08 eV in case of silicon and 0.01 eV in case of germanium) is required for an electron to leave the valence band and occupy the acceptor energy level, holes are created in the valence band by these electrons. The holes so created constitute the larger number of carriers in the semiconductor material.

Since holes can be said to have a positive charge, acceptor-doped semiconductor material is referred to as *P-type semiconductor.* The resulting P-type material is electrically neutral, for the same reasons as for the N-type material.

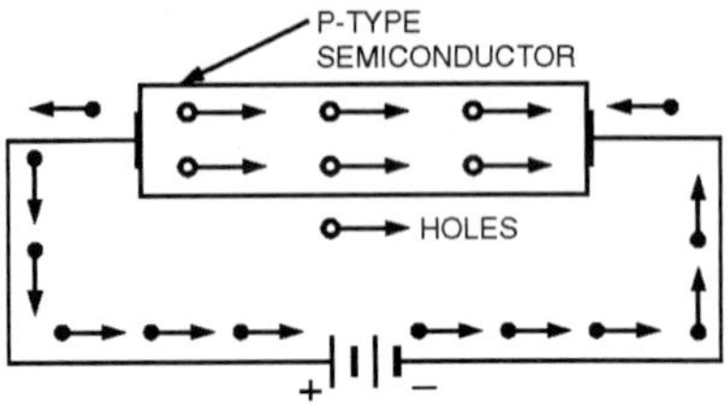

Fig. 1.21

The noteworthy points about P-type semiconductors are :

(*i*) In P-type semiconductors, the majority carriers are holes while the minority carriers are electrons.

(*ii*) P-type semiconductor remains electrically neutral as the number of mobile holes under all conditions remains equal to the number of acceptors.

(*iii*) When an electric field is applied across a P-type semiconductor, as shown in Fig. 1.21, the current conduction is primarily due to holes. Here the holes are shifted from one covalent bond to another covalent bond. As the holes are positively charged, they are directed towards the negative terminal and constitute the hole current. The hole current flows more slowly than electron current in N-type semiconductor.

(*iv*) In P-type conductors, the valence electrons move from one covalent bond to another covalent bond unlike N-type where current conduction is by free electrons.

Since the current flowing through the crystal is primarily due to holes, which have positive charge, such a semiconductor is called the P-type semiconductor and the conductivity is called the P-type conductivity. The impurity making P-type semiconductor is also called the P-type impurity.

Note : In the manufacturing process of an extrinsic semiconductor, before doping a pure crystal of silicon or germanium is produced. Then by controlling the amount of impurity, the properties of the semiconductor can be precisely controlled. Historically, pure germanium crystals were easier to produce than pure silicon crystals. This is why the earliest semiconductor devices were made of germanium. Eventually, manufacturing techniques improved and pure silicon crystals became available. Because of its advantages, silicon has become the most popular and useful semiconductor material.

1.12.1. Terms Used in Extrinsic Semiconductors

Various terms used in reference to extrinsic semiconductors are explained below.

1. *Doping.* Addition of impurity atoms to intrinsic semiconductor crystal is called the *doping* and the impurity used for doping is called the *dopant.*
2. *Diffusion.* The process of doping a semiconductor with impurities is known as *diffusion.*
3. *Donor.* Arsenic, antimony or phosphorus or any other pentavalent impurity used as dopant to produce an N-type semiconductor is called the *donor type impurity.*
4. *Acceptor.* Boron, gallium, indium or any other trivalent impurity used as dopant to produce a P-type semiconductor is called the *acceptor type impurity.*

1.12.2. Effect of Temperature on Extrinsic Semiconductors

Addition of a small amount of impurity (donor or acceptor) produces a large number of charge carriers in an extrinsic semiconductor, as already discussed above. In fact, this number is so large that the conductivity of an extrinsic semiconductor is many times that of an intrinsic semiconductor at room temperature.

If the temperature of an extrinsic semiconductor (say N-type) is raised, the additional thermal energy increases the thermally generated carriers. As a result, the concentration of minority carriers increases. Eventually, a temperature is reached, called the *critical temperature*, (85°C in case of Ge and 200°C in case of Si) when the number of covalent bonds, that are broken is very large, so that the number of holes becomes approximately the same as the number of electrons. The extrinsic semiconductor now behaves like an intrinsic semiconductor but with higher conductivity.

1.12.3. Charge Carriers

As discussed above, current conduction in an N-type semiconductor is due to excess of free electrons whereas in a P-type semiconductor it is due to excess of holes. One may think that N-type semiconductor has a net negative charge and P-type has a net positive charge. But this conclusion is wrong. Even though the N-type semiconductor has excess free electrons but these extra electrons were supplied by the atoms of donor impurity and each atom of donor impurity is electrically neutral. When the donor impurity is added, the term "*excess electrons*" refers to the excess with reference to the number of electrons required to fill the covalent bonds in the semiconductor crystal. The extra free electrons are there to increase the conductivity of the semiconductor, however, N-type semiconductor is electrically neutral. Similar situation occurs in P-type semiconductor which is also electrically neutral. Free electrons in N-type semiconductors and holes in P-type semiconductors are *charge carriers*.

1.12.4. Majority and Minority Charge Carriers

In the intrinsic state, the number of free electrons in silicon or germanium is due only to those few electrons in the valence band that have acquired sufficient energy from heat or light sources to break the covalent bond or to the few impurities that could not be removed. The vacancies left behind in the covalent bonding structure represent very limited supply of holes. In an N-type material, the number of holes has not changed significantly from this intrinsic level and the number of electrons (both added and thermally generated) far outweighs the number of thermally generated holes. Hence in an N-type material electrons are majority carriers and responsible for flow of current.

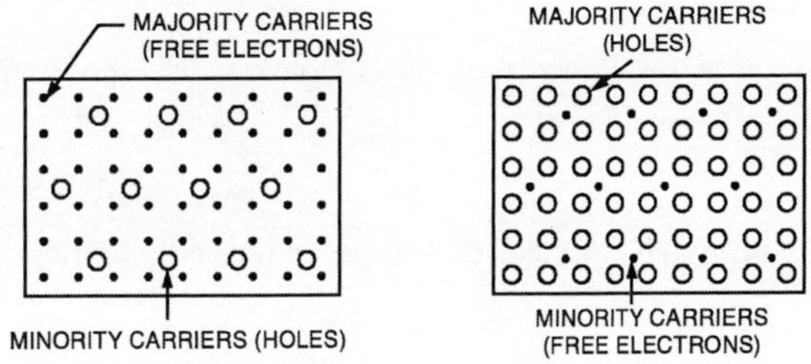

(a) N-Type Material *(b) P-Type Material*

Fig. 1.22

The reverse is true for the P-type material *i.e.* P-type materials have holes as majority carriers and free electrons as minority carriers, as illustrated in Fig. 1.22 (*b*).

1.13. MASS ACTION LAW

As already mentioned in Art. 1.11.1, concentration of free electrons and holes is always equal in an intrinsic semiconductor. When N-type impurity is added to an intrinsic semiconductor, the concentration of free electrons is increased (or the concentration of holes is reduced below the intrinsic value). Similarly addition of P-type impurity results in reduction of concentration of free electrons below the intrinsic value. Theoretical analysis reveals that under thermal equilibrium the product of concentration of free electrons and concentration of holes is constant and is independent of the amount of doping by donor and acceptor impurities. This is known as *mass action law*. Thus,

$$np = n_i^2 \qquad ...(1.11)$$

where n_i is the intrinsic concentration and is a function of temperature. For an intrinsic semiconductor $n = p = n_i$.

The above discussion reveals that the doping of impurity of either type in an intrinsic semiconductor causes increase in the conductivity and results in either free electrons or holes to dominate the situation.

This law is very important from the point of view that it enables us to determine minority carriers concentration. According to this law the addition of impurities to an intrinsic semiconductor increases the concentration of one type of carriers, which consequently become majority carriers and simultaneously decreases the concentration of the other carriers, which as a result become minority carriers.

1.14. CHARGE DENSITIES IN AN EXTRINSIC SEMICONDUCTOR

Equation (1.11) gives relationship between the electron and hole concentrations n and p respectively.

If N_D is the concentration of donor atoms, there will be N_D immobile positive charges per unit volume contributed by donor ions as the donor atoms are ionized. The positive charge density will be $N_D + p$. Similarly as the acceptor atoms are ionized at room temperature there will be N_A immobile negative charges per unit volume and total negative charge density will be $N_A + n$, where N_A is the concentration of acceptor atoms. Since the semiconductor is electrically neutral, the magnitude of positive charge-density will be equal to the magnitude of negative charge-density.

i.e. $$N_D + p = N_A + n \quad \text{...(1.12)}$$

In N-type semiconductors there is no acceptor doping *i.e.* $N_A = 0$. Also the number of electrons is much greater than the number of holes *i.e.* $n >> p$.

$$\text{So } n \simeq N_D \quad \text{...(1.13)}$$

i.e. in N-type semiconductors, the free electron concentration is approximately equal to the density of donor atoms.

Now the concentration of holes in N-type semiconductors will be given by the equation

$$p = \frac{n_i^2}{n} = \frac{n_i^2}{N_D} \quad \text{...(1.14)}$$

Similarly in case of P-type semiconductors

$$p \simeq N_A \quad \text{...(1.15)}$$

$$\text{and } n = \frac{n_i^2}{N_A} \quad \text{...(1.16)}$$

Example 1.2. What is the concentration of holes in Si crystals having donor concentration of $1.4 \times 10^{24}/m^3$ when the intrinsic carrier concentration is $1.4 \times 10^{18}/m^3$? Find the ratio of electron to hole concentration.

[*G.G.S.I.P. Univ. Analog Electronics-I December 2004*]

Solution :

Intrinsic carrier concentration, $n_i = 1.4 \times 10^{18}/m^3$

Donor concentration, $N_D = 1.4 \times 10^{24}/m^3$

Concentration of electrons, $n \simeq N_D = 1.4 \times 10^{24}/m^3$

$$\text{Concentration of holes, } p = \frac{n_i^2}{n} = \frac{(1.4 \times 10^{18})^2}{1.4 \times 10^{24}} = 1.4 \times 10^{12}/m^3$$

$$\text{Ratio of electron to hole concentration} = \frac{n}{p} = \frac{1.4 \times 10^{24}}{1.4 \times 10^{12}} = 1 \times 10^{12} \textbf{ Ans.}$$

1.15. CONDUCTIVITY OF METALS

Consider a conductor of length l metres and cross-sectional area A square metres, as shown in Fig. 1.23. Let the number of free electrons available per m^3 of the conductor material *i.e.*, electron density be n and E be the applied electric field. Due to the applied electric field, let axial drift velocity of the electrons be v metres/second.

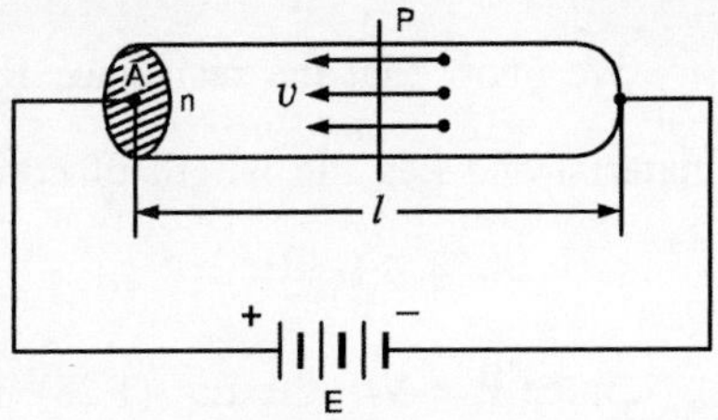

Fig. 1.23

The force F on the particle of charge q when an electric field E is applied is given as $F = qE$, so the force acting on an electron having charge e on it is given as

$$F = eE \quad ...(1.17)$$

Also $$F = ma \quad ...(1.18)$$

where a is acceleration of electrons due to applied field E

Equating Eqs. (1.17) and (1.18) we have

Acceleration, $$a = \frac{eE}{m} \quad ...(1.19)$$

Because of collisions of electrons during motion, the electrons will not get accelerated indefinitely. In fact, their velocity will fall to zero. If τ is the *relaxation* or *collision time*, the average velocity of electrons, known as *drift velocity* is given as

$$v = a \times \tau = \frac{eE}{m} \times \tau \quad ...(1.20)$$

Let I be the current flowing through the conductor on application of electric field E corresponding to drift velocity v. In time dt, the electrons will travel a distance vdt and the number of electrons crossing the cross-sectional area A in time dt will be contained in volume $Avdt$. Thus, total charge flowing through the section in time dt

$$dq = envAdt \quad ...(1.21)$$

Since the current in a conductor is defined as the rate of change of charge per unit time, current I is given as

$$I = \frac{dq}{dt} = envA \quad ...(1.22)$$

The current density J*, defined as the current per unit area is given as

$$J = \frac{I}{A} = env \quad ...(1.23)$$

The above equation may also be written as

$J = \rho v$ ...(1.24) where ρ is the charge density in coulombs/m^3 = en

Substituting the value of v from Eq. (1.20) in above Eq. (1.23), we have

$$J = en \times \frac{eE}{m} \times \tau = \frac{ne^2 E\tau}{m} \quad ...(1.25)$$

For a particular material, the quantity $\frac{ne^2\tau}{m}$ in above equation is constant at particular temperature and is known as *electrical conductivity* 'σ' of the material and thus

Current density, $$J = \sigma E \quad ...(1.26)$$

So, the current density J is proportional to the conductivity of the material σ and the magnitude of applied electric field E. In terms of resistivity ρ, current density J is given as

$$J = \frac{E}{\rho} \quad ...(1.27) \qquad \because \sigma = \frac{1}{\rho}$$

* Current density is a vector quantity and is denoted by symbol $\bar{J}$ and, therefore, in vector notation, $I = \bar{J}\bar{A}$ where $\bar{A}$ is the vector notation for area 'a'. For area of any shape we may write the expression for current I as $I = \int \bar{J}\bar{A}$.

We know that the resistance R of a conductor is given as $R = \rho\frac{l}{A}$ where ρ is the resistivity of material and l is the length of conductor. Thus

$$I = JA = \frac{EA}{\rho} = \frac{El}{R} \quad ...(1.28) \quad \because \; R = \rho\frac{l}{A}$$

Since $E = V/l$, the Eq. (1.28) may be written as

$$I = \frac{V}{l}\times\frac{l}{R} = \frac{V}{R} \quad ...(1.29)$$ which is nothing but 'Ohms' law.

Electrical conductivity of the material is also given as

$$\sigma = \frac{ne^2\tau}{m} = ne\mu_e \quad ...(1.30)$$

where $\mu_e = \frac{e\tau}{m}$ is the mobility acquired by electrons due to presence of electric field.

Using Eq. (1.20), the mobility of electrons can also be expressed as

$$\mu_e = \frac{e\tau}{m} = \frac{v}{E} \quad ...(1.31)$$

Thus, the mobility of electron in the metal is defined as the steady-state drift velocity per unit electric field.

Example 1.3. A conductor material has a free electron density of 10^{24} electrons per m^3. When a voltage is applied, a constant drift velocity of 1.5×10^{-2} m/s is attained by the electrons. If the x-sectional area of the material is 1 cm^2, calculate the magnitude of current. Electronic charge is 1.6×10^{-19} coulomb.

Solution : Electron density, $n = 10^{24}$

Electron charge, $e = 1.6 \times 10^{-19}$ C

Drift velocity, $v = 1.5 \times 10^{-2}$ m/s $= 0.015$ m/s

Cross-sectional area, $A = 1\text{ cm}^2 = 1 \times 10^{-4}\text{ m}^2$

Magnitude of current, $I = e\,n\,v\,A = 1.6 \times 10^{-19} \times 10^{24} \times 0.015 \times 1 \times 10^{-4} = 0.24$ A **Ans.**

Example 1.4. A specimen of germanium at 300 K for which the density of carriers is $2.5 \times 10^{13}/cm^3$, is doped with impurity atoms such that there is one impurity atom for 10^6 germanium atoms. All the impurity atoms may be assumed ionized. The resistivity of doped material is 0.039 Ω-cm. Carrier mobility for germanium at 300 K is 3,600 cm^2/V-s. For the doped material, find the electron and hole densities. $e = 1.602 \times 10^{-19}$ C.

Solution : Resistivity of doped material, $\rho = 0.039$ Ω-cm

Conductivity of doped material, $\sigma_n = \frac{1}{\rho} = \frac{1}{0.039} = 25.64$ mho/cm

Carrier mobility, $\mu_e = 3{,}600\text{ cm}^2/\text{V-s}$

Since $\sigma_n = n\,e\,\mu_e = N_D\,e\,\mu_e$

So concentration of donor atoms, $N_D = \frac{\sigma_n}{e\,\mu_e} = \frac{25.64}{1.602 \times 10^{-19} \times 3{,}600} = 4.45 \times 10^{16}/\text{cm}^3$

Concentration of electrons $n \simeq N_D = 4.45 \times 10^{16}/\text{cm}^3$ **Ans.**

Concentration of holes, $p = \frac{n_i^2}{n} = \frac{(2.5 \times 10^{13})^2}{4.45 \times 10^{16}} = 1.4 \times 10^{10}/\text{cm}^3$ **Ans.**

Example 1.5. A copper wire of 2 mm diameter with conductivity of 5.8×10^7 S/m and electron mobility of 0.0032 m^2/V-s is subjected to an electric field of 20 mV/m. Find (*i*) the charge density of free electrons (*ii*) the current density (*iii*) the current flowing in the wire (*iv*) the electron drift velocity. [*I.E.S. 2002*]

Solution : Diameter of copper wire, d = 2 mm = 0.002 m

Conductivity of copper, $\sigma = 5.8 \times 10^7$ S/m

Electron mobility, $\mu_e = 0.0032$ m^2/V-s

Applied electric field, E = 20 mV/m = 0.02 V/m

(*i*) From Eq. (1.30), Charge density, $n = \dfrac{\sigma}{e\mu_e} = \dfrac{5.8 \times 10^7}{1.6 \times 10^{-19} \times 0.0032} = 1.133 \times 10^{29}/m^3$ **Ans.**

(*ii*) From Eq. (1.26), Current density, $J = \sigma E = 5.8 \times 10^7 \times 0.02 = 1.16 \times 10^6$ A/m^2 **Ans.**

(*iii*) Current flowing in the wire, I = J × area of x-section of wire

$$= J \times \frac{\pi d^2}{4} = 1.16 \times 10^6 \times \frac{\pi}{4} \times (0.002)^2 = 3.644 \text{ A}$$ **Ans.**

(*iv*) From Eq. (1.31)

Electron drift velocity, $v = \mu_e \times E = 0.0032 \times 0.02 = 6.4 \times 10^{-5}$ m/s **Ans.**

1.16. CONDUCTIVITY OF SEMICONDUCTORS

As already mentioned in Art. 1.9, conduction through a semiconductor is due to two mechanisms viz. drift and diffusion.

The movement of charge carriers (electrons and holes) under the influence of an applied electric field E is termed as the *drift current*.

The conductivity of σ_e of the semiconductor due to electrons in conduction band is given as

$$\sigma_e = ne\mu_e \qquad ...(1.32(a))$$

where n is the number of electrons per unit volume of the conductor (*i.e.* electron density or concentration), e is the electron charge and μ_e is the electron mobility.

Similarly, the conductivity σ_h of the semiconductor due to holes is given as

$$\sigma_h = pe\mu_h \qquad ...(1.32(b))$$

where p is the number of holes per unit volume and μ_h is the hole mobility.

Since in semiconductors the conduction is by free electrons as well as by holes, total conductivity σ is given as

$$\sigma = \sigma_e + \sigma_h = ne\mu_e + pe\mu_h = e\,(n\mu_e + p\mu_h) \qquad ...(1.33)$$

Since in **intrinsic semiconductors** $n = p = n_i$, the intrinsic concentration of holes or free electrons in the semiconductor, **intrinsic semiconductor conductivity**

$$\sigma_i = \sigma_n + \sigma_h = e\,(n_i\,\mu_e + n_i\,\mu_h) = n_i\,e\,(\mu_e + \mu_h) \qquad ...(1.34)$$

$$\text{Current density, } J = n_i\,e\,(\mu_e + \mu_h)\,E = \sigma\,E \qquad ...(1.35)$$

$$\text{Current, } I = JA = n_i e\,(\mu_e + \mu_h)\,EA = n_i\,e\,(\mu_e + \mu_h)\,A\,V/l \qquad ...(1.36)$$

where E is applied electric field, V is the applied potential difference across the two ends of the conductor and A is the cross-sectional area of the conductor.

In case of N-type semiconductor hole concentration p_h is negligible and electron concentration $n_n = N_D$. So conductivity in N-type semiconductor

$$\sigma_n = eN_D\mu_e \qquad ...(1.37)$$

Similarly in case of P-type semiconductor electron concentration is negligible and hole concentration $p_h = N_A$. So conductivity in P-type semiconductor

$$\sigma_h = eN_A\mu_h \quad ...(1.38)$$

Total current density due to drift of electrons and holes on application of electric field E,

$$\begin{aligned} J &= \text{Current density due to hole drift } J_h + \text{current density due to electron drift, } J_e \\ &= Epe\mu_h + Ene\mu_e = Ee(p\mu_h + n\mu_e). \end{aligned} \quad ...(1.39)$$

Intrinsic Concentration. It has been established that the intrinsic concentration n_i varies with temperature according to the relation

$$n_i^2 = A_0 T^3 e^{-E_{GO}/kT} \quad ...(1.40)$$

where A_0 is a constant and is independent of temperature, k is a Boltzmann constant in eV per K and E_{GO} is the energy gap in electron volts at zero absolute temperature, T is temperature in degrees absolute and n_i is concentration in number per cubic metre. A_0 is taken as 2.33×10^{43}.

Mobility. The mobility μ of charge carriers (electrons and holes) varies as T^{-m} over a temperature range of 100 K and 400 K. For silicon $m = 2.5$ for electrons and 2.7 for holes. For germanium $m = 1.66$ for electrons and 2.33 for holes.

The carrier currents are also due to concentration gradients in the doped material which leads to **diffusion of carriers** from high concentration region to low concentration region.

A concentration gradient exists if either the number of holes or electrons is greater in one region as compared to the rest of the region. As already mentioned in Art. 1.9.2, when a concentration gradient of carriers exists in a material, the charge carriers tend to move from the region of high concentration to the region of lower concentration. The process is called the *diffusion* and the current due to this process is called the *diffusion current*.

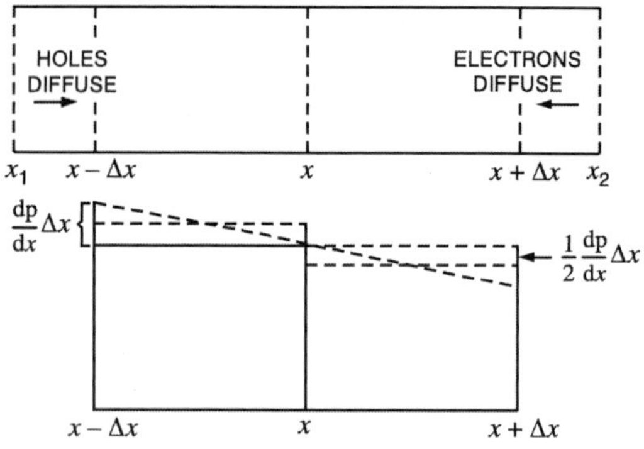

Fig. 1.24

Let us assume that at reference plane x, the density of holes is p per m^3. Let the carriers have a mean free path of Δx and mean free time of Δt between collisions. Assuming uniform density gradient of holes as $\frac{-dp}{dx}$.

The average density of holes in the region $x - \Delta x$ and x

$$= \frac{1}{2}\left[p + p - \frac{dp}{dx}\cdot\Delta x\right] = p - \frac{1}{2}\frac{dp}{dx}\cdot\Delta x$$

Flow of holes to the right per unit time,

$$p_1 = \frac{1}{2}\left[p - \frac{1}{2}\frac{dp}{dx}\Delta x\right]\frac{\Delta x}{\Delta t}$$

Left directed hole flow across reference plane x per unit time,

$$p_2 = \frac{1}{2}\left[p + \frac{1}{2}\frac{dp}{dx}\Delta x\right]\frac{\Delta x}{\Delta t}$$

The net rate of flow of holes is to the right and is given as

$$p = p_1 - p_2 = \frac{1}{2}\left[p - \frac{1}{2}\frac{dp}{dx}\Delta x\right]\frac{\Delta x}{\Delta t} - \frac{1}{2}\left[p + \frac{1}{2}\frac{dp}{dx}\Delta x\right]\frac{\Delta x}{\Delta t}$$

$$= -\frac{1}{2}\Delta x \cdot \frac{\Delta x}{\Delta t} \cdot \frac{dp}{dx} = -D_h \frac{dp}{dx} \quad \text{...(1.41)}$$

where $D_h = \frac{1}{2}\Delta x \cdot \frac{\Delta x}{\Delta t} = \frac{1}{2}\frac{\Delta x^2}{\Delta t}$ m^2/s, called the *diffusion constant.*

The current density due to diffusion of holes is given as

$$J_h = -eD_h \frac{dp}{dx} \quad \text{...(1.42)}$$

Similarly current density due to diffusion of electrons is given as

$$J_e = eD_e \frac{dn}{dx} \quad \text{...(1.43)}$$

Total Current in a Semiconductor. It is possible that a potential gradient and a concentration gradient may exist simultaneously within a semiconductor. In such a case, the total current is the sum of drift current due to potential gradient and the diffusion current due to charge carrier concentration gradient.

Thus $$J_h = Eep\mu_h - eD_h \frac{dp}{dx} \quad \text{...(1.44)}$$

and $$J_e = Een\mu_e + eD_e \frac{dn}{dx} \quad \text{...(1.45)}$$

The constants E_{G0}, μ_e, μ_h, D_e, D_h and many other important physical quantities for silicon and germanium are given below in tabular form (Table 1.1).

TABLE 1.1. ***Properties of Silicon and Germanium***

Property	*Silicon*	*Germanium*
Atomic number	14	32
Atomic weight	28.1	72.6
Density	2,330 kg/m^3	5,320 kg/m^3
Dielectric constant	12	16
Atoms/m^3	5.0×10^{28}	4.4×10^{28}
E_{G0} at 0 K	1.21 *eV*	0.785 *eV*
E_G at 300 K	1.12 *eV*	0.72 *eV*
Intrinsic concentration n_i at 300 K	$1.5 \times 10^{16}/m^3$	$2.5 \times 10^{19}/m^3$
Intrinsic resistivity ρ_i at 300 K	2,300 Ω-m	0.45 Ω-m
Electron mobility, μ_e	0.13 m^2/ V-s	0.38 m^2/ V-s
Hole mobility, μ_h	0.05 m^2/V-s	0.18 m^2/V-s
Diffusion constant, D_e	0.0034 m^2/s	0.0099 m^2/s
Diffusion constant, D_h	0.0013 m^2/s	0.0047 m^2/s

Einstein Relationship. Diffusion constant D and mobility μ are statistical thermodynamic phenomena. Hence they are not independent of each other. The diffusion constants D_h and D_e and mobilities μ_h and μ_e are related by Einstein's equation given as

$$\frac{D_h}{\mu_h} = \frac{D_e}{\mu_e} = V_T \quad ...(1.46)$$

where V_T is the *volt-equivalent of temperature* and is defined as

$$V_T = \frac{k'T}{e} = kT = \frac{T}{11{,}600} \quad ...(1.47)$$

where k′ is Boltzmann constant in joule/K and k is Boltzmann constant in eV/ K.

At room temperature (300 K)

$$V_T = \frac{300}{11{,}600} = 0.026 \text{ V or } 26 \text{ mV}$$

$$\text{and } \mu = \frac{11{,}600}{300} D = 39 D$$

Example 1.6. A 1 kΩ resistors is to be fabricated as a narrow strip of P-type silicon, 4 mm thick. If the strip is 400 μm long and 20 μm wide, what concentration of acceptor atoms is required ? Given hole mobility at room temperature as 480 cm² V⁻¹ s⁻¹. [*G.G.S.I.P. University Analog Electronics December 2010*]

Solution : Intrinsic resistivity, $\rho_i = \frac{R \times a}{l}$ $\quad \because R = \rho\frac{l}{a}$

$$= \frac{1 \times 1{,}000 \times 400 \times 10^{-6} \times 20 \times 10^{-6}}{4 \times 10^{-3}} = 0.002 \ \Omega\text{-m}$$

Intrinsic conductivity, $\sigma_i = \frac{1}{\rho_i} = \frac{1}{0.002} = 500$ S/m

From Eq. (1.32(*b*))

Intrinsic conductivity, $\sigma_i = pe\,\mu_h$

or concentration of acceptor atoms.

$$p = \frac{\sigma_i}{e\mu_h} = \frac{500}{1.6 \times 10^{-19} \times 480 \times 10^{-4}} = 6.51 \times 10^{22}/\text{m}^3 \quad \textbf{Ans.}$$

Example 1.7. The mobility of electrons and holes in a sample of intrinsic germanium at room temperature is 0.36 m²/V-s and 0.17 m²/V-s respectively. Determine the conductivity if electron and hole densities are each equal to 2.5 × 10¹⁹/m³.

Solution : Conductivity, $\sigma_i = n_i e(\mu_e + \mu_h) = 2.5 \times 10^{19} \times 1.6 \times 10^{-19}(0.36 + 0.17) = 2.12$ S/m **Ans.**

Example 1.8. A bar of intrinsic Ge 6 cm long is subjected to a potential difference of 12 V, if the velocity of electron in bar is 73 m/s, determine the mobility of electron.

Solution : Applied field, $E = \frac{V}{l} = \frac{12}{6} = 2$ V/cm = 200 V/m

Electron drift velocity, v_e = 73 m/s

Electron mobility, $\mu_e = \frac{v_e}{E} = \frac{73}{200} = 0.365 \text{ m}^2 \text{ V}^{-1}\text{s}^{-1}$ **Ans.**

Example 1.9. The mobility of free electrons and holes in pure germanium are 3,800 and 1,800 cm²/V-s respectively. The corresponding values for pure silicon are 1,300 and 500 cm²/V-s respectively. Determine the values of intrinsic conductivity for both. Assume n_i = 2.5 × 10¹³ cm⁻³ for germanium and n_i = 1.5 × 10¹⁰ cm⁻³ for silicon at room temperature.

Solution : For germanium

Mobility of free electrons, μ_e = 3,800 cm^2/V-s

Mobility of holes, μ_h = 1,800 cm^2/V-s

Intrinsic concentration, n_i = 2.5 × 10^{13}/cm^3

Intrinsic conductivity, $\sigma_i = n_i\, e(\mu_e + \mu_h) = 2.5 \times 10^{13} \times 1.6 \times 10^{-19}\ (3,800 + 1,800)$

$= 0.0224\ (\text{ohm-cm})^{-1}$ **Ans.**

For silicon :

Mobility of free electrons, μ_e = 1,300 cm^2/V-s

Mobility of holes, μ_h = 500 cm^2/V-s

Intrinsic concentration, n_i = 1.5 × 10^{10}/cm^3

Intrinsic conductivity, $\sigma_i = n_i e(\mu_e + \mu_h) = 1.5 \times 10^{10} \times 1.6 \times 10^{-19}\ (1,300 + 500)$

$= 0.4320 \times 10^{-5}\ (\text{ohm-cm})^{-1}$ **Ans.**

Example 1.10. A *n*-type germanium crystal has a current density of 100 A/m^2. The crystal has a resistivity of 0.5 Ω-m and electron mobility of 0.4 m^2/V-s. Calculate the drift velocity and the time taken by the electron to travel 10 micrometre in the crystal. Assume e = 1.6 × 10^{-19} C.

Solution : Crystal resistivity, ρ = 0.5 Ω-m

Current density, J = 100 A/m^2

Electron mobility, μ_e = 0.4 m^2/V-s

Electron drift velocity, $v = \mu_e \times E = \mu_e \times \frac{J}{\sigma} = \mu_e \times J \times \rho = 0.4 \times 100 \times 0.5 = 20$ m/s **Ans.**

Time taken by the electron to travel 10 × 10^{-6} m in crystal $= \frac{10 \times 10^{-6}}{20} = 0.5 \times 10^{-6}$ second **Ans.**

Example 1.11. The intrinsic resistivity of germanium at room temperature is 0.47 Ω-m. The electron and hole mobilities at room temperature are 0.39 and 0.19 m^2/V-s respectively. Calculate the density of electrons in the intrinsic semiconductor. Also calculate the drift velocities of these charge carriers for a field of 10 kV/m.

Solution : Intrinsic resistivity, ρ_i = 0.47 Ω-m

Intrinsic conductivity, $\sigma_i = \frac{1}{\rho_i} = \frac{1}{0.47} = 2.12766$ S/m

Electron mobility, μ_e = 0.39 m^2/V-s

Hole mobility, μ_h = 0.19 m^2/V-s

Since $\sigma_i = n_i\, e\, (\mu_e + \mu_h)$

∴ Density of electrons = Intrinsic concentration n_i

$$= \frac{\sigma_i}{e\,(\mu_e + \mu_h)} = \frac{2.12766}{1.6 \times 10^{-19}\ (0.39 + 0.19)} = 2.293 \times 10^{19}/\text{m}^3$$ **Ans.**

Drift velocity for electrons, $v_n = \mu_e E = 0.39 \times 10^4 = 3,900$ m/s **Ans.**

Drift velocity for holes, $v_h = \mu_h E = 0.19 \times 10^4 = 1,900$ m/s **Ans.**

Example 1.12. A pd of 10 V is applied longitudinally to a rectangular specimen of intrinsic germanium of length 25 mm, width 4 mm and thickness 1.5 mm. Determine at room temperature (*i*) electron and hole drift velocities (*ii*) the conductivity of intrinsic germanium if intrinsic carrier density is 2.5 × 10^{19}/m^3 and (*iii*) the total current.

Given that e = 1.6 × 10^{-19} C, μ_e = 0.38 m^2 V^{-1} s^{-1}. μ_h = 0.18 m^2 V^{-1} s^{-1}.

Solution : Applied electric field, $E = V/l = \frac{10}{0.025} = 400$ V/m

(i) Electron drift velocity, $v_e = \mu_e \times E = 0.38 \times 400 = 152$ m/s **Ans.**

Hole drift velocity, $v_h = \mu_h \times E = 0.18 \times 400 = 72$ m/s **Ans.**

Intrinsic carrier concentration, $n_i = 2.5 \times 10^{19}$ /m^2

(ii) Intrinsic conductivity of Ge, $\sigma_i = n_i e (\mu_e + \mu_h) = 2.5 \times 10^{19} \times 1.6 \times 10^{-19} (0.38 + 0.18)$

$= 2.24$ (ohm-m)$^{-1}$ **Ans.**

(iii) Total current, $I = \sigma_i EA = 2.24 \times 400 \times 4 \times 10^{-3} \times 1.5 \times 10^{-3} = 5.376$ mA **Ans.**

1.17. FERMI LEVEL IN AN EXTRINSIC SEMICONDUCTOR

The equation for f(E) specifying the fraction of all states at energy E (electron volts) occupied under conditions of thermal equilibrium is called the *Fermi-Dirac probability function* and is given as

$$f(E) = \frac{1}{1 + e^{(E - E_F)/kT}} \qquad ...(1.48)$$

where k is Boltzmann constant in eV/K, T is temperature in K and E_F is the *Fermi level* or *characteristic energy* for crystal in eV.

Concentrations of free electrons n and concentration of holes p are given as

$$n = n_C e^{-(E_C - E_F)/kT} \qquad ...(1.49)$$

and $$p = n_V e^{-(E_F - E_V)/kT} \qquad ...(1.50)$$

where n_C is the number of electrons in conduction band, n_V is the number of electrons in the valence band, E_C is the lowest energy in the conduction band in eV and E_V is the maximum energy of valence band in eV.

The only parameter that changes with the addition of impurity in Eqs. (1.49) and (1.50) is the Fermi level.

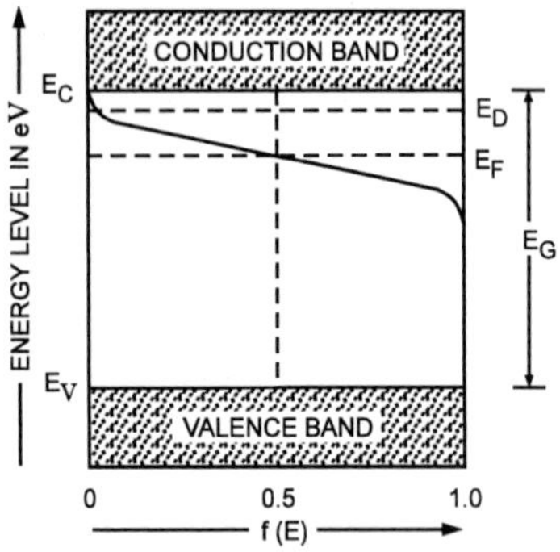

Fig. 1.25. ***Position of Fermi Level in an N-Type Semiconductor***

Fermi Level in N-type Extrinsic Semiconductor. In case of an intrinsic semiconductor, E_F lies in the middle of the energy gap, indicating equal concentrations of free electrons and holes (refer to Art. 1.11.5).

When a donor type impurity is added to the crystal, then, at a given temperature and assuming all donor atoms are ionized, the first N_D states in the conduction band will be filled. Hence it will be more difficult for the electrons from the valence band to bridge the energy gap by the thermal agitation. Consequently the number of electron-hole pairs thermally generated for that temperature will be decreased. Since the Fermi level is a measure of the probability of occupancy of the allowed energy states, it is clear that E_F must be more closer to the conduction band to indicate that many of the energy states in that band are filled by the donor electrons, and fewer holes exist in the valence band. This situation is illustrated in Fig. 1.25, for an N-type extrinsic semiconductor.

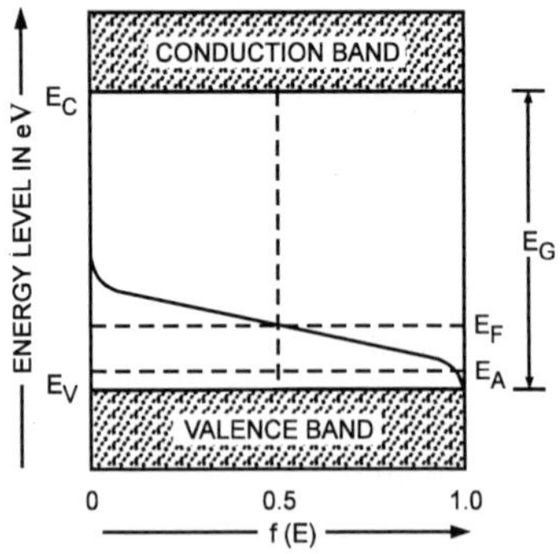

Fig. 1.26. ***Position of Fermi Level in an P-Type Semiconductor***

The same kind of argument leads to the conclusion that E_F must move from the centre of the forbidden gap closer to the valence band for a *P-type extrinsic semiconductor*.

Exact position of the Fermi level in an N-type extrinsic semiconductor can be determined by substituting $n = N_D$ in Eq. (1.49) and thus we have

$$N_D = n_C\, e^{-(E_C - E_F)/kT} \qquad ...(1.51)$$

Taking logarithm we have,

$$\log_e N_D = \log_e n_C - \frac{(E_C - E_F)}{kT}$$

Solving the above equation we have,

$$E_F = E_C - kT \log_e \frac{n_C}{N_D} \qquad ...(1.52)$$

Similarly for the P-type extrinsic semiconductor,

$$E_F = E_V + kT \log_e \frac{n_V}{N_A} \qquad ...(1.53)$$

Example 1.13. In an N-type semiconductor, the Fermi level is 0.24 eV below the conduction band at a room temperature of 300 K. If the temperature is increased to 350 K, determine the new position of the Fermi level.

Solution : The Fermi level in an N-type material is given by

$$E_F = E_C - kT \log \frac{n_C}{N_D} \qquad \text{Refer to Eq. (1.52)}$$

$$\text{or } E_C - E_F = kT \log \frac{n_C}{N_D}$$

At room temperature, T = 300 K

$$0.24 = 300 \text{ K} \log \frac{n_C}{N_D} \qquad ...(i)$$

$$\text{At temperature of 350 K} \quad E_C - E_{F_1} = 350 \text{ K} \log \frac{n_C}{N_D} \qquad ...(ii)$$

Dividing Eq. (*ii*) by Eq. (*i*), we have

$$\frac{E_C - E_{F_1}}{0.24} = \frac{350}{300}$$

$$\text{or } E_C - E_{F_1} = \frac{350}{300} \times 0.24 = 0.28 \text{ eV}$$

i.e. the new position of the Fermi level lies 0.28 eV below the conduction band.

Example 1.14. In a P-type semiconductor, the Fermi level lies 0.39 eV above the valence band. Determine the new position of the Fermi level if the concentration of acceptor atoms is trippled. Assume kT = 0.026 eV.

Solution : In a P-type semiconductor material, the concentration of acceptor atoms is given by

$$N_A = n_V\, e^{-(E_F - E_V)/kT}$$

Let initial acceptor atom concentration and Fermi level be denoted by N_{A_1} and E_{F_1} respectively. Then

$$N_{A_1} = n_V e^{-0.39/0.026} = n_V e^{-15} \; ...(i) \;\because\; E_F - E_V = 0.39 \text{ eV}$$

When acceptor concentration, $N_{A_2} = 3N_{A_1}$ and Fermi level is E_{F_2} then

$$N_{A_2} = n_V e^{-(E_{F_2} - E_V)/0.026}$$

$$\text{or } 3N_{A_1} = n_V e^{-(E_{F_2} - E_V)/0.026} \quad ...(ii)$$

Comparing Eqs. (*i*) and (*ii*), we have

$$n_V e^{-(E_{F_2} - E_V)/0.026} = 3n_V e^{-15}$$

$$\text{or } e^{-(E_{F_2} - E_V)/0.026 + 15} = 3$$

$$\text{or } \frac{-(E_{F_2} - E_V)}{0.026} + 15 = \log_e 3$$

$$\text{or } E_{F_2} - E_V = 0.026(15 - \log_e 3) = 0.36 \text{ eV} \quad \textbf{Ans.}$$

1.18. BASIC STRUCTURE OF P-N JUNCTION

Most semiconductor devices employ one or more P-N junctions. The P-N junction is the control element for the performance of all semiconductor devices such as rectifiers, amplifiers, switching devices, linear and digital integrated circuits. The P-N junction is produced* by placing a layer of P-type semiconductor next to the layer of N-type semiconductor. The interface separating the N and P regions is referred to as the *metallurgical junction.*

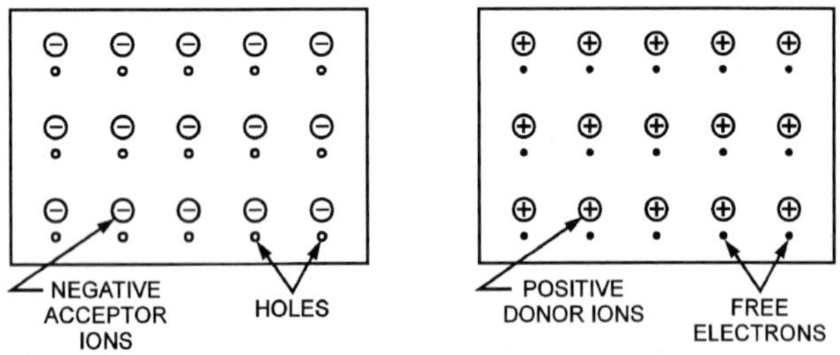

(*a*) *P-Type Semiconductor Material* (*b*) *N-Type Semiconductor Material*

Fig. 1.27

Figure 1.27 represents two blocks of semiconductor material, one P-type, and the other N-type. The P-type semiconductor block has mobile holes (shown by small circles) and the same number of fixed negative acceptor ions (shown by encircled minus sign). Similarly the N-type semiconductor block has mobile or free electrons (shown by dots) and the same number of fixed donor positive ions. Normally the holes, which are the majority charge carriers in P-type of material, are uniformly distributed throughout the volume of that material. Similarly the electrons, which are the majority charge carriers in N-type of material, are uniformly distributed throughout the volume of that material. Each region is electrically neutral because each of them carries equal positive and negative charges.

On the formation of P-N junction some of the holes from P-type material tend to diffuse across the boundary into N-type material and some of the free electrons similarly diffuse into the P-type material, as illustrated in Fig. 1.28. This happens due to density gradient (as concentration of holes is higher on P-side than that on N-side and concentration of electrons is higher on N-side than that on P-side). This process is known as *diffusion.*

The doping profile of an ideal uniformly doped P-N junction is depicted in Fig. 1.29.

As a result of the displacement of the charges, an electric field appears across the junction. Equilibrium is established when the field becomes large enough to restrain the process of diffusion. The general shape of the charge distribution may be as illustrated in Fig. 1.28 (*b*). The electric charges are confined to the neighbourhood of the junction, and consists of immobile ions.

* It is to be noted that a P-N junction is not formed by merging two separate P-type and N-type materials. If this were done, a polycrystalline structure would have resulted in, which would have not performed very well. The P-N junction found in the semiconductor devices is actually formed in a single monocrystalline structure by adding carefully controlled amounts of donor and acceptor impurities.

We see that the free electrons crossing the junction create negative ions on the P-side by giving some atoms one more electron than their total number of protons. The electrons also leave positive ions (atoms with one fewer electron than the number of protons) behind them on the N-side. As negative ions are created on the P-side of the junction, the P-side acquires a negative potential, as illustrated in Fig. 1.28 (*e*). Similarly, the positive ions are created on the N-side and the N-side acquires a positive potential as illustrated in Fig. 1.28 (*d*). The negative potential on the P-side prevents the migration of any more electrons from the N-type material to the P-type material. Similarly, the positive potential on the N-side prevents any further migration of holes across the boundary. Thus, the initial diffusion of charge carriers creates a *barrier potential* at the junction.

The region around the junction is completely ionised. As a result, there are no free electrons on the N-side, nor holes on the P-side. Since the region around the junction is depleted of mobile charges it is called the *depletion region*, the *space-charge region*, or the *transition region*. The thickness of the depletion region (or layer) is of the order of 1 micron (10^{-6} m).

The electric field intensity in the neighbourhood of the junction is illustrated in Fig. 1.28 (*c*). The electrostatic potential variation in the depletion region is shown in Fig. 1.28 (*d*), and is the negative integral of the function E of Fig. 1.28 (*c*). The variation constitutes a potential-energy barrier against the further diffusion of holes across the barrier. The form of potential-energy barrier against the flow of electrons from the N-side across the junction is shown in Fig. 1.28 (*e*). It is similar to that shown in Fig. 1.28 (*d*), except that it is inverted, since the charge on electron is negative.

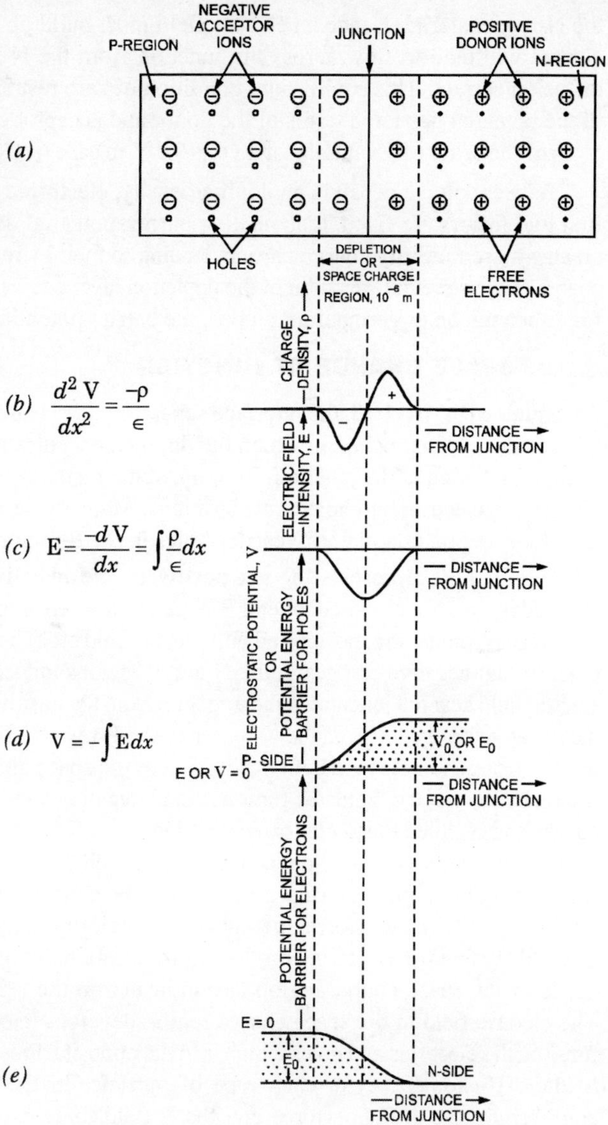

Fig. 1.28. ***A Schematic Diagram of a P-N Junction, Including The Charge Density, Electric Field Intensity, and Potential-Energy Barriers at The Junction (Not To Scale)***

The necessity for the existence of a potential barrier called the *contact*, or *diffusion, potential* is now considered further. Under open-circuited conditions the net hole current must be zero. If this statement were not true, the hole density at one end of the semiconductor would continue to increase indefinitely with time, a situation which is obviously physically impossible. Since the concentration of holes in the P-side is much greater than that in the N-side, a very large diffusion current tends to flow across the junction from the P-type material to

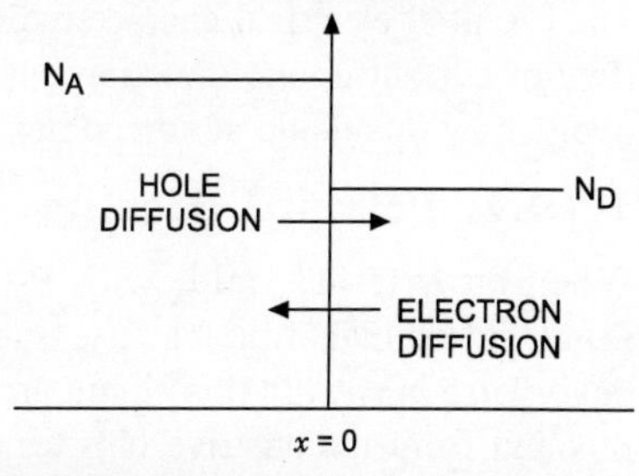

Fig. 1.29. ***Doping Profile of an Ideal Uniformly Doped P-N Junction***

the N-type material. Hence an electric field must build up across the junction in such a direction that a drift current will tend to flow across the junction from the N-side to the P-side in order to counterbalance the diffusion current. This equilibrium condition of zero resultant hole current allows us to determine the height of the potential barrier in terms of the donor and acceptor concentrations. The magnitude of barrier potential V_0 is of the order of few tenths of a volt (0.3 V in case of Ge and 0.7 V for Si).

Barrier voltage depends on doping density, electronic charge and temperature. For a given junction, the first two factors are fixed, thus making barrier potential dependent on temperature. Increase in temperature creates more minority charge carriers leading to their increased drift across the junction. These extra minority charge carriers reduce the width of the depletion layer equivalent to reduction in potential barrier. It is found that for either silicon or germanium diodes, the barrier potential decreases 2 mV for each Celsius degree rise.

1.19. SPACE CHARGE AT JUNCTION

As explained in Art. 1.18, the interface separating the N and P-regions is called the metallurgical junction. Let us consider a *step junction* in which the doping concentration is uniform in each region but there is an abrupt change in doping at the junction. Initially, at the metallurgical junction, there is a very large density gradient in both the electron and hole concentrations. Majority carrier electrons in the N-region will begin diffusing into the P-region and majority carrier holes in the P-region will begin diffusing into the N-region.

Figure 1.30 represents the net positively and negatively charged regions for a semiconductor P-N junction where there is no external connection, no applied voltage, no field etc. The net positive and negative charges in the N and P regions induce an electric field near the junction in the direction from the positive to the negative charge, or from the N-region to the P-region. Since the electrons and holes are both pushed out of the space charge region by the electric field, the region is depleted of any mobile charge and is called the *depletion region.* Density gradients still exist in the majority carrier concentrations at each edge of the space charge region. A density gradient may be considered as producing a "diffusion force" acting on the majority carriers. These diffusion forces, acting on the electrons and holes at the edges of the space charge region, are indicated in the figure. The electric field in the space charge region develops another force on the electrons and holes which is in the opposite direction to the diffusion force for each type of particle. In thermal equilibrium, the diffusion force and the E-field force exactly balance each other.

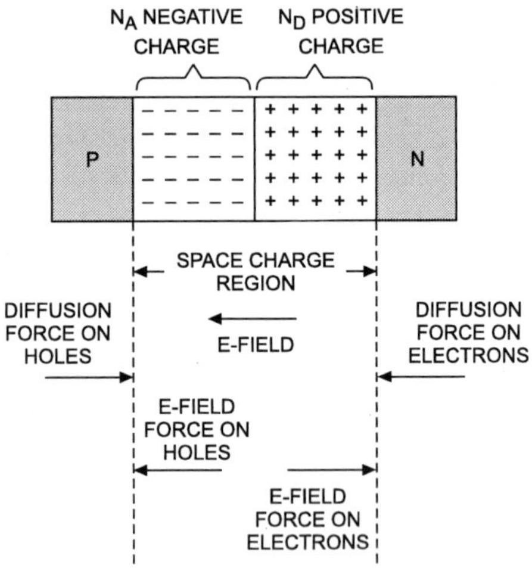

Fig. 1.30. ***Space Charge Region, Electric Field and Forces Acting on the Charged Carriers***

1.20. P-N JUNCTION AS A DIODE

The essential electrical characteristic of a P-N junction is that it constitutes a diode which permits the easy flow of current in one direction but restrains the flow in the opposite direction. We consider now, qualitatively, how this diode action comes about.

1.20.1. Forward Biasing

When an external field, with P-region connected to positive terminal and N-region connected to negative terminal of the battery, is applied across the junction, as shown in Fig. 1.31, the junction is said to be *forward biased.* In this circuit arrangement, the holes on the P-side being positively charged particles are repelled from the positive bias terminal and driven towards the junction. Similarly, the electrons on the N-side are repelled from the negative bias terminal and driven towards the junction. The result is that the depletion region is reduced in width, and the barrier potential is also reduced. If the applied bias voltage is increased from zero, the barrier potential gets progressively smaller and smaller until it effectively disappears and charge carriers can easily flow across the junction. Electrons from the N-side are then attracted across to the positive bias terminal on the P-side, and holes from the P-side flow across to the negative bias terminal on the N-side. Thus a majority carrier current flows. Since barrier potential is very small (0.3 V for Ge and 0.7 V for Si), a small forward voltage is sufficient to eliminate the barrier completely. Once the barrier is eliminated by the application of forward voltage, junction resistance becomes almost zero and a low resistance path is established in the entire circuit. The current, called the *forward current*, flows in the circuit.

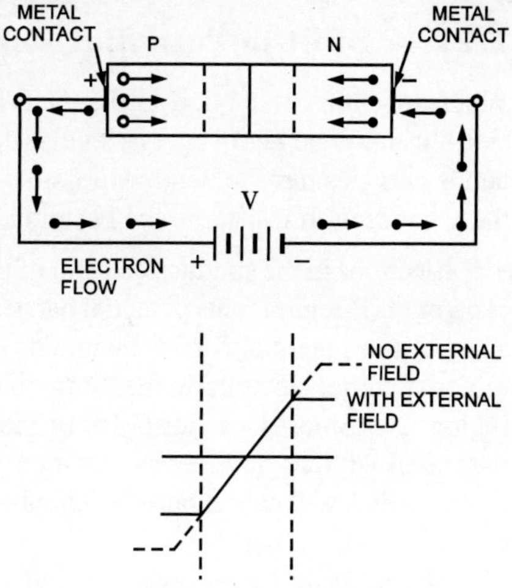

Fig. 1.31. ***Forward Biasing***

In brief it can be said that when the junction is forward biased (P-region connected to the positive terminal and N-region connected to the negative terminal of the battery), (*i*) barrier is reduced and at a forward voltage of 0.3 V in case of germanium (0.7 V in case of silicon) it is eliminated altogether, (*ii*) the junction offers low resistance, called the *forward resistance* r_f, to the flow of current and (*iii*) current flows in the circuit due to establishment of low resistance path and the magnitude of current depends upon the magnitude of applied forward voltage.

The mechanism of current flow in a forward biased P-N junction is summed up as follows :

(*i*) The electrons from the negative bias terminal continue to arrive into the N-region while the free electrons in the N-region move towards the P-N junction.

(*ii*) The electrons travel through the N-region as free electrons *i.e.* the current in N-region is by free electrons.

(*iii*) These free electrons on reaching P-N junction, combine with holes and become valence electrons. Since a hole is in the covalent bond and, therefore, when a free electron combines with a hole, it becomes a valence electron.

(*iv*) The electrons travel through P-region as valence electrons and current in this P-region is, therefore, by holes.

(*v*) These valence electrons, on reaching the left end of the crystal, flow into the positive terminal of the battery.

Thus we see that the current in the N-type region is due to movement of electrons whereas in the P-type region it is carried by the holes. However, in the external circuit (*i.e.* in connecting wires) the current is carried by electrons.

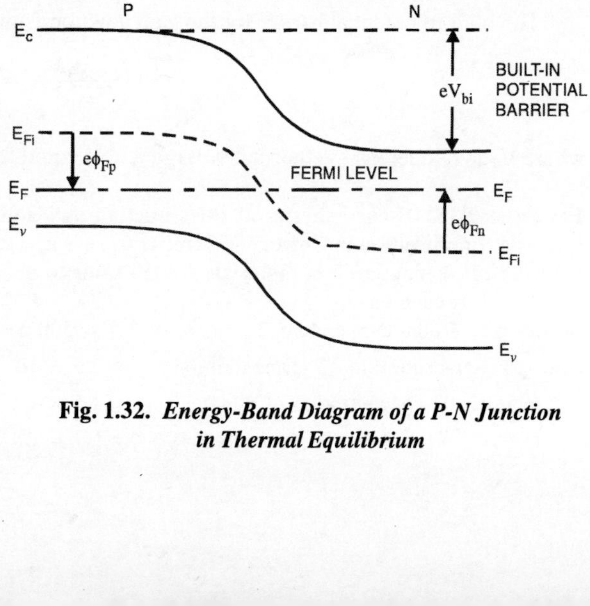

Fig. 1.32. ***Energy-Band Diagram of a P-N Junction in Thermal Equilibrium***

1.21. ZERO APPLIED BIAS

Here we will examine the properties of the step junction in thermal equilibrium, where no currents exist and no external excitation is applied. We will determine built-in potential barrier through the depletion or space charge region, electric field and width of the space charge region.

1.21.1. Built-in Potential Barrier

When no voltage is applied across the P-N junction (*i.e.* with zero applied bias), the junction is in the thermal equilibrium—the Fermi energy level is constant throughout the entire specimen. The valence and conduction bands energies have to bend while moving through the space charge region because the relative position of these bands with respect to the Fermi energy changes between P and N regions.

Electrons in the conduction band of the N-region face a potential barrier when moving into the conduction band of the P-region. This potential barrier is known as the *built-in potential barrier* and is denoted by V_{bi} or V_0. The built-in potential barrier maintains equilibrium between majority carrier electrons in the N-region and minority carrier electrons in the P-region. It also maintains equilibrium between majority carrier holes in the P-region and minority carrier holes in the N-region. This potential difference across the junction cannot be measured with a voltmeter as new potential barriers are being formed between the semiconductor and the probes which will cancel built-in potential barrier V_{bi}. As the condition pertains to equilibrium so no current is caused by this voltage.

The intrinsic Fermi level is equidistant from the conduction band edge through the junction, as illustrated in figure. The built-in potential barrier V_{bi} is the difference between the intrinsic Fermi levels of the N and P-regions.

$$V_{bi} = \phi_{Fn} + \phi_{Fp} \qquad \text{...(1.54)}$$

The electron concentration in the N-region conduction band,

$$n_0 = n_C e^{-(E_C - E_F)/k'T} = n_i e^{(E_F - E_{Fi})/k'T} = n_i e^{-e\phi_{Fn}/k'T} \qquad \text{...(1.55)}$$

where E_{Fi} and n_i are the intrinsic Fermi energy and intrinsic carrier concentration, respectively.

Potential ϕ_{Fn} in the N-region is defined as

$$e\,\phi_{Fn} = E_{Fi} - E_F \qquad \text{...(1.56)}$$

Taking natural log of both sides of Eq. (1.55), setting $n_o = N_D$ we have

$$\phi_{Fn} = \frac{-k'T}{e} \log_e \frac{N_D}{n_i} \qquad \text{...(1.57)}$$

Similarly for the P-region,

$$\phi_{Fp} = \frac{+k'T}{e} \log_e \frac{N_A}{n_i} \qquad \text{...(1.58)}$$

The built-in potential barrier for the step junction is then

$$V_{bi} = \frac{k'T}{e} \log_e \frac{N_A N_D}{n_i^2} = V_T \log_e \frac{N_A N_D}{n_i^2} \qquad \text{...(1.59)}$$

where V_T is defined as the thermal voltage and is equal to $\frac{k'T}{e}$.

Example 1.15. Discuss physics of P-N junction and analyse it when it is reverse and forward biased. The intrinsic concentration in Ge at room temperature is $n_i = 2.5 \times 10^{19}/m^3$ and of Si at room temperature is $n_i = 1.5 \times 10^{16}/m^3$. If doping levels of materials are the same to the extent $N_D = N_A = 10^{21}/m^3$, obtain the values of built-in voltage in each case.

Solution : Room temperature, T = (273 + 27) = 300 K — Assuming room temperature to be 27°

Intrinsic concentration of germanium, $n_{i(Ge)} = 2.5 \times 10^{19}/m^3$

Intrinsic concentration of silicon, $n_{i(Si)} = 1.5 \times 10^{16}/m^3$

Doping level, $N_D = N_A = 10^{21}/m^3$

Built-in voltage for germanium,

$$V_{0(Ge)} = V_T \log_e \frac{N_D N_A}{n_i^2} = \frac{T}{11,600} \log_e \frac{N_D N_A}{n_i^2} \qquad \because V_T = \frac{T}{11,600}$$

$$= \frac{300}{11,600} \log_e \frac{(1 \times 10^{21})^2}{(2.5 \times 10^{19})^2} = 0.191 \text{ V} \ \textbf{Ans.}$$

Built-in voltage of silicon, $V_{0(Si)} = \frac{300}{11,600} \log_e \frac{(1 \times 10^{21})^2}{(1.5 \times 10^{16})^2} = 0.5745 \text{ V}$ **Ans.**

Example 1.16. The resistivities of the P-region and N-region of a Ge diode are 6 Ω-cm and 4 Ω-cm respectively. Calculate the contact potential V_0. Given that $q = 1.6 \times 10^{-19}$ C, $n_i = 2.5 \times 10^{13}/cm^3$; $\mu_p = 1800$ cm²/V-s; $\mu_n = 3,800$ cm²/V-s and $V_T = 0.026$ V at 300 K.

Solution : Density of electrons, $n = N_D = \frac{\sigma_i \text{ (electrons)}}{e \mu_e}$

$$= \frac{1/\text{resistivity of N-region}}{e \mu_e} = \frac{1/4}{1.6 \times 10^{-19} \times 3,800} = 4.1 \times 10^{14}/cm^3$$

Density of holes, $p = N_A = \frac{1/6}{1.6 \times 10^{-19} \times 1,800} = 5.78 \times 10^{14}/cm^3$

Contact potential, $V = KT \log_e \frac{N_D N_A}{n_i^2} = V_T \log_e \frac{N_D N_A}{n_i^2} = 0.026 \times \log_e \frac{4.1 \times 10^{14} \times 5.78 \times 10^{14}}{(2.5 \times 10^{13})^2}$

$$= 0.154 \text{ V} \ \textbf{Ans.}$$

1.22. REVERSE APPLIED BIAS

If an external bias voltage is applied with positive terminal to N-side and negative terminal to the P-side of a P-N junction, as shown in Fig. 1.33, the junction is said to *be reverse biased.* In this arrangement, electrons from the N-side are attracted to the positive bias terminal, and holes from P-side are attracted to the negative bias terminal. Thus, as shown in Fig. 1.33, holes from the impurity atoms on the P-side of the junction are attracted away from the junction, and the electrons are attracted away from the atoms on the N-side of the junction. Thus, the depletion region is widened, and the barrier potential is increased by the magnitude of the applied bias. With the increased barrier potential, there is no possibility of majority carrier current flow across the junction. Thus the P-N junction is in a non-conductive state.

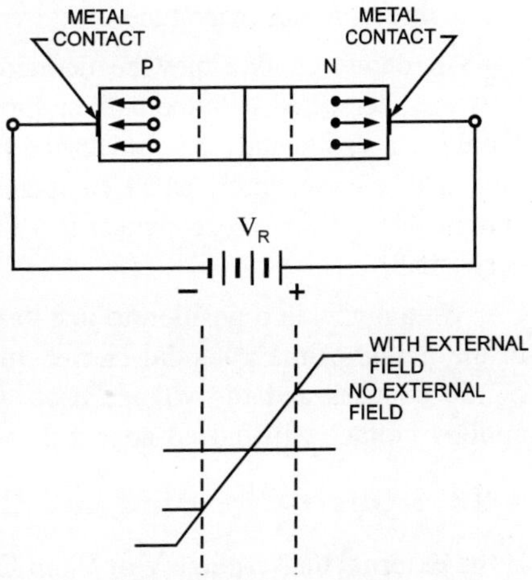

Fig. 1.33. ***Reverse Biasing***

In brief it can be said that when the P-N junction is reverse biased (*i*) potential barrier is increased and width of depletion is also increased (*ii*) the junction offers high resistance, called the reverse resistance r_r, to the flow of current (*iii*) owing to establishment of high resistance path, very small current, called the *leakage current* or *reverse current*, flows through the circuit. The reverse or leakage current is not affected by barrier and depends on temperature, material and construction of the junction.

Thus the P-N junction allows flow of large current only when forward biased and only small reverse (or leakage) current when reverse biased.

1.23. METAL-SEMICONDUCTOR JUNCTIONS

Metal-semiconductor junctions are very common in all semiconductor devices and are very important. Depending upon the doping concentration, materials, and the characteristics of the interface, the metal semiconductor junctions can act as either an ohmic contact or as a Schottky barrier.

1.23.1. Structure of Metal-Semiconductor Junction

A metal-semiconductor junction, as its name indicates, consists of a metal in contact with a piece of semiconductor, as shown in Fig. 1.34. The active junction is the interface between the metal, which acts as an anode, and the semiconductor. The other interface between the semiconductor and the metal, which acts as a cathode, is an Ohmic contact and there is no potential drop at this junction.

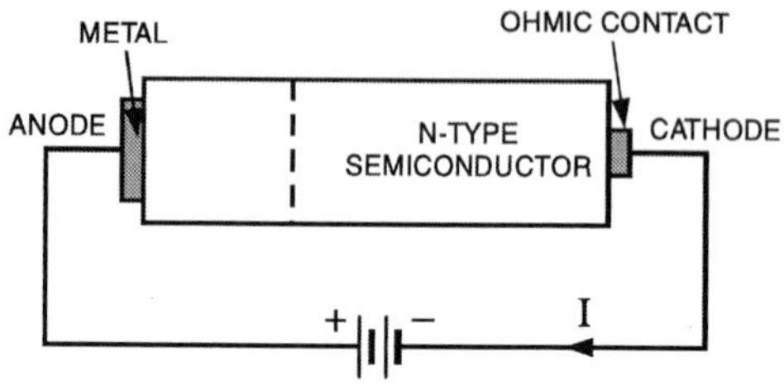

Fig. 1.34. ***Structure of Metal-Semiconductor Junction***

1.23.2. Ohmic Contacts

It has been assumed that the external bias voltage (forward or reverse) appears directly across the junction and has the effect of lowering or raising the electrostatic potential across the junction. In order to justify the above assumption it should be specified how electrical contact is made to the semiconductor from the external bias circuit. In Figs. 1.31 and 1.33 we indicate metal contacts with which the homogeneous P-type and N-type materials are provided. Thus two metal-semiconductor junctions, one at each end of the diode, have been introduced. So a contact potential is expected to be developed across these additional junctions. However, it can be assumed that metal-semiconductor contacts shown in Figs. 1.31 and 1.33 have been manufactured in such a way that they are nonrectifying *i.e.*, the contact potential across these junctions is almost independent of direction and magnitude of current. A contact of this type is referred to as an *ohmic contact.*

An ohmic contact may be defined as a metal-semiconductor contact that has a negligible contact resistance compared with the bulk or spreading resistance of the semiconductor. A satisfactory ohmic contact should not significantly perturb device performance, and it can supply the required current with a voltage drop that is considerably small compared with the drop across the active region of the device. Ideally, the current through the ohmic contact is a linear function of applied voltage, and the applied voltage should be very small.

We are now in a position to justify our assumption that the entire applied voltage appears as a *change* in the *height* of the potential barrier. In as much as the metal-semiconductor contacts are low-resistance ohmic contacts and the voltage drop across bulk of the crystal is negligible, approximately the entire applied voltage will indeed appear as a change in the height of the potential barrier at the P-N junction.

1.23.3. Short-Circuited and Open-Circuited P-N Junction

If the external bias voltage V or V_R in Figs. 1.31 and 1.33 were set equal to zero, the P-N junction would be short circuited. Under these conditions, as explained below, no current can flow *i.e.*, $I = 0$ and electrostatic potential V_0 remains unchanged and equal to the value under open-circuited conditions. If there were a current *i.e.* $I \neq 0$, the metal would become heated. Since there is no external source of energy available, the energy required to heat the metal wire would have to be supplied by the P-N bar. The semiconductor bar, therefore, would have to cool off. Obviously, under thermal equilibrium the simultaneous heating of metal and cooling of semiconductor bar is impossible, and so it is concluded that $I = 0$. Since under short-circuited conditions, the sum of the voltages around the closed loop must be zero the junction potential V_0 must be exactly compensated by the metal-to-semiconductor contact potentials at ohmic contacts. Since the current I is zero, the wire can be cut without changing the situation, and the voltage drop across the cut must remain zero. If in an attempt to measure junction potential V_0 a voltmeter is connected across the cut, the voltmeter will indicate zero *i.e.*, contact difference of potential cannot be measured directly with a voltmeter.

1.23.4. Large Forward Voltages

Suppose forward bias voltage V, in Fig. 1.31, is increased until V approaches junction potential V_0. If V were equal to V_0, the barrier would disappear and the current could be arbitrarily large, exceeding the diode rating. In practice, barrier cannot be reduced to zero because, as the current increases without limit, the bulk resistance of the crystal and the resistance of the ohmic contacts, will limit the current. Thus it is no longer possible to assume that all the voltage V appears as a change across the P-N junction. It is concluded that, as the forward bias voltage V becomes comparable with V_0, the current through a P-N junction diode will be governed by the ohmic-contact resistance and the crystal bulk resistance. Thus the V-I characteristic becomes almost a straight line.

1.23.5. Bulk Resistance

After overcoming the barrier potential (or offset voltage), all that impedes the diode current is the resistance of the P and N regions, symbolized by the r_P and r_N in Fig. 1.35, if ohmic contact resistance is neglected. The sum of these resistances is called the bulk resistance r_B of the diode *i.e.*

$$r_B = r_P + r_N \qquad \text{...(1.60)}$$

The bulk resistance of the diode depends upon the doping level, and the size of the P and N regions. Typically, the bulk resistance of rectifier diode is less than 1 Ω.

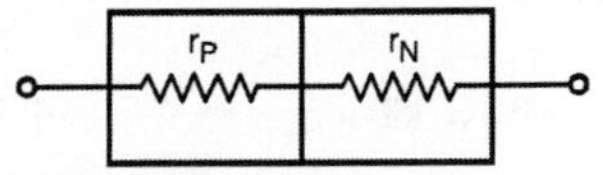

Fig. 1.35. ***Bulk Resistance***

1.23.6. Energy Band Structure of an Open-Circuited P-N Junction

Let us consider that a P-N junction is formed by placing P and N-type materials in intimate contact on an atomic scale. Under such conditions the Fermi level must be constant throughout the specimen at equilibrium. If this were not true, electrons on one side of the junction would have an average energy higher than those on the other side and there would be a transfer of electrons and energy until the Fermi level on the other side equalize. In Art. 1.17 we have seen that Fermi level E_F is closer to the conduction band edge E_{CN} in the N-type material and closer to the valence band edge E_{VP} in the P-type material. Obviously, the conduction band edge E_{CP} in the P-type material cannot be at the same level as E_{CN}, nor can the valence band edge E_{VN} in the N-side line up with E_{VP}. Thus on formation of a P-N junction, conduction band edge E_{CP} in P-type material is higher than E_{CN} in N-type material and valence band edge E_{VP} in P-type material is higher than E_{VN} in N-type material. This is illustrated in Fig. 1.36, where a shift in energy level E_0 is shown.

The total shift in the energy level E_0 is given as

$$E_0 = E_{CP} - E_{CN} + E_{VP} - E_{VN}$$

$$= E_1 + E_2 \qquad \text{...(1.61)}$$

where E_1 and E_2 represent the shifts in the Fermi level from the intrinsic condition in P- and N-type materials. The energy E_0 represents the potential energy of the electrons at the junction, as shown in Fig. 1.28 (*e*). From Fig. 1.36, we have

$$E_F - E_{VP} = \frac{1}{2} E_G - E_1 \qquad \text{...(1.62)}$$

and $$E_{CN} - E_F = \frac{1}{2} E_G - E_2 \qquad \text{...(1.63)}$$

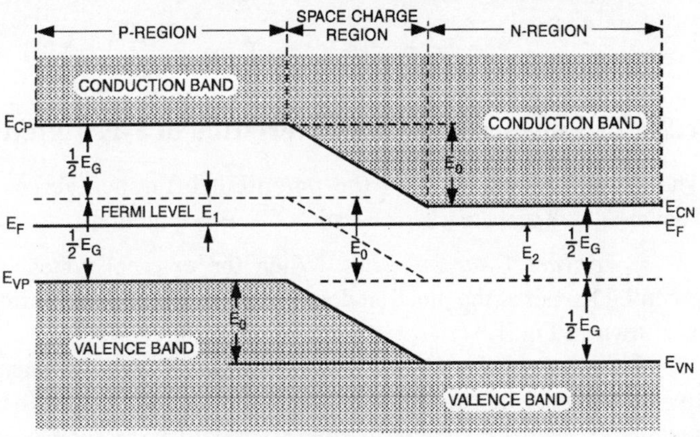

Fig. 1.36. ***Energy Band Diagram For a P-N Junction Under Open-Circuited Conditions***

Adding Eqs. (1.62) and (1.63) we have

$$E_1 + E_2 = E_G - (E_{CN} - E_F) - (E_F - E_{VP})$$

$$\text{or} \quad E_0 = E_G - (E_{CN} - E_F) - (E_F - E_{VP}) \qquad ...(1.64)$$

From Eqs. (1.49) and (1.50) we have

$$n_i^2 = np = n_C n_V \; e^{-[(E_C - E_F) + (E_F - E_V)]/kT} = n_C n_V e^{-(E_C - E_V)/kT} = n_C n_V \; e^{-E_G/kT}$$

or $$E_G = kT \log_e \frac{n_C n_V}{n_i^2} \qquad ...(1.65)$$

From Eqs. (1.52) and (1.53) we have

$$E_{CN} - E_F = kT \log_e \frac{n_C}{N_D} \qquad ...(1.66)$$

and $$E_F - E_{VP} = kT \log_e \frac{n_V}{N_A} \qquad ...(1.67)$$

Substituting values of E_G, $(E_{CN} - E_F)$ and $(E_F - E_{VP})$ from Eqs. (1.65), (1.66) and (1.67) in Eq. (1.64) we have

$$E_0 = kT \log_e \frac{n_C n_V}{n_i^2} - kT \log_e \frac{n_C}{N_D} - kT \log_e \frac{n_V}{N_A} = kT \log_e \left[\frac{n_C n_V}{n_i^2} \cdot \frac{N_D}{n_C} \cdot \frac{N_A}{n_V} \right] = kT \log_e \frac{N_D N_A}{n_i^2} \qquad ...(1.68)$$

It is to be noted that in above equations, the E's are expressed in electron volts and k has the dimensions of electron volts per Kelvin. The contact difference of potential V_0 is expressed in volts and is numerically equal to E_0. It is obvious from Eq. (1.68) that E_0 depends upon the equilibrium concentrations and not upon the charge density in transition region.

Example 1.17. Consider a Ge diode with $N_D = N_A$ and with impurity concentration of $8 \times 10^{14}/cm^3$. Assume $n_i = 2 \times 10^{13}/cm^3$. At room temperature of 300 K calculate the height of the potential barrier under open-circuit conditions. Assume Boltzmann's constant $k = 8.61 \times 10^{-5}$ eV/K and electron charge, $e = 1.6 \times 10^{-19}$ C.

Solution : Impurity concentration, $N_D = N_A = 8 \times 10^{14}/cm^3$

Intrinsic carrier concentration, $n_i = 2 \times 10^{13}/cm^3$

Boltzmann's constant, $k = 8.61 \times 10^{-5}$ eV/K

Temperature, T = 300 K

Potential barrier, $V_0 = kT \log_e \frac{N_D N_A}{n_i^2} = 300 \times 8.61 \times 10^{-5} \log_e \frac{(8 \times 10^{14})^2}{(2 \times 10^{13})^2} = 0.19$ V **Ans.**

1.23.7. Volt-Ampere Characteristics of P-N Junction

The graph plotted between the potential difference across the P-N junction and the circuit current is known as volt-ampere characteristic of a P-N junction.

1. *Forward Characteristics.* When the external voltage is zero *i.e.* when the circuit is open, the potential barrier at the junction does not allow the flow of current and, therefore, the circuit current is zero, as shown in Fig. 1.37.

With forward bias to the P-N junction (*i.e.* P-region connected to +ve terminal and N-region connected to –ve terminal of the battery) very little current, called the *forward current*, flows until the *forward voltage exceeds* the junction barrier potential (0.3 V for Ge and 0.7 V for Si). The characteristics follow an exponential law. As the forward voltage is increased to the knee of the characteristic, the barrier potential is progressively reduced to zero, allowing more and more majority charge carriers to flow across the junction.

Beyond the knee of the characteristic, the potential barrier is completely eliminated, forward current increases almost linearly with the increase in forward voltage and the P-N junction starts behaving as a resistor. If the forward voltage is increased beyond a certain value, extremely large current will flow and the P-N junction may get destroyed due to overheating.

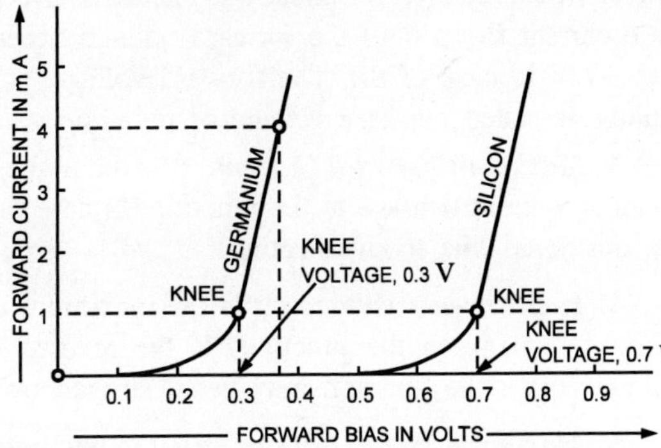

Fig. 1.37. ***Forward Characteristics of P-N Junction***

Because of the nonlinear knee of the curve, if a very small signal voltage is applied to the diode so that it only operates around the knee, the signal will be distorted. The signal must, therefore, be large enough to operate the diode over the linear part of the characteristic. Forward resistance of P-N junction is low (of the order of 75 Ω in case of Ge and 150 Ω in case of Si).

2. *Reverse Characteristics.* When a reverse bias is applied (*i.e.* when P-region is connected to the negative bias terminal and N-region is connected to the positive bias terminal), as shown in Fig. 1.33, the potential barrier at junction is increased, therefore, the junction resistance becomes very high and there is no possibility of majority carriers flowing across a reverse biased junction. But minority carriers generated on each side can still cross the junction. Electrons on the P-side are attracted across the junction to the positive bias terminal on the N-side and holes on the N-side may travel across the junction to the negative bias terminal on the P-side. This results in a very small current, as illustrated in Fig. 1.38. This current is called *reverse current* and is due to minority carriers.

For silicon the reverse current* is typically less than 1 μA and may be as low as 1 *n*A. In case of germanium, the reverse current may exceed 10 μA. Only a very small reverse voltage is necessary to direct all available minority carriers across the junction, and when all minority carriers are flowing across, further increase in bias voltage will not increase the current. This current is referred to as a *reverse saturation current* and is designated as I_0. Reverse resistance is very high (of the order of MΩ).

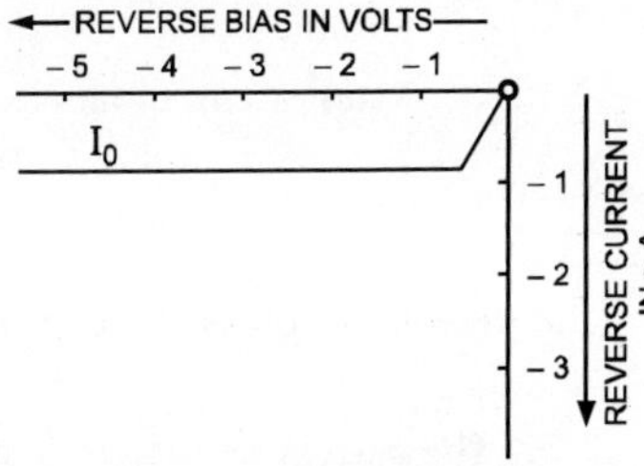

Fig. 1.38. ***Reverse Characteristics of P-N Junction***

As discussed above, the reverse saturation current I_0 is made up of minority charge carriers crossing the junction. When the temperature of the semiconductor material is increased, the additional thermal energy causes more electrons to break away from atoms. This creates more hole-electron pairs and generates more minority charge carriers. So reverse saturation current I_0 increases with the increase in the junction temperature.** It has been found that I_0 approximately doubles for each 10°C rise in temperature in case of germanium and for each 8°C rise in temperature in case of silicon.

1.23.8. Important Terms Often Used in P-N Junction

1. *Breakdown Voltage.* Under normal reverse voltage, a very little reverse current flows through a P-N junction. However, on increasing the reverse voltage, a point may reach at which the junction breaks down with a sudden rise in reverse current. Breakdown voltage is defined as the reverse voltage at which P-N junction breaks down with a sudden rise in reverse current.

* There are more minority charge carriers available in germanium than in silicon, since charge carriers are more easily detached from germanium atoms.

* It can be shown that reverse saturation current I_0 depends upon electronic charge, doping density and junction area, as well as temperature. With the exception of temperature, all other factors (*i.e.* electronic charge, doping density and junction area) are constant for a given junction, I_0 is affected only by variation in temperature.

2. *Knee Voltage.* As already explained above in Art. 1.23.7, with forward bias to P-N junction very little current flows until the forward voltage exceeds the junction barrier potential (0.3 V in case of Ge and 0.7 V in case of Si). The forward voltage, at which the current through the junction starts increasing rapidly, is called the *knee voltage* or the *cutin voltage*.

3. *Maximum Forward Current.* It is the highest instantaneous forward current that a P-N junction can conduct without damage to the junction. In case the forward current exceeds this rating, the junction will get destroyed due to overheating.

4. *Peak Inverse Voltage.* It is the maximum reverse voltage that can be applied to the P-N junction without damage to the junction. If the reverse voltage across the junction exceeds its peak inverse voltage (PIV), the junction may get destroyed owing to excessive heat.

5. *Maximum Power Rating.* It is the maximum power that can be dissipated at the junction without any damage to it.

Note : Every P-N junction has limiting values of maximum forward current, peak inverse voltage and maximum power rating. The P-N junction gives satisfactory performance when operated within these limiting values. The P-N junction may get destroyed due to excessive heat if any of these values is exceeded.

HIGHLIGHTS

1. An electron is a negatively charged particle having negligible mass. Some of the important properties of an electron are

$$\text{Charge on an electron, } e = 1.602 \times 10^{-19} \text{ C}$$
$$\text{Mass of an electron, } m = 9.107 \times 10^{-31} \text{ kg}$$
$$\text{Radius of an electron, } r = 1.9 \times 10^{-15} \text{ m}$$
$$\text{Ratio of e/m of an electron} = 1.759 \times 10^{11} \text{ C/kg}$$
$$\text{Total energy of an electron} = \frac{-e^2}{8\pi\varepsilon_0 r}$$

2. Electron volt is the unit of work or energy and

$$1 \text{ eV} = 1.602 \times 10^{-19} \text{ joule}$$

3. The current density in a conducting medium is given by

$$J = nev = \rho v$$

where $\rho = ne$, the charge density in C/m^3 and v is the velocity in m/s.

4. A semiconductor is a material that has a resistivity somewhere between that of a good conductor and that of an insulator. The resistivity of different semiconductor materials lies in a very wide range from 10^{-4} to about 0.5 Ω-m whereas the resistivities of conductors and insulators are of the order of 10^{-8} Ω-m and 10^{12} Ω-m respectively. Most semiconductor materials used in electronics industry have negative temperature coefficients.

   The semiconductors have another very important property—*the resistivity (or conductivity) of semiconductors changes considerably when even minute amounts of certain other substances, called the impurities, are added to them.* Semiconductor materials are nonlinear resistors.

5. Within any given material there are two distinct energy bands in which electrons may exist. These two energy bands are valence energy band (or valence band) and conduction energy band (or conduction band) and are separated by an energy gap in which no electron can normally exist. This energy gap is termed as forbidden energy gap (or forbidden gap).

   The distinction between conductors, insulators and semiconductors is largely concerned with the relative widths of the forbidden energy gaps (no forbidden gap in case of conductors, about 1 eV in case of semiconductors and more than 5 eV in case of insulators).

6. The two basic charge transport mechanisms are drift, due to an applied electric field, and diffusion, due to a density gradient.

The steady flow of electrons in one direction caused by the applied electric field constitutes an electric current, called the drift current. The process whereby particles flow from a region of high concentration to a region of low concentration is known as diffusion process and the current that results from the diffusion of charged particles is termed as diffusion current.

7. A bonding of atoms, strengthened by sharing of electrons between neighbouring atoms, is called covalent bonding. The covalent bonding arrangement exists in silicon and germanium.

8. Intrinsic materials are those semiconductors that have very low level of impurities, whereas extrinsic materials are semiconductors that have been exposed to a doping process.

Silicon (Si) and germanium (Ge) are the two most widely used semiconductor materials in electronics industry.

At absolute zero temperature E_G for silicon and germanium are 1.21 eV and 0.785 eV respectively.

Hole, strictly speaking, is not a particle; but for all practical purposes it is viewed as a positive charged particle capable of conducting current. The mobility of hole is about half that of an **electron.**

The resistance of a semiconductor depends largely on (*a*) *temperature* — decreases with the increase in temperature, (*b*) *illumination* — decreases in brighter surroundings and (*c*) *magnitude of the electric field* — the current does not increase in proportion to the applied potential, as in case of conductors obeying Ohm's law, but increases by far more than the voltage *i.e., the semiconductors are nonlinear.*

In intrinsic semiconductors, the Fermi level lies between the conduction and valence bands.

Conduction in a pure semiconductor consists of electron flow in the conduction band and hole flow in the valence band.

A germanium device is seldom used at a temperature higher than 75°C, a silicon seldom at higher than 175°C.

9. Depending on the type of impurity added, the extrinsic semiconductors can be divided into two classes viz. (*i*) N-type semiconductors and (*ii*) P-type semiconductors.

An N-type material is formed by adding donor atoms (such as arsenic, antimony, bismuth or phosphorus) that have five valence electrons to establish a high level of relatively free electrons. In an N-type material, the electron is the majority carrier and the hole is the minority carrier.

A P-type material is formed by adding acceptor atoms (such as boron, gallium, indium or aluminium) that have three valence electrons to establish a high level of holes in the material. In a P-type material, the hole is the majority charge carrier and the electron is the minority charge carrier.

10. Under thermal equilibrium the product of concentration of free electrons and concentration of holes is constant and is independent of the amount of doping by donor and acceptor impurities. This is known as **mass action law.** Thus

$$np = n_i^2$$

11. Concentration of holes in N-type semiconductors is given by

$$p = \frac{n_i^2}{n} = \frac{n_i^2}{N_D}$$

where N_D is the concentration of donor atoms.

Similarly in case of P-type semiconductors

$$n = \frac{n_i^2}{N_A}$$

where N_A is the concentration of acceptor atoms.

12. Carrier mobility is the ratio of the average drift velocity and applied electric field *i.e.*

$$\text{Drift velocity, } v = \mu_e E$$

13. If τ is the relaxation or collision time, the average velocity of electrons, known as drift velocity, is given by

$$v = \frac{eE}{m} \times \tau$$

The current density J, defined as the current per unit area, is given as

$$J = env = \frac{ne^2E\tau}{m} = \sigma E$$

Electrical conductivity, $\sigma = \frac{ne^2\tau}{m} = ne\mu_e$

14. Total current density due to drift of electrons and holes on application of electric field is given by

$$J = Ee(n\mu_e + p\mu_h)$$

The current due to diffusion of electrons and holes is given by

$$J_e = eD_e\frac{dn}{dx} \text{ and } J_h = -eD_h\frac{dp}{dx}$$

The diffusion coefficient and mobility are related through the Einstein relationship, given by

$$\frac{D_h}{\mu_h} = \frac{D_e}{\mu_e} = V_T = \frac{T}{11,600}$$

15. Exact position of Fermi level in N-type extrinsic semiconductor is given by

$$N_D = n_C e^{-(E_C - E_F)/kT}$$

or

$$E_F = E_C - kT \log\frac{n_C}{N_D}$$

Similarly in case of P-type semiconductor

$$E_F = E_V + kT \log\frac{n_V}{N_A}$$

16. A space charge region or depletion region is formed on either of the metallurgical junction separating the P and N junctions. This region is essentially depleted of any mobile electrons or holes. A net positive charge density, due to the positively charged donor impurity ions, exists in the N region and a negative charge density, due to the negatively charged acceptor impurity ions, exists in the P region. An electric field exists in the depletion region due to the net space charge density. The direction of the electric field is from the N region to the P region.

A pd exists across the space-charge region. Under zero applied bias, the pd, known as built-in potential barrier, maintains thermal equilibrium and holds back majority carrier electrons in the N-region and majority carrier holes in the P region.

The pd existing across the space charge region is known as the contact potential. The built-in potential barrier or the contact potential for the step junction is given by

$$V_{bi} \text{ or } V_0 = V_T \log_e \frac{N_A N_D}{n_i^2}$$

17. A forward-biased PN junction has the positive terminal applied to the P side and the negative to the N side. The reverse is true for a reverse-biased junction.

When a forward bias voltage is applied across a PN junction, the potential barrier is lowered so that holes from the P-region and electrons from the N region can flow across the junction.

18. An applied reverse bias voltage (N region positive w.r.t. the P region) increases the potential barrier, increases the space-charge width, and increases the magnitude of the electric field.

19. Metal-semiconductor junction can also form ohmic contacts, which are low-resistance junctions providing conduction in both directions with very little voltage drop across the junction.

EXERCISES

1. What is an electron volt ? How is it related with energy unit joule ?
2. Define current density and derive its expression in terms of charge density and velocity of movement of charge.
3. Show the energy level diagrams in insulators, semiconductor and conductor materials.

4. Explain why the temperature coefficient of resistance of a semiconductor is negative ?
   Draw the energy band diagrams of insulator, semiconductor and metal, and explain them.
5. Giving relevant defining equations, explain the phenomenon of drift and diffusion associated with carrier movement in semiconductors.
6. What do you understand by drift current and diffusion current ? Write an equation for the net electron current in a semiconductor. **[G.G.S.I.P. Univ. Analog Electronics December 2010]**
7. Explain the difference between (*i*) drift current and (*ii*) diffusion current.
   [G.G.S.I.P. Univ. Analog Electronics December 2010]
8. Explain the transport phenomenon of mobile charge carriers in a semiconductor.
9. Explain with suitable examples the ionic, covalent and metallic bondings in crystals.
10. Define an intrinsic material, a negative temperature coefficient and covalent bonding. List three materials that have negative temperature coefficient.
11. Is hole a fundamental particle in an atom ? Between electron and hole which one has greater mobility ? Give a reason for the difference.
12. Differentiate among conductor, insulator and semiconductor using energy band concept. Also differentiate between N type and P type semiconductor materials.
13. Draw the atomic structure of a P type material and explain the effect of acceptor impurities on the energy band structure.
14. Distinguish between intrinsic and extrinsic semiconductors. What is the effect of temperature on the conductivity of a semiconductor ?
15. Sketch the energy-band picture for (*i*) an intrinsic (*ii*) an N-type and (*iii*) a P-type semiconductor. Indicate the positions of the Fermi, the donor and the acceptor levels.
16. Distinguish between 'majority' and 'minority' carriers in a semiconductor. Define mobility of charge carriers.
17. Explain what do you mean by donor and acceptor impurities ?
18. Differentiate between
    (*i*) Donor and acceptor impurities.
    (*ii*) Intrinsic and extrinsic impurites.
19. How are the carrier mobilities related with resistivity of a semiconductor ? Does a hole in a semiconductor contribute to a flow of current ? If yes, how and if no, why ?
20. What is the difference between an intrinsic and an extrinsic semiconductor ? Define a hole in a semiconductor.
21. Define :
    (*i*) Donor and acceptor imputies.
    (*ii*) Mobility and conductivity.
22. Describe the difference between majority and minority carriers.
23. Prove that the conductivity of semiconductor is given by $\sigma = q(n\mu_A + p\mu_h)$.
    [G.G.S.I.P. Univ. Analog Electronics December 2008]
24. Explain the current conduction mechanism in a semiconductor. Explain how doping changes the conductivity of a semiconductor ? Show that the Fermi level lies half way between the valence and conduction bands of an intrinsic semiconductor.
25. Draw the energy band diagram of an N-type semiconductor. State the factors on which the position of Fermi level depends.
26. What is P-N junction diode ? How potential barrier is formed in a P-N junction diode ?
27. What is a P-N junction ? What happens when P-N junction is biased in forward direction ?
28. Explain the behaviour of P-N junction at no bias, reverse bias and forward bias. Sketch V-I characteristics of P-N junction diode.

29. Draw the characteristics in forward and reverse bias and explain them.

30. Explain the following terms in context with a semiconductor diode.

(*i*) Potential barrier, (*ii*) Depletion layer, (*iii*) Breakdown voltage.

(*iv*) Peak inverse voltage, (*v*) Knee voltage.

SHORT ANSWER TYPE QUESTIONS WITH ANSWERS

Q. 1. What is Fermi level ?

Ans. The maximum energy that an electron in a metal has at the absolute zero temperature is called the Fermi level of energy and is designated by E_F.

Q. 2. Name two semiconductors except Si and Ge.

Ans. Semiconductor materials, in addition to Si and Ge, are grey (crystalline) tin, selenium (Se), tellurium (Te), boron (B) etc.

Q. 3. What is the basis for classifying a material as a conductor, semiconductor or a dielectric ? What is the conductivity of perfect dielectric ?

Ans. Conductors possess high conductivity whereas the characteristic property of insulating materials (or dielectrics) is poor conductivity. Semiconductors occupy an intermediate position between conductors and insulators. Though there is no rigid line separating the conductors from semiconductors and semiconductors from insulators, but still according to resistivity the materials of resistivity of the order from 10^{-8} to 10^{-3}, 10^{-3} to 10^{6} and 10^{6} to 10^{18} ohm-metres may be classified as conductors, semiconductors and dielectrics respectively.

Another classification is based on temperature coefficient of resistivity. Metals have positive temperature coefficient of resistivity; semiconductors have small negative temperature coefficient of resistivity and insulators have large negative temperature coefficient of resistivity.

Q. 4. Differentiate semiconductors, conductors and insulators on the basis of band gap.

Ans. The distinction between conductors, insulators and semiconductors is largely concerned with the relative width of the forbidden energy gaps in their energy band structures. As illustrated in Figs. 1.5 (*a*), (*b*) and (*c*), there is a wide forbidden gap (more than 5 eV) for insulators, narrow forbidden gap (about 1 eV) in case of semiconductors and no forbidden gap in case of conductors. For details refer to Art. 1.7.

Q. 5. What is the importance of valence shell and valence electrons ?

Ans. The outermost shell of an atom is called the valence shell and electrons in this shell are called the valence electrons. Formation of energy bands occur owing to overlapping of energy levels of these valence electrons in valence shells. With the decrease in interatomic distance between the atoms in a crystal, the energy levels of electrons in outermost shells of atoms overlap to form energy bands.

Q. 6. What is the forbidden energy gap ? What is its magnitude for Ge and Si ?

Ans. The energy gap between the valence band and conduction band is known as forbidden energy gap. It is a region in which no electron can stay as there is no allowed energy state. Magnitude of forbidden energy gap in germanium and silicon is 0.72 eV and 1.12 eV respectively at 300 K and 0.785 eV and 1.21 eV respectively at absolute zero temperature.

Q. 7. Is hole a fundamental particle in an atom ?

Ans. Hole is not a fundamental particle in an atom. Holes may be thought of as positive particles, and as such, they move through an electric field in a direction opposite to that of electrons.

Q. 8. Define a hole in a semiconductor.

Ans. When an energy is supplied to a semiconductor, a valence electron is lifted to a higher energy level, the departing electron leaves a vacancy in the valence band. This vacancy is called a hole. Thus, a vacancy left in the valence band because of lifting of an electron from valence band to conduction band is known as hole.

Q. 9. What is hole current ?

Ans. The movement of the hole (positively charged vacancy in the valence band) from positive terminal of the supply to the negative terminal through semiconductor constitutes hole current.

Q. 10. What is intrinsic semiconductor ?

Ans. An intrinsic semiconductor is one which is made of the semiconductor material in the extremely pure form (impurity content not exceeding one part in 100 million parts of semiconductors).

Q. 11. Why silicon and germanium are the two widely used semiconductor materials ?

Ans. Because the energy required to release an electron from their valence band (*i.e.*, to break their covalent bonds) is very small (1.12 eV for Si and 0.72 eV for Ge).

Q. 12. Which of the two semiconductor materials Si or Ge has larger conductivity at room temperature ? Why ?

Ans. Since energy required in transferring electrons from valence band to conduction band is more in case of silicon than that in case of germanium, the conductivity of Ge will be more than that of Si at room temperature.

Q. 13. Why does a pure semiconductor behave like an insulator at absolute zero temperature ?

Ans. For a pure semiconductor at a temperature of absolute zero (– 273.15°C) the valence band is usually full and there may be no electron in the conduction band and it is difficult to provide additional energy required for lifting electron from valence band to conduction band by applying electric field. Hence, the conductivity for a pure semiconductor at absolute zero temperature is zero and it behaves like an insulator.

Q. 14. What is the main factor for controlling the thermal generation and recombination ?

Ans. Temperature, because with the increase in temperature, concentrations of free electrons and holes increase and the rate of recombination is proportional to the product of concentrations of free electrons and holes and also the rate of production of electron-hole pairs (thermal generation) increases with the rise in temperature.

Q. 15. Define mean lifetime of a carrier.

Ans. The amount of time between the creation and disappearance of a free electron is called the lifetime. It varies from a few nanoseconds to several microseconds depending how perfect the crystal is and other factors.

Q. 16. In which bands do the movement of electrons and holes take place ?

Ans. Free electrons move in conduction band while holes in valence band.

Q. 17. What is the mechanism by which conduction takes place inside the semiconductor.

Ans. Conduction occurs in any given material when an applied electric field causes electrons to move in a desired direction within the material. This may be due to one or both of two processes, electron motion and hole transfer. In case of former process, free electrons in the conduction band move under the influence of the applied electric field. Hole transfer involves electrons which are still attached to the atoms *i.e.* those in valence band. For details refer to Art. 1.8.

Q. 18. What do you mean by mobility of a carrier ? How does it depend on temperature and doping concentrations ?

Ans. Mobility of charge carriers is essentially a measure of the ease with which the carriers can move under the influence of an electric field. The drift velocity per unit electric field is known as mobility of charge carriers.

The mobility μ of charge carriers varies as T^{-m} over a temperature range of 100 K and 400 K. For silicon m = 2.5 for electrons and 2.7 for holes. For germanium m = 1.66 for electrons and 2.33 for holes.

With the increase in cocentration of dopant, the carrier mobility at a given temperature increases.

Q. 19. What do you mean by drift velocity and mobility of a free electron ?

Ans. The average velocity of a free electron is known as drift velocity whereas mobility of an electron is defined as the drift velocity per unit electric field.

Q. 20. Define mobility of a carrier. Show that the mobility constant of electron is larger than that of a hole.

Ans. Mobility is defined as the average particle drift velocity per unit electric field.

The mobility of electrons is more than that of holes because the probability of an electron having the energy required to move to an empty state in the conduction band is much greater than the probability of an electron having the energy required to move to the empty state in valence band. The mobility of electron is about double that of an hole.

Q. 21. Define diffusion current in a semiconductor.

Ans. The diffusion of charge carriers is as a result of a gradient of carrier concentration (*i.e.*, the difference of carrier concentration from one region to another). In this case concentration of charge carriers (either electrons or holes) tends to distribute them uniformly throughout the semiconductor crystal. This movement continues until all the carriers are evenly distributed throughout the material. This type of movement of charge carriers is called the diffusion current.

Q. 22. What happens to the conductivity of the semiconductor with the rise in temperature? Compare with the conductivity of metals.

Ans. With the increase in temperature, the concentration of charge carriers increases resulting in increase of conductivity of semiconductors. The conductivity of metal decreases with the increase in temperature.

Q. 23. Why the temperature coefficient of resistance of a semiconductor is negative?

Ans. With the increase in temperature, the concentration of charge carriers (electrons and holes) increases. As more charge carriers are made available, the conductivity of a pure semiconductor increases *i.e.* resistivity of a pure semiconductor decreases with the rise in temperature *i.e.* semiconductors have negative temperature coefficient of resistance.

Q. 24. What is meant by Fermi level in semiconductor ? Where does the Fermi level lie in an intrinsic semiconductor ?

Ans. Fermi level in a semiconductor can be defined as the maximum energy that an electron in a semiconductor have at absolute zero temperature.

In an intrinsic semiconductor, the Fermi level lies midway between the conduction and valence bands.

Q. 25. Can the Fermi level be pushed into the conduction band ? If yes, explain how.

Ans. If a diode is heavily doped, then Fermi level can be pushed inside the conduction band. Fermi level for N-type extrinsic semiconductor is given by an expression

$$E_F = E_C - kT \log_e \frac{n_C}{N_D}$$

Since $n_C \simeq 10^{19}/\text{cm}^3$, then for donor concentration in excess of this amount ($N_D > 10^{19}/\text{cm}^3$), $\log_e \frac{n_C}{N_D}$ is negative. Thus $E_F > E_C$ and the Fermi level in the N-type material will be pushed into the conduction band.

Q. 26. Differentiate between intrinsic semiconductors and extrinsic semiconductors ? Give examples.

Ans. An intrinsic semiconductor is one which is made of the semiconductor material in its extremely pure form.

When a small amount of impurity is added to a pure semiconductor crystal during the crystal growth in order to increase its conductivity, the resulting crystal is called the extrinsic semiconductor.

Q. 27. Describe the difference between P-type and N-type semiconductor materials.

Ans. When a small amount of trivalent impurity (such as boron, gallium, indium or aluminium) is added to a pure semiconductor crystal during crystal growth, the resulting crystal is called the P-type semiconductor.

When a small amount of pentavalent impurity (such as arsenic, antimony, bismuth or phosphorus) is added to a pure semiconductor crystal during crystal growth, the resulting crystal is called the N-type semiconductor.

Q. 28. What do you mean by donor and acceptor impurities?

Ans. Donor impurities (such as arsenic, antimony, bismuth or phosphorus), when added to a pure semiconductor lattice, form N-type extrinsic semiconductor. The pentavalent impurities are called the donor type impurities as such impurities donate electrons to the lattice.

Acceptor impurities (such as boron, gallium, indium or aluminium), when added to a pure semiconductor lattice, form P-type extrinsic semiconductor. The trivalent impurities are called the acceptor type impurities because such impurities accept electrons from the lattice.

Q. 29. Explain the term doping and its need.

Ans. The electrical conductivity of intrinsic semiconductor, which has little current conduction capability at ordinary room temperature and so is of little use, can be increased many times by adding very small amount of impurity (of the order of one atom per million atoms of pure semiconductor) to it in the process of crystallization. This process is called the doping.

Q. 30. "An N-type semiconductor has more number of electrons than holes, hence it has negative charge". Justify or nullify this statement.

Ans. Although N-type semiconductor has excess of electrons but it is electrically neutral. This is due to the fact that electrons are created by the addition of pentavalent impurity atoms to the pure semiconductor *i.e.* there is no addition of either negative or positive charge.

Q. 31. What is the effect of temperature on extrinsic semiconductors ?

Ans. With the increase in temperature of an extrinsic semiconductor, the number of thermally generated carriers is increased resulting in increase of concentration of minority carriers. At temperature exceeding critical temperature the extrinsic semiconductor behaves like an intrinsic semiconductor but with higher conductivity.

Q. 32. What are the charge carriers in N-type and P-type semiconductors ?

Ans. Free electrons in N-type semiconductors and holes in P-type semiconductors are the charge carriers.

Q. 33. For the same order of doping, why does N-type semiconductor exhibit larger conductivity than P-type semiconductor ?

Ans. Since the mobility of electrons is greater than that of holes, for same level of doping, N-type semiconductor exhibits larger conductivity.

Q. 34. State law of mass action.

Ans. Under thermal equilibrium conditions, the product of concentration n of free electrons and concentration p of holes is constant and is independent of the amount of doping by donor and acceptor impurities. *i.e.* $n\,p = n_i^2$. This is known as mass action law.

Q. 35. Define the term intrinsic concentration.

Ans. Intrinsic concentration of holes or free electrons is called the intrinsic concentration, n_i, which varies with temperature according to the relation $n_i = A_0\, T^3\, e^{-E_{G_0}/kT}$.

where A_0 is a constant and independent of temperature, k is a Boltzmann constant in eV per K and E_{Go} is the energy gap in eV at absolute zero temperature, T is the temperature in degrees absolute and n_i is the concentration in number/m^3. A_0 is taken as 2.33×10^{43}.

Q. 36. Define the volt equivalent of temperature. What is its magnitude at room temperature ?

[G.G.S.I.P. Univ. December 2009]

Ans. Volt-equivalent of temperature, V_T is defined as

$$V_T = \frac{k'T}{e} = kT = \frac{T}{11{,}600}$$

where k' is Boltzmann constant in Joules/K, k is Boltzmann constant in eV/K and T is absolute temperatrue.

At room temperature, $V_T = \dfrac{300}{11{,}600} = 0.026$ V or 26 mV **Ans.**

Q. 37. Define diffusion constant of an electron and hole.

Ans. Diffusion constant of any charge carrier (electron or hole) may be defind from the Einstein relationship

$$\frac{D_h}{\mu_h} = \frac{D_e}{\mu_e} = V_T$$

where D_h and D_e are diffusion constants, μ_h and μ_e are mobilities and V_T is the volt-equivalent of temperature.

Q. 38. What is a P-N junction ?

Ans. The contact surface between the layers of P-type and N-type semiconductor pieces placed together so as to form a P-N junction is called the P-N junction.

Q. 39. What is potential barrier in a diode ? How it gets established ?

Ans. The free electrons crossing the P-N junction create negative ions on the P-side by giving some atoms one more electron than their total number of protons. The electrons also leave positive ions behind them on the N-side. As negative ions are created on the P-side of the junction, the P-side acquires a negative potential. Similarly, the positive ions are created on N-side and the N-side acquires a positive potential. The negative potential on the P-side prevents the migration of any more electrons from the N-type material to the P-type material. Similarly, the positive potential on the N-side prevents any further migration of holes across the boundary. Thus, the initial diffusion of charge carriers results in a barrier potential at the junction.

Q. 40. How do the transition region width and contact pontential across a P-N junction vary with the applied bias voltage ?

Ans. When the P-N junction is forward biased, the transition region width is reduced and the contact potential is also reduced with the increase in applied bias voltage.

When the P-N junction is reverse biased, the transition region is widened, and the contact potential is increased with the increase in applied bias voltage.

Q. 41. Which type of charges are present on the two opposite faces of the junction ?

Ans. Positive charge on N-side and negative charge on P-side of the junction.

Q. 42. What types of carriers are present in space charge region ?

Ans. No mobile carrier is present in the space charge region.

Q. 43. Why is space region called the depletion region ?

Ans. The region around the junction is completely ionised on formation of P-N junction. As a result, there are neither free electrons on the N-side nor the holes on the P-side. Since the region around the junction is depleted of mobile charges, it is called the depletion region.

Q. 44. Why an electric field is produced in a depletion region of a P-N junction ?

Ans. The separation of positive and negative space charge densities in a P-N junction results in an electric field.

Q. 45. What is space charge width ?

Ans. The space charge region extends into the N and P-regions from the metallurgical junction. The distance is known as the space charge width.

Q. 46. The electric field in the space charge region decreases with forward bias and increases with reverse bias. Why ?

Ans. Because applied electric field opposes built-in field.

Q. 47. Define cutin voltage of a P-N junction diode ?

Ans. The forward voltage, at which the current through the P-N junction starts increasing rapidly, is called the cut-in voltage.

Q. 48. What do you understand by reverse saturation current of a diode ?

Ans. Reverse saturation current of a diode is due to minority carriers and is caused when the diode is reverse biased. Only a very small voltage is required to direct all available minority carriers across the junction, and when all minority carriers are flowing across, further increase in bias voltage will not cause increase in current. This current is referred to as a reverse saturation current.

Q. 49. Why is silicon preferred to germanium in the manufacturing of semiconductor devices ?

Ans. Silicon is preferred to germanium in the manufacturing of semiconductor devices because such devices have higher peak inverse voltage and current ratings and wider temperature range than germanium one.

Q. 50. Define peak inverse voltage ?

Ans. Peak inverse voltage is the maximum voltage that can be applied to the P-N junction without damage to the junction. If the reverse voltage across the junction exceeds its peak inverse voltage (PIV), the junction may get destroyed owing to excessive heat.

Q. 51. Define the limitations in the operating conditions of P-N junction.

Ans. Every P-N junction has limiting values of (*i*) maximum forward current (*ii*) peak inverse voltage (PIV) and (*iii*) maximum power rating. The P-N junction provides satisfactory performance when operated within these limiting values. The P-N junction diode may get destroyed due to excessive heat if any of these values is exceeded.

PROBLEMS

1. For silicon the intrinsic concentration is approximately 10^{16} carriers/m^3. If an impurity concentration of 10^{22} donor atoms/m^3 is doped, determine the electron and hole concentration. **[Ans.** 10^{22}; 10^{10}]
2. A silicon wafer (intrinsic carrier concentration 1×10^{16} m^{-3}) is doped with 2×10^{22} aluminium atoms/m^3 and 1×10^{22} arsenic atoms/m^3. Determine the minority carrier concentration. Assume complete dopant ionization. **[Ans.** 1×10^{10}/m^3]
3. The electron and hole concentration in a P-type semiconductor is 10^{12}/m^3 and 10^{22}/m^3 respectively. If the mobility of electrons and holes are 0.04 and 0.06 m^2/V-s respectively, determine the conductivity of the material. **[Ans.** 96 (ohm-metre)$^{-1}$]
4. Determine the number of electrons per m^3 if the resistivity of the conductor material is 0.01531 $\times 10^{-6}$ Ω-m and mobility is 7.04×10^{-3} m^2/V-s. Also determine the relaxation time. Given that $e = 1.6 \times 10^{-19}$ C and $m = 9.1 \times 10^{-31}$ kg. **[Ans.** 5.8×10^{28}; 4.004×10^{-14} second]
5. Find the concentration of holes and electrons in an N-type silicon at 300 K if the conductivity is 0.1 (Ω-cm)$^{-1}$. Given that n_i at 300 K for silicon is 1.5×10^{10}/cm^3. μ_e at 300 K for silicon is 1,300 cm^2/V-s. **[Ans.** $n = 4.8 \times 10^{14}$ electrons/cm^3, $p = 4.688 \times 10^5$ holes/cm^3]
6. A sample of pure germanium at room temperature has a density of charge carriers of 2.5×10^{19}/m^3. It is doped with donor impurity atoms at the rate of one impurity atom for every 10^6 atoms of germanium. The density of germanium atoms is 4.2×10^{28} atoms/m^3. Determine hole density. **[Ans.** 1.42×10^{16}/m^3]
7. The intrinsic resistivity of silicon at 27°C is 2.8×10^3 Ω-m. The electron and hole mobilities are 0.38 and 0.18 m^2/V-s respectively. Calculate intrinsic carrier density at the given temperature. **[Ans.** 3.986×10^{15}/m^3]

□□□

CHAPTER 2

Semiconductor Diodes

INSIDE THIS CHAPTER

2.1. INTRODUCTION

One of the first uses of semiconductor materials was to join a P-type material with an N-type to form a simple but important semiconductor device called a *diode.*

The term *diode* denotes a two-electrode device. A *semiconductor diode* is simply a P-N junction with connecting leads or terminals on the two sides of the P-N junction. P-N junction is also called the crystal diode because the P-N junction is grown out of crystal. A diode is a unidirectional device permitting the easy flow of current in one direction but restraining the flow in opposite direction. A major application of diodes is in *rectification i.e.* in conversion of ac into dc. Semiconductor diode is becoming more and more popular these days due to its smaller size, cheapness, robustness and higher operating efficiency.

The P-type material of the diode is the positive electrode and is called the *anode* whereas the N-type material is the negative electrode and is called the *cathode.* The arrow symbol used for the anode represents the *direction of flow of conventional current.* The *electron current* flows in a direction *opposite* to the arrow.

The symbol for semiconductor diode is shown in Fig. 2.1. The Diodes have *polarity* and, therefore, must be connected in circuits in the correct direction. Small diodes are marked with a band to indicate the cathode end. Larger diodes use the diode symbol to indicate their polarity. Diodes are available in many package configurations, depending on their ratings. Small diodes, used for low current application in circuits up to 1 A of average current, are generally packaged in a simple glass or plastic tube. Diodes for handling larger currents are packaged in metal containers and are often mounted by studs on metal plates to dissipate heat.

Diodes are rated in terms of their peak and average forward current-carrying capability and in terms of their reverse voltage or peak inverse voltage ratings. None of these values can be exceeded, otherwise the diode will get damaged. For the P-N diode junction, the reverse voltage rating applies to the instantaneous peak reverse voltage across the diode.

The diode junction can be viewed as an automatic switch that is closed when the P-type material is relatively positive and open when the P-type material is relatively negative. Remember, it is the *voltage across the diode* that is important *not* the voltage to another part of the circuit. Because the depletion region is never completely absent, it takes a small voltage across a diode to make current flow, even when the polarity is correct for flow of larger current.

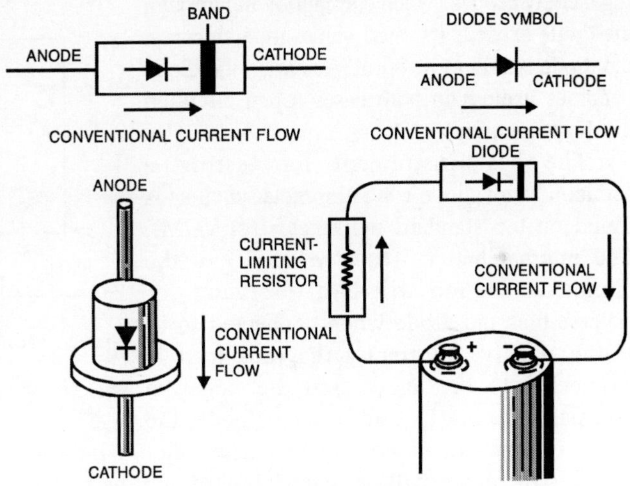

Fig. 2.1. ***Semiconductor Diode***

Most diodes in use today make use of silicon semiconductor material; however, for some applications, germanium is also used.

The silicon junction diode is similar in appearance to the germanium diode but differs in properties. Silicon diodes have, in general, higher peak inverse voltage (PIV) and current ratings and wider temperature ranges than germanium diode.

(*i*) The peak inverse voltage ratings for the silicon can be as high as 1,000 V whereas for germanium it is near about 400 V.

(*ii*) Silicon can be operated at temperatures up to about 200°C compared with about 75–80°C for germanium.

(*iii*) The forward voltage drop of silicon diode is approximately double of that of the corresponding germanium diode. It is typically of the order of 0.7 V for commercially available silicon diodes and 0.3 V for germanium diodes.

(*iv*) The reverse current at a given temperature and voltage is approximately a hundredth of that of the corresponding germanium diode, but there is little sign of current saturation as in case of germanium. In fact, the reverse current of a silicon diode is approximately proportional to the square root of the voltage until breakdown is approached.

(*v*) The reverse current of a silicon diode, for a given voltage, practically doubles for every 8°C rise of temperature, compared with 10°C for germanium.

2.2. DIODE FAILURE MODES AND OHMMETER CHECK

We would certainly like for solid-state components to be 100% reliable but this is unfortunately not true. All solid-state devices have a certain probability of failure. To be good circuit designers and/or good circuit trouble shooters, we should have some idea of the most likely failure modes for solid-state devices. This knowledge can help us anticipate failures in new circuit designs, and locate failures in a malfunctioning electronic circuit.

A diode or P-N junction may get open circuited or short circuited. The probability of the diodes becoming open circuited when they fail is not as great. In normal service usually, diode is short circuited.*

* As a rule of thumb, small-signal, rectifiers, Schottky, and varactor diodes will become shorts, or exhibit excessive leakage current, 75% of the time when they fail, zener and avalanche diodes will become short circuited 60% of the time when they fail.

It generally happens when current flowing through the diode exceeds its rated value. In such cases, the diode is said to be burnt. Any way, the diode becomes useless in both cases (open circuited or short circuited).

The best instrument for testing a semiconductor diode is an ohmmeter such as is found on the standard multimeter (or VOM). The internal battery (usually 1.5 V) of the ohmmeter section will either forward- or reverse-bias the diode when connected to it. If the positive (normally the red) lead is connected to the anode and the negative (normally the black) lead to the cathode, the diode is forward biased and the meter should indicate a low value of resistance. With the reverse polarity, the diode will be reverse biased and the meter should indicate very large value of resistance. A large forward-bias resistance means an open circuit, while small reverse-bias resistance means a short circuit condition. The basic connections for the tests are shown in Fig. 2.2.

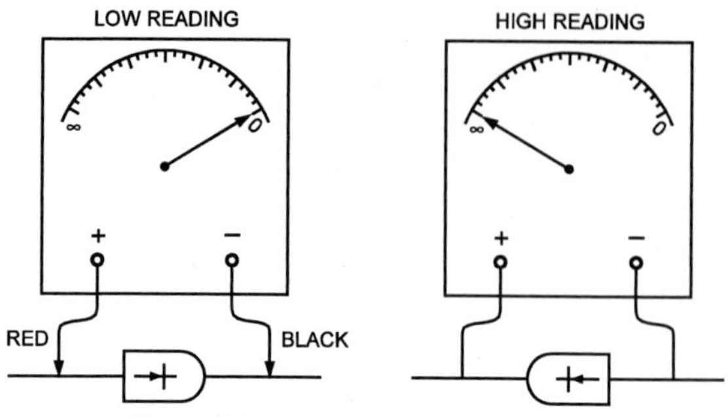

(a) Forward Bias *(b) Reverse Bias*

Fig. 2.2. ***Ohmmeter Testing of a Semiconductor Diode***

Precautions To Be Taken During Testing of Diodes 1. An ohmmeter scale that has a high internal current should not be used because high current will damage the diode under test.

2. One side of the diode should be disconnected from the circuit to eliminate any parallel resistance path.

Tunnel diodes should not be checked with an ohmmeter.

2.3. IDEAL DIODE

It is a two terminal device and permits only *unidirectional conduction.* It conducts well in the forward direction and poorly in the reverse direction. It would have been ideal if it acted as a perfect conductor (with zero resistance or zero voltage drop across it) when forward biased and as a perfect insulator (with an infinite resistance or with no current through it) when reverse biased. The volt-ampere characteristics of such an ideal diode have been shown in Fig. 2.3 (*b*). An ideal diode acts like an *automatic switch.* The switch is closed when the diode is forward biased and is opened when reverse biased.

No diode can act as an ideal diode. An actual diode does not behave as a perfect conductor when forward biased and as a perfect insulator when reverse biased. Neither it offers zero resistance when forward biased nor infinite

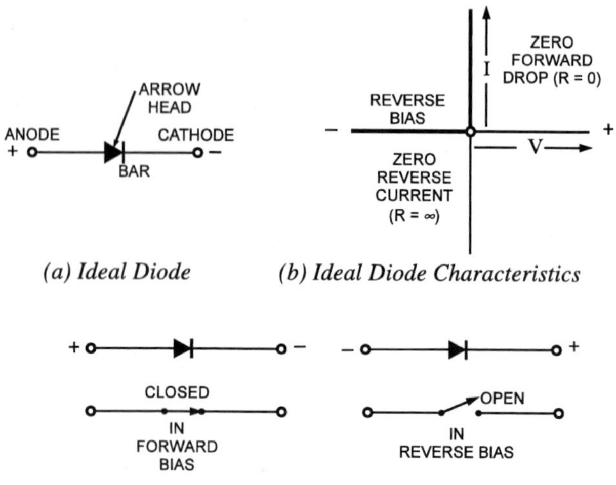

(a) Ideal Diode *(b) Ideal Diode Characteristics*

(c) Switch Analogy

Fig. 2.3

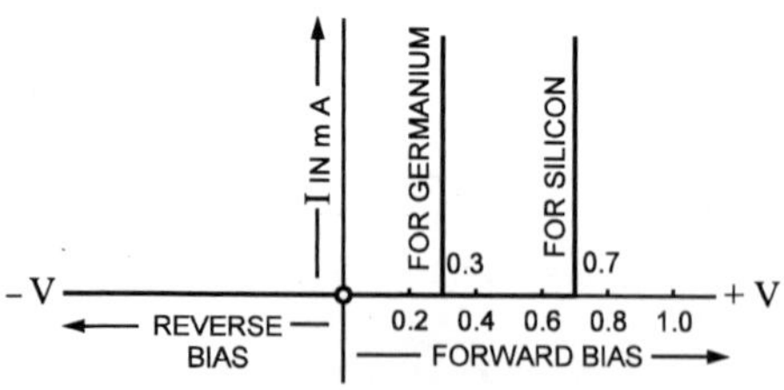

Fig. 2.4. ***Approximate Characteristics For Ge and Si Diodes***

resistance when reverse biased. However, there are many applications in which diodes can be assumed to be nearly ideal devices, if the voltage drop across the diode when it is forward biased *i.e.*, V_D is taken into account. In cases when the circuit supply voltage V_S is much larger than the forward voltage drop V_D, V_D can be assumed constant without introducing any serious error. Also, the diode forward current I is usually so much larger than the reverse saturation current I_0 that the I_0 can just be ignored. These assumptions lead to a nearly ideal, or approximate, characteristics for germanium and silicon diodes, as illustrated in Fig. 2.4.

2.4. CURRENT COMPONENTS IN A P-N DIODE

A P-N junction with a forward bias with an external voltage V is shown in Fig. 2.5 (*a*). Due to forward bias, there exists a potential gradient in P and N materials. Now the holes from P-region and electrons from N-region are injected into the N-region and P-region respectively. The holes injected into the N-region and electrons injected into the P-region are the minority carriers. The minority carriers diffuse away from the junction exponentially with the distance, as illustrated in Fig. 2.5 (*b*). Their concentration falls steadily due to recombination with electrons and holes respectively. Since the diffusion current due to minority carriers is proportional to the concentration gradient [refer to Eqs. (1.42) and (1.43)], this current must also vary exponentially with distance. There are two minority (diffusion) currents, I_{ep} and I_{hn} as shown in Fig. 2.5 (*b*). Let the symbols $I_{ep}(x)$ and $I_{hn}(x)$ represent the electron current in P-region and hole current in N-region respectively.

Electrons crossing the junction at $x = 0$ from right to left constitute a current in the same direction as the holes crossing the junction from left to right. Thus the total current I at junction (*i.e.*, $x = 0$) is given as

$$I = I_{hn}(0) + I_{ep}(0) \qquad ...(2.1)$$

Since the current is the same throughout a series circuit, total current I is independent of x, and is shown as a horizontal line in Fig. 2.5 (*b*). So, in the P-region there must be a second component of current I_{hp} which combining with I_{ep} gives the total current I. So the majority carrier (or hole) current, I_{hp} in the P-region is given as

$$I_{hp}(x) = I - I_{ep}(x) \qquad ...(2.2)$$

Similarly the majority (or electron) current, I_{en} in the N-region is given as

$$I_{en}(x) = I - I_{hn}(x) \qquad ...(2.3)$$

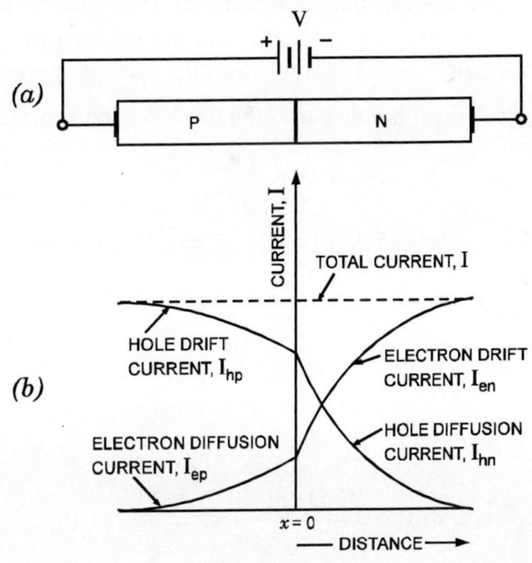

Fig. 2.5. ***Current Components in a Forward Biased P-N Junction Diode***

The currents I_{hp} and I_{en} are plotted as a function of distance in Fig. 2.5 (*b*). The Fig. 2.5 (*b*) is drawn for unsymmetrically doped diode, so that $I_{ep} \neq I_{hn}$.

It may be noted that far away from the junction in P-region the current is drift current I_{hp} of holes sustained by small electric field in the semiconductor. As the holes approach the junction, some of them recombine with the electrons, which have been injected into the P-region from the N-region. Thus part of the current I_{hp} becomes a negative current just equal in magnitude to the diffusion current I_{ep}. The drift current I_{hp}, therefore, decreases toward the junction. The rate of decrease is such that the total current remains constant, independent of distance. The reduced hole current at the junction now enters the N-region and becomes hole diffusion current I_{hn}. Similar remarks can be made with respect to electron drift current in N-region, I_{en}. Thus in a forward-biased P-N diode, the current enters the P-region as a hole current and leaves the N-region as an electron current of the same magnitude.

From the above discussion it is concluded that the current in a P-N diode is bipolar in character (because it is made up of both positive and negative charge carriers) and the total current is constant throughout the device, but the proportion of current due to holes and that due to electrons varies with the distance, as illustrated in Fig. 2.5 (*b*).

2.5. QUANTITATIVE THEORY OF P-N DIODE CURRENTS

Here an expression for the total current will be derived as a function of the applied voltage with the assumption that depletion layer thickness is negligible (*i.e.*, *barrier* width is zero).

When a P-N diode is forward biased, holes are injected from P-region into the N-region. The concentration p_n of the holes in the N-region is increased above its thermal equilibrium value p_{no}. The hole concentration in N-region is given as

$$p_n(x) = p_{no} + p'_n(0)\, e^{-x/L_h} \qquad \text{...(2.4)}$$

where p_{no} is the hole concentration in thermal-equilibrium condition, L_h is the diffusion length for holes in N-type material and x is the distance from the junction where concentration is considered.

The injected, or excess, concentration of holes at $x = (0)$, $p'_n(0)$ is given as

$$p'_n(0) = p_n(0) - p_{no} \qquad \text{...(2.5)}$$

These several hole concentration components are shown in Fig. 2.6, which shows that the concentration $p_n(x)$ falls exponentially with distance x into the N-type material.

The diffusion hole current in the N-region is given as

$$I_{hn} = -AeD_h \frac{dp_n}{dx} \qquad \text{...(2.6)}$$

Taking the derivative of Eq. (2.4) and substituting the value of $\frac{dp_n}{dx}$ in above Eq. (2.6) we have

$$I_{hn}(x) = \frac{AeD_h\, p'_n(0)}{L_h} e^{-x/L_h} \qquad \text{...(2.7)}$$

At junction *i.e.*, $x = 0$

$$I_{hn}(0) = \frac{AeD_h\, p'_n(0)}{L_h} = \frac{AeD_h}{L_h}[p_n(0) - p_{no}] \qquad \text{...(2.8)}$$

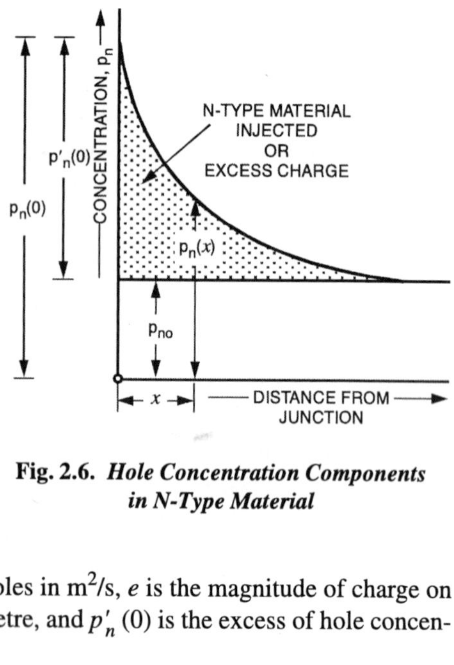

Fig. 2.6. ***Hole Concentration Components in N-Type Material***

where A is area of material in m^2, D_h is diffusion constant for holes in m^2/s, e is the magnitude of charge on holes, L_h is the diffusion length of holes in N-type material in metre, and $p'_n(0)$ is the excess of hole concentration at junction.

Using Boltzmann relationship of kinetic theory of gases, we have

$$p_n(0) = p_{no}\, e^{V/V_T} \qquad \text{...(2.9)}$$

where V is the voltage applied across the P-N diode and V_T is the volt-equivalent of temperature and from Eq. (1.47) it is given as

$$V_T = \frac{k'T}{e} = \frac{T}{11,600}$$

where k′ is the Boltzmann constant.

The Eq. (2.9) is known as the *law of junction.*

2.5.1. Diffusion Length

Diffusion length is defined as the distance travelled by free charge carriers (electrons or holes) before recombination. It may also be defined as the average distance covered by an excess charge carrier during its lifetime τ.

The diffusion length L_h and lifetime τ_h are related to each other as :

$$L_h = \sqrt{D_h \tau_h} \quad ...(2.10)$$

Similarly in case of P type semiconductor

$$L_e = \sqrt{D_e \tau_e} \quad ...(2.11)$$

2.5.2. Forward Currents

The total diode current I at $x = 0$ is given as

$$I = I_{hn}(0) + I_{ep}(0) \quad ...(2.12)$$ Refer to Eq. (2.1)

where $I_{hn}(0)$ is the current caused by holes entering the N-region and $I_{ep}(0)$ is the current caused by electrons entering the P-region.

From Eqs. (2.8) and (2.9)

$$I_{hn}(0) = \frac{AeD_h}{L_h}[p_{no}\, e^{V/V_T} - p_{no}] = \frac{AeD_h p_{no}}{L_h}[e^{V/V_T} - 1] \quad ...(2.13)$$

$$\text{Similarly } I_{ep}(0) = \frac{AeD_e n_{po}}{L_e}[e^{V/V_T} - 1] \quad ...(2.14)$$

From Eqs. (2.12), (2.13) and (2.14) we have total diode current

$$I = \left[\frac{AeD_h p_{no}}{L_h} + \frac{AeD_e n_{po}}{L_e}\right](e^{V/V_T} - 1)$$

$$\text{or } I = I_0(e^{V/V_T} - 1) \quad ...(2.15)$$

$$\text{where } I_0 = \frac{AeD_h p_{no}}{L_h} + \frac{AeD_e n_{po}}{L_e} \quad ...(2.16)$$

2.5.3. Reverse Saturation Current

In the foregoing discussion a positive value of V indicates a forward bias. Equation (2.15) is equally valid for reverse bias [*i.e.* for negative values of applied voltage V]. For a reverse bias voltage exceeding V_T (*i.e.* 26 mV) at room temperature (27 °C or 300 K), current $I \to -I_0$. Thus I_0 is called the *reverse saturation current.*

From Eq. (1.14), concentration of holes in N-type region,

$$p_{no} = \frac{n_i^2}{N_D}$$

and from Eq. (1.16), concentration of electrons in P-type region,

$$n_{po} = \frac{n_i^2}{N_A}$$

where N_D and N_A are the concentrations of donor and acceptor atoms respectively.

Substituting $p_{no} = \dfrac{n_i^2}{N_D}$ and $n_{po} = \dfrac{n_i^2}{N_A}$ in Eq. (2.16) we have

$$I_0 = Ae\left[\frac{D_h}{L_h N_D} + \frac{D_e}{L_e N_A}\right] n_i^2 \qquad ...(2.17)$$

From Eq. (1.40) $$n_i^2 = A_0 T^3 e^{-E_{GO}/kT} = A_0 T^3 e^{-V_{GO}/V_T} \qquad ...(2.18)$$

where V_{GO} is a voltage numerically equal to the forbidden energy gap E_{GO} in electron volts, and V_T is the volt equivalent of temperature and is equal to T/11,600.

For germanium the diffusion constants D_h and D_e vary approximately inversely proportional to temperature T. Hence I_0 may be given as

$$I_0 = K_1 T^2 e^{-V_{GO}/V_T} \qquad ...(2.19)$$

where K_1 is a constant, independent of temperature T.

In the discussion made above, the effect of carrier generation and recombination in the space-charge region has been ignored. Such an assumption is valid for a germanium diode (not for a silicon diode). For the silicon diode, a diffusion current is negligible in comparison with the transition-layer charge-generation current, which is given approximately by

$$I = I_0 (e^{V/\eta V_T} - 1) \qquad ...(2.20)$$

where $\eta \approx 2$ for small (rated) currents and $\eta \approx 1$ for large currents. Also I_0 is found to be proportional to n_i instead of n_i^2 so

$$I_0 = K_2 T^{1.5} e^{-V_{GO}/2V_T} \qquad ...(2.21)$$

where K_2 is a constant, independent of temperature T.

2.6. TERMINAL CHARACTERISTICS OF JUNCTION DIODES

The volt-ampere characteristic of a silicon junction diode is depicted in Fig. 2.7. The same characteristic is depicted in Fig. 2.8 with some scales expanded and others compressed to reveal details—the changes in scale have resulted in the apparent discontinuity at the origin.

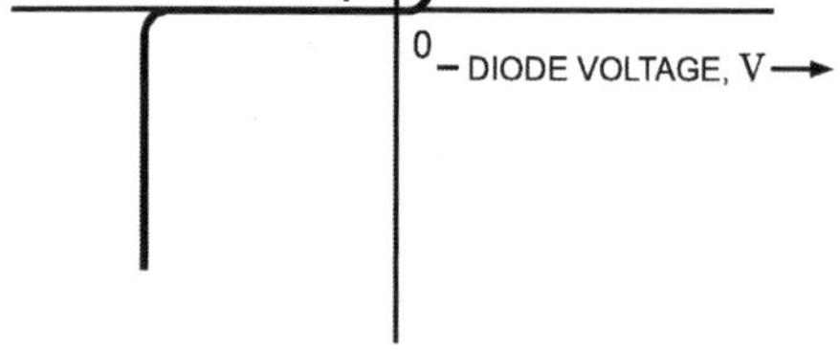

Fig. 2.7. ***Volt-Ampere Characteristic of a Silicon Junction Diode***

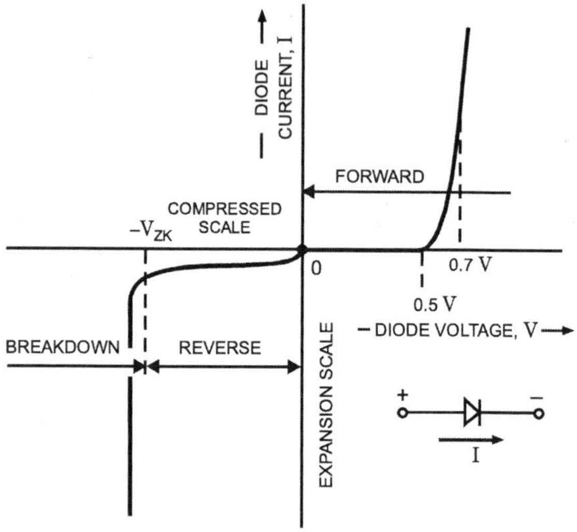

Fig. 2.8. ***V-I Characteristics of the Diode With Some Scales Expanded and Others Compressed In Order To Reveal Details***

As shown in the figure, the characteristic curve consists of three distinct regions :

1. The forward-bias region, determined by $V > 0$.
2. The reverse-bias region, determined by $V < 0$.
3. The breakdown region, determined by $V < -V_{ZK}$.

The above three regions of operation are described below :

2.6.1. Forward-Bias Region

A positive value of current I means the current flowing from the P-region to the N-region. The diode is forward biased if the applied voltage V is positive *i.e.* P-side is connected to the positive terminal and N-side is connected to negative terminal as shown in Fig. 1.31.

The current flowing through the junction diode is given by Eq. (2.20).

In Eq. (2.20) I_0 is a constant for a given diode at a given temperature. A formula for I_0 in terms of the temperature will be given in Art. 2.8. The current I_0 is usually called the saturation current, and sometimes denoted by I_s. Another name for the saturation current is the *scale current*. This name arises from the fact that I_0 is directly proportional to the cross-sectional area of the diode. Thus doubling of junction area results in a diode with double the value of I_0 and, as the diode equation indicates, double the value of current I for a given forward voltage V. For "small-signal" diodes (small size diodes intended for low-power applications), I_0 is of the order of 10^{-15} A. The value of I_0 is, however, a very strong function of temperature. As a thumb rule, I_0 approximately doubles for every 5°C rise in temperature.

Voltage V_T in Eq. (2.20) is the volt equivalent of temperature and is equal to $\frac{T}{11,600}$. At room temperature V_T is taken equal to be 2.6 mV.*

In Eq. (2.20) the constant η has a value between 1 and 2, depending on the material and physical structure of the diode. Diodes made using the standard IC fabrication process exhibit $\eta = 1$ when operated under normal conditions. Diodes available as discrete two-terminal components generally exhibit $\eta = 2$.

For appreciable current I in the forward direction, specifically for $I >> I_0$, Equation (2.20) can be approximated by the exponential relationship

$$I \simeq I_0 e^{V/\eta V_T} \qquad \text{...(2.22)}$$

This relationship can be expressed alternatively in the logarithmic form

$$V = \eta V_T \log_e \frac{I}{I_0} \qquad \text{...(2.23)}$$

The exponential relationship of the current I to the voltage V holds over many decades of current (a span of as many as seven decades—*i.e.*, a factor of 10^7—can be found). This is quite a remarkable property of junction diodes, one that is also found in bipolar junction transistors and that has been exploited in many interesting applications.

From Eq. (2.22), the currents I_1 and I_2 for voltages V_1 and V_2 are given respectively by equations

$$I_1 = I_0 e^{V_1/\eta V_T} \qquad \text{...(2.24)}$$

$$I_2 = I_0 e^{V_2/\eta V_T} \qquad \text{...(2.25)}$$

* A slightly higher ambient temperature (25°C or so) is usually assumed for electronic equipment operating inside a cabinet. At this temperature, $V_T = 25.8$ mV $\simeq$ 26 mV. Nevertheless, for the sake of simplicity, and to promote rapid circuit analysis, sometimes more arithmetically convenient value of 25 mV is used.

Dividing Eq. (2.25) by Eq. (2.24), we have

$$\frac{I_2}{I_1} = e^{(V_2 - V_1)/\eta V_T}$$

$$\text{or } V_2 - V_1 = \eta V_T \log_e \frac{I_2}{I_1}$$

$$\text{or } V_2 - V_1 = 2.3\eta V_T \log_{10} \frac{I_2}{I_1} \quad ...(2.26)$$

The above equation reveals that for a decade (factor of 10) change in diode current, the diode voltage drop changes by $2.3\eta V_T$, which is about 60 mV for $\eta = 1$ and 120 mV for $\eta = 2$. This also suggests that the diode volt-ampere characteristic is most conveniently drawn on semilog paper. Using the vertical, linear axis for V and horizontal, log axis for current I, we have a straight line with a slope of $2.3\eta V_T$ per decade of current. Finally, it is noteworthy that not knowing the exact value of η, circuit designer use the convenient approximate number of 0.1 V/decade for the slope of the diode logarithmic characteristic.

Voltage-current characteristic in the forward region (Fig. 2.8) reveals that the current is negligibly small for voltage smaller than about 0.5 V. This value is usually referred to as the *cutin* voltage. It needs to be emphasized, however, that this apparent threshold in the characteristic is simply a consequence of the exponential relationship. Another consequence of this exponential relationship is the rapid increase of current I. Thus, for a "fully conducting" diode, the average drop lies in a narrow range, approximately 0.6 V to 0.8 V. This gives rise to a simple "model" for the diode where it is assumed that a conducting diode has approximately 0.7 V drop across the diode. Diodes with different current ratings (*i.e.* different areas and correspondingly different I_0) will exhibit the 0.7 V drop at different currents.

2.6.2. Reverse-Bias Region

When the diode is reverse biased and V is several times V_T in magnitude, current, $I = -I_0$. The reverse current, is, therefore, constant, independent of the applied reverse bias, as shown in Fig. 2.8. Consequently, I_0 is referred to as the *reverse saturation current.*

Real diodes exhibit currents, that though quite small, are much larger than I_0. For instance, a small-signal diode whose current I_0 is of the order of 10^{-14} to 10^{-15} A could show a reverse current of the order of 1 nA. The reverse current also increases somewhat with the increase in magnitude of the reverse voltage. Because of the very small magnitude of the current, these details are not clearly evident on the diode V-I characteristic depicted in Fig. 2.8.

A large component of the reverse current is due to leakage effects. These leakage currents are proportional to junction area, just as I_0 is. Their dependence on temperature, however, is different from that of I_0. Thus, whereas I_0 doubles for every 5°C rise in temperature, the corresponding rule of thumb for the temperature dependence of the reverse current is that it doubles for every 10°C rise in temperature.

2.6.3. Breakdown Region

The third distinct region of diode operation is the breakdown region, which can be easily identified on the volt-ampere characteristic depicted in Fig. 2.8. The breakdown region is entered when the magnitude of the reverse voltage exceeds a threshold value that is specific to the particular diode, called the *breakdown voltage.* This is the voltage at the "knee" of the V-I characteristic curve in Fig. 2.8, and is denoted by V_{ZK}, where the subscript Z stands for zener (to be explained shortly) and K denotes knee.

As can be seen from Fig. 2.8, in the breakdown region the reverse current increases rapidly, with the associated increase in voltage drop being very small. Diode breakdown is normally not destructive provided that the power dissipated in diode is limited by external circuitry to a "safe" level. This safe value is normally specified on the data sheets of the device. It is, therefore, necessary to limit the reverse current in the breakdown region to a value consistent with the permissible power dissipation.

The fact that the diode V-I characteristic in breakdown is almost a vertical line enables it to be used in voltage regulation.

2.7. DETERMINATION OF DIODE CHARACTERISTICS

The forward characteristics of a diode can be obtained by use of the circuit depicted in Fig. 2.9 (*a*). The diode voltage is set at a series of convenient levels, and the corresponding current levels are measured and recorded. The characteristics are then plotted from the values obtained. The reverse characteristics can be obtained in the same way, except that a very sensitive microammeter is required for measuring the diode reverse current. Also, the microammeter must be connected directly in series with the diode as shown in Fig. 2.9 (*b*); otherwise the voltmeter current may introduce a serious error.

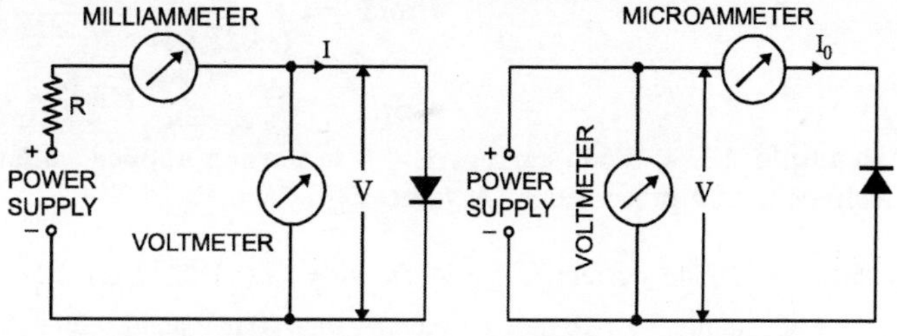

(a) For Forward Characteristics *(b) For Reverse Characteristics*

Fig. 2.9. ***Circuits For Determination of Diode Characteristics***

The forward characteristics of a diode can also be determined with the use of an XY recorder as shown in Fig. 2.10. The voltage across resistance R is directly proportional to the diode current I, while the voltage across the diode is of course V_D. The voltage across the resistor R *i.e.* V_R is applied to the vertical input terminals of the X-Y recorder, and voltage across the diode, V_D is applied to the horizontal terminals. When the power supply voltage is slowly increased from zero, the diode forward characteristics is traced out by the pen. If resistance R is of 1 kΩ, there is a voltage drop of 1 V across it for every 1 mA of diode current. Thus, with the scale of the XY recorder set to 1 V/cm, the (vertical) current ordinate of the graph is 1 mA/cm. A convenient scale for the (horizontal) voltage ordinate is 0.1 V/cm.

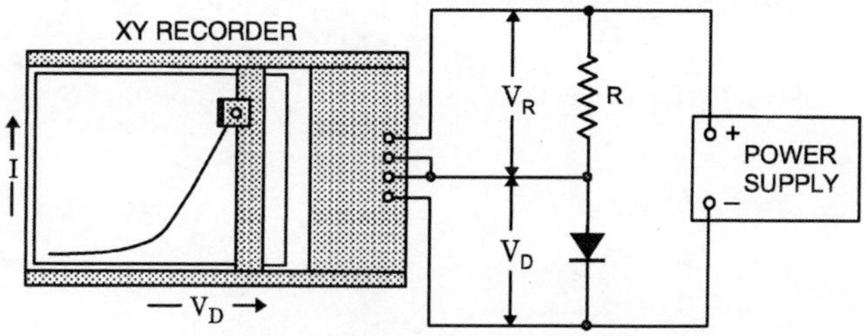

Fig. 2.10. ***Determination of Diode Forward Characteristics With The Use of XY Recorder***

Example 2.1. The current flowing through a silicon diode at room temperature is 9×10^{-7} A, when a large reverse voltage is applied. Calculate the current through the diode when 0.4 V forward bias is applied. Consider η = 2, where η is ideality factor. *[G.G.S.I.P. University Analog Electronics, December 2010]*

Solution : Diode current is given as $I = I_0(e^{V/\eta V_T} - 1)$ ...(*i*)

Reverse saturation current, $I_0 = 9 \times 10^{-7}$ A

Forward bias voltage, V = 0.4 V

For silicon, η = 2

At room temperature, V_T = 26 mV = 0.026 V

Substituting the values in Eq. (*i*), we have

$$\text{Diode current, } I = 9 \times 10^{-7} (e^{0.4/2 \times 0.026} - 1) = 0.00197 \text{ A} = 1.97 \text{ mA} \textbf{ Ans.}$$

Example 2.2. A germanium diode carries a current of 1 mA at room temperature when a forward bias of 0.15 V is applied. Estimate the reverse saturation current at room temperature.

Solution : Applied forward bias voltage, V = 0.15 V

Forward current, I = 1 mA

Volt equivalent of room temperature, V_T = 26 mV = 0.026 V

If reverse saturation current = I_0

then, diode current is given as, $I = I_0(e^{V/\eta V_T} - 1)$

$$\text{or } I_0 = \frac{I}{(e^{V/V_T} - 1)} = \frac{1 \times 10^{-3}}{(e^{0.15/0.026} - 1)} \qquad \because \text{ for Ge, } \eta = 1$$

$$= 3.12 \times 10^{-6}\text{ A} = \mathbf{3.12\ \mu A} \quad \textbf{Ans.}$$

Example 2.3. A diode current is 0.6 mA when applied voltage is 400 mV and 20 mA when the applied voltage is 500 mV. Find η. Assume V_T = 26 mV.

Solution : Diode current is given as $I = I_0\ (e^{V/\eta V_T} - 1) \simeq I_0 e^{V/\eta V_T}$

Substituting V_T = 26 mA, I = 0.6 mA for 400 mV and I = 20 mA for 500 mV, we have

$$0.6\text{ mA} = I_0\ e^{\frac{400}{26\eta}} \qquad ...(i)$$

$$20\text{ mA} = I_0\ e^{\frac{500}{26\eta}} \qquad ...(ii)$$

Dividing Eq. (*ii*) by Eq. (*i*), we have

$$\frac{e^{\frac{500}{26\eta}}}{e^{\frac{400}{26\eta}}} = \frac{20}{0.6} = 33.333$$

$$\text{or} \quad e^{\frac{100}{26\eta}} = 33.333$$

$$\text{or } \eta = \frac{100}{26 \log_e 33.333} = 1.097 \quad \textbf{Ans.}$$

Example 2.4. A silicon diode has reverse saturation current of 2.5 μA at 300 K. Find forward voltage for a forward current of 10 mA.

Solution : From Eq. (2.20) $\qquad I = I_0\ (e^{V/\eta V_T} - 1)$

Substituting forward current I = 10 mA (or 0.01 A), reverse saturation current $I_0 = 2.5 \times 10^{-6}$ A in above equation, we have

$$0.01 = 2.5 \times 10^{-6}\ (e^{V/\eta V_T} - 1)$$

$$\text{or } e^{V/\eta V_T} = \frac{0.01}{2.5 \times 10^{-6}} + 1 = 4 \times 10^3 + 1 = 4001$$

$$\text{or } \frac{V}{\eta V_T} = \log_e 4001$$

$$\text{or } V = \eta \times V_T \log_e 4001 = 2 \times 0.026 \times \log_e 4001 = 0.43\text{ V} \quad \textbf{Ans.}$$

$$\because \eta = 2 \text{ for silicon and } V_T \text{ at 300 K} = 26\text{ mV or } 0.026\text{ V}$$

Example 2.5. What is the ratio of current for forward biased mode with voltage 0.05 V to the current with same magnitude of reverse bias ?

Solution : From Eq. (2.20) $\qquad I = I_0(e^{V/\eta V_T} - 1)$

$$\text{So } I_F = I_0(e^{0.05/\eta V_T} - 1) \qquad ...(i)$$

$$I_R = I_0(e^{-0.05/\eta V_T} - 1) \quad ...(ii)$$

Dividing Eq. (*i*) by Eq. (*ii*), we have

$$\frac{I_F}{I_R} = \frac{e^{0.05/\eta V_T} - 1}{e^{-0.05/\eta V_T} - 1} = \frac{e^{0.05/0.026} - 1}{e^{-0.05/0.026} - 1} = \frac{5.842}{-0.8538} = -6.84 \textbf{ Ans.}$$

Example 2.6. Find the reverse saturation current density in an abrupt silicon junction. Given the following data :

$$\mathbf{N_D = 10^{21}\ m^{-3};\ N_A = 10^{22}\ m^{-3};\ D_e = 3.4 \times 10^{-3}\ m^2\ s^{-1};}$$
$$\mathbf{D_h = 1.2 \times 10^{-3}\ m^2\ s^{-1};\ L_e = 7.1 \times 10^{-4}\ m}$$
$$\mathbf{L_h = 3.5 \times 10^{-4}\ m;\ n_i = 1.6 \times 10^{16}\ m^{-3}.}$$

Solution : From Eq. (2.17) reverse saturation current.

$$I_0 = Ae\left[\frac{D_h}{L_h N_D} + \frac{D_e}{L_e N_A}\right]n_i^2$$

or reverse saturation current density $= \dfrac{I_0}{A} = \left[\dfrac{D_h}{L_h N_D} + \dfrac{D_e}{L_e N_A}\right]en_i^2$

$$= \left[\frac{1.2 \times 10^{-3}}{3.5 \times 10^{-4} \times 10^{21}} + \frac{3.4 \times 10^{-3}}{7.1 \times 10^{-4} \times 10^{22}}\right] \times 1.602 \times 10^{-19} \times (1.6 \times 10^{16})^2$$

$$= 1.6 \times 10^{-7}\ \text{A} \quad \text{or} \quad 0.16\ \mu\text{A} \quad \textbf{Ans.}$$

2.8. TEMPERATURE DEPENDENCE OF P-N DIODES

Temperature is very important consideration in the analysis or design of electronic devices or systems. It affects virtually all of the characteristics of a semiconductor device.

The diode current at a given voltage depends upon the operating temperature. The variation may be determined from Eq. (2.20), where the temperature is contained implicitly in V_T and also in the reverse saturation current I_0.

From Eq. (2.20) $\quad I = I_0(e^{V/\eta V_T} - 1)$

At room temperature $V_T = 26\ \text{mV} = 0.026\ \text{V}$

For germanium $\eta = 1$ and for silicon $\eta = 2$

So substituting the above values in the expression for diode current we have

$$\text{For germanium, } I = I_0(e^{V/0.026} - 1) \simeq I_0 e^{40V} \quad ...(2.27)$$

$$\text{and for silicon, } I = I_0(e^{V/2\times 0.026} - 1) \simeq I_0 e^{20V} \quad ...(2.28)$$

The dependence of I_0 on temperature T is, from Eqs. (2.19) and (2.21), given approximately as

$$I_0 = KT^m e^{-V_{GO}/\eta V_T} \quad ...(2.29)$$

where K is a constant and eV_{GO} is the forbidden energy gap in joules

For silicon : $\eta = 2;\ m = 1.5;\ V_{GO} = 1.21\ \text{V}$

For germanium : $\eta = 1;\ m = 2;\ V_{GO} = 0.785\ \text{V}$

Taking the logarithm on both sides of Eq. (2.29), we have

$$\log_e I_0 = mK\log_e T - \frac{V_{GO}}{\eta V_T}$$

Taking the derivative of the above equation we have

$$\frac{d(\log_e I_0)}{dT} = \frac{1}{I_0}\frac{dI_0}{dT} = \frac{m}{T} + \frac{V_{GO}}{\eta T V_T} \quad ...(2.30)$$

At room temperature (300 K), from Eq. (2.30)

$$\frac{d(\log_e I_0)}{dT} = 0.08 \text{ per °C for silicon and } 0.11 \text{ per °C for germanium.}$$

The performance of commercial diodes is only consistent with above result. The reason for the discrepancy is that, in a practical diode, there is a component of the reverse saturation current due to leakage over the surface that has not been taken into account in Eq. (2.29). Since leakage component is independent of temperature, it may be expected to find a smaller rate of change of I_0 with temperature than that predicted above.

From experimental data it is found that reverse saturation current I_0 increases 7 per cent per °C for both silicon and germanium. Since $(1.07)^{10} \simeq 2.0$, the reverse saturation current approximately doubles for every 10°C rise in temperature.

Germanium is more temperature dependent than silicon because its reverse saturation current is approximately 1,000 times larger.

The change in V-I characteristics of a semiconductor diode due to variations in temperature above and below room temperature is shown in Fig. 2.11.

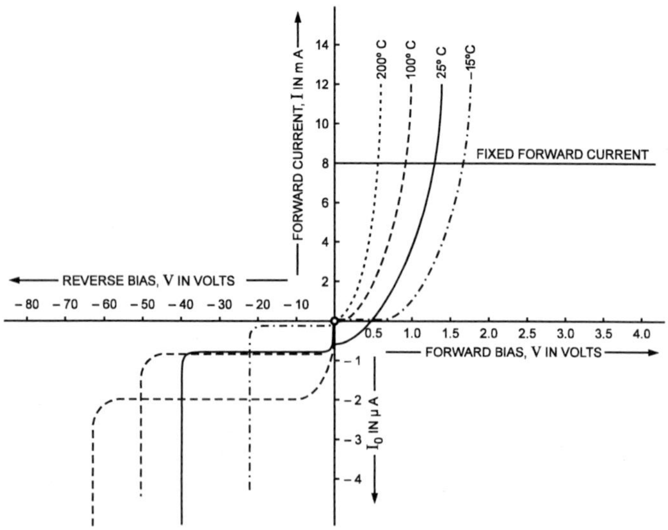

Fig. 2.11. ***Variation in Diode Characteristics With Variations in Temperature***

From Fig. 2.11 we see that for a fixed level of forward voltage, forward current increases with the increase in temperature. If the forward current is measured, at the increased temperature, for several levels of forward voltage and the results are plotted, it is seen that current-voltage characteristic is moved to the left. The horizontal line on Fig. 2.11 shows that for a fixed forward current, forward voltage drops as the junction temperature increases *i.e.* forward voltage has a negative temperature coefficient. It is found that the temperature coefficient for the forward voltage of a P-N junction is approximately – 1.8 mV/°C for Si and – 2.02 mV/ °C for Ge. Zener potential is also experiencing a pronounced change in level.

2.9. DIODE RESISTANCE

As already discussed, a forward-biased diode conducts easily whereas a reverse-biased diode restrains the flow of current, so a diode has a low forward resistance as compared to high blocking or reverse resistance.

Forward Resistance. The resistance offered by a diode in the circuit, when forward biased, is known as the *forward resistance.* This resistance is not the same for dc as well as changing current. Thus, forward resistance is of two types viz. dc or *static resistance* and ac or *dynamic resistance*

DC or **static resistance**, R is the resistance offered by a diode to the direct current. It is simply the ratio of the dc voltage across the diode to the direct current flowing through it. It can be determined by determining the ratio of voltage drop across the diode to the current flowing through it. At any point P on the V-I characteristic of the diode (Fig. 2.12), the voltage across the diode is OA and the corresponding current is OB.

So dc or static resistance, $R = \frac{V}{I} = \frac{OA}{OB}$

Thus at any point on the V-I characteristic of the diode, the dc or static resistance R is equal to the reciprocal of the slope of the line joining the operating point to the origin.

DC or static resistance is not constant but depends on the operating point on the V-I characteristic of the diode. It is different at different points, as illustrated in Fig. 2.12. From Fig. 2.12 it is obvious that the dc resistance of a diode in the forward-bias region decreases as the region of higher voltages and currents is approached.

In the reverse-bias region, the current $I = I_0$, constant in magnitude, the resistance of the ideal diode is essentially infinite resistance (open-circuit equivalent).

Any way dc or static resistance does not form a useful parameter of the diode.

AC or **dynamic resistance** *r* is resistance offered by a diode to the changing forward current. It may also be defined as the reciprocal of the slope of the forward characteristic of the diode. It forms an important device parameter for small signal operation. However, it varies with the operating voltage.

From Eq. (2.20) $r = \frac{dV}{dI} = \frac{\eta V_T}{I_0 e^{V/\eta V_T}} = \frac{\eta V_T}{I + I_0}$ ...(2.31)

For a forward bias greater than a few tenths of a volt, $\left|\frac{V}{\eta V_T}\right| >> 1$ and according to Eq. (2.20) $I >> I_0$, and dynamic resistance *r* is given approximately by

$$r = \frac{\eta V_T}{I} \quad \text{...(2.32}a\text{)}$$

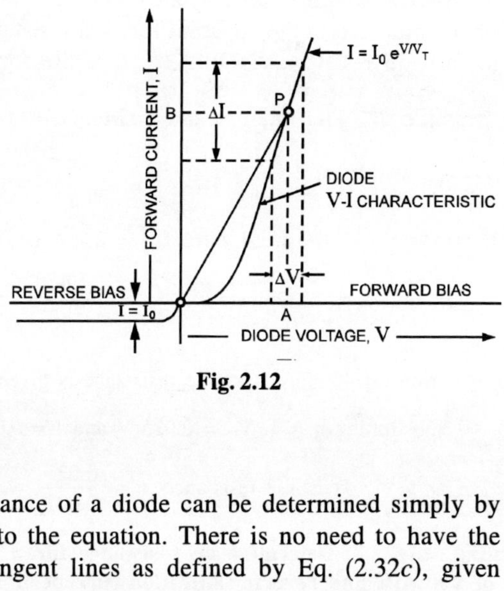

Fig. 2.12

From above equation it is obvious that for forward bias *r* varies inversely with the current I. At room temperature (27°C or 300 K) V_T = 0.026 V or 26 mV.

Substituting η = 1 and V_T = 26 mV in above Eq. (2.32*a*), we have

$$r = \frac{26 \text{ mV}}{I \text{ (mA)}} \quad \text{...(2.32}b\text{)}$$

The above equation implies that the dynamic resistance of a diode can be determined simply by substituting the quiescent value of the diode current into the equation. There is no need to have the characteristic available or to worry about sketching tangent lines as defined by Eq. (2.32*c*), given below.

The *ac* or *dynamic resistance* can be determined from the forward characteristic, as illustrated in Fig. 2.12.

$$\text{AC or dynamic resistance, } r = \frac{\text{Small change in forward voltage}}{\text{Small change in forward current}} = \frac{\Delta V}{\Delta I} \quad \text{...(2.32}c\text{)}$$

Equation (2.32*b*) indicates that the incremental forward resistance of the P-N junction diode is quite low. For example, when diode current is 26 mA, the diode resistance is 1 Ω.

Actually diode ac or dynamic resistance for forward bias is slightly different from that obtained from above equation. This is because of bulk and contact resistance of the semiconductor device. This additional resistance level can be included in Eq. (2.32*b*), by adding a factor denoted r_B as appearing in Eq. (2.32*d*)

$$r' = \frac{26 \text{ mV}}{\text{I (mA)}} + r_B \qquad \text{...(2.32}d\text{)}$$

For large values of forward current the first term $\frac{26 \text{ mV}}{\text{I (mA)}}$ giving apparent forward or junction resistance r is negligible so $r' = r_B$. For small values of forward current, r_B is negligible so $r' = \frac{26 \text{ mV}}{\text{I (mA)}}$.

The incremental backward resistance (*i.e.*, resistance with reverse bias), is ideally infinite because of the zero slope of the V-I characteristic in this region.

This conclusion also follows from Eq. (2.31), upon inserting $I = -I_0$ for negative values of bias voltage. Thus diode resistance for reverse bias

$$= \frac{\eta V_T}{I + I_0} = \frac{\eta V_T}{-I_0 + I_0} = \infty$$

However, the apparent backward resistance is not infinite because for any value of reverse bias there does exist a small current I_0. For germanium and silicon junction diodes this resistance is quite large compared to the apparent forward resistance. In germanium diodes the ratio of apparent backward to forward resistance is 400,000 : 1 while for silicon diodes this ratio is 1,000,000 : 1.

Example 2.7. Find the dynamic resistance of a P-N junction diode at a forward current of 2 mA. Assume $\frac{k'T}{e}$ = 25 mV.

Solution : Forward current, I = 2 mA or 0.002 A

Volt equivalent of temperature, $V_T = \frac{k'T}{e}$ = 25 mV or 0.025 V

From Eq. [2.32*a*], dynamic resistance is given approximately by $r = \frac{\eta V_T}{I}$

Substituting $\eta = 1$, $V_T = 0.025$ V and I = 0.002 A in above equation we have

$$\text{Dynamic resistance, } r = \frac{0.025}{0.002} = 12.5\ \Omega \quad \textbf{Ans.}$$

Example 2.8. Determine ac resistance for a germanium semiconductor diode having a forward bias of 200 mV and reverse saturation current of 1 μA at room temperature.

Solution : At room temperature (300 K),

V_T = 26 mV = 0.026 V

Applied forward voltage, V = 200 mV = 0.2 V

Reverse saturation current, $I_0 = 1\ \mu A = 1 \times 10^{-6}$ A

From Eq. (2.31), ac resistance for the diode,

$$r = \frac{\eta V_T}{I_0 e^{V/\eta V_T}} = \frac{0.026}{1 \times 10^{-6} \times e^{0.2/0.026}} = 11.86\ \Omega \quad \textbf{Ans.}$$

Taking $\eta = 1$ for Ge diode

Example 2.9. Explain the static and dynamic resistances in P-N junction diode. Determine dc resistance levels for the diode in following figure at (*i*) I_D = 2 mA, (*ii*) I_D = 20 mA, (*iii*) V_D = –10 V.

Solution : (*i*) When $I_D = 2$ mA, $V_D = 0.5$ V

$$\text{So resistance, } R_1 = \frac{V_D}{I_D} = \frac{0.5}{2 \times 10^{-3}} = 250\ \Omega \textbf{ Ans.}$$

(*ii*) When $I_D = 20$ mA $= 0.02$ A, $V_D = 0.8$ V

$$\text{So resistance, } R_2 = \frac{0.8}{0.02} = 40\ \Omega \textbf{ Ans.}$$

(*iii*) When $V_D = -10$ V,

$$I_D = -1\ \mu A = -1 \times 10^{-6} \text{ A}$$

$$\text{So resistance, } R_3 = \frac{-10}{-1 \times 10^{-6}} = 10\ M\Omega \textbf{ Ans.}$$

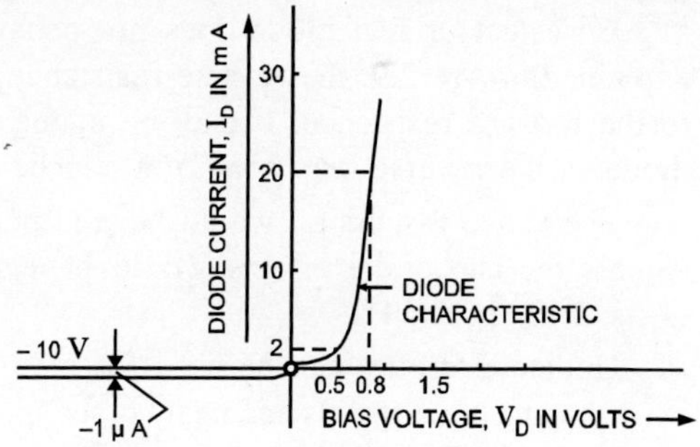

Fig. 2.13

2.10. REAL DIODE

As already mentioned in Art. 2.3, an actual diode neither behaves as a perfect conductor when forward biased nor as a perfect insulator when reverse biased.

When a diode is forward biased, it will not conduct till the potential barrier V_0 (0.7 V for Si and 0.3 V for Ge) at the junction is overcome. When the applied voltage exceeds the potential barrier V_0, the diode starts conducting. The forward current, I flowing through the diode causes a voltage drop in its internal resistance r'. Thus the voltage V applied across the diode is to overcome (*i*) the potential barrier V_0 and (*ii*) internal drop Ir'

$$\text{So } V = V_0 + Ir' \quad ...(2.33)$$

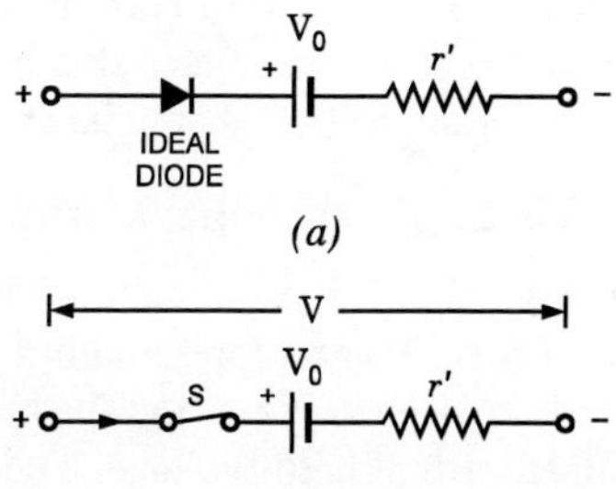

Fig. 2.14. ***Real Diode Representation Under Forward Bias Condition***

Hence, a real diode may be considered as equivalent to an ideal diode in series with a barrier potential of V_0 volts and a forward bias ac resistance r', as illustrated in Fig. 2.14 (*a*). The equivalent circuit of a real diode under forward bias condition is shown in Fig. 2.14 (*b*). This circuit shows that a real diode still acts as a switch when forward biased but the voltage required to operate this switch is V given by Eq. (2.33).

Voltage-current characteristic of a real diode under forward bias condition is shown in Fig. 2.15 which follows the Eq. (2.33). Here, the reciprocal of the slope of this characteristic represents the ac or dynamic resistance r' of the diode.

For *large signal sources* whose voltage is much greater than the diode barrier potential V_0 (or knee voltage V_k), forward current I would be large and the ac or dynamic resistance r' of the diode will be equal to bulk resistance r_B of the diode, as already explained in Art. 2.9.

For *small signal sources* whose voltage is much smaller than V_0 or V_k, forward current I would be small and the ac or dynamic resistance of r' of the diode will be equal to apparent forward or junction resistance r given by Eq. (2.32*b*) because for smaller values of forward current r_B is negligible.

For most applications, the ac or dynamic resistance r' of a diode can be neglected, being very small as compared to other elements and the real diode can be represented as shown in Fig. 2.16. The V-I characteristics for a real diode, neglecting its dynamic resistance become as illustrated in Fig. 2.4.

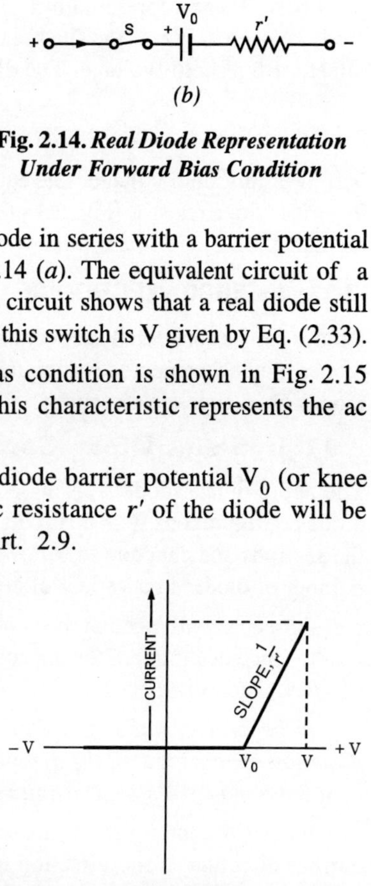

Fig. 2.15

An actual or real diode does not behave as a perfect insulator when reverse biased. As already explained in Art. 2.9, the reverse resistance of silicon and germanium diodes is quite large compared to the forward resistance. It may be of the order of many thousands of mega ohms. Hence an actual diode in the reverse bias condition can be thought of as equivalent to a resistor of high resistance.

The above discussion would be true only in case of small frequency signals. For large frequency signals the capacitive effects would have to be taken into account and will be discussed later on (Arts. 2.13 and 2.14).

Unilateral diodes can be used for rectification of alternating currents, waveshaping of ac signals and many other applications. Waveshaping involves alteration of ac signals by passing only certain portions of the signal while blocking other portions.

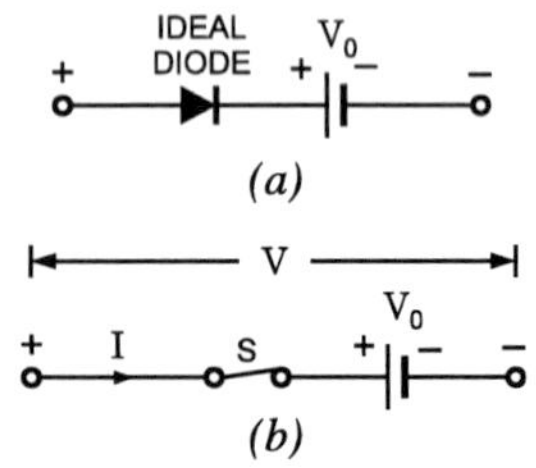

Fig. 2.16

Junction diodes designed for use as rectifiers must have volt-ampere characteristics close to those of an ideal diode. The reverse current should be negligible and forward current should exhibit little voltage dependence. The reverse breakdown voltage should be large and the offset or barrier potential V_0 in forward direction should be small. No diode can offer all these properties.

2.10.1. Piecewise Linear Diode Characteristic

A large-signal approximation which usually leads to a sufficiently accurate engineering solution is the **piecewise linear** representation. For instance, the piecewise linear approximation for a semiconductor diode is given in Fig. 2.15. The break point is not at the origin, and therefore, V_0 (or V_γ) is also known as the offset, or threshold, voltage. The diode behaves like an open circuit when the applied voltage is less than potential barrier V_0, and has a constant incremental resistance $r' = \frac{dV}{dI}$ when applied voltage exceeds potential barrier or threshold voltage. The equivalent circuit for this linear approximation is a constant voltage source in series with a resistor [Fig. 2.14 (*b*)]. This method models the diode with segments of straight lines : thus the name *piecewise linear model.*

2.11. INTERCONNECTION OF DIODES

Diodes can be connected in several ways such as series, parallel or series-parallel. In series arrangement, the diodes may be connected with polarities in the same direction (in forward-biased mode or in reverse-biased mode) or in opposite directions.

2.11.1. Series Diode Configurations

The approximate models discussed in preceding section are used for investigation of a number of series diode configurations with dc inputs. This establishes a foundation in diode analysis that will carry over into the sections and chapters to follow. The procedure described can, in fact, be applied to networks with any number of diodes in a variety of configurations.

For each configuration the state of each diode is first determined which diodes are "on" and which are "off"? Once determined, the appropriate equivalent can be substituted and the remaining parameters of the network determined.

In general, a diode is in the "on" state if the current caused by the applied sources is such that its direction matches that of the arrow in the diode symbol, and the diode voltage V_D exceeds V_0 (0.7 V in case of silicon and 0.3 V for germanium diode).

For each configuration, mentally replace the diodes with resistive elements and note the resulting current direction. If the direction of resulting current "matches" with the arrow in the diode symbol, the diode is in conducting or "on" state. If the diode is in the "on state", diode can be replaced by a 0.7 V drop across the element or the network is redrawn with the V_0 or V_K equivalent circuit. It is preferable to include the 0.7 V drop across each "on" diode and draw a line through each diode in the "off" or open state.

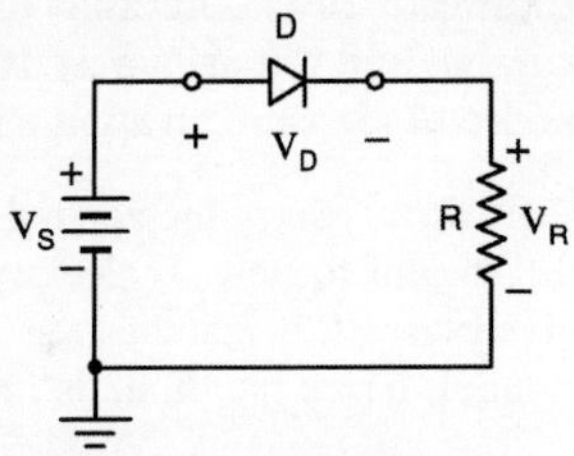

Fig. 2.17. ***Series Diode Configuration***

Let us consider a series circuit, shown in Fig. 2.17, to understand the approach described above.

The state of the diode is first determined by mentally replacing the diode with a resistive element as shown in Fig. 2.18 (*a*). The resulting direction of current I matches with the arrow in the diode symbol, and since V_S exceeds V_0 or V_K, the diode is in the conducting (or "on") state. The network is redrawn, as depicted in Fig. 2.18 (*b*), with the appropriate equivalent model for the forward-biased semiconductor diode. Note for future reference that the polarity of V_D is the same as would result if in fact the diode were a resistive element. The various relations between different voltages and current are given below :

Voltage drop across diode,

$$V_D = V_0 (\text{or} V_K) \quad ...(2.34)$$

Voltage drop across resistor R,

$$V_R = V_S - V_D = V_S - V_0 \quad ...(2.35)$$

Diode current,
$$I_D = I_R = \frac{V_R}{R} \quad ...(2.36)$$

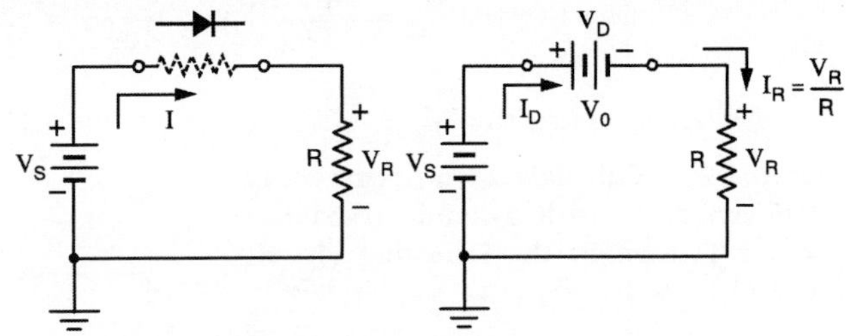

Fig. 2.18

Now let us consider the diode connected in reverse-biased mode, as shown in Fig. 2.19. If the diode is replaced mentally with a resistive element as depicted in Fig. 2.20 (*a*), it is revealed that the direction of flow of resulting current does not match the arrow in the diode symbol. It means that the diode is in "off" state and the resultant equivalent circuit will be as shown in Fig. 2.20 (*b*). Open circuit will not allow the current to flow in the circuit *i.e.* $I_D = I_R = 0$.

Voltage drop across resistor, $V_R = I_R R = 0 \times R = 0\,V$ ...(2.37)

and Voltage drop across diode, $V_D = V_S$ volts ...(2.38)

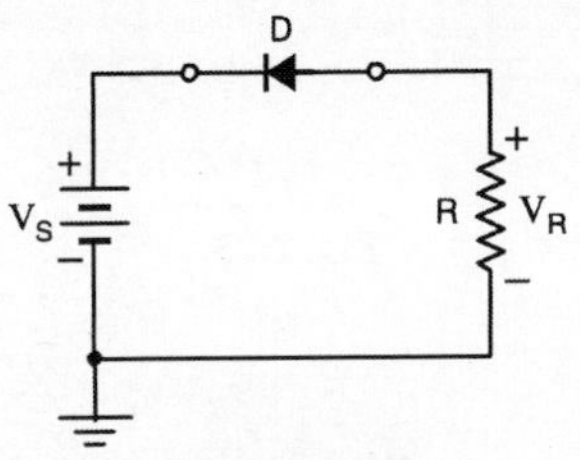

Fig. 2.19

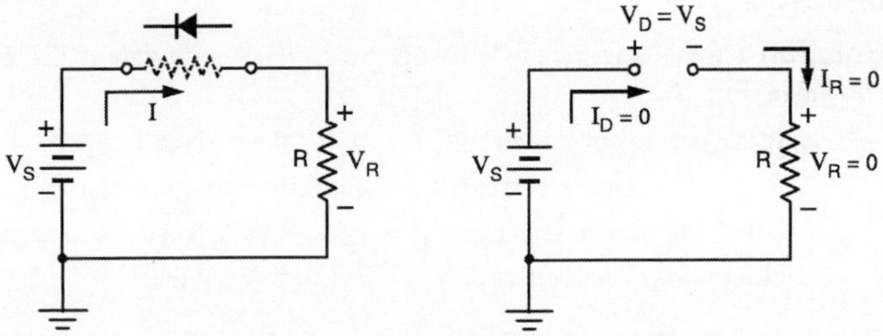

Fig. 2.20

2.11.2. Parallel and Series-Parallel Configurations

The method described in the preceding section can be extended to the analysis of parallel and series-parallel configurations. For each area of application, simply the sequential series of steps applied to series diode configurations are to be matched.

Example 2.10. Determine diode voltage, diode current and voltage across 100 Ω resistor in the series diode configuration shown in Fig. 2.21.

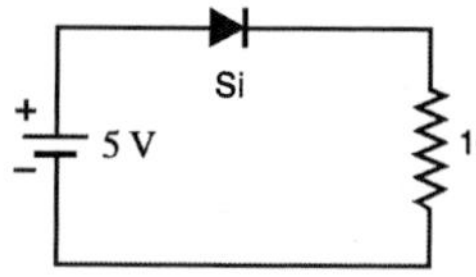

Fig. 2.21

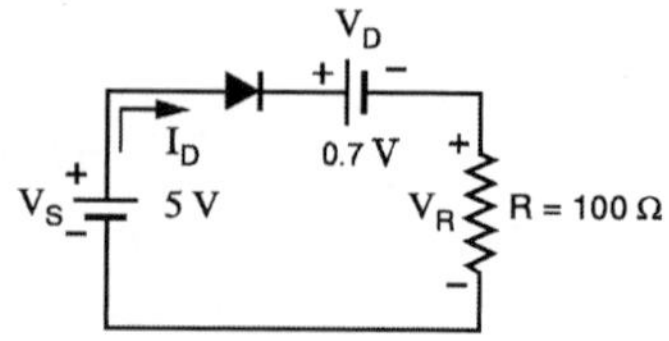

Fig. 2.22

Solution : Since the applied voltage establishes the current to flow in clockwise direction in the given circuit, it matches the arrow of the diode symbol, hence the diode is in conducting state.

Since the diode is silicon one, and the barrier potential for silicon diode is 0.7 V, so V_0 = 0.7 V.

The barrier potential acts in the direction opposite to that of the applied voltage and the equivalent circuit becomes as shown in Fig. 2.22. So

$$\text{Diode voltage, } V_D = V_0 = 0.7 \text{ V}$$

$$\text{Diode current, } I_D = \frac{V_S - V_D}{R} = \frac{5 - 0.7}{100} = 0.043 \text{ A or } 43 \text{ mA} \quad \textbf{Ans.}$$

neglecting diode forward resistance

$$\text{Voltage across resistor, } V_R = V_S - V_D = 5 - 0.7 = 4.3 \text{ V} \quad \textbf{Ans.}$$

Example 2.11. Calculate forward current I_F for silicon diode with dynamic resistance r_d = 0.25 Ω used in the following circuit (Fig. 2.23).

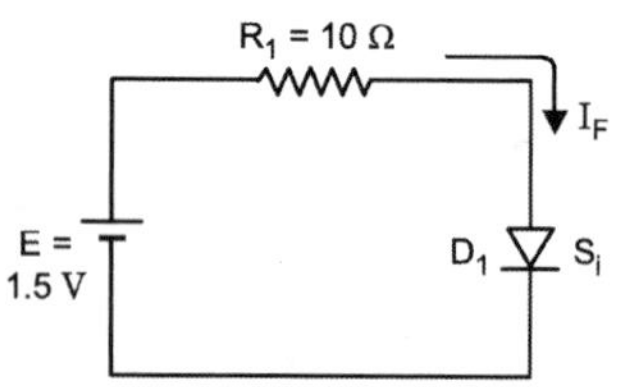

Fig. 2.23

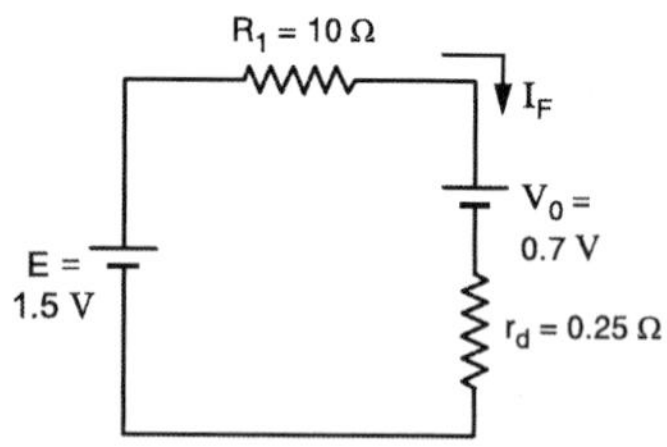

Fig. 2.24

Solution : Supply voltage, E = 1.5 V

Dynamic resistance, r_d = 0.25 Ω

Resistance, R_1 = 10 Ω

Potential barrier of silicon diode,

$$V_0 = 0.7 \text{ V}$$

The equivalent circuit is shown in Fig. 2.24.

$$\text{Forward current, } I_F = \frac{E - V_0}{R_1 + r_d} = \frac{1.5 - 0.7}{10 + 0.25} = 0.078 \text{ A or } 78 \text{ mA} \ \textbf{Ans.}$$

Example 2.12. Determine diode voltage, diode current and voltage across 100 Ω resistor in the series diode configuration shown in Fig. 2.21 with the diode polarity reversed.

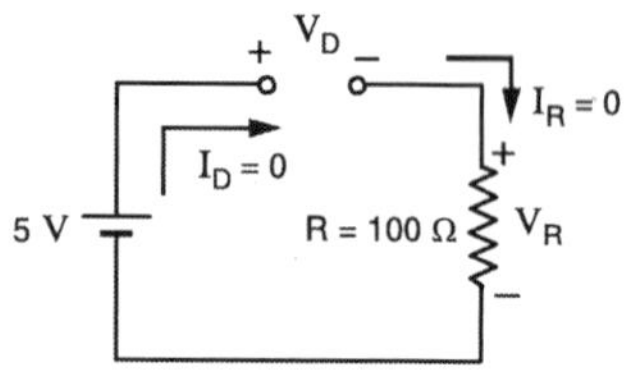

Fig. 2.25

Solution : Replacing the diode by open circuit, the equivalent circuit becomes as shown in Fig. 2.25.

Resulting current in the circuit, I = 0 ∵ the circuit is open

Diode current, I_D = Current through resistor, I_R = I = 0 A **Ans.**

Voltage across resistor, $V_R = I_R \times R = 0 \times 100 = 0$ V **Ans.**

Voltage drop across diode, $V_D = V_S = 5$ V **Ans.**

Example 2.13. Determine diode current, diode voltage and voltage across 500 Ω resistor, for the series diode configuration shown in Fig. 2.26.

Fig. 2.26

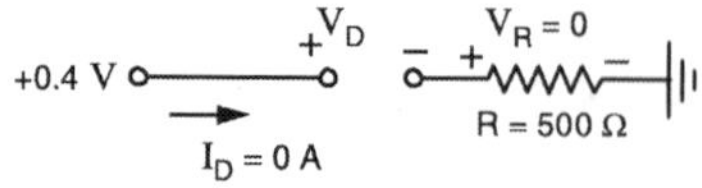

Fig. 2.27

Solution : Although the applied voltage establishes a current in the same direction as that of diode arrow symbol, but the magnitude of applied voltage is insufficient, being less than potential barrier of silicon diode, to turn the silicon diode "on". The equivalent circuit is shown in Fig. 2.27.

Diode current, I_D = 0 A

Resistor current, I_R = 0 A

Voltage drop across resistor, $V_R = I_R \times R = 0 \times 500 = 0$ V **Ans.**

Diode voltage, $V_D = V_S = 0.4$ V **Ans.**

Example 2.14. Find the voltage V_A of the circuit given in Fig. 2.28.

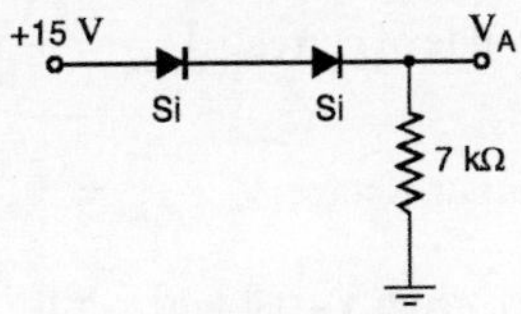

Fig. 2.28

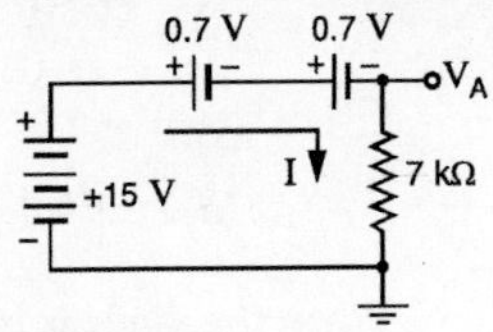

Fig. 2.29

Solution : Assuming diodes of negligible forward resistance, the given circuit is redrawn, as given in Fig. 2.29.

Current through the circuit,

$$I = \frac{V - 2V_0}{R} = \frac{15 - 2 \times 0.7}{7\ k\Omega} = 1.943\ mA$$

$\because$ barrier potential of silicon diode, $V_0 = 0.7$ V

Potential V_A = Voltage drop across 7 kΩ = 1.943 mA × 7 kΩ = 13.6 V **Ans.**

Example 2.15. Determine I, V_1, V_2 and V_0 for the series configuration of the following figure (Fig. 2.30).

Solution : Mentally replacing the diode with a resistive element, it is revealed that the resulting current direction matches the arrow of diode symbol and the diode is in conducting state. The equivalent circuit is shown in Fig. 2.31.

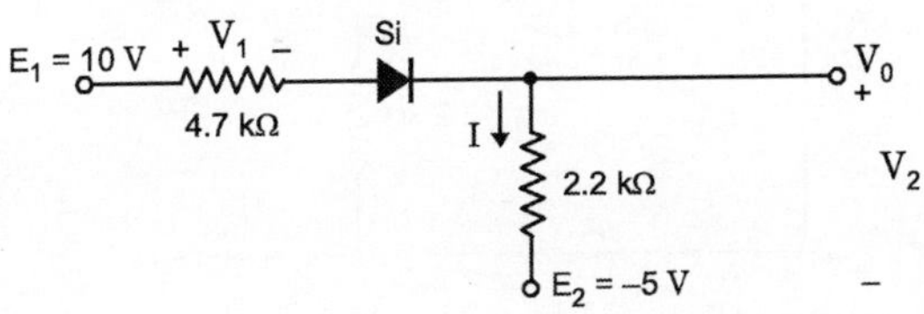

Fig. 2.30

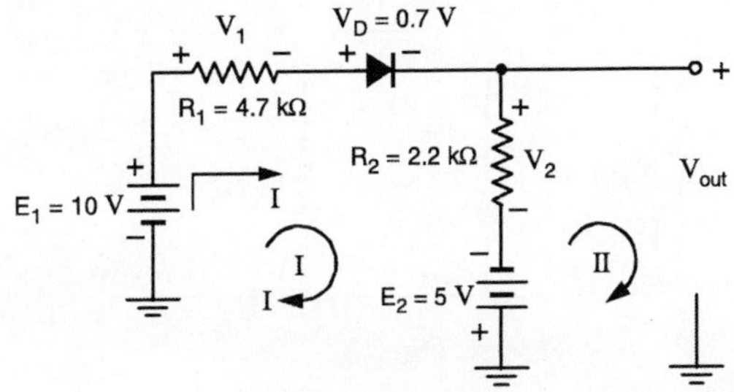

Fig. 2.31

Applying KVL to loop I, we have

$$E_1 - IR_1 - V_D - IR_2 + E_2 = 0$$

$$\text{or } I = \frac{E_1 + E_2 - V_D}{R_1 + R_2} = \frac{10 + 5 - 0.7}{(4.7 + 2.2) \times 10^3} = 2.072\ mA\ \textbf{Ans.}$$

Voltage drop across resistor R_1, $V_1 = IR_1 = 2.072 \times 10^{-3} \times 4.7 \times 10^3 = 9.74$ V **Ans.**

Voltage drop across resistor R_2, $V_2 = IR_2 = 2.072 \times 10^3 \times 2.2 \times 10^3 = 4.56$ V **Ans.**

Applying KVL to output loop, we have

$$-E_2 + V_2 - V_{out} = 0$$

or Output voltage, $V_{out} = V_2 - E_2 = 4.56 - 5 = -0.44$ V **Ans.**

Example 2.16. An ac voltage of peak value 25 V is connected in series with a silicon diode and a load resistance of 600 Ω. If the forward resistance of the diode is 15 Ω determine (*i*) the peak value of current through the diode (*ii*) peak value of the output voltage.

What will be the peak values of current through the diode and output voltage if the diode is assumed to be ideal ?

Solution :

Peak value of applied voltage, V_{max} = 25 V

Forward resistance of diode, r_f = 15 Ω

Load resistance, R_L = 600 Ω

Potential barrier voltage, V_0 = 0.7 V

The current through the diode will be maximum (peak) at the instant the input or applied voltage attains positive peak *i.e.* V_{in} = 25 V (*i*) Peak value of current, $I_{max} = \frac{V_{max} - V_0}{R_L + r_f} = \frac{25 - 0.7}{600 + 15} = 0.0395$ A or 39.5 mA **Ans.**

(*ii*) Peak value of output voltage, $V_{out\,max} = I_{max} \times R_L = 0.0395 \times 600 = 23.7$ V **Ans.**

When the diode is ideal *i.e.* when $V_0 = 0$ and $r_f = 0$

Peak value of current, $I'_{max} = \frac{V_{max}}{R_L} = \frac{25}{600} = 0.04167$ A or 41.67 mA **Ans.**

Peak value of output voltage, $V_{out\,max} = I'_{max} \times R_L = 0.04167 \times 600 = 25$ V **Ans.**

Example 2.17. Calculate V_0 and I_D for the following (Fig. 2.32).

Solution : Converting 10 mA current source having a 2.2 kΩ resistor connected across it into an equivalent voltage source of 10 mA × 2.2 kΩ *i.e.* 22 V in series with a resistance of 2.2 kΩ as shown in Fig. 2.33, we have

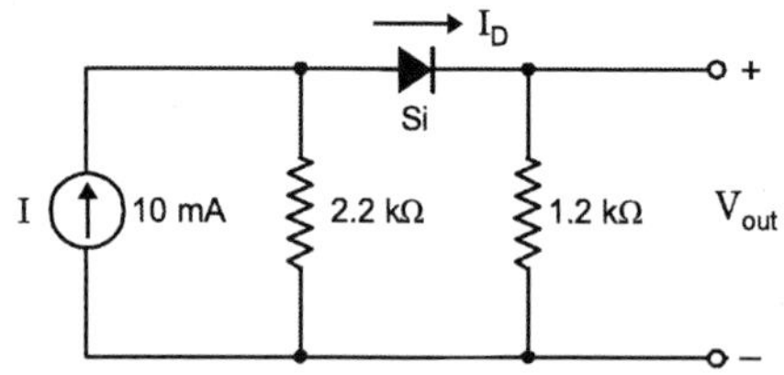

Fig. 2.32

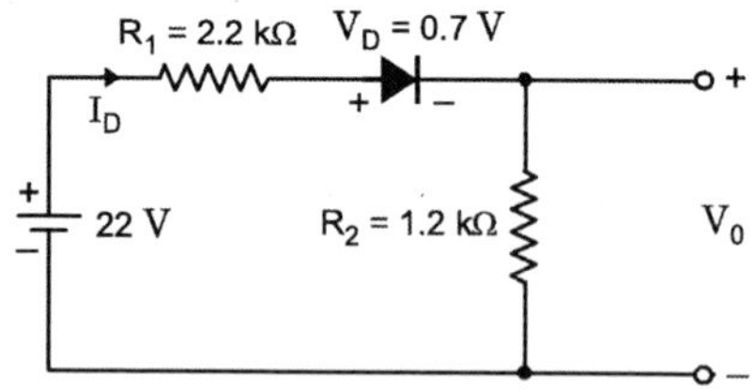

Fig. 2.33

Diode current, $I_D = \frac{V_S - V_D}{R_1 + R_2} = \frac{22 - 0.7}{(2.2 + 1.2) \times 10^3} = 6.265 \times 10^{-3}$ A or 6.265 mA **Ans.**

Output voltage, V_{out} = Voltage drop across 1.2 kΩ resistor

$= I_D \times 1.2 \times 10^3 = 6.265 \times 10^{-3} \times 1.2 \times 10^3 = 7.52$ V **Ans.**

Example 2.18. Determine diode current, diode voltage and output voltage for the circuit shown in Fig. 2.34.

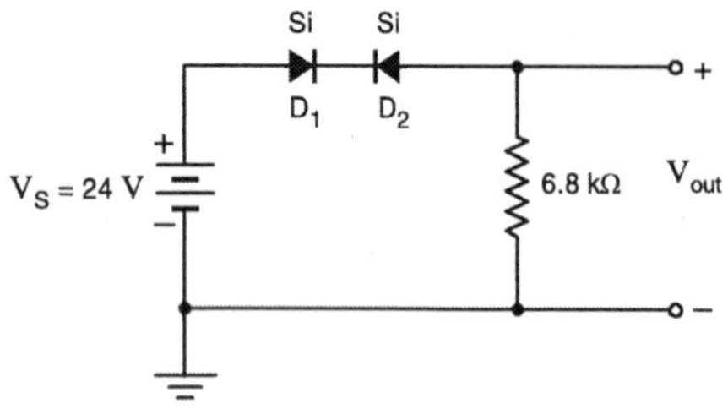

Fig. 2.34

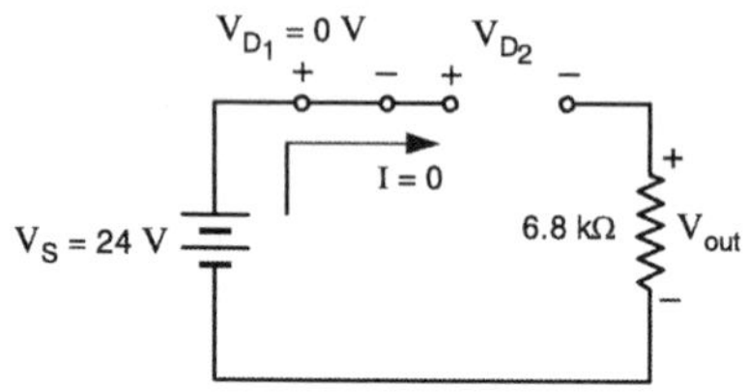

Fig. 2.35

Solution : Although the current established by the applied voltages matches the direction of symbol arrow of diode D_1 but not that of diode D_2. Thus there is an open circuit and the equivalent circuit is shown in Fig. 2.35.

Since the equivalent circuit shown in Fig. 2.35 is open one,

Diode current, I_0 = Output current, I_R

= Resulting circuit current, I = 0 A **Ans.**

Diode voltage, $V_{D_1} = 0$ because there is no bias

Output voltage, $V_{out} = IR = 0 \times 6.8$ kΩ $= 0$ V **Ans.**

Diode voltage, $V_{D_2} = V_S - V_{D_1} - V_{out} = 24 - 0 - 0 = 24$ V **Ans.**

Example 2.19. Find the voltage drop across each of the silicon junction diodes shown in Fig. 2.36 at room temperature. Assume that no reverse saturation current flows in the circuit and the magnitude of the reverse breakdown voltage is greater than 5 V.

Solution : Diode D_1 is forward biased and diode D_2 is reverse biased. Let the diode currents be I_1 and I_2 respectively, as shown in Fig. 2.37. Diode currents are given by the expressions

$$I_1 = I_0 (e^{V_{D_1}/\eta V_T} - 1) \quad ...(i)$$

and

$$I_2 = I_0 (e^{-V_{D_2}/\eta V_T} - 1) \quad ...(ii)$$

where I_0 is the reverse saturation current.

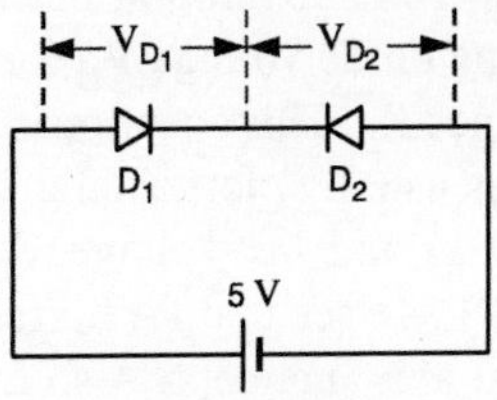

Fig. 2.36

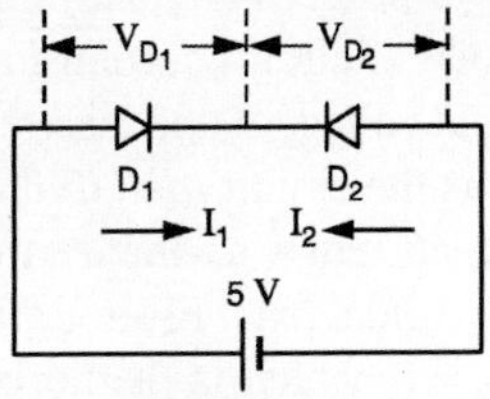

Fig. 2.37

Now from circuit shown in Fig. 2.37

$$I_1 + I_2 = 0$$

$$\text{or } I_0 (e^{V_{D_1}/\eta V_T} - 1) + I_0 (e^{-V_{D_2}/\eta V_T} - 1) = 0$$

$$\text{or } e^{V_{D_1}/\eta V_T} + e^{-V_{D_2}/\eta V_T} = 2$$

$$\text{or } e^{V_{D_1}/\eta V_T} + e^{-(5 - V_{D_1})/\eta V_T} = 2 \qquad \because V_{D_1} + V_{D_2} = V_S = 5 \text{ V}$$

$$\text{or } e^{V_{D_1}/\eta V_T} (1 + e^{-5/\eta V_T}) = 2$$

$$e^{V_{D_1}/2 \times 0.026} (1 + e^{-5/2 \times 0.026}) = 2 \qquad \because \eta \text{ for Si} = 2 \text{ and } V_T = 0.026 \text{ V}$$

$$\text{or } e^{V_{D_1}/0.052} (1 + 1.74 \times 10^{-42}) = 2$$

$$\text{or } V_{D_1} = 0.052 \log_e 2 = 0.036 \text{ V} \quad \textbf{Ans.}$$

$$V_{D_2} = V_S - V_{D_1} = 5 - 0.036 = 4.964 \text{ V} \quad \textbf{Ans.}$$

Example 2.20. Determine the voltage V_A in the circuit shown in Fig. 2.38.

Solution : It appears that on switching on the supply both the diodes turn 'on', but this is not true. When the supply voltage is switched on, germanium diode whose barrier potential is 0.3 V will turn on first and a level of 0.3 V will be maintained across the parallel branch. The silicon diode will never get opportunity to have 0.7 V across it and, therefore, remain in open-circuit state, as illustrated in Fig. 2.39.

So, $V_A = 15 - 0.3 = 14.7$ V **Ans.**

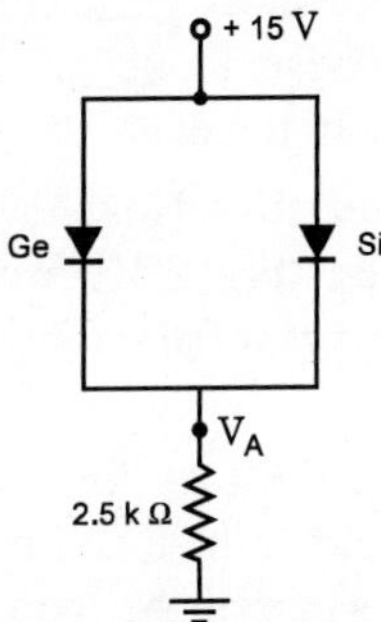

Fig. 2.38

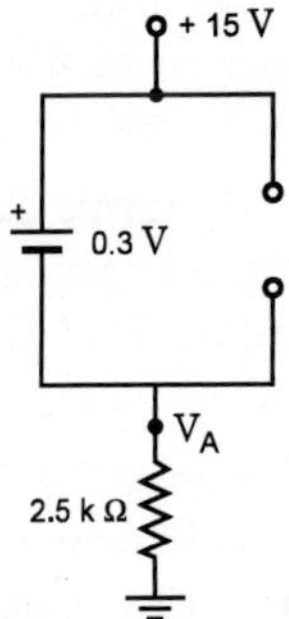

Fig. 2.39

2.12. DIODE AS A CIRCUIT ELEMENT

The P-N junction diode is considered as a circuit element. One of the most important application of diode is in rectification (*i.e.* in conversion of sinusoidal ac supply into dc supply.

The basic diode circuit, shown in Fig. 2.40, consists of a diode in series with a load resistane R_L across an input voltage of V volts.

2.12.1. Load Line Concept

Kirchhoff's voltage law is applicable both to linear and nonlinear circuits, so we may write

$$V = V_D + V_{out}$$

$$= V_D + IR_L \quad ...(2.39)$$ where I is the circuit current.

$$\text{or } I = \frac{V}{R_L} - \frac{V_D}{R_L} \quad ...(2.40)$$

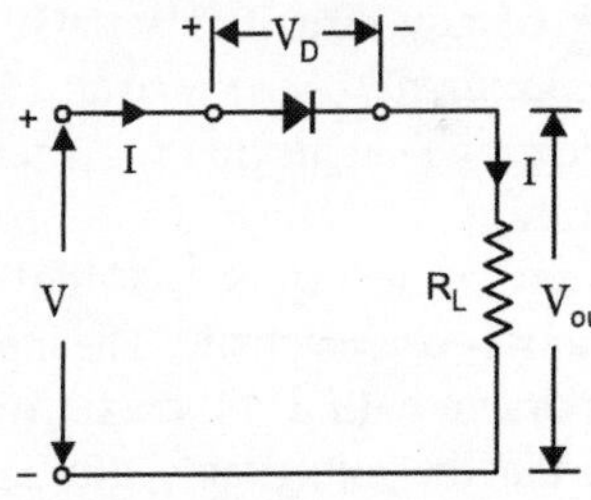

Fig. 2.40. ***Basic Diode Circuit***

The above equation gives a linear relation between the diode or circuit current I and the diode voltage V_D for a given input voltage V and load resistance R_L. This equation is referred to as the circuit *load line,* and is usually plotted on graph with the current I as the vertical axis and the voltage V_D as the horizontal axis. From Eq. (2.40), we see that for current I = 0, $V_D = V$ which is the horizontal axis intercept. Also from this equation, if $V_D = 0$, then current $I = \frac{V}{R_L}$ which is the vertical axis intercept. The load line can be drawn between these two points. The slope of the load line is determined by R_L, the negative value of the slope is equal to $\frac{1}{R_L}$.

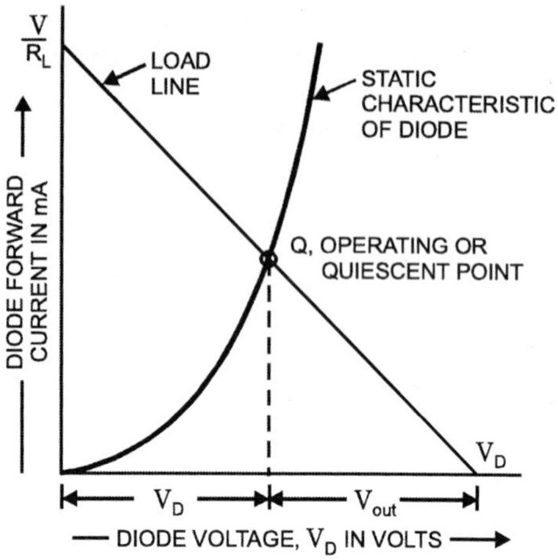

Fig. 2.41. ***The Diode and Load Line Characteristics For The Circuit Given in Fig. 2.40***

The intersection of load line with the static characteristic of diode provides the operating point, sometimes referred to as quiescent or Q point. The dc load illustrates all dc conditions that could exist within the circuit for given supply voltage V and load resistance R_L.

The graphical analysis method can yield accurate results, but it is somewhat cumbersome. However, the concept of load line and the graphical approach are useful for "visualizing" the response of the circuit and the load line is used extensively in the evaluation of electronic circuits.

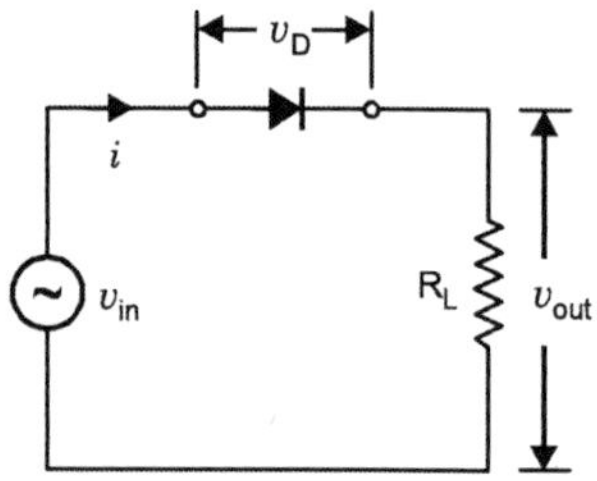

Fig. 2.42. ***Basic Diode Circuit***

Dynamic Characteristics. Now let a variable voltage v_{in} be applied to a diode circuit, depicted in Fig. 2.42. A plot of current i vs input voltage v_{in} is called the *dynamic characteristic.* The curve is drawn for different values of input voltage v_{in}.

Figure 2.43 depicts the load lines for two voltages v_{in} and v'_{in}. It is observed from Fig. 2.43 that the load lines drawn for the two input voltages v_{in} and v'_{in} are parallel to each other. This is because slope of the load line does not vary with the change in input voltage since R_L is constant. The load lines intersect the static characteristics at points A and A′ respectively. Now, vertical lines are drawn from v_{in} and v'_{in} and horizontal lines are drawn from points A and A′ respectively. A curve drawn passing through B′BO is known as dynamic characteristic.

Transfer Characteristic. The curve that relates the output voltage v_{out} to the input voltage v_{in} of any circuit is known as the *transfer* or *transmission characteristic.* Figure 2.44 illustrates the graphical method for determination of output voltage waveform for a given input voltage waveform (say, sinusoidal waveform) using transfer characteristic. The input signal waveform is drawn with its time axis vertically downward, so that the voltage axis is horizontal. The different input voltages are at different time instants t_1, t_2, t_3 etc. and corresponding vertical lines are drawn

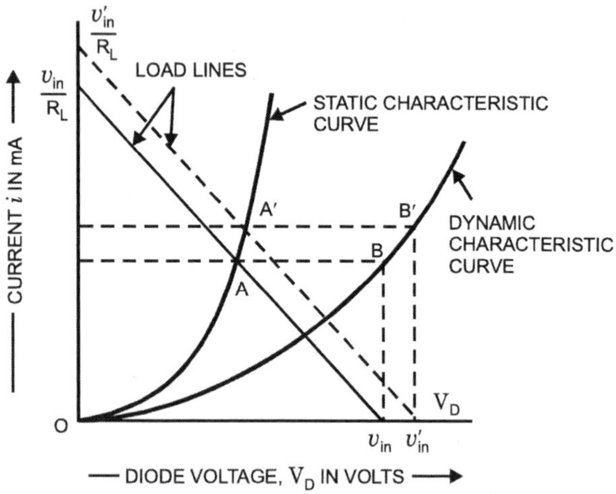

Fig. 2.43. ***Dynamic Characteristic***

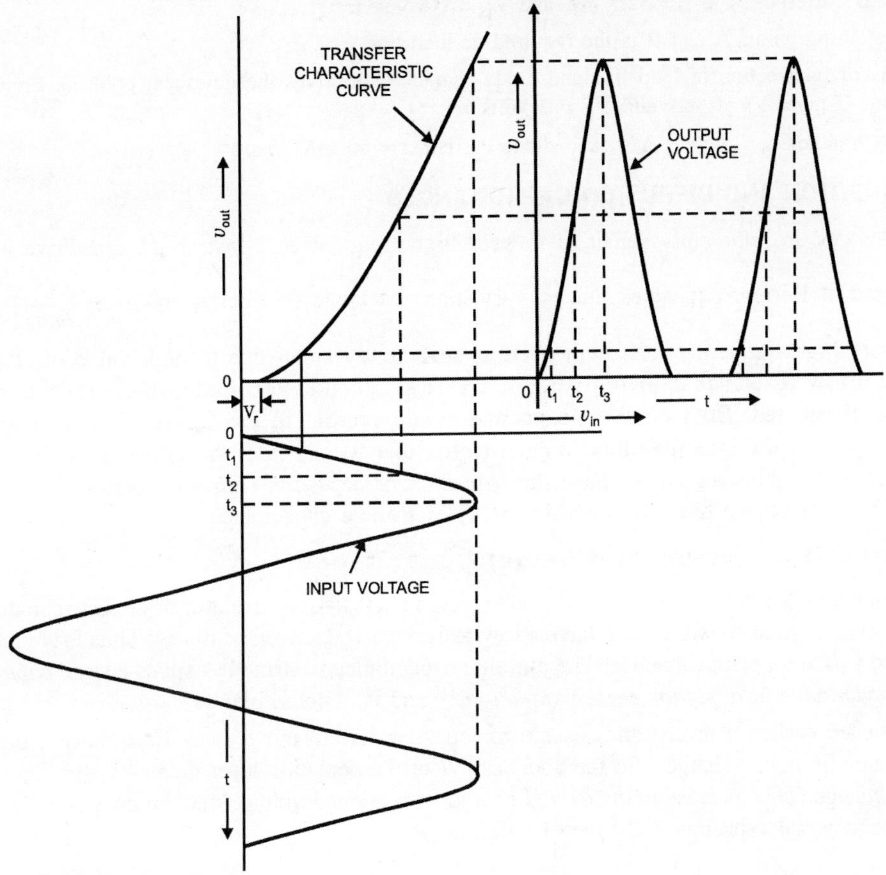

Fig. 2.44. *Graphical Method For Obtaining Output Voltage Waveform Using Transfer Curve*

to meet the transfer curve. Now horizontal lines are drawn from the intersection points on transfer curve to provide the output voltage corresponding to time instants t_1, t_2, t_3 etc. Thus output voltage waveform can be obtained by joining these points. From Fig. 2.44, it can be visualized that (*i*) the output voltage is distorted at the start due to nonlinear transfer characteristic and (*ii*) for $v_{in} < V_r$, the diode acts as a clipper and a portion of the input signal does not appear at the output.

Example 2.21. A diode is connected in series with a load resistance of 160 Ω across a dc supply voltage of 10 V. Draw a dc load line and determine the voltage drop across the diode and diode current. Diode characteristic is given.

Solution : From circuit shown in Fig. 2.40.

Supply voltage, $V = V_D + IR_L$

When diode current I is zero, the voltage across the diode, $V_D = V = 10$ V because voltage drop across the load resistance $R_L = 0$ volts.

So point A corresponds to I = 0 and V_D = 10 volts.

When diode voltage V_D is zero, voltage across the load resistance $R_L = V = 10$ V and, therefore, diode current,

$$I = \frac{V}{R_L} = \frac{10}{160} = 0.0625 \text{ A or } 62.5 \text{ mA}$$

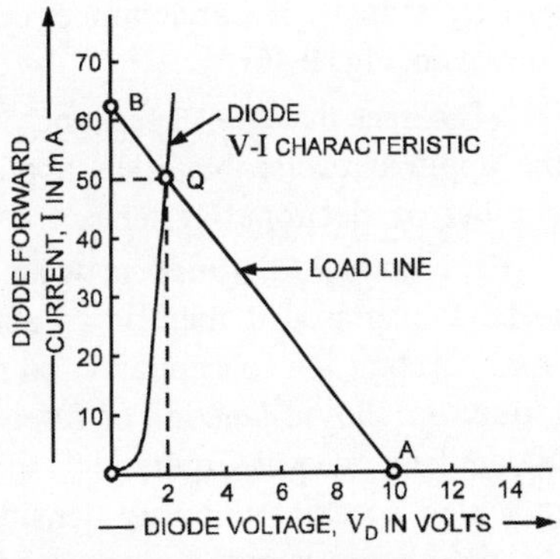

Fig. 2.45

So point B corresponds to I = 62.5 mA and V_D = 0 volts.

The line joining points A and B is the required dc load line.

The point of intersection of load line and diode characteristic gives the quiescent point Q. From Fig. 2.45, quiescent point Q gives I = 50 mA and V_D = 2 volts.

So diode voltage V_D = 2 volts **Ans.** and diode current, I = 50 mA **Ans.**

2.13. TRANSITION AND DIFFUSION CAPACITANCES

Electronic devices are inherently sensitive to very high frequencies. Most shunt capacitive effects that can be ignored at lower frequencies due to very high reactance (X_C being equal to $\frac{1}{2\pi f C}$) cannot be ignored at very high frequencies. X_C will become sufficiently small due to high value of frequency f to introduce a low reactance *shorting* path. In a P-N semiconductor diode, there are two capacitive effects to be considered. Both types of capacitances are present in the forward-bias and reverse-bias regions, but one so outweighs the other in each region that we consider the effects of only one in each region. In the reverse-bias region we have the *transition* or *depletion-region capacitance* (C_T), while in the forward bias region we have the diffusion (C_D) or storage capacitance.

2.13.1. Transition (or Space-Charge) Capacitance

When a P-N junction is reverse biased, the depletion region acts like an insulator or dielectric material while the P- and N-type regions on either side have a low resistance and act as the plates. Thus P-N junction may be considered a parallel plate capacitor. The junction capacitance is termed as space charge *capacitance* or *transition capacitance* or *depletion region capacitance* and is denoted by C_T.

As mentioned earlier, a reverse bias causes majority carriers to move away from the junction, thereby uncovering more immobile charges. So the thickness W of the depletion layer increases with the increase in reverse bias voltage. This increase in uncovered charge with applied voltage may be considered a *capacitive effect*. The incremental capacitance C_T may be defined as

$$C_T = \left|\frac{dQ}{dV}\right| \qquad ...(2.41)$$

where dQ is the increase in charge due to increase in voltage, dV.

$$\text{But } i = \frac{dQ}{dt} = C_T \frac{dV}{dt} \qquad ...(2.42)$$

Since the depletion region increases with the increase in reverse-bias potential, the resulting transition capacitance decreases as shown in Fig. 2.46.

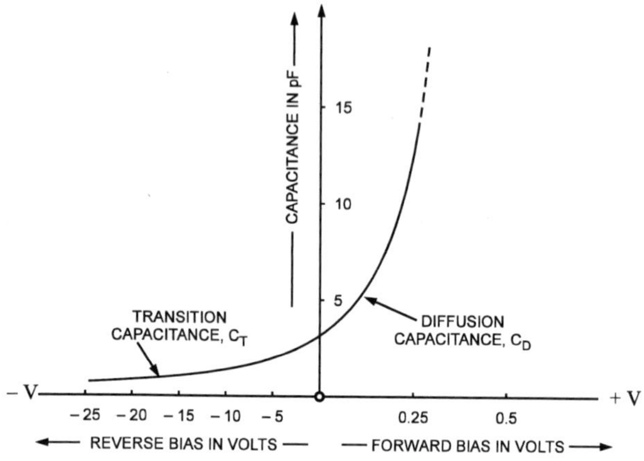

Fig. 2.46

The fact that the capacitance depends on the applied reverse bias, has application in a number of electronic systems.

1. *Step Graded Junction.* A junction is said to be step graded if there is an abrupt change from acceptor ion concentration on the P-side to donor ion concentration on the N-side. Such a junction gets formed in alloyed junction or fused junction diode. Usually the acceptor density N_A and the donor density N_D are kept unequal. Figure 2.47 shows the charge density as a function of distance from an alloy junction in which acceptor impurity density is assumed to be much larger than the donor concentration. Since the net charge is zero,

$$eN_AW_p = eN_DW_n \quad ...(2.43)$$

When N_A is much larger than N_D, the $W_p << W_n$ and $W_n \simeq W$.

The relationship between potential and charge density is given by the Poisson's equation,

$$\frac{d^2V}{dx^2} = \frac{-eN_D}{\in} \quad ...(2.44)$$

where $\in$ is the permittivity of the semiconductor and is given as $\in = \in_r \in_0$ where $\in_r$ is the relative permittivity of the dielectric and $\in_0$ is the permittivity of free space.

The electric lines of flux start on the positive donor ions and terminate on the negative acceptor ions. So there are no flux lines to the right of the boundary $x = W_n$ in Fig. 2.47 and field intensity

$$\in = -\frac{dV}{dx} = 0$$

at $x = W_n \simeq W$

Integrating Eq. (2.44) subject to these boundary conditions, we have

$$\frac{dV}{dx} = \frac{-eN_D}{\in}(x - W)$$

$$= -\varepsilon \quad ...(2.45)$$

Neglecting the small voltage drop across W_p, we may arbitrarily choose $V = 0$ at $x = 0$

Integrating Eq. (2.45) subject to these conditions

$$V = \frac{-eN_D}{2\in}(x^2 - 2Wx) \quad ...(2.46)$$

The linear variation in field intensity and the quadratic dependence of potential upon distance are plotted in Fig. 2.47 (*c*) and (*d*). These graphs should be compared with the corresponding curves of Fig. 1.28.

At $x = W$, $V = V_B$, the barrier height. Thus

$$V_B = \frac{eN_DW^2}{2\in} \quad ...(2.47)$$

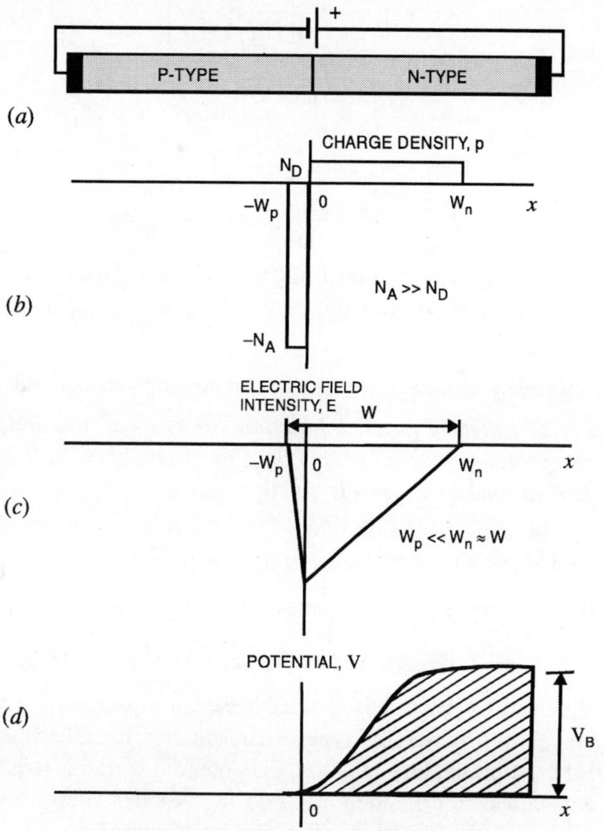

Fig. 2.47. ***Charge Density, Field Intensity and Potential Variation With Distance x***

Here $V_B = V_0 - V$ where V is the negative number for an applied reverse bias and V_0 is the contact potential.

Hence, the width of depletion layer increases with applied reverse bias and varies as $\sqrt{V_B}$.

The total charge density of N-type material with area of junction A is given by

$$Q = eN_DWA$$

The transition capacitance C_T, given by Eq. (2.41), is

$$C_T = \left|\frac{dQ}{dV}\right| = eN_DA\left|\frac{dW}{dV_B}\right|$$

From Eq. (2.47), $\left|\dfrac{dW}{dV_B}\right| = \dfrac{\in}{eN_D W}$, and therefore,

$$C_T = \frac{\in A}{W}^* \qquad ...(2.48)$$

where $\in$ is the absolute permittivity of the semiconductor medium, A is the cross-sectional area of the junction and W is the width of the depletion layer and is given as

$$W^2 = \left[\frac{2 \in V_B}{e}\right]\left[\frac{1}{N_A} + \frac{1}{N_D}\right] \qquad ...(2.49)$$

$$\text{or } W = \sqrt{\frac{2 \in V_B}{e N_D}} \quad \text{when } N_A >> N_D \qquad ...(2.50)$$

$$\text{So } C_T = \in A \sqrt{\frac{eN_D}{2 \in V_B}} = A\sqrt{\frac{N_D}{V_B}} \cdot \sqrt{\frac{e\in}{2}} \qquad ...(2.51)$$

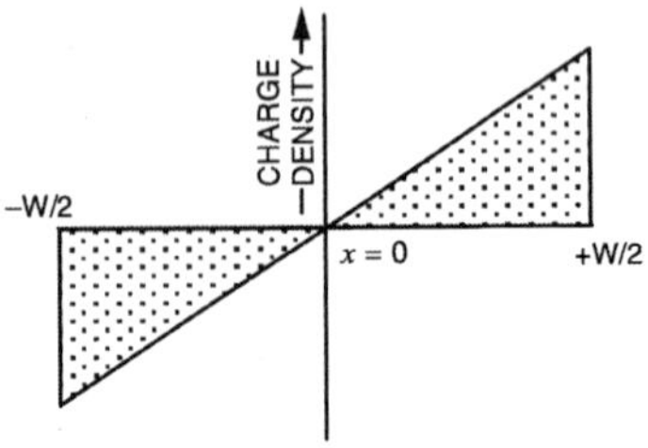

Fig. 2.48. ***Charge-Density Variation With Distance in a Grown P-N Junction***

Thus in a step graded junction, C_T is inversely proportional to the square root of the junction voltage V_B where V_B is given as

$$V_B = V_0 - V_R \qquad ...(2.52)$$

where V_R is a negative number indicating the applied reverse bias and V_0 is the contact potential.

2. *Linearly Graded Junction (or Grown Junction).* A junction is said to be linearly graded if the charge density varies gradually (almost nearly) with the distance in the transition region. Such a junction gets formed in a growth junction diode.

In such a junction diode also, C_T is given by Eq. (2.48). Thus the expression for C_T for a linearly graded junction is the same as for a step graded junction. But here W is the total width of the space-charge layer. However, it now comes out that W varies as $V_B^{1/3}$ instead of $V_B^{1/2}$.

2.13.2. Diffusion (or Storage) Capacitance

When a P-N junction is forward biased, a capacitance, which is much larger than the transition capacitance, comes into play. This type of capacitance is called the *diffusion capacitance*, C_D. We know that for a forward-biased P-N junction, the potential barrier is reduced. Now the electrons from N-side enter the P-side and holes from P-side enter into the N-side. These charge carriers diffuse away from the junction and progressively recombine. The density of charge carriers is high near the junction and decays exponentially with distance. Thus a charge is stored on both sides of the junction when forward biased. It is observed that the amount of stored charge varies with the applied potential as for a true capacitor. It is convenient to introduce an incremental capacitance, called *diffusion* or *storage capacitance* C_D given by the equation

$C_D = \dfrac{dQ}{dV}$ where dQ represents the change in number of minority carriers stored outside the depletion region when a change in voltage across diode, dV, is applied.

If τ is the mean lifetime of charge carriers, then a flow of charge Q yields a diode current I given as

$$I = \frac{Q}{\tau} \quad \text{or} \quad Q = \tau I$$

* It is interesting to note that this relationship is exactly the expression derived for a parallel plate capacitor of area A (m^2) and plate separation W (m), containing a dielectric of permittivity $\in$.

and from Eq. (2.20), $I = I_0 (e^{V/\eta V_T} - 1)$

$$\text{So } Q = \tau I_0 (e^{V/\eta V_T} - 1) = \tau I_0 e^{V/\eta V_T} \qquad \because e^{V/\eta V_T} >> 1$$

So diffusion capacitance

$$C_D = \frac{dQ}{dV} = \frac{d\tau I_0 e^{V/\eta V_T}}{dV} = \frac{\tau I_0}{\eta V_T} e^{V/\eta V_T} = \frac{\tau(I + I_0)}{\eta V_T}$$

For a forward bias greater than a few tenths of a volt, $\left[\frac{V}{\eta V_T}\right] >> 1$ and according to Eq. (2.20) $I >> I_0$ and so the diffusion capacitance is given as

$$C_D = \frac{\tau I}{\eta V_T} \qquad \text{...(2.53)}$$

Thus the diffusion capacitance is directly proportional to the forward current through the diode. This relationship occurs since the number of majority carriers injected into the depletion region is directly proportional to the forward current of the diode. The variation in diffusion capacitance with the variation in forward bias applied across a P-N junction is shown in Fig. 2.46.

This capacitance assumes great significance in the operation of devices which are required to switch rapidly from forward bias to reverse bias. Large value of C_D, delays both of the switching-on and the switching-off.

In fact in a forward-biased diode, both the diffusion capacitance C_D and the transition capacitance C_T are present, but $C_D >> C_T$. Typically C_D is more than a million times greater than C_T. So in a forward-biased diode C_T may be neglected and only C_D may be considered. Typical value of C_D is 0.02 μF.

Similarly in a reverse-biased diode both C_D and C_T are present but $C_T >> C_D$. Hence in a reverse-biased diode C_D is neglected and only C_T is considered.

The capacitive effects described above may be represented by a capacitor connected in parallel with an ideal diode, as illustrated in Fig. 2.49.

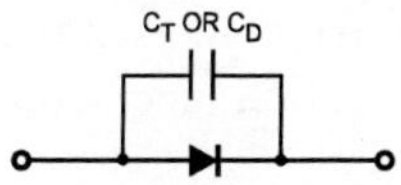

Fig. 2.49

2.14. EQUIVALENT CIRCUIT OF A DIODE

It is usually advantageous to replace a device or system by its equivalent circuit. An equivalent circuit of a device or a system is a combination of electric elements properly chosen to best represent the actual terminal characteristics of the device or the system. Once the equivalent circuit is determined, the device can be replaced by its equivalent circuit without severely affecting the behaviour of the overall system.

We have seen that a forward-biased junction offers a dynamic or ac resistance $r' = (r + r_B)$ and possesses diffusion capacitance C_D which comes into picture only when frequency of the applied voltage is very high. Hence, it can be represented by the equivalent circuit shown in Fig. 2.50 (*a*). An opposing battery has been connected in series with r' to account for the junction barrier potential V_0. The complete equivalent circuit for a forward biased diode is shown in Fig. 2.50 (*a*).

The equivalent circuit for the forward-biased diode may be modified to form a *small-signal ac equivalent circuit.* This circuit is employed for diodes which are maintained in a forward-bias condition, but which are subjected to small variations in voltage V and current I. The small-signal ac equivalent circuit is drawn [Fig. 2.50 (*b*)] by dropping the opposing battery representing the junction barrier potential V_0 from the circuit shown in Fig. 2.50 (*a*).

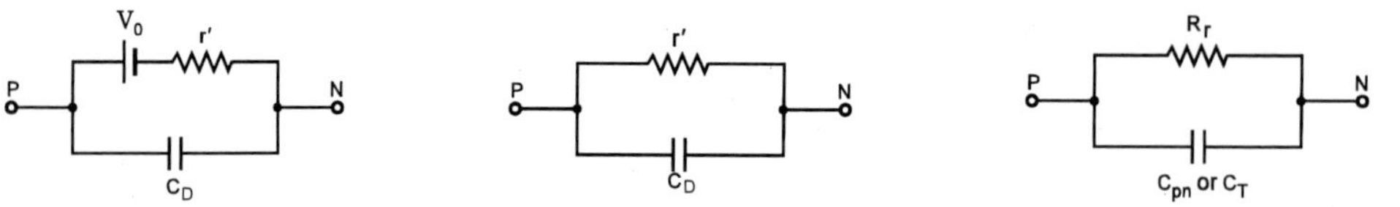

(a) Equivalent Circuit For Forward Biased Diode
(b) AC Equivalent Circuit For Forward Biased Diode
(c) Equivalent Circuit For Reverse Biased Diode

Fig. 2.50

A reverse-biased diode can be simply represented by the reverse resistance R_r in parallel with the depletion layer capacitance C_{pn} or C_T. This is illustrated in Fig. 2.50 (*c*).

2.15. P-N DIODE SWITCHING TIMES

A P-N junction diode may be used as an electrical switch. The electrical circuit can be made 'on' and 'off' by forward biasing or reverse biasing the diode. In both the cases, the diode response is accompanied by transient, and an interval of time elapses before the diode recovers to its steady state. The duration of time elapsed before the recovery of the steady state of the diode is called the recovery time, t_r. On switching over of a diode from reverse-bias condition to forward-bias condition, the diode takes certain time interval, called the *forward recovery time*, t_{fr}. The forward recovery time t_{fr} is defined as the time interval between the instant of 10% diode voltage and the instant this voltage reaches 90% of its final value. It is found that t_{fr} does not usually constitute a serious practical problem and so only the more important situation of reverse recovery will be discussed here.

2.15.1. Diode Reverse Recovery Time

As already discussed, when a P-N junction is forward biased a large number of electrons cross over from N-region to P-region and similarly a large number of holes cross over from P-region to N-region. The electrons in the P-type region and the holes in the N-type region establish a large number of minority carriers in each region. The variation of steady-state minority carrier density with distance x from the junction in a forward biased P-N junction is shown in Fig. 2.51 (*a*). The number of minority carriers is very

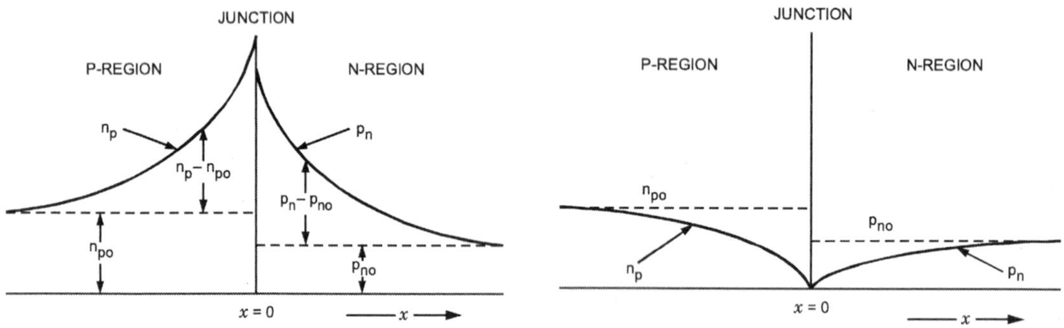

(a) Forward Biased Junction
(b) Reverse Biased Junction

Fig. 2.51. ***Minority-Carrier Density Distribution as a Function of Distance x from The Junction***

large. These minority carriers have, in each case, been supplied from the other side of the junction, where, being majority carriers, they are in plentiful supply. The concentration of holes in N-region, p_n is increased above its thermal-equilibrium value p_{no}. Similarly the concentration of electrons in P-region, n_p is increased above its thermal-equilibrium value n_{po}. The difference of excess of holes *i.e.*, $(P_n - P_{no})$ and excess of electrons *i.e.* $(n_p - n_{po})$ is also shown in the figure.

When the junction is reverse biased, the steady-state density of minority carriers is as shown in Fig. 2.51 (*b*). Far from the junction the minority carriers are equal to their thermal-equilibrium values p_{no} and n_{po} as is also the situation in Fig. 2.51 (*a*). As the minority carriers approach the junction, they are rapidly swept across, and the density of minority carriers diminishes to zero at this junction. The reverse saturation current I_0 is very small because the density of the thermally generated minority carriers is very small.

If the external voltage is suddenly reversed in a diode circuit which has been carrying current in the forward direction, the diode current will not immediately fall to its steady-state reverse voltage value, as illustrated in Fig. 2.52. However, because of the large number of minority carriers in both of the regions, the diode current will simply reverse, as illustrated in Fig. 2.52 and stay at this measurable level required for the minority carriers to reduce to the distribution shown in Fig. 2.51 (*b*) from the minority-carrier distribution shown in Fig. 2.51 (*a*). The diode will continue to conduct for a time, known as *diode reverse recovery time,* t_{rr}, until excess minority-carrier density $P_n - P_{no}$ or $(n_p - n_{po})$ has dropped normally to zero. In this situation, the current is determined by the external resistance in the diode circuit.

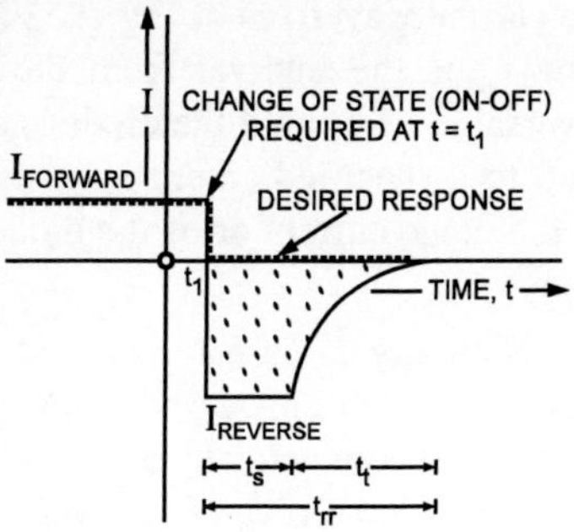

Fig. 2.52

2.15.2. Storage and Transition Times

The sequence of events occurring on reverse biasing of a conducting diode is shown in Fig. 2.53.

Let the input voltage v_{in} applied to a diode-resistance circuit shown in Fig. 2.53 (*a*) be reversed abruptly at time instant t_1. Up to a time t_1, the diode is conducting in forward direction and input voltage $v_{in} = V_F$. Assuming load resistance R_L to be large enough to cause voltage drop across it large in comparison with the voltage drop across the diode, the current flowing through the circuit, $i \simeq \frac{V_F}{R_L} \simeq I_F$.

At instant of time t_1, the input voltage reverses abruptly to a value $-V_R$. Now the current does not drop to zero, but instead reverses and remains at a value $i \simeq \frac{-V_R}{R_L} \simeq -I_R$ for the time duration t_s (storage time interval) required for the minority carriers to return to their majority-carrier state in the opposite region. At instant of time t_2, as seen in Fig. 2.53 (*c*), the injected minority-carrier density at $x = 0$ has reached the equilibrium state.

Now if we consider the diode voltage v, we find that at time instant t_1 the diode voltage falls slightly by $(I_F + I_R) R_d$, where R_d is the diode ohmic resistance, but does not reverse. At time instant t_2 the process of sweeping of the excess minority carriers in the vicinity of junction back has completed. So the diode voltage begins to reverse and the magnitude of diode current begins to decrease. During the time interval t_1 to t_2, the stored minority charge has remained stored and, therefore, this time interval is called the *storage time* t_s. This is illustrated in Fig. 2.53 (*e*).

After all excess minority carriers are swept out or recombined in the neighbourhood of junction, the diode voltage begins to reverse and the magnitude of diode current begins to fall exponentially to the normally steady-state reverse saturation current value. The time duration which elapses between t_2 and the time instant t_3 when the diode has nominally recovered, is called the *transition time* t_t. This recovery time interval will be completed when the minority carriers, which are at some distance from the junction, have diffused to the junction and crossed it and when, in addition, the junction transition capacitance across the reverse-biased junction has charged through load resistance R_L to the voltage $-V_R$.

The sum of the storage time t_s and transition time t_t is called the diode *reverse recovery time* t_{rr} *i.e.*, $t_s + t_t = t_{rr}$.

Manufacturers normally specify the reverse recovery time of a diode t_{rr} in a typical operating condition in terms of the current waveform of Fig. 2.53 (*d*). The time t_{rr} is the interval from the current reversal at $t = t_1$ until the diode has recovered to a specified extent in terms either of the diode current or of the diode resistance.

Naturally t_{rr} is an important consideration in high speed switching applications. Most commercially available switching diodes have reverse recovery time t_{rr} in the range of a few nanoseconds to 1 μs. Units are available, however, with a t_{rr} of only a few hundred picoseconds.

2.16. SWITCHING DIODES

In discussing the rectifier diodes, the importance of reducing the reverse-bias current and the power losses under forward bias to the minimum was emphasized. But in case a junction diode is to be employed for high speed switching applications, special consideration is required to be given to its charge control properties. A diode with fast switching properties must either store very little charge in the neutral regions for steady forward currents, or have a very short carrier lifetime, or both.

Switching speed of a diode can be improved by adding efficient recombination centers to the bulk material. For silicon diodes, Au (gold) doping can serve this purpose. The carrier lifetime varies approximately with the reciprocal of the recombination center concentration. For example, a P^+-N silicon diode may have τ_h = 1 μs and a reverse recovery time t_{rr} of 0.1 μs before Au doping. In case of addition of 10^{14} Au atoms/cm^3 reduces the lifetime to 0.1 microsecond and storage delay time t_{sd} (time required for the stored charge to become zero) to 0.01 μs. 10^{15} cm^{-3} Au atoms could reduce τ_h to 0.01 μs and t_{sd} to 1 ns. This process cannot be continued indefinitely, however, the reverse current due to creation of carriers from the Au centres in the depletion region becomes appreciable with large Au concentration. In addition, as the Au

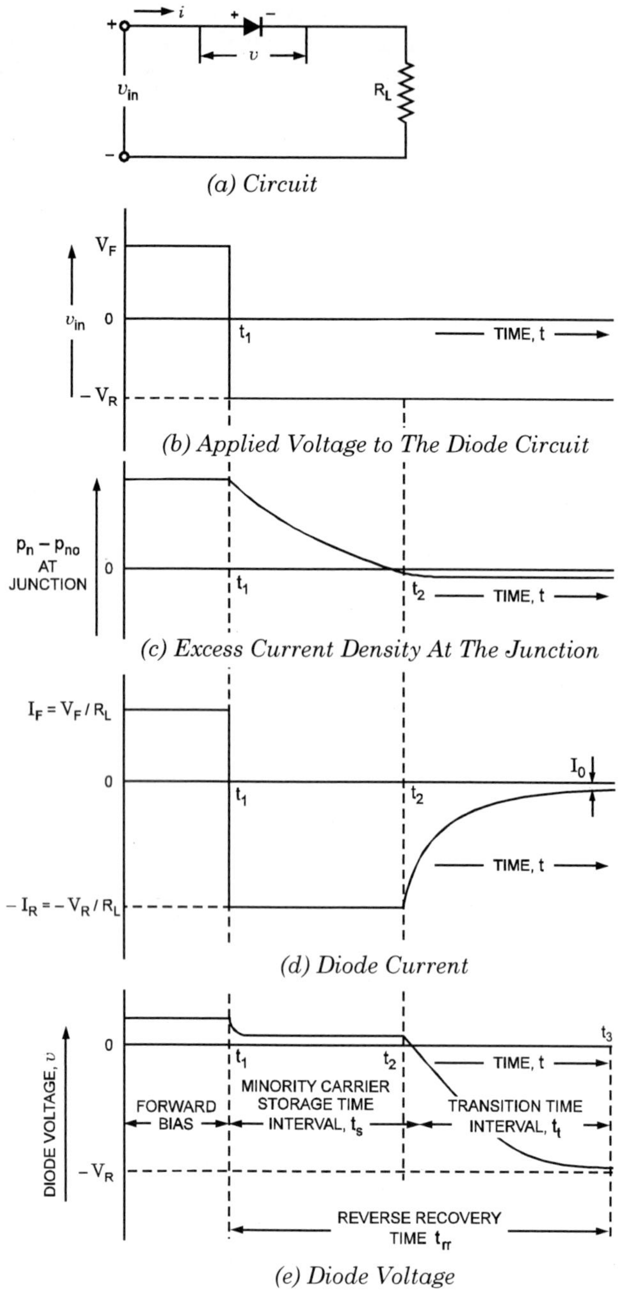

Fig. 2.53. ***Waveform in a Diode On Reversing of Bias***

concentration approaches the lightest doping of the junction, the equilibrium carrier concentration of that region can be affected.

A diode switching time can also be improved by making the lightly doped neutral region shorter than a minority carrier diffusion length. This is the *narrow base diode.* In this case the stored charge for forward conduction is very small because most of the injected carriers diffuse through the lightly doped region to the end contact. When such a diode is switched to reverse conduction, very little time is required for eliminating the stored charge in the narrow neutral region.

2.17. P-N DIODE APPLICATIONS

An ideal P-N junction diode is a two terminal polarity sensitive device that has zero resistance (*i.e.* diode conducts) when it is forward biased and infinite resistance (*i.e.* diode does not conduct) when reverse biased. Because of this property the diode finds use in many applications as enumerated below :

(*i*) As rectifiers in dc power supplies.

(*ii*) In demodulation or detector circuits.

(*iii*) In clamping networks employed as dc restorers in TV receivers and voltage multipliers.

(*iv*) In clipping circuits used as waveshaping circuits in computers, radars, radio and TV receivers.

(*v*) As switches in digital logic circuits.

The same P-N junction with different doping concentrations finds special applications as follows :

(*i*) As zener diodes in voltage regulators, peak clippers, and in switching operation.

(*ii*) As tunnel diodes as a relaxation oscillator at microwave frequencies.

(*iii*) As light-emitting diodes (LEDs) in digital displays.

(*iv*) As LASER diodes in optical communications.

(*v*) As varactor diodes in tuning sections of radio and TV receivers.

(*vi*) As detectors (PIN photodiode) in optical communication circuits.

2.18. DIODE DATA SHEETS

For selection of a diode for a particular application, the data sheets provided by the device manufacturers must be consulted.

Most data sheets start off with the device type number at the top of the sheet, a short descriptive title and a list of major applications of the device. The information is usually followed by mechanical data in the form of illustration showing the package shape and dimensions. The absolute maximum ratings of the diode at 25°C are listed next. These are maximum voltages, currents, temperatures etc., that the device can bear/carry without breaking down. It is very important, for the safe operation of the device, the maximum ratings given in the data sheets are never exceeded. For reliability the maximum ratings should not even be approached. Also, the maximum ratings must be adjusted downward for operation at temperatures exceeding 25°C. Following the absolute maximum ratings, there is normally a complete list of electrical characteristics for the device. Again, these are specified at 25°C, and allowances are necessary for variation of temperature.

Some of the important parameters are considered below :

1. *Peak reverse voltage or peak inverse voltage, V_R or V_{RM}*. This is the absolute peak of voltage that must be applied in reverse across the diode.

2. *Steady-state forward current, I_0 or I_F.* This is the maximum current that may be passed continuously through the diode.

3. *Peak forward surge current, $I_{FM\,(surge)}$.* The surge current is very much higher than the normal maximum forward current. This rating defines the maximum value and the time interval for such surges in current level. It is a current that may flow briefly when a circuit is first switched on.

4. *Peak repetitive forward current* is the maximum instantaneous value of repetitive forward current. Note that since it is at this level for a brief period of time, its level can be higher than the continuous level.

5. *Average rectified current.* A half-wave-rectified signal has an average value defined by I_{av} = 0.318 I_{peak}. The *average current rating* is lower than the *continuous* or *peak repetitive* forward currents because a half-wave current waveform will have instantaneous values much higher than the average value.

6. *Static maximum voltage drop, $V_{F\,(max)}$.* This is the maximum forward voltage drop for a given forward current and device temperature.

7. *Continuous power dissipation at 25°C, P.* This is the maximum power that the device can safely dissipate on a continuous basis in free air. This rating must be downgraded for operation at higher temperatures, and may be upgraded when the device is mounted on a heat sink.

8. *Reverse recovery time, t_{rr}.* This is the maximum time for the device to switch from on to off.

Depending on the type of diode considered, additional data may also be provided, such as frequency range, noise level, switching time, thermal resistance levels, and peak repetitive values.

Appendix 1.1 shows a data sheet for rectifier diodes, which indicates that these devices can pass an average forward current of 1 A and have maximum reverse voltages ranging from 50 V to 1,000 V. The noteworthy point is that the data sheet does not bear the reverse recovery time t_{rr}. The reason is that rectifier diodes are generally intended for low frequency (50 Hz to perhaps 400 Hz) applications in which switching time is not important. Obviously, these devices can be described as relatively high-current diodes, and some are also high-voltage diodes.

HIGHLIGHTS

1. A diode is a unidirectional device permitting the easy flow of current in one direction but restraining the flow in opposite direction. A semiconductor diode is simply a P-N junction with connecting leads or terminals on its two sides. P-N junction is also called the crystal diode because it is grown out of crystal.
2. A P-N junction may get open circuited or short circuited and the best instrument for testing a P-N junction is an ohmmeter.
3. An ideal diode is a two terminal polarity sensitive device that has zero resistance when it is forward biased and infinite resistance when reverse biased.
4. The current in a P-N diode is bipolar in character (because it is made up of both positive and negative charge carriers) and the total current is constant throughout the device, but the proportion of current due to holes and that due to electrons varies with the distance.
5. In the forward-bias condition the diode current increases exponentially with the increase in voltage across the diode. Diode current, when the diode is forward biased, is given by

$$I = I_0(e^{V/V_T} - 1) \text{ where } V_T = \frac{T}{11{,}600}$$

and $I_0 = Ae\left[\frac{D_h}{L_h N_D} + \frac{D_e}{L_e N_A}\right] n_i^2$

Reverse saturation current may be also given by

$$I_0 = A_0 T^3 e^{-V_{GO}/V_T}$$

6. The reverse saturation current I_0 will just double in magnitude for every 10-fold increase in temperature.
   The temperature coefficient for the forward voltage of a P-N junction is approximately –1.8 mV/°C for silicon diode and –2.02 mV/°C for germanium diode.
7. The static and dynamic resistances of a diode may be determined from the characteristic curves and represent the dc and ac opposition to the flow of current offered by the diode.
   The dc resistance of a diode is determined by the ratio of the diode voltage and current at the point of interest and is not sensitive to the shape of the curve. The dc resistance decreases with the increase in diode current or voltage.
   The ac resistance of a diode is sensitive to the shape of V-I characteristic curve in the region of interest and decreases for higher level of diode current or voltage.
8. An equivalent circuit of a diode is a linearization of the diode's characteristic curve and typically consists of an ideal diode in series with a battery (representing the cutin, offset or threshold voltage) and a resistance equal to the dynamic resistance of the diode over its linear portion.
   The cutin or threshold voltage is about 0.7 V for silicon diodes and 0.3 V for germanium diodes.
9. A P-N junction exhibits a transition capacitive effect of several picofarads in reverse bias, but have many microfarads of diffusion capacitance in the forward-biased state.
   The capacitance of diode increases exponentially with the increase in the forward-bias voltage. Its lowest levels are in the reverse-bias region.
   In fact in a forward biased diode, both the diffusion capacitance C_D and the transition capacitance C_T are present, but $C_D >> C_T$. Typically C_D is more than a million times greater than C_T. So in a forward biased diode C_T may be neglected and only C_D may be considered. Typical value of C_D is 0.02 µF.
   Similarly in a reverse biased diode both C_D and C_T are present but $C_T >> C_D$. Hence in a reverse biased diode C_D is neglected and only C_T is considered.
   The capacitive effects described above may be represented by a capacitor connected in parallel with an ideal diode, as illustrated in Fig. 2.49.
10. When a P-N junction is switched from forward bias to reverse bias, the stored excess minority carrier charge must be removed from the junction. The time required to remove this charge is called the storage time and is a limiting factor in the switching speed of a diode.
11. P-N junction diodes find use in many applications like rectifiers, demodulation circuits, clipping circuits, clamping circuits and switching circuits.

EXERCISES

1. What is a crystal diode ? Why is silicon preferred over germanium in the manufacture of semiconductor devices ?
2. Explain how a P-N junction diode can be tested.
3. Draw and explain V-I characteristics of diode with the help of diode equation of current.
4. Explain how diode works as on and off switch. Also derive diode current equation.
5. What is an ideal diode ? How does an actual diode differ from an ideal diode ?
6. Plot the hole current, the electron current, and the total current as a function of distance on the both sides of a P-N junction. Indicate the transition region.
7. Sketch typical forward and reverse characteristics for a germanium diode and for a silicon diode. Compare the characteristics and explain why the reverse saturation current in a silicon diode is much smaller than that in a comparable germanium diode.
8. Explain the effect of temperature on V-I characteristics of a P-N junction diode.
   [G.G.S.I.P. Univ. Analog Electronics December 2009]
9. Explain the P-N junction diode characteristics in forward and reverse biased conditions. What is the effect of temperature on these characteristics ? **[G.G.S.I.P. Univ. Analog Electronics December 2010]**

10. Explain clearly the difference between static and dynamic resistance of a P-N junction diode.
[G.G.S.I.P. Univ. Analog Electronics-I, December 2004]
11. Using diode equation show that the dynamic resistance of an ideal P-N junction diode under forward bias condition is inversely proportional to the forward current.
[G.G.S.I.P. Univ. Analog Electronics-I, December 2004]
12. Define the static and dynamic resistance of the diode, how these resistances are measured ?
13. What is meant by dc and ac resistance of a crystal diode ? How the diode resistances can be determined from its V-I characteristics ?
14. How does dynamic resistance of a diode vary with current ? Draw the piecewise linear characteristic of a diode.
15. Sketch the piecewise linear characteristics of a diode. What are the approximate cutin voltages for silicon and germanium ?
16. Consider a circuit consisting of diode D, a resistance R and a signal source v_i in series. Define
(1) Static characteristics, (2) Dynamic characteristics (3) Transfer characteristics.
17. Define diffusion capacitance of a p-n junction diode. Obtain an expression for the same. Why is the diffusion capacitance negligible for a reverse-biased diode ?
18. Show that the transition capacitance of a step graded P-N junction diode is given by $C_T = A\left(\frac{q\varepsilon N_D}{2(V_0 - V_D)}\right)^{1/2}$. Assume $N_A >> N_D$. All the symbols have their usual meaning.
[G.G.S.I.P. Univ. Analog Electronics December 2009]
19. Derive an expression for transition capacitance of a linearly graded P-N junction. Explain the significance of each symbol in the expression. **[G.G.S.I.P. Univ. Analog Electronics December 2010]**
20. Explain the significance of diffusion capacitance and derive expression for it.
[G.G.S.I.P. Univ. Analog Electronics December 2010]
21. What is the difference between diffusion and transition capacitance ? How will you represent the capacitive effect of a practical diode on an ideal diode ?
22. What are the two types of capacitances across P-N junction ? Which of these is more important in case of forward bias ? Why ?
23. Draw the equivalent circuit of a junction diode.
24. Draw the large signal model of diode for both forward and reverse bias and explain the model.
[G.G.S.I.P. Univ. Analog Electronics December 2010]
25. Explain the large signal model and small signal model of a PN junction diode in both forward bias and reverse bias condition. **[G.G.S.I.P. Univ. Analog Electronics December 2010]**
26. Draw switching waveform of diode when bias is suddenly changed from forward to reverse. Indicate diode reverse recovery time, storage time and transition time.
27. Discuss "junction diode switching time" to justify diode reverse recovery time, storage time and transition time in brief. Elaborate these definitions by labelling an appropriate waveform for diode.
28. Explain (*i*) storage time (*ii*) transition time (*iii*) reverse recovery time in a PN junction diode with the help of waveforms. How the storage time can be reduced ?
[G.G.S.I.P. Univ. Analog Electronics December 2010]

SHORT ANSWER TYPE QUESTIONS WITH ANSWERS

Q. 1. What is a P-N junction ?
Ans. The contact surface between the layers of P-type and N-type semiconductor pieces placed together so as to form a P-N junction is called the P-N junction.
Q. 2. Why is silicon preferred over germanium in the manufacture of semiconductor devices ?

Ans. The silicon semiconductor devices have, in general, higher PIV and current ratings and wider temperature range than germanium semiconductor devices and, therefore, silicon is preferred over germanium in the manufacture of semiconductor devices.

Q. 3. What does the arrowhead represent in a schematic symbol of a P-N junction ?

Ans. The arrowhead in the schematic symbol of a P-N junction indicates the direction of conventional current flow when the diode is forward biased.

Q. 4. How diode acts as a rectifier ?

Ans. Rectifier is a device which converts the sinusoidal ac voltage into unidirectional pulsating voltage. Diode is an electronic device consisting of two elements known as cathode and anode. Since in a diode electrons can flow in only one direction *i.e.* from cathode to anode, the diode provides the unilateral conduction necessary for rectification. This is true for diodes of all types—vacuum, gas-filled, crystal or semiconductor diodes.

Q. 5. In a semiconductor diode P-side is grounded and N-side is applied a potential of –3 V. Will the diode conduct ? Explain.

Ans. The *p-n* diode connected with *p*-side grounded and with *n*-side to –ve potential of 3 V, the diode will conduct, because *n*-side is at negative potential w.r.t. positive side.

Q. 6. What is a junction diode ? Draw ideal diode characteristics.

Ans. Diode is a two terminal device which conducts exponentially in one direction and has a poor conduction in reverse direction. It will be an ideal case if its conduction in one direction is like a perfect conductor and in other direction as perfect insulator. Ideal diode characteristic is shown in Fig. 2.3 (*b*).

Q. 7. What is the main difference between the characteristic of a simple switch and those of an ideal diode ?

Ans. The important difference between the characteristic of a simple switch and those of an ideal diode is that an ideal diode acts like an automatic switch—the switch is closed when the diode is forward biased and is opened when reverse biased.

Q. 8. What is reverse saturation current ?

Ans. Reverse current of a diode is due to minority carriers and is caused when the diode is reverse biased. The reverse current is constant, independent of the applied reverse bias, consequently it is called the reverse saturation current.

Q. 9. Is reverse saturation current of a junction diode independent of reverse bias voltage ?

Ans. Yes.

Q. 10. Germanium is more temperature dependent than silicon. Why ?

Ans. Because the reverse saturation current in case of a germanium diode is approximately 1,000 times larger.

Q. 11. What is the effect of temperature on the reverse saturation current of a diode ?

Ans. Reverse saturation current, theoretically, increases by 8 percent per °C for silicon and 11 percent per °C for germanium. But from experimental data it is found that the reverse saturation current increases 7 percent per °C for both silicon and germanium. This is because a surface leakage current component of reverse saturation current is independent of temperature. Since $(1.07)^{10} \simeq 2.0$, the reverse saturation current approximately doubles for every 10°C rise in temperature.

Q. 12. Explain the effect of temperature on the V-I characteristics of the diode.

Ans. Reverse saturation current I_0 and volt equivalent of temperature V_T increase with the increase in temperature. I_0 doubles for every 10°C rise in temperature. This causes lowering of threshold or cutin voltages. With the increase in temperature, forward characteristic becomes more ideal. In reverse bias, the breakdown voltage increases with the increase in temperature. Refer to Fig. 2.11.

Q. 13. What is static resistance of a diode ?

Ans. The static or dc resistance of a diode is the resistance offered by it to the direct current. It is defined as the ratio of the diode voltage and current at the point of interest and is not sensitive to the shape of V-I characteristic curve. It decreases with the increase in diode current or voltage.

Q. 14. Define dynamic resistance of a P-N junction diode in forward biased condition.

Ans. The resistance offered by a P-N junction diode to the changing forward current is defined as the dynamic resistance.

It may be given as

$$\text{AC or dynamic resistance, } r = \frac{\text{Small change in forward voltage}}{\text{Small change is forward current}}$$

Thus ac resistance of a diode is sensitive to the shape of *v-i* characteristic curve in the region of interest. It decreases for higher level of diode current or voltage.

Q. 15. Explain what do you understand by the expression "linearization of a nonlinear device".

Ans. A nonlinear device, such as a diode, has a nonlinear voltage-current characteristic. In a diode V and I are related by equation

$$I = I_0 (e^{V/\eta V_T} - 1)$$

Diode shows nonlinearity until the threshold or cutin voltage V_T is reached. Voltage-current characteristics of the diode follow a linear relationship after it has attained the cutin voltage, just as for a resistance of very low magnitude. It means that diode exhibits nonlinearity for lower voltages ($V < V_T$) but linearity for voltages exceeding V_T.

The equivalent circuit for the linear approximation is a constant-voltage source in series with a resistor [Fig. 2.14 (*b*)].

Q. 16. What is step graded junction ?

Ans. A junction is said to be step graded if there is an abrupt change from acceptor ion concentration on the P-side to donor ion concentration on the N-side such as alloyed or fused junctions.

Q. 17. What is linear graded junction ?

Ans. A junction is said to be linearly graded if the charge concentration varies gradually with the distance in its transition region such as a growth junction.

Q. 18. What are the two types of capacitances across P-N junction ? Which of these is more important in case of forward bias ?

Ans. In a P-N junction, there are two capacitive effects which are to be considered. Thus, there are two types of capacitances namely transition or depletion region capacitance and diffusion or storage capacitance. In the case of a forward biased diffusion capacitance is much larger than transition capacitance (more than a million times).

Q. 19. Define transition and diffusion capacitances.

Ans. When a P-N junction is reverse biased, the depletion region acts like an insulator or dielectric material while the P- and N-type regions on either side have a low resistance and act as the plates. Thus P-N junction may be considered a parallel plate capacitor. The junction capacitance is termed as space charge *capacitance* or *transition capacitance* or *depletion region capacitance* and is denoted by C_T.

When a P-N junction is forward biased, a capacitance, which is much larger than the transition capacitance, comes into play. This type of capacitance is called the diffusion capacitance C_D.

Q. 20. How does diffusion capacitance C_D vary with dc diode current?

Ans. In a P-N junction diode, the diffusion capacitance is directly proportional to the diode forward current.

Q. 21. How does transition capacitance vary with junction voltage in a P-N junction diode ?

Ans. Transition capacitance is inversely proportional to the square root of the junction voltage.

Q. 22. What is the relation between the transition capacitance and reverse bias voltage in a PN diode ?

Ans. Transition capacitance $C_T = \frac{dQ}{dV}$ where dQ is the increase in charge due to increase in voltage dV.

Since depletion region increases with the increase in reverse-bias potential, the resulting transition capacitance C_T decreases. Transition capacitance C_T is inversely proportional to the square root of the junction voltage.

Q. 23. What is reverse recovery time ?

Ans. If the external voltage is suddenly reversed in a diode circuit which has been carrying current in the forward direction the diode current will not fall to its steady-state reverse value. The diode will continue to conduct for a time, known as diode reverse recovery time t_{rr}, until excess minority-carrier density has dropped down to zero.

Reverse recovery time is the maximum time for the device to switch from *on* to *off.*

Q. 24. Is 'reverse recovery time' due to majority carriers or the minority carriers ?

Ans. Reverse recovery time is due to large number of minority carriers in both of the P- and N-regions.

Q. 25. Define storage time in conducting diode.

Ans. In a conducting diode during the time interval t_1 to t_2 (Refer to Fig. 2.53), the stored minority charge has remained stored and, therefore, this time interval is called the storage time t_s.

Q. 26. How can the switching speed of a diode be improved ?

Ans. Switching speed of a diode can be improved by adding efficient recombination centres to the bulk material. For silicon diodes Au doping can serve this purpose.

A diode switching time can also be improved by making the lightly doped neutral region shorter than a minority carrier diffusion length.

PROBLEMS

1. A germanium diode draws 50 mA with a forward bias of 0.27 V. The junction is at room temperature of 27°C. Determine the reverse saturation current of the diode. **[Ans.** 1.545 μA]
2. Determine the reverse saturation current density in an abrupt silicon diode with following data. $N_D = 1.25 \times 10^{21}$ m^{-3}; $N_A = 1.05 \times 10^{22}$ m^{-3}; $D_e = 3.4 \times 10^{-3}$ m^2 s^{-1}; $D_h = 1.2 \times 10^{-3}$ m^2 s^{-1}; $L_e = 6.8 \times 10^{-4}$ m ; $L_h = 2.8 \times 10^{-4}$ m; $n_i = 1.6 \times 10^{16}$ m^{-3}; $e = 1.6 \times 10^{-19}$ C. **[Ans.** 0.16 μ A]
3. Determine the voltage at which the reverse current in a germanium P-N junction diode attains a value of 95% of its saturation value at room temperature of 300 K. Assume V_T = 2.6 mV. **[Ans.** –0.078 V]
4. Determine the forward resistance of a silicon diode, when the forward current is 4 mA at T = 300 K. **[Ans.** 12.93 Ω]
5. Determine ac resistance for a semiconductor diode with a forward bias of 0.25 V. Reverse saturation current at room temperature is of 1.2 μA. **[Ans.** 1.445 Ω]
6. Determine the current flowing in the circuit shown in Fig. 1. **[Ans.** 1.94 mA]

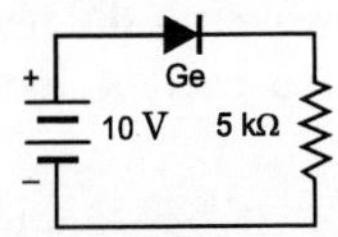

Fig. 1

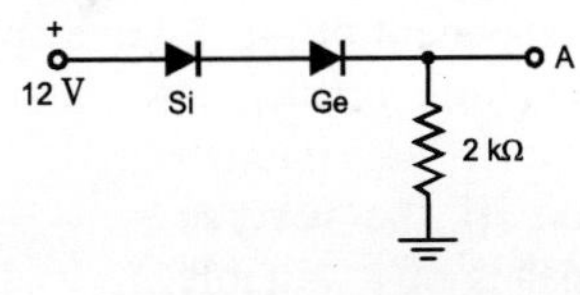

Fig. 2

7. Determine the current flowing in the circuit shown in Fig. 2. Also determine the potential of point A. **[Ans.** 5.5 mA, 11 V]
8. A silicon diode has a saturation current of 5 μA at room temperature of 300 K. Determine its value at 400 K. **[Ans.** 5.12 mA]

❑❑❑

CHAPTER 3

Zener and Other Two-Terminal Devices

3.1. INTRODUCTION

INSIDE THIS CHAPTER

There are many two-terminal devices which have a single P-N junction, as an ordinary semiconductor diode have, but operate somewhat in different ways, giving different terminal characteristics and have different applications. Few of such devices like zener, tunnel, varactor, Schottky, power diodes, photodiode, light-emitting diode will be introduced in this chapter.

3.2. JUNCTION BREAKDOWN

When an ordinary P-N junction diode is reverse biased, normally only very small reverse saturation current flows. This current is due to movement of minority carriers. It is almost independent of the voltage applied. However, if the reverse bias is increased, a point is reached when the junction breaks down and the reverse current increases abruptly, as shown in Fig. 3.1. This current could be large enough to destroy the junction. If the reverse current is limited by means of a suitable series resistor, the power dissipation at the junction will not be excessive, and the device may be operated continuously in its breakdown region to its normal (reverse saturation) level. It is found that for a suitably designed diode, the breakdown voltage is very stable over a wide range of reverse currents. This quality gives the *breakdown diode* many useful applications as a *voltage reference source.*

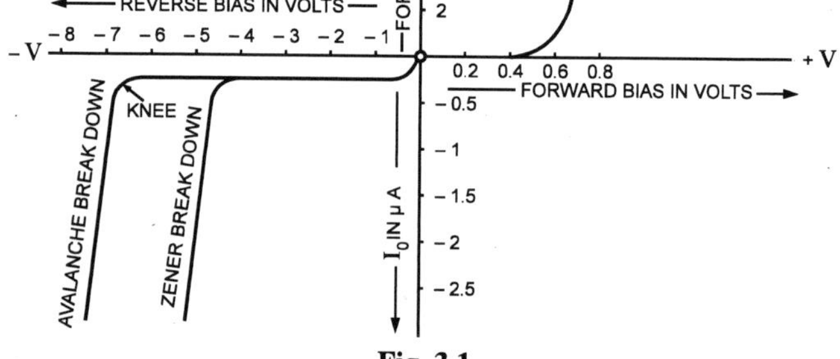

Fig. 3.1

The critical value of the voltage, at which the breakdown of a P-N junction diode occurs is called the *breakdown voltage.* The breakdown voltage depends on the width of the depletion region, which, in turn, depends on the doping level. The junction offers almost zero resistance at the breakdown point.

There are two mechanisms by which breakdown can occur at a reverse biased P-N junction : *avalanche breakdown* and *zener breakdown.*

The minority carriers, under reverse biased conditions, flowing through the junction acquire a kinetic energy which increases with the increase in reverse voltage. At a sufficiently high reverse voltage (say 5 V or more), the kinetic energy of minority carriers becomes so large that they knock out electrons from the covalent bonds of the semiconductor material. As a result of collision, the liberated electrons in turn liberate more electrons and the current becomes very large leading to the breakdown of the crystal structure itself. This phenomenon is called the *avalanche breakdown.* The breakdown region is the *'knee'* of the characteristic curve. Now the current is not controlled by the junction voltage but rather by the external circuit.

Under a very high reverse voltage, the depletion region expands and the potential barrier increases leading to a very high electric field across the junction. The electric field will break some of the covalent bonds of the semiconductor atoms leading to a large number of free minority carriers, which suddenly increase the reverse current. This is called the *zener effect.* The breakdown occurs at a particular and constant value of reverse voltage called the *breakdown voltage.* It is found that zener breakdown occurs at electric field intensity of about 3×10^7 V/m.*

Either of the two (zener breakdown or avalanche breakdown) may occur independently, or both of these may occur simultaneously. Diode junctions that breakdown below 5 V are caused by *zener effect.* Junctions that experience breakdown above 5 V are caused by *avalanche effect*. Junctions that break down around 5 V are usually caused by combination of two effects. The zener breakdown occurs in heavily doped junctions (P-type semiconductor moderately doped and N-type heavily doped), which produce narrow depletion layers. The avalanche breakdown occurs in lightly doped junctions, which produce wide depletion layers.

With the increase in junction temperature zener breakdown voltage is reduced while the avalanche breakdown voltage increases. The zener diodes have a negative temperature coefficient while avalanche diodes have a positive temperature coefficient. Diodes that have breakdown voltages around 5 V have zero temperature coefficient.

The breakdown phenomenon is reversible and harmless so long as the safe operating temperature is maintained.

Despite the fact that avalanche and zener constitute two types of breakdown diode, the name zener is commonly applied to both types of diodes.

3.3. ZENER DIODES

Zener diode, also sometimes called the *breakdown diode*, is a P-N junction diode specially designed for operation in the breakdown region in reverse bias condition. The diode may use either zener breakdown mechanism or avalanche breakdown mechanism, already discussed in detail in Art. 3.2. Despite the fact that avalanche and zener constitute two types of breakdown diodes, the name zener is commonly applied to both types of diodes. The breakdown diode may be silicon or germanium one but silicon is preferred over germanium because of higher operating temperature and current capability. The knee point is also more sharp in case of a silicon diode. These diodes are analogous to gas diodes in which large current appears on reaching the breakdown potential.

Zener diodes are connected into circuits with their *cathodes* as the *positive* terminal because they

---

* From basic circuit theory electric field intensity E is given as $\frac{V}{d}$, where V is the applied reverse voltage in volts and *d* is the width of depletion region in metres. The narrower the depletion region, the smaller the required bias to cause zener breakdown. A small reverse bias can produce a sufficiently strong electric field in a narrow depletion region.

operate in the reverse-bias mode. Zener diodes are packaged in the same way as conventional semiconductor diodes. Small zener diodes are marked the same way as conventional diodes with a cathode band. The larger diodes are marked with the zener diode symbol shown in the illustration (Fig. 3.2). In addition to have a voltage rating (reverse breakdown), zener diodes have a power rating that must not be exceeded.

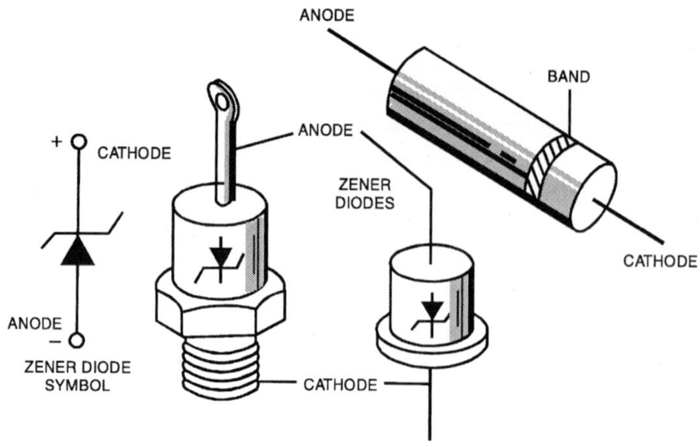

Fig. 3.2. *Zener Diode*

The symbol of a zener diode is given in Fig. 3.2. This is similar to that of a normal diode except that the line representing the cathode is bent at both ends *i.e.* the bar is turned into Z-shape; stands for zener.

3.3.1. Construction

Zener diode is like an ordinary P-N junction diode except that it is properly doped so as to have a sharp breakdown voltage. There are several methods used in the manufacture of zener diodes, such as diffused structure, diffused and passivated structure and alloy diffused structure.

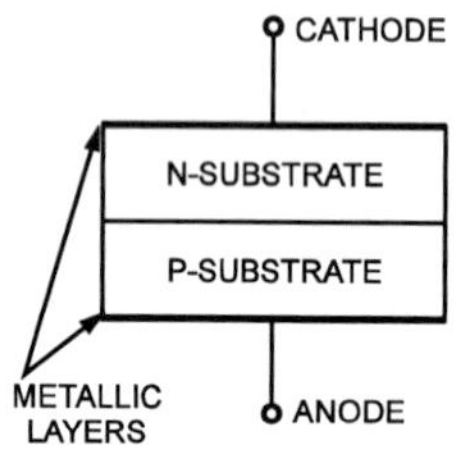

Fig. 3.3. *Diffused Structure of a Zener Diode*

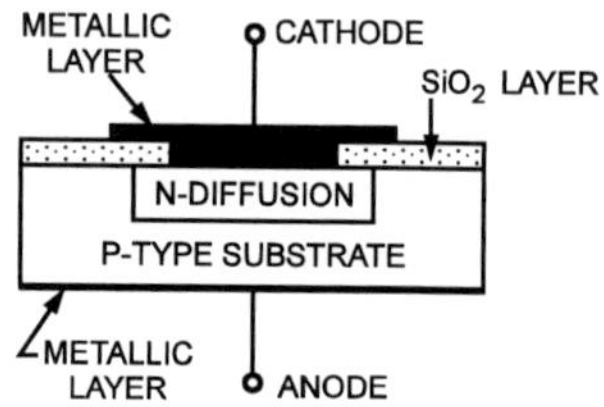

Fig. 3.4. *Diffused and Passivated Structure of a Zener Diode*

Diffused structure zener diode consists of two N and P substrates diffused together and has metallic layers deposited on both sides to connect anode and cathode terminals, as illustrated in Fig. 3.3. The drawback of such a construction is that the edges of the junction are exposed to contamination.

In passivated structure of a zener diode, the edges of the junction are covered by layer of SiO_2, as illustrated in Fig. 3.4.

Similarly, all junctions are protected in alloy diffused structures, as shown in Fig. 3.5.

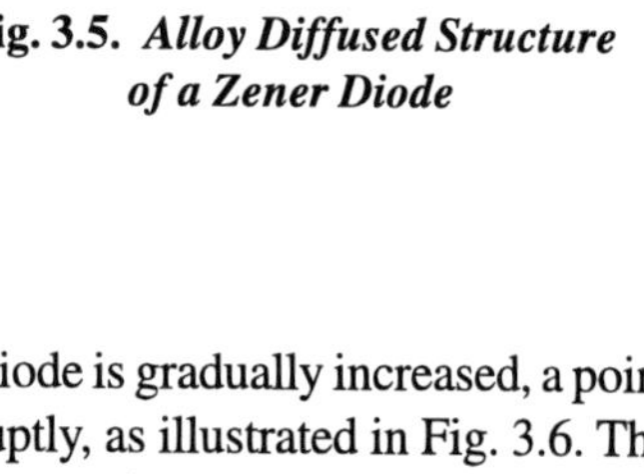

Fig. 3.5. *Alloy Diffused Structure of a Zener Diode*

In general, the alloy diffused structure gives better zener characteristics at lower zener voltages. On the other hand, diffused and passivated zener diode is used at higher zener voltages.

3.3.2. Volt-Ampere Characteristics

It has already been discussed in Art. 3.2 that when the reverse bias on a crystal diode is gradually increased, a point is reached when the junction breaks down and a reverse current increases abruptly, as illustrated in Fig. 3.6. The breakdown region is the knee of the reverse characteristic, as illustrated in Fig. 3.6. This breakdown of junction was first explained by the American scientist, C. Zener. So, the breakdown voltage is sometimes called, *zener voltage* V_Z and the sharply increased current is called the *zener current* I_Z.

Volt-ampere characteristics of typical silicon and germanium diodes are illustrated in Fig. 3.6. Breakdown diode is operated over the breakdown region and the current is limited by an external resistance. The breakdown or zener voltage V_Z depends upon the amount of doping. A heavily doped diode will have very thin depletion layer, so zener voltage will be very low. By controlling the thickness of depletion layer and doping concentration it is possible to make the zener diode to break down at a specified voltage, called the zener voltage (ranging from 2.4 V to 200 V). The forward characteristics of a zener diode are similar to those of an ordinary P-N junction diode, as illustrated.

During the operation in the breakdown region, it does not burn out immediately. As long as the current through the diode is limited by the external circuit within permissible values, it does not burn out.

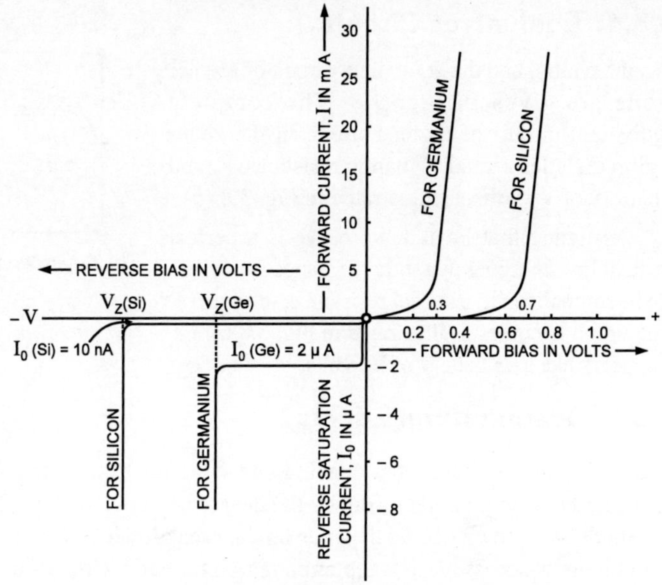

Fig. 3.6. *Volt-Ampere Characteristics of Typical Silicon and Germanium Zener Diodes*

3.3.3. Zener Impedance

This is very important parameter of a zener diode. Zener impedance Z_Z is essentially the dynamic resistance of a zener diode (similar to dynamic resistance of an ordinary P-N diode). It is defined as the reciprocal of the slope of the zener curve

$$i.e.\ r_Z = \frac{\Delta V_Z}{\Delta I_Z} \qquad ...(3.1)$$

where ΔV_Z and ΔI_Z are the small variations in voltage and current respectively.

It is usually measured just above the knee in zener region and is named as Z_{ZK}. When measured at test point, it is designated Z_{ZT}. It decreases with increase in zener current. For this reason, the diode should be operated with as much reverse current as possible, consistent with rating limitations.

Small value of zener impedance is usually desirable as it indicates steep curve where relatively large change in current, ΔI_Z occurs with a small change in voltage ΔV_Z.

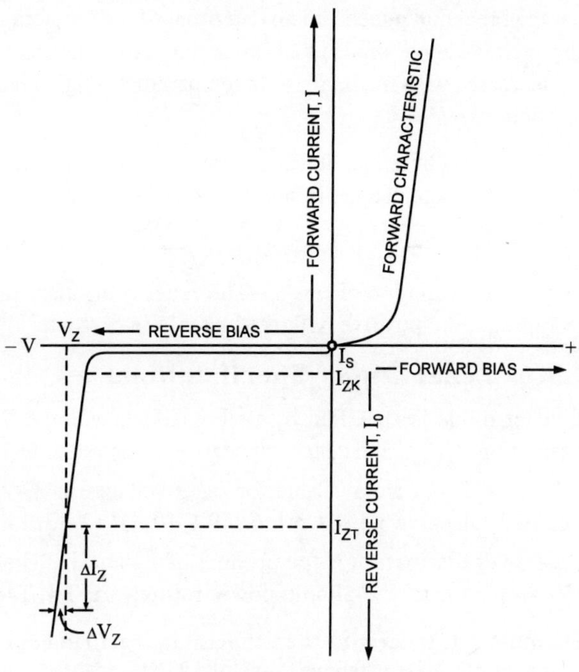

Fig. 3.7. *Zener Diode Characteristic With Enlarged View of Breakdown Region*

Ideally the breakdown curve should be a perfect vertical line and r_Z should be zero. In practice, however, r_Z has a small value ranging from a few ohms to several hundred ohms. There is also a variation in impedance of zener diodes among those having different values V_Z. Zener diodes with breakdown voltage near 7 V have the smallest impedances.

3.3.4. Equivalent Circuit

Circuit symbol and the equivalent circuit of a zener diode are given in Fig. 3.8. The complete equivalent circuit of a zener diode in the zener region includes a small dynamic resistance r_Z and a battery of V_Z volts, as illustrated in Fig. 3.8 (*b*).

Assuming that breakdown curve is a perfect vertical line *i.e.* zener dynamic resistance is negligible in comparison to external resistance, the equivalent circuit becomes as illustrated in Fig. 3.8 (*c*) *i.e.* it appears like a *dc* battery of V_Z volts.

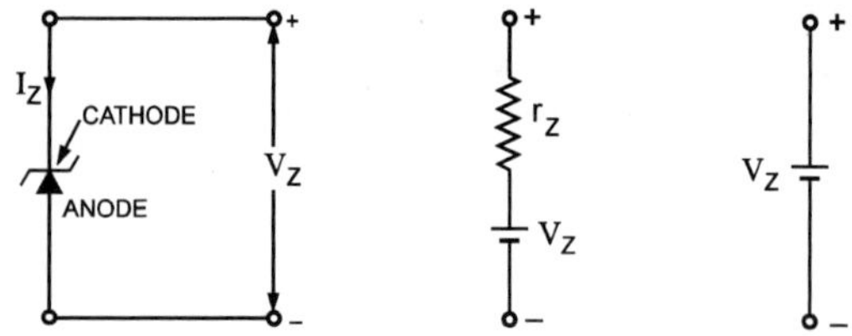

(*a*) *Zener Diode Symbol* (*b*) *Complete* (*c*) *Approximate*

Fig. 3.8. ***Equivalent Circuits of Zener Diode***

3.3.5. Temperature Effects

The breakdown voltage of a zener diode is a function of the width of its depletion region*, which is controlled during manufacturing by the degree of impurity doping. As discussed in Art. 3.2, the mechanism by which breakdown occurs depends on the breakdown voltage itself. For V_Z less than 5 V, the high electric field intensity across narrow depletion region (about 3×10^7 V/m) strips carriers directly from their bonds, a phenomenon usually known as zener breakdown. For V_Z exceeding 8 V, breakdown occurs as a result of collision between high-energy carriers, the mechanism known as *avalanching*. Between 5 V and 8 V, both the avalanching and zener mechanisms contribute to breakdown. The practical significance of these facts is that the breakdown mechanism determines how temperature variations affect the value of zener voltage V_Z. Low-voltage zener diodes that break down by zener mechanism have negative temperature coefficient *i.e.* V_Z decreases with the increase in temperature. Higher-voltage avalanche zeners have positive temperature coefficients.

The effect of temperature on zener voltage is given in terms of temperature coefficient, which is defined as the percentage change in nominal zener voltage for each degree Celsius of change in junction temperature. Mathematically it is given as $\frac{\Delta V_Z}{V_Z(t - t_0)} \times 100$ where ΔV_Z is the change in zener voltage due to temperature variation of $(t - t_0)$. The zener coefficient may be positive or negative, depending on current, becoming more positive with the increase in current.

3.3.6. Zener Diode Specifications

A zener diode is specified by its breakdown voltage V_Z, breakdown current, I_{ZK}, the maximum power dissipation $P_{Z\,(max)}$ and zener impedance measured at test point, Z_{ZT}.

Zener diodes are available for zener voltages of 2.4 to 200 V with accuracies between 5 and 20%. The preferred values are 3.3, 4.7, 5.1, 6.1, 9.1, 10, 11, 12, 13, 15, to 200 V.

Power dissipation is the product of V_Z and I_Z. The maximum power rating varies from 150 mW to 250 W. Commonly available power ratings are 0.4, 1, 3, 10, and 50 watts.

Example 3.1. Determine the temperature coefficient of a 5 V zener diode at 25°C, if the nominal voltage drops to 4.8 V at a temperature of 100°C.

Solution : Change in zener voltage, $\Delta V_Z = 4.8 - 5 = -0.2$ V

Change in temperature, $(t - t_0) = 100 - 25 = 75$°C.

$$\text{Temperature coefficient of zener diode} = \frac{\Delta V_Z}{V_Z(t - t_0)} \times 100 = \frac{-0.2}{5 \times 75} \times 100 = -0.0533\% \ \textbf{Ans.}$$

* Heavy doping increases conductivity, which narrows the depletion region and, therefore, reduces the voltage at which breakdown occurs.

Example 3.2. A zener diode has an impedance of 40 Ω in the range for I_Z = 1 mA to 10 mA. The voltage corresponding to I_Z of 1 mA is 9 V. For constant impedance over the given range, what minimum and maximum zener voltages can be expected if the diode is used in an application where the zener current varies from 2 mA to 8 mA ?

Solution : The voltage change between I_Z = 1 mA and I_Z = 2 mA,

$$\Delta V_Z = \Delta I_Z r_Z = (2\text{ mA} - 1\text{ mA}) \times 40\ \Omega = 0.04\text{ V}$$

So minimum zener voltage, $V_{Z\,min} = V + \Delta V_Z = 9 + 0.04 = 9.04$ V **Ans.**

The voltage change between I_Z = 2 mA and 8 mA

$$\Delta V_Z = \Delta I_Z r_Z = (8\text{ mA} - 2\text{ mA}) \times 40\ \Omega = 0.24\text{ V}$$

So maximum zener voltage, $V_{Z\,max} = V_{Z\,min} + \Delta V_Z = 9.04 + 0.24 = 9.28$ V **Ans.**

3.4. ZENER DIODE APPLICATIONS

Zener diodes have numerous applications in transistor circuitry. Some of their common applications are explained below.

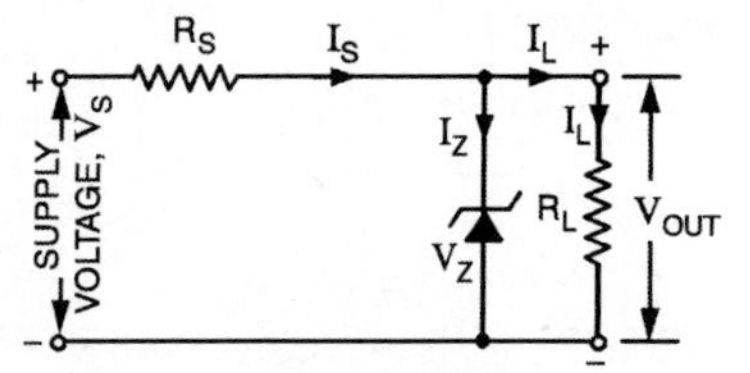

Fig. 3.9

1. Zener Diode Shunt Regulator (or Voltage Regulation). Voltage regulation is a measure of a circuit's ability to maintain a constant output voltage even when either input voltage or load current varies.

Figure 3.9 shows how a zener diode can be used as a voltage regulator to provide a constant voltage from a source whose voltage may vary appreciably. A resistor R_S is necessary to limit the reverse current through the diode to safer value. The voltage source V_S and the resistor R_S are so selected that the diode operates in the breakdown region. The diode voltage in this region, which is also the voltage across the load R_L, is called *zener voltage* V_Z and the diode current is called the *zener current* I_Z. The series resistor R_S absorbs the output voltage fluctuations so as to maintain voltage across the load constant. Zener diode is reverse connected across the input voltage whose variations are to be regulated.

As long as voltage across the load resistor R_L is less than the breakdown voltage V_Z, the zener diode does not conduct, the resistors R_S and R_L constitute a potential divider across V_S. At an increased supply voltage V_S, the voltage drop across load resistor (or zener diode) becomes greater than the zener breakdown voltage. It then operates in its breakdown region. The series resistor R_S limits the zener current I_Z from exceeding its rated maximum $I_{Z\,max}$. Current through resistor R_S is given as

$$I_S = \frac{V_S - V_Z}{R_S} \quad ...(3.2)$$

The current from the power supply splits at the junction of zener diode and the load resistor R_L.

$$\text{So } I_S = I_Z + I_L \quad ...(3.3)$$

When the zener diode operates in its breakdown region, the voltage across it, V_Z remains fairly constant even though the current I_Z flowing through it may vary considerably.

When the supply voltage V_S increases, the current through both the zener diode and load resistance R_L increases. At the same time, however, the zener diode resistance decreases and the current through the diode increases more than proportionately. As a result, a greater voltage drop will occur across the series resistor R_S and the output voltage V_{OUT} (voltage across the diode or load resistance R_L) will become very close to the original value. The reverse is also true. Thus a zener diode can maintain the output voltage V_{OUT} within a fraction of a volt when the supply or input voltage V_S may vary over a range of several volts. The zener diode will maintain a constant voltage across the load as long as the supply voltage is more than the zener voltage.

Let us examine the other cause of the output voltage variation. When the load resistance R_L decreases for constant input voltage V_S, load current I_L increases. This additional current is not supplied from the source of supply but the demand of additional load current is met by decrease in zener current I_Z. This keeps the voltage drop across series resistance R_S constant and so the output voltage V_{OUT}. The worst case occurs for minimum source voltage and maximum load current because the zener current reduces to a minimum. In such a case,

$$I_{S\,min} = \frac{V_{S\,min} - V_Z}{R_{S\,max}}$$

$$\text{or} \quad R_{S\,max} = \frac{V_{S\,min} - V_Z}{I_{S\,min}} \qquad \text{...(3.4)}$$

From Eq. (3.3) $\quad I_Z = I_S - I_L$

In worst case, this may be written as

$$I_{Z\,min} = I_{S\,min} - I_{L\,max}$$

The critical point occurs when $I_{L\,max} = I_{S\,min}$. At this point, the zener current, I_Z reduces to zero, and regulation is lost.

By substituting $I_{L\,max}$ for $I_{S\,min}$ in Eq. (3.4), we have

$$R_{S\,max} = \frac{V_{S\,min} - V_Z}{I_{L\,max}} \qquad \text{...(3.5)}$$

where $R_{S\,max}$ is the critical value of series resistance, $V_{S\,min}$ is the minimum source voltage, V_Z is zener voltage and $I_{L\,max}$ is maximum load current.

The *critical resistance,* $R_{S\,max}$ is the maximum allowable series resistance. The series resistance R_S must always be less than the critical value; otherwise; breakdown operation is lost, and the regulator stops its operation.

The advantages of zener diode voltage regulators over other voltage regulators are that they are smaller, lighter, more rugged and have a longer life. They are also simpler and inherently cheaper.

Such regulators have the following drawbacks :

1. Their efficiency is low for heavy load currents because there is a considerable power loss in series resistor R_S and the diode.
2. The output voltage varies slightly due to zener impedance r_Z.
3. The output voltage cannot be choosen independently but depends upon the breakdown voltage of the zener diode.

This regulator is employed when there are small variations in load current and supply voltage.

2. Meter Protection. Figure 3.10 shows how does a zener diode protect a meter (AVO meter or multimeter) movement or circuit against damage from an accidental application of excessive voltage. In this arrangement zener diode is connected in parallel with the meter, as shown in Fig. 3.10. In case of accidental overload most of the current will pass through the zener diode and so the meter movement will be saved from damage. Two zener diodes connected, as shown in Fig. 3.11, can provide overload protection regardless of the applied polarity.

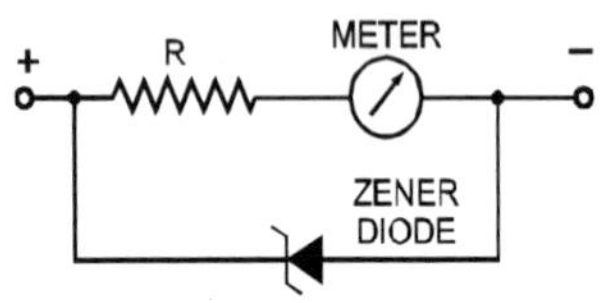

Fig. 3.10

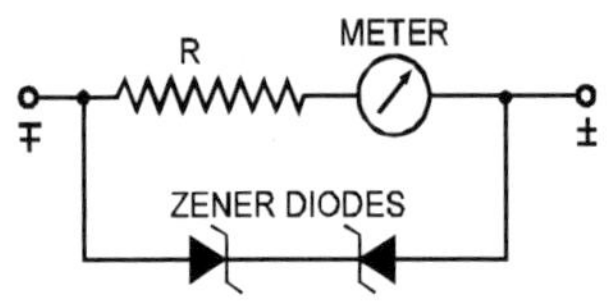

Fig. 3.11

3. Switching Operation. Zener diode can produce a sudden change from low current to high current, so it is useful for switching operations. The use of zener diode in switching operations has made possible an extremely fast performance in computer applications.

4. Zener Diode as a Reference Element. There are many applications in which it is desirable that a constant voltage is maintained between two points in a circuit, and use this voltage as reference for comparing other voltages against it. The difference between the two voltages (the voltage under comparison and reference voltage) is usually amplified and then used to perform some control function. Such an arrangement is employed in power supply voltage regulator circuits, measurement circuits and servomechanism circuits. The constant voltage characteristic of a zener diode in its breakdown region makes it desirable for such an application.

A circuit employing zener diode as a reference element is given in Fig. 3.12. The value of R is so chosen that the diode operates well within its breakdown region. V_{REF} equals the zener breakdown voltage V_Z. The difference of input voltage V_{IN} and reference voltage V_{REF} (*i.e.* $V_{IN} - V_{REF}$) gives the control output. Obviously V_{IN} should be more than V_{REF}.

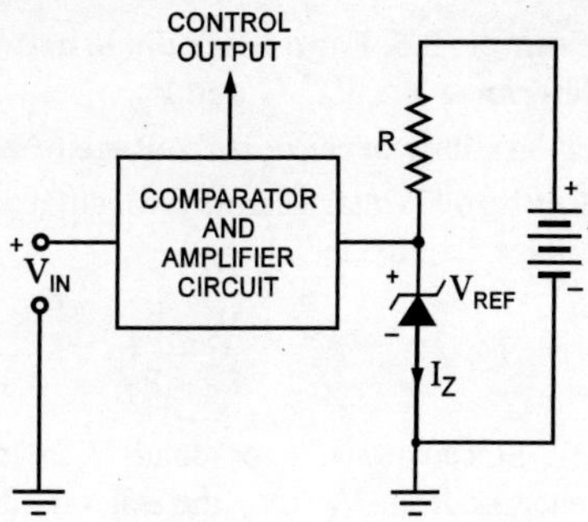

Fig. 3.12. ***Zener Diode As Reference Element***

Example 3.3. For the circuit shown in Fig. 3.13, find

(*a*) the output voltage,

(*b*) the voltage drop across R_S,

(*c*) the current through zener

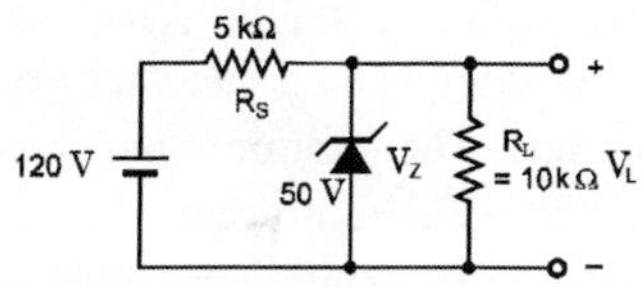

Fig. 3.13

Solution : From the figure shown in Fig. 3.14 we have

(*a*) Output voltage, $V_L = V_Z = 50$ V **Ans.**

(*b*) Voltage drop across $R_S = I_S \times R_S$

$= V_S - V_Z = 120 - 50 = 70$ V **Ans.**

(*c*) Load current, $I_L = \dfrac{V_L}{R_L} = \dfrac{50\text{ V}}{10\text{ k}\Omega} = 5$ mA

Current through R_S, $I_S = \dfrac{70\text{ V}}{5\text{ k}\Omega} = 14$ mA

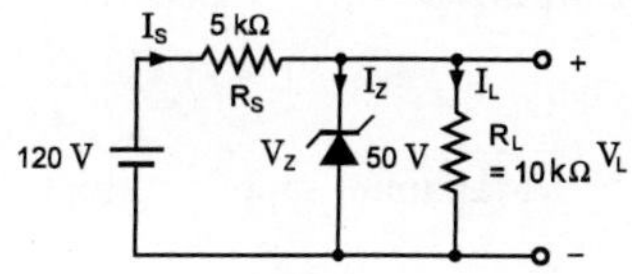

Fig. 3.14

Hence, current through zener diode is given by

$$I_Z = I_S - I_L = 14\text{ mA} - 5\text{ mA} = 9\text{ mA} \quad \textbf{Ans.}$$

Example 3.4. For the zener regulator shown in Fig. 3.15, find the maximum and minimum current flowing through the zener diode. [*G.G.S.I.P. Univ. Analog Electronics, December 2009*]

Solution : Load current, $I_L = \dfrac{V_{out}}{R_L} = \dfrac{V_Z}{R_L} = \dfrac{50}{10\text{ k}\Omega} = 5$ mA

Zener current will be maximum when input voltage is maximum *i.e.* 125 V.

Corresponding current through series resistance of 5 kΩ,

$$I_{S\,max} = \frac{V_{S\,max} - V_{out}}{R_S} = \frac{125 - 50}{5\text{ k}\Omega} = 15\text{ mA}$$

Corresponding zener current,

$$I_{Z\,max} = I_{S\,max} - I_L = 15 - 5 = 10\text{ mA} \textbf{ Ans.}$$

The zener current will be minimum when input voltage is minimum *i.e.* 85 V. Corresponding current through series resistance (5 kΩ),

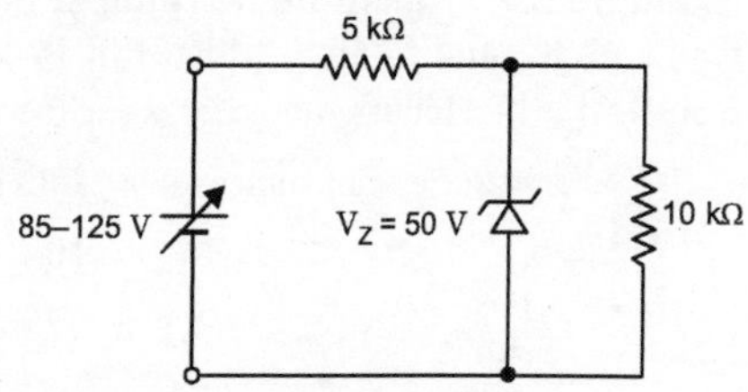

Fig. 3.15

$$I_{S\,min} = \frac{85 - 50}{5\text{ k}\Omega} = 7\text{ mA}$$

Corresponding zener current,

$$I_{Z\,min} = I_{S\,min} - I_L = 7 - 5 = 2\text{ mA} \textbf{ Ans.}$$

Example 3.5. For a zener diode network shown below in Fig. 3.16, determine V_L, V_R, I_Z and P_Z.

Assume breakdown voltage of zener diode 10 V.

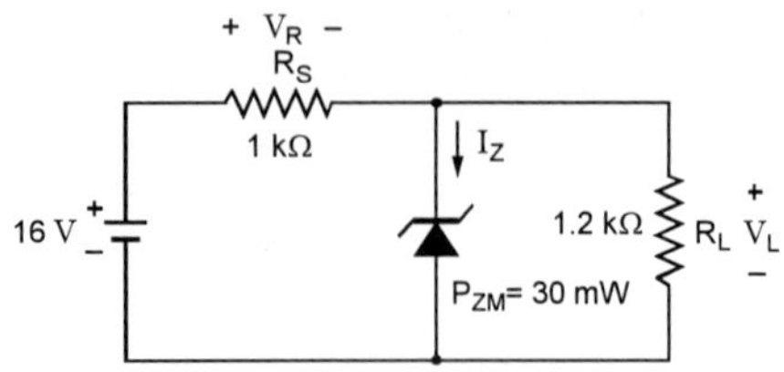

Fig. 3.16

Solution : Voltage across the load, if zener diode is open circuited, by voltage divide rule,

$$V_L = \frac{V_S \times R_L}{R_S + R_L} = \frac{16 \times 1.2}{1 + 1.2} = 8.73 \text{ V}$$

Since voltage across load, V_L is less than breakdown voltage of zener diode *i.e.* $V_L < V_Z$, the zener diode will not conduct and

$$V_L = 8.73 \text{ V}$$

$$V_R = V_S - V_L = 16 = 8.73 = 7.27 \text{ V} \quad \textbf{Ans.}$$

$$I_Z = 0$$

$$P_Z = V_Z I_Z = V_L I_Z = 8.73 \times 0 = 0 \text{ watt} \quad \textbf{Ans.}$$

Example 3.6. For the circuit shown in Fig. 3.17, R_S = 20 Ω, V_Z = 18 V and R_L = 200 Ω. If V_{in} can vary from 20 V to 30 V, find

(*i*) The minimum and maximum currents in zener diode.

(*ii*) The minimum and maximum power dissipated in the diode.

(*iii*) The maximum rated power dissipation that R_S should have.

[*G.G.S.I.P. Univ. Analog Electronics December 2009*]

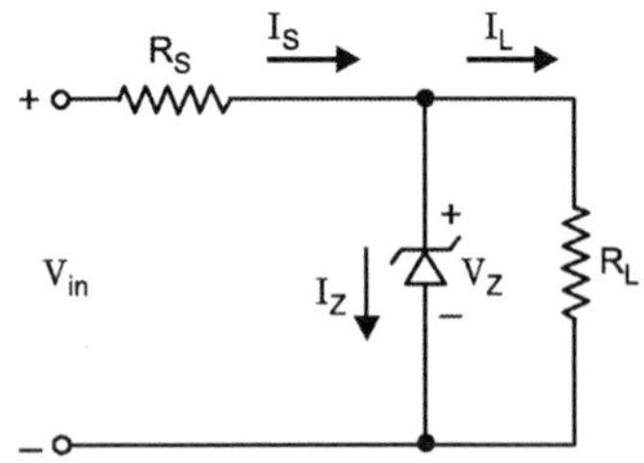

Fig. 3.17

Solution : Load current, $I_L = \frac{V_{out}}{R_L} = \frac{V_Z}{R_L} = \frac{18}{200} = 0.09 \text{ A}$

$$\text{Maximum supply current, } I_{S\,max} = \frac{V_{in\,max} - V_Z}{R_S} = \frac{30 - 18}{20} = 0.6 \text{ A}$$

(*i*) Corresponding zener current, $I_{Z\,max} = I_{S\,max} - I_L = 0.6 - 0.09 = 0.51$ A **Ans.**

$$\text{Minimum supply current, } I_{S\,min} = \frac{V_{in\,min} - V_Z}{R_S} = \frac{20 - 18}{20} = 0.1 \text{ A}$$

Corresponding zener current, $I_{Z\,min} = I_{S\,min} - I_L = 0.1 - 0.09 = 0.01$ A or 10 mA **Ans.**

(*ii*) Minimum power dissipation in diode $= V_Z \times I_{Z\,min} = 18 \times 0.01 = 0.18$ W **Ans.**

Maximum power dissipation in diode $= V_Z \times I_{Z\,max} = 18 \times 0.51 = 9.18$ W **Ans.**

(*iii*) Maximum power dissipation rating of $R_S = I^2_{Smax} R_S = 0.6^2 \times 20 = 7.2$ W **Ans.**

Example 3.7. Explain the working of diode as a shunt regulator. For the circuit of Fig. 3.18, determine the range of R_L and I_L that will result in V_{R_L} being maintained at 10 V.

Solution : Input voltage, $V_{in} = 50$ V

Series resistance, $R = 1$ kΩ

$V_Z = 10$ V

$I_{Zm} = 32$ mA

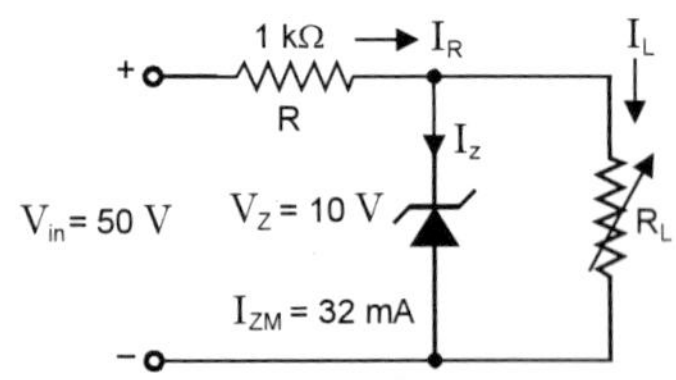

Fig. 3.18

$$\text{Supply current, } I_R = \frac{V_{in} - V_L}{R} = \frac{50 - 10}{1,000} = 40 \text{ mA}$$

Load current I_L will be maximum when zener current $I_Z = 0$

$$\text{So } I_{L\,max} = I_R - I_{Z\,min} = 40 - 0 = 40 \text{ mA}$$

Corresponding load resistance will be minimum and

$$R_{L\,min} = \frac{V_L}{I_{L\,max}} = \frac{V_Z}{I_{L\,max}} = \frac{10}{40 \times 10^{-3}} = 250\ \Omega$$

Load current I_L will be minimum when I_Z is maximum *i.e.* 32 mA.

$$\text{So } I_{L\,min} = I_R - I_{Z\,max} = 40 - 32 = 8\text{ mA}$$

Corresponding load resistance will be maximum and

$$R_{L\,max} = \frac{V_L}{I_{L\,min}} = \frac{10}{8 \times 10^{-3}} = 1.25\text{ k}\Omega$$

Hence range of R_L : From 250 Ω to 1.25 kΩ **Ans.** and range of I_L : From 8 mA to 40 mA **Ans.**

Example 3.8. In the circuit shown in Fig. 3.19, a 5 V zener diode is used which provides regulation for 50 mA ≤ I_z ≤ 1 A. Determine the range of load currents for which regulation is achieved if the unregulated voltage, V_s varies 7.5 V and 10 V. The resistance R_s = 4.75 Ω. *[G.G.S.I.P. Univ. Analog Electronics December 2010]*

Solution : Maximum supply voltage, $V_{S\,max}$ = 10 V

Minimum supply voltage, $V_{S\,min}$ = 7.5 V

$I_{Z\,max}$ = 1 A

$I_{Z\,min}$ = 50 mA = 0.05 A

Load voltage, $V_L = V_Z = 5$ V

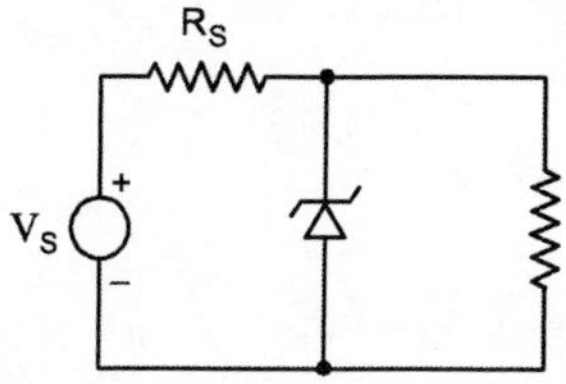

Fig. 3.19

$$\text{Maximum supply current, } I_{S\,max} = \frac{V_{S\,max} - V_Z}{R_S} = \frac{10-5}{4.75} = 1.053\text{ A}$$

$$\text{Minimum supply current, } I_{S\,min} = \frac{V_{S\,min} - V_S}{R_S} = \frac{7.5-5}{4.75} = 0.526\text{ A}$$

Maximum load current $I_{L\,max} = I_{S\,min} - I_{Z\,min} = 0.526 - 0.05 = 0.476$ A or 476 mA **Ans.**

Minimum load current, $I_{L\,min} = I_{S\,max} - I_{Z\,max} = 1.053 - 1 = 0.053$ A or 53 mA **Ans.**

Example 3.9. A zener voltage regulator has variable load R_L requiring load current to vary 10 mA to 85 mA. It is given that V_Z = 10 V. $I_{Z\,min}$ = 15 mA; I_{Zmax} = 100 mA and the series resistance R_S = 40 Ω. Calculate the range of dc variation permissible and zener power dissipation.

[G.G.S.I.P. Univ. Analog Electronics December 2010]

Solution : Minimum supply current, $I_{S\,min} = I_{L\,max} + I_{Zmin} = 85 + 15 = 100$ mA

Maximum supply current, $I_{S\,max} = I_{L\,min} + I_{Zmax} = 10 + 100 = 110$ mA

Maximum supply voltage, $V_{S\,max} = I_{S\,max} R_S + V_Z = 110 \times 10^{-3} \times 40 + 10 = 14.4$ V

Minimum supply voltage, $V_{S\,min} = I_{S\,min} R_S + V_Z = 100 \times 10^{-3} \times 40 + 10 = 14$ V

i.e. supply voltage variation is from 14 V to 14.4 V **Ans.**

Example 3.10. Determine V_L, I_L, I_Z and I_R for the following circuit.

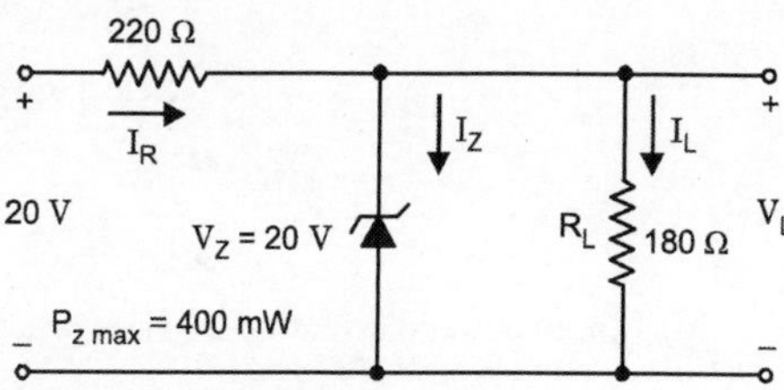

Fig. 3.20

Solution : Voltage across load if zener diode is open circuited.

$$V_L = V_{in} \times \frac{R_L}{R + R_L} = 20 \times \frac{180}{220 + 180} = 9\text{ V}$$

Since voltage across load is less than the breakdown voltage of zener diode *i.e.* $V_L < V_Z$, the zener diode will not conduct and voltage drop across load, V_L = 9 V **Ans.**

Zener diode current, I_Z = 0 ∵ Zener diode is open circuited

$$\text{Load current, } I_L = \frac{V_{in}}{R + R_L} = \frac{20}{220 + 180} = 0.05 \text{ A} \quad \textbf{Ans.}$$

Current through resistance R, $I_R = I_L = 0.05$ A **Ans.**

Example 3.11. Describe the physical mechanism for zener breakdown. For the circuit shown in Fig. 3.21, find the voltage drop across the resistance of 5 kΩ.

Solution : Removing the zener diode from the circuit, we have

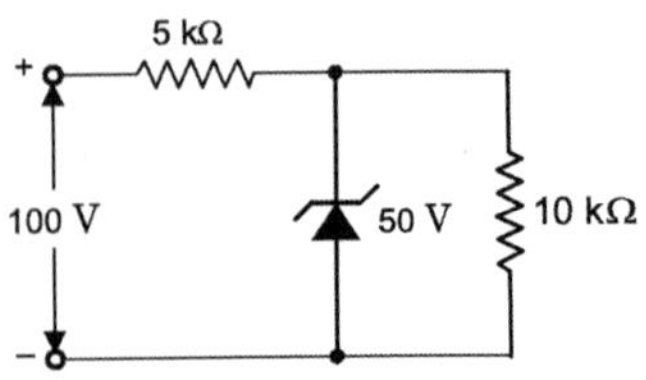

Fig. 3.21

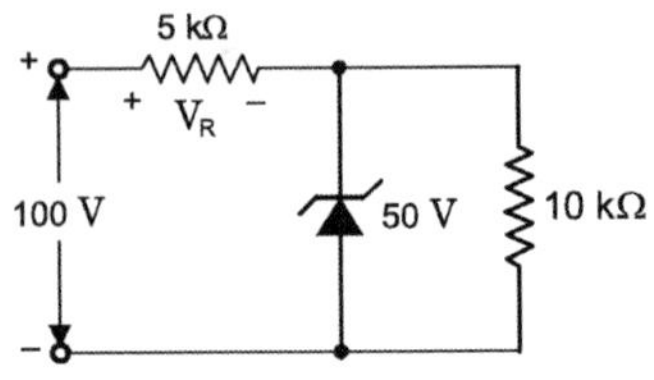

Fig. 3.22

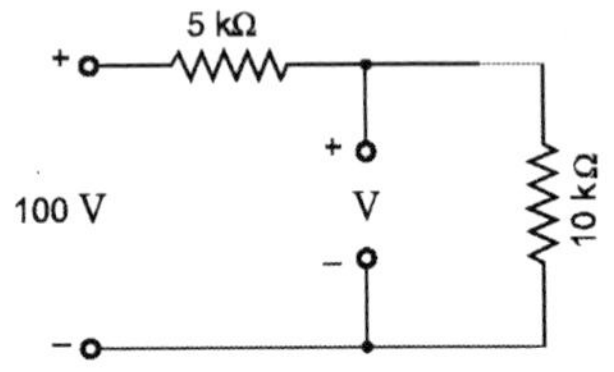

Fig. 3.23

$$V_L = \frac{10}{10+5} \times 100 = 66.67 \text{ V} \qquad \therefore \quad V_L > V_Z \ (i.e. \ 50 \text{ V})$$

So zener diode is in ON state.

Thus the voltage drop across the 10 kΩ load resistor

$$V_L = V_Z = 50 \text{ V}$$

Voltage drop across 5 kΩ resistance $= V_S - V_L = 100 - 50 = 50$ V **Ans.**

Example 3.12. A zener diode regulator has V_Z = 12 V. The input voltage may vary from 20 to 35 V and load current from 15 to 75 mA. To hold voltage constant under all conditions, what should be the value of series resistance ?

Solution : Maximum value of required series resistance,

$$R_{S\,max} = \frac{V_{S\,min} - V_Z}{I_{L\,max}} = \frac{20 - 12}{75 \times 10^{-3}} = 106.7 \ \Omega$$

Select R_S = 100 Ω, standard value **Ans.**

Example 3.13. Draw and design a zener voltage regulator circuit to provide output of 12 V for a maximum load current of 0.5 A when the input voltage variation is 15 to 18 V.

Solution : Assume $I_{Z\,min}$ = 5% of $I_{L\,max} = 0.05 \times 0.5 = 0.025$ A

$$\text{Source resistance (maximum value), } R_{S\,max} = \frac{V_{S\,min} - V_Z}{I_{S\,min}} = \frac{V_{S\,min} - V_Z}{I_{L\,max} + I_{Z\,min}} = \frac{15 - 12}{0.5 + 0.025} = 5.7 \ \Omega$$

Select R_S = 5.6 Ω (standard value)

$$I_{Z\,max} = \frac{V_{S\,max} - V_Z}{R_S} - I_L = \frac{18 - 12}{5.6} - 0.5 = 0.57 \text{ A}$$

Maximum power dissipation in series resistance,

$$P_{R_{S\,max}} = \frac{(V_{S\,max} - V_Z)^2}{R_S} = \frac{(18 - 12)^2}{5.6} = 6.4 \text{ W}$$

Maximum power dissipated in zener diode $P_{Z\,max} = I_{Z\,max} \times V_Z = 0.57 \times 12 = 6.84$ W **Ans.**

3.5. TUNNEL DIODE

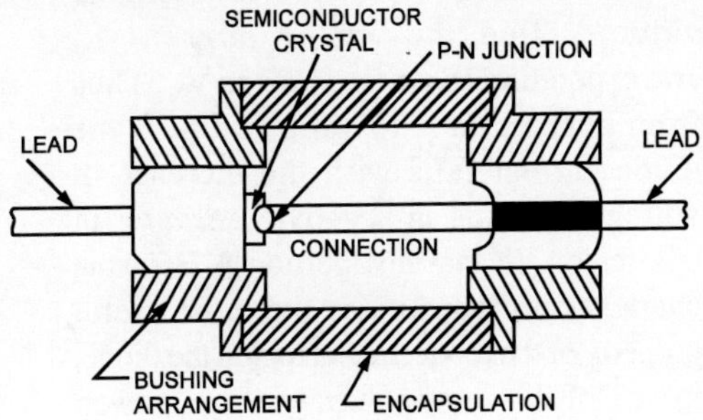

Fig. 3.24

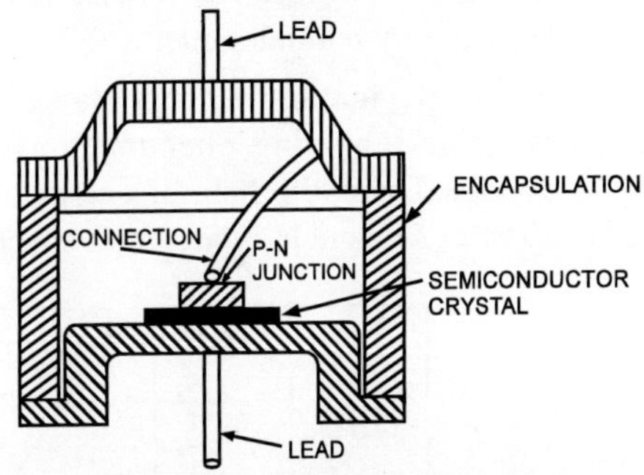

Fig. 3.25. ***Pellet Tunnel Diode***

The tunnel diode (sometimes called the *Esaki diode* after its inventer, Dr. Leo Esaki) is a high conductivity, heavily doped two terminal P-N junction. This diode differs from the normal P-N junction diode in the amount of impurity concentration. The impurity concentration in a normal P-N junction diode is about 1 part in 10^8, whereas in the tunnel diode the impurity concentration is about 1 part in 10^3. Because of heavy doping depletion layer width is reduced* to an extremely small value of the order of 10^{-5} mm : reverse breakdown voltage is also reduced to a very small value (approaching zero) resulting in appearance of the diode to be broken for any reverse voltage, and a –ve resistance section is produced in volt-ampere characteristic of the diode. It is found that the reduced depletion layer can result in carriers '*punching through*' the junction with the velocity of light even when they do not possess enough energy to overcome the potential barrier. The result is that large forward current is produced at relatively low forward voltage (less than 100 mV). Such a mechanism of conduction in which charge carriers (possessing very little energy) punch through a barrier directly instead of climbing over it is called *tunneling*. That is why such diodes are called the *tunnel diodes*. Because of heavy doping the tunnel diode can conduct in reverse as well as in forward direction but it is usually used in forward-biased mode.

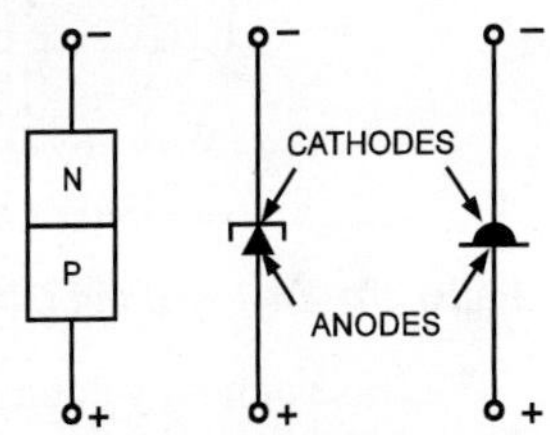

Fig. 3.26. ***Tunnel Diode Symbols***

Construction. Tunnel diodes are usually fabricated from germanium, gallium arsenide or gallium antimonide. Silicon is not used in the construction of tunnel diodes because the ratio of peak value of forward current to the value of valley current (*i.e.* I_P/I_V) is maximum in case of gallium arsenide (approximately 20) and comparatively smaller for germanium (roughly 10) but very small in case of silicon (about 3).

Tunnel diodes are ordinarily manufactured by alloying from gallium arsenide. The source materials are highly doped semiconductor crystals with an impurity concentration of the order of 10^{25} per cubic metre or more. Arrangement is shown in Fig. 3.24.

An alternative arrangement is shown in Fig. 3.25. Such a tunnel diode is called the *pellet tunnel diode.*

It is to be noted that the tunnel diode, being a low power device, can be easily damaged by heat and static electricity.

The commonly used schematic symbols for the tunnel diodes are shown in Fig. 3.26.

Volt-Ampere Characteristic. Forward V-I characteristic of a tunnel diode is illustrated in Fig. 3.27. Forward bias produces immediate conduction—as soon as forward bias is applied, significant current is produced. The current attains quickly its peak value, I_P when the applied forward bias attains a value of V_P volts. The current variation in the vicinity of origin is due to quantum mechanical tunneling of electrons through narrow space charge region of the junction. With the further increase in forward voltage, the diode current

* The width of junction barrier varies inversely as the square root of impurity concentration.

starts decreasing till it attains its minimum value, called the *valley current* I_V corresponding to *valley voltage* V_V. Thus from peak point A to valley point B the diode current falls with the increase in voltage resulting in *negative resistance* in this region. In fact this portion A B of the characteristic constitutes the most useful property of the diode. In this region the diode, instead of absorbing power, produces power. So the tunnel diode can be used as a very high-frequency oscillator. For voltages higher than valley voltage V_V current starts increasing as in any normal diode.

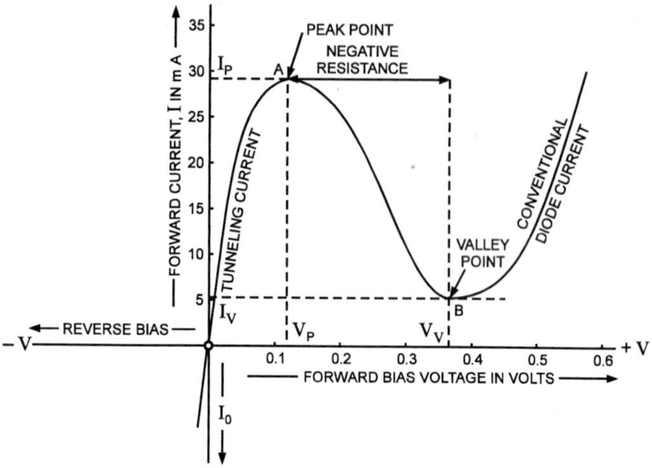

Fig. 3.27. ***V-I Characteristic of Tunnel Diode***

Depletion Region and Energy Band Diagrams. The tunneling phenomenon can be explained with the help of energy band diagrams of P- and N-type semiconductor materials.

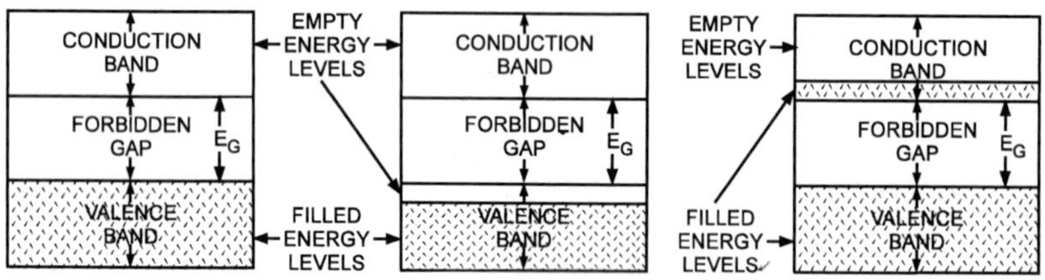

(a) Normally Doped P and N-Type Materials *(b) Very Heavily Doped P-Type Material* *(c) Very Heavily Doped N-Type Material*

Fig. 3.28. ***Energy Band Diagrams For Normally Doped and For Very Heavily Doped Semiconductor Materials***

Because of heavy doping in case of tunnel diode, the depletion region is extremely narrow and, therefore, it does not constitute much of a barrier to the electron flow. As a result, a small forward or reverse bias (not large enough to overcome the barrier potential) can provide charge carriers sufficient energy to cross the depletion region. When this occurs, the charge carriers are said to be *tunneling* through the barrier.

When the semiconductor material is normally doped (either N-type or P-type), electrons fill all the holes in the valence band of energy level and conduction band is empty as illustrated in Fig. 3.28 (*a*). In case of a heavily doped P-type semiconductor material there is a deficiency of electrons and the valence band cannot be taken as filled. The result is that at the top of valence band there is a layer of empty energy levels, as illustrated in Fig. 3.28 (*b*). With very heavily doped N-type semiconductor material, there is an abundance of electrons. The result is that electrons fill the valence band and create a layer of filled energy level at the bottom of the conduction band, as illustrated in Fig. 3.28 (*c*).

The energy band diagram for a heavily doped unbiased P-N junction is shown in Fig. 3.29. Note that the depletion region is very narrow and the filled levels on the N-side are exactly opposite those on the P-side. In this condition, no tunneling occurs as there are no empty lower energy levels to which electrons from either side might cross the depletion region. It is also to be noted that the conduction and valence bands on P-side are (negatively) higher than those on the N-side. This is due to formation of depletion region and barrier potential as a result of electrons crossing from the N-region to the P-region. The N-region has lost negative charges and P-region has gained them.

When the junction is *reverse biased* the P-region moves up with respect to the N-region. As a result, filled energy levels on the P-side become opposite empty energy levels on the N-side. At this stage electrons tunnel through the narrow space charge region from the higher energy levels on the P-side to the lower energy levels on the N-side. Despite the fact that the junction is reverse biased, significant current flows. With the increase in reverse bias, more and more electrons tunnel from the P-side to N-side and a larger current flows. Thus the reverse characteristic of a tunnel diode is linear, just like that of a resistor (Fig. 3.27).

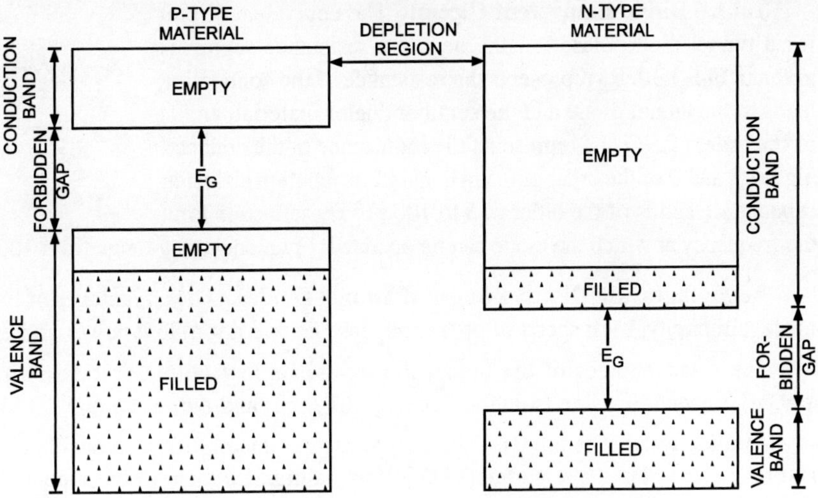

Fig. 3.29. ***Energy Band Diagrams For a Tunnel Diode***

When the tunnel diode is forward biased, its initial behaviour is similar to that when it is reverse biased. Now, some of the filled energy levels on the N-side shift to a high energy level than empty level on the P-side (Fig. 3.29). Electron tunneling now occurs from the N-side to the P-side. With the increase in forward bias, more and more electrons tunnel from the N-side to the P-side. Eventually, however, a maximum level of tunneling is reached when a band of filled energy levels at the bottom of conduction band on the N-side is directly opposite the band of empty energy levels at the top of the valence band on the P-side. At this stage diode current attains its peak value I_P and the forward bias is V_P ($\simeq$ 100 mV). With the further increase in forward bias, part of the band of filled energy levels on the N-side is raised to an energy level corresponding to forbidden energy gap on the P-side. Electrons cannot tunnel to a forbidden energy level; thus the current flow due to tunneling is reduced. With continued increase in forward bias, tunneling goes on reducing. When all of the band of filled levels at the bottom of the conduction band on the N-side is raised to a level corresponding to the forbidden energy gap on the P-side, the current flow due to tunneling is reduced to the minimum level, called the *valley current* I_V. Now, however, the normal process of current flow across the forward biased junction begins to take over, as the bias becomes large enough to overcome the barrier potential. For voltages exceeding valley voltage V_V, the current starts increasing as in a normal diode.

Diode Parameters. We have seen that between I_P and I_V (points A and B on the V-I characteristic of tunnel diode illustrated in Fig. 3.27) the diode current decreases with the increase in voltage applied across the diode. This region of the characteristic is called the *negative resistance region*, and the negative resistance R_N of the tunnel diode is its most important property. Its value may be determined by computing the slope of the characteristic in the negative resistance region. The value of negative resistance slightly differs from point to point on the negative resistance portion of the characteristic. So R_N is usually specified at the centre of the negative resistance region.

Parameters for a typical tunnel diode are given below :

Peak current I_P : 1 – 100 mA; Peak voltage V_P : 50 – 200 mV; Valley current I_V : 0.1 – 5 mA; Valley voltage V_V : 350 – 500 mV; Forward bias V : 0.5 – 1 V; Negative resistance R_N : – 10 to – 200 Ω.

Tunnel Diode Equivalent Circuit. The equivalent circuit for a tunnel diode biased in the negative resistance region is given in Fig. 3.30. R_S represents the resistance of the connecting leads to the tunnel diode and the semiconductor material, and is of the order of 5 Ω; L_S represents the inductance of the connecting leads and is of the order of 0.5 nH; and C_D is junction diffusion capacitance and is of the order of 5 to 100 pF. These factors limit the frequency at which the diode can be operated. Frequency may range from 100 MHz to 10,000 MHz.

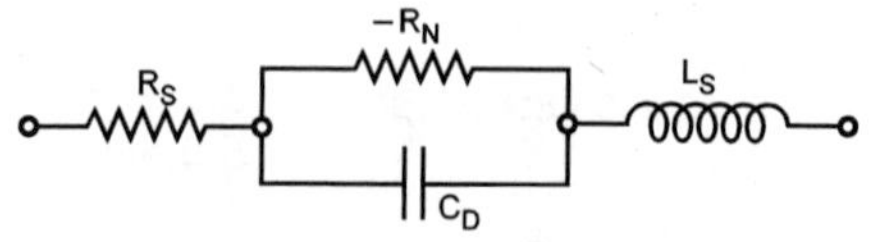

Fig. 3.30. ***Equivalent Circuit For a Tunnel Diode***

Advantages and Disadvantages of Tunnel Diodes. The advantages of tunnel diodes are low cost, low noise, simplicity, high speed of operation, ease of operation, low power and environmental immunity.

The disadvantages of the tunnel diodes are its low output-voltage swing and being a two-terminal device, it cannot provide isolation between the input and output circuits.

Applications. The tunnel diode can be used as an amplifier, an oscillator, or a switch. Because of very fast response to inputs, it is almost exclusively a high frequency component. The tunnel diode is, however, not widely used because it is a low current device and also that other better devices are available for the operations listed above.

3.6. Varactor Diode

The *varactor* [also called the *varicap, voltcap, epicap,* VVC (voltage-variable capacitance) or *tunning*] diode is a semiconductor, voltage-dependent, variable capacitor. Basically, it is just a reverse-biased junction diode whose mode of operation is based on its transition capacitance. The P- and N-regions (away from the space-charge region) are essentially low resistance areas because of high concentration of majority carriers. The space-charge or depletion region, which is depleted of majority carriers, serves as a *dielectric*. The P- and N-regions act as the plates of the capacitor and the depletion region acts as an insulating dielectric. The reverse-biased P-N junction thus possesses junction capacitance, called the *transition capacitance* C_T. The value of junction capacitance is given as

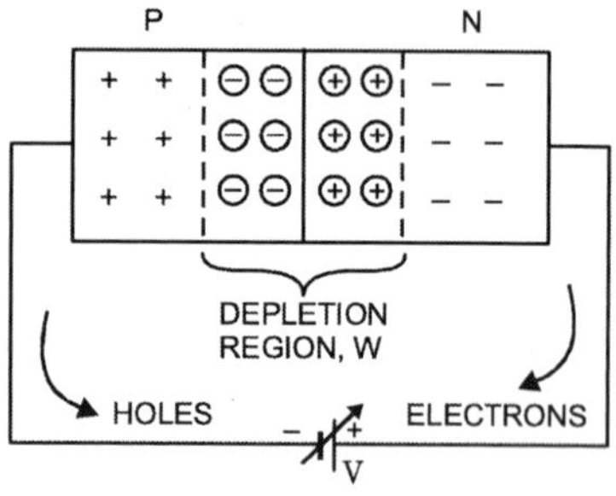

Fig. 3.31. ***Depletion Region in a Reverse Biased P-N Junction***

$$C_T = \frac{\epsilon A}{W} \quad ...(3.6)$$

where ∈ is the permittivity of the semiconductor material, A is the area of the P-N junction and W is the width of the space-charge region.

With the increase in reverse-bias potential, the width of the space-charge region increases, which in turn reduces the transition capacitance C_T. Figure 3.32 illustrates graphically how the capacitance of a varactor diode varies with the reverse bias voltage. The normal range of voltage V_R for varactor diodes is limited to about 20 V. In terms of applied reverse bias V_R, the transition capacitance is given approximately by the following equation

$$C_T = \frac{K}{(V_K + V_R)^n} \quad ...(3.7)$$

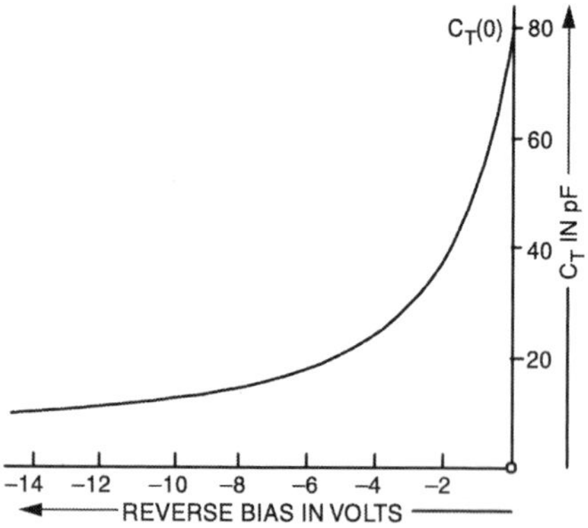

Fig. 3.32

where K is a constant determined by the semiconductor material and fabrication technique, V_K is the diode knee voltage or barrier potential, V_R is the reverse bias applied in volts and $n = \frac{1}{2}$ for alloyed junctions and $\frac{1}{3}$ for diffused junctions.

In terms of the capacitance at zero-bias condition $C_T(0)$, the capacitance as a function of reverse bias voltage is given as

$$C_T = \frac{C_T(0)}{\left(1 + \frac{V_R}{V_k}\right)^n} \qquad ...(3.8)$$

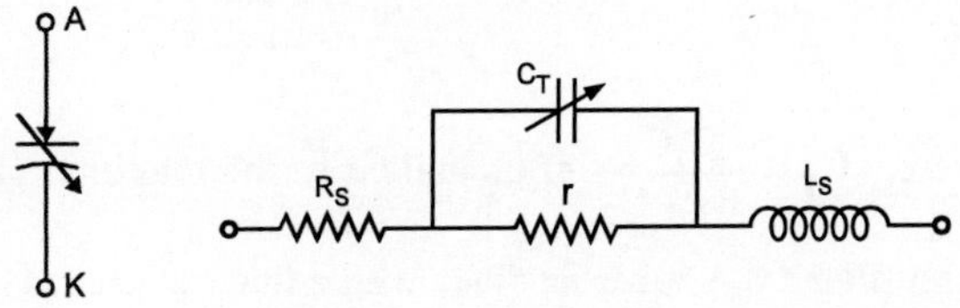

(*a*) *Symbol* (*b*) *Equivalent Circuit in Reverse-Bias Mode*

Fig. 3.33. ***Varactor Diode***

where V_k is the diode barrier potential (*e.g.* 0.7 V for Si).

The schematic symbol and a simple equivalent circuit for a varactor diode are given in Fig. 3.33. The resistance R_S represents the body (ohmic) series resistance of the diode. Typical values of C_T and R_S are 20 pF and 8.5 Ω, respectively at a reverse bias of 4 V. The reverse diode resistance *r* shunting C_T is very large (exceeding 1 MΩ). To ensure that *r* is as large (for minimum leakage current) as possible, silicon is normally used in varactor diodes. The fact that the device is used at very high frequencies, series inductance L_S is required to be included in the equivalent circuit, even though it is measured in nanohenries.

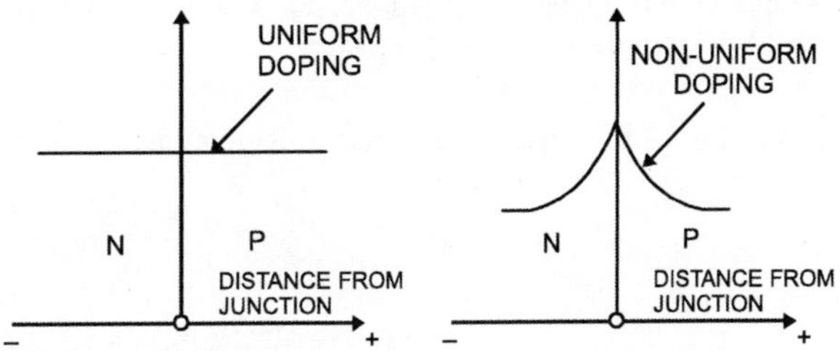

(*a*) *Abrupt Doping Profile* (*b*) *Hyper-abrupt Doping Profile*

Fig. 3.34

The tunning range of a varactor diode depends on the doping level. Fig. 3.34 (*a*) shows the doping profile for an abrupt junction diode (ordinary type of diode). Note that the doping is uniform on both sides of the junction *i.e.* the number of free electrons and holes is equally distributed. The tuning range of an abrupt-junction diode is between 3:1 and 4:1.

To get larger tuning ranges, some varactors have a hyper abrupt junction, whose doping profile appears like that shown in Fig. 3.34 (*b*). In this case concentration of charge carriers increases as the distance from the junction reduces. The heavier concentration leads to a narrower depletion region and a larger capacitance. Such a varactor diode has a tuning range of about 10:1, enough to tune an AM radio through its frequency range (535 to 1,605 kHz).

Varactor diodes possessing voltage controlled capacitance have replaced mechanically tuned capacitors in many applications such as television receivers and automobile radios. A varactor diode in parallel with an inductor gives a resonant tank circuit, whose resonant frequency can easily be changed by varying the reverse bias voltage applied to the diode. These diodes find wide use in the high-frequency (as defined by the small capacitance levels) areas which may include FM modulators, automatic frequency control devices, adjustable bandpass filters and parametric amplifiers.

3.6.1. Varactor Diode in Tuning Circuit

Figure 3.35 depicts two varactor diodes D_1 and D_2 which provide total variable capacitances in a parallel resonant circuit. Here, V_C is a variable dc voltage that controls the reverse bias and thus the capacitance of varactor diodes

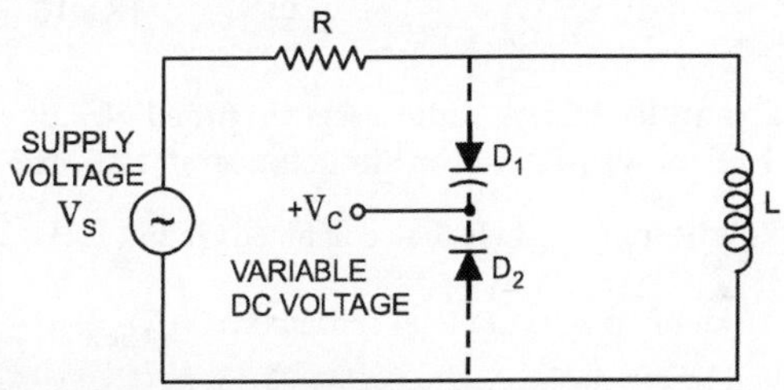

Fig. 3.35. ***Varactor Diode in Tuning Circuit***

and L is the inductance of the resonant circuit in henry. The resonant frequency of the circuit shown in Fig. 3.35 is

$$f_0 = \frac{1}{2\pi\sqrt{LC_T}}$$

where $C_T = \frac{C_1C_2}{C_1 + C_2}$ if C_1 and C_2 be the maximum and minimum values of diode capacitances.

Example 3.14. A varactor diode with a linearly graded doping profile has a capacitance of 50 pF when no bias is applied to the diode. Determine the junction capacitance for the silicon diode when the reverse bias applied to the diode is of 8 V.

Solution : Capacitance of diode with zero bias, $C_T(0)$ = 50 pF

Reverse bias applied to the diode, V_R = 8 V

Barrier potential for Si, V_k = 0.7 V

n = 1/3

From Eq. (3.8), junction capacitance is given as

$$C_T = \frac{C_T(0)}{\left(1+\frac{V_R}{V_k}\right)^n} = \frac{50}{\left(1+\frac{8}{0.7}\right)^{1/3}} = 21.59 \text{ pF } \textbf{Ans.}$$

Example 3.15. A varactor diode has a capacitance of 18 pF when the reverse bias voltage applied across it is 4 V. Determine the capacitance if the diode bias voltage is increased to 8 V.

[G.G.S.I.P. Univ. Analog Electronics, December 2009]

Solution : From Eq. (3.7)

Transition capacitance, $C_T = \frac{K}{(V_K + V_R)^n} = \frac{K}{(0.7 + V_R)^{1/3}}$ $\because n = \frac{1}{3}$ for diffused junction and V_K = 0.7 V

When reverse bias voltage, V_R is 4 V, C_T =18 pF = 18 × 10^{-12} F, so

$$18 \times 10^{-12} = \frac{K}{(4.7)^{1/3}} \quad ...(i)$$

When reverse bias voltage V_R is 8 V, let the capacitance be C, then

$$C = \frac{K}{(8.7)^{1/3}} \quad ...(ii)$$

Dividing Eq. (*ii*) by Eq. (*i*), we have

$$\frac{C}{18 \times 10^{-12}} = \left(\frac{4.7}{8.7}\right)^{1/3}$$

$$\text{or } C = 18 \times 10^{-12} \times \left(\frac{4.7}{8.7}\right)^{1/3} = 14.66 \text{ pF } \textbf{Ans.}$$

Example 3.16. A radio receiver tuned circuit uses varactor diodes, whose capacitance can be varied from 4 pF to 40 pF, and an inductance of 12.5 mH. Determine the tuning range of the circuit.

Solution : Inductance in tuned circuit, L = 12.5 mH = 12.5 × 10^{-3} H

Minimum values of total capacitance, $(C_T)_{min} = \frac{C_1C_2}{C_1 + C_2} = \frac{4 \times 4}{4 + 4} = 2$ pF

Maximum values of total capacitance, $(C_T)_{max} = \frac{40 \times 40}{40 + 40} = 20$ pF

Resonant frequency will be maximum when C_T is minimum and therefore,

$$(f_0)_{max} = \frac{1}{2\pi\sqrt{12.5 \times 10^{-3} \times 2 \times 10^{-12}}} = 1 \times 10^6 \text{ Hz or } 1 \text{ MHz}$$

Resonant frequency will be minimum when C_T is maximum and therefore

$$(f_0)_{min} = \frac{1}{2\pi\sqrt{12.5 \times 10^{-3} \times 20 \times 10^{-12}}} = 318.31 \text{ kHz}$$

Thus, the tuning range of the circuit lies between 318.31 kHz and 1 MHz. **Ans.**

3.7. SCHOTTKY DIODE

At lower frequencies, an ordinary diode can easily turn off when the bias changes from forward to reverse. But with the increase in frequency the diode reaches a point where it cannot turn off fast enough to prevent noticeable current during part of the reverse half cycle. The effect is called the *charge storage.* It restricts the useful frequency of ordinary rectifier diode. How does it happen, it is explained below :

When a diode is forward biased, some of the carriers in the depletion region have not yet recombined. If the diode is suddenly reverse biased, the carriers can flow in the reverse for a while. The greater the lifetime, the longer these charges can contribute to reverse current.

The reverse recovery time* is so short in small-signal diodes that its effect cannot be noticed at frequencies below 10 MHz or so. It becomes very important well above 10 MHz.

The solution is a special-purpose device called a *Schottky diode*. Such a diode has no depletion layer eliminating the stored charges at the junction. Because of lack of charge storage the Schottky diode can switch off faster than an ordinary diode. In fact, a Schottky diode can easily rectify frequencies exceeding 300 MHz.

Its construction is very different from the normal P-N junction in that a metal semiconductor junction is developed, as illustrated in Fig. 3.36 (*a*). On one side of the junction a metal (such as gold, silver, platinum, molybdenum, chrome or tungsten) is used and on the other side of the junction N-type doped silicon (but gallium is often used, especially at high frequencies) is used. Different construction techniques result in a different set of characteristics for the device, such as increased frequency range, lower forward bias, and so on. In general, Schottky diode construction results in a more uniform junction region and increased ruggedness compared to a point contact diode–its main rival.

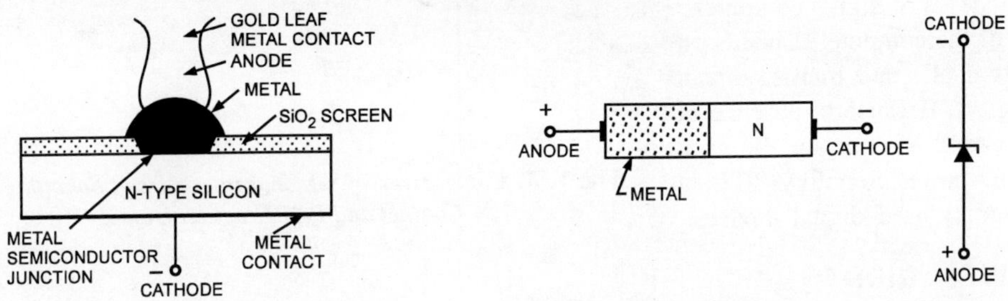

(*a*) *Passivated Hot Carrier Diode* (*b*) *Schematic Symbols of Schottky Diode*

Fig. 3.36

In both materials, the electrons are the majority carriers. In the metal, the level of minority carriers (holes) is insignificant. When the diode is unbiased, electrons on the N-side have low energy levels than electrons in the metal and so the electrons cannot cross the junction barrier, called the *Schottky barrier.* But when the diode is forward biased, the electrons on the N-side gain enough energy to cross the junction

* Already explained in Art. 2.15.1.

and enter the metal. Since these electrons plunge into the metal with very large energy, they are usually called the *hot carriers* and the diode is called the *hot-carrier diode.*

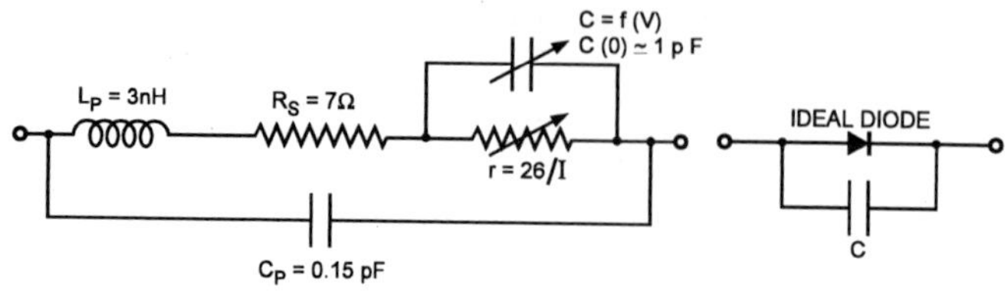

(*a*) *Equivalent Circuit* (*b*) *Approximate Equivalent Circuit*

Fig. 3.37. ***Schottky Diode***

The equivalent circuit for the device (with typical values) is given in Fig. 3.37 (*a*). The inductance L_P and capacitance C_p are the package values, and R_S is the body (ohmic) series resistance of the diode. The *r* is the diode resistance and C is the junction capacitance. For many applications, an approximate equivalent circuit, simply including an ideal diode in parallel with the junction capacitance C, illustrated in Fig. 3.37 (*b*), is used.

Schottky diode possesses two unique features over an ordinary P-N junction diode viz. (*i*) It is a unipolar device because of absence of minority carriers in reverse direction (no significant current from metal to the semiconductor in reverse-biased condition), while P-N junction diode is a bipolar device as it has both electrons and holes as majority carriers. (*ii*) Because of absence of holes in the metal, there is no depletion layer or stored charges to worry about. So Schottky diode can switch off faster than a bipolar diode. Because of larger contact area between semiconductor and metal its forward resistance is lower (*i.e.* noise is comparatively lower). The Schottky diode also has a lower barrier potential (about 0.2 – 0.25 V) where normal silicon P-N junction diode has a barrier potential of 0.7 V. This makes it more efficient in high power applications than silicon rectifiers. However, Schottky diode do exhibit higher leakage currents and lower reverse breakdown voltages. Consequently, they are only used when reverse recovery time or efficiency are the main concerns.

A comparison of V-I characteristics of Schottky, point contact and P-N junction diodes is given in Fig. 3.38.

The most important application of Schottky diodes is in digital computers. The speed of the computer depends on how fast its diodes and transistors can turn on and off. Because of absence of charge storage, it has become the back bone of low-power Schottky TTL, a group of widely used digital devices.

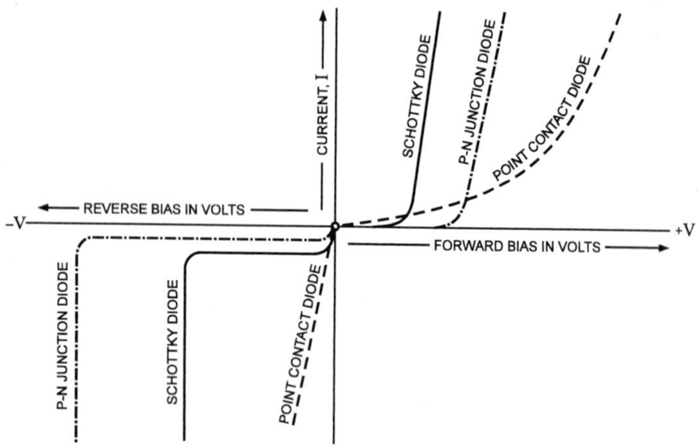

Fig. 3.38. ***Comparison of V-I Characteristics of Schottky, Point Contact and P-N Junction Diodes***

3.8. POWER DIODES

Diodes find many applications in electronics and electrical engineering circuits. Power diodes play a significant role in power electronics for conversion of electric power. A diode acts as a switch to perform various functions, such as switching in rectifiers, freewheeling in switching regulators, charge reversal of capacitor and energy transfer between components, voltage isolation, energy feedback from load to the power source, and trapped energy recovery.

Power diodes can be assumed as ideal switches for most applications but practical diodes differ from the ideal characteristics and have certain limitations. The power diodes are similar to P-N junction signal or low-power diodes. However, the power diodes have large power–, voltage–, and current-handling capabilities than those of ordinary diodes. The frequency response (or switching speed) is low in comparison to that of ordinary diodes.

Power diodes are of three types: General purpose, high speed (or fast-recovery), and Schottky. General-purpose diodes are available up to 6000 V, 4500 A, and the rating of fast-recovery diodes can go up to 6000 V, 1100 A. The reverse recovery time varies between 0.1 and 5 μs. The fast-recovery diodes are essentially required for high-frequency switching of power converters. Schottky diodes have low on-state voltage and very small recovery time, typically nanoseconds. The leakage current increases with the voltage rating and their ratings are limited to 100 V, 300 A. A diode conducts when its anode voltage is higher than that of the cathode; and the forward voltage drop of a power diode is very small, typically 0.5 and 1.2 V.

3.8.1. Basic Structure

The structure of power diode is slightly more complex than that of a signal diode. In a power diode a lightly doped N^- expitaxial layer is grown on an heavily doped N^+ substrate. The P-N junction is developed by alloying N^- layer with heavily doped P^+ region as shown in Fig. 3.39. The N^- layer in the figure is termed as drift region which is not the part of the signal diodes. The function of drift region is to absorb depletion layer of the reversed biased P^+ N^- junction J_1. The width of this region determines the reverse voltage that can be applied across the device under safe operation. The drawback of N^- layer is to add significant ohmic resistance to the diode when it is conducting in forward mode. This leads to large power dissipation in the diode, requiring proper cooling arrangements in large rating diodes.

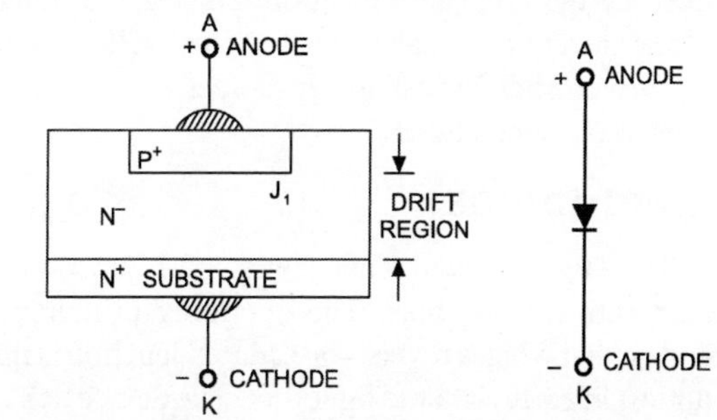

(*a*) *Basic Structure* (*b*) *Circuit Symbol*

Fig. 3.39

The circuit symbol of a power diode is the same as that for a signal diode [Fig. 3.39 (*b*)].

The modifications in the context of diode, explained above, make them appropriate for high-power applications. As diode, or P-N junction, is the basic building block of all other power semiconductor devices; same basic modifications are required to be made in all low-power semiconductor devices; in the order to enhance their power-handling capabilities.

3.8.2. Types of Power Diodes

Ideally, a diode should have no reverse recovery time. However, the manufacturing cost of such a diode may increase. In many applications, the effects of reverse recovery time is not significant, and cheaper diodes can be used. Depending on the recovery characteristics and manufacturing techniques, the power diodes may be classified as (*i*) Standard or general purpose diodes (*ii*) Fast-recovery diodes (*iii*) Schottky diodes.

The characteristics and practical limitations of these types restrict their applications.

1. General Purpose Diodes. These diodes have relatively high reverse recovery time, typically 25 μs; and are employed in low-speed applications, where recovery time is not critical (*e.g.*, diode rectifiers and converters for a low-input frequency up to 1 kHz applications and line-commutated converters). Their ratings cover a very wide range from less than 1 A to several thousands of amperes with voltage rating from 50 V to around 5 kV. These diodes are generally manufactured by diffusion process. However, alloyed types of rectifiers, used in welding power supplies, are most cost effective and rugged, and their ratings can go up to 1500 V, 400 A.

2. Fast-Recovery Diodes. As the name indicates their recovery time is very low, generally less than 5 μs. Their current ratings are from about 1 A to several thousands of amperes and voltage ratings from about

50 V to about 3 kV. For high voltage ratings (more than 400 V), they are manufactured by diffusion process and the recovery time is controlled by platinum or gold diffusion. For less than 400 V rating epitaxial diodes provide very fast recovery time. Epitaxial diodes have a very narrow base width resulting in recovery times of about 50 ns.

3.8.3. Power Diode Ratings

1. Voltage Rating. Power diodes have two voltage ratings viz., repetitive peak inverse voltage (V_{RRM}) and non-repetitive peak inverse voltage (V_{RM}). The non-repetitive peak inverse voltage determines the capability of the diode to withstand a reverse voltage which may occur occasionally. On the other hand, repetitive peak inverse voltage is applied across a diode in every cycle.

2. Current Ratings. Current rating in every device is based on its capacity to dissipate heat so that the temperature does not rise beyond permissible limit. In all semiconductor devices the current rating is determined by the allowable junction temperature. Power diodes have three current ratings, *i.e.,* average current, rms current and peak current. Actual value of each of these currents must be below the rated value. For any given current waveshape the actual values of average, rms and peak current can be found either by measurement or calculation.

3.9. PHOTODIODE

Photodiode is a two-terminal semiconductor P-N junction device and is designed to operate with reverse bias. The operating principle of a semiconductor photodiode is that when a reverse-biased P-N junction is illuminated, the current varies almost linearly with the light flux. The device consists of a P-N junction embedded in a clear plastic, as shown in Fig. 3.40. Radiation is allowed to fall upon one surface across the junction. The remaining sides of the plastic are either painted black or enclosed in a metallic case. The entire unit is extremely small and has dimensions of the order of 25 mm. The basic biasing arrangement and symbols for the device are given in Fig. 3.41.

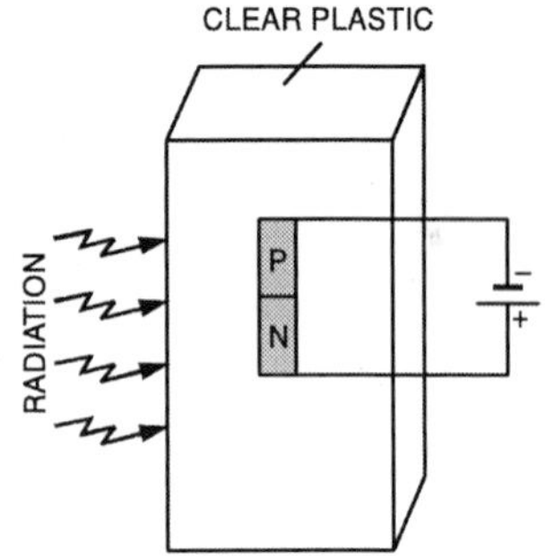

Fig. 3.40. ***Semiconductor Diode Construction***

The output voltage is taken from across a series-connected load resistor R. This resistance may be connected between the diode and ground or between the diode and the positive terminal of the supply, as illustrated in Fig. 3.41.

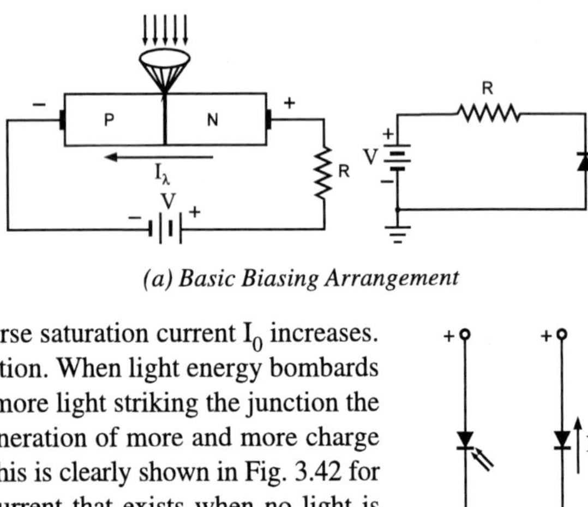

(a) Basic Biasing Arrangement

When the P-N junction is reverse biased, a reverse saturation current flows due to thermally generated holes and electrons being swept across the junction as the minority carriers. With the increase in temperature of the junction more and more hole-electron pairs are created and so the reverse saturation current I_0 increases. The same effect can be had by illuminating the junction. When light energy bombards a P-N junction, it dislodges valence electrons. The more light striking the junction the larger the reverse current in a diode. It is due to generation of more and more charge carriers with the increase in level of illumination. This is clearly shown in Fig. 3.42 for different intensity levels. The *dark current* is the current that exists when no light is incident. It is to be noted here that current becomes zero only with a positive applied bias equals to V_0. The almost equal spacing between the curves for the same increment in luminous flux reveals that the reverse saturation current I_0 increases linearly with the luminous flux as shown in Fig. 3.43. Increase in reverse voltage does not increase the

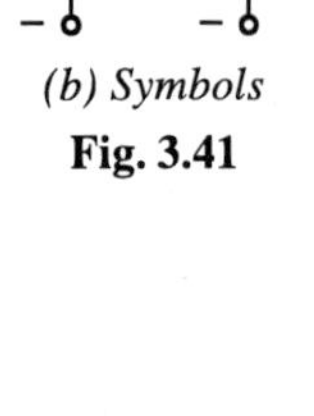

(b) Symbols

Fig. 3.41

reverse current significantly, because all available charge carriers are already being swept across the junction. For reducing the reverse saturation current I_0 to zero, it is necessary to forward bias the junction by an amount equal to barrier potential. Thus the photodiode can be used as a *photoconductive device.*

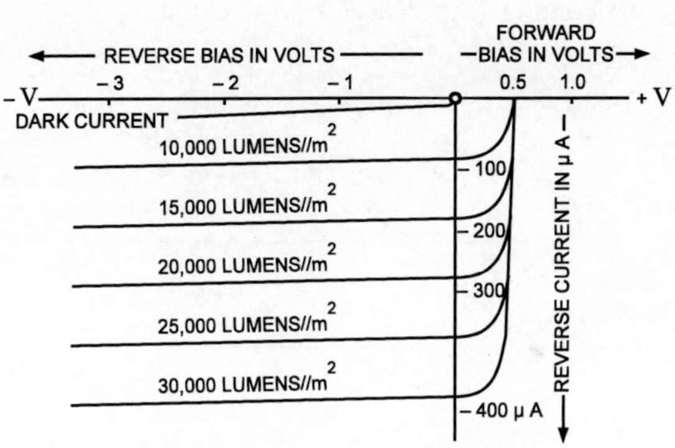

Fig. 3.42. ***Photodiode Characteristics***

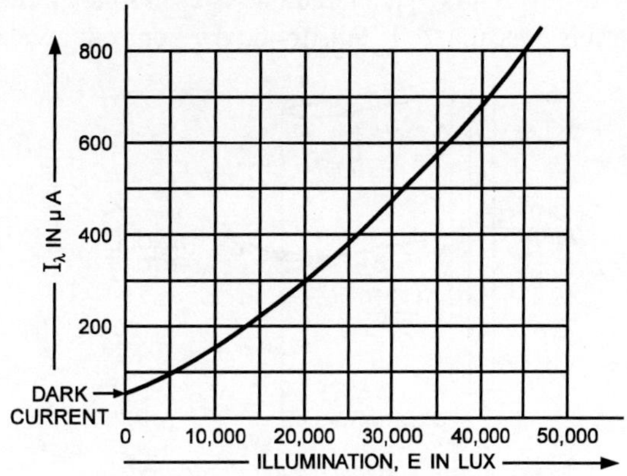

Fig. 3.43

On removal of reverse bias applied across the photodiode, minority charge carriers continue to be swept across the junction while the diode is illuminated. This has the effect of increasing the concentration of holes in the P-side and that of electrons in the N-side. But the barrier potential is negative on the P-side and positive on the N-side, and was created by holes flowing from P to N-side and electrons from N to P-side during fabrication of junction. Thus the flow of minority carriers tends to reduce the barrier potential.

When an external circuit is connected across the diode terminals, the minority carriers return to the original side via the external circuit. The electrons which crossed the junction from P to N-side now flow out through the N-terminal and into the P-terminal. This means that the device is behaving as a voltage cell with the N-side being the negative terminal and the P-side the positive terminal. Thus, the photodiode is a *photovoltaic* device as well as *photoconductive* device. When the device is operated with a reverse bias, it operates as a photoconductive device and when operated without the reverse bias, it operates as a photovoltaic device.

The reverse saturation current I_0 in a P-N diode is proportional to the concentration p_{no} and n_{po} of minority carriers in the N and P regions, respectively, when a reverse-biased P-N junction is illuminated, the number of hole-electron pairs is proportional to the number of incident photons. So, the current under large reverse bias is $I = I_0 + I_{sc}$, where I_{sc}, the short-circuit current, is proportional to the light intensity. Hence the V-I characteristic is given by

$$I = I_{sc} + I_0(1 - e^{V/\eta V_T}) \qquad \text{...(3.9)}$$

where I, I_{sc} and I_0 represent the magnitude of the reverse current, and V is positive for a forward voltage and negative for a reverse bias. The parameter η is unity for germanium and 2 for silicon, and V_T is the volt equivalent of temperature.

With typical photodiodes, the reverse current is in the tens of microamperes.

Since the rise and fall times (change of state parameters) are very small for this device (in the nanosecond range), it can be employed for high speed counting or switching applications.

Photodiodes have a far lower light sensitivity than cadmium sulphide LDRs, but give a fair quicker response in light level. Generally LDRs are ideal for use in slow acting direct coupled light-level sensing applications, while photodiodes are ideal for use in fast acting ac coupled signalling applications. Typical photodiode applications include detection (both visible and invisible), demodulation, switching, logic circuits that need stability and high speed, character recognition, optical communication equipment, IR remote control circuits, encoders etc.

An alarm system using a photodiode is shown in Fig. 3.44 (*a*). The reverse current I_λ will continue to flow as long as the light beam is not interrupted. When the light beam gets interrupted, I_λ falls to the dark current level and sounds the alarm.

In Fig. 3.44 (*b*) a photodiode is used to count items on a conveyor belt. As each item passes, the light beam is interrupted, I_λ falls to the dark current level and the counter is increased by one count.

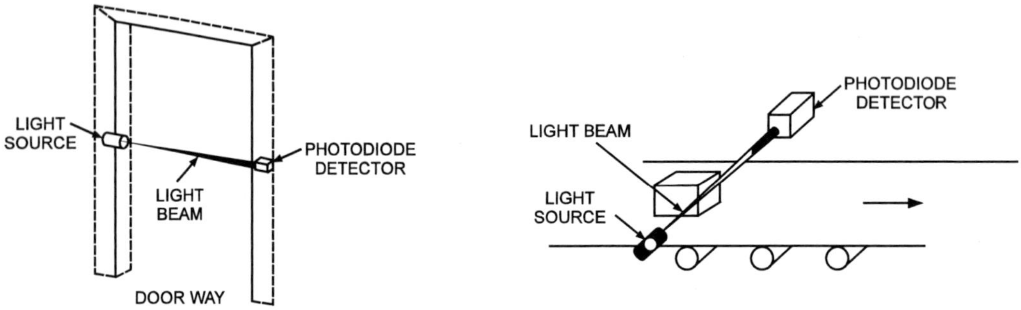

(*a*) *Alarm System Using a Photodiode* (*b*) *Counter Operation Using a Photodiode*

Fig. 3.44

Example 3.17. A photodiode is connected as shown in Fig. 3.45. When the device is illuminated with light intensity of 100 lumens/cm², a current of 1.8 mA flows in the circuit. Calculate the resistance offered by the photodiode. If the illumination is doubled, what happens to the resistance value ?

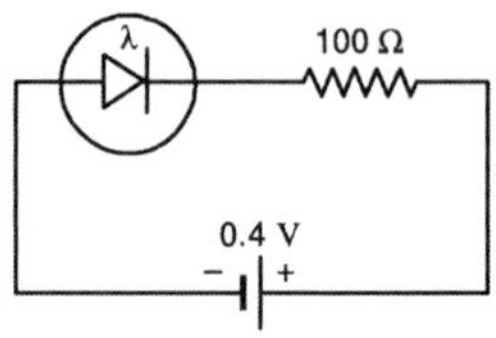

Fig. 3.45

Solution : Photodiode current, $I_d = 1.8 \text{ mA} = 1.8 \times 10^{-3}$ A

Voltage applied, $V_s = 0.4$ V

Series resistor, $R_1 = 100\ \Omega$

If R is the resistance offered by the photodiode in ohms then

$$I_d = \frac{V_s}{R + R_1}$$

$$\text{or } R = \frac{V_s}{I_d} - R_1 = \frac{0.4}{1.8 \times 10^{-3}} - 100 = 122.22\ \Omega \ \textbf{Ans.}$$

With double illumination, photodiode current I_d will be doubled *i.e.* $2 \times 1.8 \times 10^{-3} = 3.6 \times 10^{-3}$ A and now resistance offered by photodiode is given as

$$R' = \frac{V_s}{2I_d} - R_1 = \frac{0.4}{3.6 \times 10^{-3}} - 100 = 11.11\ \Omega \ \textbf{Ans.}$$

Example 3.18. A photodiode is connected in a circuit shown in Fig. 3.46 (*a*). The supply polarity reverse biases the device. The characteristics of the photodiode are shown in Fig. 3.46 (*b*). Draw the 250 Ω dc load line for the circuit and determine the diode currents and voltages at 2,000 and 25,000 lumens/m² illumination.

Solution : When diode current, $I_d = 0$

Voltage across resistance R_1 (250 Ω) $= V_{R1} = I_d R_1 = 0$

and diode voltage, $V_d = V_S$,

Supply voltage $= -0.75$ V

When diode voltage, $V_d = 0$

Voltage across 250 Ω resistance, $V_{R1} = V_S = -0.75$ V

$$\text{and diode current, } I_d = \frac{V_{R1}}{R_1} = \frac{-0.75}{250} = -3.0 \text{ mA}$$

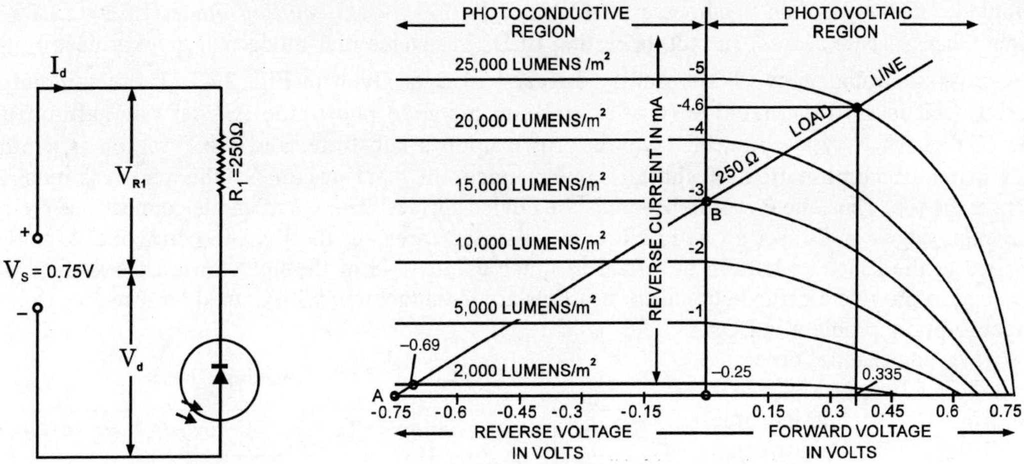

(a) Circuit Diagram *(b) Characteristics*

Fig. 3.46

The two points ($V_d = -0.75$ V, $I_d = 0$) and ($V_d = 0$, $I_d = -3.0$ mA) are plotted as points A and B respectively. The straight line joining points A and B is 250 Ω load line. **Ans.**

From the load line we get,

At 2,000 lumens/m^2 : $I_d = -0.25$ mA; $V_d = -0.69$ V **Ans.**

At 25,000 lumens/m^2 : $I_d = -4.6$ mA; $V_d = 0.35$ V **Ans.**

It is to be noted that the polarity of the diode, voltage V_d, changes from – ve to + ve at the highest level of illumination.

3.10. LIGHT-EMITTING DIODES

The best known of all optoelectronic devices is the so called LED (lights-emitting diode), which emits a fairly narrow bandwidth of visible (usually red, orange, yellow or green) or invisible (infrared) light when its internal diode junction is stimulated by a forward electric current/voltage (power). LEDs have typical power-to-light energy conversion efficiencies some 10 to 50 times greater than of a simple tungsten lamp and have very fast response times (approximately 0.1 μs compared with tens or hundreds of milliseconds for a tungsten lamp), and are thus widely employed as visual indicators and as moving light displays.

The operation of light-emitting diode (LED) is based on the phenomenon of electroluminance, which is the emission of light from a semiconductor under the influence of an electrical field. The recombination of charge carriers take place in a forward P-N junction as the electrons cross from the N-region and recombine with holes existing in P-region. Free electrons are in the conduction band of energy levels, while holes are in the valence energy band. Therefore, the electrons are at high energy levels than the holes. For the electrons to recombine with the holes, they must give some of their energy. Typically, these electrons give up energy in the form of heat and light.

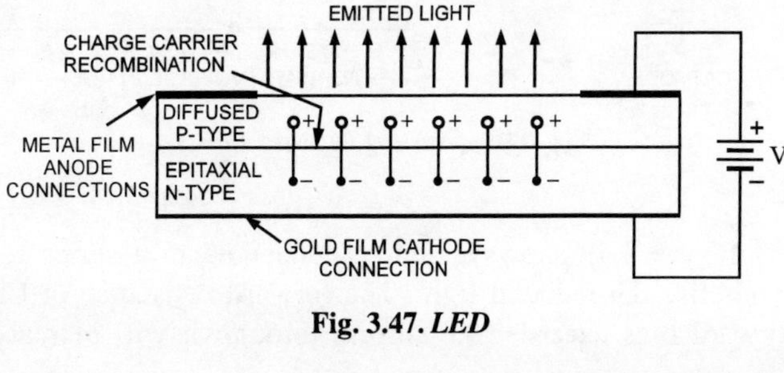

Fig. 3.47. ***LED***

In silicon and germanium diodes, most of the electrons give up their energy in the form of heat. However, with GaAsP, and GaP semiconductors, the electrons give up their energy by emitting photons. If the semiconductor is transluscent, the light will be emitted and the junction becomes a source of light, *i.e.* a *light-emitting diode* (LED). LEDs emit no light when *reverse biased*. In fact, operating LEDs in reverse bias mode will quickly destroy them.

A cross-sectional view of a typically diffused LED is given in Fig. 3.47. The semiconductor material used is gallium arsenide (GaAs), gallium arsenide phosphide (GaAsP) or gallium phosphide (GaP). An N-type epitaxial layer is grown upon a substrate, and the P-region is produced by diffusion. Recombination of charge carriers occur in the P-region so this region is required to be kept at the top. Thus, the P-region becomes the device surface. The metal anode connections are made at the outer edges of P-layer so as to allow more surface area for the light to come out. A gold film is applied to the substrate bottom to reflect as much as possible of the light toward the surface of the device and to provide a cathode connection. LEDs are manufactured with domed lenses so as to reduce the reabsorption problem. LEDs are always encased in order to protect their delicate wires. LEDs made from GaAs emit invisible infrared light, LEDs constructed of GaAsP tend to emit either red or yellow light. The GaP LEDs give either red or green light. LEDs can be manufactured to produce light of virtually any colour. Table 3.1 provides a list of common compound semiconductors and the light they produce. In addition, the typical range of forward bias voltage for each is listed.

TABLE 3.1. Light-Emitting Diodes

Colour	Construction	Typical Forward Voltage in Volts
Amber	AllnGaP	2.1
Blue	GaN	5.0
Green	GaP	2.2
Orange	GaAsP	2.2
Red	GaAsP	1.8
White	GaN	4.1
Yellow	AllnGaP	2.1

Figure 3.48 (*a*) shows a circuit symbol for a LED. The arrows indicate radiation emitted by the diode. Figs. 3.48 (*b*) and 3.48 (*c*) give two curves used to determine LED operating characteristics. Figure 3.48 (*b*) is forward bias V-I curve for a typical LED employed in burglar alarms. Forward bias of approximately one volt is required to give significant forward current. Figure 3.48 (*c*) provides radiant power-forward current curve. The radiant output power is rather small and indicates a very low efficiency of electrical to radiant energy conversion.

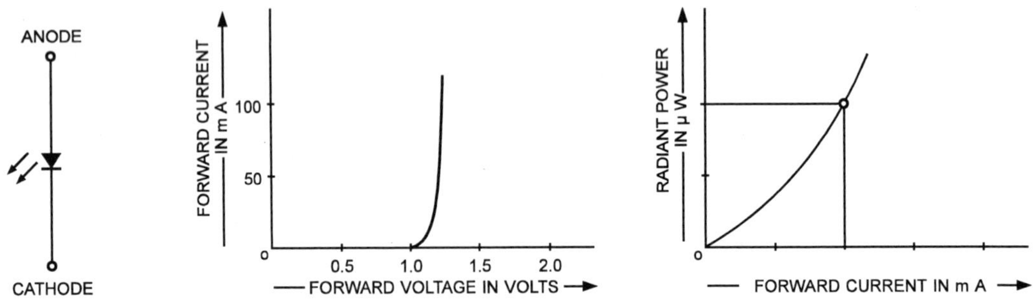

(*a*) (*b*) *Forward Bias V-I Curve* (*c*) *Output Characteristic Curve*

Fig. 3.48. ***LED***

Figure 3.49 shows a source connected to a series resistor R_S and a LED. The outward arrows symbolise the radiated light. The forward resistance of LEDs is very low which means that once the forward bias exceeds, the current through it will increase rapidly for only a very small increase in voltage. Thus it becomes necessary to use an external series current limiting resistor. The value of series resistor R_S can be determined from the following equation

$$R_S = \frac{V_S - V_D}{I} \qquad ...(3.10)$$

where V_S is the supply voltage, V_D is LED forward bias voltage and I is the desired current.

For most of the commercially available LEDs, the typical voltage drop (diode voltage) is from 1.5–2.5 V for currents between 10 and 50 mA. The exact voltage drop depends on the LED current, colour, tolerance, etc.

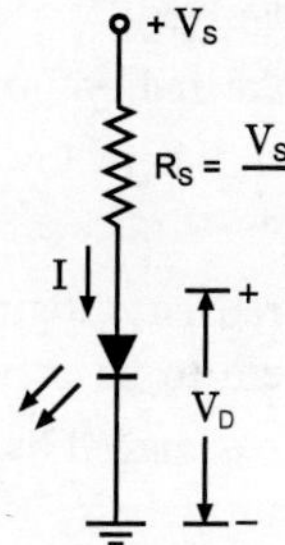

Fig. 3.49. ***LED Circuit***

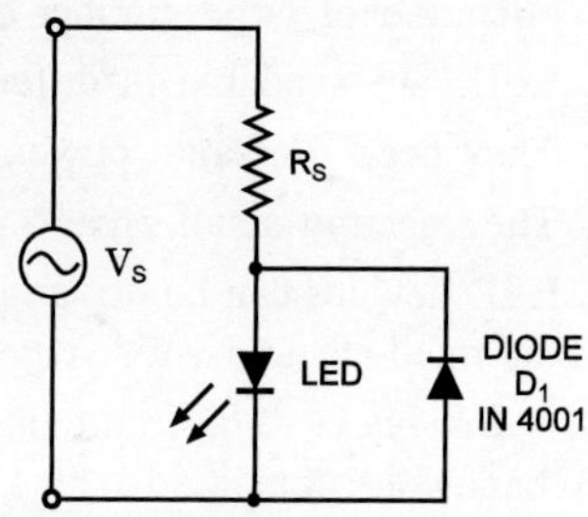

Fig. 3.50. ***LED As An Indicator in AC Circuit***

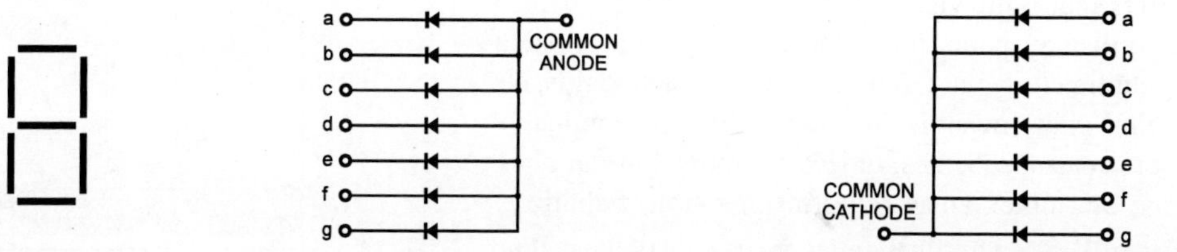

(a) Display Format *(b) Common Anode Configuration* *(c) Common Cathode Configuration*

Fig. 3.51

A LED can be used as an indicator in ac circuit by wiring it in inverse parallel with a normal diode, as shown in Fig. 3.50, to prevent the LED being reverse biased; for a given brightness, the value of R_s should be halved relative to that of the dc circuit.

LED displays are available in many different sizes and shapes. The light-emitting region is available in lengths from 2.5 to 25 mm. Numbers can be created by such segments. Typical arrangement of a seven segment LED numerical display is shown in Fig. 3.51 (*a*). Any desired numeral from 0 to 9 can be displayed by passing current through the appropriate segments. The LEDs in a seven segment display may be connected in common anode or in common cathode configuration, as illustrated in Figs. 3.51 (*b*) and 3.51 (*c*) respectively. LEDs are usually switched *on* and *off* by transistor circuit.

Transistor circuit for switching on and off LEDs is given in Fig. 3.52.

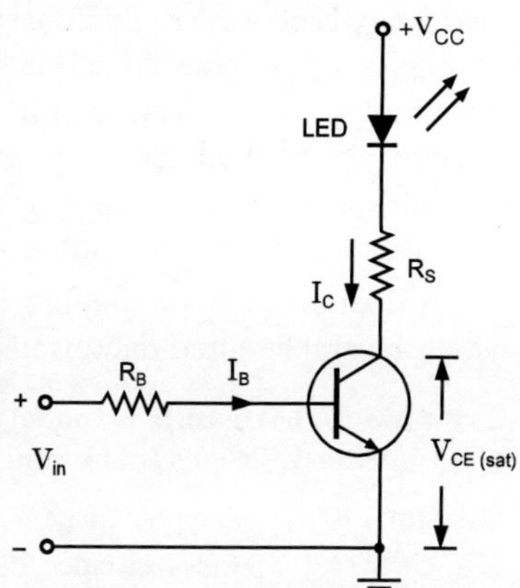

Fig. 3.52. ***Transistor Circuit For Switching On and OFF a LED***

Advantages of LEDs. 1. LEDs are miniature in size and can be stacked together to form numeric and alphanumeric displays in high density matrix.

2. Smooth control of light intensity from an LED because light output from an LED is a function of current flowing through it.
3. LEDs are economical and have a high degree of reliability because they are manufactured with the same type of technology as is used for transistors and ICs.
4. LEDs are rugged and can, therefore, withstand shocks and vibrations.

5. LEDs can be operated over a wide temperature range (0–70°C).
6. The switching time (both on and off) is less than 1 ns. So they are very suitable where dynamic operation of large number of arrays is involved.
7. LEDs are available in different colours like red, yellow, green and amber.
8. They need moderate power. They are used where low dc power is available.
9. They occupy small area.
10. LED devices can be driven by transistor-transistor logic (TTL) whereas gas discharge devices need intermediate transistor stages to allow them to be driven by 5 V TTL.

The *drawbacks of LEDs* are that they may get damaged by overvoltage or overcurrent. They have wide optical bandwidth compared to the LASER (≃ 10 nm). Their temperature depends on the radiant output power and wavelength. LEDs are not suited for large area displays, primarily because of their high cost. For large displays, devices using gas filled plasma are used.

LEDs that emit visible light are widely used in instrument display panel indicator, digital watches, calculators, multimeters, intercom, telephone switchboards, etc. LEDs that emit invisible infrared light find application in remote control schemes, object detectors, burglar alarm systems; and other areas requiring invisible radiation.

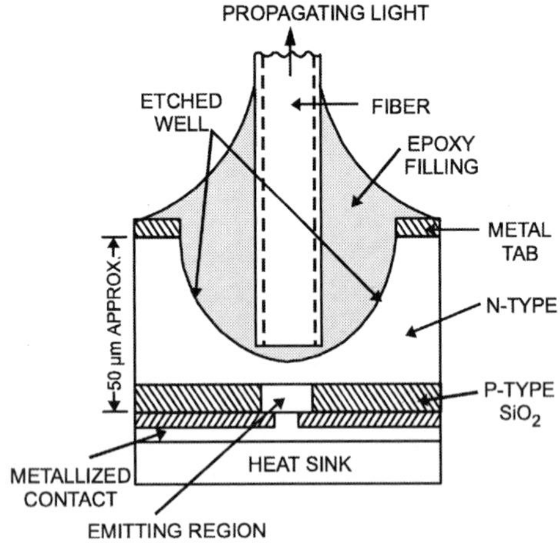

Fig. 3.53. ***LEDs as Source For Fiber Optic System***

The LEDs used for fiber optics are usually of the gallium arsenide (GaAs) type, with various dopants added to shift the center wavelength of the radiation spectrum. Dopants used are phosphorus (P), indium (In), and aluminium (Al). Gallium arsenide phosphide (GaAsP) diodes can be made with bandgaps in the range of 1.5 to 2.0 eV (0.62 – 0.83 μm wavelength). Gallium indium arsenide (GaInAs), indium arsenide phosphide (InAsP), or aluminium indium arsenide (AlInAs) diodes cover the range from 0.5 to 1.5 eV (0.83 to 2.5 μm). Some of these structures may also be used to form laser diodes.

A fiber to the diode coupling is shown in Fig. 3.53. Currently, diode fiber combinations are available in which several hundred microwatts at the most can be produced and injected into an optical fiber.

Example 3.19. An LED is connected across a voltage source of +10 V through a series resistance of 820 Ω. Calculate the LED current. Assume the voltage drop across an LED of 1.5 V.

Solution :

Supply voltage, $V_S = 10$ V

Series resistance, $R_S = 820\ \Omega$

Voltage drop across LED, $V_D = 1.5$ V

$$\text{LED current, } I = \frac{V_S - V_D}{R_S} = \frac{10 - 1.5}{820} = 10.4 \text{ mA} \quad \textbf{Ans.}$$

HIGHLIGHTS

1. There are two mechanisms by which breakdown can occur at a reverse biased P-N junction : avalanche breakdown and zener breakdown. The breakdown phenomenon is reversible and harmless so long as the safe operating temperature is maintained.
2. A zener diode is a P-N junction operated in the reverse-biased mode to take advantage of its

sharply defined breakdown voltage. The direction of conduction for a zener diode is opposite to that of the arrow in the symbol, and the zener voltage has a polarity opposite to that of a forward-biased diode.

3. The zener voltage V_Z is specified at some test value of current I_{ZT}, at which the diode will exhibit some dynamic impedance Z_Z known as zener impedance and given by $Z_Z = \frac{\Delta V_Z}{\Delta I_Z}$ where ΔV_Z and ΔI_Z are the small variations in voltage and current respectively.

4. Zener diodes have a temperature coefficient, which generally is negative for V_Z below 5 V but positive above 6 V, and is expressed in percent of V_Z per °C with the change in zener voltage given by E_q.

$\alpha_Z = \frac{\Delta V_Z}{V_Z(t - t_0)} \times 100$. Reference zener diodes are available with α_Z as low as 0.0005% per °C.

5. A zener diode may be used to regulate the load voltage at the value V_Z by acting as a "bypass value" to counteract line voltage or load current variations. Zener diodes are also useful in voltage reference circuits. The percent regulation, a figure of merit for the circuit, is a function of the range of input voltage and load resistance values, and of the individual device parameters.

6. The tunnel diode (sometimes called the *Esaki diode* after its inventer, Dr. Leo Esaki) is a high conductivity, heavily doped two terminal P-N junction. This diode differs from the normal P-N junction diode in the amount of impurity concentration. The impurity concentration in a normal P-N junction diode is about 1 part in 10^8, whereas in the tunnel diode the impurity concentration is about 1 part in 10^3. It can conduct in reverse as well as forward direction but is usually used in forward biased mode.

   A tunnel diode's characteristics exhibit a negative resistance region and it may be used as an ultra-fast switching device or in ultrahigh frequency oscillators.

7. The varactor [also called the *varicap, voltcap, epicap* (voltage variable capacitance) or *tuning*] diode is a semiconductor, voltage-dependent, variable capacitor. Basically, it is just a reverse-biased junction diode whose mode of operation is based on its transition capacitance.

8. A metal on a lightly doped semiconductor can produce a rectifying contact that is known as a Schottky barrier diode. The ideal voltage-current relationship of the Schottky barrier diode is the same as that of the P-N junction diode. However, since the current mechanism is different from that of the P-N junction diode, the switching speed of the Schottky diode is faster. In addition, its reverse saturation current is larger than that of the P-N junction diode, so it needs less forward bias voltage to achieve a given current compared to a P-N junction diode.

9. The power diodes are similar to P-N junction signal or low-power diodes but have large power-, voltage- and current-handling capabilities than those of conventional P-N junction diodes. Power diodes find many applications in electronics and electrical engineering circuits.

10. Photodiode is a two-terminal semiconductor P-N junction and is designed to operate with reverse bias. The photodiode is a photovoltaic device as well as photoconductive device. When the device is operated with a reverse bias, it operates as a photoconductive device and when operated without the reverse bias, it operates as a photovoltaic device.

11. Light-emitting diodes (LEDs) emit light under forward bias conditions but need 2 V to 4 V for good emission.

    LEDs may emit visible radiation when the PN junction is diffused in GaAsP, infrared if constructed from GaAs, and a laser beam if sufficiently high pulsed current is applied to a GaAs junction.

EXERCISES

1. Explain the breakdown mechanism in a diode. **[G.G.S.I.P. Univ. Analog Electronics December 2009]**
2. Explain the difference between zener breakdown and Avalanche breakdown.
3. Explain various types of breakdown mechanisms in P-N junction.
4. What are the various breakdown mechanisms in junction diode ? Give example of, at least one type, for each of these mechanisms and elaborate their applications.

5. Breakdown diodes with a breakdown voltage greater than 5 V generally have a positive temp. coefficient but those with lower breakdown voltage have a negative temp. coefficient. Do you agree with this statement. Justify.
6. Draw the V-I characteristics of zener diode and explain the phenomenon of zener breakdown and avalanche breakdown. **[G.G.S.I.P. Univ. Analog Electronics, December 2009]**
7. Explain with reference to the zener diode characteristic, zener voltage, breakdown current and dynamic impedance. How does the dynamic impedance vary with zener current ? Draw the equivalent circuit of an ideal zener diode in the breakdown region.
8. Explain zener diode as voltage regulator.
9. Briefly describe the working of zener diode shunt regulator.
10. Explain the working of a zener diode and its use as a reference voltage device.
11. How does a tunnel diode differ from a conventional P-N junction diode ? With the help of its energy-band diagram explain its principle of operation in forward and reverse bias conditions. Sketch its V-I characteristics and mark negative resistance on it.
12. Draw and explain V-I characteristics of a tunnel diode.
13. Draw and explain the construction and operation of a varactor diode.
14. Which device produces voltage variable capacitor ? How the voltage variable capacitance varies with change in voltage across it ? Enumerate its special features.
15. Explain the working principle of (*i*) Schottky diode (*ii*) Tunnel diode. Give their characteristics and typical applications. **[G.G.S.I.P. Univ. Analog Electronics December 2010]**
16. What do you mean by hot carrier diode ? Compare the characteristics of hot carrier and P-N junction diodes.
17. How does a power diode differ from a signal diode ? What are the types of power diodes ? Give applications of power diodes. Explain the construction and characteristics of Schottky diode.
18. Explain how light signal is converted into electrical signal in a photodiode.
19. What is a photodiode ? Give its biasing arrangement and construction. Give its one or two applications.
20. What is LED ? Give its principle of working, construction, circuit symbol, merits, demerits and applications.
21. Give the basic principle of LED. **[G.G.S.I.P. Univ. Analog Electronics December 2010]**
22. Write short notes on the following :
    (*i*) Junction breakdown. (*ii*) Zener diode and its applications.
    (*iii*) Tunnel diode. (*iv*) Varactor diode. (*v*) Photodiode.

SHORT ANSWER TYPE QUESTIONS WITH ANSWERS

Q. 1. What are the two mechanisms of breakdown in a P-N junction ?
Ans. Avalanche and zener breakdowns.
Q. 2. Name the breakdown mechanism in a lightly doped P-N junction under reverse biased condition.
Ans. Avalanche breakdown.
Q. 3. Name the breakdown mechanism in a heavily doped P-N junction under reverse biased condition.
Ans. Zener breakdown.
Q. 4. What is a zener diode ?
Ans. Zener diode is a P-N junction diode specially designed for operation in its breakdown region.
Q. 5. What is zener voltage ?
Ans. The voltage at which the zener diode breaks down is called the zener voltage.
Q. 6. What is meant by the temperature coefficient ?
Ans. The effect of temperature on zener voltage is given in terms of temperature coefficient which is defined as the percentage change in nominal zener voltage for each degree centigrade of change in junction temperature.

Mathematically it is given as $\frac{\Delta V_Z}{V_Z(t-t_0)} \times 100$ where ΔV_Z is the change in zener voltage due to temperature variation of $(t - t_0)$.
Q. 7. What happens to the series current, load current and zener current when the dc input voltage of a zener regulator increases ?
Ans. Zener current and series current increase while the load current remains unchanged.

Q. 8. Why is zener diode used as voltage regulator ?

Ans. Zener diode has the property of behaving like a dc battery in 'on' state (*i.e.* when the voltage across zener diode exceeds its zener voltage rating V_Z). In 'on state', the voltage across zener diode remains constant until the voltage across it drops less than V_Z. This property of zener diode makes its use as voltage regulator.

Q. 9. Explain how zener diode maintains constant voltage across the load ?

Ans. Zener diode has the property of behaving like a dc battery in 'on' state. If the zener diode is shunted across the load R_L and the voltage across zener diode is more than zener voltage V_Z then zener diode is in 'on' state, and any variation in voltage across the zener diode due to variations either in supply voltage or in load resistance is not able to change the output voltage. Thus zener diode maintains voltage constant across the load.

Q. 10. What is tunnel diode ?

Ans. The tunnel diode (sometimes called the *Esaki diode* after its inventer, Dr. Leo Esaki) is a high conductivity, heavily doped two terminal P-N junction. This diode differs from the normal P-N junction diode in the amount of impurity concentration. The impurity concentration in a normal P-N junction diode is about 1 part in 10^8, whereas in the tunnel diode the impurity concentration is about 1 part in 10^3. It can conduct in reverse as well as forward direction but is usually used in forward biased mode.

A tunnel diode's characteristics exhibit a negative resistance region and it may be used as an ultra-fast switching device or in ultrahigh frequency oscillators.

Q. 11. What is tunneling ?

Ans. The mechanism of conduction in a semiconductor diode in which charge carriers (possessing very little energy) punch through a barrier directly instead of climbing over it is called tunneling.

Q. 12. What are the applications of tunnel diodes ?

Ans. Tunnel diodes are used as amplifiers, oscillators or switching devices, being an exclusive high-frequency component because of its very fast response of inputs.

Q. 13. What is a varactor diode ?

Ans. A varactor diode is a specially fabricated P-N junction with proper impurity concentration profile and operated under reverse-biased mode so as to give a variable junction capacitance.

Q. 14. Which device produces voltage variable capacitor ? How the voltage variable capacitance varies with the change in voltage across it ?

Ans. The varactor diode produces voltage variable capacitor. The junction capacitance of a varactor diode varies inversely as the square root of the reverse bias voltage in case of alloyed junction and varies inversely as the cube root of the reverse bias voltage for diffuse junction.

Q. 15. What is Schottky diode ?

Ans. Schottky diode is quite different in construction from the normal P-N junction diode. It has metal (such as gold, silver, platinum, molybdenum, chrome or tungsten) on one side and N-type doped silicon on the other side of the junction. It has no storage charge. The junction barrier is called the Schottky barrier.

Q. 16. Why is Schottky diode called hot-carrier diode ?

Ans. Since in forward bias operation of the Schottky diode, the electrons on the N-side gain enough energy to cross the junction and plunge into the metal with very large energy, they are usually called hot carriers and the diode is called the hot-carrier diode.

Q. 17. What are power diodes ?

Ans. The power diodes are similar to P-N junction signal or low-power diodes but have large power-, voltage- and current-handling capabilities than those of conventional P-N junction diodes. Power diodes find many applications in electronics and electrical engineering circuits.

Q. 18. Why the current in power diodes varies linearly rather than exponentially with voltage ?

Ans. The large magnitude of current in power diodes leads to ohmic drop that hides the exponential part of the V-I characteristic curve.

Q. 19. What is photodiode ?

Ans. Photodiode is a two-terminal semiconductor P-N junction device and is designed to operate with reverse bias.

Q. 20. State the working principle of photodiode.

Ans. Photodiode is a two-terminal semiconductor P-N junction device and is designed to operate with reverse bias. When light energy bombards a P-N junction, it dislodges valence electrons. The more light striking the junction, the larger the reverse current in a diode. It is due to generation of more and more charge carriers with the increase in level of illumination. This is the working principle of photodiode.

Q. 21. How the current is reduced to zero in a photodiode ?

Ans. The reverse saturation current I_0 flowing through a photodiode is reduced to zero by applying a forward bias voltage of magnitude equal to barrier potential.

Q. 22. What is radiative recombination in LED ?

Ans. Whenever LED is forward biased, a forward current starts flowing through it. This current flows because of electrons supplied by the battery. The electrons supplied by the battery are free electrons and they reside in the conduction band. The energy level of conduction band is higher than that of valence band. These electrons recombine with holes to become valence electrons and come to a lower energy level. While coming back to the valence band, they give up the excess heat in the form of light (refer to Fig. 3.47). This process is called recombination in LED because of which light emits in LED.

Q. 23. What precautions are required to be observed in the use of LEDs ?

Ans. LEDs should neither be reverse biased nor operated near their maximum current rating. The leads of a LED should never be bent closer than about 2 mm from the encapsulation.

Q. 24. State differences between LED and photodiodes.

Ans. Photodiode is a two-terminal semiconductor P-N device and is designed to operate with reverse bias. It works when light falls on depletion layer of P-N junction of photodiode. Photodiode is a photovoltaic device as well as photoconductive device. When it is operated with a reverse bias, it is photoconductive device and when operated without the reverse bias, it is a photovoltaic device.

LED stands for light-emitting diode. It emits a fairly narrow bandwidth of visible (usually red, orange, yellow or green) or invisible (infrared) light when its internal diode junction is stimulated by a forward electric current/voltage (power). The operation of LED is based on the phenomenon of electroluminence, which is the emission of light from a semiconductor under the influence of electric field.

PROBLEMS

1. A zener diode shunt regulator is shown in Fig. 1.
Determine
(*i*) load voltage,
(*ii*) voltage drop across series resistance,
(*iii*) current through the diode.
[**Ans.** (*i*) 12 V (*ii*) 18 V (*iii*) 51 mA]

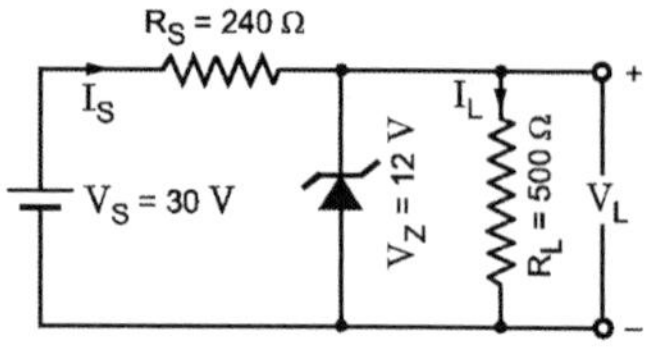

Fig. 1

2. Determine the maximum and minimum values of zener current if in the circuit shown in Fig. 3.9 the load resistance R_L = 5 kΩ, series resistance is 2 kΩ, input voltage varies between 80 V and 100 V and the output voltage is 50 V. [**Ans.** 15 mA and 5 mA]

3. In the circuit shown in Fig. 3.9 the series resistance is 2 kΩ, load resistance is 5 kΩ, input voltage is 60 V and output voltage is 35 V. Determine (*i*) voltage drop across series resistance (*ii*) load current (*iii*) current through series resistance and (*iv*) zener diode current. [**Ans.** (*i*) 25 V (*ii*) 7.0 mA (*iii*) 12.5 mA (*iv*) 5.5 mA]

4. Determine the temperature coefficient of a 7.5 V zener diode at 25°C if the nominal voltage increases to 7.8 V at a temperature of 105°C. [**Ans.** + 0.05%]

5. For the circuit of the following Fig. 2, find (*a*) the output voltage (*b*) voltage across R_S and (*c*) the current through zener diode. [**Ans.** (*a*) 8 V (*b*) 4 V (*c*) 0 A]

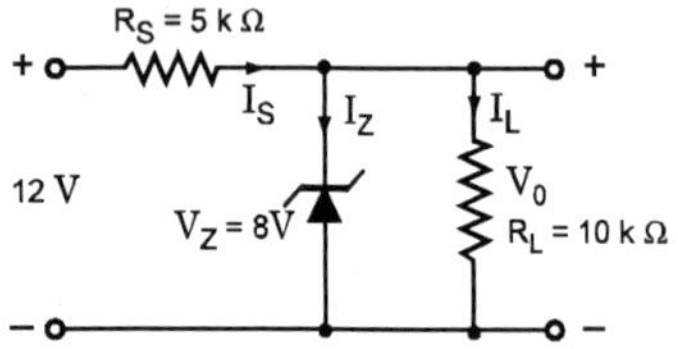

Fig. 2

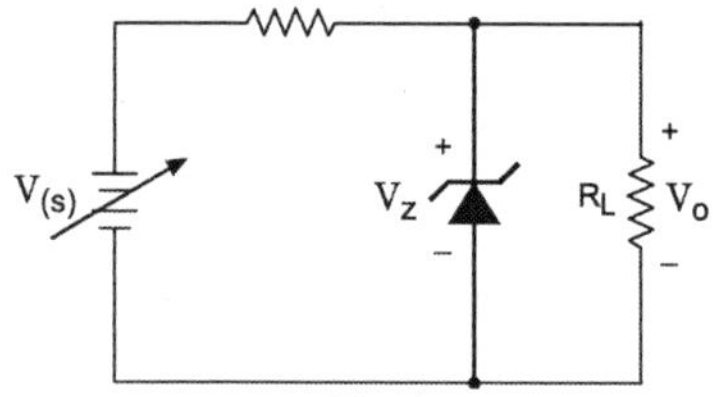

Fig. 3

6. Explain 'Avalanche' breakdown. Draw the *v-i* characteristics of zener diode and explain how does a zener regulate a voltage ? The input voltage for the Fig. 3 varies from 35 V to 45 V, V_Z = 20 V, r_Z = 5 Ω, I_L (min) = 0 mA, I_L (max) = 100 mA, I_Z (min) = 10 mA, I_Z (max) = 400 mA. Find the values of R and P_z (max). [**Ans.** 136.4 Ω, 1.67 W]

7. A varactor diode with a linearly graded doping profile has a capacitance of 40 pF when no bias is applied to the diode. Determine the junction capacitance for the silicon diode when the reverse bias applied to the diode is of 7 V. [**Ans.** 18 μF]

❑❑❑

CHAPTER 4

Rectifiers

4.1. INTRODUCTION

INSIDE THIS CHAPTER

The electrical power is almost exclusively generated, transmitted and distributed in the form of ac because of economical consideration but for operation of most of the electronic devices and circuits, dc supply is required. Dry cells and batteries can be used for this purpose. No doubt, they have the advantages of being portable and ripple free but their voltages are low, they need frequent replacement and are expensive in comparison to conventional dc power supplies. Nowadays, almost all electronic equipment include a circuit that converts ac supply into dc supply. The part of equipment that converts ac into dc is called the *dc power supply*. In general, at the input of the power supply, there is a power transformer. It is followed by a rectifier (a diode circuit), a *smoothing filter* and then by a *voltage* regulator circuit. A block diagram of such a power supply is given in Fig. 4.1.

As obvious from the block diagram shown in Fig. 4.1, the basic power supply is constituted by four elements viz. a *transformer*, a *rectifier*, a *filter*, and a *regulator* put together. The output of the dc power supply is used to provide a constant dc voltage across the load. Let us briefly outline the function of each of the elements of the dc power supply.

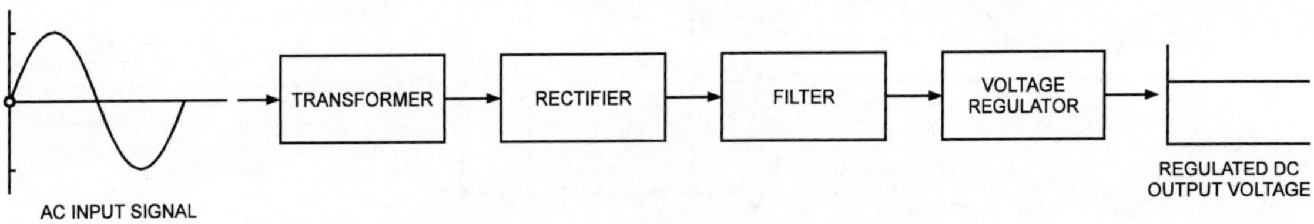

Fig. 4.1. ***Block Diagram of a DC Power Supply***

Transformer is used to step-up or step-down (usually to step-down) the supply voltage as per need of the solid-state electronic devices and circuits to be supplied by the dc power supply. It can provide isolation from the supply line—an important safety consideration. It may also include internal shielding to prevent unwanted electrical noise signal on the power line from getting into the power supply and possibly disturbing the load.

Rectifier is a device which converts the sinusoidal ac voltage into either positive or negative pulsating dc. P-N junction diode, which conducts when forward biased and practically does not conduct when reverse biased, can be used for rectification *i.e.* for conversion of ac into dc. The rectifier typically needs one, two or four diodes. Rectifiers may be either half-wave or full-wave (centre-tap or bridge) type.

The output voltage from a rectifier circuit has a pulsating character *i.e.*, it contains unwanted ac components (components of supply frequency *f* and its harmonics) along with dc component. For most supply purposes, constant direct voltage is required than that furnished by a rectifier. To reduce ac components from the rectifier output voltage a *filter circuit* is required. Thus *filter* is a device which passes dc component to the load and blocks ac components of the rectifier output. Filter is typically formed from reactive circuit elements such as capacitors and/or inductors and resistors.

The magnitude of output dc voltage may vary with the variation of either the input ac voltage or the magnitude of load current. So at the output of a rectifier-filter combination a voltage regulator is required, as shown in Fig. 4.1, to provide an almost constant dc voltage at the output of the regulator. The voltage regulator may be constructed from a zener diode, and/or discrete transistors, and/or integrated circuits (ICs). Its main function is to maintain a constant dc output voltage. However, it also rejects any ac ripple voltage that is not removed by the filter. The regulator may also include protective devices such as short-circuit protection, current limiting, thermal shutdown, or overvoltage protection.

4.2. HALF-WAVE RECTIFIERS

When a single rectifier unit is placed in series with the load across an ac supply, as illustrated in Fig. 4.2, it converts alternating voltage into unidirectional pulsating voltage, using one half cycles of the applied voltage, the other half cycles being suppressed because it conducts only in one direction. Unless there is an inductance or battery in the circuit, the current will be zero, therefore, for half the time. This is called *half-wave rectification.*

As already discussed, diode is an electronic device consisting of two elements known as cathode and anode. Since in a diode electrons can flow in one direction only *i.e.* from cathode to anode so the diode provides the unilateral conduction necessary for rectification. This is true for diodes of all types–vacuum, gas-filled, crystal, semiconductor or metallic (copper oxide and selenium types) diodes. Semiconductor diodes, because of their inherent advantages (already mentioned in Art. 1.6) are usually used as a rectifying device. However, for very high voltages, vacuum diodes may be employed.

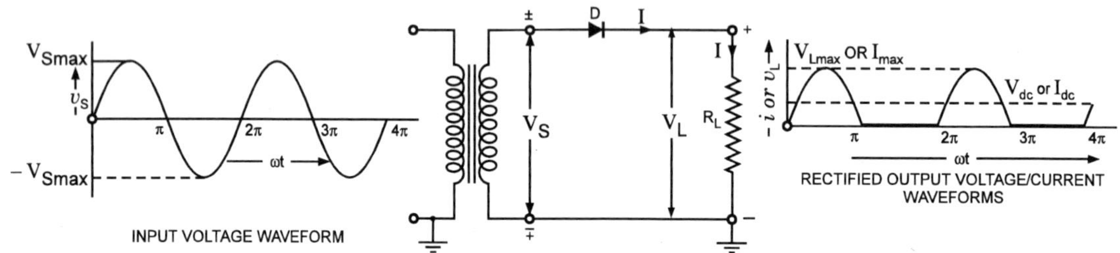

Fig. 4.2. ***Half-Wave Rectifiers***

Circuit. The half-wave rectifier circuit using a semiconductor diode with a load resistance R_L but no smoothing filter is given in Fig. 4.2. The diode is connected in series with the secondary of the transformer and the load resistance R_L, the primary of the transformer is being connected to the ac supply mains.

Working. The ac voltage across the secondary winding changes polarity after every half cycle. During the positive half cycles of the input ac voltage *i.e.* when upper end of the secondary winding is positive w.r.t. its lower end, the diode is forward biased and, therefore, conducts current. If the forward resistance of the diode is assumed to be zero (in practice, however, a small resistance exists) the input voltage during the positive half cycles is directly applied to the load resistance R_L, making its upper end positive w.r.t. its lower end. The waveforms of the output current and output voltage are of the same shape as that of the input ac voltage.

During the negative half cycles of the input ac voltage *i.e.* when the lower end of the secondary winding is positive w.r.t. its upper end, the diode is reverse biased and so does not conduct. Thus during the negative half cycles of the input ac voltage the current through and voltage across the load remains zero if the reverse current, being very small in magnitude, is neglected. Thus for the negative half cycles no power is delivered to the load.

Thus the output voltage developed across load resistance R_L (V_L) is a series of positive half cycles of alternating voltage, with intervening very small constant negative voltage levels, as shown in the Fig. 4.2. It is obvious from the figure that the output is not a steady dc, but only a pulsating dc wave. Since only half cycles of the input wave are used, it is called a *half-wave* rectifier.

Power Supply Specifications. The most important characteristics which are required to be specified for a power supply are given below :

1. The required output dc voltage.
2. The average and peak currents in the diode.
3. The peak inverse voltage (PIV) of each diode.
4. The regulation.
5. The ripple factor.

The above factors will be considered in the analysis of the circuit given below.

Analysis of Half-Wave Rectifier. Under this sub-article, the following items will be considered.

(*i*) Peak inverse voltage
(*ii*) Average and peak currents in the diode
(*iii*) DC output voltage
(*iv*) RMS current and voltage
(*v*) Rectification efficiency
(*vi*) Ripple factor, and
(*vii*) Regulation.

1. *Peak Inverse Voltage.* It is the maximum voltage that the rectifying diode has to withstand, when it is reverse biased. The peak inverse voltage (PIV) rating of a diode is of the primary importance in the design of rectification systems.

During negative half cycles of the input voltage, the diode is reverse biased, no current flows through the load resistance R_L and so causes no voltage drop across load resistance R_L and consequently the whole of the input voltage appears across the diode. Thus the maximum voltage, that appears across the diode, is equal to the peak value of the secondary voltage *i.e.* $V_{S\,max}$. Thus for a half-wave rectifier

$$PIV = V_{S\,max} \qquad ...(4.1)$$

2. *Peak Current.* Assuming the voltage across the transformer secondary sinusoidal of peak value $V_{S\,max}$, instantaneous value of the voltage applied to the rectifier is given as

$$v_s = V_{S\,max} \sin \omega t \qquad ...(4.2)$$

neglecting diode cutin voltage V_K, being very small as compared with $V_{S\,max}$

If the diode is assumed to have a forward resistance of R_F ohms and reverse resistance equal to infinity, then the current flowing through the diode (or load resistance R_L) is given as

$$i = I_{max} \sin \omega t \text{ for } 0 \leq \omega t \leq \pi \qquad ...(4.3)$$

$$i = 0 \text{ for } \pi \leq \omega t \leq 2\pi \qquad ...(4.4)$$

where peak value of current flowing through diode (or load resistance R_L) is given as

$$I_{max} = \frac{V_{S\,max}}{R_F + R_L} \qquad ...(4.5)$$

3. ***DC Output Current.*** The average or dc value of output current is given as

$$I_{dc} = \frac{1}{2\pi}\int_0^{2\pi} i\,d(\omega t) = \frac{1}{2\pi}\left[\int_0^{\pi} I_{max} \sin \omega t\, d(\omega t) + \int_{\pi}^{2\pi} 0\, d(\omega t)\right]$$

$$= \frac{1}{2\pi}\left[I_{max}\{-\cos \omega t\}_0^{\pi}\right] = \frac{I_{max}}{\pi} = 0.318\ I_{max} \qquad ...(4.6)$$

Substituting the value of I_{max} from Eq. (4.5) we have

$$I_{dc} = \frac{V_{S\,max}}{\pi(R_L + R_F)} = \frac{V_{S\,max}}{\pi R_L} \text{ if } R_L >> R_F \qquad ...(4.7)$$

4. ***DC Output Voltage.*** Average or dc value of voltage across the load is given as

$$V_{dc} = I_{dc} R_L = \frac{V_{S\,max}}{\pi(R_L + R_F)} \cdot R_L = \frac{V_{S\,max}}{\pi\left(1 + \frac{R_F}{R_L}\right)} \qquad ...(4.8)$$

If $R_L >> R_F$, then $$V_{dc} \simeq \frac{V_{S\,max}}{\pi} \qquad ...(4.9)$$

5. ***RMS Value of Current.*** RMS value of current flowing through the diode (or load resistance R_L) is given as

$$I_{rms}^2 = \frac{1}{2\pi}\int_0^{2\pi} i^2\, d(\omega t) = \frac{1}{2\pi}\left[\int_0^{\pi} I_{max}^2 \sin^2 \omega t\, d(\omega t) + \int_{\pi}^{2\pi} 0\, d(\omega t)\right] = \frac{I_{max}^2}{4}$$

or $$I_{rms} = \frac{I_{max}}{2} \qquad ...(4.10)$$

Substituting the value of I_{max} from Eq. (4.5) we have

$$I_{rms} = \frac{V_{S\,max}}{2(R_F + R_L)} \qquad ...(4.11)$$

6. ***RMS Value of Output Voltage.*** RMS value of voltage across the load is given as

$$V_{L\,rms} = I_{rms} R_L = \frac{V_{S\,max} \times R_L}{2(R_F + R_L)} = \frac{V_{S\,max}}{2\left[1 + \frac{R_F}{R_L}\right]} \qquad ...(4.12)$$

If $R_L >> R_F$, then

$$V_{L\,rms} = \frac{V_{S\,max}}{2} \qquad ...(4.13)$$

7. *Form Factor and Peak Factor.* The form factor, defined as the ratio of rms value to average value, is given by

$$K_f = \frac{\text{RMS value}}{\text{Average value}} = \frac{I_{rms}}{I_{dc}} = \frac{V_{Smax}/2(R_F + R_L)}{V_{Smax}/\pi(R_F + R_L)} = \frac{\pi}{2} = 1.57 \qquad ...(4.14)$$

The peak factor, defined as the ratio of peak value to rms value, is given by

$$K_p = \frac{\text{Peak value}}{\text{RMS value}} = \frac{V_{Smax}/(R_F + R_L)}{V_{Smax}/2(R_F + R_L)} = \frac{V_{Smax}}{V_{Smax}/2} = 2 \qquad ...(4.15)$$

8. *Output Frequency.* The output frequency is the same as the input frequency. This can be realized by comparing output waveform with the input waveform in Fig. 4.2. Thus for an half-wave rectifier

$$f_{out} = f_{in} \qquad ...(4.16)$$

9. *Rectification Efficiency.* Rectification efficiency, which is defined as the ratio of dc output power to the ac input power, is given as

$$\eta = \frac{\text{DC power delivered to the load}}{\text{AC input power from the transformer}} = \frac{P_{dc}}{P_{ac}}$$

Now dc power delivered to the load,

$$P_{dc} = I_{dc}^2 R_L = \left(\frac{I_{max}}{\pi}\right)^2 R_L \qquad ...(4.17)$$

AC power input from the transformer,

$$P_{ac} = \text{Power dissipated in diode junction} + \text{power dissipated in load resistance } R_L$$

$$= I_{rms}^2 R_F + I_{rms}^2 R_L$$

$$= \left(\frac{I_{max}}{2}\right)^2 R_F + \left(\frac{I_{max}}{2}\right)^2 R_L = \frac{I_{max}^2}{4}(R_F + R_L) \qquad ...(4.18)$$

So rectification efficiency, $\eta = \frac{P_{dc}}{P_{ac}} = \frac{I_{max}^2 R_L/\pi^2}{I_{max}^2 (R_F + R_L)/4}$

$$= \frac{4}{\pi^2}\frac{R_L}{(R_F + R_L)} = \frac{0.406}{1 + \frac{R_F}{R_L}} \qquad ...(4.19)$$

If R_F is neglected, η becomes equal to 0.406 or 40.6 per cent. Obviously it is the maximum possible efficiency for a half-wave rectifier.

10. *Ripple Factor.* The pulsating output of a rectifier can be considered to contain a dc component and ac components called the *ripples*. The ripple current is undesirable and its value should be the smallest possible in order to make the rectifier effective.

The ripple voltage or current is measured in terms of the ripple factor which is defined as the ratio of the effective value of the ac components of voltage (or current) present in the output from the rectifier to the diode or average value of the output voltage (or current).

The effective value of the load current is given as

$$I^2 = I_{dc}^2 + I_1^2 + I_2^2 + I_4^2 + \ldots = I_{dc}^2 + I_{ac}^2 \qquad ...(4.20)$$

where I_1, I_2, I_4 etc. are the rms values of fundamental, second, fourth etc. harmonics and I_{ac}^2 is the sum of the squares of the rms values of the ac components.

So ripple factor, $$\gamma = \frac{I_{ac}}{I_{dc}} = \frac{\sqrt{I^2 - I_{dc}^2}}{I_{dc}} = \sqrt{\left[\frac{I_{rms}}{I_{dc}}\right]^2 - 1} = \sqrt{K_f^2 - 1} \qquad ...(4.21)$$

where K_f is the form factor of the input voltage. For half-wave rectifier, form factor is given as

$$K_f = \frac{I_{rms}}{I_{av}} = \frac{I_{max}/2}{I_{max}/\pi} = \frac{\pi}{2} = 1.57$$

So ripple factor, $\gamma = \sqrt{(1.57)^2 - 1} = 1.21$

11. *Transformer Utilisation Factor.* Determination of transformer rating becomes necessary while designing a power supply. For this transformer utilisation factor (TUF), which is defined as the ratio of power delivered to the load and ac rating of the transformer secondary, should be known.

$$\text{So TUF} = \frac{P_{dc}}{P_{ac\,(rated)}} = \frac{I_{dc}^2 R_L}{V_{S\,rms}\, I_{rms}}$$

$$= \frac{(I_{max}/\pi)^2 R_L}{\frac{V_{S\,max}}{\sqrt{2}} \cdot \frac{I_{max}}{2}} = \frac{2\sqrt{2}}{\pi^2} \cdot \frac{I_{max} R_L}{V_{S\,max}} \qquad ...(4.22)$$

But $V_{S\,max} = I_{max} (R_F + R_L)$

$$\text{So TUF} = \frac{2\sqrt{2}}{\pi^2} \cdot \frac{I_{max} R_L}{I_{max}(R_F + R_L)} = \frac{0.286\, R_L}{R_L + R_F} \qquad ...(4.23)$$

Neglecting forward resistance R_F of the diode we have TUF = 0.286.

12. *Regulation.* The variation of dc output voltage as a function of dc load current is called *regulation*. Percentage regulation is given as

$$\%\ \text{regulation} = \frac{V_{NL} - V_{FL}}{V_{FL}} \times 100$$

For an ideal power supply, output voltage should be independent of load current and the percentage regulation should be equal to zero.

For half-wave rectifier, from Eq. (4.8)

$$V_{dc} = \frac{V_{S\,max} R_L}{\pi (R_F + R_L)}$$

$$= \frac{V_{S\,max}}{\pi}\left[1 - \frac{R_F}{R_F + R_L}\right]$$

$$= \frac{V_{S\,max}}{\pi} - I_{dc} R_F \qquad ...(4.24)$$

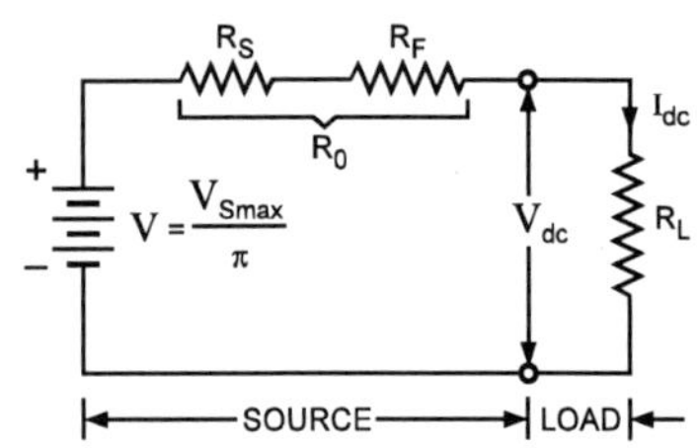

Fig. 4.3. *Thevenin's Model For a Half-Wave Rectifier*

From above equation it seems that the half-wave rectifier functions as if it were a constant voltage source of $\frac{V_{S\,max}}{\pi}$ volts in series with an internal resistance being equal to R_F. Thus V_{dc} equals $\frac{V_{S\,max}}{\pi}$ at no load and dc output voltage decreases linearly with the increase in output dc current I_{dc}. Clearly, the effective internal resistance of the power supply is R_F. In practice, the resistance R_S of the transformer secondary is in series with the diode, and in Eq. (4.24) R_S is to be added in R_F. Thevenin's model for a half-wave rectifier is given in Fig. 4.3.

Effect of Diode Cutin Voltage on Rectifier Output. We do not get a perfect half-wave voltage across the load resistor. Because of the barrier potential (or cutin voltage) of diode, the diode does not turn on until the ac source voltage attains approximately 0.7 V in case of silicon diode. When the peak source voltage is much greater than barrier potential (V_K) of diode, the load voltage will resemble a half-wave signal. If the peak source voltage is not much greater than diode barrier potential,

Peak value of voltage across load resistor R_L,

$$V_{out\,(max)} = V_{S\max} - V_K \qquad \text{...(4.25)}$$

DC output voltage, $$V_{dc} = \frac{V_{S\max} - V_K}{\pi} \qquad \text{...(4.26)}$$

DC output current, $$I_{dc} = \frac{V_{dc}}{R_L} = \frac{V_{S\max} - V_K}{\pi R_L} \qquad \text{...(4.27)}$$

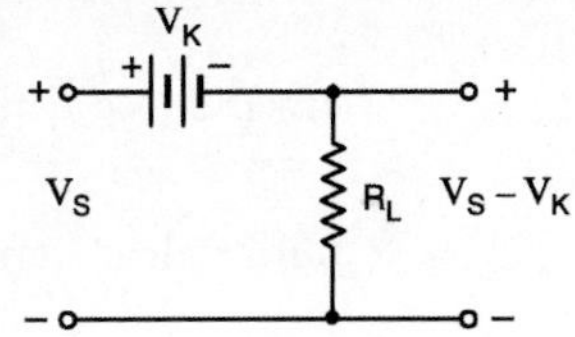

Fig. 4.4

Condition For Maximum DC Output Power in a Half-Wave Rectifier. DC output power in a half-wave single phase rectifier is given by

$$P_{dc} = I_{dc}^2 R_L = \left(\frac{I_{max}}{\pi}\right)^2 R_L = \frac{V_{S\max}^2}{\pi^2 (R_L + R_F)^2} R_L \qquad \because I_{max} = \frac{V_{S\max}}{R_L + R_F}$$

Differentiating both sides of above equation w.r.t. R_L we have

$$\frac{dP_C}{dR_L} = \frac{V_{S\max}^2}{\pi^2}\left[\frac{(R_L^2 + R_F^2 + 2R_L R_F) - R_L(2R_L + 2R_F)}{(R_L^2 + R_F^2 + 2R_L R_F)^2}\right] = \frac{V_{S\max}^2 (R_F^2 - R_L^2)}{\pi^2 (R_L^2 + R_F^2 + 2R_L R_F)^2}$$

Output will be maximum if $\dfrac{dP_C}{dR_L} = 0$

or $R_F^2 - R_L^2 = 0$

or $R_L = R_F$ *i.e.* when load resistance equals diode resistance.

The advantages and disadvantages of a half-wave rectifier are given below.

Advantages. Simple circuit and low cost.

Disadvantages. 1. The output current in the load contains, in addition to dc component, ac components of basic frequency equal to that of the input voltage frequency. Ripple factor is high and an elaborate filtering is, therefore, required to give steady dc output.

2. The power output and, therefore, rectification efficiency is quite low. This is due to the fact that power is delivered only half the time.

3. Transformer utilization factor is low.

4. DC saturation of transformer core resulting in magnetising current and hysteresis losses and generation of harmonics.

The type of supply available from a half-wave rectifier is not satisfactory for general power supply. This type of supply can be satisfactory for some particular purposes such as battery charging.

Example 4.1. A diode whose internal resistance is 20 Ω, has to supply power to a 200 Ω load from a 110 V (rms) source of supply. Calculate the (*i*) Peak load current (*ii*) DC load current (*iii*) AC load current (*iv*) Rectification efficiency (*v*) Peak inverse voltage. [*G.G.S.I.P. Univ. Analog Electronics December 2010*]

Solution :

$$V_{rms} = 110 \text{ V}$$
$$R_L = 200 \ \Omega$$
$$R_F = 20 \ \Omega$$

Maximum value of supply voltage, $V_{Smax} = \sqrt{2} \times V_{rms} = 110\sqrt{2}$ V

(*i*) Peak load current, $I_{max} = \dfrac{V_{Smax}}{R_L + R_F} = \dfrac{110\sqrt{2}}{200+20} = 0.707 \text{ A}$ **Ans.**

(*ii*) DC load current, $I_{dc} = \dfrac{I_{max}}{\pi} = \dfrac{0.707}{\pi} = 0.225 \text{ A}$ **Ans.**

(*iii*) AC load current, $I_{ac} = \dfrac{I_{max}}{2} = \dfrac{0.707}{2} = 0.3535 \text{ A}$ **Ans.**

(*iv*) Rectification efficiency, $\eta = \dfrac{0.406}{1+\dfrac{R_F}{R_L}} = \dfrac{0.406}{1+\dfrac{20}{200}} = 0.369 \text{ or } 36.9\%$ **Ans.**

(*v*) Peak inverse voltage, PIV $= V_{Smax} = 110\sqrt{2} \text{ V} = 155.56 \text{ V}$ **Ans.**

Example 4.2. A sinusoidal voltage of 40 V and frequency 50 Hz is applied to a half-wave rectifier. R_L = 200 Ω, V_γ = 0, R_F = 20 Ω, R_r = ∝. Find V_{dc}, I_{dc}, I_{max}, I_{rms}, P_{dc}, η, ripple factor, % regulation.

Solution :

$$V_{s\,rms} = 40 \text{ V}$$
$$\text{Frequency, } f = 50 \text{ Hz}$$
$$R_L = 200 \ \Omega$$
$$R_F = 20 \ \Omega$$
$$R_r = \propto$$

$$I_{max} = \frac{V_{s\,max}}{R_L + R_F} = \frac{40\sqrt{2}}{200+20} = 0.257 \text{ A or } 257 \text{ mA} \ \textbf{Ans.}$$

$$I_{rms} = \frac{I_{max}}{2} = \frac{257}{2} = 128.5 \text{ mA} \ \textbf{Ans.}$$

$$I_{dc} = \frac{I_{max}}{\pi} = \frac{257}{\pi} = 81.8 \text{ mA} \ \textbf{Ans.}$$

$$V_{dc} = I_{dc}R_L = 81.8 \times 10^{-3} \times 200 = 16.36 \text{ V} \ \textbf{Ans.}$$

$$P_{dc} = I_{dc}^2 R_L = (81.8 \times 10^{-3})^2 \times 200 = 1.338 \text{ W} \ \textbf{Ans.}$$

$$P_{ac} = \frac{I_{max}^2}{4}(R_F + R_L) = \frac{(257 \times 10^{-3})^2}{4} \times (20 + 200) = 3.63 \text{ W}$$

$$\eta = \frac{P_{dc}}{P_{ac}} \times 100 = \frac{1.338}{3.63} \times 100 = 36.86\% \ \textbf{Ans.}$$

$$\text{Ripple factor, } \gamma = \left(\frac{I_{rms}^2}{I_{dc}^2} - 1\right)^{1/2} = \sqrt{\left(\frac{128.5}{81.8}\right)^2 - 1} = 1.21 \ \textbf{Ans.}$$

$$V_{dc} = \frac{V_{max}}{\pi} - I_{dc}R_F = \frac{40\sqrt{2}}{\pi} - 81.8 \times 10^{-3} \times 20 = 18 - 1.636 = 16.364 \text{ V}$$

$$\text{Regulation} = \frac{V_{NL} - V_{FL}}{V_{FL}} \times 100 = \frac{\dfrac{V_{max}}{\pi} - I_{dc}R_F}{I_{dc}\,R_F} \times 100 = \frac{18 - 16.364}{16.364} \times 100$$

$$= 9.998\% \ \textbf{Ans.}$$

Example 4.3. A half-wave rectifier uses a diode with a forward resistance of 100 Ω. If the input ac voltage is 220 V and load resistance is 2 kΩ, determine the transformer utilisation factor.

Solution : Diode forward resistance, $R_F = 100\ \Omega$ or $0.1\ k\Omega$

Load resistance, $R_L = 2\ k\Omega$

$$\text{Transformer utilisation factor, TUF} = \frac{0.286}{1+\frac{R_F}{R_L}} = \frac{0.286}{1+\frac{0.1}{2}} = 0.2724 \text{ Ans.}$$

Example 4.4. What is the ripple 2 V on average of 50 V ?

Solution : RMS value of ac component, $V_{rms} = 2$ V

Average value of output voltage, $V_{dc} = 50$ V

$$\text{Ripple factor, } \gamma = \frac{V_{rms}}{V_{dc}} = \frac{2}{50} = 0.04 \text{ Ans.}$$

Example 4.5. A dc voltage supply provides 60 V when the output is unloaded. When connected to a load the output drops to 56 V. Calculate the value of voltage regulation.

Solution : No-load voltage, $V_{NL} = 60$ V

Full-load voltage, $V_{FL} = 56$ V

$$\text{Voltage regulation} = \frac{V_{NL} - V_{FL}}{V_{FL}} \times 100 = \frac{60-56}{56} \times 100 = 7.14\% \text{ Ans.}$$

Example 4.6. A half wave rectifier is used to supply 10 V dc to a resistive load of 400 Ω. If the crystal diode has a forward resistance of 20 Ω, determine the value of ac voltage supplied to the circuit.

Solution : Maximum value of supply voltage

$$V_{s\,max} = V_{dc} \times \pi \left(1+\frac{R_F}{R_L}\right) = 10 \times \pi \left(1+\frac{20}{400}\right) = 33 \text{ V} \qquad \text{Refer to Eq. (4.8)}$$

RMS value of ac voltage applied to the circuit

$$V_s = \frac{V_{s\,max}}{\sqrt{2}} = \frac{33}{\sqrt{2}} = 23.3 \text{ V Ans.}$$

4.3. FULL-WAVE RECTIFIERS

In half-wave rectifiers only one half cycles of the input are utilized but in full-wave rectifiers both half cycles of the input are utilized. Alternate half cycles are inverted to give unidirectional load current. There are two types of full-wave rectifier circuits namely 1. *centre-tap rectifier* and 2. *bridge rectifier*.

4.3.1. Centre-Tap Full-Wave Rectifier

In such a rectifier, the ac input is applied through a transformer, the anodes of the two diodes D_1 and D_2 (having similar characteristics) are connected to the opposite ends of the centre-tapped secondary winding and two cathodes are connected to each other and are connected also through the load resistance R_L and back to the centre of the transformer, as shown in Fig. 4.5.

When the top of the transformer secondary winding is positive, say during the first half cycle of the supply, the anode of diode D_1 is positive w.r.t. cathode, and anode of diode D_2 is negative w.r.t. cathode. Thus only diode D_1 conducts, being forward biased and current flows from cathode to anode of diode D_1, through load resistance R_L and top half of the transformer secondary making cathode end of load resistance R_L positive. During the second half cycle of the input voltage the polarity is reversed, making the bottom of the secondary winding positive w.r.t. centre tap and thus diode D_2 is forward biased and diode D_1 is reverse biased. Consequently during this half cycle of the input only

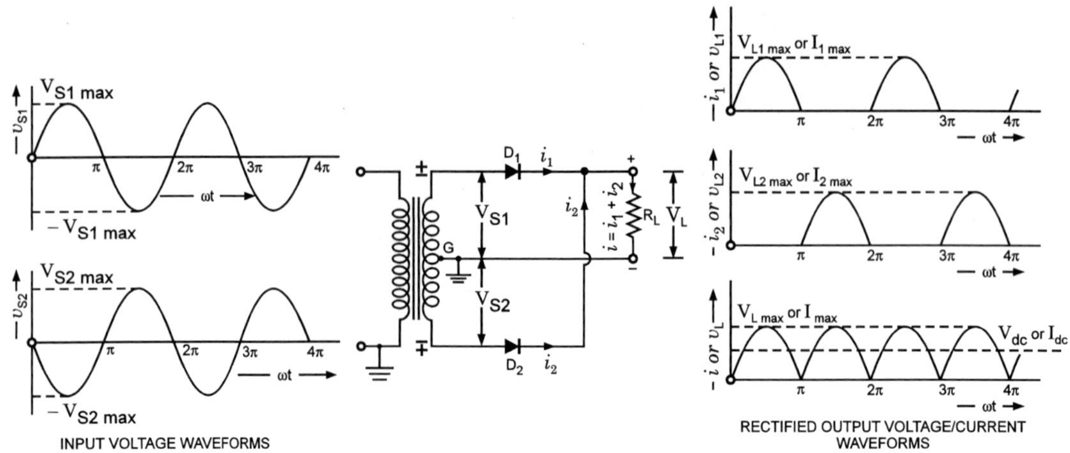

Fig. 4.5. ***Centre-Tap Full-Wave Rectifier***

the diode D_2 conducts and current flows through the load resistance R_L and bottom of the transformer secondary making the cathode end of the load resistance R_L positive. Thus the direction of flow of current through the load resistance R_L remains the same during both halves of the input supply voltage. Thus the circuit shown in Fig. 4.5 acts as a full-wave rectifier.

Peak Inverse Voltage. As already mentioned before, peak inverse voltage (PIV) is the maximum possible voltage across a diode when it is reverse biased. During first half cycle of the supply *i.e.* when the top of the transformer secondary winding is positive, diode D_1 conducts and offers almost zero resistance. So whole of the voltage $V_{S\,max}$ of the upper half winding is developed across the load resistance R_L. Now voltage across the non-conducting diode D_2 is the sum of voltage across the lower half of the transformer secondary and the voltage across the load resistance R_L.

Thus PIV of diode, $D_2 = V_{S\,max} + V_{S\,max} = 2V_{S\,max}$

Similarly PIV of diode $D_1 = 2\,V_{S\,max}$ ...(4.28)

4.3.2. Bridge Rectifier

In the bridge circuit four diodes are connected in the form of a Wheatstone bridge, two diametrically opposite junctions of the bridge are connected to the secondary of a transformer and the other two are connected to the load, as shown in Fig. 4.6.

When the upper end of the transformer secondary winding is positive, say during first half cycles of the input supply, diodes D_1 and D_3 are forward biased and current flows through arm AB, enters the load at positive terminal, leaves the load at negative terminal, and returns back flowing through arm DC. During this half of each input cycle, the diodes D_2 and D_4 are reverse biased and so the current is not allowed to flow in arms AD and BC. The flow of current is

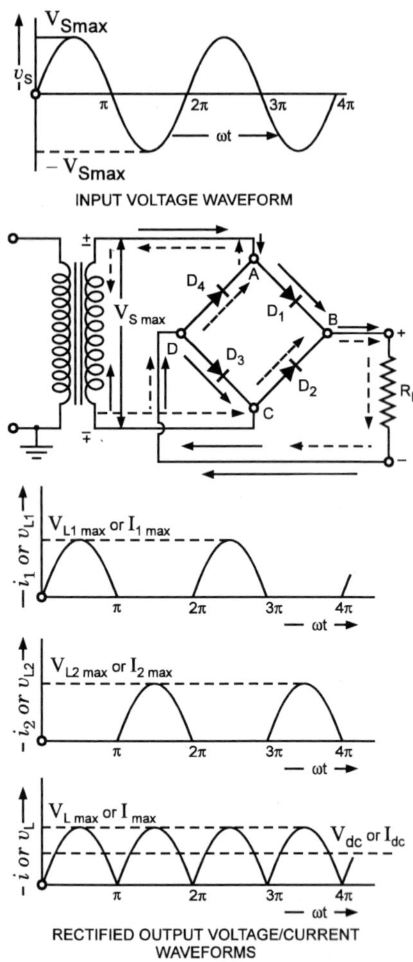

Fig. 4.6. ***Bridge Rectifier***

indicated by solid arrows in the figure. In the second half of the input cycle the lower end of ac supply becomes positive, diodes D_2 and D_4 become forward biased and current flows through arm CB, enters the load at the positive terminal, leaves the load at negative terminal and returns back flowing through arm DA. Flow of current has been shown by dotted arrows in the figure. Thus the direction of flow of current through the load resistance R_L remains the same during both half cycles of the input supply voltage.

Peak Inverse Voltage (PIV). Figure 4.6 shows a bridge rectifier circuit. Let us consider the instant the secondary voltage attains its positive peak value $V_{S\,max}$. Now diodes D_1 and D_3 are conducting whereas diodes D_2 and D_4 are non-conducting being reverse biased. The conducting diodes D_1 and D_3 have almost zero resistance (*i.e.* zero voltage drop across them). Point B has the same potential as point A and similarly point D has the same potential as point C. The entire voltage of the transformer secondary winding, V_{Smax} is developed across the load resistance R_L. The same voltage *i.e.* V_{Smax} acts across each of the non-conducting diodes D_2 and D_4. Thus

$$PIV = V_{Smax} \qquad ...(4.29)$$

4.3.3. Circuit Analysis

The analysis of both of the full-wave rectifier circuits (*i.e.* centre-tap and bridge type) is the same except that (*i*) in a bridge rectifier circuit two diodes conduct during each half cycle and forward resistance becomes double *i.e.* 2 R_F and (*ii*) in a bridge rectifier circuit $V_{S\,max}$ is the maximum voltage across the transformer secondary winding whereas in a centre tap rectifier circuit $V_{S\,max}$ represents the maximum voltage across each half of the secondary winding.

1. ***Peak Current.*** Instantaneous value of voltage applied to the rectifier is given as

$$v_s = V_{S\,max} \sin \omega t \qquad ...(4.30)$$

If the diode is assumed to have a forward resistance of R_F ohms and reverse resistance equal to infinity, then current flowing through the load resistance is given as

$$i_1 = I_{max} \sin \omega t \text{ and } i_2 = 0 \text{ for first half cycle} \qquad ...(4.31)$$

$$\text{and } i_1 = 0 \text{ and } i_2 = I_{max} \sin \omega t \text{ for second half cycle} \qquad ...(4.32)$$

The total current flowing through the load resistance R_L, being the sum of currents i_1 and i_2, is given as

$$i = i_1 + i_2 = I_{max} \sin \omega t \text{ for the whole cycle} \qquad ...(4.33)$$

where peak value of current flowing through the load resistance R_L is given as

$$I_{max} = \frac{V_{Smax}}{R_F + R_L} \qquad ...(4.34) \text{ in case of centre-tap rectifier}$$

$$\text{and } I_{max} = \frac{V_{Smax}}{2R_F + R_L} \qquad ...(4.35) \text{ in case of bridge rectifier}$$

2. ***Output Current.*** Since the current is the same through the load resistance R_L in the two halves of the ac cycle, magnitude of direct current I_{dc}, which is equal to the average value of alternating current, can be obtained by integrating the current i_1 between 0 and π or current i_2 between π and 2π.

$$\text{So } I_{dc} = \frac{1}{\pi}\int_0^{\pi} i_1\, d(\omega t) = \frac{1}{\pi}\int_0^{\pi} I_{max} \sin \omega t\, d(\omega t) = \frac{2I_{max}}{\pi} \qquad ...(4.36)$$

3. ***DC Output Voltage.*** Average or dc value of voltage across the load is given as

$$V_{dc} = I_{dc}R_L = \frac{2}{\pi} I_{max} R_L \qquad ...(4.37)$$

4. *RMS Value of Current.* RMS or effective value of current flowing through the load resistance R_L is given as

$$I_{rms}^2 = \frac{1}{\pi}\int_0^{\pi} i_1^2 \, d(\omega t) = \frac{1}{\pi}\int_0^{\pi} I_{max}^2 \sin^2 \omega t \, d(\omega t) = \frac{I_{max}^2}{2}$$

$$\text{or} \quad I_{rms} = \frac{I_{max}}{\sqrt{2}} \qquad \text{...(4.38)}$$

5. *RMS Value of Output Voltage.* RMS value of voltage across the load is given as

$$V_{L\,rms} = I_{rms} R_L = \frac{I_{max}}{\sqrt{2}} R_L \qquad \text{...(4.39)}$$

6. *Form Factor and Peak Factor.*

$$\text{Form factor, } K_f = \frac{\text{RMS value}}{\text{Average value}} = \frac{I_{rms}}{I_{dc}} = \frac{I_{max}/\sqrt{2}}{2I_{max}/\pi} = \frac{\pi}{2\sqrt{2}} = 1.11 \qquad \text{...(4.40)}$$

$$\text{Peak factor, } K_p = \frac{\text{Peak value}}{\text{RMS value}} = \frac{I_{max}}{I_{max}/\sqrt{2}} = \sqrt{2} \qquad \text{...(4.41)}$$

7. *Output Frequency.* With a half-wave rectifier, the output frequency equals the input frequency. But the frequency of the full-wave signal is double the input frequency. A full-wave output has twice as many cycles as the sine-wave input has. The full-wave rectifier inverts each negative half cycle, so that we get the number of positive half cycles. The effect is to double the frequency. Thus, for full-wave rectifier

$$f_{out} = 2 f_{in} \qquad \text{...(4.42)}$$

8. *Rectification Efficiency.* Power delivered to load,

$$P_{dc} = I_{dc}^2 R_L = \left(\frac{2}{\pi} I_{max}\right)^2 R_L = \frac{4}{\pi^2} I_{max}^2 R_L \qquad \text{...(4.43)}$$

$$\text{AC input power, } P_{ac} = I_{rms}^2 (R_L + R_F) = \frac{I_{max}^2}{2} (R_L + R_F) \qquad \text{...(4.44)}$$

$$\text{Rectification efficiency, } \eta = \frac{P_{dc}}{P_{ac}} = \frac{\frac{4}{\pi^2} I_{max}^2 R_L}{\frac{1}{2} I_{max}^2 (R_L + R_F)} = \frac{8}{\pi^2} \frac{1}{\left(1 + \frac{R_F}{R_L}\right)} = \frac{0.812}{1 + \frac{R_F}{R_L}} \qquad \text{...(4.45)}$$

In case of bridge rectifier, rectification efficiency is given as

$$\eta = \frac{0.812}{1 + \frac{2 R_F}{R_L}} \qquad \text{...(4.46)}$$

9. *Ripple Factor.* Form factor of the rectified output voltage of a full-wave rectifier is given as

$$K_f = \frac{I_{rms}}{I_{av}} = \frac{I_{max}/\sqrt{2}}{2 I_{max}/\pi} = \frac{\pi}{2\sqrt{2}} = 1.11$$

Ripple factor is given as $\gamma = \sqrt{K_f^2 - 1} = \sqrt{(1.11)^2 - 1} = 0.482$ ...(4.47)

10. *Regulation.* From Eq. (4.37).

$$V_{dc} = \frac{2}{\pi} I_{max} R_L = \frac{2V_{S\,max} R_L}{\pi (R_F + R_L)} = \frac{2V_{S\,max}}{\pi}\left[1 - \frac{R_F}{R_F + R_L}\right]$$

$$= \frac{2V_{S\,max}}{\pi} - I_{dc} R_F \qquad ...(4.48)$$

In case of a bridge rectifier, $V_{dc} = \frac{2V_{S\,max}}{\pi} - 2I_{dc}R_F$ ...(4.49)

11. *Transformer Utilization Factor For Centre-Tap Transformer.* In a centre-tap full-wave rectifier, there is no problem due to saturation of core because the dc currents in the two halves of the transformer secondary flow in opposite directions. The average transformer utilisation factor is found by considering the primary and secondary windings.

Transformer utilisation factor of primary

$$= \frac{P_{dc}}{\text{VA rating of primary}} = \frac{I_{dc}^2 R_L}{V_{S\,rms} I_{rms}} = \frac{\left(\frac{2I_{max}}{\pi}\right)^2 R_L}{\frac{V_{S\,max}}{\sqrt{2}} \times \frac{I_{max}}{\sqrt{2}}}$$

$$= \frac{4I_{max}^2 R_L / \pi^2}{\frac{I_{max}}{\sqrt{2}}(R_F + R_L) \times \frac{I_{max}}{\sqrt{2}}}$$

$$= \frac{8}{\pi^2} \times \frac{R_L}{(R_F + R_L)} = \frac{8}{\pi^2} \times \frac{1}{\left(1 + \frac{R_F}{R_L}\right)} \simeq 0.812$$

Centre-tap transformer can be thought of as equivalent to two half-wave rectifiers feeding to a common load. Hence, TUF of two half secondaries can be written as

$$TUF_{(\text{Full secondary})} = 2 \times TUF_{(\text{Half-wave})} = 2 \times 0.286 = 0.572$$

The average transformer utilisation factor,

$$(TUF)_{av} = \frac{TUF_{Primary} + TUF_{Secondary}}{2} = \frac{0.812 + 0.572}{2} = 0.692$$

12. *Transformer Utilisation Factor For Bridge Rectifier.* The current flow through both of primary and secondary windings are sinusoidal. Due to this TUF of both the primary and secondary are 0.812 and the overall TUF is 0.812. Since in case of bridge rectifier TUF is 0.812 whereas the TUF for centre-tap transformer is 0.692 *i.e.* TUF of bridge rectifier is larger than that of centre-tap rectifier, for the same dc output power, a smaller transformer can be used in the bridge circuit.

4.3.4. Effect of Diode Cutin Voltage on Rectifier Output and PIV of Diodes

1. *Centre-Tap Rectifier.* Peak value of voltage across load resistor R_L.

$$V_{out\,(max)} = V_{max} - V_K \qquad ...(4.50) \text{ as in case of half-wave rectifier}$$

DC output voltage, $V_{dc} = \frac{2}{\pi}(V_{S\,max} - V_K)$ ...(4.51)

DC output current, $I_{dc} = \frac{V_{dc}}{R_L} = \frac{2}{\pi}\frac{(V_{S\,max} - V_K)}{R_L}$ ...(4.52)

Peak inverse voltage (PIV) of diode

$$= V_{max} + \text{voltage developed across load resistor } R_L$$

$$= V_{max} + V_{out\,(max)} = V_{max} + (V_{max} - V_K)$$

$$= 2V_{max} - V_K \quad ...(4.53)$$

2. ***Bridge Rectifier.*** Voltage developed across load resistor R_L,

$$V_{out\,(max)} = V_{S\,max} - 2V_K \quad ...(4.54)$$

$$\text{DC output voltage, } V_{dc} = \frac{2}{\pi}(V_{Smax} - 2V_K) \quad ...(4.55)$$

$$\text{DC output current, } I_{dc} = \frac{V_{dc}}{R_L} = \frac{2}{\pi}\frac{(V_{Smax} - 2V_K)}{R_L} \quad ...(4.56)$$

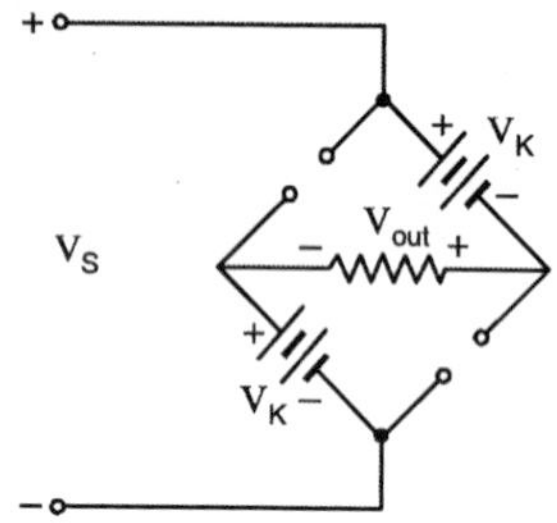

Fig. 4.7

Peak inverse voltage (PIV) of diode

$$= \frac{1}{2}(V_{Smax} + \text{voltage developed across load resistor } R_L)$$

$$= \frac{1}{2}[V_{Smax} + (V_{Smax} - 2V_K)] = V_{S\,max} - V_K \quad ...(4.57)$$

4.3.5. Merits and Demerits of Full-Wave Rectifiers Over Half-Wave Rectifiers

Merits. 1. The rectification efficiency of full-wave rectifier is double of that of a half-wave rectifier.

2. The ripple voltage is low and of higher frequency in case of a full-wave rectifier, so simple filtering circuit is required.
3. Higher output voltage, higher output power and higher TUF in case of a full-wave rectifier.
4. In a full-wave rectifier, there is no problem due to dc saturation of the core because the dc currents in the two halves of the transformer secondary flow in opposite directions.

Demerits. Full-wave rectifier needs more circuit elements and is costlier.

4.3.6. Merits and Demerits of Bridge Rectifiers Over Centre-Tap Rectifiers

With the availability of low-cost, highly reliable and small-sized silicon diodes bridge rectifier is becoming more and more popular in comparison to centre-tap rectifier. It has many advantages over a centre-tap rectifier, as given below.

1. No centre tap is required in the transformer secondary so in case of a bridge rectifier the transformer required is simpler. If stepping up or stepping down of voltage is not required, transformer can be eliminated even.
2. The PIV is one-half that of centre-tap rectifier. Hence bridge rectifier is highly suited for high-voltage applications.
3. Transformer utilisation factor, in case of a bridge rectifier, is higher than that of a centre-tap rectifier.
4. For a given power output, power transformer of smaller size can be used in case of the bridge rectifier because current in both (primary and secondary) windings of the supply transformer flow for the entire ac cycle.

The main drawback of a bridge rectifier is that it needs four diodes, two of which conduct in alternate half cycles. Because of this the total voltage drop in diodes becomes double of that in case of centre-tap rectifier, losses are increased and rectification efficiency is somewhat reduced. This poses a problem when low voltages are required. Another drawback of bridge rectifier is that the load resistor R_L and the supply source have no common point which may be earthed.

4.3.7. Comparison Between Half-Wave, Centre-Tap and Bridge Rectifiers

TABLE 4.1.

S.No.	*Particulars*	*Half-Wave Rectifiers*	*Full-Wave Rectifiers*	
			Centre-Tap	*Bridge*
1.	Number of diodes required	1	2	4
2.	Transformer requirement	not essential	essential by centre tap	not essential
3.	Average value of current, I_{dc}	I_{max}/π	$2\,I_{max}/\pi$	$2\,I_{max}/\pi$
4.	RMS value of current, I_{rms}	$I_{max}/2$	$I_{max}/\sqrt{2}$	$I_{max}/\sqrt{2}$
5.	Peak inverse voltage (PIV)	$V_{S\,max}$	$2\,V_{S\,max}$	$V_{S\,max}$
6.	Peak load current, I_{max}	$\frac{V_{S\,max}}{R_L + R_F}$	$\frac{V_{S\,max}}{R_L + R_F}$	$\frac{V_{S\,max}}{R_L + 2\,R_F}$
7.	DC output voltage, V_{dc}	$\frac{I_{max}}{\pi} R_L$	$\frac{2}{\pi} I_{max}\, R_L$	$\frac{2}{\pi} I_{max}\, R_L$
8.	Rectification efficiency (max)	40.6%	81.2%	81.2%
9.	Ripple factor	1.21	0.482	0.482
10.	Fundamental frequency of ripple	Supply frequency, f	$2f$	$2f$
11.	Voltage regulation	good	better	good
12.	Transformer utilisation factor	0.286	0.692	0.812
13.	Form factor	1.57	1.11	1.11
14.	Peak factor	2	$\sqrt{2}$	$\sqrt{2}$

Example 4.7. A sine wave of RMS value 70.7 volts is applied to the primary of full-wave centre-tapped rectifier with a transformer turns-ratio of 1:2. Assuming diode drop of 0.7 V, determine the maximum PIV rating of each diode.

Solution : Primary voltage, $V_1 = 70.7$ V

$$\text{Secondary voltage, } V_2 = V_{S_1} + V_{S_2} = 2V_s = 70.7 \times \frac{2}{1} = 141.4 \text{ V}$$

Supply voltage across one-half of secondary winding,

$$V_S = \frac{141.4}{2} = 70.7 \text{ V}$$

$$\text{PIV of each diode} = 2V_{Smax} - \text{Voltage drop across diode} = 2 \times \sqrt{2} \times 70.7 - 0.7 = 199.3 \text{ V} \quad \textbf{Ans.}$$

Example 4.8. In a bridge rectifier circuit the peak value of secondary voltage is $240\sqrt{2}$ V and frequency is 50 Hz. Determine no-load dc voltage, PIV and output frequency.

Solution : Maximum supply voltage, $V_{max} = 240\sqrt{2}$ V

$$\text{No-load dc output voltage} = \frac{2\,V_{max}}{\pi} = \frac{2 \times 240\sqrt{2}}{\pi} = 216 \text{ V} \quad \textbf{Ans.}$$

$$\text{PIV rating of diodes, PIV} = V_{smax} = 240\sqrt{2} = 339.4 \text{ V} \quad \textbf{Ans.}$$

$$\text{Output frequency, } f_{out} = 2f_{in} = 2 \times 50 = 100 \text{ Hz} \quad \textbf{Ans.}$$

Example 4.9. A full-wave rectifier shown in Fig. 4.8 uses ideal silicon diodes. Determine (*i*) V_{dc} at the load (*ii*) PIV rating of each diode (*iii*) I_{max} through each diode (*iv*) power rating of each diode (*v*) efficiency. Given n = 10 : 1, $R_L = 5\ \Omega$.

[*G.G.S.I.P. Univ. Analog Electronics December 2010*]

Solution : Supply voltage, $V_{rms} = \frac{120}{10} = 12$ V

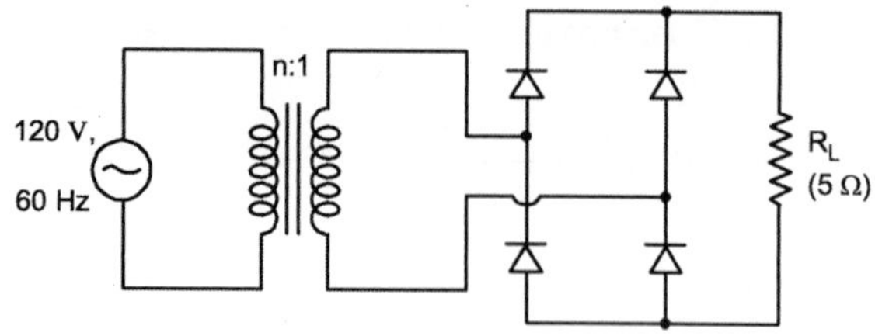

Fig. 4.8

Maximum value of supply voltage, $V_{S\,max} = \sqrt{2}\, V_{rms}$

$= 12\sqrt{2}$ V

Diode barrier potential, $V_K = 0$

Diode forward resistance, $R_F = 0$ } ∵ Diodes are ideal ones

Load resistance, $R_L = 5\ \Omega$

Output dc current, $I_{dc} = \frac{2I_{max}}{\pi} = \frac{2}{\pi}\frac{V_{S\,max}}{R_L} = \frac{2}{\pi} \times \frac{12\sqrt{2}}{5} = 2.16$ A

(*i*) V_{dc} at load $= I_{dc} \times R_L = 2.16 \times 5 = 10.8$ V **Ans.**

(*ii*) PIV rating of each diode $= V_{S\,max} = 12\sqrt{2}$ V $= 16.97$ V **Ans.**

(*iii*) I_{max} through each diode $= \frac{V_{max}}{R_L} = \frac{12\sqrt{2}}{5} = 3.394$ V **Ans.**

(*iv*) Power rating of each diode $= I_{dc}^2\, R_L = (2.16)^2 \times 5 = 23.328$ W **Ans.**

(*v*) Rectification efficiency, $\eta = \frac{0.812}{1 + \frac{R_F}{R_L}} = 0.812$ or 81.2% **Ans.** $\because R_F = 0$

Example 4.10. A full-wave bridge rectifier with 120 V_{rms} sinusoidal input has a load resistor of 1 kΩ.

(*i*) If silicon diodes are applied, what is dc voltage available at load.

(*ii*) Determine required PIV rating of each diode.

(*iii*) Find maximum current through each diode during conduction.

(*iv*) What is required power rating of each diode ?

Solution : RMS value of supply voltage to the bridge rectifier,

$V_{S\,rms}$ = RMS value of output voltage of transformer = 120 V

$V_{S\,max} = \sqrt{2}\, V_{S\,rms} = \sqrt{2} \times 120\text{ V} = 169.7$ V

(*ii*) PIV rating of each diode $= V_{S\,max} - V_K = 169.7 - 0.7 = 169$ V **Ans.**

(*iii*) Peak current, $I_{max} = \frac{V_{max} - 2V_K}{R_L} = \frac{169.7 - 2 \times 0.7}{1 \times 10^3} = 168.3$ mA **Ans.**

Average current, $I_{dc} = \frac{2I_{max}}{\pi} = \frac{2 \times 168.3}{\pi} = 107.14$ mA

(*i*) DC voltage available at load, $V_{dc} = I_{dc} R_L = 107.14 \times 10^{-3} \times 1 \times 10^3 = 107.14$ V **Ans.**

(*iv*) Power rating of each diode $= I_{dc}^2 R_L = (107.14 \times 10^{-3})^2 \times 1 \times 10^3 = 11.48$ W **Ans.**

Example 4.11. What is transformer utilisation factor ? Determine the rating of a transformer to deliver a 100 W of dc power to a load under full wave rectifier.

Solution : Transformer rating $= \frac{P_{dc}}{\text{Transformer utilisation factor}}$

$= \frac{100}{0.692} = 144.5$ VA **Ans.** in case of centre-tap transformer

and transformer rating $= \frac{100}{0.812} = 123$ VA **Ans.** in case of bridge rectifier

Example 4.12. For the circuit shown in Fig. 4.9, find and sketch the output voltage $v_0(t)$ for $0 \le t \le 5$ ms. Assume that the diode is ideal. [*G.G.S.I.P. Univ. Analog Electronics December 2010*]

Solution : Let the input voltage be expressed as

$$v_{in}(t) = Kt \quad \text{where } t \text{ is in ms}$$

Substituting $v_{in}(t) = 10$ V and $t = 5$ ms, we have

$$k = \frac{10}{5} = 2 \text{ V/ms}$$

$$\text{So } v_{in}(t) = 2t \qquad ...(i)$$

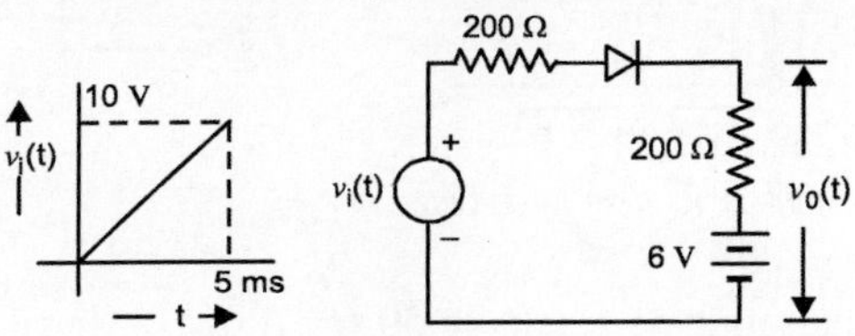

Fig. 4.9

The semiconductor diode remains reverse biased as long as input voltage does not exceed 6 V, so

$$\text{Output voltage, } v_{out}(t) = 6 \text{ V for } v_{in}(t) \le 6 \text{ V or } t \le \frac{6}{2} \text{ ms} \le 3 \text{ ms}$$

When the input voltage exceeds 6 V, the diode is forward biased and current supplied is

$$= \frac{v_{in}(t) - 6}{200 + 200} = \frac{v_{in}(t) - 6}{400}$$

$$\text{Voltage drop in 200 } \Omega \text{ resistor} = \frac{v_{in}(t) - 6}{400} \times 200 = \frac{1}{2} v_{in}(t) - 3$$

$$\text{So } v_{out}(t) = v_{in}(t) - \left[\frac{1}{2} v_{in}(t) - 3\right] = \frac{1}{2} v_{in}(t) + 3 = \frac{1}{2} \times 2t + 3 = t + 3 \text{ volts}$$

$$\text{So output voltage } v_{out}(t) = 6 \text{ V} \qquad \text{for } t \le 3 \text{ ms}$$

$$\text{and } v_{out}(t) = t + 3 \text{ volts} \qquad \text{for } 3 \text{ ms} \le t \le 5 \text{ ms}$$

Sketch of output waveform is shown in Fig. 4.10.

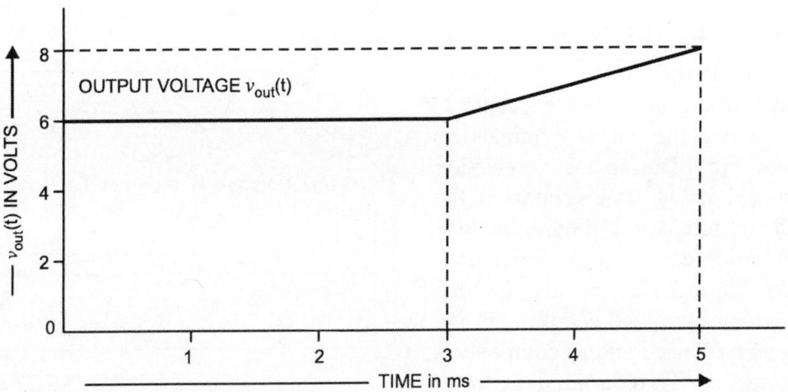

Fig. 4.10

4.4. FILTER CIRCUITS

The output from any of the rectifier circuits just discussed is not purely dc but also has some ac components, called *ripples*, along it. The ripples are maximum in the single phase half-wave rectifier and being reduced in full-wave rectifier and being reduced further with the increase in number of phases. Such supply is not useful for driving sophisticated electronic devices/circuits. Of course, for a circuit such as battery charger the pulsating nature of supply available from a rectifier is no great detriment as long as the dc level provided results in charging of battery. But for supply circuits to radio or tape recorder the

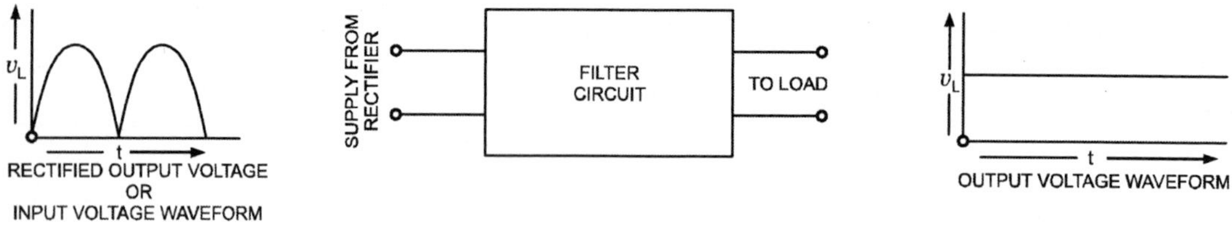

Fig. 4.11

pulsating dc results in 50 (or 100) Hz signal appearing in the output, thereby making the operation of the overall circuit poor. For such applications, as well as for many more, the output dc developed will have to be much steady or smoother than that of the pulsating dc obtained directly from half-wave or full-wave rectifier circuits.

Hence, it becomes essential to reduce the ripples from the pulsating dc supply available from rectifier circuits to the minimum. This is achieved by using a filter or smoothing circuit which removes (or filters out) the ac components and allows only the dc component to reach the load. Obviously, a filter circuit should be placed between the rectifier and the load, as shown in Fig. 4.11.

A filter circuit is a device that converts pulsating output of a rectifier into a steady dc level.

A filter circuit is generally a combination of inductors L and capacitors C. The filtering action of L and C depends upon the facts that an inductor allows dc only and capacitor allows ac only to pass. So a suitable L and C network can effectively filter out (or remove) the ac components from the rectified output.

Commonly used filter circuits are (*a*) series inductor filter (*b*) shunt capacitor filter (*c*) choke input filter (*d*) capacitor input or π filter and (*e*) R-C filter. Only shunt capacitor filter will be discussed here.

4.4.1. Shunt Capacitor Filter

This is the most simple form of the filter circuit and in this arrangement a high value capacitor C is placed directly across the output terminals, as shown in Fig. 4.12 (*a*). During the conduction period it gets charged and stores up energy in the electrostatic field and discharges through the load resistance R_L delivering energy to it during non-conduction period. Through this process, the time duration during which current flows through the load resistance gets prolonged and ac components or ripples get considerably reduced. It is to be noted here that the capacitor C gets charged to the peak value of input voltage quickly because charging time constant is almost zero. It is so because there is no resistance (except the negligible forward resistance of diode) in the charging path. But the discharging time is quite large (roughly 100 times more than the charging time depending upon the value of R_L) because it discharges through load resistance R_L.

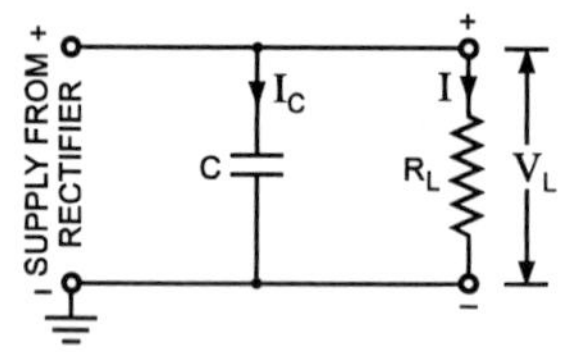

(a) Circuit Diagram

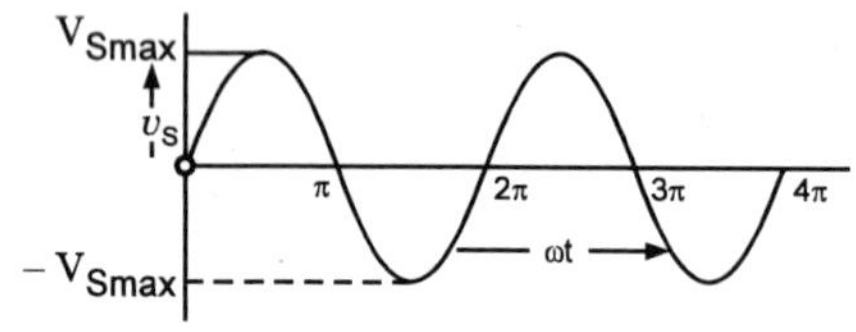

(b) Input Voltage Waveform To Rectifier

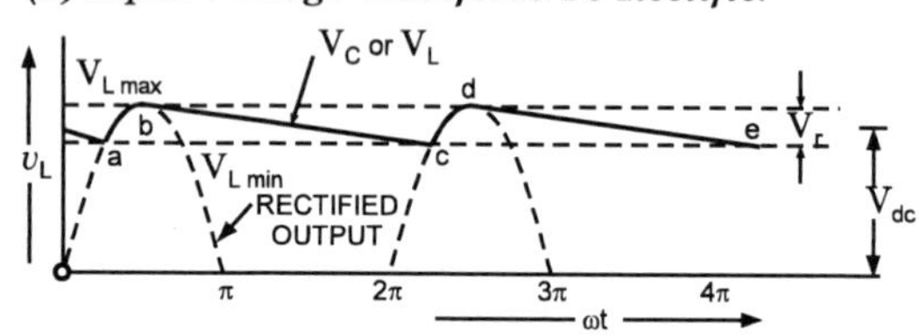

(c) Rectified and Filtered Output Voltage Waveforms

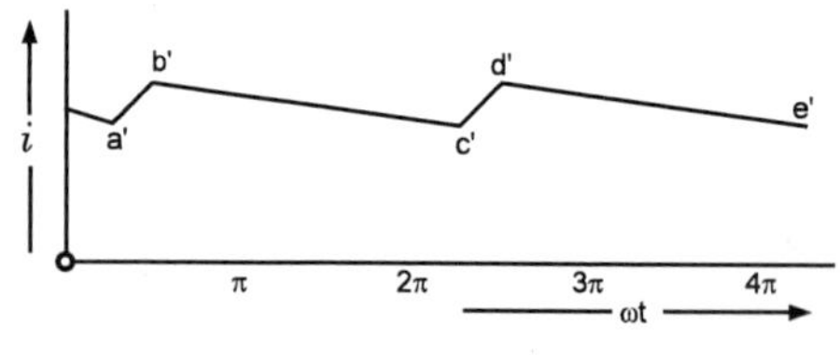

(d) Load Current Waveform

Fig. 4.12. ***Half-Wave Rectifier With Shunt Capacitor Filter***

The function of the capacitor filter may be viewed in terms of impedances. The large value capacitor C offers a low impedance shunt path to the ac components or ripples but offers high impedance to the dc component. Thus ripples get bypassed through capacitor C and only dc component flows through the load resistance R_L.

Capacitor filter is very popular because of its low cost, small size, light weight and good characteristics. It is quite useful for load up to 50 mA as in transistor radio, battery eliminators.

1. *Half-Wave Rectifier With Shunt Capacitor Filter.* The waveforms of ac input voltage, rectified and filtered output voltages and load current are shown in Figs. 4.12 (*b*), (*c*) and (*d*) respectively.

During the positive half cycle of the ac input, the diode of the rectifier (Fig. 4.2) is forward biased and so it conducts. This quickly charges the capacitor C to peak value of the supply voltage V_{Smax} because of almost zero charging time constant. This is shown by point *b* in Fig. 4.12 (*c*). After being fully charged, the capacitor holds the charge till input ac supply to the rectifier goes negative. During the negative half cycle, the diode gets reverse biased and so stops conduction. So the capacitor C discharges through load resistance R_L and loses charge. Voltage across R_L (v_L) or across C (v_C), both being equal, decreases exponentially with time constant CR_L along the curve bc, as illustrated in Fig. 4.12 (*c*). Because of large discharge time constant CR_L, the capacitor does not have sufficient time to discharge appreciably. Due to this fact the capacitor maintains a sufficiently large voltage across R_L, even during the negative half cycle of the input supply. During the next positive half cycle, when the rectified voltage exceeds the capacitor voltage V_C represented by point *c* in Fig. 4.12 (*c*), the capacitor C again gets charged quickly to $V_{S\,max}$ (or $V_{L\,max}$) as represented by point *d* in the figure. This process of charging and discharging of capacitor C is repeated for each cycle of input supply voltage. It is seen, from the figure, that nearly constant dc voltage appears across load resistance R_L at all times and also the dc component of the output voltage is increased considerably.

In case of a purely resistive load, the *output current waveform* is of the same nature as the waveform of output voltage. This is illustrated in Fig. 4.12 (*d*). During the periods *a′ b′* and *c′ d′* the current is supplied by the diode and during periods *b′ c′* and *d′ e′* etc. by capacitor C.

Diode current flows during short intervals of time *ab* and *cd* etc. in Fig. 4.12 (*c*). During these time intervals, diode output voltage exceeds the capacitor (or load) voltage. Hence diode current is short-duration pulses (*i.e.* surging current). A small resistor, called the *surge limiting resistor* is, therefore, always connected in series with the diode to limit this surge current.

The noteworthy points about shunt capacitor filter are :

1. For a fixed-value filter capacitance larger the load resistance R_L larger will be the discharge time constant CR_L and, therefore, lower the ripples and more the output voltage. On the other hand lower the load resistance (or more the load current), lower will be the discharge time constant and, therefore, more the ripples and lower the output voltage.

2. Similarly smaller the filter capacitor, the less charge it can hold and more it will discharge. Thus the peak-to-peak value of the ripple will increase, and the average dc level will decrease. Larger the filter capacitor, the more charge it can hold and the less it will discharge. Hence the peak-to-peak value of the ripple will be less, and the average dc level will increase. But, the maximum value of the capacitance that can be employed is limited by another factor. The larger the capacitance value, the greater is the current required to charge the capacitor to a given voltage. The maximum current that can be handled by a diode is limited by the figure quoted by the manufacturer. Thus the maximum value of the capacitance, that can be used in the shunt filter capacitor is limited.

Approximate Analysis. Total change in output voltage is equal to V_r, the ripple component of output voltage. The average or dc value of output voltage, V_{dc} is almost between the peak value $V_{L\,max}$ and the minimum value given by points *b* and *c* respectively in Fig. 4.12 (*c*). Thus

$$V_{dc} = V_{L\,max} - \frac{V_r}{2} \quad ...(4.58)$$

The total charge lost during non-conduction (or discharge) duration T_2 through load is given as

$$Q_{discharge} = I_{dc}\,T_2 \quad ...(4.59)$$

This charge is replenished during time interval T_1, in which voltage across the capacitor increases by V_r volts. So charge gained by capacitor,

$$Q_{charge} = CV_r \quad ...(4.60)$$

In steady-state, $Q_{charge} = Q_{discharge}$

or $C\,V_r = I_{dc}\,T_2$

or $V_r = \frac{I_{dc}\,T_2}{C}$

Assuming $T_1 << T_2$, $\quad T_2 = T = \frac{1}{f} \quad$ So $V_r = \frac{I_{dc}}{fC}$ ...(4.61)

Hence from Eq. (4.58) $\quad V_{dc} = V_{L\,max} - \frac{I_{dc}}{2fC}$ ...(4.62)

From Eq. (4.61) it is revealed that ripple voltage varies directly with the load current I_{dc} and inversely with the capacitance C.

Ripple Factor. The rms value of the ripple component of almost triangular wave is independent of the slope or the length of the almost straight lines *bc* and *cd* but depends only on the peak value V_r. The rms value is given as

$$V_{ac\,rms} = \frac{V_r}{2\sqrt{3}}$$

Hence ripple factor, $\gamma = \frac{V_{ac\,rms}}{V_{dc}} = \frac{V_r}{2\sqrt{3}I_{dc}R_L} = \frac{I_{dc}}{2\sqrt{3}\,I_{dc}\,R_L\,fC} = \frac{1}{2\sqrt{3}\,fC\,R_L}$ ...(4.63)

2. *Full-Wave Rectifier With Shunt Capacitor Filter.* The filtering action of shunt capacitor filter on a full wave rectifier is shown in Fig. 4.13. In this case capacitance C discharges twice during one cycle. Because both the diodes conduct, non-conduction period has reduced. The result is that ripple voltage V_r has been reduced to half and V_{dc} has been increased relative to half-wave rectifier.

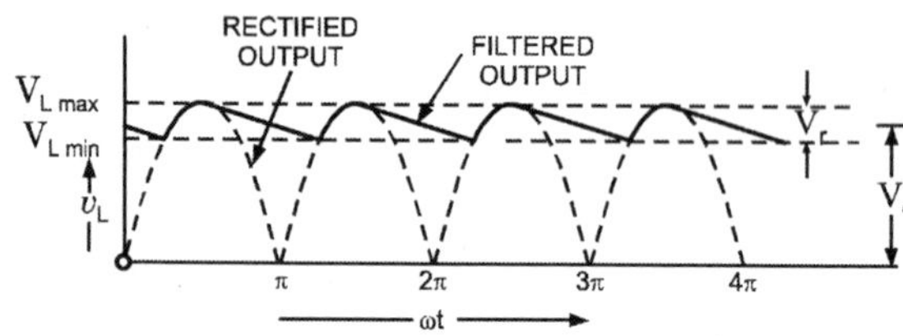

Fig. 4.13. ***Rectified and Filtered Output Voltage Waveform From Full-Wave Rectifier With Shunt Filter Capacitor***

In this case because $T_2 = \frac{T}{2} = 1/2f$

So $V_r = \frac{I_{dc}}{2fC}$ ...(4.64)

$$V_{dc} = V_{L\,max} - \frac{V_r}{2} = V_{L\,max} - \frac{I_{dc}}{4fC} \quad ...(4.65)$$

$$V_{ac\,rms} = \frac{V_r}{2\sqrt{3}}$$

So ripple factor, $\gamma = \frac{V_{ac\,rms}}{V_{dc}} = \frac{V_r}{2\sqrt{3}I_{dc}R_L} = \frac{1}{4\sqrt{3}\,fCR_L}$ ...(4.66)

Voltage regulation in this case is better than that in half-wave rectifier.

HIGHLIGHTS

1. Half-wave and full-wave rectifier circuits convert an ac signal to an approximate dc signal. A dc power supply, which is used to bias electronic circuits and systems, utilizes such circuits.
2. A half-wave rectifier is characterized by an excessive ripple, small values of rectification efficiency and transformer utilisation factor (TUF) and dc saturation of transformer core.
3. The centre-tapped and bridge type full-wave rectifiers improve on the half-wave characteristics but the three-phase, full-wave rectifier is even better.
4. Ripple factor is defined as the ratio of the effective value of the ac components of voltage (or current) present in the output from the rectifier to the direct or average value of the output voltage (or current). It is an indication of how successful is a circuit in converting ac to dc.

   Ripple factor for half-wave is 1.21 whereas for full-wave rectifier it is 0.482.
5. Rectification efficiency is defined as the ratio of dc output power to the ac input power from the transformer secondary $\left(i.e.\ \eta = \frac{P_{dc}}{P_{ac}}\right)$ and is a figure of merit to compare rectifiers.

   Rectification efficiency is 40.6% in case of half-wave rectifiers and 81.2% in case of full-wave rectifiers.
6. Transformer utilisation factor is defined as the ratio of power delivered to the load and ac rating of transformer secondary

   $$i.e.\ \text{TUF} = \frac{P_{dc}}{P_{ac\,(rated)}} = \frac{I_{dc}^2 R_L}{V_{S\,rms} I_{rms}}.$$

   It is used in determination of the necessary rating of a transformer for a given dc load.

   TUF for half-wave, centre-tap and bridge rectifiers are 0.286, 0.692 and 0.812 respectively.
7. The variation of dc output voltage as a function of dc load current is called regulation. Percentage regulation is given as

   $$\%\ \text{voltage regulation} = \frac{V_{NL} - V_{FL}}{V_{FL}} \times 100$$

   It is an indication of how the output voltage of a power supply varies with load.
8. The internal resistance of a power supply, given by

   $$R_{in} = \frac{V_{NL} - V_{FL}}{I_{FL}}$$

   is due to the resistance of the diodes and transformer winding.
9. It is very important that the PIV rating of a diode be checked for a particular application. Simply determine the maximum voltage across the diode under reverse-bias conditions, and compare it to the name plate rating. For the typical half-wave and full-wave bridge rectifiers, it is the peak value of the applied signal. For the centre-tapped transformer full-wave rectifier, it is twice the peak value.
10. The filtering action of a capacitor filter depends upon its property of opposing any change in voltage across it.

    The ripple from a capacitor filter decreases as the load resistance increases (or load current decreases) and provides effecive filtering only for light loads of 50 mA and less.

    The capacitor filter provides poor voltage regulation because of the increase in ripple with load current causing a decrease in average voltage.

    A capacitor filter is characterized by the large peak current that the diodes must handle due to the capacitor recharging in a very short interval, and this current increases with the value of capacitance.

    Ripple factor, $\gamma = \frac{1}{2\sqrt{3} f C R_L}$ in case of an half-wave rectifier.

    Ripple factor, $\gamma = \frac{1}{4\sqrt{3} f C R_L}$ in case of a full-wave rectifier.

    A diode's PIV in a half-wave rectifier using a capacitor filter is $2V_{max}$.

EXERCISES

1. Derive an expression for the dc current in a half-wave rectifier if the diode used is not ideal, *i.e.* $V_r \neq 0$, $R_f \neq 0$. **[G.G.S.I.P. Univ. Analog Electronics December 2009]**
2. What is rectification ? Draw the circuit diagram of a half-wave rectifier and explain its working giving input and output waveforms. Derive an expression for the efficiency of half-wave rectifier.
3. With reference to a rectifier circuit define and explain the following : (*i*) ripple factor, (*ii*) rectification efficiency. **[G.G.S.I.P. Univ. Analog Electronics-I December 2004]**
4. Derive the expressions for (*i*) ripple factor (*ii*) rectification efficiency. **[G.G.S.I.P. Univ. Analog Electronics December 2010]**
5. Draw the circuit diagram of a full-wave rectifier using two diodes and calculate (*i*) I_{dc}, (*ii*) I_{rms}, (*iii*) PIV rating of diode.
6. Sketch the circuit for a full-wave rectifier using two diodes only. Derive the expression for (*i*) the dc current, (*ii*) the dc load voltage, (*iii*) the dc diode voltage, (*iv*) the rms current.
7. Draw the circuit of a bridge rectifier and explain its working with the help of input and output waveshapes. **[G.G.S.I.P. Univ. Analog Electronics-I December 2004]**
8. For a full-wave rectifier derive expression for average current, rms current, ripple factor and rectification efficiency. **[G.G.S.I.P. Univ. Analog Electronics December 2010]**
9. Compare half-wave, full-wave and bridge rectifier circuits. What is the advantage of each ?
10. Compare center tap full wave rectifier and bridge rectifier. **[G.G.S.I.P. Univ. Analog Electronics December 2009]**
11. Draw the circuit diagram of a half-wave rectifier and its output waveform and then show the effect on this waveform of connecting a capacitor across the load resistance.
12. Draw a neat sketch of simple capacitor filter for a half-wave rectifier and show the output voltage across the load. Making suitable approximations, determine the dc output voltage as well as the ripple factor of the half-wave rectifier with capacitor filter.

SHORT ANSWER TYPE QUESTIONS WITH ANSWERS

Q. 1. What is a dc power supply ?

Ans. The part of the equipment that converts ac into dc is called the dc power supply.

Q. 2. What is meant by rectifier ?

Ans. A rectifier is a device which converts alternating current (or voltage) into unidirectional current (or voltage).

Q. 3. Enumerate the properties that the junction diodes designed for use as rectifiers should possess.

Ans. Junction diodes designed for use in rectifiers should have the properties : (*i*) V-I characteristics as close as possible to that of an ideal diode (*ii*) Negligible forward resistance (*iii*) Negligible reverse current *i.e.* large reverse resistance (*iv*) Large reverse breakdown voltage and (*v*) Small offset voltage V_0 in the forward direction.

Q. 4. Define peak inverse voltage (PIV) of a diode. What is the difference between PIV of a center tapped FWR and bridge rectifier ? **[G.G.S.I.P. Univ. December 2009]**

Ans. Peak inverse voltage (PIV) is the maximum possible voltage that occurs across a diode when it is reverse biased.

PIV of a center-tapped FWR is double of that of bridge rectifier.

Q. 5. What is the importance of peak inverse voltage ?

Ans. If the applied voltage in reverse biased condition exceeds peak inverse voltage (PIV) rating of the diode, then the diode may get damaged.

Q. 6. Why half-wave rectifiers are generally not used in dc power supply ?

Ans. The type of supply available from half-wave rectifier is not satisfactory for general power supply. That is why it is generally not used in dc power supply.

Q. 7. Why diodes are not operated in the breakdown region in rectifiers ?

Ans. In breakdown region, a diode has a risk of getting damaged or burnt because the magnitude of current flowing through it increases in an uncontrollable manner. That is why diodes are not operated in the breakdown region in rectifiers.

Q. 8. What is rectification efficiency of a full-wave rectifier ?

Ans. $\eta = \dfrac{0.812}{1 + \dfrac{2R_F}{R_L}}$ *i.e.* maximum value is 81.2%

Q. 9. What is ripple factor ?

Ans. The pulsating output of a rectifier can be considered to contain a dc component and ac components called the *ripples*. The ripple current is undesirable and its value should be the smallest possible in order to make the rectifier effective.

The ripple voltage or current is measured in terms of the ripple factor which is defined as the ratio of the effective value of the ac components of voltage (or current) present in the output from the rectifier to the diode or average value of the output voltage (or current).

Q. 10. What is transformer utilisation factor ?

Ans. Transformer utilisation factor is defined as the ratio of power delivered to the load and ac rating of secondary of supply power transformer.

Q. 11. The output of a 60 Hz full-wave bridge rectifier has a 60 Hz ripple. Is the circuit working properly ?

Ans. A full-wave rectifier with 60 Hz input must have lowest ripple frequency equal to twice the input frequency *i.e.* 120 Hz. If the ripple frequency is 60 Hz, it means some diodes in the circuit are not working.

Q. 12. Why is it necessary to include voltage stabilizer in a power supply ?

Ans. The magnitude of output dc voltage may vary with the variation of either the input ac voltage or the magnitude of load current. So at the output of a rectifier-filter combination a voltage stabilizer is required.

PROBLEMS

1. A half-wave rectifier has a load resistance of 3.5 kΩ. If the diode and secondary of the transformer have a total resistance of 800 Ω and the input voltage has an ac signal of 240 V (peak value), determine (*i*) peak, average and rms values of current flowing (*ii*) dc power output (*iii*) ac power input (*iv*) rectification efficiency and (*v*) ripple factor.
   [**Ans.** (*i*) 55.81 mA, 17.766 mA; 27.91 mA (*ii*) 1.1 W (*iii*) 3.35 W (*iv*) 32.85% (*v*) 1.21]
2. A half-wave rectifier employs a diode having a forward resistance of 10 Ω. If the input voltage to the rectifier circuit is 12 V rms, find the dc output voltage at a load of 100 mA and PIV. (Fig. 1) [**Ans.** 4.4 V; 17 V]

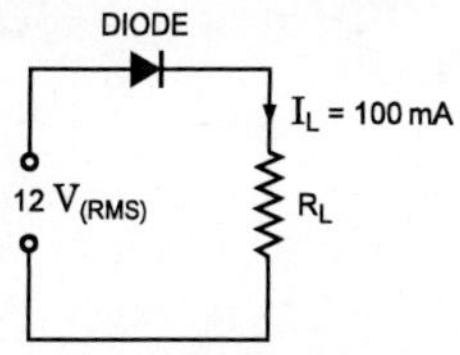

Fig. 1

3. A full-wave rectifier uses a double diode with each element having a constant forward resistance of 500 Ω. The transformer rms secondary voltage from the centre tap to each plate is 300 V and the load has a resistance of 2.5 kΩ. Determine (*i*) dc output power (*ii*) ac input power (*iii*) the rectification efficiency and (*iv*) voltage regulation from no-load to full-load. [**Ans.** (*i*) 20.3 W (*ii*) 30 W (*iii*) 67.67% (*iv*) 20%]
4. A single phase full-wave rectifier uses two diodes, the internal resistance of each being 20 Ω. The transformer rms secondary voltage from centre tap to each end of secondary is 50 V and load resistance is 980 Ω. Find (*i*) the mean load current (*ii*) rms load current and output efficiency.
   [**Ans.** (*i*) 45 mA (*ii*) 50 mA; (*iii*) 79.38%]
5. In a bridge rectifier circuit the peak value of secondary voltage is $240\sqrt{2}$ V and frequency is 50 Hz. Determine the no-load output dc voltage (*ii*) PIV and (*iii*) output frequency.
   [**Ans.** (*i*) 216 V (*ii*) 339.4 V (*iii*) 100 Hz]
6. What is transformer utilization factor ? Determine the rating of a transformer to deliver a 100 watts of dc power to a load under full-wave rectifier.
   [**Ans.** 144.5 W in case of centre-tap transformer rectifier, and 123.2 W in case of bridge rectifier]
7. A centre-tapped transformer has a 220 V primary winding and a secondary winding rated at 12-0-12 V and is used in a full-wave rectifier circuit with a load of 100 Ω. What is the dc output voltage, dc load current and the PIV rating required for diodes ? [**Ans.** 10.8 V; 108 mA; 33.94 V]
8. A 220 V, 50 Hz ac voltage is applied to the primary of 4 : 1 stepdown transformer, which is used in bridge rectifier, having a load resistance of 1 kΩ. Assuming the diodes to be an ideal, determine the following :
   (*i*) DC output voltage (*ii*) DC power delivered to load
   (*iii*) PIV of each diode (*iv*) Output frequency.
   [**Ans.** (*i*) 49.52 V; (*ii*) 2.452 W; (*iii*) 77.78 V; (*iv*) 100 Hz]

❑❑❑

UNIT II

BIPOLAR JUNCTION TRANSISTOR AND BIAS STABILIZATION

Contents in This Unit

BIPOLAR JUNCTION TRANSISTOR : Introduction; Transistor Construction; Transistor Operations; BJT Characteristics; Load Line; Operation Point; Leakage Currents; Saturation and Cutoff Mode of Operations; Ebers-Moll Model.

BIAS STABILIZATION : Need for Stabilization; Fixed Bias; Emitter Bias; Self Bias; Bias Stability with Respect to Variations in I_{CO}, V_{be} and β; Stabilization Factors; Thermal Stability.

SMALL SIGNAL AMPLIFIERS : CB, CE, CC Configurations; Hybrid Model For Transistor at Low Frequencies; R-C Coupled Amplifiers; Midband Model; Gain and Impedance; Comparison of Different Configurations; Darlington Pair; Hybrid π-Model at High Frequencies.

CHAPTER 5

Bipolar Junction Transistor (BJT)

INSIDE THIS CHAPTER

5.1. INTRODUCTION

The transistor is a solid state device, whose operation depends upon the flow of electric charge carriers within the solid. Transistor is capable of amplification and in most respect it is analogous to a vacuum triode. The main difference between the two is that the transistor is a current controlled device whereas vacuum triode is a voltage controlled device. The transistor is only about 6 decade old, yet it is replacing vacuum tubes in almost all applications. The reasons are obviously its advantages over vacuum tubes such as compact size, light weight, rugged construction, more resistive to shocks and vibrations, instantaneous operation (no heating required), low operating voltage, high operating efficiency (no heat loss) and long life with essentially no ageing effect if operated within permissible limits of temperature and frequency. However, transistors, in comparison to vacuum triodes, have some drawbacks also such as loud hum noise, restricted operating temperature (up to 75°C) and operating frequency (up to a few MHz only).

It consists of a silicon or germanium (preferably silicon because of its smaller cutoff current I_{CBO}, smaller variations in I_{CBO} due to variations in temperature and higher operating temperature) crystal in which a layer of N-type material is sandwiched between two layers of P-type material, as shown in Fig. 5.1 (*a*). Alternatively a transistor may consist of a layer of P-type material sandwiched between two layers of N-type material, as shown in Fig. 5.1 (*b*). In the former case the transistor is referred to as a P-N-P *transistor* and in latter case, as an N-P-N *transistor*. Each type of transistor has two P-N junctions—one junction between the emitter and base, called the *emitter-base junction* or simply the *emitter junction* and the other junction between the base and collector, called the *collector-base junction* or simply the *collector junction.* Thus a transistor is like two P-N junction diodes connected back to back. Each of these P-N junction diodes have all the characteristics discussed in earlier chapters. The two junctions give rise to three regions provided with three terminals called the *emitter*, *base* and *collector*, as shown in Fig. 5.1. The emitter, base and collector correspond in a general way to the cathode, grid and plate or anode of a vacuum triode.

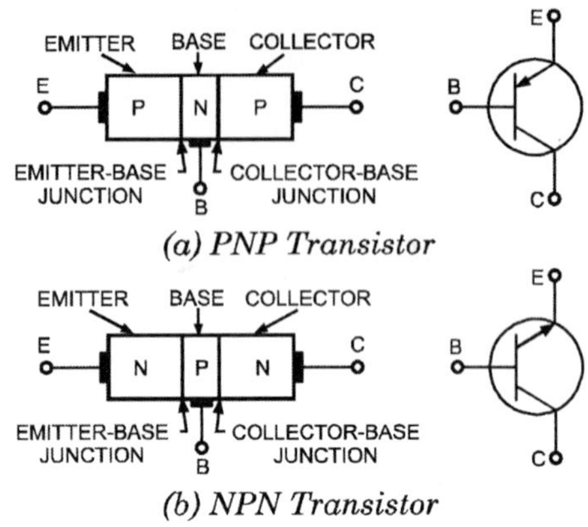

Fig. 5.1

5.2. TRANSISTOR TERMINALS

As mentioned above, transistor is a single crystal in which there are two P-N junctions, as shown in Fig. 5.1. The idea behind is to have first section to supply the charges (either holes or electrons) to be collected by the third section through the middle section. One side section supplying free charges is called the *emitter*, other side section collecting these charges is called the *collector* and the middle section which is formed between the emitter and collector is called the *base*.

1. ***Emitter.*** It is left hand section (or region) of the transistor and its main function is to supply majority charge carriers (electrons in case of NPN transistors and holes in case of P-N-P transistors) to the base. The emitter is always forward biased w.r.t. base so that it is able to supply majority charge carriers to the base. The emitter is heavily doped so that it may be able to inject a large number of charge carriers. It is of moderate size in order to maintain heavy doping without diluting it or mesh formation in it.

2. ***Collector.*** It is the right hand section of the transistor and its main function is to collect majority charge carriers. Collector is always reverse biased so as to remove the charge carriers away from its junction with the base. It is moderately doped to avoid the chances of mesh formation even after taking the carriers from the emitter. Large in size to withstand the temperature generated at the collector.

3. ***Base.*** It is the middle section of the transistor and is very lightly doped to reduce the recombination within the base so as to increase collector current and is very thin (of the order of μm) in comparison to either emitter or collector so that it may pass most of the injected charge carriers to the collector.

Base forms two junctions *i.e.* emitter junction and collector junction, each having its own barrier voltage. The forward-biased emitter-base junction offers low resistance to the emitter current whereas the collector-base junction, which is reverse biased, offers high resistance to the collector current. As the resistance of emitter-base junction is very small as compared to that of collector-base junction, therefore, the forward bias applied to the emitter-base junction is usually very small whereas the reverse bias on the collector-base junction is much large.

In most of the transistors, the collector region is made physically larger than emitter region because collector is to dissipate much power.* Due to this difference, collector and emitter are not interchangeable. However, for sake of convenience, it is customary to show emitter and collector to be of equal size.

As regards the symbols, arrowhead is always at the emitter. The direction indicates the conventional direction of current flow (from emitter to base in case of P-N-P transistor and from base to emitter in case of NPN transistor).** Generally, no arrow head is marked for collector since its reverse leakage current is always opposite to the direction of emitter current.

5.3. TRANSISTOR ACTION

Transistors of both types (P-N-P and N-P-N) behave exactly in the same way except change in biasing and majority carriers. In P-N-P transistors the conduction is by holes whereas in N-P-N transistors

* During operation of a transistor, much heat is generated at the collector junction.

** In a P-N-P transistor emitter is positive w.r.t. base while in case of an N-P-N transistor base is positive w.r.t. emitter.

conduction is by electrons. However, N-P-N transistors are preferred due to their better high frequency response.

5.3.1. Unbiased Transistor

When no battery is connected between the different terminals of a transistor, the transistor is said to be in an unbiased state or in an open-circuit state.

We know that when a semiconductor is doped with donor or pentavalent impurities, it becomes an N-type semiconductor. The additional free electrons tend to raise the Fermi (average electron energy) level within the crystal. Similarly, when a semiconductor is doped with acceptor or trivalent impurities, it becomes a P-type semiconductor. The additional holes tend to lower the Fermi level. This is illustrated in Fig. 5.2 (*b*).

We also know that when an N-type region is formed next to a P-type region, a barrier potential (0.7 V in case of Si and 0.3 V in case of Ge at room temperature) is produced at the P-N junction. The free electrons in the N-region diffuse into the adjacent P-region to annihilate holes. Consequently, a layer of positive ions is formed on the N side, and a layer of negative ions is formed on the P-side. This serves to create the barrier potential.

The same phenomenon occurs at both the P-N junctions found in the N-P-N transistor structure [Fig. 5.2 (*a*)]. The depletion regions (ionization layers), which are formed, extend into the semiconductor crystal as a function of the doping levels. Specifically, it extends deeply into the lightly doped regions, and slightly into the heavily doped regions. Since emitter is heavily doped than base and collector, the emitter-base depletion region width is smaller than that of collector-base depletion region width. Note that the barrier potentials are negative on the base and positive on the emitter and collector for the N-P-N device. In case of P-N-P device it would be just reverse *i.e.*, positive on the base and negative on emitter and collector. This is illustrated in Fig. 5.2 (*a*).

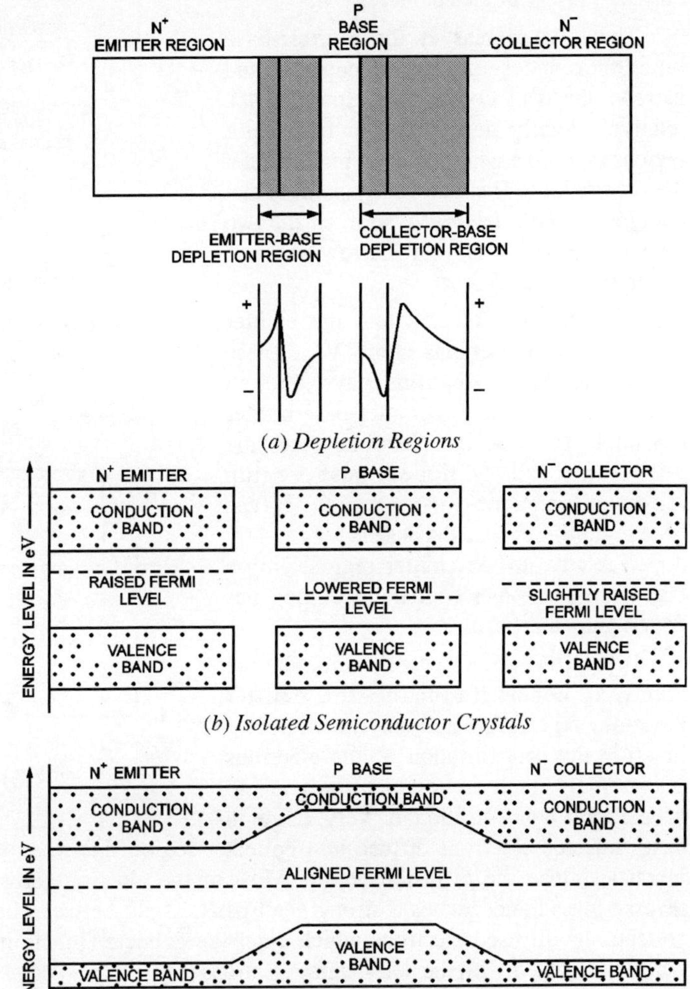

(*a*) *Depletion Regions*

(*b*) *Isolated Semiconductor Crystals*

(*c*) *Energy Diagram of An N-P-N Transistor*

Fig. 5.2. ***Unbiased Transistor***

The formation of barrier potential corresponds to the alignment of the Fermi levels throughout the crystals. Fermi level alignment produces a distortion in the valence and conduction energy bands, as illustrated in Figs. 5.2 (*b*) and 5.2 (*c*).

5.3.2. Operation of Transistor

The process of applying dc voltages across the different terminals of a transistor is called *biasing*. For normal operation the emitter-base junction is always forward biased while the collector-base junction is always reverse biased. This is depicted in Fig. 5.3.

The forward bias at the emitter-base junction reduces the barrier potential and narrows the depletion region. However, the relatively lightly doped base and collector regions produce a wide depletion region under the reverse bias. The noteworthy point is that the *effective base width* W_b between the two depletion regions is very narrow. This is illustrated in Fig. 5.3 (*a*).

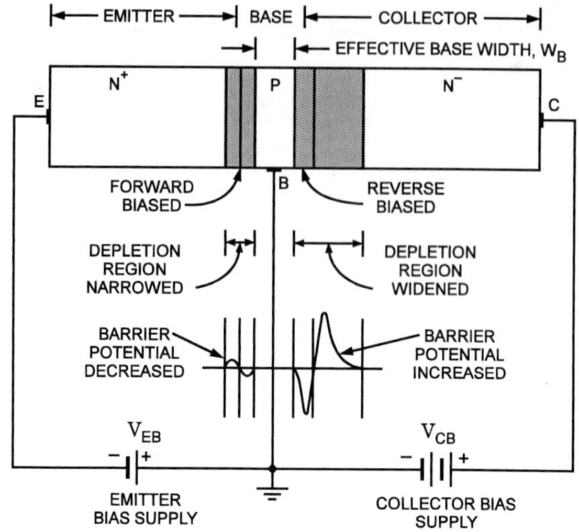

(a) Depletion Regions, Barrier Potentials and W_B

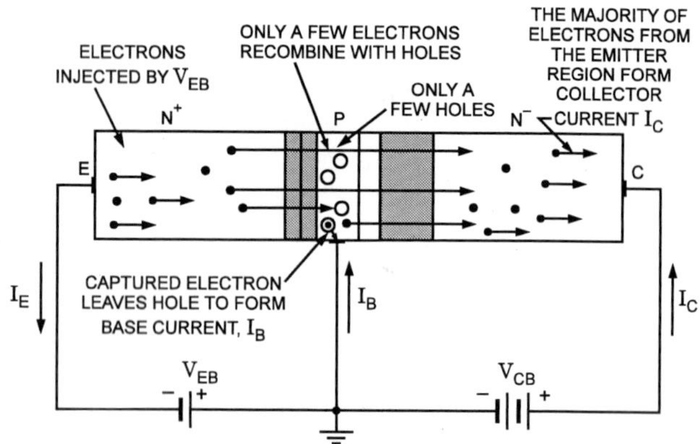

(b) Charge Carrier Movement

Fig. 5.3

Electrons are injected into the emitter region by the emitter bias supply V_{EB}. These conduction band electrons have enough energy to overcome the emitter-base barrier potential. The injected electrons enter the very thin, lightly doped base region (electrons are minority carriers in the P-type base region). Because the base is very lightly doped relative to the emitter region, only a few of the electrons recombine with the holes doped into the base.

Injection of electrons makes the electrons concentration on the emitter junction very large and on the collector junction the concentration of the electrons is extremely small and as such the gradient of electron concentration is very large in base. Injected electrons diffuse into collector region due to extremely small thickness of base which is much less than the diffusion length. Most of the electrons cross into the collector region. Collector is reverse biased and creates a strong electrostatic field between base and collector. The field immediately collects the diffused electrons which enter the collector junction. Flow of electrons into the base region when confronted with the holes, a few electrons (say 1 to 5%) combine and neutralize and rest of the electrons (say 95 to 99%) of the injected electrons diffuse into the collector region and collected by the collector electrode. For each electron that combines with a hole, an electron leaves the base region via the base terminal. This action creates a very small base current (1 to 5% of the electron current from emitter to collector). Thus the emitter current I_E is equal to the sum of collector current I_C and base current I_B. The ratio of collector current I_C to emitter current I_E is called the α and it is a measure of the possible current amplification. In a transistor α cannot be higher than unity, but practical values of 0.95 to 0.99 are attained in commercial transistors.

The operation of the transistor may also be examined by means of its energy band diagram. Since the forward bias tends to align the valence and conduction energy bands, while reverse bias tends to increase their misalignment, the diagram shown in Fig. 5.4 is arrived at. As can be seen from Fig. 5.4, electrons must lose energy to travel from the emitter region to the base region. Further, the electrons must also lose energy in moving from base region into collector region. Only a few of the electrons in the base region lose energy to recombine with the (valence band) hole. Typically, at least 95% of the electrons have a long enough lifetime to diffuse into the collector region.

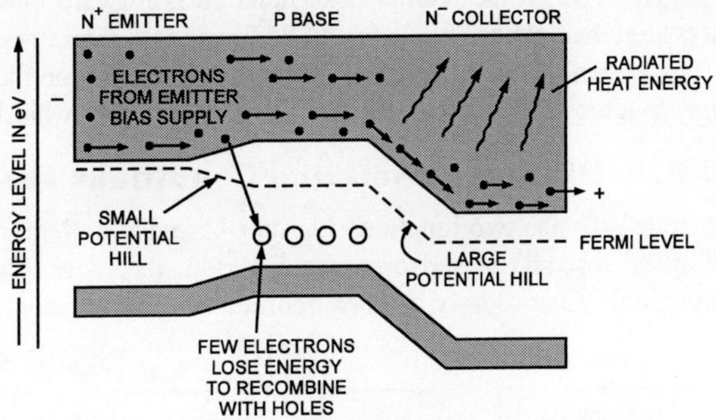

Fig. 5.4. ***Energy Band Diagram For an N-P-N Transistor in its Active Region***

It is observed from Fig. 5.4 that the potential hill (bend in the Fermi level) between the base and the collector region is very steep. Specifically, the electrons must lose a considerable amount of energy as they move from the base region to the collector region. Typically, the electrons give up their energy in the form of heat. Consequently, the collector region must be able to dissipate this heat energy. It is for this reason that it is typically the largest of the three regions.

Operation of P-N-P Transistor. The P-N-P transistor behaves exactly the same way as an N-P-N device, with the exception that the majority charge carriers are holes. As illustrated in Fig. 5.5, holes are emitted from the P-type emitter across the forward-biased emitter-base junction into the base. In the lightly doped N-type base, the holes find few electrons to recombine with. Some of the holes flow out via the base terminal, but most are drawn across the collector by the positive-negative electric field at the reverse biased collector-base junction. As in the case of NPN device, the forward bias at the emitter base junction controls the collector and emitter currents.

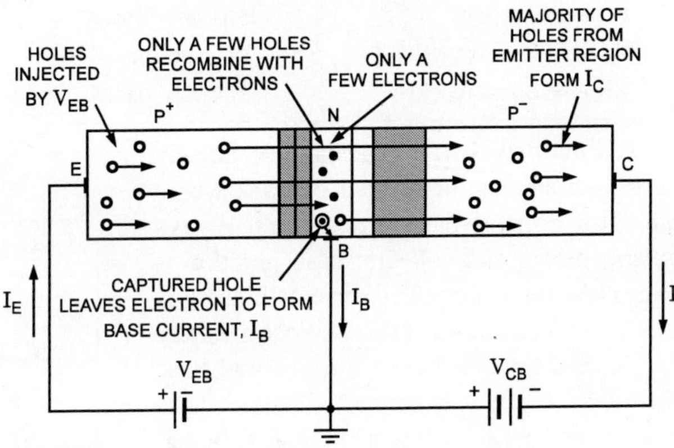

Fig. 5.5. ***Charge Carrier Movement in a P-N-P Transistor***

5.4. TRANSISTOR BIASING

As already mentioned above, for normal operation of a transistor emitter-base junction is always forward biased and collector-base junction is always reverse biased.

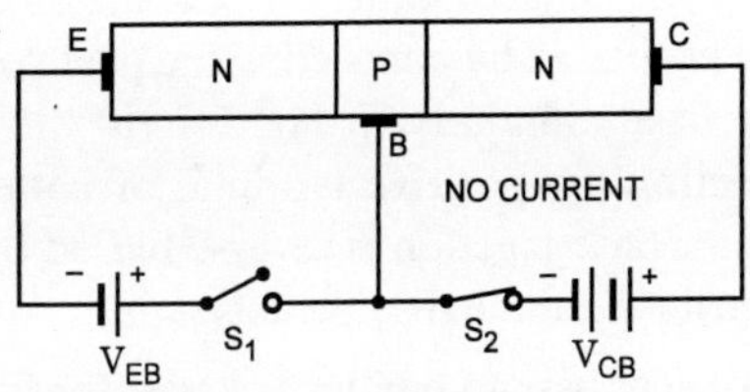

(*a*)

For a moment if the biasing of emitter-base junction is ignored and only collector-base junction, which is reverse biased [Fig. 5.6 (*a*)] is considered, then due to high resistance of reverse biased collector diode, practically* no current would flow in the collector circuit.

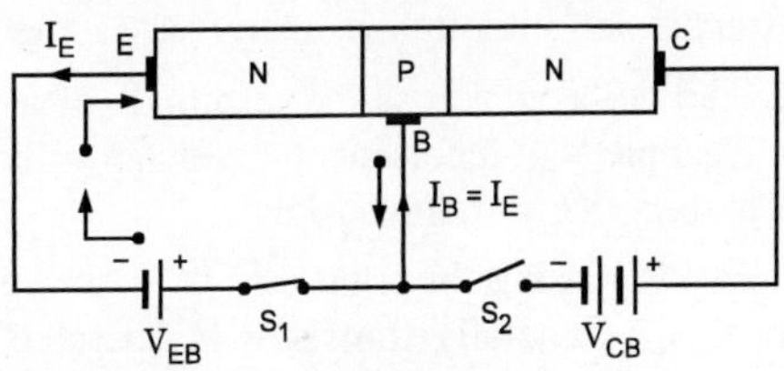

(*b*)

Fig. 5.6

Now if the biasing of collector-base junction is ignored and only emitter-base junction, which is forward biased [Fig. 5.6 (*b*)] is considered then practically whole of the emitter current flows through the base *i.e.* $I_B = I_E$.

* In actual practice a very small reverse saturation current would flow in the collector circuit. This is called collector cut-off current and is because of minority carriers.

However, if the emitter-base junction is forward biased and the collector-base junction is reverse biased it is seen that almost whole (about 95%) of the current constituted by the emitter flows through collector circuit. Thus the magnitude of collector current I_C depends upon the magnitude of emitter current I_E. No emitter current means almost no collector current. This is precisely what happens in a transistor.

5.4.1. Different Operating Conditions For a Transistor

A transistor has two junctions (emitter-base and collector-base junctions), and each of these two junctions, may be forward biased or reverse biased, therefore, there are four possible ways of biasing these two junctions. Accordingly it may operate in different conditions as listed in Table 5.1.

TABLE 5.1

Condition	*Emitter-Base (EB) Junction*	*Collector-Base (CB) Junction*	*Region of Operation*
1. Forward-Reverse (FR)	Forward-Biased	Reverse-Biased	Active
2. Forward-Forward (FF)	Forward-Biased	Forward-Biased	Saturation
3. Reverse-Reverse (RR)	Reverse-Biased	Reverse-Biased	Cutoff
4. Reverse-Forward (RF)	Reverse-Biased	Forward-Biased	Inverted

1. Forward-Reverse Biasing. As already mentioned above, for normal operation of a transistor emitter-base junction is always forward biased and collector-base junction is always reverse biased. This type of biasing is known as *FR biasing*. For this purpose a battery V_{EB} is connected between emitter and base while a battery V_{CB} is connected between the collector and base, as shown in Fig. 5.7. In Fig. 5.7 (*a*), the emitter-base junction of N-P-N transistor is forward biased by connecting the negative terminal of battery V_{EB} to the emitter and positive terminal to the base while collector-base junction is reverse biased by connecting the positive terminal of the battery V_{CB} to the collector and negative terminal to the base. Similarly in Fig. 5.7 (*b*) the emitter-base junction of P-N-P transistor is forward biased by connecting the positive terminal of battery V_{EB} to the emitter and negative terminal to the base while collector-base junction is reverse biased by connecting the negative terminal of battery V_{CB} to the collector and positive terminal to the base.

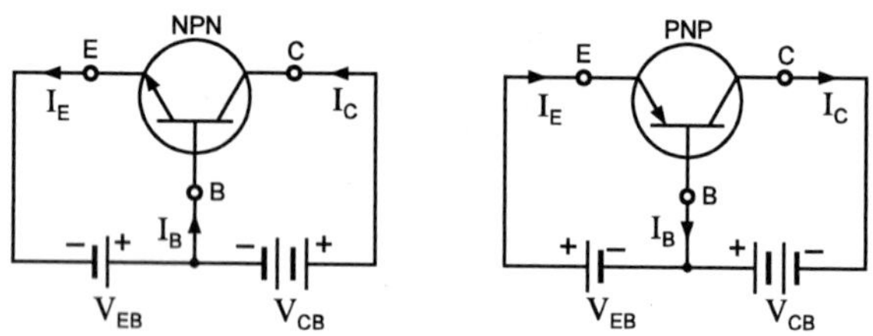

(*a*) *NPN Transistor* (*b*) *PNP Transistor*

Fig. 5.7. ***Active Region***

Typical base-to-emitter voltages for both N-P-N and P-N-P transistors are 0.7 V for silicon and 0.3 V for germanium. Typical collector-to-base voltages might be anything between 3 V and 20 V for most types of transistors, although in many cases the collector-to-base voltage may exceed 20 volts even.

The voltages across the terminals of a transistor follow the usual double subscript voltage convention. For example V_{BE} represents the voltage of base with respect to the emitter. If V_{BE} is equal to 0.7 V it means the base is 0.7 V more positive.

In FR biasing the transistor is in active region and the collector current I_C, depends upon the emitter current I_E. Generally transistor is operated in this region for amplification.

For connecting batteries properly in the transistor circuits the following scheme may be remembered.

The battery polarities to the emitter and base are the same as the letter of the emitter and base impurity designations. Thus, in the P-N-P type, the emitter receives the P (positive) terminal, and the base receives the N (negative) terminal. Similarly, in N-P-N type, the emitter receives the N (negative) terminal, the base receives the P (positive) terminal. The collector receives a battery polarity opposite to its impurity type designation. Thus P-type collector receives the N (negative) terminal of the battery and the N-type collector receives the P (positive) terminal.

2. Forward-Forward Biasing. In this biasing, both collector and emitter are forward biased, as illustrated in Fig. 5.8. The transistor operates in saturation region. In this region, the collector current becomes independent of the base current. Thus the transistor acts like a close switch.

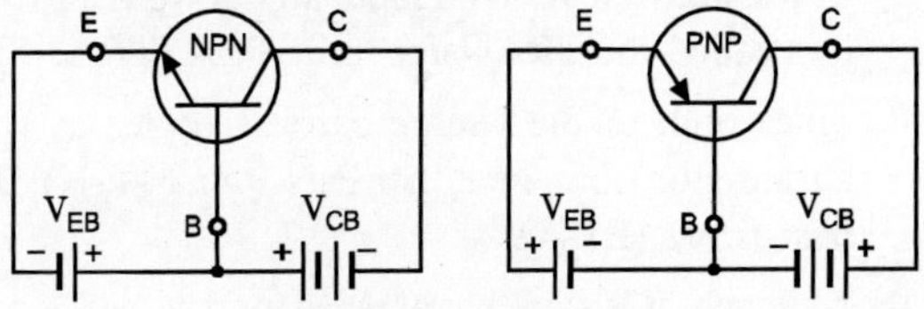

Fig. 5.8. ***Saturation Region***

3. Reverse-Reverse Biasing. In this biasing, both the junctions are reverse biased. In this biasing, the transistor has practically zero current because the emitter does not emit charge carriers into the base and no charge carriers are collected by the collector except a few thermally generated minority carriers. Thus, the transistor acts like an open switch.

4. Reverse-Forward Biasing. In this biasing, the emitter junction is reverse biased while the collector junction is forward biased. As the collector is not doped to the extent as the emitter is doped, therefore, it cannot inject the majority charge carriers to the base. So, in this region, the action of transistor is poor. This region is of little importance.

The *active mode* which is also called forward active mode, is the one used if the transistor is to operate as an amplifier. Switching operations (*e.g.*, logic circuits) utilize both the *cutoff mode* and the *saturation mode*. The reverse active (or inverse active) mode has very limited application but is conceptually significant.

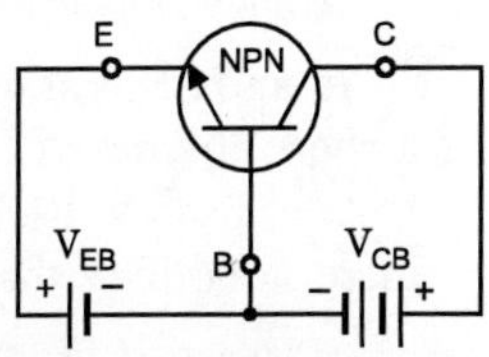

Fig. 5.9. ***Cutoff Region***

5.5. IMPORTANT POINTS REGARDING WORKING OF TRANSISTORS

The noteworthy points regarding working of transistors are summarized below.

1. Current conduction in N-P-N transistor is by electrons and the conventional current flow will be in the opposite direction. Current conduction in P-N-P transistor is by holes but in external leads, the current conduction will be by flow of electrons.
2. The collector current I_C is always less than emitter current I_E, being the difference of emitter current I_E and base current I_B.
3. The base current is only a small fraction (usually 5%) of emitter current.
4. As a standard convention, all the currents entering into the transistor are considered to be positive. Conversely currents that flow out of the transistor are taken to be negative. It means that if the actual conventional current flows in the outward direction, a negative sign is included along with its magnitude. Thus in an N-P-N transistor, the emitter current I_E is to be taken negative because it flows out of the transistor [Figs. 5.3 (*b*) and 5.7 (*a*)] while both the base current I_B and collector current I_C are to be taken positive because they flow into the transistor. Similarly in a P-N-P transistor I_E is to be taken positive and both base current I_B and collector current I_C are to be taken negative [Figs. 5.5 and 5.7(*b*)]. However, to avoid confusion, the actual direction of flow of current is indicated in the diagrams.
5. Emitter-base junction is always forward biased and collector-base junction is always reverse biased for active region of operation.
6. The input circuit (*i.e.* emitter-base junction), because of forward bias, offers low resistance and so needs usually very small bias (approximately 0.7 V for Si and 0.3 V for Ge). The output circuit (*i.e.*

collector-base junction), because of reverse bias, offers high resistance and, therefore, needs much higher bias (3 to 20 V).

7. Transistor transfers the input signal from a low resistance circuit to a high resistance circuit, therefore, it is called the TRANsfer reSISTOR (TRANSISTOR).
8. Since both of the charge carriers (holes as well as electrons) are involved in current flow through a transistor (may be either P-N-P or N-P-N), these devices are sometimes called the *bipolar junction transistors* (BJTs).
9. Although, it is said that a transistor is a device with two PN diodes connected back to back, but it does not mean that two discrete diodes connected back to back, as shown in Fig. 5.10, can work as a transistor. In such a case, each diode has two equally doped regions, so that the overall circuit has four equally doped regions. This would not work as base region is not the same as in a transistor.

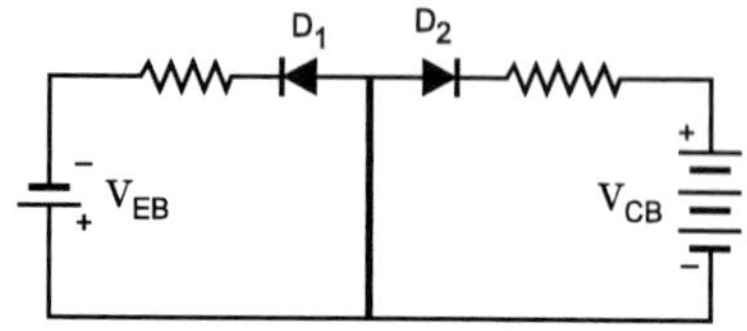

Fig. 5.10. ***Two Diodes Connected Back to Back***

The key to the transistor action is the lightly doped thin base between the heavily doped emitter and moderately doped collector.

In an N-P-N transistor, the free electrons passing through the base to the collector region have a short lifetime. As long as the base is thin, the free electrons can reach the collector. But in case of two discrete back to back connected diodes there are four doped regions instead of three and there is nothing that resembles a thin base region between an emitter and a collector.

Hence two discrete diodes connected back to back can never work as a transistor.

10. The choice of N-P-N transistor is made more often because majority charge carriers are electrons whose mobility is much more (almost double) than that of holes.

5.6. TRANSISTOR CURRENT COMPONENTS

The various current components which flow across the forward-biased emitter junction and reverse biased collector junction are indicated in Fig. 5.11. The current flowing into the emitter is referred to as the *emitter current* and designated as I_E. The emitter current I_E consists of two constituents (hole current I_{hE} due to crossing of holes from emitter into the base and electron current I_{eE} due to crossing of electrons from base into the emitter).

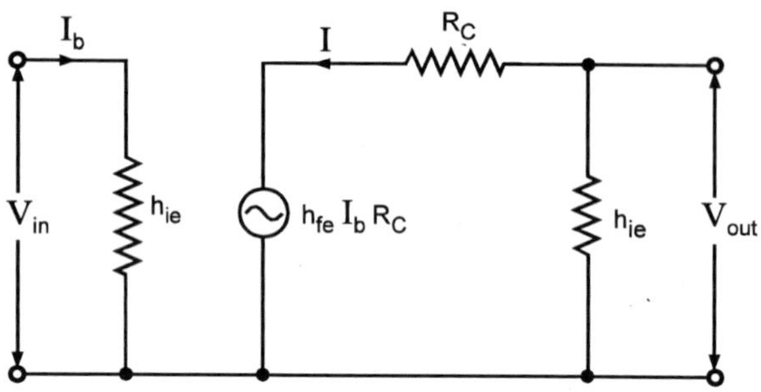

Fig. 5.11. ***Current Components in a PNP Transistor For a Forward Biased Emitter Junction and a Reverse Biased Collector Junction***

i.e. $I_E = I_{hE} + I_{eE}$ ...(5.1)

In a commercial transistor the emitter is heavily doped and the base is very lightly doped and, therefore, electron current I_{eE} is negligibly small in comparison to hole current I_{hE}. Thus the entire emitter current I_E in a P-N-P transistor is due to flow of holes from emitter to the base. It is to be noted that the indicated direction of I_E external to the transistor is the conventional current direction.

All the holes crossing the emitter junction J_E do not reach the collector junction J_C because some of them combine with the electrons in the N-type base. If I_{hC} is the hole current at collector junction J_C, there must be a bulk recombination current I_{hE} less I_{hC} leaving the base, as shown in Fig. 5.11. In fact electrons enter the base region through the base lead to replenish those electrons which have been lost by recombination with the holes injected into the base across the emitter junction J_E. The holes on reaching the collector junction J_C cross this junction and enter the collector P-region.

If the emitter were open circuited, I_E would have been zero and then I_{hC} also would have been zero. Under these conditions, the base and collector would act as a reverse biased diode and the collector current I_C would be equal to that reverse saturation current I_{CBO} (or simply I_{CO}).

When the emitter circuit is closed *i.e.* when $I_E \neq 0$, then from Fig. 5.11

$$\text{Collector current, } I_C = I_{hC} + I_{CO} \quad ...(5.2)$$

For a P-N-P transistor, I_{CO} consists of holes moving across junction J_C from base region to collector region and electrons crossing junction J_C in the opposite direction.

$$\text{Thus } I_{CO} = I_{hCO} + I_{eCO} \quad ...(5.3)$$

In the active region operation with emitter junction J_E forward biased, the collector current is given as

$$I_C = \alpha I_E + I_{CO} \quad ...(5.4)$$

where α is the fraction of the total current I_E which constitutes I_{hC}.

It is to be noted that the indicated direction of emitter current I_E external to the transistor is the conventional current direction. Base current I_B and collector current I_C are also shown external to the transistor conventional current direction. According to the Kirchhoff's first law total current flowing into the transistor must be equal to the total current flowing out of it. So emitter current I_E is equal to the sum of collector and base currents I_C and I_B.

$$\text{i.e. } I_E = I_C + I_B \quad ...(5.5)^*$$

Let us now define different parameters which relate the current components discussed above.

Emitter Efficiency. The emitter or injection efficiency denoted by γ is defined as the ratio of current of injected carriers at J_E to the total emitter current

$$\text{i.e. } \gamma = \frac{\text{Current of injected carriers at } J_E}{\text{Total emitter current}} = \frac{I_{hE}}{I_{hE} + I_{eE}} = \frac{I_{hE}}{I_E} \quad ...(5.6)$$

where I_{hE} is the injected hole diffusion current at emitter junction and I_{eE} is the injected electron diffusion current at emitter junction.

Transport Factor. Tansport factor, denoted by β, is defined as the ratio of injected carrier current reaching J_C to the injected carrier current at J_E

$$\text{i.e. } \beta = \frac{\text{Injected carrier current reaching } J_C}{\text{Injected carrier current at } J_E} = \frac{I_{hC}}{I_{hE}} \quad ...(5.7)$$

Large Signal Current Gain α. The term α has already been defined above. However, α can alternatively be defined from Eq. (5.4). From Eq. (5.4) α may be defined as the ratio of collector current increment from cutoff value ($I_C = I_{CO}$) to I_C and the emitter current increment from cutoff value ($I_E = 0$) to I_E. Thus, we may write

$$\alpha = \frac{I_C - I_{CO}}{I_E - 0} = \frac{I_C - I_{CO}}{I_E} \quad ...(5.8)$$

Term α is called the *large signal current gain* of a common base transistor. The parameter α is extremely important in transistor theory. It is to be pointed here that α is not constant, but varies with emitter current I_E, collector-base voltage V_{CB}, and the temperature. Typically α lies in the range 0.95 to 0.99.

* Some authors taking standard transistor convention, write this equation as $I_E = -(I_C + I_B)$

From Eqs. (5.2) and (5.8) we have

$$\alpha = \frac{I_{hC}}{I_E} = \frac{I_{hC}}{I_{hE}} \times \frac{I_{hE}}{I_E} = \beta\gamma \quad ...(5.9) \quad \because \text{From Eq. (5.7) } \frac{I_{hC}}{I_{hE}} = \beta \text{ and from Eq. (5.6) } \frac{I_{hE}}{I_E} = \gamma$$

For an efficient transistor we would like β and γ to be very near unity *i.e.* the emitter current should be due mostly to holes ($\gamma = 1$) and most of the injected holes should eventually participate in the collector current ($\beta \simeq 1$).

The product βγ is defined as the factor α, called the *current gain ratio,* which represents the emitter-to-collector current amplification. There is no real amplification between these currents, since α is lesser than unity. On the other hand, the relation between collector current and base current is more promising for amplification.

DC Current Gain. If I_{CO} is negligibly small in comparison to I_C, α approximately equals I_C/I_E. This is referred to as the *dc current gain* of the common base (CB) transistor and is denoted by α_{dc}

Thus $$\alpha_{dc} = \frac{I_C}{I_E} \quad ...(5.10)$$

α_{dc} is always positive and less than unity.

Small Signal Current Gain α_{ac}. Small signal current gain α_{ac} is defined as the ratio of change in collector current to change in emitter current

$$\text{i.e. } \alpha_{ac} = \frac{\Delta I_C}{\Delta I_E} \text{ for constant collector-base voltage } V_{CB} \quad ...(5.11)$$

α_{ac} is always positive but less than unity (very close to unity). It also varies with I_E, V_{CB} and temperature.

Generalized Expression For Collector Current. We know that a small reverse current flows across a P-N junction diode due to *thermally generated minority carriers* that are propelled by the barrier potential. When the junction is reverse biased, this reverse current increases slightly. For moderate reverse-bias voltages, the reverse current attains its saturation value, I_s (or I_0). Because the collector-base junction of a transistor is reverse biased, there is likewise a reverse current due to thermally generated carriers. Of course this "reverse" current, in the context of a transistor, is in the same direction as the main (collector) current flowing through the device due to the injection of minority carriers into the base. The total collector current is, therefore, the sum of these two components: the injected minority carriers and the thermally generated minority carriers

$$\text{i.e. } I_C = \alpha I_E + I_{CO}$$

The reverse current I_{CBO} is given by the volt-ampere relationship of Eq. (2.20) with I_0 replaced by $-I_{CO}$ and V by V_C where symbol V_C represents the voltage drop across junction J_C from the P-side to the N-side and the complete expression for collector current I_C may be written as

$$I_C = \alpha I_E + I_{CO}(1 - e^{V_C/\eta V_T}) \quad ...(5.12)$$

If V_C is negative and is of large magnitude in comparison with that of V_T, Eq. (5.12) reduces to Eq. (5.4).

The physical interpretation of Eq. (5.12) is that the P-N junction diode current crossing the collector junction J_C is supplemented by a fraction α of the current I_E flowing in the emitter.

Example 5.1. For a transistor if I_E = 1 mA and α = 0.98, determine the value of I_B.

Solution : Emitter current, I_E = 1 mA

Current amplification factor, α = 0.98

Collector current, $I_C = \alpha I_E = 0.98 \times 1$ mA = 0.98 mA

Base current, $I_B = I_E - I_C = 1.0 - 0.98 = 0.02$ mA **Ans.**

5.7. CURRENT AMPLIFICATION FACTORS

There are two current amplification factors, the alpha factor (α) and the beta factor (β) defined below :

The alpha factor (α) is also called the *current amplification factor* and is the ratio of collector current (I_C) to emitter current (I_E).

$$\text{In general, } \alpha = \frac{\Delta I_C}{\Delta I_E} \; V_{CB} \text{ remaining constant} \qquad ...(5.13)$$

The beta factor (β) is the current gain factor (*also called the transport factor*) of a common emitter circuit and is defined as the ratio of collector current (I_C) and base current (I_B).

$$\text{In general, } \beta = \frac{\Delta I_C}{\Delta I_B} \qquad ...(5.14)$$

5.8. RELATIONSHIP BETWEEN α AND β

$$\text{As defined above } \beta = \frac{\Delta I_C}{\Delta I_B}$$

$$\text{and } \alpha = \frac{\Delta I_C}{\Delta I_E} = \frac{\Delta I_C}{\Delta I_C + \Delta I_B} \qquad \because \Delta I_E = \Delta I_B + \Delta I_C$$

$$\text{or } \frac{1}{\alpha} = \frac{\Delta I_C + \Delta I_B}{\Delta I_C} = 1 + \frac{\Delta I_B}{\Delta I_C} = 1 + \frac{1}{\beta} = \frac{1+\beta}{\beta}$$

$$\text{Hence } \alpha = \frac{\beta}{1+\beta} \quad \text{and} \quad \beta = \frac{\alpha}{1-\alpha} \qquad ...(5.15)$$

$$\text{Thus, if } \alpha = 0.98 \qquad \beta = \frac{0.98}{0.02} = 49$$

$$\text{and if } \alpha = 0.99 \qquad \beta = \frac{0.99}{0.01} = 99$$

i.e. a small variation in α corresponds to a large variation in β. It is, therefore, better to determine β experimentally and calculate therefrom the corresponding value of α by means of expression

$$\alpha = \frac{\beta}{1+\beta} \qquad ...(5.16)$$

Example 5.2. In a common base configuration of a PNP transistor the current amplification factor is 0.988. Determine the base current if emitter current is 1.2 mA. Neglect collector leakage current.

Solution :

Current amplification factor, $\alpha = 0.988$

Emitter current, $I_E = 1.2$ mA

Collector leakage current, $I_{CO} \simeq 0$

From Eq. (5.4) Collector current, $I_C = \alpha I_E + I_{CO} = 0.988 \times 1.2 + 0 = 1.1856$ mA

Base current, $I_B = I_E - I_C = 1.2 - 1.1856 = 0.0144$ mA $= 14.4\ \mu$A **Ans.**

Example 5.3. For a given transistor $\alpha_{dc} = 0.98$, $I_E = 2$ mA. Calculate β_{dc} and I_B.

Solution :

Emitter current, $I_E = 2$ mA

$\alpha_{dc} = 0.98$

$$\beta_{dc} = \frac{\alpha_{dc}}{1-\alpha_{dc}} = \frac{0.98}{1-0.98} = 49 \text{ Ans.}$$

$$\text{and base current, } I_B = \frac{I_C}{\beta_{dc}} = \frac{\alpha_{dc} I_E}{\beta_{dc}} = \frac{0.98 \times 2}{49} = 0.04 \text{ mA or } 40\ \mu\text{A Ans.}$$

5.9. BASE SPREADING RESISTANCE

With two depletion layers penetrating the base, the base holes are confined to the thin region of P-type semiconductor illustrated in Fig. 5.12. The resistance of this thin region (or section) is called the *base spreading resistance* $r_{bb'}$. With the increase in reverse bias V_{CB} on the collector diode, the width of the P-section is reduced, which is equivalent to an increase in $r_{bb'}$.

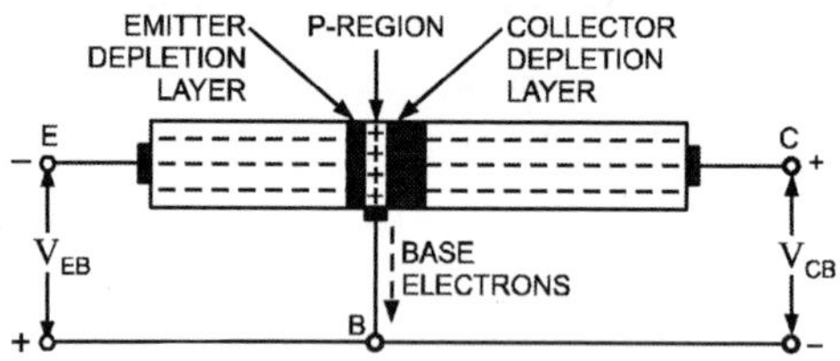

Fig. 5.12. ***Base Spreading Resistance***

The recombination current in the base I_B must flow through $r_{bb'}$. When it does so it causes a voltage drop. The value of base spreading resistance $r_{bb'}$ depends upon the width of the P-section as well as doping of the base. In rare case $r_{bb'}$ may be as high as 1,000 Ω. Typically, it is in the range of 50 – 150 Ω. The effects of $r_{bb'}$ are important in high-frequency circuits. At low frequencies $r_{bb'}$ usually has little effect and so it is ignored.

5.10. EBERS-MOLL MODEL

Device modelling aims at relating physical device parameters to device terminal characteristics. Device modelling is especially important for integrated circuits, since simple and accurate device models are required to perdict the performance of a circuit. Generally, by making a model more accurate, we make it more complex. Thus, there is a trade-off between accuracy and complexity.

Ebers-Moll model is a transistor model which describes the operating states of a transistor. This model is useful to obtain information about the dc characteristics of a transistor. The transistor operates in active region when emitter junction is forward biased and collector junction is reverse biased. This model generalizes the behaviour of a transistor by taking into account the inverted mode of operation of transistor.

The dependence of the currents in a transistor upon the junction voltages, or vice versa may be obtained by starting with Eq. (5.12), reproduced here

$$i.e. \quad I_C = -\alpha_N I_E + I_{CO}\left[1 - e^{V_C/V_T}\right]$$

$$\text{or} \quad I_C = -\alpha_N I_E - I_{CO}\left[e^{V_C/V_T} - 1\right] \qquad ...(5.17)$$

Subscript N to α has been added in order to indicate the normal operation of the transistor. Minus sign has been introduced with $\alpha_N I_E$ to take into account its direction.

In inverted mode of operation, equation corresponding to Eq. (5.17) may be written as

$$I_E = -\alpha_I I_C - I_{EO}\left[e^{V_E/V_T} - 1\right] \qquad ...(5.18)$$

Here α_I is the inverted common-base current gain, just as α_N in Eq. (5.17) is the current gain in normal operation. I_{EO} is the emitter-junction reverse saturation current, and V_E is the voltage drop from P-side to N-side at the emitter junction and is positive for a forward-biased emitter.

Ebers-Moll model interpreting Eqs. (5.17) and (5.18) in terms of a circuit is given in Fig. 5.13 for a P-N-P transistor. It involves two ideal diodes connected back to back with reverse saturation currents $-I_{EO}$ and $-I_{CO}$ and two dependent current-controlled current sources shunting the ideal diodes. The current sources account for the minority-carrier transport across the base. By applying KCL to the collector node (Fig. 5.13), we have

$$\text{Collector current, } I_C = -\alpha_N I_E + I = -\alpha_N I_E + I_0 (e^{V_C/V_T} - 1)$$

where I is the diode current. As I_0 is the magnitude of reverse saturation current, then $I_0 = -I_{CO}$. Substituting $I_0 = I_{CO}$ in above equation we have

Collector current, $I_C = -\alpha_N I_E - I_{CO}(e^{V_C/V_T} - 1)$

which is nothing but the general expression for collector current of a transistor given in Eq. (5.17). Thus, this model is valid for both forward- and reverse-biased static voltages applied across the transistor junctions.

Here, base spreading resistance has been omitted and the difference between I_{CBO} and I_{CO} have also been neglected.

From Fig. 5.13 it is evident that dependent current sources can be eliminated from the figure provided $\alpha_N = \alpha_I = 0$. For example, by making the width of the base much larger than the diffusion length of the minority carriers in the base, all the minority carriers will recombine in the base and none will survive to reach the collector. In such a condition transport factor β and hence also current gain factor α will be zero. Thus under these conditions, the transistor action ceases, and we simply have two diodes placed back to back.

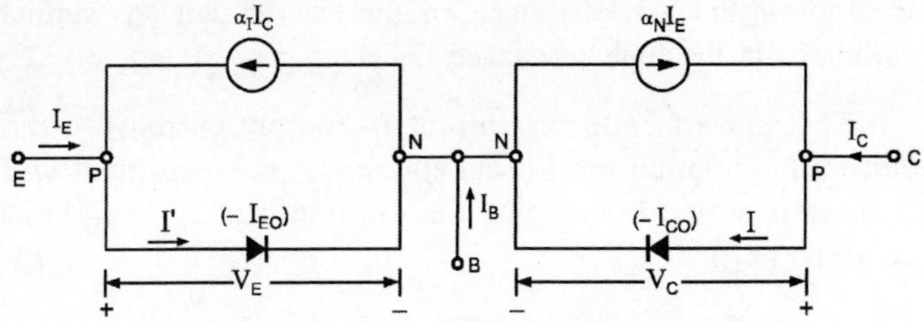

Fig. 5.13. ***Ebers-Moll Model For a PNP Transistor***

From the above discussion it is obvious that a transistor cannot be constructed by simply connecting two separate (isolated) diodes back to back.

5.11. TRANSISTOR CIRCUIT CONFIGURATIONS

As already mentioned, a transistor is a three-terminal device (having three terminals namely emitter, base and collector) but we require four terminals—two for the input and two for the output for connecting it in a circuit. Hence one of the terminals of the transistor is made common to the input and output circuits. Thus there are three types of configurations for operation of a transistor. These configurations are *common base* (CB), *common emitter* (CE) and *common collector* (CC) configurations. The term 'common' is used to denote the electrode that is common to input and output circuits. Since the common terminal is usually grounded, these configurations are also referred to as *grounded base*, *grounded emitter* and *grounded collector* configurations, as shown in Figs. 5.14 (*a*), 5.14 (*b*) and 5.14 (*c*) respectively. Each configuration has its own merits and demerits. It is to be remembered that regardless of the circuit configurations, the emitter is always forward biased and the collector is always reverse biased.

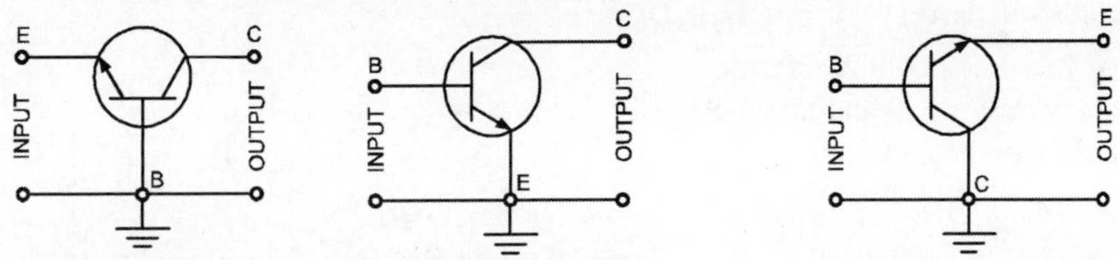

(*a*) *Common Base Configuration* (*b*) *Common Emitter Configuration* (*c*) *Common Collector Configuration*

Fig. 5.14. ***Different Circuit Configurations For NPN Transistor***

5.12. COMMON BASE (CB) CONFIGURATION

In common base (CB) configuration, input is connected between emitter and base and output is taken across collector and base, as illustrated in Fig. 5.15 for an N-P-N transistor. Thus base is common between input and output circuits.

In this arrangement (Fig. 5.15) the emitter-base junction is forward biased and collector-base junction reverse biased. The emitter current I_E flows in the input circuit and the collector current I_C in the output circuit. A change in emitter current produces a similar change in the collector current due to change in diffusion of electrons or holes across the junctions. The impedances of these two circuits are different and thus some voltage and power gains can be achieved since a current variation in the low resistance emitter-base circuit has similar current variations in the high resistance collector-base circuit.

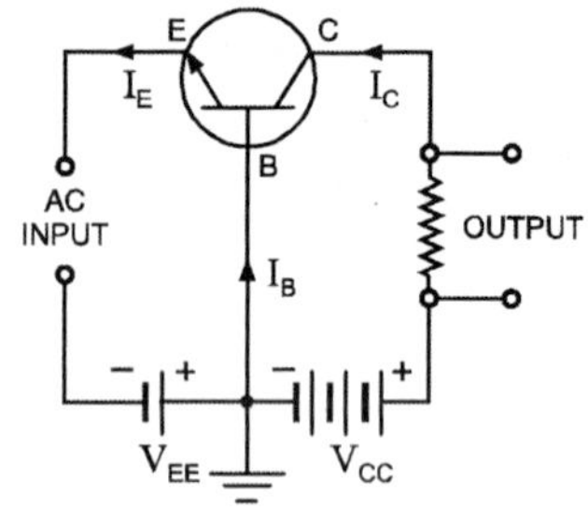

Fig. 5.15. ***Common Base NPN Transistor***

The ratio of collector current I_C (output current) to the emitter current I_E (input current) is called the *current amplification* factor α. If there is no input ac signal then the ratio of I_C to I_E is called the dc alpha (α_{dc}).

i.e. $\alpha_{dc} = \frac{-I_C}{I_E}$ (negative sign signifies that I_E flows out of transistor while I_C into transistor).*

If we write α_{dc} simply by α and ignore –ve sign,

$$\alpha = \frac{I_C}{I_E} \quad ...(5.19)$$

Incidently, there is also an ac alpha (α_{ac}) for a transistor and it refers to the ratio of change in collector current to change in emitter current

i.e. $\alpha_{ac} = \frac{-\Delta I_C}{\Delta I_E}$ for constant collector-base voltage V_{CB}. ...(5.20)

α** of a transistor is a measure of its quality–higher the value of α better is the transistor in the sense that I_C approaches I_E, its value ranges from 0.95 to 0.99. The value of α can be increased by making base thin and lightly doped.

The collector current consists of two parts (*i*) the current produced by normal transistor action *i.e.* component depending upon emitter current (α I_E) which is produced by majority carriers and (*ii*) the leakage current due to movement of minority carriers across base-collector junction on account of it being reverse biased. This leakage current flows in the same direction as that of I_C and is abbreviated as I_{CBO}* (collector-base current with emitter open).

∴ Total collector current, $I_C = \alpha I_E + I_{CBO}$ ...(5.21)

The current I_{CBO}, sometimes referred to as I_{CO}, is usually very small and may be neglected in transistor calculations.

From Eq. (5.21)

$$\alpha = \frac{I_C - I_{CO}}{I_E} \simeq \frac{I_C}{I_E} \quad ...(5.22)$$

neglecting I_{CO} in comparison with I_C

$$I_B = I_E - I_C = I_E - (\alpha I_E + I_{CO})$$
$$= (I - \alpha) I_E - I_{CO} \quad ...(5.23)$$

∵ from Eq. (5.21) $I_C = \alpha I_E + I_{CO}$

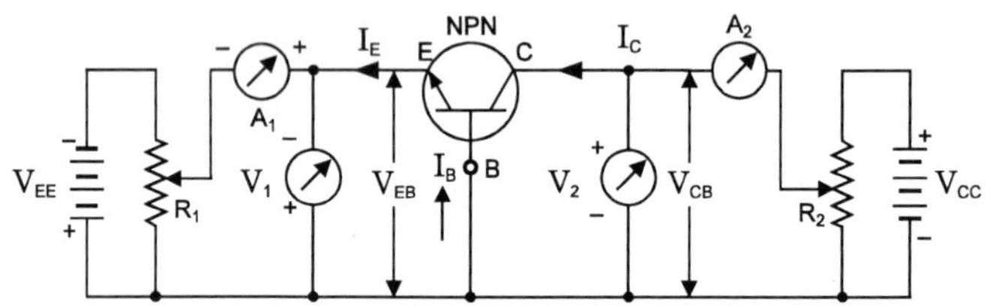

Fig. 5.16. ***Circuit Arrangement For Determination of Input and Output Characteristic Curves of a Transistor in CB Configuration***

* In case of a P-N-P transistor emitter current I_E flows into the transistor and collector current I_C flows out of it.

** For all practical purposes α_{dc}, α_{ac} and α have the same values.

5.12.1. Characteristics of Common Base Configuration

The performance of transistors, when connected in a circuit, may be determined from their characteristic curves that relate different dc currents and voltages of a transistor. Such curves are known as *static characteristic curves.* There are two important characteristics of a transistor viz. input characteristics and output characteristics.

Both of the input and output characteristics can be determined with the same circuit arrangement shown in Fig. 5.16.

Input Characteristics. The curve drawn between emitter current I_E and emitter-base voltage V_{EB} for a given value of collector-base voltage V_{CB} is known as *input characteristic.* For determination of the input characteristics, the output voltage V_{CB} is maintained constant, and the input voltage V_{EB} is set at several convenient levels. For each level of input voltage, the input current I_E is noted. Then the curve is plotted between emitter current I_E and emitter base voltage V_{EB}, as shown in Fig. 5.17.

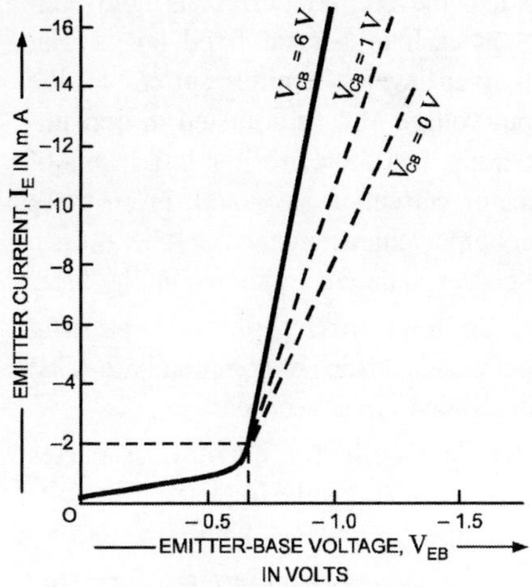

Fig. 5.17. ***Input Characteristics For Common Base NPN Transistor***

For a given value of V_{CB} the curve is just like that of a forward biased P-N junction because emitter-base junction in forward bias mode is similar to a P-N junction. This characteristic may be used to determine input resistance of the transistor.

From Fig. 5.17 it is seen that with the increase in V_{CB}, it conducts better, although the effect is not very significant. This is because large collector-base (reverse bias) voltages cause the depletion layer at the collector-base junction to penetrate deeper into the base of the transistor, thus reducing the distance and the resistance between the emitter-base and collector-base regions.

There exists a *cutin*, *offset* or *threshold* voltage V_γ below which the emitter current is very small. Substantial emitter current flows only when emitter-base junction is forward biased beyond the knee of the characteristic (approximately 0.7 V for Si and 0.3 V for Ge).

The emitter current I_E increases rapidly with the small increase in V_{EB} which means low dynamic input resistance because input resistance of a transistor is given as the reciprocal of the slope of its input characteristic *i.e.* mathematically

$$r_{in} = \frac{\Delta V_{EB}}{\Delta I_E} \quad ...V_{CB} \text{ constant} \qquad ...(5.24)$$

Since a small increase in V_{EB} (ΔV_{EB}) causes large change in I_E (ΔI_E), r_{in} is quite low.

The value of dynamic input resistance r_{in} varies from point to point in the initial part of the characteristic because it is nonlinear. Its value over the linear part of the characteristic is about 100 Ω but for lower values of V_{EB}, it is considerably larger. This dependence of r_{in} on V_{EB} causes distortion of the signals handled by the transistors.

Output Characteristics. The curve drawn between collector current I_C and collector-base voltage V_{CB} for a given value of emitter current I_E is known as *output characteristic.*

---

* I_{CBO} is exactly like the reverse saturation current I_S or I_0 of a reverse biased diode discussed earlier and is extremely temperature dependent as it is made up of thermally generated minority carriers.

For determination of output characteristics, the emitter current is held constant at each of several fixed levels. For each fixed level of emitter current I_E, the output voltage V_{CB} is adjusted in convenient steps, and the corresponding levels of collector current I_C are noted. From these data output characteristics can be drawn. The curves so drawn are shown in Fig. 5.18.

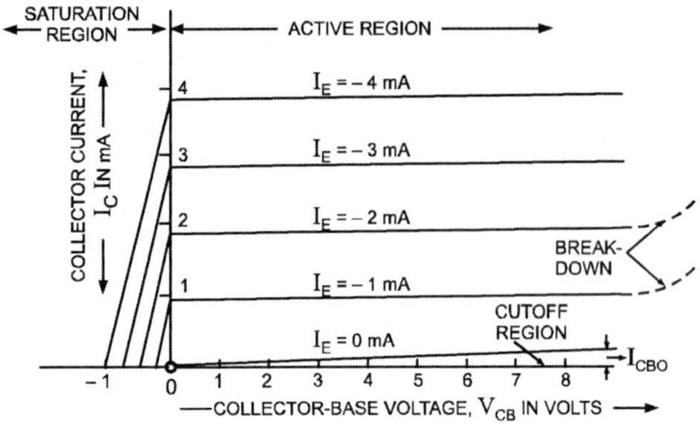

Fig. 5.18. ***Output Characteristics For Common Base NPN Transistor***

The noteworthy points regarding output characteristics of common base (CB) configuration are given below :

(*i*) The collector current I_C varies with V_{CB} only for very low voltage (below 1 V) but transistor is never operated in this region.

(*ii*) In active region (emitter forward biased and collector reverse biased) collector current I_C is almost equal to I_E and appears to remain constant when V_{CB} is increased. In fact, there is very small increase in I_C with increase in V_{CB}. This is because the increase in V_{CB} expands the collector-base depletion region and thus shortens the distance between the two depletion regions. With emitter current I_E held constant, however, the increase in I_C is so small that it is usually noticeable only for large variations in V_{CB}. I_C is slightly lesser than I_E in magnitude.

Transistor is normally operated in active region.

(*iii*) Although the collector current I_C is practically independent of V_{CB} over the transistor operating range, however, if V_{CB} is increased beyond a certain value, I_C eventually increases rapidly because of avalanche or zener (or both) effects. This condition is known as *punch-through* or *reach-through*. When it occurs large currents can flow, possibly destroying the device. The extension of the depletion region is, of course, the direct consequence of the increase in V_{CB}. Thus it is very essential to maintain V_{CB} below the maximum safe limit specified by the manufacturer of the device.

(*iv*) A very large change in collector voltage causes a very small change in collector current *i.e.* output resistance of CB configuration is very high (of the order of few hundred kΩ)—the dynamic output resistance r_{out} being given as the ratio of ΔV_{CB} and ΔI_C for a given value of I_E.

(*v*) In *cutoff* region (emitter and collector junctions both reverse biased) small collector current I_C flows even when emitter current $I_E = 0$. This is the collector leakage current I_{CBO} or I_{CO}.

(*vi*) In saturation region (both emitter and collector junctions forward biased) collector current I_C flows even when $V_{CB} \simeq 0$. Even when the externally applied bias voltage is reduced to zero, there is still a barrier potential existing at the collector-base junction, and this assists in the flow of I_C. To stop it the collector-base junction has to be forward biased. Consequently, as illustrated in Fig. 5.18, collector current I_C is reduced to zero when V_{CB} is increased negatively.

We have already determined that a small signal voltage impressed in the low-resistance input (emitter) circuit of a transistor causes a relatively large emitter current. Almost the same amount of current will flow in the high-resistance output (collector) circuit of transistor, where the voltage may be very high. Evidently, then, both the output voltage and power can be quite large, as compared to the tiny input voltage and power present at the emitter.

CB configuration is rarely used in audio-frequency (AF) circuits because its current gain is less than unity and its input and output resistances are quite different.

Current Gain Characteristics. The current gain characteristics, also referred to as *forward transfer characteristics,* can be determined experimentally (by keeping V_{CB} constant at a convenient value, recording I_C value measured for different settings of I_E and then plotting I_C versus I_E) or from the common-base output characteristics as shown in Fig. 5.19. A vertical line is drawn through a selected value of V_{CB} and corresponding values of I_E and I_C are read off along this line. Then a graph is plotted between I_C versus I_E, and the characteristic is labelled with the V_{CB} used.

Note that for I_E = 2 mA, this line (representing I_C) is always less than 2 mA. Similarly, for all other values of I_E, $I_C < I_E$.

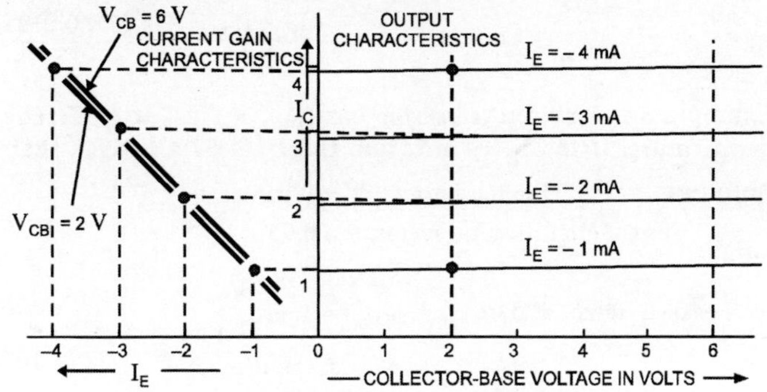

Fig. 5.19. ***Current Gain Characteristics For Common Base NPN Transistor***

Example 5.4. A transistor is connnected in CB configuration. When the emitter voltage is changed by 200 mV, the emitter current changes by 5 mA. During this variation, the collector-to-base voltage is kept fixed. Calculate the dynamic input resistance of transistor.

Solution : The dynamic input resistance for CB configuration is given as

$$r_{in} = \left.\frac{\Delta V_{EB}}{\Delta I_E}\right|_{V_{CB} = \text{ constant}} = \frac{200 \text{ mV}}{5 \text{ mA}} = 40 \ \Omega \ \textbf{Ans.}$$

Example 5.5. In a grounded base configuration (Fig. 5.15) the voltage drop across load resistance of 4 kΩ is 3 V. Determine base current. α = 0.96.

Solution : Load resistance $R_L = 4 \text{ k}\Omega$

Voltage drop across load resistance, $I_C R_L = 3$ V

$$\text{Collector current, } I_C = \frac{I_C R_L}{R_L} = \frac{3 \text{ V}}{4 \text{k}\Omega} = 0.75 \text{ mA}$$

$$\text{Emitter current, } I_E = \frac{I_C}{\alpha} = \frac{0.75}{0.96} = 0.78 \text{ mA}$$

$$\text{Base current, } I_B = I_E - I_C = 0.78 - 0.75 = 0.03 \text{ mA } \textbf{Ans.}$$

Example 5.6. In a common base configuration, the emitter current is 1.6 mA. The collector current with emitter circuit open is 10 μA. Determine the total collector current.

Solution : Emitter current, I_E = 1.6 mA

Collector current with emitter circuit open, I_{CO} = 10 μA = 0.01 mA

Assume α = 0.98

Collector current, $I_C = \alpha I_E + I_{CO}$ Refer to Eq. (5.4)

$= 0.98 \times 1.6 + 0.01 = 1.578$ mA **Ans.**

Example 5.7. The reverse saturation current of an N-P-N transistor in common base circuit is 12.5 μA. For an emitter current of 2 mA, collector current is 1.97 mA. Determine the current gain and base current.

Solution : Emitter current, $I_E = 2$ mA

Collector current, $I_C = 1.97$ mA

Base current, $I_B = I_E - I_C = 2 - 1.97 = 0.03$ mA **Ans.**

Reverse saturation current, $I_{CO} = 12.5\ \mu A = 0.0125$ mA

$$\text{Current gain, } \alpha = \frac{I_C - I_{CO}}{I_E} = \frac{1.97 - 0.0125}{2} = 0.979 \quad \textbf{Ans.}$$

Example 5.8. A silicon transistor has I_{CBO} = 1 μA at 27°C. The leakage current doubles every 6°C rise in temperature. If the I_E = 1 mA, find the I_B at 57°C. Given that α = 0.95.

Solution : Emitter current, $I_E = 1$ mA

Current amplification factor, $\alpha = 0.95$

Leakage current at 57°C, $I_{CBO_{57}} = I_{CBO_{27}} \times 2^{\frac{57-27}{6}}$ $\quad \because I_{CBO}$ doubles every 6°C rise in temperature

$$= 1 \times 10^{-6} \times 2^5 = 32 \times 10^{-6} \text{ A} = 0.032 \text{ mA}$$

Collector current, $I_C = \alpha I_E + I_{CBO} = 0.95 \times 1 + 0.032 = 0.982$ mA

Base current at 57°C, $I_B = I_E - I_C = 1 - 0.982 = 0.018$ mA $= 18\ \mu A$ **Ans.**

5.13. EARLY EFFECT AND BASE-WIDTH MODULATION

In active region, the emitter-base (EB) junction is forward biased and collector-base (CB) junction is reverse biased, so the barrier width at EB junction is negligible in comparison with the space-charge width at CB junction, as illustrated in Fig. 5.20.

The transition or space-charge region at a junction is the region of uncovered charges on both sides of the junction. As the voltage applied across the junction increases, the transition region penetrates deeper into the collector and base. Because neutrality of the charge is to be maintained, the number of uncovered charges on each side remains equal. Since the doping in the base is substantially smaller than that of collector, the penetration of the transition region into the base is much larger than that into the collector region. Hence the collector depletion region is neglected and all the immobile charge is indicated in the base region.

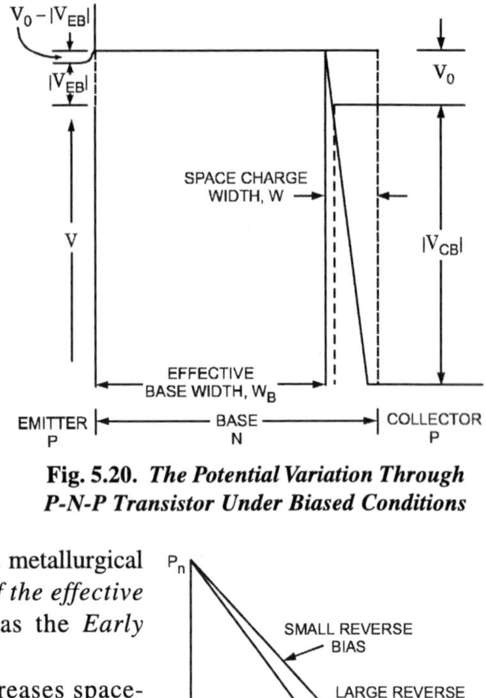

Fig. 5.20. ***The Potential Variation Through P-N-P Transistor Under Biased Conditions***

As obvious from Fig. 5.20 the effective base width is metallurgical base width less space charge width W. This *modulation of the effective base width* by the collector voltage V_{CB} is known as the *Early effect*.

An increase in magnitude of collector voltage increases space-charge width W [$\because$ from Eq. (2.50) $V_B = \dfrac{eN_D W^2}{2\epsilon}$] at the output junction diode and this, in turn, causes the effective base width W_B to decrease.

The decrease in effective base width W_B with increasing reverse bias has three consequences.

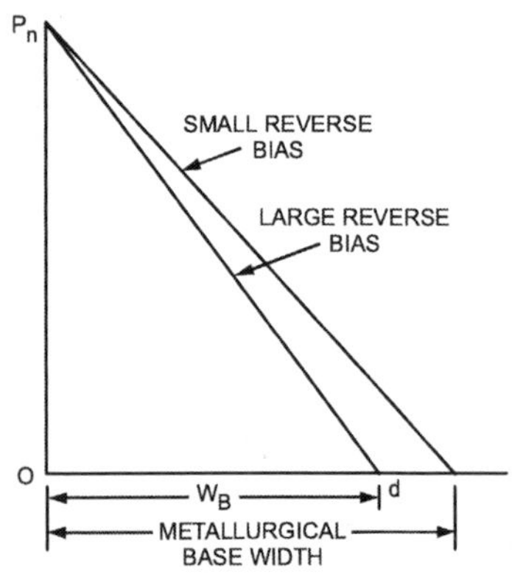

Fig. 5.21. ***Minority Carrier Charge Density Within The Base***

(*i*) There is less chance for recombination within the base region. Hence the transport factor β* and also α, increase with an increase in the magnitude of CB junction voltage.

(*ii*) The concentration gradient of minority carriers is increased within the base, and consequently, the current of minority carriers injected across the emitter junction increases, with increasing reverse collector voltage.

(*iii*) For extremely large voltages, W_B may be reduced to zero causing voltage breakdown. When effective base width W_B is reduced to zero, the emitter barrier becomes V', which is smaller than normal value $V_0 - |V_{EB}|$ because the collector voltage has reached through the base region. This lowering of the emitter-junction voltage causes an excessively large emitter current, thus placing an upper limit on the magnitude of the collector voltage. This phenomenon is known as *punch through.*

The punch-through (or reach-through) phenomenon may be defined as the mechanism by which a transistor usefulness may be terminated by increasing the collector voltage.

5.14. COMMON EMITTER (CE) CONFIGURATION

In common emitter configuration input is connected between base and emitter while the output is taken between collector and emitter, as shown in Fig. 5.22. Thus emitter is common to input and output circuits.

In this configuration, the bias voltages are applied between collector and emitter and base and emitter. Emitter-base junction is forward biased and the base is made more positive than the emitter by V_{BB}, collector-emitter junction is reverse biased and the collector is made more positive than emitter by V_{CC}. The value of V_{CC} must be greater than that of V_{BB}.

In this circuit arrangement (Fig. 5.22) the base current I_B flows in the input circuit and collector current I_C flows in the output circuit. In this arrangement current gain between the input and output sides is obtained; and since the input resistance is again less than the output resistance (though difference is not as much as in case of common base arrangement) there will be high voltage and power gains like those in equivalent vacuum tube circuits. The common emitter produces a reversal between input and output signals. Common emitter (CE) configuration is commonly used because its current, voltage and power gains are quite high and output to input impedance ratio is moderate.

The ratio of change in collector current (output current) and change in base current (input current) is called the *base current amplification factor* β

$$i.e.\ \beta = \frac{\Delta I_C}{\Delta I_B} \qquad ...(5.25)$$

Almost in all transistors, the base current is less than 5% of emitter current I_E so β is usually higher than 20. Usually β ranges from 20 to 100.

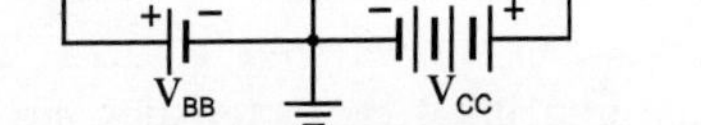

Fig. 5.22. ***Common Emitter NPN Transistor***

$$\text{Also } \beta = \frac{I_C}{I_B} = \frac{I_C}{I_E - I_C} = \frac{I_C/I_E}{1 - \frac{I_C}{I_E}} = \frac{\alpha}{1-\alpha} \qquad ...(5.26)$$

In CE configuration, a small collector current flows even when base current is zero (*i.e.* base lead is open). This is the collector *cutoff* current and denoted by I_{CEO}. I_{CEO} must be larger than I_{CBO}.

The current equations are $I_E = I_B + I_C$ ...(5.27)

and $I_C = \alpha\, I_E + I_{CBO} = \alpha\, (I_B + I_C) + I_{CBO}$

* When no signal is applied, then the ratio of I_C to I_B is called the dc beta (β_{dc}) of a transistor. With signal applied, the ratio of ΔI_C to ΔI_B is called the ac beta (β_{ac}).

$$\text{or}\quad I_C\,(1-\alpha) = \alpha\, I_B + I_{CBO}$$

$$\text{or}\quad I_C = \frac{\alpha}{1-\alpha} I_B + \frac{I_{CBO}}{1-\alpha} \qquad ...(5.28)$$

and total collector current, $I_C = \beta\, I_B + I_{CEO}$ ...(5.29)

Comparing Eqs. (5.28) and (5.29) we have

$$\beta = \frac{\alpha}{1-\alpha}$$ as already derived in Eq. (5.26)

$$\text{and}\quad I_{CEO} = \frac{1}{1-\alpha} I_{CBO} \qquad ...(5.30)$$

Substituting the value of $I_{CEO} = \frac{1}{1-\alpha} I_{CBO}$ in Eq. (5.29) we have

$$I_C = \beta I_B + \frac{1}{1-\alpha} I_{CBO} = \beta I_B + (\beta + 1)\, I_{CBO} \qquad ...(5.31)$$

$\because$ from Eq. (5.26) $\frac{1}{1-\alpha} = \beta + 1$

5.14.1. Characteristics of Common Emitter Configuration

The circuit used for determination of transistor common emitter characteristics is shown in Fig. 5.23. The input voltage is applied between base and emitter terminals, and the output is taken from collector and emitter terminals. Input and output voltages and currents are measured by voltmeters and ammeters as shown.

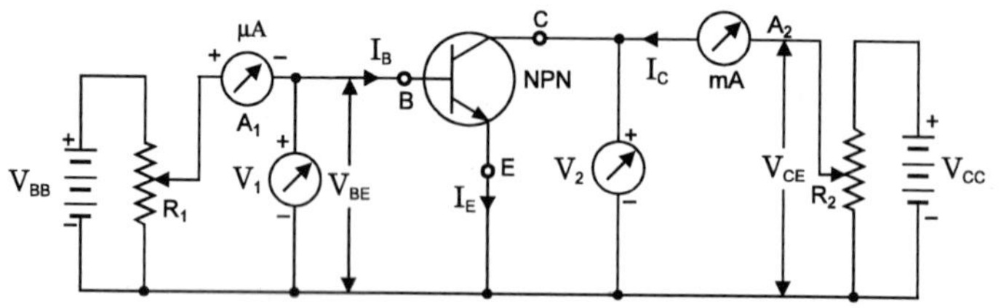

Fig. 5.23. ***Circuit Arrangement For Determination of Input and Output Characteristics of a Transistor in CE Configuration***

Input Characteristics. The curve drawn between base current I_B and base-emitter voltage V_{BE} for a given value of collector-emitter voltage V_{CE} is known as the *input characteristic*. For determination of input characteristics, collector-emitter voltage V_{CE} is held constant and base current I_B is recorded for different values of base-emitter voltage V_{BE}. Now the curves are drawn between base current I_B and base-emitter voltage V_{BE} for different levels of V_{CE}, as shown in Fig. 5.24. The noteworthy points regarding input characteristics are given below :

(*i*) The input characteristics of CE transistors are quite similar to those of a forward biased diode because the base-emitter region of the transistor is a diode and it is forward biased.

(*ii*) In comparison to common base (CB) arrangement base (or input) current increases less rapidly with the increase in base-emitter voltage, V_{BE}. This indicates that input resistance is larger in common emitter configuration than that in common base configuration. Due to initial nonlinearity of

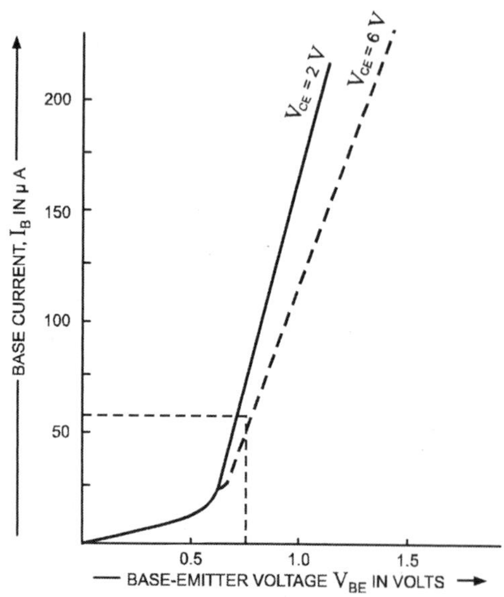

Fig. 5.24. ***Input Characteristics For Common Emitter NPN Transistor***

the curve, input resistance varies from point to point in the initial part of the characteristic. Its value over the linear part of the curve is of the order of few hundred ohms.

(*iii*) An increment in value of V_{CE} causes the input current I_B to be lower for a given level of V_{BE}. This is because the higher levels of V_{CE} provide greater collector-base junction reverse bias, causing greater depletion region penetration into the base and thus reducing the distance between the collector-base and emitter-base regions. As a result more of the charge carriers from the emitter flows across the collector-base junction, and few flow out through the base lead. Usually effect of change in V_{CE} on the input characteristic is ignored.

The ratio of change in base-emitter voltage (ΔV_{BE}) to the resulting change in base current (ΔI_B) at constant collector-emitter voltage (V_{CE}) is known as *dynamic input resistance* (r_{in}). It is calculated as the reciprocal of the slope of the input characteristic at a given V_{BE}.

$$i.e.,\ r_{in} = \left.\frac{\Delta V_{BE}}{\Delta I_B}\right|_{V_{CE}\,=\,\text{constant}} \qquad ...(5.32)$$

Output Characteristics. Output characteristic for a common emitter transistor is the curve drawn between collector current I_C and collector-emitter voltage V_{CE} for a given value of base current I_B. For determination of common emitter output characteristics, base current I_B is maintained constant at several convenient levels. At each fixed level of I_B, collector-emitter voltage V_{CE} is adjusted in steps, and the corresponding values of collector current I_C are noted. For each level of I_B, I_C is plotted versus V_{CE}. Thus a family of characteristics is obtained which are typically as illustrated in Fig. 5.25. The noteworthy points regarding output characteristics are given below :

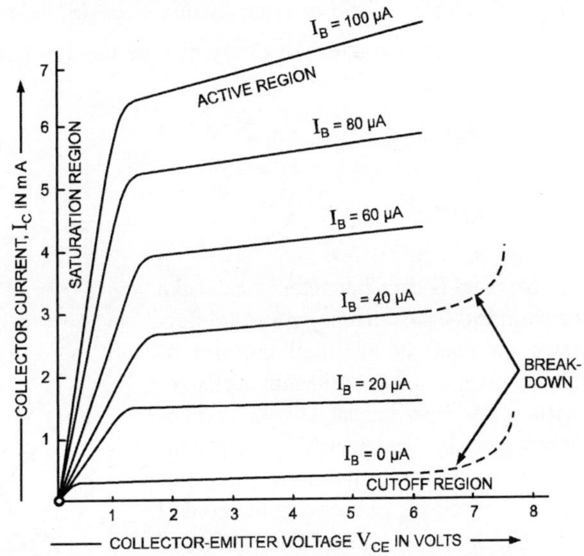

Fig. 5.25. ***Output Characteristics For Common Emitter NPN Transistor***

(*i*) The collector current I_C varies with V_{CE} for V_{CE} between 0 and 1 V and then becomes almost constant and independent of V_{CE}. The transistors are always operated above 1 V.

(*ii*) Output characteristic in CE configuration has some slope while CB configuration has almost horizontal characteristics. This indicates that output resistance in case of CE configuration is less than that in CB configuration.

(*iii*) In active region (collector junction reverse biased and emitter junction forward biased), for small values of base current I_B the effect of collector voltage V_C over I_C is small but for large values of I_B this effect increases.

The shape of the characteristic is same as in CB configuration but with the difference that collector current I_C is larger than input current (base current I_B). Thus the current gain for this configuration is larger than unity. The transistor must always be operated in the active region, when employed as an amplifying device.

(*iv*) With low values (ideally zero) of V_{CE} the transistor is said to be operated in saturation region and in this region base current I_B does not cause a corresponding change in collector current I_C.

(*v*) With much higher V_{CE}, the collector-base junction completely breaks down and because of this avalanche breakdown collector current I_C increases rapidly and the transistor gets damaged.

(*vi*) In cutoff region, small amount of collector current I_C flows even when base current $I_B = 0$. This is called I_{CEO}. Since main current I_C is zero, the transistor is said to be cutoff.

(*vii*) Moderate output to input impedance ratio makes this configuration an ideal one for coupling between various transistor stages.

The ratio of change in collector-emitter voltage (ΔV_{CE}) to the change in collector current (ΔI_C) at constant base current is known as *dynamic* ouput resistance.

It is calculated as the reciprocal of the slope of output characteristic at a given V_{CE}.

$$i.e.,\ r_{out} = \left.\frac{\Delta V_{CE}}{\Delta I_C}\right|_{I_B = \text{constant}} \quad ...(5.33)$$

The output resistance of CE configuration is less than that of CB configuration as the slope of the ouput characteristic is more in this case. Its value is of the order of 50 kΩ.

The output characteristics may also be used to determine the dc current gain β and ac current gain β_0 as follows :

$$\text{DC current gain, } \beta = \frac{I_C}{I_B} \quad ...(5.34)$$

$$\text{AC current gain, } \beta_0 = \left.\frac{\Delta I_C}{\Delta I_B}\right|_{V_{CE} = \text{constant}} \quad ...(5.35)$$

Current Gain Characteristics. Like the common-base current gain characteristics, they may be obtained experimentally, or determined from the output characteristics. For experimental determination of current gain characteristics, V_{CE} is kept constant at a convenient value and value of collector current I_C is observed and recorded for different values of base current I_B and then a graph is plotted between I_C and I_B.

The current gain characteristics can be obtained from the output characteristics as shown in Fig. 5.26, and detailed in Art. 5.12 except that I_C is plotted against I_B rather than I_C against I_E.

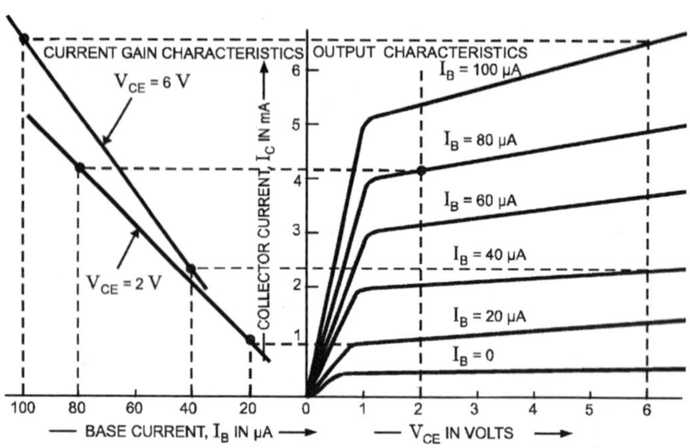

Fig. 5.26. ***Current Gain Characteristics For Common Emitter NPN Transistor***

Example 5.9. A change of 250 mV in base-emitter voltage causes a change of 1 mA in the base current. Determine the dynamic input resistance.

Solution : Change in base-emitter voltage, ΔV_{BE} = 250 mV

Change in base current, ΔI_B = 1 mA

$$\text{Dynamic input resistance, } r_{in} = \frac{\Delta V_{BE}}{\Delta I_B} = \frac{250 \times 10^{-3}}{1 \times 10^{-3}} = 250\ \Omega \ \textbf{Ans.}$$

Example 5.10. Increase in collector-emitter voltage from 5 V to 10 V causes increase in collector current from 5 mA to 5.8 mA. Determine the dynamic output resistance.

Solution : Change in collector-emitter voltage, ΔV_{CE} = 10 – 5 = 5 V

Change in collector current, ΔI_C = 5.8 – 5.0 = 0.8 mA

$$\text{Dynamic output resistance, } r_{out} = \frac{\Delta V_{CE}}{\Delta I_C} = \frac{5}{0.8 \times 10^{-3}} = 6.25\ \text{k}\Omega \ \textbf{Ans.}$$

Example 5.11. In a CE configuration, the collector supply voltage is 10 V. When a resistor $R_C = 1\ k\Omega$ is connected in the collector circuit, the voltage drop across it is 0.5 V. For $\alpha = 0.98$, determine the collector-emitter voltage and the base current.

Solution : Collector supply voltage, $V_{CC} = 10$ V

Voltage drop across load resistance $= I_C R_C = 0.5$ V

Collector-emitter voltage, $V_{CE} = V_{CC} - I_C R_C = 10 - 0.5 = 9.5$ V **Ans.**

$$\text{Current gain factor, } \beta = \frac{\alpha}{1-\alpha} = \frac{0.98}{1-0.98} = 49$$

$$\text{Collector current, } I_C = \frac{I_C R_C}{R_C} = \frac{0.5}{1\times 10^3} = 0.5\ \text{mA}$$

$$\text{Base current, } I_B = \frac{I_C}{\beta} = \frac{0.5\ \text{mA}}{49} = 10.2\ \mu\text{A}\ \textbf{Ans.}$$

Example 5.12. $I_E = 5$ mA, $I_C = 4.95$ mA, $I_{CEO} = 200\ \mu$A. Calculate β_{dc} and leakage current I_{CBO}.

Solution : Base current, $I_B = I_E - I_C = 5 - 4.95 = 0.05$ mA $= 50\ \mu$A **Ans.**

$$\text{Current gain factor, } \beta_{dc} = \frac{I_C}{I_B} = \frac{4.95}{0.05} = 99\ \textbf{Ans.}$$

$$\text{Current amplification factor, } \alpha_{dc} = \frac{\beta_{dc}}{1+\beta_{dc}} = \frac{99}{1+99} = 0.99$$

Collector-to-base leakage current, $I_{CBO} = I_{CEO}(1 - \alpha_{dc}) = 200(1 - 0.99) = 2\ \mu$A **Ans.**

Example 5.13. In an NPN transistor, $\alpha = 0.995$, $I_E = 10$ mA and $I_{CO} = 0.5\ \mu$A. Determine the values of I_C, I_B, β and I_{CEO}.

Solution : Emitter current, $I_E = 10$ mA

Collector-to-base leakage current, $I_{CO} = 0.5\ \mu\text{A} = 0.0005$ mA

Collector current, $I_C = \alpha I_E + I_{CO} = 0.995 \times 10 + 0.0005 = 9.9505$ mA **Ans.**

Base current, $I_B = I_E - I_C = 10 - 9.9505 = 0.0495$ mA $= 49.5\ \mu$A **Ans.**

$$\text{Current gain factor, } \beta = \frac{\alpha}{1-\alpha} = \frac{0.995}{1-0.995} = 199\ \textbf{Ans.}$$

$$\text{Collector current with base lead open, } I_{CEO} = \frac{I_{CBO}}{1-\alpha} = \frac{0.5\times 10^{-6}}{1-0.995} = 100\ \mu\text{A}\ \textbf{Ans.}$$ Refer to Eq. (5.30)

Example 5.14. The reverse leakage current of a transistor when connected in CB configuration is 0.2 μA and it is 18 μA when the same transistor is connected in CE configuration. Calculate α_{dc} and β_{dc} of the transistor. Assume $I_B = 30$ mA. *[G.G.S.I.P. Univ. Analog Electronics December 2009]*

Solution : $I_{CBO} = 0.2\ \mu$A

$I_{CEO} = 18\ \mu$A

$$\alpha_{dc} = \frac{I_{CEO} - I_{CBO}}{I_{CEO}} = \frac{18-0.2}{18} = 0.989\ \textbf{Ans.}$$

$$\beta = \frac{\alpha}{1-\alpha} = \frac{0.989}{1-0.989} = 89\ \textbf{Ans.}$$

Example 5.15. The collector leakage current in a transistor is 300 μA in CE arrangement. If the transistor is now connected in CB arrangement, what will be the leakage current. Given that $\beta = 100$.

Solution : I_{CEO} = Collector leakage current in CE arrangement = 300 μA

Current gain factor, $\beta = 100$

Current amplification factor, $\alpha = \frac{\beta}{1+\beta} = \frac{100}{101} = 0.99$

Leakage current in CB arrangement $= I_{CBO} = I_{CEO}\ (1 - \alpha) = 300(1 - 0.99) = 3\ \mu A$ **Ans.**

Example 5.16. The collector and base current of an N-P-N transistor are measured as $I_C = 5$ mA, $I_B = 50\ \mu A$ and $I_{CBO} = 1\ \mu A$.

(*i*) Determine α, β and I_E

(*ii*) Determine the new level of I_B required to produce $I_C = 10$ mA.

Solution : (*i*) Given that

$$I_C = 5 \text{ mA}$$
$$I_B = 50\ \mu A$$
$$I_{CBO} = 1\ \mu A$$

Since $I_C = \beta I_B + (\beta + 1)\, I_{CBO}$

$$\text{or } \beta = \frac{I_C - I_{CBO}}{I_B + I_{CBO}} = \frac{5\times 10^{-3} - 1\times 10^{-6}}{50\times 10^{-6} + 1\times 10^{-6}} = \mathbf{98} \quad \textbf{Ans.}$$

$$\alpha = \frac{\beta}{1+\beta} = \frac{98}{99} = \mathbf{0.99} \quad \textbf{Ans.}$$

$$I_E = I_B + I_C = 50\times 10^{-6} + 5\times 10^{-3} = \mathbf{5.05\ mA} \quad \textbf{Ans.}$$

$$(ii)\ I_B = \frac{I_C - (\beta+1)\, I_{CBO}}{\beta} = \frac{10\times 10^{-3} - (98+1)\times 1\times 10^{-6}}{98} = \mathbf{101\ \mu A} \quad \textbf{Ans.}$$

5.15. COMMON COLLECTOR (CC) CONFIGURATION

In this circuit arrangement, input is applied between base and collector while the output is taken across emitter and collector. Thus the collector forms the terminal common to both the input and output circuits. The circuit is essentially the same as that of CE configuration except that load resistor R_L is now placed in the emitter circuit instead of collector circuit. The operation of the circuit in terms of the currents that flow is basically the same as that for the CE configuration. In this arrangement base current I_B flows in the input circuit and emitter current I_E flows in the output circuit. So the change in emitter current ΔI_E to the change in base current ΔI_B gives the current amplification factor γ. The circuit provides nearly the same current gain as in case of CE arrangement because $\Delta I_E \simeq \Delta I_C$.

With base current I_B equal to I_{CO}, the emitter current I_E is zero so no current flows in the load resistor R_L. With the increase in input current I_B, the transistor passes through the active region and finally reaches saturation. Almost the complete supply voltage, except for a small voltage drop across the transistor appears across the load resistor R_L.

The current equations are

$$I_C = \alpha\, I_E + I_{CBO}$$

$$\text{and } I_E = I_B + I_C = I_B + \alpha\, I_E + I_{CBO}$$

$$\text{or } I_E\,(1 - \alpha) = I_B + I_{CBO}$$

$$\text{or } I_E = \frac{I_B}{1-\alpha} + \frac{I_{CBO}}{1-\alpha} = (\beta + 1)\, I_B + (\beta + 1)\, I_{CBO} \quad \ldots(5.36) \quad \because \frac{1}{1-\alpha} = \beta + 1$$

$$\text{Current gain, } \gamma = \frac{\Delta I_E}{\Delta I_B}$$

$$= \frac{\Delta I_E}{\Delta I_E - \Delta I_C}$$

$$= \frac{\Delta I_E / \Delta I_C}{\frac{\Delta I_E}{\Delta I_C} - 1}$$

$$= \frac{1/\alpha}{1/\alpha - 1}$$

$$= \frac{1}{1-\alpha} = \beta + 1 \qquad ...(5.37)$$

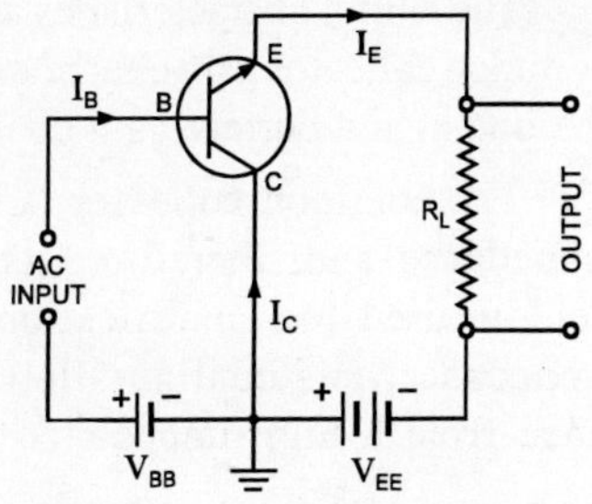

Fig. 5.27. ***Common Collector NPN Transistor***

5.15.1. Characteristics of Common Collector Configuration

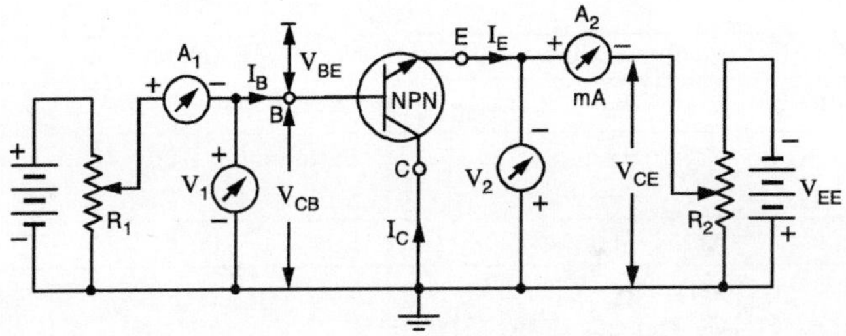

Fig. 5.28. ***Circuit Arrangement For Determination of Input and Output Characteristics of a Transistor in CC Configuration***

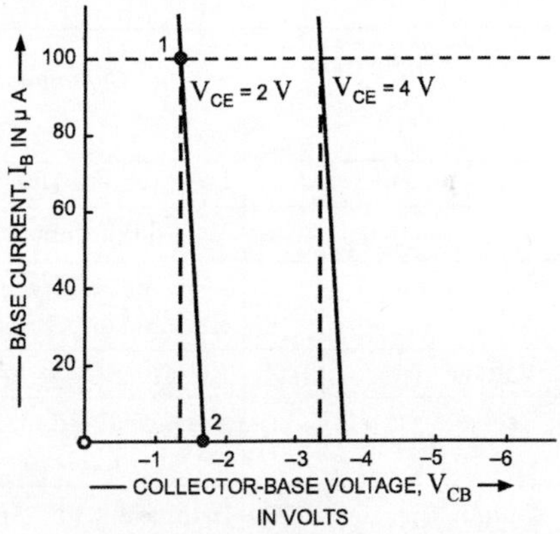

Fig. 5.29. ***Input Characteristics For Common Collector NPN Transistor***

The circuit used for determination of transistor common collector configuration is shown in Fig. 5.28. For determination of input characteristics, V_{CE} is kept at a suitable fixed level and base current I_B is recorded for different values of collector-base voltage V_{CB}. Now the curves are drawn between base current and collector-base voltage, as shown in Fig. 5.29.

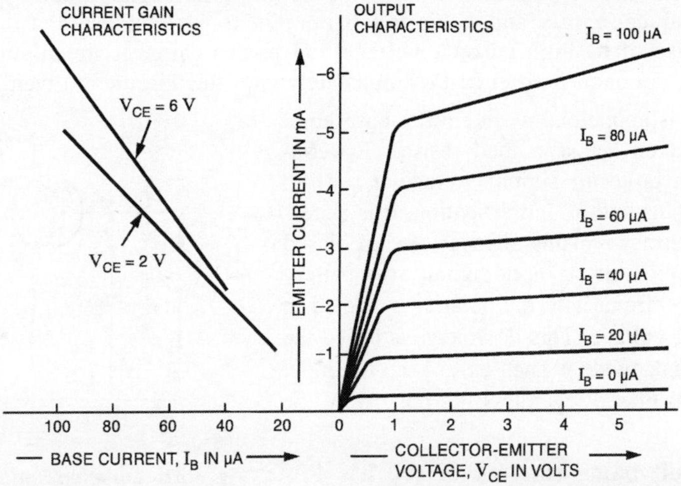

Fig. 5.30. ***Output and Current Gain Characteristics of Transistor in CC Configuration***

The output characteristics are plotted between emitter current I_E and collector-emitter voltage V_{CE}. The common collector gain characteristics are plotted between emitter-current and base current. Fig. 5.30 depicts the output and current gain characteristics.

The common collector (CC) arrangement gives very high input impedance and very low output impedance and, therefore, its voltage gain is always less than unity. Hence this configuration is seldom used for amplification. However, owing to relatively high input impedance and low output impedance, this configuration is primarily used for impedance matching *i.e.* for driving low impedance from a high impedance source. This configuration is also called *emitter follower.*

5.16. COMPARISON OF CHARACTERISTICS OF TRANSISTORS IN DIFFERENT CONFIGURATIONS

The essential characteristics of each of basic configurations are tabulated below :

Configuration / *Characteristics*	*Common Base*	*Common Emitter*	*Common Collector*
Input impedance	Low (about 100 Ω)	Medium (about 800 Ω)	Very high (about 750 kΩ)
Output impedance	Very high (about 500 kΩ)	High (about 50 kΩ)	Low (about 50 Ω)
Current gain	Less than unity but usually more than 0.9 (about 0.98)	High (about 80)	High (about 100)
Voltage gain	About 150	About 500	Less than unity
Leakage current	Very small (5 μA for Ge and 1 μA for Si)	Very large (500 μA for Ge and 20 μA for Si)	Very large
Output Signal phase	In phase with input	Reverse	In phase with input
Applications	For high frequency applications	For AF applications	For impedance matching

5.17. TRANSISTOR AS AN AMPLIFIER

The main utility of a transistor lies in its ability of amplifying weak signals. The weak signal is applied at the input terminals and the amplified output is obtained across the output terminals.

A transistor alone cannot perform the function of amplification and some passive components such as resistors and capacitors and a biasing battery is to be connected. Common emitter (CE) configuration, because of its high current, voltage and power gains, is much suited for most of the amplifier circuits. A common emitter N-P-N transistor amplifier circuit is given in Fig. 5.31.

The weak signal is applied between emitter-base junction and output is taken across a load resistor R_L connected in series with collector supply voltage V_{CC} (Fig. 5.31). In order to obtain faithful amplification, it is necessary that the input circuit remains always forward biased regardless the polarity of the ac input signal. So a battery V_{BB} is inserted in the circuit with the polarity indicated in addition to the signal voltage. This dc voltage is called the *bias voltage* and its magnitude is such as to keep the input circuit always forward biased regardless the polarity of the input signal.

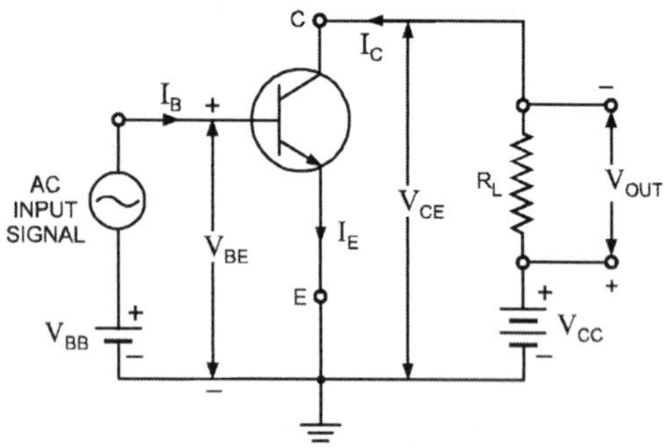

Fig. 5.31. ***Common Emitter NPN Transistor Amplifier Circuit***

The input circuit being forward biased has low resistance and a small change ΔV_{in} in input signal voltage causes a relatively large change ΔI_E in emitter current. This causes almost the same change in collector current because of transistor action. The collector current flowing through a high load resistance R_L develops a large voltage across it. The change in output voltage across load resistance R_L may be many times the change in input signal voltage. Thus voltage amplification $A = \frac{\Delta V_{out}}{\Delta V_{in}}$ will be greater than unity, and the transistor acts as an amplifier. This is further illustrated below by considering typical circuit values.

Let the load resistance R_L be of 10 kΩ and a change of 0.1 V in input signal voltage cause a change of 0.5 mA in emitter current. This change of 0.5 mA in emitter current will also change collector current I_C by approximately 0.5 mA. This change of 0.5 mA in collector current will produce a change of $0.5 \times 10^{-3} \times 10 \times 10^3$ *i.e.* 5 V in output voltage appearing across the load resistor R_L of 10 kΩ. Thus the change of 0.1 V in input signal voltage causes a change of 5 V in the output voltage giving a voltage amplification of $\frac{5}{0.1} = 50$.

Necessity of Biasing. For most of the applications, transistors are required to operate as *linear amplifiers* (*i.e.* to amplify output voltage as a linear function of the input voltage). To achieve this, it is necessary to operate the transistor over region of its characteristic curves which are linear, parallel and equi-spaced for equal increments of the parameters. Such an operation can be ensured by proper selection of zero signal operating point and limiting the operation of the transistor over the linear portion of the characteristics. For proper selection of zero signal operating point, proper biasing *i.e.* application of dc voltages at emitter-to-base junction and collector-to-base junction is required.

If the transistor is not biased properly, it would work inefficiently and produce distortion in the output signal.

5.18. TRANSISTOR LOAD LINES

The concept of load line is very important in understanding the working of a transistor. It is defined as the locus of operating point on the output characteristic of the transistor. It is the line on which the operating point moves when ac signal is applied to the transistor.

5.18.1. DC Load Line

In the circuit shown in Fig. 5.31, V_{CC} is the supply voltage to collector, R_C (or R_L) is the collector resistance (or load resistance) and V_{CE} is the collector-to-emitter voltage. Applying Kirchhoff's second law to the output or collector circuit we have,

$$V_{CC} = V_{CE} + I_C R_C \qquad ...(5.38)$$

$$\text{or} \quad I_C = \frac{V_{CC} - V_{CE}}{R_C}$$

$$\text{or} \quad I_C = \frac{-V_{CE}}{R_C} + \frac{V_{CC}}{R_C} \qquad ...(5.39)$$

This equation is to be plotted on the output characteristic of the transistor, V_{CE} and I_C are variables.

Identify this equation as $y = mx + c$ where $m = \frac{-1}{R_C}$ as the slope of the line and $I_C = \frac{V_{CC}}{R_C}$ as intercept of the line on vertical current axis (Fig. 5.32).

Consider the following two particular situations :

(*i*) When $V_{CE} = 0$, $I_C = \frac{V_{CC}}{R_C}$ ...saturation point A

(*ii*) When $I_C = 0$, $V_{CE} = V_{CC}$ ...cutoff point B

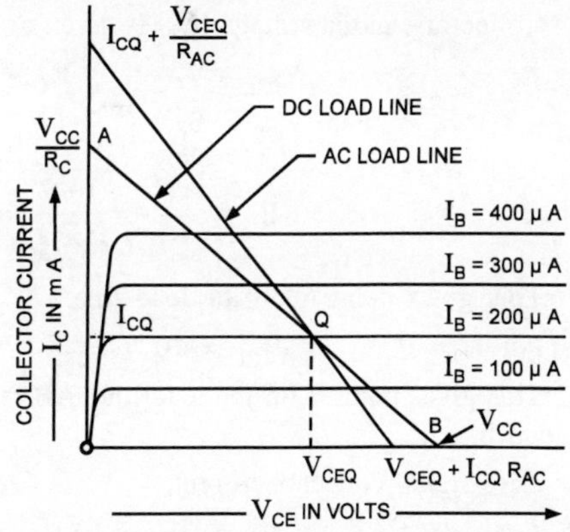

Fig. 5.32. ***Common Emitter Transistor Output Characteristics With DC and AC Load Lines***

By joining these two points A and B, *dc load line* is obtained. The dc load line represents the dynamic characteristic of the device. The dc load line gives the values of collector current I_C and collector-emitter voltage V_{CE} corresponding to zero signal conditions.

5.18.2. Quiescent Point

It is a point on the dc load line which represents dc collector-emitter voltage V_{CE} and collector current I_C in the absence of ac signal. It is also called the *operating point* because the variations in V_{CE} and I_C take place at about this point when signal is applied. The best position for this point is midway between cutoff and saturation points where $V_{CE} = \frac{1}{2} V_{CC}$. Point Q is quiescent point marked on the output characteristics shown in Fig. 5.32. Selection of the operating point is done as per application for which the device is to be used. For example in case of a small signal amplifier, in which power is conserved, operating point is selected so as to give lowest quiescent value of I_C, while for an amplifier operated to deliver small amount of power, operating point is selected so that available quiescent current is about one-half of the maximum permissible collector current I_C.

5.18.3. AC Load Line

When an ac signal is applied, the transistor voltage V_{CE} and collector current I_C vary above and below the quiescent point Q. So point Q is common to both dc and ac load lines. The *ac load line* gives the values of V_{CE} and I_C when an ac signal is applied. For drawing ac load line, take a convenient collector current change ΔI_C and compute the corresponding collector-emitter voltage change $\Delta V_{CE} = -\Delta I_C . R_C$ to obtain another point lying on the ac load line. Now the ac load can be drawn by joining this point and point Q. AC load is steeper than dc load line but the two lines intersect at the quiescent point Q determined by the biasing dc voltages and currents. AC load line takes into account the ac load resistance while the dc load line considers only the dc load resistance.

Example 5.17. In CE configuration (Fig. 5.31) collector supply voltage V_{CC} = 10 V, load resistance R_C is 8 kΩ. Draw dc load line. Determine the operating point Q for zero signal if base current is 15 μA and β is 40.

Solution :

Collector supply voltage, V_{CC} = 10 V

Load resistance, R_C = 8 kΩ

Zero signal base current, I_B = 15 μA

Current gain factor, β = 40

Collector-emitter voltage V_{CE} is given as

$$V_{CE} = V_{CC} - I_C R_C$$

For $V_{CE} = 0$,

$$I_C = \frac{V_{CC}}{R_C} = \frac{10 \text{ V}}{8 \text{k}\Omega} = 1.25 \text{ mA}$$

This gives point A on the load line.

For $I_C = 0$, $V_{CE} = V_{CC} = 10$ V

This gives point B on the load line. AB is the required dc load line.

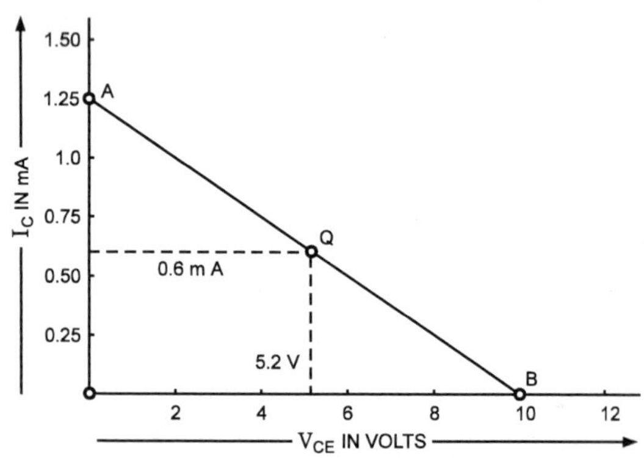

Fig. 5.33

Zero signal collector current,

$$I_C = \beta I_B = 40 \times 15 \times 10^{-6} = 0.6 \text{ mA}$$

Zero signal collector-emitter voltage,

$$V_{CE} = V_{CC} - I_C R_C = 10 - 0.6 \text{ mA} \times 8 \text{ k}\Omega = 5.2 \text{ V}$$

∴ Operating point Q is (5.2 V, 0.6 mA) **Ans.**

Example 5.18. Draw the load line for the following figure (Fig. 5.34). What is I_C at saturation point ? Find V_{CE} at cutoff point.

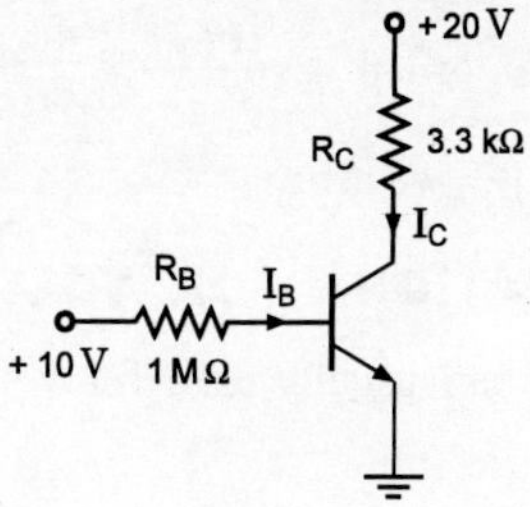

Fig. 5.34

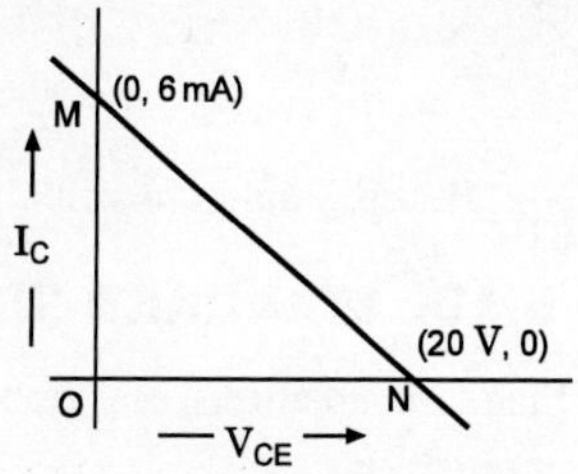

Fig. 5.35

Solution : (*i*) Applying Kirchhoff 's voltage law to the output loop we have

$$V_{CC} = I_C R_C + V_{CE}$$

For cutoff point, $I_C = 0$

and, therefore, $V_{CE} = V_{CC}$

$= 20$ V **Ans.**

$\therefore$ For saturation point, $V_{CE} = 0$

$$\text{So } I_C = \frac{V_{CC}}{R_C} = \frac{20}{3.3 \times 10^3} = 6 \text{ mA } \textbf{Ans.}$$

Therefore, load line coordinates are given as (20 V, 0 mA) and (0 V, 6 mA) as shown in Fig. 5.35.

Example 5.19. Determine V_C and V_B for the network given below (Fig. 5.36) with $\beta = 45$ and $V_{BE} = 0.7$ volt.

Solution : $\beta = 45$ and $V_{BE} = 0.7$ volt

Applying Kirchhoff's voltage law in input loop we have

$$I_B R_B + V_{BE} + V_{EE} = 0$$

Putting values in above equation we have

$$I_B \times 100 \times 10^3 + 0.7 - 9 = 0$$

$$\text{or } I_B = \frac{9 - 0.7}{10^5} = 83 \ \mu\text{A}$$

$$I_C = \beta \times I_B = 45 \times 83 \times 10^{-6} \text{ A} = 3.735 \text{ mA}$$

Collector voltage, $V_C = V_{CC} - I_C R_C$

$= 0 - 3.735 \times 10^{-3} \times 1.2 \times 10^3$

$= -4.482$ V **Ans.**

$$V_B = V_{BE} + V_{EE} = 0.7 - 9 = -8.3 \text{ volts } \textbf{Ans.}$$

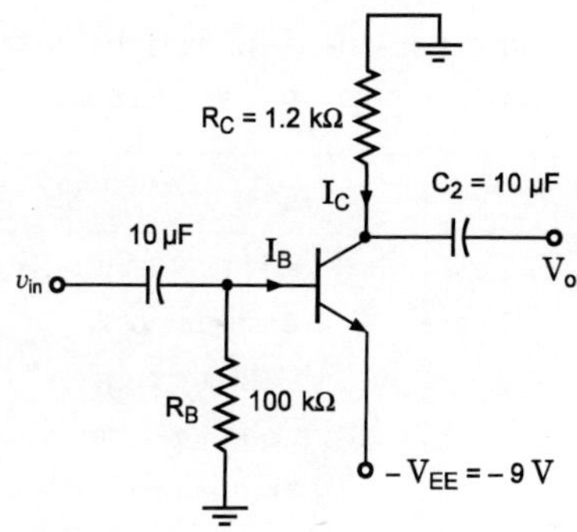

Fig. 5.36

Example 5.20. In the circuit shown in Fig. 5.37, what should be the minimum value of β such that the transistor is in saturation ? Assume $V_{CE\ Sat} = 0.2$ V.

Solution : In the given circuit,

Base current, $I_B = 0.1$ mA

Base resistance, $R_B = 10$ kΩ

Collector resistance, $R_C = 1$ kΩ

Supply voltage, $V_{CC} = 5$ V

Saturation collector-emitter voltage,

$$V_{CE\,(sat)} = 0.2 \text{ V}$$

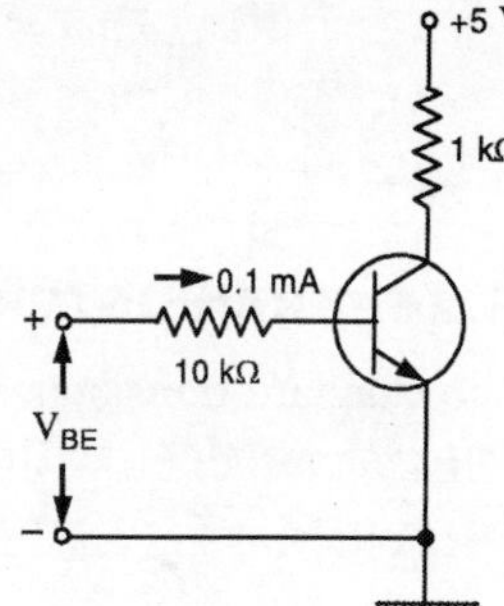

Fig. 5.37

Applying Kirchhoff's voltage law to the output loop, we have

$$V_{CC} - I_{C\,(sat)} R_C - V_{CE\,(sat)} = 0$$

$$\text{or } I_{C\,(sat)} = \frac{V_{CC} - V_{CE(sat)}}{R_C} = \frac{5 - 0.2}{1 \times 10^3} = 4.8 \text{ mA}$$

For transistor to be in saturation

$$\text{Base current, } I_B \geq \frac{I_{C\,(sat)}}{\beta}$$

$$\text{or} \quad \beta \geq \frac{I_{C(sat)}}{I_B} \geq \frac{4.8}{0.1} \geq 48$$

Hence, minimum value of $\beta = 48$ **Ans.**

5.19. STANDARD NOTATIONS FOR VOLTAGES AND CURRENTS

Standard notations or symbols are usually used for voltages and currents in transistor circuit analysis and applications.

1. Subscripts E (or *e*), B (or *b*) and C (or *c*) usually refer to emitter, base and collector quantities respectively.
2. Upper case (capital) letters with capital subscripts of the proper electrode symbol are used for dc or average values of voltages and currents.
3. Upper case (capital) letters with lower subscripts of the proper electrode symbol are used for rms values of ac voltages and currents.
4. Lower case (small) letters with lower subscripts of proper electrode symbol are used for instantaneous values of ac voltages and currents.

Notations used for various quantities (voltages and currents) in transistor circuit analysis and applications are given below in tabular form.

S.No.	*Particular*	*DC or average*	*RMS ac*	*Instantaneous ac*
1.	Emitter current	I_E	I_e	i_e
2.	Base current	I_B	I_b	i_b
3.	Collector current	I_C	I_c	i_c
4.	Collector-emitter voltage	V_{CE}	V_{ce}	v_{ce}
5.	Emitter-base voltage	V_{EB}	V_{eb}	v_{eb}
6.	Collector-base voltage	V_{CB}	V_{cb}	v_{cb}
7.	Emitter supply voltage	V_{EE}	—	—
8.	Base supply voltage	V_{BB}	—	—
9.	Collector supply voltage	V_{CC}	—	—
10.	Emitter voltage to ground	V_E	V_e	v_e
11.	Base voltage to ground	V_B	V_b	v_b
12.	Collector voltage to ground	V_C	V_c	v_c

5.20. TRANSISTOR MAXIMUM RATINGS

The standard transistor data sheet will include at least three maximum ratings—*collector dissipation, collector voltage* and *collector current.*

Following are maximum ratings, for the 2N3903 and 2N3904 transistors, at 25°C free air temperature.

$$P_{C\,max} = 310 \text{ mW}$$

$$I_{C\,max} = 200 \text{ mA}$$

$$V_{CE\,max} = 40 \text{ V}$$

The *power* or *dissipation rating* of a transistor is defined as the power that a transistor can handle without destruction. It is given as the product of collector voltage and collector current. For CE configuration

$$P_{C\,max} = V_{CE}\, I_C \qquad ...(5.40)$$

Small-signal transistors can dissipate half a watt or less whereas *power transistors* can dissipate more than half a watt.

Specifying of total device dissipation as 310 mW at a maximum free air temperature of 25°C means that ($V_{CE} \times I_C$) must not exceed 310 mW. For example, if the maximum collector current ($I_{C\,max}$) of 200 mA is to be used, then $V_{CE\,max}$ should be $\frac{310\text{ mW}}{200\text{ mA}}$ *i.e.*, 1.55 V.

It is important to note that if the free air temperature exceeds 25°C, the device power or dissipation rating must be derated. The derating factor for 2N3903 – 2N3904 transistors is 2.81 mW/°C.

The nonlinear curve indicated by Eq. (5.40) can be obtained by choosing different values of V_{CE} (or I_C) and finding the other variables using Eq. (5.40).

This has been illustrated in Example 5.22.

Example 5.21. A 2N3904 transistor has a maximum power dissipation of 310 mW at free air temperature of 25°C and derating factor of 2.81 mW/°C. Determine the maximum level of collector current I_C if the transistor is to be used in a circuit to be operated at a free air temperature of 105°C. V_{CE} = 20 V.

Solution : Free air temperature for circuit = 105°C

Temperature in excess of 25°C = 105°C – 25°C = 80°C

Derated power dissipation value, $P_{C\,max}$ = Original maximum power dissipation value – deration

= 310 – 80 × 2.81 = 85.2 mW

The device dissipation is $V_{CE} \times I_C$.

Thus, $20 \times I_{C\,max} = 85.2$ mW

$$I_{C\,max} = \frac{85.2}{20} = 42.6 \text{ mA}$$ **Ans.**

Example 5.22. A 2N3055 transistor has a maximum power dissipation of 115 watts at free air temperature of 25°C and derating factor of 0.66 W per °C. The output characteristic of the device may be assumed, as illustrated in Fig. 5.38. Plot a maximum power dissipation curve for a case temperature of 85°C.

Solution : Maximum power dissipation at 25°C,

$P_{C\,max}$ = 115 watts

Derating factor = 0.66 W/°C

Case temperature for device = 85°C

Temperature in excess of 25°C = 85°C – 25°C

= 60°C

Device must be derated by 0.66 W per °C × 60 = 39.6 watts

Maximum power dissipation at 85°C = 115 – 39.6

= 75.4 watts

From Eq. (5.40) $$I_C = \frac{P_{C\,max}}{V_{CE}} = \frac{75.4}{V_{CE}}$$

Let us determine value of I_C for V_{CE} to be 10, 20, 30, 40, 50 and 60 V. Thus,

Collector currents, I_C for V_{CE} of 10 V = $\frac{75.4}{10}$ = 7.54 A

I_C for V_{CE} of 20 V = $\frac{75.4}{20}$ = 3.77 A

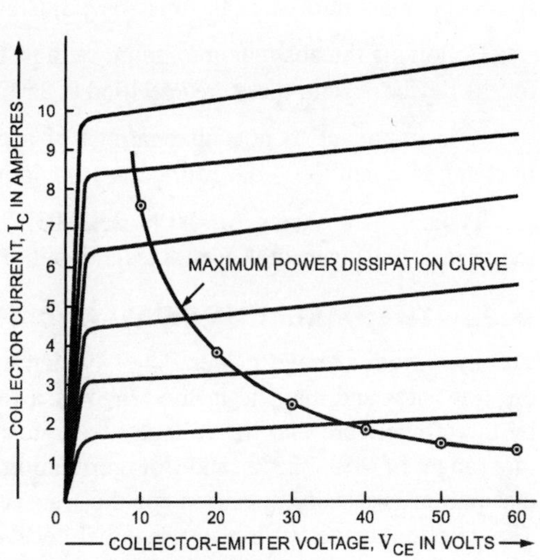

Fig. 5.38

$$I_C \text{ for } V_{CE} \text{ of } 30 \text{ V} = \frac{75.4}{30} = 2.51 \text{ A}$$

$$I_C \text{ for } V_{CE} \text{ of } 40 \text{ V} = \frac{75.4}{40} = 1.885 \text{ A}$$

$$I_C \text{ for } V_{CE} \text{ of } 50 \text{ V} = \frac{75.4}{50} = 1.51 \text{ A}$$

$$I_C \text{ for } V_{CE} \text{ of } 60 \text{ V} = \frac{75.4}{60} = 1.26 \text{ A}$$

Now the curve drawn between points (10 V, 7.54 A), (20 V, 3.77 A), (30 V, 2.51 A), (40 V, 1.885 A), (50 V, 1.51 A) and (60 V, 1.26 A) is the required maximum power dissipation curve.

The transistor voltage and current conditions must at all times be maintained in the portion of characteristics below the maximum power dissipation curve.

5.21. TRANSISTOR DATA SHEETS

For selection of a transistor for a particular application, the data sheets provided by the device manufacturers must be consulted.

Most data sheets start off with the device type number at the top of the sheet, a descriptive title, and a list of major applications of the device. This information is usually followed by mechanical data in the form of illustration showing the package shape and dimensions. The lead orientation is also identified here.

The absolute maximum ratings of the transistor at a temperature of 25°C are listed next. These are the maximum voltages, currents etc. that the device can bear/carry without breaking down. The standard data sheet includes at least three maximum ratings (collector power dissipation, collector voltage and collector current). The power or dissipation rating is the product of the collector voltage and collector current. In case of a common-emitter configuration the maximum collector dissipation rating, $P_{C\ max} = V_{CE}\ I_C$.

It is very important, for the safe operation of the device, that the maximum ratings given in the data sheets are never exceeded. For reliability, the maximum ratings should not even be approached. Also, the maximum ratings must be adjusted for operation at temperatures exceeding 25°C.

Following the absolute maximum ratings, there is normally a complete list of electrical characteristics for the device. Again, these are specified at 25°C, and allowances are necessary for variations of temperature.

It is important to note that ratings of a given transistor are stated for specified circuit conditions. In changed conditions, the ratings are no longer valid.

Typical data sheets for NPN 2N3903, 2N3904 (silicon) transistors and 2N3055 (silicon) power transistors are given at Appendixes 1.3 and 1.4 respectively.

5.22. THERMAL RUNAWAY AND HEAT SINK

The maximum average power $P_{C\ max}$ which a transistor can dissipate depends upon the construction of the transistor and may lie in the range of a few milliwatts and 200 W. This maximum power is limited by the temperature that the collector-base junction withstand. For silicon transistor this temperature is in the range of 150–225°C and for germanium it is between 60°C and 100°C. The maximum power dissipation is usually specified for the transistor enclosure (or casing) or ambient temperature of 25°C. The junction temperature may rise either because of rise in ambient temperature or because of self heating. The problem of self heating arises due to dissipation of power at the collector junction.

The leakage current I_{CBO} is extremely temperature dependent and increases with the rise in temperature of collector-base junction. An increase in I_{CBO} causes an increase in collector current I_C considerably.* With the increase in collector current I_C, collector power dissipation increases which raises the junction temperature that leads to further increase in collector current I_C. The process is cumulative and may lead to the eventual destruction of the transistor. This phenomenon is known as *thermal runaway* of the transistor. In practice, thermal runaway is prevented in a well designed circuit by using stabilization circuitry.

When the power dissipation at the collector-base junction of a transistor is small, as in case of a small-signal transistor, the surface area of the transistor case is normally large enough to allow all of the heat to escape. But for the large power dissipation that can occur in large-power transistor the transistor surface area is not large enough and junction temperature may rise to a dangerous level. However, power handling capacity of a transistor can be increased by making suitable provision for rapid conduction of heat away from the transistor junction. This is achieved by using a sheet of metal called the *heat sink* which increases the area of contact with the atmosphere.

Heat sink is a direct practical means of combating the undesirable thermal effects *e.g.*, thermal runaway. It may be noted that the ability of any heat sink to transfer heat to the surroundings depends upon its material, volume, area, shape, contact between case and sink and movement of air around the sink.

For achieving efficient heat dissipation heat sinks of different types are employed. A few of these are shown in Fig. 5.39.

One type of heat sink, called the *push-on heat sink*, is shown in Fig. 5.39 (*a*). When this is pushed on to the transistor case, heat radiates more quickly because of the increased surface area of the fins.

Power tab transistor outline is shown in Fig. 5.39 (*b*). A metal tab provides a path out of the transistor for heat. This metal tab can be fastened to the chassis of electronic equipment. Because the chassis is a massive heat sink, heat can easily escape out from the transistor to the chassis.

Large power transistors like shown in Fig. 5.39 (*c*) have the collector connected to the case to dissipate heat as easily as possible. The transistor case is then fastened to the chassis. To prevent the collector from shorting to chassis ground, a thin mica washer is used between the transistor case and the chassis. The important idea here is that heat can escape out from transistor at a faster rate, which means that the transistor has a higher power rating at the given ambient temperature. Sometimes, the transistor is fastened to a large heat sink with fins; this is even more efficient in heat dissipation from transistor.

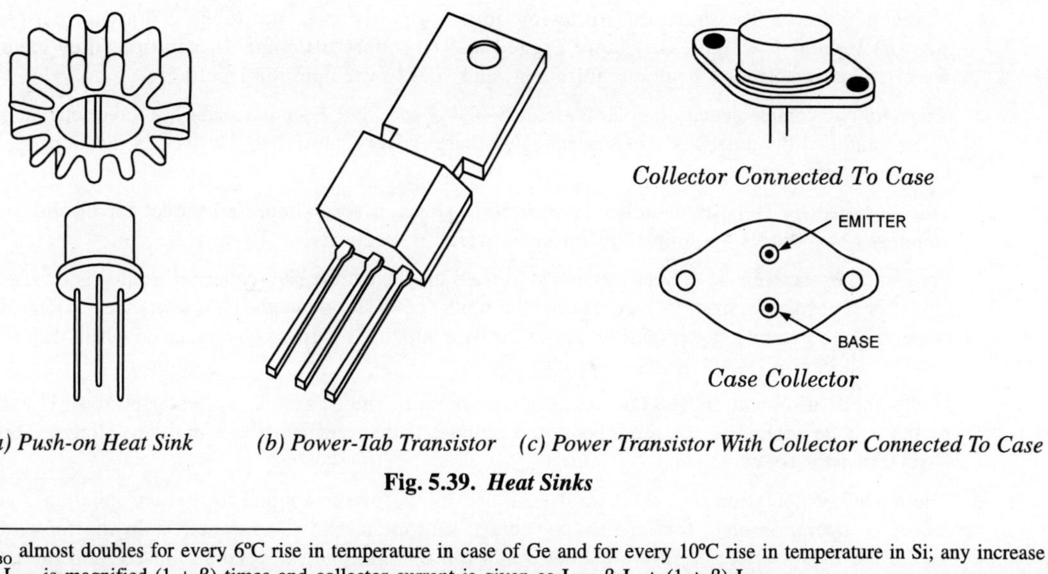

(a) Push-on Heat Sink *(b) Power-Tab Transistor* *(c) Power Transistor With Collector Connected To Case*

Fig. 5.39. ***Heat Sinks***

* I_{CBO} almost doubles for every 6°C rise in temperature in case of Ge and for every 10°C rise in temperature in Si; any increase in I_{CBO} is magnified $(1 + \beta)$ times and collector current is given as $I_C = \beta\, I_B + (1 + \beta)\, I_{CBO}$.

Finned aluminium heat sinks yield the best heat transfer per unit cost.

For maximum efficiency a heat sink (*i*) is kept in good thermal contact with the transistor case (*ii*) is made with the largest possible surface area (*iii*) is painted black and (*iv*) is mounted in such a position that free air can flow past it.

The noteworthy point here is that the use of heat sink may not be sufficient to avoid thermal runaway under all conditions. In designing a transistor circuit, consideration should also be given to the choice of (*i*) operating point (*ii*) ambient temperatures which are likely to be encountered and (*iii*) the type of transistor e.g. metal case transistors are more readily cooled by conduction than plastic ones. Circuits may be designed to compensate automatically for temperature variations and thus stabilise the operation of the transistor components.

5.23. IMPORTANCE OF COLLECTOR-EMITTER VOLTAGE, V_{CE}

The collector-emitter voltage, V_{CE} plays a significant role in indication of region of operation of transistor *i.e.* whether the transistor is operating in cutoff, saturation or well into saturation regions.

Equality of V_{CE} and V_{CC} (*i.e.* $V_{CE} = V_{CC}$) indicates that the transistor is operating in *cutoff mode* (*i.e.* the transistor is turned *off*). Zero value of V_{CE} indicates that the transistor is operating in saturation mode (*i.e.* the transistor is turned *on* fully). Negative value (less than zero value) of V_{CE} indicates that the transistor is *well in saturation.* In practice, both of these conditions are avoided, except in switching circuits.

In practice, the transistor is operated with $V_{CE} = \frac{1}{2} V_{CC}$ (*i.e.* at approximately $\frac{1}{2}$ *on*). Thus, variation in base current I_B in either direction controls collector current I_C well in both directions. It means that collector current I_C increases or decreases with the increase or decrease of base current I_B. However, with I_B *off* ($I_B = 0$), I_C is also *off* ($I_C = 0$). On the other hand, if the collector has been turned fully *on* (*i.e.* transistor is operating in saturation), the collector current is maximum, and the further increase in I_B will not reflect in I_C.

HIGHLIGHTS

1. Semiconductor devices have the following advantages over vacuum tubes : They are (*i*) smaller in size (*ii*) lighter in weight, (*iii*) more rugged and (*iv*) more efficient. In addition, they have (*i*) no warm up period, (*ii*) no heater requirement and (*iii*) lower operating voltages.
2. There are two complementary bipolar transistors—NPN and PNP. Each transistor has three separately doped regions and two PN junctions. The center region (base) is very narrow, so the two PN junctions are said to be interacting junctions.
3. Since a transistor consists of either PNP or NPN slices, a very simplified model for biasing purposes comprises two diodes "pointing inward or outward" respectively.
4. For normal operation, base-emitter junction is forward biased and the base-collector junction is reverse biased. Under these conditions, majority carriers from the emitter are injected into the base where they become minority carriers. These minority carriers diffuse across the base into the B-C space charge region where they are swept into the collector.

   In this mode of operation, the current at one terminal of the transistor (collector current) is controlled by the voltage applied across the other two terminals of the transistor (base-emitter voltage). This is the basic transistor action.
5. The modes of operation of a BJT are determined by the biases applied to the two junctions. The four modes of operation are : forward active, cutoff, saturation, and inverse active. In the forward-active mode, the B-E junction is forward biased and the B-C junction is reverse biased, and the collector and base currents are related by the common-emitter current gain $\beta(I_C = \beta I_B)$. The relationship is

the same for both NPN and PNP transistors, as long as the conventional current directions are maintained. In the cutoff mode of operation, the B-E and B-C junctions are both reverse biased and all currents are zero. In saturation mode, the B-E and B-C junctions are both forward biased and the collector current is no longer a function of base current.

6. The dc emitter current I_E is always the largest one of a transistor, whereas the base current I_B is always the smallest. The collector current I_C is almost as large as emitter current I_E (*i.e.* $I_C \simeq I_E$). The emitter current is always equal to the sum of the other two (*i.e.* $I_E = I_B + I_C$).

7. The collector current I_C is made up of two components : the majority component and the minority component (also known as the leakage current).

   i.e., $I_C = I_{C_{majority}} + I_{CO_{minority}}$

8. The arrow in the transistor symbol defines the direction of conventional current flow for the emitter current and thereby defines the directions for the other currents in the device.

   The arrow in the symbol of an NPN transistor points out of the device (not pointing in), whereas the arrow points to the centre of the symbol for a PNP transistor (pointing in).

9. On an average basis, as a second approximation, the base-to-emitter voltage of an operating transistor can be assumed to be 0.7 V in case of silicon device and 0.3 V in germanium device.

10. The quantity α relates the collector and emitter currents $\left(\alpha = \frac{I_C}{I_E}\right)$ and is always approximately equal to one (never greater than 1).

11. The impedance between terminals of a forward-biased junction is always relatively small, whereas the impedance between terminals of a reverse-biased junction is usually quite large.

12. The quantity beta β provides an important relationship between the base and collector currents and is usually between 50 and 400.

13. The dc beta is defined by a simple ratio of dc currents at an operating point $\left(i.e.\ \beta_{dc} = \frac{I_C}{I_B}\right)$, whereas the ac beta β_{ac} is sensitive to the characteristics in the region of interest. For most applications, however, the two are considered equivalent as a first approximation.

14. A three-terminal device needs two sets of characteristics to completely define its characteristics.

    The transistor input characteristic is essentially that of a forward biased junction, with some dependence upon the transistor's output voltage.

15. For a transistor used in CB configuration, the collector or output characteristics consist of practically horizontal lines of I_C versus V_{CB} with I_E as the controlling parameter.

16. The CB ac current gain, $\alpha_{ac} = \left.\frac{\Delta I_C}{\Delta I_E}\right|_{V_{CB} = \text{Constant}}$ while $\alpha_{dc} = \frac{I_C}{I_E}$ and is less than unity.

17. The reverse saturation current in a CB configuration = $I_{CBO} = I_{CO}$, while in a CE configuration it is $I_{CEO} = (\beta + 1)I_{CBO}$.

18. The collector or output characteristics for a transistor used in the CE configuration have the collector current I_C versus collector-emitter voltage curves sloping upward slightly with base current as the controlling parameter.

19. The CE ac current gain, $\beta_{ac} = \left.\frac{\Delta I_C}{\Delta I_B}\right|_{V_{CE} = \text{Constant}}$, while $\beta_{dc} = \frac{I_C}{I_B}$.

20. The permissible operating area of a transistor's characteristics is defined by $I_{C\,max}$, the maximum dissipation hyperbola, and $V_{(BR)CEO}$.

21. In either CB or CE configuration, majority carriers are swept across the narrow, lightly doped base to the collector. The collector current is controlled by a relatively small base input current that is the result of the forward-biased base-emitter junction, which accounts for the amplifying action in a transistor.
22. To ensure that a transistor is operating within its maximum power level rating, simply find the product of the V_{CE} and I_C, and compare it to the rated value.
23. Transistors have maximum ratings of voltages, currents and powers. Small-signal transistors can dissipate 1 W or less. Power transistors can dissipate more than 1 W. Temperature can change the value of the transistor characteristics. Maximum power decreases with the increase in temperature. Also current gain varies greatly with temperature.

EXERCISES

1. Draw NPN and PNP transistors. Label all the currents and show the direction of flow. How are all the currents of a transistor related ?
2. What is a transistor ? Show the transistor action for the N-P-N and P-N-P transistors.
3. Explain the working of a PNP transistor with the help of its energy band diagram. Explain 'emitter injection efficiency' and base 'transport factor'. How do they influence the operation of a transistor ?
4. If a transistor has its collector-base junction forward biased and the other junction reverse biased, will it work ? Explain.
5. If two P-N junction diodes are connected back to back as shown in Fig. 1, will it work as a transistor with the terminals designated in the figure ? Explain.

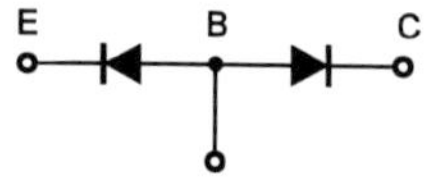

Fig. 1

6. Define with respect to BJT the following :

   (*i*) I_{CO}, (*ii*) α, (*iii*) β, (*iv*) I_{CEO}, (*v*) I_{CBO}
7. Differentiate between I_{CO} and I_{CBO}. What is the effect of temperature on I_{CBO} ?
8. In how many modes the BJT works ? Also explain the biasing pattern for each of them.
9. Explain the following regions in a transistor : (*i*) Active, (*ii*) Saturation, (*iii*) Cutoff.
10. Explain the saturation and cutoff condition in a CE connected transistor.
11. Explain different components of currents flow through the structure of a N-P-N transistor. How the emitter injection efficiency and base transport factor influences the amplification factor ?
12. Define base spreading resistance of BJT.
13. Give the Ebers-Moll model for an NPN transistor and give the equations for I_E and I_C. With the help of this model explain why it is not possible to construct a BJT by using two diodes connected back to back. **[G.G.S.I.P. Univ. Analog Electronics-I December 2004]**
14. Explain the Ebers-Moll model for a PNP transistor. **[G.G.S.I.P. Univ. Analog Electronics December 2009]**
15. Draw the Ebers-Moll model for a PNP transistor and give equations for emitter current and collector current. **[G.G.S.I.P. Univ. Analog Electronics December 2009]**
16. Explain why a transistor action cannot be achieved by connecting two back-to-back diodes. In a transistor explain, why emitter region is heavily doped, base width is small and collector area is large.
17. Write Ebers Moll equations for an NPN transistor. Based on these equations, draw and explain the output characteristics of NPN transistor in CE configuration. **[G.G.S.I.P. Univ. Analog Electronics December 2010]**
18. Differentiate between CE, CC and CB amplifiers.
19. Sketch a family of CB output characteristics for a transistor. Indicate the active, cutoff and saturation regions. Explain the shapes of the curves qualitatively.
20. Sketch typical output characteristics for a PNP transistor in CB configuration. Label all variables and indicate active, cutoff and saturation regions. **[G.G.S.I.P. Univ. Analog Electronics December 2009]**
21. Explain Early effect. **[G.G.S.I.P. Univ. Analog Electronics December 2010]**
22. Explain the operation of NPN transistor in CB configuration. Show that $\alpha = \dfrac{\beta}{1+\beta}$. Also give the reason why the input impedance of CB is less than input impedance of CE configuration.

23. Explain the output characteristics of an NPN transistor in (*i*) CB configuration (*ii*) CE configuration. **[G.G.S.I.P. Univ. Analog Electronics December 2010]**

24. For a transistor in CE configuration, (*i*) Derive the expression between I_C, I_B and I_{CO}, (*ii*) Explain why slope of the output characteristics in CE configuration is more than in CB configuration (*iii*) cutoff condition. **[G.G.S.I.P. Univ. Analog Electronics December 2009]**

25. Draw the output characteristics of a transistor in CE configuration and show the total region when transistor can be used as an amplifier. Explain why the slope of characteristic is more in CE mode compared to CB mode. **[G.G.S.I.P. Univ. Analog Electronics December 2010]**

26. Compare the CB, CE and CC configurations of a transistor. For CE configuration, prove that $I_C = \beta I_B + (\beta + 1) I_{CO}$.

27. Draw the circuits of various transistor configurations. List their important features. Why CE configuration is mainly used ? What are uses of other configurations ?

28. Draw a simple circuit of a common-collector amplifier and explain its operation. Obtain an expression for its voltage gain.

29. Compare the characteristics of transistor amplifiers in the three possible configurations.]

30. Explain the difference in ac and dc load lines, which is more steeper—ac load or dc load ? **[G.G.S.I.P. Univ. Analog Electronics December 2009]**

31. Explain the significance of ac and dc load lines. **[G.G.S.I.P. Univ. Analog Electronics December 2010]**

32. Explain heat dissipation curve and its importance, while selecting heat sink.

33. Explain the ratings that limit the operation of a transistor as an amplifier. Show the region on the output characteristics of the transistor. **[G.G.S.I.P. Univ. Analog Electronics December 2009]**

34. What do you understand by thermal runaway in BJT circuits. **[G.G.S.I.P. Univ. Analog Electronics December 2010]**

SHORT ANSWER TYPE QUESTIONS WITH ANSWERS

Q. 1. Explain why an ordinary transistor is called bipolar ?

Ans. Because the transistor operation is carried out by two types of charge carriers (majority and minority carriers), an ordinary transistor is called bipolar.

Q. 2. Why transistor is called current controlled device ?

Ans. The output voltage, current or power is controlled by the input current in a transistor so it is called the current controlled device.

Q. 3. What is the significance of arrowhead in the transistor symbol ?

Ans. Arrowhead is always marked on the emitter. The direction indicates the conventional direction of current flow (from emitter-to-base in case of P-N-P transistor and from base-to-emitter in case of N-P-N transistor). Generally, no arrowhead is marked for collector since its reverse leakage current is always opposite to the direction of emitter current.

Q. 4. What do you understand by transistor biasing ? Why is it necessary to bias a transistor ?

Ans. For normal operation, base-emitter junction should be forward biased and the collector-base junction reverse biased. The proper flow of zero signal collector current and the maintenance of proper collector-emitter voltage during the passage of signal is called the transistor biasing. The amount of bias required is significant for establishment of the operating or Q-point which is dictated by the mode of operation desired.

In case the transistor is not biased properly, it would (*i*) work inefficiently, (*ii*) produce distortion in the output signal and (*iii*) with the change in transistor parameters or temperature rise, the operating point may shift and the amplifier output will be unstable.

Q. 5. What are 'emitter injection efficiency' and 'base transport factor' and how do they influence the transistor operation ?

Ans. The ratio of current of injected carriers at emitter junction to the total emitter current is called the emitter injection efficiency. The ratio of injected carrier current reaching collector junction J_C to the injected carrier current at junction J_E is known as transport factor, β

$$i.e.\ \beta = \frac{I_{hC}}{I_{hE}}$$

The larger the value of emitter injection efficiency, the larger the injected carriers at emitter junction and this increases the collector current. The larger the β value larger the injected carriers across collector junction and hence collector current increases.

Q. 6. Which of the transistor currents is always the largest ? Which is always the smallest ? Which two currents are relatively close in magnitude ?

Ans. The emitter current I_E is always the largest one. The base current I_B is always the smallest. The collector current I_C and emitter current I_E are relatively close in magnitude *i.e.* $I_C \simeq I_E$.

Q. 7. Why silicon type transistors are more often used than germanium type ?

Ans. Because silicon transistor has smaller cutoff current I_{CBO}, small variations in I_{CBO} due to variations in temperature and high operating temperature as compared to those in case of germanium type.

Q. 8. Why is size of collector kept maximum in BJT ?

Ans. The potential hill (bend in the Fermi level) between the base and collector region is very steep. Specifically the electrons must lose a considerable energy as they move from the base region to the collector region. Typically the electrons give up their energy in the form of heat, therefore, the collector region must be capable of dissipating this heat energy. It is for this reason that the size of the collector is kept maximum in BJT.

Q. 9. Why the width of the base region of a transistor is kept very small compared to other regions ?

Ans. Base region of a transistor is kept very small and very lightly doped so as to pass most of the injected charge carriers to the collector.

Q. 10. Why emitter is always forward biased ?

Ans. Emitter is always forward biased w.r.t. base so as to supply majority charge carriers to the base.

Q. 11. Why collector is always reverse biased w.r.t. base ?

Ans. Collector is always reverse biased w.r.t. base so as to remove the charge carriers away from the base-collector junction.

Q. 12. What are three regions of operation of a transistor ?

Ans. The three regions of operation of a transistor are :

(*i*) Cutoff region, (*ii*) Active region and (*iii*) Saturation region.

Q. 13. Can a transistor be obtained by connecting two semiconductor diodes back to back ?

Ans. No. Because in case of two discrete back to back connected diodes there are four doped regions instead of three and there is nothing that resembles a thin base region between an emitter and a collector.

Q. 14. How α and β are related to each other ?

Ans. α and β are related as below :

$$\alpha = \frac{\beta}{1+\beta} \quad \text{or} \quad \beta = \frac{\alpha}{1-\alpha}$$

Q. 15. Define beta of a transistor.

Ans. The β factor of a transistor is the common-emitter current gain of that transistor and is defined as the ratio of collector current to base current.

$$i.e., \ \beta = \frac{I_C}{I_B}.$$

Q. 16. Indicate whether the β-value of a BJT increases or decreases with the increase in the values of the following parameters : (*i*) base width (*ii*) minority carrier lifetime in the base region, (*iii*) temperature (*iv*) collector current (*v*) collector voltage.

Ans. The β-value of a BJT

(*i*) decreases with the increase in base width.

(*ii*) increases with the increase in minority carrier lifetime in the base region.

(*iii*) increases with the increase in temperature.

(*iv*) increases with the increase in collector current.

(*v*) increases with the increase in collector voltage.

Q. 17. Why there is a maximum limit of collector supply voltage for a transistor ?

Ans. Although collector current is practically independent of collector supply voltage over the transistor operating range, but if V_{CB} is increased beyond a certain value collector current I_C eventually increases rapidly and possibly destroys the device.

Q. 18. Explain why $I_{CEO} >> I_{CBO}$?

Ans. The collector cutoff current denoted by I_{CEO} is much larger than I_{CBO}. I_{CEO} is given as

$$I_{CEO} = \frac{1}{1-\alpha} I_{CBO}$$

Because α is nearly equal to unity (slightly less than unity), $I_{CEO} >> I_{CBO}$.

Q. 19. Give reasons why common emitter (CE) configuration is widely used in amplifier circuits.

Ans. CE configuration is mainly used because its current, voltage and power gains are quite high and the ratio of output impedance and input impedance are quite moderate.

Q. 20. Why is CE configuration preferred for cascaded amplifiers ?

Ans. CE configuration, because of its moderate output to input impedance ratio, is preferred for cascaded amplifiers.

Q. 21. Why CC configuration is called a voltage buffer ? What is other name ?

Ans. Because of its high input impedance and low output impedance, the common collector circuit finds wide application as a buffer amplifier between a high impedance source and low impedance load, that is why it is called a voltage buffer. Its other name is emitter follower.

Q. 22. What is Early effect ?

Ans. The modulation of the effective base width by the collector voltage is known as Early effect. An increase in collector voltage increases the space charge width at the output junction diode and thus the effective base width W_b is reduced.

Q. 23. Elucidate three consequences of the Early effect in bipolar junction transistor.

Ans. Three consequences of the Early effect in BJT are as follows :

1. There is less chance for recombination within the base region. Hence α and β increase with increasing collector-base voltage $|V_{CB}|$.
2. The concentration gradient of minority carriers is increased within the base, and consequently, the current of minority carriers injected across the emitter junction increases with increasing reverse collector voltage.
3. For extremely large collector-base voltage, effective base width W_b may be reduced to zero resulting in a phenomenon known as punch through.

Q. 24. What is punch through effect ?

Ans. When on increasing collector voltage effective base width is reduced to zero, the emitter barrier becomes smaller than normal value $V_0 - |V_{EB}|$ because the collector voltage has reached through the base region. This lowering of emitter-junction voltage causes an excessively large emitter current, thus placing an upper limit on the magnitude of collector voltage. This phenomenon is known as punch through.

Q. 25. What do you understand by collector reverse saturation ? In which configuration does it have a greater value ? **[G.G.S.I.P. Univ. December 2009]**

Ans. When input current (I_E in case of CB configuration and I_B in case of CE configuration) is zero, collector current I_C is not zero although it is very small. In fact this is the reverse leakage current or collector reverse saturation current (I_{CBO} or simply I_{CO} in CB configuration and I_{CEO} in CE configuration). In case of CE configuration it is much more than that in case of CB configuration.

Q. 26. What do you mean by operating point ?

Ans. Operating point or Quiescent point is a point on the *dc* load line which represents V_{CE} and I_C in the absence of *ac* signal and variations in V_{CE} and I_C take place around this point when ac signal is applied.

Q. 27. What limits the maximum current a transistor is capable of handling ?

Ans. The junction area of the collector region limits the maximum current a transistor can handle. Large power ratings are obtained from larger junction areas. More is the surface area, more is the effective heat dissipation and more is the magnitude of current, a transistor can handle.

Q. 28. Why heat sinks are used in power amplifiers ?

Ans. Heat sink is a specially designed metal sheet over which the power transistor is mounted so that it dissipates heat more effectively and protects the power transistor from overheating. The idea behind the heat sink is to increase the area of contact with the atmosphere. It is to be noted that the ability of any heat sink to transfer heat to the surroundings depends upon its material, volume, area, shape, contact between case and sink and movement of air around the sink.

PROBLEMS

1. Determine emitter current I_E, collector current I_C for a transistor with $\alpha_{dc} = 0.97$ and collector to base leakage current 10 μA. Base current is 50 μA. [**Ans.** 2.00 mA, 1.95 mA]
2. For a certain transistor collector current is 6.25 mA, base current is 100 μA and collector-to-base leakage current 5 μA. Determine emitter current and α_{dc}. [**Ans.** 6.35 m A, 0.983]
3. In a particular transistor the collector current is 5.6 mA and the emitter current is 5.75 mA. Determine α_{dc}. [**Ans.** 0.974]
4. A transistor has $\alpha = 0.98$. If emitter current of the transistor is 1 mA, determine the base current and gain factor β. [**Ans.** 20 μA; 49]
5. Determine β if $\alpha = 0.99$. [**Ans.** 99]
6. A bipolar junction transistor has a collector current of 4 mA and base current of 20 μA. Determine its β. [**Ans.** 200]
7. A BJT has a base current of 200 μA and emiter current of 20 mA. Determine collector current and β. [**Ans.** 19.8 mA; 99]
8. In a transistor $I_C = 0.95$ mA, $I_E = 1$ mA. Determine base current, α and β in CB configuration. [**Ans.** 50 μA; 0.95; 19]
9. In a transistor CE configuration the voltage across 5 kΩ load resistance is 5 V. Determine collector current, base current and emitter current if $\beta = 50$. [**Ans.** 1.0 mA; 20 μA; 1.02 mA]
10. In a grounded base configuration the voltage drop across load resistance of 4 kΩ is 3.2 V. Determine collector current, emitter current and base current if $\alpha = 0.96$. [**Ans.** 0.8 mA; 0.8333 mA; 0.0333 mA]
11. A CE transistor amplifier has input resistance of 100 Ω and output resistance of 6 kΩ. Determine the voltage gain and power gain if current gain factor is 90. [**Ans.** 5,400; 486,000]
12. Increase in collector-emitter voltage from 4 V to 8 V causes increase in collector current from 5 mA to 6 mA. Determine the output resistance. [**Ans.** 4 kΩ]
13. A change of 200 mV in base-emitter voltage causes a change of 60 μA in the base current. Determine its input resistance. [**Ans.** 3.33 kΩ]
14. A transistor is connected in CB configuration. When the emitter voltage is changed by 200 mV, the emitter current changes by 5 mA. During this variation, the collector to base voltage is kept fixed. Calculate the dynamic input resistance of transistor. [**Ans.** 40 Ω]

❑❑❑

CHAPTER **6**

Bias Stabilization

INSIDE THIS CHAPTER

6.1. TRANSISTOR BIASING

The basic function of a transistor is amplification. The process of raising the strength of a weak signal without any change in its general shape is referred to as *faithful amplification.* For faithful amplification it is essential that (*i*) emitter-base junction is forward biased; (*ii*) collector-base junction is reverse biased; and (*iii*) proper zero signal collector current.

For achieving faithful amplification, fulfilment of the following basic conditions is essential.

1. ***Proper Zero Signal Collector Current.*** Zero signal collector current should be at least equal to the maximum collector current due to signal alone *i.e.* zero signal current $\geq$ maximum collector current due to signal alone.
2. ***Minimum Proper Base-Emitter Voltage At Any Instant.*** The base-emitter voltage V_{BE} should not fall below 0.3 V for germanium transistors and 0.7 V for silicon transistors at any instant. If the base-emitter voltage V_{BE} falls below these values (potential barrier values) during any part of the signal, that part will be amplified to smaller extent due to small collector current and, therefore, faithful amplification will not be available.
3. ***Minimum Proper Collector-Emitter Voltage At Any Instant.*** The collector-emitter voltage V_{CE} should not fall below knee voltage (0.5 V for germanium and 1.0 V for silicon transistors). With collector-emitter voltage V_{CE} less than knee voltage, the collector-base junction is not properly reverse biased and, therefore, the collector cannot attract the charge carriers emitted by the emitter and hence a larger portion of them goes to the base. Consequently, collector current decreases and base current increases. Thus when V_{CE} falls below knee voltage V_{knee} during any part of the signal, that part will be amplified to lesser extent due to reduced value of β and result in unfaithful amplification. For collector-emitter voltage exceeding knee voltage ($V_{CE} > V_{knee}$), the collector-base junction is properly reverse biased and the value of β remains constant, resulting in faithful amplification.

The conditions 1 and 2 ensure that emitter-base junction remains properly forward biased during all parts of the signal and condition 3 ensures that collector-base junction remains properly reverse biased at all times. By fulfilling these conditions the transistor is made to operate over region of its characteristic curves which are linear, parallel and equi-spaced for equal increments of the parameter.

The proper flow of zero signal collector current and the maintenance of proper collector-emitter voltage during the passage of signal is called the *transistor biasing*.

If the transistor is not biased properly, it would work inefficiently and produce distortion in the output signal. A transistor is biased either with the help of battery or associating a circuit with the transistor. The latter method is more efficient and is frequently used. The circuit used for transistor biasing is called the *biasing circuit*. It may be noted that transistor biasing is very essential for the proper operation of transistor in any circuit.

Example 6.1. An N-P-N silicon transistor has V_{CC} = 10 V and the collector load resistance R_C = 3 kΩ. Determine (*i*) the maximum collector current i_C that can be allowed to flow during the application of signal for faithful amplification and (*ii*) the minimum zero signal collector current required.

What is the maximum input signal if β = 100 and a change of 1 V in base-emitter voltage V_{BE} causes a change of 4 mA in collector current i_C? Determine also the base current (maximum value).

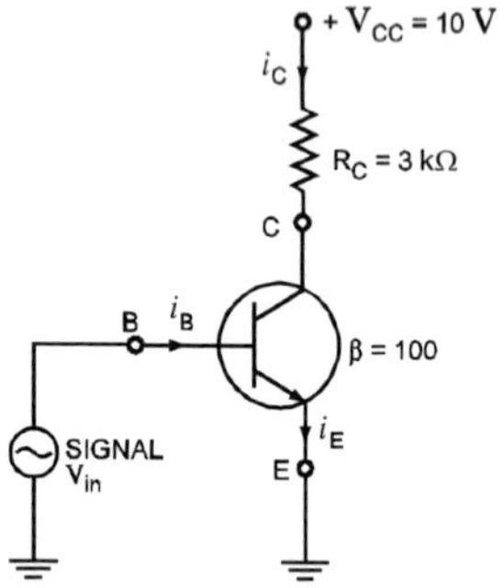

(a) Circuit Diagram

Solution :

$$\text{Collector supply voltage, } V_{CC} = 10 \text{ V}$$

$$\text{Collector load resistance, } R_C = 3 \text{ k}\Omega$$

$$\text{Knee voltage for silicon transistor, } V_{knee} = 1 \text{ V}$$

So maximum voltage drop across resistance R_C that can be allowed,

$$i_{C\,(max)}\, R_C = V_{CC} - V_{knee} = 10 - 1 = 9 \text{ V}$$

$$\text{Maximum allowable collector current, } i_{C\,(max)} = \frac{9}{R_C} = \frac{9}{3\text{ k}\Omega} = 3 \text{ mA } \textbf{Ans.}$$

(*i*) Thus, the maximum collector current that can be allowed during any part of the input signal is 3 mA. If the collector current i_C is allowed to exceed this value, collector-emitter voltage V_{ce} will fall below knee voltage of 1 V and available amplification will not be faithful one.

(*ii*) During negative peak of the input signal, i_C can at the most be allowed to drop to zero value. As the negative and positive half cycles are equal, so the change in collector current i_C due to these will also be equal but in opposite direction.

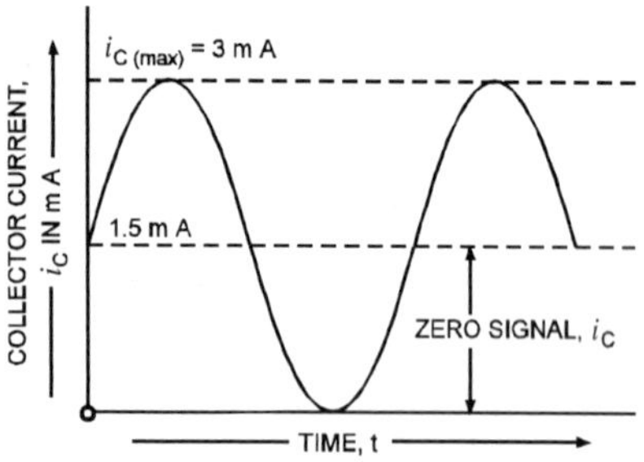

(b)

Fig. 6.1

$$\therefore \text{ Zero signal collector current} = \frac{3\text{ mA}}{2} = 1.5 \text{ mA } \textbf{Ans.}$$

$$\text{Maximum base current, } i_{B\,(max)} = \frac{i_{C\,(max)}}{\beta} = \frac{3\text{ mA}}{100} = 30\ \mu\text{A}$$

$$\text{Signal voltage (base-voltage)} = \frac{\text{Collector current}}{\text{Change in } i_C \text{ per volt change in } V_{BE}}$$

$$= \frac{3\text{ mA}}{4\text{ mA/V}} = 0.75 \text{ V or } 750 \text{ mV } \textbf{Ans.}$$

6.2. SELECTION OF OPERATING POINT

To study the effects of biasing conditions on the performance of a transistor, it is necessary to draw a dc load line on the output characteristics of the transistor. Consider the common emitter transistor amplifier circuit shown in Fig. 6.2. The input terminals are base and the emitter and the output is taken from the collector and emitter. Note that the polarity of the transistor terminal voltage in Fig. 6.2 is such that the base-emitter junction is forward biased and collector-base junction is reverse biased. These are the normal bias polarities for the device.

The output characteristics of the transistor are shown in Fig. 6.3. DC load line is drawn on the output characteristics, as mentioned in Art. 5.18.1.

In designing a circuit a point on the load line is selected as the dc *bias point* or *quiescent point*. This bias point specifies the collector current I_C and collector-emitter voltage V_{CE} that exist when no input signal is applied. When an input signal is applied base current varies according to the amplitude of the input signal and causes collector current to vary, consequently producing an output voltage variation.

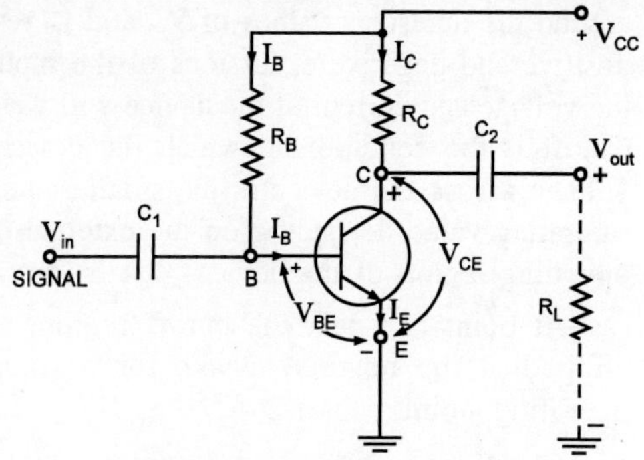

Fig. 6.2. *Common Emitter Transistor Amplifier*

For faithful amplification of the input signal a judicious selection of the operating or quiescent point is of utmost importance. From our discussion of transistor characteristics, it is clear that the transistor operates linearly when it is constrained to operate in the active region. Once an operating point Q is established, time-varying excursions of the input signal should cause an output signal (collector current or collector-emitter voltage) of the same waveform. If the output signal is not a faithful reproduction of the input signal, for example, if it is clipped on one side, the operating point is unsatisfactory and needs relocation on the output characteristics. The question now naturally arises as how to choose the operating point.

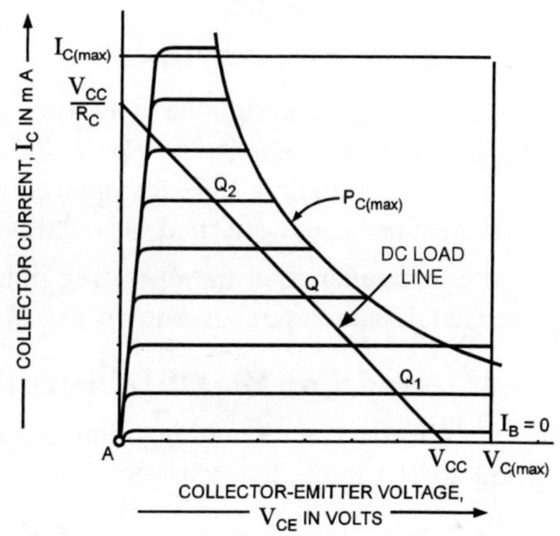

Fig. 6.3

Figure 6.3 gives the output characteristics, with four indicated operating points, of the transistor used in Fig. 6.2. The biasing circuit may be designed to set the device to operate at any of these points or others within the *operating region*. The operating region is the area of current or voltage within the maximum limits for the particular device. The maximum ratings, as mentioned in the manufacturer's specification sheet, are maximum collector current $I_{C\,(max)}$, maximum collector voltage $V_{C\,(max)}$, maximum collector dissipation $P_{C\,(max)}$, and maximum emitter-to-base voltage $V_{BE\,(max)}$. Figure 6.3 shows three of these bounds on typical output characteristics.

The BJT device could be biased to operate outside these maximum limit points but such an operation would result in either a considerable shortening of lifetime of the device or destruction to the device. Confining ourselves to the safe operating region, we may select several operating points, such as Q, Q_1 and Q_2. The exact point often depends on the intended use of the circuit. Let us consider some differences between operation at various points shown in Fig. 6.3 to present some basic ideas about the operating point and, thereby, the bias circuit.

If no bias were used, the device would initially be completely cutoff, which would result in zero current through the device and zero voltage across it (*i.e.* $I_C = 0$ and $V_{CE} = 0$, as represented by point A). It is, therefore, necessary to bias the device so that it can respond or cause change in current and voltage for the entire range of an input signal.

Now let us consider point Q, located at the centre of the load line. When a signal is applied to the circuit, *in addition to the bias level*, collector voltage and current will vary approximately symmetrically around the quiescent values of V_C and I_C respectively allowing the device to react and amplify both the positive and negative excursions of the input signal. If, as could be the case, the input signal is small, the voltage and current of the device will vary but not enough to drive the device into *cutoff* or *saturation.* Cutoff is the condition in which the device no longer conducts. Saturation is the condition in which voltage across the device is the smallest possible with the current in the device reaching a limiting or saturating value depending on the external circuit. The usual amplifier action desired occurs within the operating region of the device, that is, between saturation and cutoff.

At point Q_1, near the cutoff region, output current and voltage would be allowed to vary but clipped at the negative peaks for a sinusoidally varying input signal. So this is not a suitable operating point.

At point Q_2, near the saturation region, the output signal would be clipped at the positive peaks for a sinusoidally varying input signal. Thus this is also not suitable operating point.

Thus point Q, located at the centre of load line, seems to be the best operating point in terms of linear gain or largest possible voltage and current swing.

6.3. BIAS STABILIZATION

Only the fixing of a suitable operating point is not sufficient but it is also to be ensured that the operating point remains stable *i.e.* it does not shift due to change in temperature or due to variations in transistor parameters (due to replacement of transistor). Unfortunately it is not possible in practice unless special efforts are made to achieve it.

The maintenance of the operating point stable (independent of temperature variations or variations in transistor parameters) is known as *stabilization.*

6.3.1. Need For Bias Stabilization

The stabilization of operating point is essential because of (*i*) temperature dependence of collector current I_C (*ii*) individual variations and (*iii*) thermal runaway.

(i) Temperature Dependence of Collector Current I_C. The instability of collector current I_C, being equal to $\beta I_B + (1 + \beta) I_{CO}$, due to variations in temperature, is caused because of the following three main factors.

(*a*) Reverse saturation current (leakage current), I_{CO}, which doubles for every 10°C rise in temperature.

(*b*) Transistor current gain, β, which increases with the increase in temperature.

(*c*) Base-emitter voltage, V_{BE}, which decreases by 2.5 mV per °C.

Any or all of the above factors can cause the bias point to shift from the values originally fixed by the circuit because of a change in temperature.

Though with the change in temperature collector-emitter voltage V_{CE} also changes but the change is very small and collector current I_C is not affected.

(ii) Individual Variations. The value of β and V_{BE} are not exactly the same for any two transistors even of the same type. For an example BC147 is a silicon transistor with β varying from 100 to 600 (for one transistor β may be 100 and for the other it may be 600). The major reason for these variations is that transistor is a new device and manufacturing techniques have not too much advanced. For an instance, it has not been possible to control the base width and it may vary, although slightly, from one transistor to another one of the same type. Such little variations result in a large change in the transistor parameters such as β, V_{BE} etc.

Thus when a transistor is replaced by another transistor of the same type, the operating point may shift. Such problems do not arise in case of vacuum tube circuits because it is possible to manufacture vacuum tubes with identical characteristics.

(iii) Thermal Runaway. The collector current I_C, being equal to $\beta I_B + (1 + \beta) I_{CO}$, increases with the increase in temperature. This leads to increased power dissipation with further increase in temperature. Being a cumulative process, it can lead to *thermal runaway* resulting in burn out of the transistor. The self destruction of an unstabilized transistor is called the *thermal runaway.*

However, if by some modification, I_B is made to fall with increase in temperature automatically, then decrease in the term βI_B can be made to neutralize the increase in the term $(1 + \beta) I_{CO}$, thereby keeping I_C almost constant. This will achieve thermal stability resulting in bias stability.

6.3.2. Means of Achieving Stability For Operating Point

Operating point stability may be achieved by adopting either stabilization techniques or compensation techniques.

Stabilization technique makes use of a resistive biasing circuit that permits such a variation of base current (biasing current) I_B as to maintain collector current I_C almost constant inspite of variations in reverse saturation current I_{CO}, base-emitter voltage V_{BE} and β.

Compensation techniques make use of temperature sensitive devices such as diodes, transistors, thermistors etc. Such devices produce compensating voltages and currents so as to make operating point stable.

Additional means are adopted to maintain the ambient and junction temperatures constant. Air-conditioning or thermostat chambers (wherever possible) may be used for maintaining the ambient or case temperature constant. Heat sink and air blast may be used to remove heat from the transistor.

6.4. STABILITY FACTOR

The degree of success achieved in stabilizing I_C in the face of variations in I_{CO} is expressed in terms of stability factor S* and it is defined as the rate of change of collector current w.r.t. I_{CO} keeping β and V_{BE} constant

$$i.e.\ S = \frac{\partial I_C}{\partial I_{CO}} \approx \frac{\Delta I_C}{\Delta I_{CO}} \simeq \frac{dI_C}{dI_{CO}} \text{ at constant } \beta \text{ and } V_{BE} \text{ (or } I_B) \qquad ...(6.1)$$

From above equation it is obvious that smaller is the value of S, higher is the stability. The ideal value of S is unity (because I_C includes I_{CO}) but it is never possible to achieve it in practice.

While defining stability factor S above, V_{BE} and β are assumed to remain constant while they vary with temperature. Hence other two stability factors S_V and S_β are defined as below :

Stability factor S_V is defined as the rate of change of collector current I_C with base-emitter voltage V_{BE}, keeping I_{CO} and β constant

$$i.e.\ S_V = \frac{\partial I_C}{\partial V_{BE}} = \frac{\Delta I_C}{\Delta V_{BE}} \text{ at constant } I_{CO} \text{ and } \beta \qquad ...(6.2)$$

Stability factor S_β is defined as the rate of change of collector current I_C with respect to β keeping I_{CO} and V_{BE} constant *i.e.*

$$S_\beta = \frac{\partial I_C}{\partial \beta} \approx \frac{\Delta I_C}{\Delta \beta} \text{ at constant } I_{CO} \text{ and } V_{BE} \qquad ...(6.3)$$

6.4.1. General Expression For Stability Factor S

In the active region the basic relationship between collector current I_C and base current I_B is given below :

$$I_C = \beta I_B + (1 + \beta) I_{CO} \qquad ...(6.4)$$

* The larger the value of S, the more likely the circuit to exhibit thermal instability, so it will be more proper to call S an instability factor.

Differentiating above Eq. (6.4) with respect to I_C considering β to be constant, we have

$$1 = \beta \frac{dI_B}{dI_C} + (1+\beta)\frac{dI_{CO}}{dI_C} = \beta \frac{dI_B}{dI_C} + \frac{1+\beta}{S}$$

So stability factor, $S = \dfrac{1+\beta}{1-\beta \dfrac{dI_B}{dI_C}}$ ...(6.5)

The value of $\dfrac{dI_B}{dI_C}$ depends upon the biasing arrangement used and for determination of the stability factor S, it is only necessary to find the relationship between I_C and I_B.

6.4.2. General Expression For Stability Factor S_b

Differentiating Eq. (6.4) w.r.t. I_C considering I_{CO} to be constant we have

$$1 = \beta \frac{dI_B}{dI_C} + I_B \frac{d\beta}{dI_C} + I_{CO} \frac{d\beta}{dI_C}$$

or $\dfrac{d\beta}{dI_C}[I_{CO} + I_B] = 1 - \beta \dfrac{dI_B}{dI_C}$

or $\dfrac{1}{S_\beta}[I_{CO} + I_B] = 1 - \beta \dfrac{dI_B}{dI_C}$ $\because$ from Eq. (6.3) $S_\beta = \dfrac{\delta I_C}{\delta \beta}$

or $S_\beta = \dfrac{I_{CO} + I_B}{1-\beta \dfrac{dI_B}{dI_C}}$ ...(6.6)

6.5. TRANSISTOR BIASING CIRCUITS

There is a large number of circuits for biasing of a transistor. Some of these are quite simple. These circuits differ so as to their ability to keep the quiescent point fixed inspite of variations in transistor characteristics and also effects of temperature variations and ageing. Before discussing different biasing circuits, it becomes imperative to know the basic requirements of a biasing circuit.

Requirements of Biasing Circuit. A biasing network associated with a transistor should fulfill the following requirements :

1. Establish the operating point in the middle of the active region of the characteristics, so that on applying the input signal the instantaneous operating point does not move either to the cutoff region or to the saturation region, even at the extreme values of the input signal.
2. Stabilize the collector current I_C against temperature variations.
3. Make the operating point independent of transistor parameters so that replacement of transistor by another transistor of the same type in the circuit does not shift the operating point.

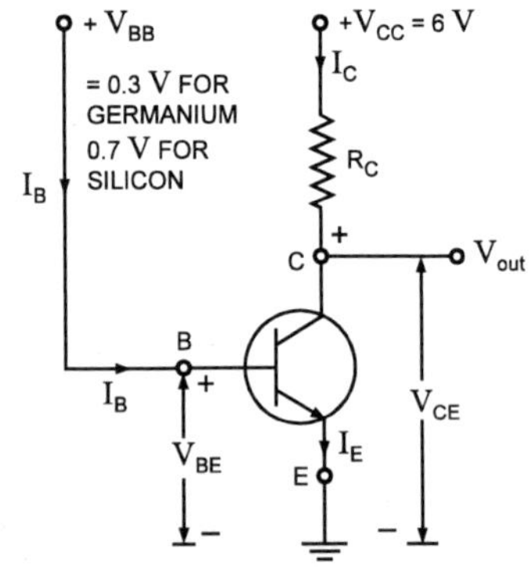

Fig. 6.4

6.5.1. Simplest Biasing Circuit

The simplest biasing circuit could be as shown in Fig. 6.4. The emitter-base junction is forward biased by the battery V_{BB} and the collector-base junction is reverse biased by the battery V_{CC}. The voltage V_{BE} across the forward biased junction is very low (0.3 V for Ge and 0.7 V for Si), therefore, the battery voltage V_{BB} must also be of the same order. The voltage V_{CC} of battery required for reverse bias junction should be of a much large value than voltage V_{BB} so as to reverse bias the collector-base junction. Since it is extremely difficult to have a battery of voltage as low as 0.3 V or 0.7 V, this is not a practical circuit.

6.5.2. Fixed Bias Circuit

The circuit shown in Fig. 6.5 uses two batteries V_{BB} and V_{CC}. The battery V_{BB} gives potential to the base terminal through resistance R_B (of the order of hundred kΩ) and battery V_{CC} supplies power to the collector through resistance R_C (of the order of kΩ).

The base current can be determined by applying Kirchhoff's second law for the branch circuit, starting from battery V_{BB} to ground.

$$V_{BB} = I_B R_B + V_{BE}$$

$$\text{or} \quad I_B = \frac{V_{BB} - V_{BE}}{R_B} \quad \text{...(6.7)}$$

Since V_{BE} is very small (0.7 V in case of silicon transistor and 0.3 V in case of Ge transistor) in comparison to V_{BB}, not much error will be committed if it is neglected. The base current is then given by the simple expression

$$I_B \simeq \frac{V_{BB}}{R_B} \quad \text{...(6.8)}$$

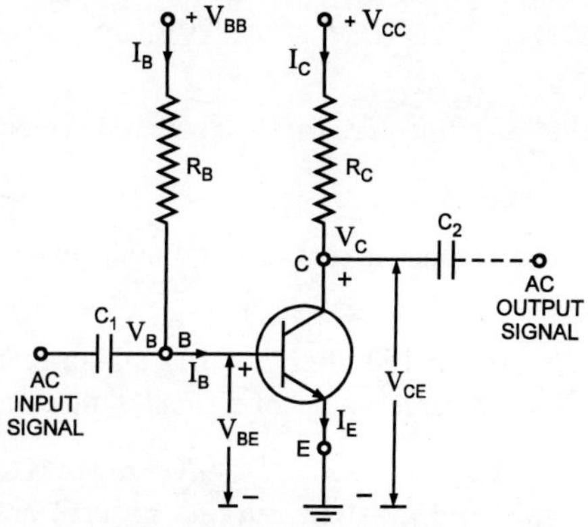

Fig. 6.5. ***Fixed Bias Circuit***

The supply voltage V_{BB} being of fixed value—once the resistance R_B is chosen—the base current I_B is also fixed. Hence the name fixed bias circuit is given.

The collector-emitter section consists of the supply battery V_{CC}, the collector resistor and the transistor collector-emitter junction. The currents through the collector and emitter are about the same since I_B is very small in comparison to either. For linear amplifier operation, the collector current will be β times base current, where β is the transistor current gain *i.e.*

$$I_C = \beta I_B \quad \text{...(6.9)}$$

The base current I_B is determined from the operation of the base-emitter section of the circuit. The collector current, as shown above, will be β times greater than the base current and not at all dependent on the resistance in the collector circuit. Thus collector current I_C is controlled by base-emitter section of the circuit.

Applying Kirchhoff's second law to the branch circuit starting from V_{CC} to ground.

$$V_{CC} = I_C R_C + V_{CE}$$

$$\text{or} \quad V_{CE} = V_{CC} - I_C R_C \quad \text{...(6.10)}$$

The quiescent point is given by I_C, V_{CE}.

The coupling capacitors C_1 and C_2 have reactances quite small at the frequency of ac signal, so that very little voltage drop of ac signal takes place across them. These capacitors are open circuit for direct currents, so that dc biasing potentials of this amplifier stage do not affect the circuits connected at the input as well as output terminals.

The fixed bias circuit is the simplest way to bias a transistor, but from stability point of view it is the worst circuit, because with the rise in temperature of transistor, β changes and also there is increase of leakage current and drop in V_{BE} value.

Since base current I_B controls the collector current I_C and I_C is equal to βI_B, with the change in the value of β, I_C changes (I_B remaining unchanged) and the operating point gets shifted.

The circuit shown in Fig. 6.5 may be modified by using a single battery, as shown in Fig. 6.2. From the practical point of view, it is always preferable to have an electronic circuit operated from a single battery.

The expressions for I_B, I_C and V_{CE} are given below

$$I_B = \frac{V_{CC} - V_{BE}}{R_B} \simeq \frac{V_{CC}}{R_B} \quad ...(6.11)$$

$$I_C = \beta I_B \quad ...(6.12)$$

$$V_{CE} = V_{CC} - I_C R_C \quad ...(6.13)$$

Stability Factor S. From Eq. (6.5) stability factor $S = \dfrac{1+\beta}{1-\beta\dfrac{dI_B}{dI_C}}$ and in fixed biasing method I_B is independent of I_C, as obvious from Eq. (6.8) or Eq. (6.11), so stability factor is given as

$$S = 1 + \beta \quad ...(6.14)$$

If $\beta = 150$, then $S = 151$ which means that collector current I_C increases 151 times as much as I_{CO}. Such a large value of S makes thermal runaway, a definite possibility with this circuit.

Advantages and Disadvantages. Though this biasing circuit is very simple, as very few components (only two resistors and one battery) are required, biasing conditions can be easily set, the calculations are simple and there is no loading of the source by the biasing circuit, as no resistor is used across base-emitter junction but this method of biasing is rarely used. This is due to the following reasons.

1. This method provides poor stability, as there is no means to stop a self increase in collector current I_C owing to temperature rise and individual variations. For example, if β increases either due to temperature rise or due to replacement of transistor, then collector current, I_C also increases by the same factor as base current I_B is constant.
2. There are good chances of thermal runaway. This is due to high stability factor S.

Example 6.2. In the fixed bias circuit of transistor, V_{CC} = 15 V, R_B = 300 kΩ, R_L = 2 kΩ. If β = 100, I_{CO} = 20 nA and V_{BE} = 0.7 V. Find the stability factor of Q-point with respect to I_{CO}.

Solution : In a fixed bias circuit, the stability factor of operating point w.r.t. I_{CO} is given by equation

$$S = 1 + \beta \qquad \text{Refer to Eq. (6.14)}$$

Substituting β = 100 in above equation, we have

$$S = 1 + 100 = 101 \quad \textbf{Ans.}$$

Example 6.3. In a fixed bias circuit using transistor with α = 0.97, the temperature changes from 30°C to 60°C producing change in I_C from 2.2 mA to 3.8 mA. Assume reverse saturation current changes 1.7 μA/°C. What is stability factor ?

Solution : Change in collector current, $\Delta I_C = 3.8 - 2.2 = 1.6$ mA or 1.6×10^{-3} A

Change in temperature, $\Delta t = 60 - 30 = 30$°C

Change in reverse saturation current for Δt of 30°C

$$\Delta I_{CO} = 30 \times 1.7 \times 10^{-6} = 51 \times 10^{-6} \text{ A}$$

$$\text{Stability factor, } S = \frac{\Delta I_C}{\Delta I_{CO}} = \frac{1.6 \times 10^{-3}}{51 \times 10^{-6}} = 31.37 \quad \textbf{Ans.}$$

Example 6.4. Draw the dc load line and locate the operating point for fixed biasing transistor circuit shown in the Fig. 6.6 (*a*). What will be its stability factor ? $V_{BE} = 0.7$ V.

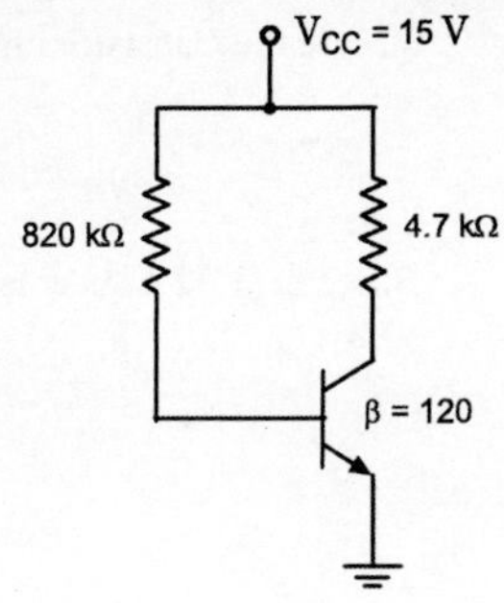

(*a*)

Solution : Base current, $I_B = \dfrac{V_{CC} - V_{BE}}{R_B}$

$$= \frac{(15 - 0.7)\text{ V}}{820\text{ k}\Omega}$$

$$= 17.44\ \mu\text{A}.$$

Collector current, $I_C = \beta I_B = 120 \times 17.44 \times 10^{-6}$

$$= 2.09\text{ mA}$$

Collector emitter voltage, $V_{CE} = V_{CC} - I_C R_C$

$$= 15 - 2.09 \times 10^{-3} \times 4.7 \times 10^3$$

$$= 5.164\text{ V}$$

Thus $I_{CQ} = 2.09$ mA and $V_{CEQ} = 5.164$ V

For fixed bias stability factor, $S = \beta + 1 = 120 + 1 =$ **121 Ans.**

For dc load line point A $\left(0, \dfrac{V_{CC}}{R_L}\right)$ and point B $(V_{CC}, 0)$

$V_{CE} = V_{CC} = 15$ V when $I_C = 0$

and $I_C = \dfrac{V_{CC}}{R_L} = \dfrac{15\text{ V}}{4.7\text{ k}\Omega} = 3.2$ mA when $V_{CE} = 0$

Load line is drawn joining points A and B, as shown in Fig. 6.6 (*b*).

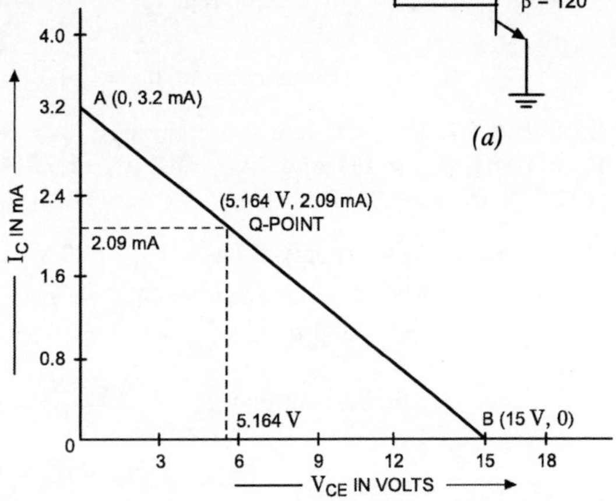

(*b*)

Fig. 6.6

Example 6.5. What is the base current in the following figure (Fig. 6.7) ?

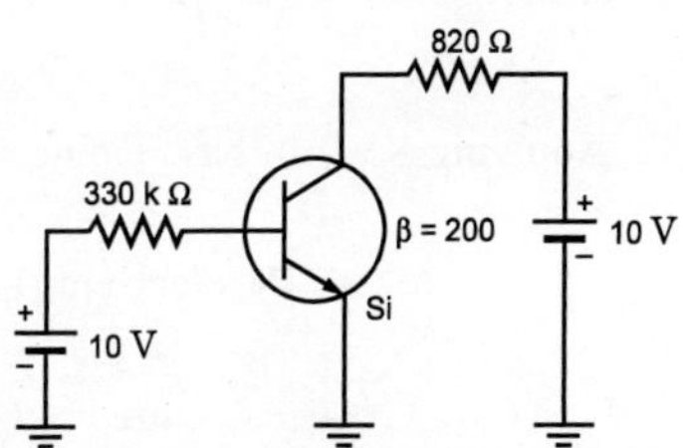

Fig. 6.7

Solution : Base supply voltage, $V_{BB} = 10$ V

Base-emitter voltage, $V_{BE} = 0.7$ V ∵ transistor is silicon one

Base current, $I_B = \dfrac{V_{BB} - V_{BE}}{R_B} = \dfrac{10 - 0.7}{330 \times 10^3} = 28.2\ \mu\text{A}$ **Ans.**

Example 6.6. Consider the transistor circuit of Fig. 6.8 which has a resistance included between emitter and ground. Show that the transistor is operating in active mode. Calculate I_C, I_E and I_B. Given : $\beta = 50$.

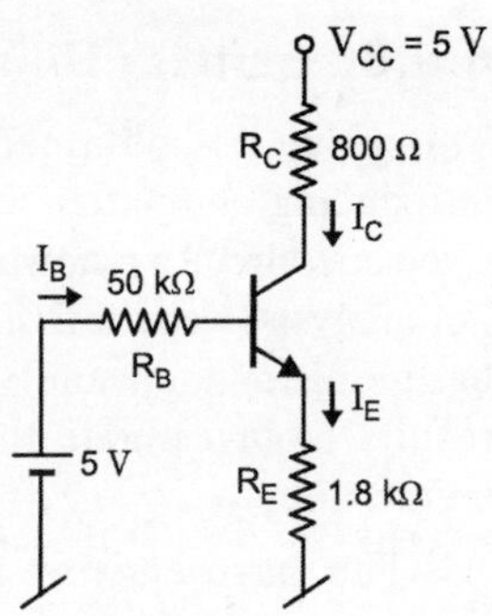

Fig. 6.8

Solution : Applying KVL to input loop, we have

$$V_{BB} = I_B R_B + V_{BE} + I_E R_E = \frac{I_C}{\beta} R_B + V_{BE} + \frac{\beta + 1}{\beta} I_C R_E$$

$$\because\ I_B = \frac{I_C}{\beta} \text{ and } I_E = \left(1 + \frac{1}{\beta}\right) I_C$$

$$\text{or } I_C = \frac{V_{BB} - V_{BE}}{\dfrac{R_B}{\beta} + \dfrac{\beta + 1}{\beta} R_E} = \frac{5 - 0.7}{\dfrac{50 \times 10^3}{50} + \dfrac{51}{50} \times 1.8 \times 10^3} = 1.52\text{ mA } \textbf{Ans.}$$

In case the transistor circuit goes into saturation, the maximum

$$\text{Collector current, } I_{C\,max} = \frac{V_{CC}}{R_C + R_E} = \frac{5}{800 + 1{,}800} = 1.92 \text{ mA}$$

Since I_C (1.52 mA) is less than $I_{C\,max}$ (1.92 mA), the circuit is in active mode.

$$\text{Base current, } I_B = \frac{I_C}{\beta} = \frac{1.52 \text{ mA}}{50} = 30.4\ \mu\text{A} \quad \textbf{Ans.}$$

$$\text{Emitter current, } I_E = I_B + I_C = 0.0304 + 1.52 = 1.55 \text{ mA} \quad \textbf{Ans.}$$

Example 6.7. In a CE transistor circuit V_{CC} = 30 V, V_{BB} = 6V, R_C = 1 kΩ, R_B = 1 MΩ, β_{dc} = 100 and I_{CBO} = 0.1 μA at 25 °C. Compute I_B and I_C at ((*i*) 25 °C (*ii*) 55 °C).

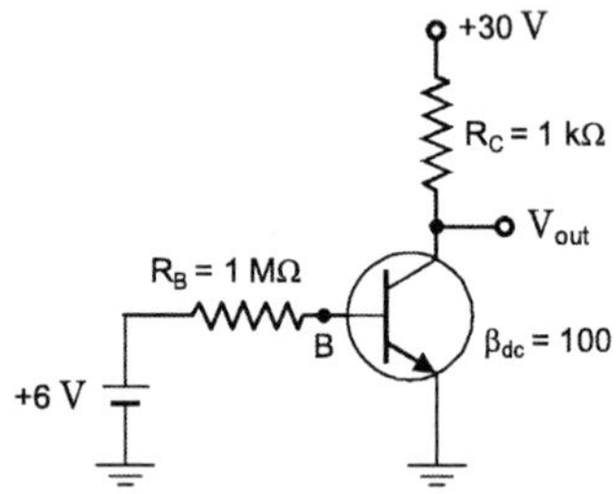

Fig. 6.9

Solution : Collector supply voltage, V_{CC} = 30 V

(*i*) Applying KVL to base-emitter loop we have

$$V_{BB} - I_B R_B - V_{BE} = 0$$

$$\text{or Base current, } I_B = \frac{V_{BB} - V_{BE}}{R_B} = \frac{6 - 0.7}{1 \times 10^6} = 5.3\ \mu\text{A} \quad \textbf{Ans.}$$

$$\text{So } I_C = \beta I_B + (\beta + 1)\, I_{CBO}$$
$$= 100 \times 5.3 \times 10^{-6} + (100 + 1) \times 0.1 \times 10^{-6} = 0.54 \text{ mA} \quad \textbf{Ans.}$$

(*ii*) At 55°C, $I_{CBO} = I_{CBO}$ at 25°C $\times\ 2^{\frac{55-25}{10}} = 0.1 \times 10^{-6} \times 2^3 = 0.8\ \mu$A

At 55°C, $V_{BE} = 0.7 - 2.5 \times 10^{-3} \times (55 - 25) = 0.625$ V

Applying KVL to base-emitter loop we have

$$\text{Base current, } I_B = \frac{V_{BB} - V_{BE}}{R_B} = \frac{6 - 0.625}{1 \times 10^6} = 5.375\ \mu\text{A} \quad \textbf{Ans.}$$

$$\text{and collector current, } I_C = \beta I_B + (\beta + 1)\, I_{CBO}$$
$$= 100 \times 5.375 \times 10^{-6} + (100 + 1) \times 0.8 \times 10^{-6} = 0.6183 \text{ mA} \quad \textbf{Ans.}$$

6.5.3. Emitter Bias Circuit

This circuit is obtained by simply introducing an emitter resistor to the fixed bias circuit as shown in Fig. 6.10. For analysis, we will first examine the base-emitter loop and then use the results to investigate the collector-emitter loop.

The base-emitter loop of the network given in Fig. 6.10 may be redrawn as shown in Fig. 6.11. Applying KVL around the indicated loop, we have

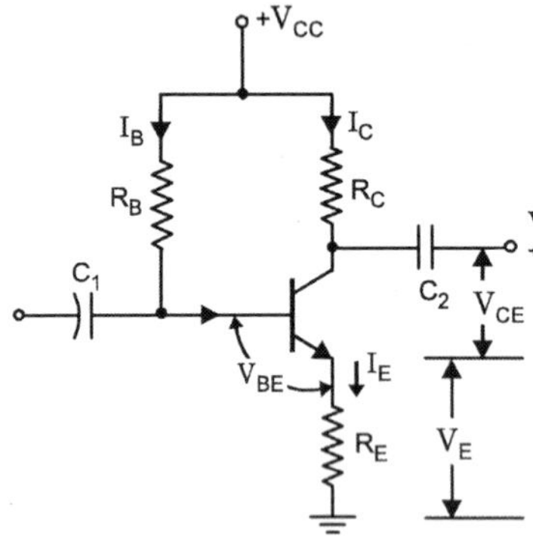

Fig. 6.10. ***Emitter Bias Circuit***

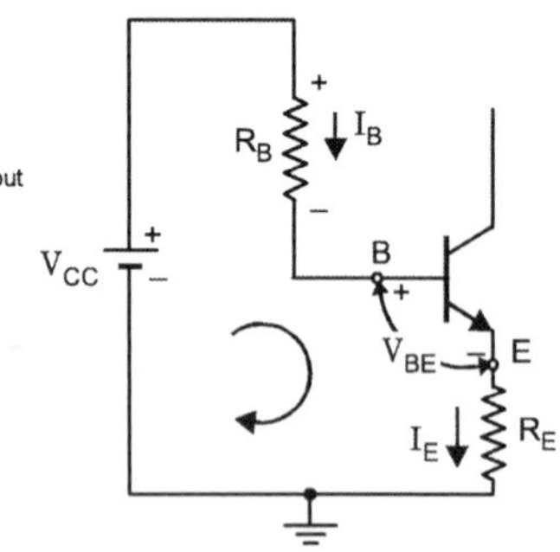

Fig. 6.11

$$V_{CC} = I_B R_B + V_{BE} + I_E R_E = I_B R_B + V_{BE} + (\beta + 1) I_B R_E \qquad \because I_E = (\beta + 1) I_B$$

$$\text{or } I_B = \frac{V_{CC} - V_{BE}}{R_B + (\beta + 1) R_E} \qquad \text{...(6.15)}$$

Note that the only difference between the above equation and equation for the fixed bias configuration $\left[I_B = \frac{V_{CC} - V_{BE}}{R_B}\right]$ is the term $(\beta + 1)R_E$.

The collector-emitter loop is redrawn in Fig. 6.12.

Applying KVL for the indicated loop, we have

$$I_C R_C + V_{CE} + I_E R_E = V_{CC}$$

$$\text{or } V_{CE} = V_{CC} - I_C R_C - I_C R_E \quad \because I_C \simeq I_E$$

$$\text{or } V_{CE} = V_{CC} - I_C(R_C + R_E) \qquad \text{...(6.16)}$$

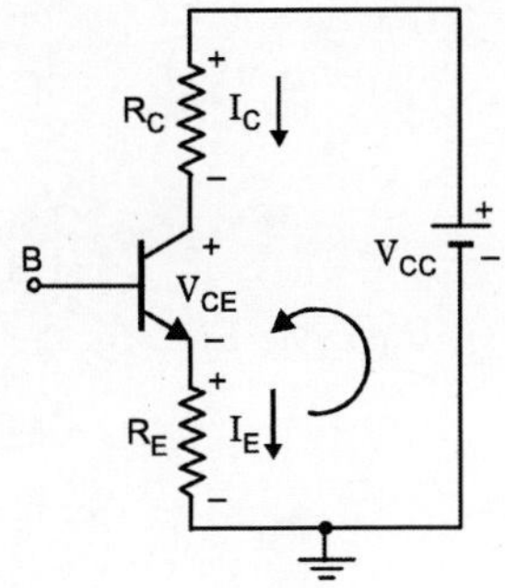

Fig. 6.12

Voltage from emitter-to-ground is

$$V_E = I_E R_E \qquad \text{...(6.17)}$$

while the voltage from collector-to-emitter is

$$V_{CE} = V_C - V_E = V_C - I_E R_E \qquad \text{...(6.18)}$$

$$\text{Collector voltage, } V_C = V_{CC} - I_C R_C \qquad \text{...(6.19)}$$

$$\text{and Base voltage, } V_B = V_{CC} - I_B R_B \qquad \text{...(6.20)}$$

$$= V_{BE} + V_E \qquad \text{...(6.21)}$$

From Eq. (6.15), we have $V_{CC} = I_B R_B + V_{BE} + (I_B + I_C)R_E$

Differentiating above equation w.r.t. I_C (considering V_{BE} to be independent of I_C), we have

$$0 = (R_B + R_E)\frac{dI_B}{dI_C} + R_E$$

$$\text{or } \frac{dI_B}{dI_C} = \frac{-R_E}{R_B + R_E} \qquad \text{...(6.22)}$$

From Eq. (6.5) stability factor S is given as

$$S = \frac{1+\beta}{1 - \beta\frac{dI_B}{dI_C}}$$

Substituting the value of $\frac{dI_B}{dI_C}$ from Eq. (6.22) in above expression, we have

$$\text{Stability factor, } S = \frac{1+\beta}{1+\beta\frac{R_E}{R_B + R_E}} = (1+\beta)\frac{1+\frac{R_B}{R_E}}{1+\beta+\frac{R_B}{R_E}} \qquad \text{...(6.23)}$$

The above equation indicates that stability factor S varies between 1 for smaller values of $\frac{R_B}{R_E}$ and $(1 + \beta)$ for larger values of $\frac{R_B}{R_E}$. For proper operation both V_{CC} and R_E should be larger and R_B smaller.

The addition of the emitter resistor to the dc bias of the BJT provides improved stability, *i.e.* the dc bias currents and voltages remain closer to where they were fixed by the circuit when the outside conditions, such as temperature, and transistor beta change.

Example 6.8. Determine I_C, V_E, V_B, V_C and I_B for the following circuit (Fig. 6.13).

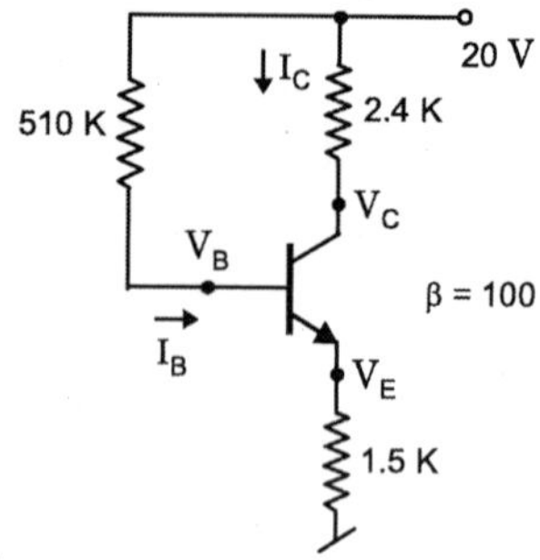

Fig. 6.13

Solution : For given emitter bias CE amplifier,

$$R_B = 510 \text{ k}\Omega$$
$$R_C = 2.4 \text{ k}\Omega$$
$$R_E = 1.5 \text{ k}\Omega$$
$$V_{CC} = 20 \text{ V}$$
$$\text{and } \beta = 100$$

Applying KVL for the input loop

$$V_{CC} = I_B R_B + V_{BE} + I_E R_E = I_B R_B + V_{BE} + (\beta + 1) I_B R_E$$

$$\text{or } I_B = \frac{V_{CC} - V_{BE}}{R_B + (\beta+1) R_E} = \frac{20 - 0.7}{510 \times 10^3 + 101 \times 1.5 \times 10^3}$$

$$= \frac{19.3}{661.5 \times 10^3} = 0.0292 \text{ mA } \textbf{Ans.}$$

So collector current, $I_C = \beta I_B = 100 \times 0.0292 = 2.92$ mA **Ans.**

Collector voltage, $V_C = V_{CC} - I_C R_C = 20 - 2.92 \times 10^{-3} \times 2.4 \times 10^3 = 13$ V **Ans.**

Emitter voltage, $V_E = I_E R_E = (\beta + 1) I_B R_E = 101 \times 0.0292 \times 10^{-3} \times 1.5 \times 10^3 = 4.42$ V **Ans.**

Base voltage, $V_B = V_{CC} - I_B R_B = 20 - 0.0292 \times 10^{-3} \times 510 \times 10^3 = 5.12$ V **Ans.**

Example 6.9. Determine the resistor values for the following network for the indicated operating point and power supply voltage.

Solution : The given circuit is emitter biased CE amplifier circuit and its operating point is

$$V_{CEQ} = 10 \text{ V and } I_{CQ} = 2 \text{ mA}$$

Applying KVL in the outer loop, we have

$$V_{CC} = I_{CQ} R_C + V_{CEQ} + I_{EQ} R_E$$

$$\text{or } V_{CC} = I_{CQ}(R_C + R_E) + V_{CEQ} \quad \because I_{CQ} \simeq I_{EQ}$$

$$\text{or } R_C + R_E = \frac{V_{CC} - V_{CEQ}}{I_{CQ}} = \frac{20 - 10}{2 \times 10^{-3}} = 5 \text{ k}\Omega$$

Choosing two practical values for R_C and R_E

$$R_C = 3.9 \text{ k}\Omega \textbf{ Ans.}$$
$$R_E = 1.0 \text{ k}\Omega \textbf{ Ans.}$$

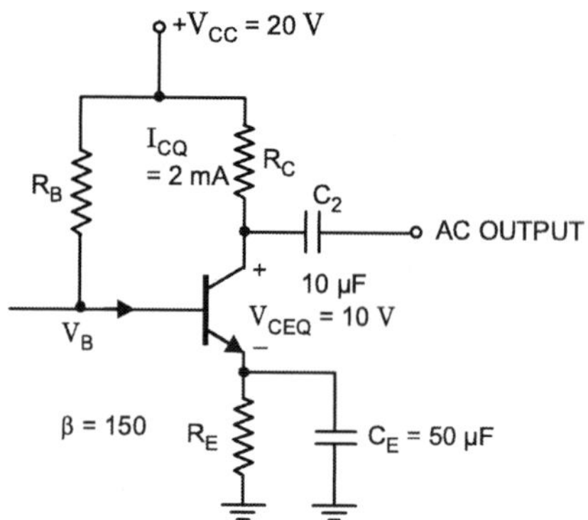

Fig. 6.14

Now applying KVL to the input loop, we have

$$V_{CC} = I_B R_B + V_{BE} + I_E R_E$$

$$\text{or } V_{CC} = \frac{I_C}{\beta} R_B + 0.7 + I_C R_E$$

$$\text{or } R_B = \frac{V_{CC} - 0.7 - I_C R_E}{I_C} \times \beta = \frac{20 - 0.7 - 2 \times 10^{-3} \times 1 \times 10^3}{2 \times 10^{-3}} \times 150$$

$$= 1{,}297 \text{ k}\Omega \simeq 1.2 \text{ M}\Omega \text{ (standard value) } \textbf{Ans.}$$

Example 6.10. For the bias circuit shown in Fig. 6.15, determine the value of R_B to yield an operating point at $I_C = 5$ mA. Assume $\beta_0 = 60$, $V_{EB} = 0.7$ V, $R_C = 500\ \Omega$, $R_E = 100\ \Omega$ and $V_{CC} = -6$ V.

What is one advantage and one limitation of this type of biasing ?

Solution : The circuit is redrawn, with dc biasing as shown in Fig. 6.16.

Applying Kirchhoff's voltage law to the collector-emitter loop, we have

$$V_{CC} = I_{CQ}R_C + V_{CEQ} + I_{EQ}R_E$$

$$-6 = -5 \times 10^{-3} \times 500 + V_{CEQ} - \frac{60+1}{60} \times 5 \times 10^{-3} \times 100 \qquad \because I_{EQ} = (\beta + 1)I_{CQ}$$

$$\text{or } V_{CEQ} = -6 + 2.5 + 0.508 = -2.992 \text{ V}$$

Applying Kirchhoff's voltage law to base-emitter loop, we have

$$R_B = \frac{V_{CC} - V_{BE} - I_{EQ}R_E}{I_{BQ}}$$

$$= \frac{V_{CC} - V_{BE} - \left(\frac{\beta+1}{\beta}\right)I_{CQ}R_E}{\frac{I_{CQ}}{\beta}}$$

$$\because I_{BQ} = \frac{I_{CQ}}{\beta} \text{ and } I_{EQ} = \frac{\beta+1}{\beta}I_{CQ}$$

$$= \frac{-6-(-0.7)-\left(\frac{60+1}{60}\right)(-5\times 10^{-3})\times 100}{\frac{-5\times 10^{-3}}{60}}$$

$$= \frac{4.792 \times 60}{5} = 57.5 \text{ k}\Omega \quad (56 \text{ k}\Omega\text{, standard value}) \textbf{ Ans.}$$

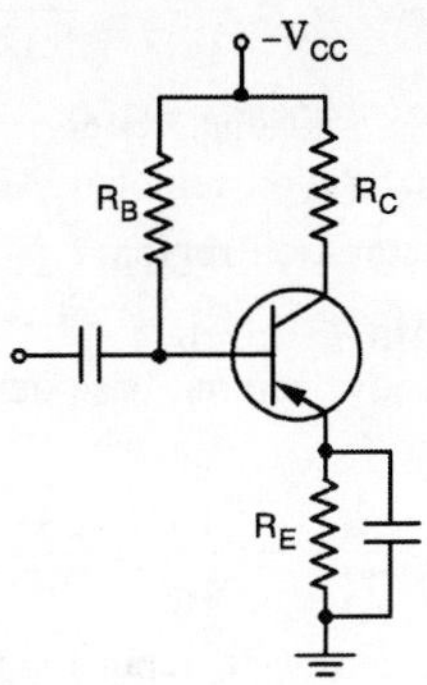

Fig. 6.15

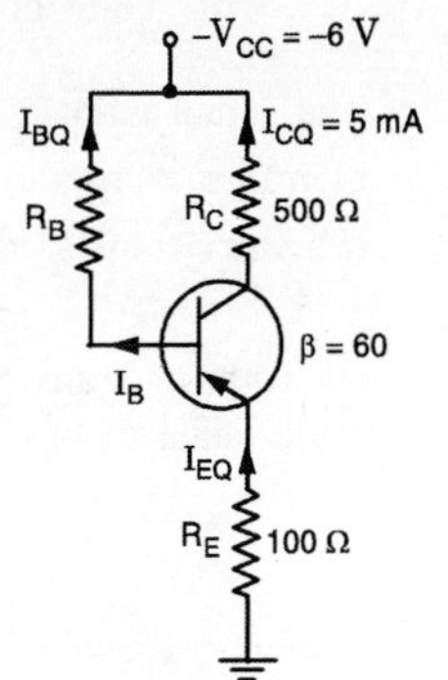

Fig. 6.16

This circuit stabilises the base current and so the operating point can be stabilized but the operating point stability also depends upon other factors like junction temperature and transistor parameters mentioned in Art. 6.3.1. So this circuit is not able to stabilize the operating point.

Example 6.11. For the circuit shown in Fig. 6.17(*a*), assume h_{FE} = 100 and V_{BE} = 0.8 V.

(*a*) Find if the silicon transistor is in cutoff, saturation or active region.

(*b*) Find V_C.

(*c*) Find the minimum value for the base resistor for which the transistor operates in active region.

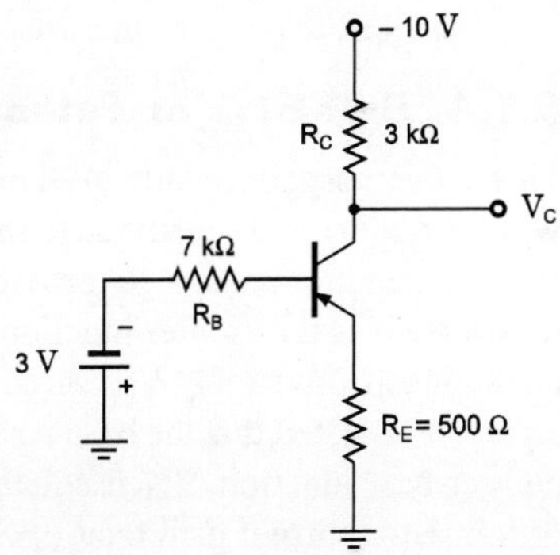

Fig. 6.17(*a*)

Solution : The circuit is redrawn as shown in Fig. 6.17 (*b*).

(*a*) Applying Kirchhoff's voltage law to the input loop we have

$$V_{BB} = R_E \times (I_B + I_C) + R_B \times I_B + V_{BE} \qquad ...(i)$$

Putting all the values in above Eq. (*i*) we have

$$3 = 500\,(I_B + I_C) + 7{,}000\,I_B + 0.8$$

$$\text{or } 7{,}500\,I_B + 500\,I_C = 2.2 \qquad ...(ii)$$

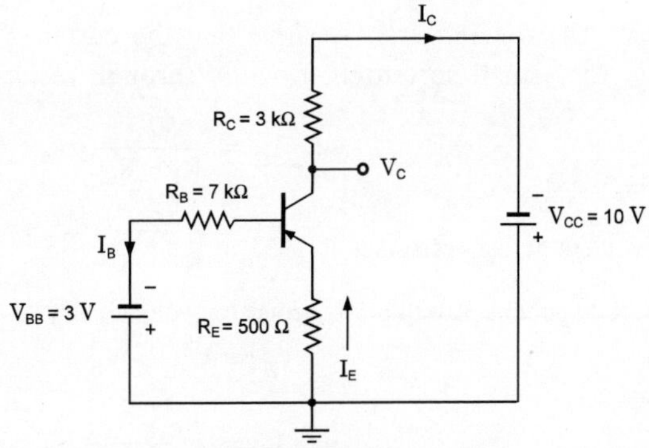

Fig. 6.17 (*b*)

Applying Kirchhoff's voltage law to output loop we have

$$V_{CC} = 500\,(I_B + I_C) + 3{,}000\,I_C + V_{CE}$$

$$\text{or } 500\,I_B + 3{,}500 I_C = 10 - V_{CE}$$

Taking limiting value of V_{CE} = 1 V for saturation we have

$$500\ I_B + 3{,}500\ I_C = 9.0 \qquad ...(iii)$$

Solving Eqs. (*ii*) and (*iii*) we have

$$I_C = 2.55 \text{ mA}$$

$$I_B = 0.123 \text{ mA}$$

$$\text{Gain of amplifier, } \beta = \frac{I_C}{I_B} = \frac{2.55}{0.123} = 20.7$$

The actual gain is very less than h_{FE} = 100,
therefore, transistor is in saturation region. **Ans.**

(*b*) Since transistor is in saturation region,

$$V_C = -V_{CE} - (I_B + I_C)\,R_E = -1 - (0.123 + 2.55) \times 10^{-3} \times 500 = -2.3365 \text{ V} \quad \textbf{Ans.}$$

(*c*) To bring transistor to active region, base current I_B should be reduced (by increasing resistance R_B). The maximum value of base current, $I_{B\,(max)}$ for which transistor remains in active region

$$I_{B\,(max)} = \frac{I_C}{h_{fE}} = \frac{2.55}{100} = 0.0255 \text{ mA}$$

Applying Kirchhoff's voltage law to input loop we have

$$V_{BB} = I_C R_E + V_{BE} + R_B I_{B\,(max)} \qquad \because I_E \simeq I_C$$

Putting the values we have

$$3 = 2.55 \times 10^{-3} \times 500 + 0.8 + 0.0255 \times 10^{-3} R_B$$

$$\text{or } R_B = \frac{3 - 1.275 - 0.8}{0.0255 \times 10^{-3}} = 36.27 \text{ k}\Omega \qquad (39 \text{ k}\Omega \text{ standard value})$$

Therefore, to bring the transistor in active region the minimum value for base resistance should be 39 kΩ **Ans.**

6.5.4. Self Bias or Potential Divider Bias Circuit

This is the most commonly used biasing arrangement. The arrangement is shown in Fig. 6.18 (*a*). The name voltage divider is derived due to the fact that the voltage divider is formed by the resistors R_1 and R_2 across V_{CC}. The emitter resistor R_E provides stabilization. The resistor R_E causes a voltage drop in a direction so as to reverse bias the emitter junction. Since the emitter-base junction is to be forward biased, the base voltage is obtained from supply V_{CC} through R_1–R_2 network. For forward biasing the emitter-base junction R_1 and R_2 are so adjusted that the base terminal becomes more positive than emitter. The net forward bias across the emitter-base junction, V_{BE} is equal to V_B minus dc voltage drop across R_E. The dc bias circuit is independent of transistor current gain factor β. In case of amplifiers, to avoid the loss of ac signal (because of feedback caused by R_E) a capacitor of large capacitance is connected across R_E. The capacitor offers a very small reactance to the ac signal and so it passes through the capacitor.

Circuit Analysis. Assume that the current flowing through resistance R_1 is I_1. As base current I_B is very small so current flowing through resistance R_2 can also be assumed to be equal to I_1

$$\text{and } I_1 = \frac{V_{CC}}{R_1 + R_2}$$

$$\text{Voltage across resistance } R_2,\ V_B = \frac{V_{CC}}{R_1 + R_2} \cdot R_2$$

Applying Kirchhoff's second (or voltage) law to the base-emitter loop [Fig. 6.18(*a*)] we have

$$V_B = V_{BE} + V_E = V_{BE} + I_E R_E$$

$$\text{or } I_E = \frac{V_B - V_{BE}}{R_E}$$

and collector current, $I_C \simeq I_E \simeq \dfrac{V_B - V_{BE}}{R_E}$ ...(6.24)

Applying Kirchhoff's second (or voltage) law to the collector-emitter loop we have

$$V_{CC} = I_C R_C + V_{CE} + I_E R_E = I_C R_C + V_{CE} + I_C R_E \quad \text{since } I_C \simeq I_E$$

$$= V_{CE} + I_C (R_C + R_E)$$

$$\text{or } V_{CE} = V_{CC} - I_C (R_C + R_E) \quad ...(6.25)$$

From Eqs. (6.24) and (6.25) the values of I_C and V_{CE} can be determined and the quiescent point Q is established.

It is clear from Eq. (6.24) that I_C does not at all depend upon β. Though collector current I_C depends upon V_{BE} but in practice V_{BE} is very small in comparison to V_B and so collector current I_C is practically independent of V_{BE}. Thus collector current I_C in this biasing circuit is almost independent of transistor parameters and hence good stabilisation is ensured.

In this biasing circuit the emitter resistance R_E provides excellent stabilisation. This is explained as below :

Now let the temperature of transistor junction rise when it is loaded. This causes increase in leakage currents and so increase in the value of β. Hence collector current I_C tends to increase. With the increase in the value of I_C, voltage drop across emitter resistance R_E increases. Since voltage drop across R_2 (*i.e.* V_B) is independent of collector current, therefore, V_{BE} decreases and so I_B, I_E and I_C.

Thus we see that the circuit has tendency to hold the Q point (I_C) stable automatically. This is due to feedback action. The increase in I_C has immediately a reaction change of feedback so as to correct the situation.

Stability Factor S. The circuit to the left of the base terminal is replaced by its Thevenin's equivalent circuit, as shown in Fig. 6.18 (*b*).

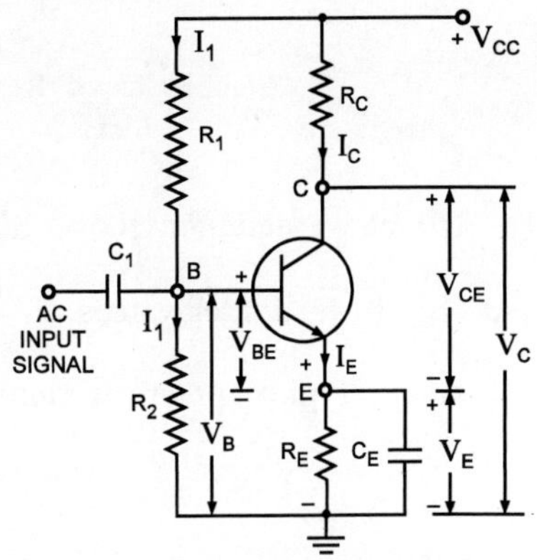

(a) Self Bias Circuit

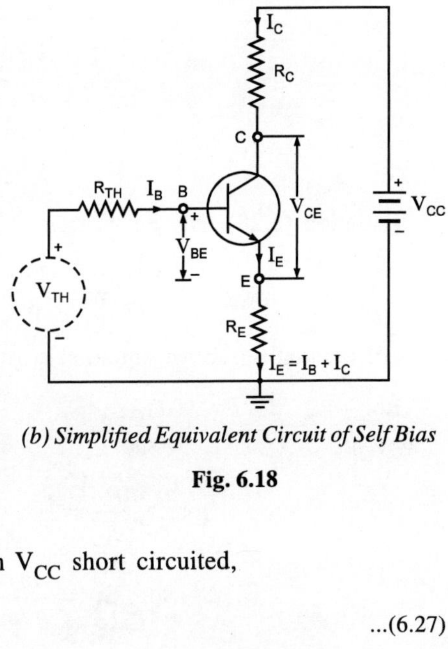

(b) Simplified Equivalent Circuit of Self Bias

Fig. 6.18

Open-circuit voltage across base and ground terminals,

$$V_{Th} = V_B = \frac{R_2}{R_1 + R_2} V_{CC} \quad ...(6.26)$$

Resistance seen into the base and ground terminals with V_{CC} short circuited,

$$R_{Th} = R_1 \parallel R_2 = \frac{R_1 R_2}{R_1 + R_2} \quad ...(6.27)$$

Applying Kirchhoff's voltage law around the closed base circuit [Fig. 6.18 (*b*)] yields

$$V_{Th} = I_B R_{Th} + V_{BE} + I_E R_E = I_B R_{Th} + V_{BE} + (I_B + I_C) R_E \quad ...(6.28)$$

Differentiating above Eq. (6.28) w.r.t. I_C (considering V_{BE} to be independent of I_C) we have

$$0 = R_E + (R_{Th} + R_E) \frac{dI_B}{dI_C}$$

$$\text{or } \frac{dI_B}{dI_C} = \frac{-R_E}{R_{Th} + R_E} \quad ...(6.29)$$

From Eq. (6.5) stability factor S is given as

$$S = \frac{1+\beta}{1-\beta\dfrac{dI_B}{dI_C}}$$

Substituting the value of $\dfrac{dI_B}{dI_C}$ from Eq. (6.29) in above equation we have

$$\text{Stability factor, } S = \frac{1+\beta}{1+\beta\dfrac{R_E}{R_{Th}+R_E}} = (1+\beta)\frac{1+R_{Th}/R_E}{1+\beta+\dfrac{R_{Th}}{R_E}} \qquad ...(6.30)$$

The above equation shows that stability factor S varies between 1 for small values of R_{Th}/R_E and $(1 + \beta)$ for larger values of $\dfrac{R_{Th}}{R_E}$. For proper operation both R_E and V_{CC} should be larger and R_{Th} small. Typical practical value of S for this type of biasing circuit is about 10.

$$S_\beta = \frac{I_{CO}+I_B}{1-\beta\dfrac{dI_B}{dI_C}} \qquad \text{...Refer to Eq. (6.6)}$$

$$= \frac{I_{CO}+I_B}{1-\beta\times\dfrac{-R_E}{R_{Th}+R_E}} \qquad \because \text{ From Eq. (6.29) } \frac{dI_B}{dI_C} = \frac{-R_E}{R_{Th}+R_E}$$

$$= \frac{I_{CO}+I_B}{1+\dfrac{\beta R_E}{R_{Th}+R_E}} \qquad ...(6.31)$$

Equation (6.28) may be written as

$$V_{Th} = \frac{I_C}{\beta}R_{Th} + V_{BE} + \left(\frac{I_C}{\beta}+I_C\right)R_E = \frac{I_C}{\beta}(R_{Th}+R_E) + V_{BE} + I_C R_E$$

Differentiating above equation w.r.t. V_{BE}, we have

$$\frac{R_{Th}+R_E}{\beta}\frac{dI_C}{dV_{BE}} + 1 + \frac{dI_C}{dV_{BE}}R_E = 0$$

$$\text{or} \quad \frac{dI_C}{dV_{BE}} = \frac{-\beta}{R_{Th}+R_E+\beta R_E} = \frac{-\beta}{R_{Th}+R_E(1+\beta)}$$

$$\text{So} \quad S_V = \frac{dI_C}{dV_{BE}} = \frac{-\beta}{R_{Th}+R_E(1+\beta)} \qquad ...(6.32)$$

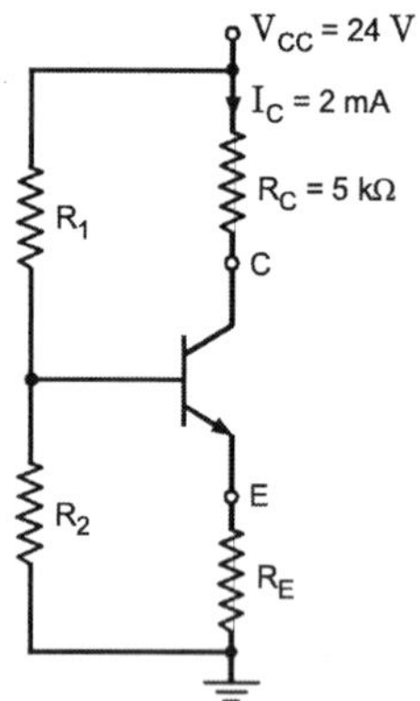

Fig. 6.19

Example 6.12. A silicon transistor with $\beta = 100$ is to be used in self biasing circuit, such that Q point corresponds to $V_{CE} = 12$ V, $I_C = 2$ mA. Find R_E if $V_{CC} = 24$ V and $R_C = 5$ kΩ.

Solution :

$$I_C = 2 \text{ mA}$$

$$\text{Emitter current, } I_E = \frac{\beta+1}{\beta}I_C = \frac{100+1}{100}\times 2 \text{ mA} = 2.02 \text{ mA}$$

Applying KVL to outer loop, we have

$$V_{CC} = I_C R_C + V_{CE} + I_E R_E$$

$$\text{or} \quad R_E = \frac{V_{CC}-I_C R_C - V_{CE}}{I_E} = \frac{24-2\times10^{-3}\times5\times10^{3}-12}{2.02} \simeq 1 \text{ k}\Omega \text{ **Ans.**}$$

Example 6.13. For a voltage divider biasing circuit shown in Fig. 6.20. Find I_C, V_{CE}, I_B, V_E and V_B.

Solution : Applying Thevenin's theorem to given potential divider network, we have

$$\text{Thevenin's resistance, } R_{Th} = R_1 \parallel R_2 = \frac{R_1 \times R_2}{R_1 + R_2} = \frac{47 \times 5.6}{47 + 5.6} \text{ k}\Omega = 5 \text{ k}\Omega$$

$$\text{Thevenin's voltage, } V_{Th} = \frac{R_2}{R_1 + R_2} \times V_{CC} = \frac{5.6}{47 + 5.6} \times 20 = 2.13 \text{ V}$$

The circuit will be redrawn as shown in Fig. 6.21.

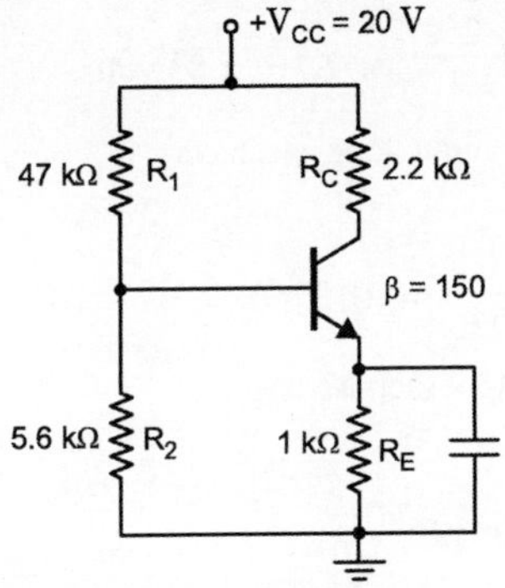

Fig. 6.20

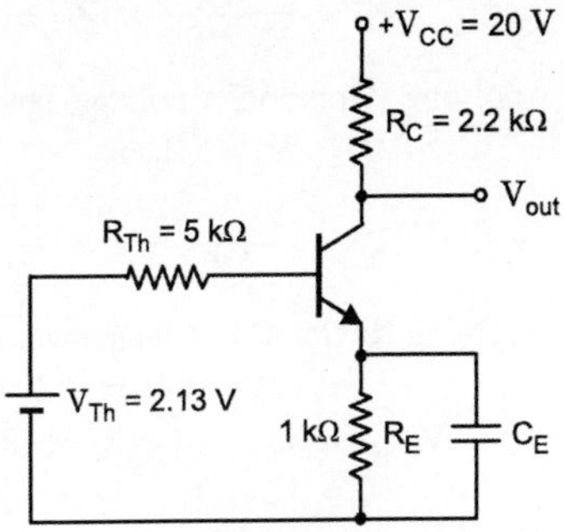

Fig. 6.21

Applying KVL in input loop, we have

$$V_{Th} = I_B R_{Th} + V_{BE} + I_E R_E$$

$$= I_B R_{Th} + V_{BE} + (1 + \beta) I_B R_E$$

$$\text{or } I_B = \frac{V_{Th} - V_{BE}}{R_{Th} + (\beta + 1) R_E}$$

$$= \frac{2.13 - 0.7}{5 \times 10^3 + (150 + 1) \times 1 \times 10^3}$$

$= 9.2\ \mu\text{A}$ **Ans.**

Collector current,

$I_C = \beta I_B = 150 \times 9.2 \times 10^{-6} = 1.375$ mA **Ans.**

Emitter current, $I_E = I_C + I_B = 1.375 + 0.0092 = 1.384$ mA **Ans.**

Emitter voltage, $V_E = I_E R_E = 1.384 \times 10^{-3} \times 1 \times 10^3 = 1.384$ V **Ans.**

Base voltage, $V_B = V_{BE} + V_E = 0.7 + 1.384 = 2.084$ V **Ans.**

Applying KVL in output loop $V_{CE} = V_{CC} - I_C R_C - I_E R_E = 20 - 1.375 \times 10^{-3} \times 2.2 \times 10^3 - 1.384 = 15.6$ V **Ans.**

Example 6.14. The potential divider biasing of a transistor is as shown in Fig. 6.22 (*a*). The transistor has $\beta = 100$ and operates at $I_C = 3$ mA, and $V_{CE} = 10$ V. Assume $V_{BE} = 0.7$ V. Find V_{CC}, R_{B1}, V_B and R_E.

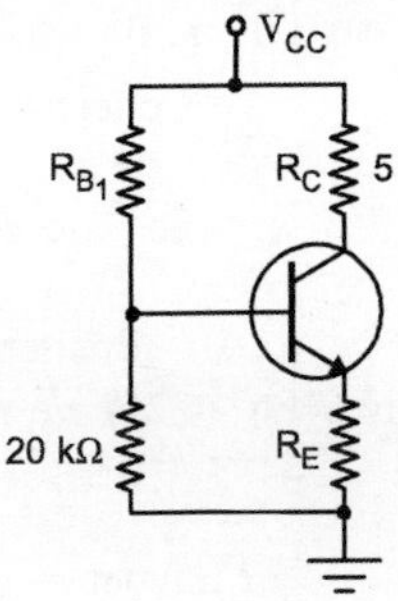

(*a*)

Solution : Base current, $I_B = \dfrac{\text{Collector current, } I_C}{\beta} = \dfrac{3 \times 10^{-3}}{100} = 0.03$ mA

$$I_1 = 10\, I_B = 10 \times 0.03 = 0.3 \text{ mA}$$

The value of R_{B_1} is so decided that the current I flowing through R_{B_1} and R_{B_2} (20 kΩ) is about 10 times of base current to achieve good stabilization.

For the circuit shown in Fig. 6.22 (*b*), we have

$$V_{CB} + V_{BE} = V_{CE} \qquad ...(i)$$

$$\text{and} \quad I_1 R_{B_1} = I_C R_C + V_{CB} \qquad ...(ii)$$

So from Eq. (*i*), we have

$$V_{CB} = V_{CE} - V_{BE} = 10 - 0.7 = 9.3 \text{ V}$$

and from Eq. (*ii*), we have

$$I_1 R_{B_1} = I_C R_C + V_{CB} = 3 \times 10^{-3} \times 5 \times 10^3 + 9.3 = 24.3 \text{ V}$$

and $R_{B_1} = \dfrac{24.3}{I_1} = \dfrac{24.3}{0.3 \times 10^{-3}} = 81$ kΩ **Ans.**

$V_B = I_1 \times R_{B_2} = 0.3 \times 10^{-3} \times 20 \times 10^3 = 6$ V **Ans.**

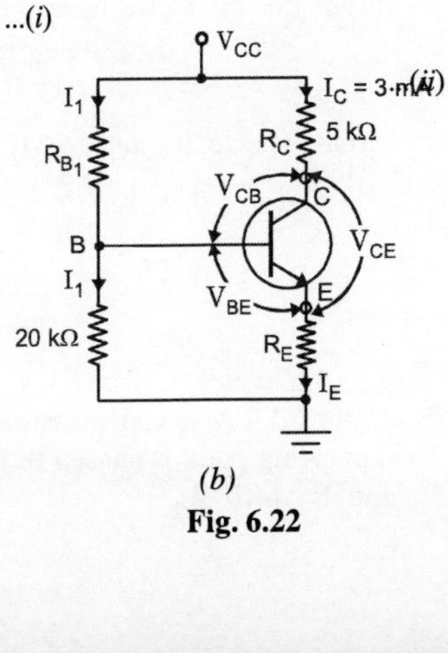

(*b*)

Fig. 6.22

$$V_{CC} = I_1(R_{B_1} + R_{B_2}) = 0.3 \times 10^{-3}(81+20) \times 10^3 = 30.3 \text{ V} \quad \textbf{Ans.}$$

Now $I_E R_E = V_B - V_{BE} = 6 - 0.7 = 5.3 = V$

or $R_E = \dfrac{I_E R_E}{I_E} = \dfrac{5.3}{3 \times 10^{-3}} = 1.766 \text{ k}\Omega$ (1.8 kΩ standard value) **Ans.** Assuming $I_E \simeq I_C$

Example 6.15. Determine V_{CEQ} and I_{CQ} for the BJT amplifier of Fig. 6.23. Also find I_B.

Solution : As we have to calculate biasing current I_{CQ}, we are not concerned with signal V_S.

Now for given potential divider network,

$$V_A = \frac{V_{CC}}{R_1 + R_2} \times R_2 = \frac{12}{15+2.7} \times 2.7 = 1.83 \text{ volts}$$

Applying Kirchhoff's voltage law to input loop we have

$$V_A = V_{BE} + I_E R_E$$

$$I_E = \frac{V_A - V_{BE}}{R_E} = \frac{1.83 - 0.7}{1 \times 10^3} = 1.13 \text{ mA} \quad ...(i)$$

Applying Kirchhoff's voltage law to output loop we have

$$V_{CC} = I_C R_C + V_{CE} + I_E R_E$$

or $V_{CEQ} = V_{CC} - I_C(R_C + R_E)$ $\quad \because I_C \simeq I_E$

$= 12 - 1.13(1 + 1) = 9.74$ V **Ans.**

Thus $V_{CEQ} = 9.74$ V and $I_{CQ} = 1.13$ mA **Ans.**

and $I_B = \dfrac{I_C}{\beta} = \dfrac{1.13 \text{ mA}}{100} = 11.3\ \mu\text{A}$ **Ans.**

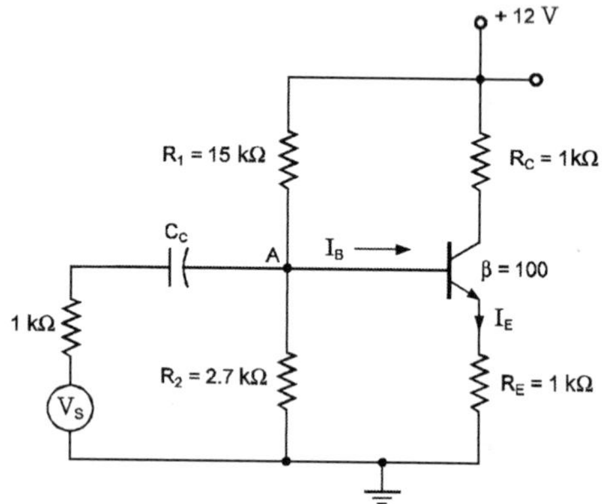

Fig. 6.23

Example 6.16. In a single stage CE amplifier, with single battery supply V_{CC} = 20 V, the device has a maximum h_{FE} value of 20 and I_{CO} = 10 μA. If R_L = 4 kΩ and the quiescent point is required to be at V_{CE} = 6 V and I_C = 2 mA, estimate the values of R_1 and R_2.

Solution : From Eq. (6.4)

Collector current, $I_C = \beta I_B + (1 + \beta) I_{CO}$

So base current, $I_B = \dfrac{I_C - (1+\beta)I_{CO}}{\beta} = \dfrac{2\text{mA} - (20+1) \times 10\mu\text{A}}{20} = \dfrac{1.79\text{mA}}{20} = 89.5\ \mu\text{A}$

Emitter current, $I_E = (\beta + 1) I_B = (20 + 1) \times 89.5\ \mu\text{A} = 1.88$ mA

From Eq. (6.25) we have

$$V_{CC} = I_C R_L + V_{CE} + I_E R_E$$

or Emitter resistance, $R_E = \dfrac{V_{CC} - I_C R_L - V_{CE}}{I_E} = \dfrac{20 - 2\text{mA} \times (4\text{k}\Omega) - 6}{1.88} = 3.2 \text{ k}\Omega$

From Eq. (6.28) we have $V_{Th} = I_B R_{Th} + V_{BE} + I_E R_E = I_B R_{Th} + 0.7 + 1.88 \text{ mA} \times 3.2 \text{ k}\Omega = I_B R_{Th} + 6.7$

Assuming $R_{Th} = 20$ kΩ

$V_{Th} = 89.5 \times 10^{-6} \times 20 \times 10^3 + 6.7 = 8.49$ V

From Eqs. (6.26) and (6.27) we have

$R_1 = \dfrac{R_{Th} V_{CC}}{V_{Th}} = \dfrac{20\text{k}\Omega \times 20\text{V}}{8.49\text{V}} = 47.1$ kΩ (47 kΩ standard value) **Ans.**

and $R_2 = \dfrac{R_1 V_{Th}}{V_{CC} - V_{Th}} = \dfrac{47.1 \times 8.49}{20 - 8.49} = 34.74$ kΩ (33 kΩ, standard value) **Ans.**

Example 6.17. A silicon transistor is used in the self biasing arrangement with V_{CC} = 16 V and R_C = 1.5 kΩ. The operating point is chosen to be V_{CE} = 8 V and I_C = 4 mA. A stability factor S = 12 is desired if β = 50, find R_1 and R_2 with R_E.

Solution : Base current, $I_B = \frac{I_C}{\beta} = \frac{4}{50} = 0.08$ mA

Emitter current, $I_E = I_B + I_C = 0.08 + 4 = 4.08$ mA

From Eq. (6.25), we have

$$\text{Emitter resistance, } R_E = \frac{V_E}{I_E} = \frac{V_{CC} - I_C R_C - V_{CE}}{I_E} = \frac{16 - 4 \times 10^{-3} \times 1.5 \times 10^3 - 8}{4.08 \times 10^{-3}}$$

$$= 0.5 \text{ k}\Omega \text{ or } 500\ \Omega \quad \textbf{Ans.}$$

From Eq. (6.30), we have
$$S = (1+\beta)\frac{1+\frac{R_{Th}}{R_E}}{1+\beta+\frac{R_{Th}}{R_E}}$$

$$\text{So } 12 = (1+50)\frac{1+\frac{R_{Th}}{0.5}}{1+50+\frac{R_{Th}}{0.5}}$$

$$\text{or } R_{Th} = 7.2 \text{ k}\Omega$$

$$V_{Th} = I_B R_{Th} + V_{BE} + I_E R_E = 0.08 \times 10^{-3} \times 7.2 \times 10^3 + 0.7 + 4.08 \times 10^{-3} \times 500$$

$$= 0.576 + 0.7 + 2.04 = 3.02 \text{ V}$$

$$\text{Also } V_{Th} = \frac{R_2}{R_1 + R_2} V_{CC} = \frac{R_{Th}}{R_1} V_{CC}$$

$$\text{or } R_1 = R_{Th} \times \frac{V_{CC}}{V_{Th}} = 7.2 \times 10^3 \times \frac{16}{3.02} = 38.14 \text{ k}\Omega \text{ (39 k}\Omega \text{ standard value)} \quad \textbf{Ans.}$$

$$\text{and since } V_{Th} = \frac{R_2}{R_1 + R_2} V_{CC}$$

$$\text{So } R_2 = \frac{V_{Th} \times R_1}{V_{CC} - V_{Th}} = \frac{3.02 \times 39 \times 10^3}{16 - 3.02} = 9.1 \text{ k}\Omega \text{ (10 k}\Omega\text{, standard value)} \quad \textbf{Ans.}$$

6.5.5. Collector-To-Base Bias (or Base Bias With Collector Feedback)

This circuit is like a fixed bias circuit except that the base resistor R_B is returned to the collector terminal instead of supply V_{CC}. It derives its name from the fact that voltage for R_B is derived from the collector. There exists a negative feedback effect which tends to stabilise I_C against changes either as a result of change in temperature or as a result of replacement of the transistor by another one.

Let us see how does this circuit operate. If the collector current I_C tends to increase (either due to rise in temperature or due to replacement of transistor), V_{CE} decreases due to larger voltage drop across collector resistor R_C. The result is that base current I_B is reduced. The reduced base current in turn reduces the original increase in collector current I_C. Thus a mechanism exists in the circuit which does not allow collector current I_C to increase rapidly.

Circuit Analysis. The required value of base resistor R_B to give the zero signal current I_C can be determined as follows :

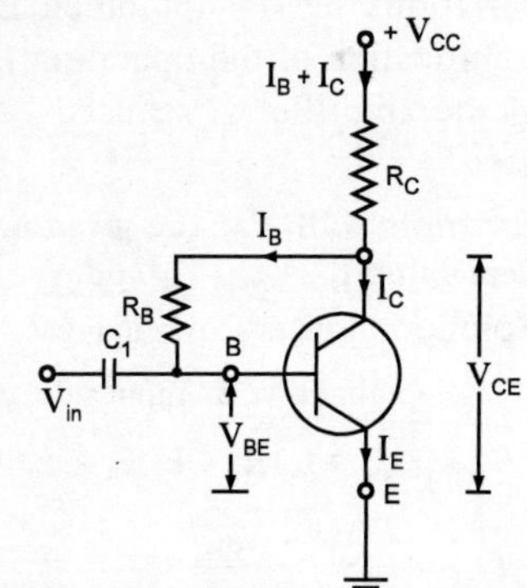

Fig. 6.24. ***Collector-To-Base Bias Circuit***

From circuit diagram shown in Fig. 6.24, applying Kirchhoff's voltage law to the input circuit, we have

$$V_{CC} = (I_B + I_C)R_C + I_B R_B + V_{BE}$$

$$\text{or } I_B (R_B + R_C) = V_{CC} - V_{BE} - I_C R_C$$

$$\text{or } R_B = \frac{(V_{CC} - V_{BE} - I_C R_C)\beta}{I_C} - R_C \quad \text{...(6.33)} \because \beta = \frac{I_C}{I_B}$$

From the output section of the circuit we have

$$V_{CE} = V_{CC} - (I_B + I_C) R_C$$

$$\text{or } R_C = \frac{V_{CC} - V_{CE}}{I_B + I_C} = \frac{V_{CC} - V_{CE}}{I_C\left(\frac{1}{\beta}+1\right)} \quad \text{...(6.34)}$$

Stability Factor S. From Eq. (6.33)

$$I_B = \frac{I_C}{\beta} = \frac{V_{CC} - V_{BE} - I_C R_C}{R_B + R_C} \quad \text{...(6.35)}$$

Since V_{BE} is almost independent of collector current, so differentiating above equation w.r.t. I_C we have

$$\frac{dI_B}{dI_C} = \frac{-R_C}{R_B + R_C}$$

Substituting $\frac{dI_B}{dI_C} = \frac{-R_C}{R_B + R_C}$ in Eq. (6.5) we have

$$\text{Stability factor, } S = \frac{1+\beta}{1+\beta\frac{R_C}{R_B + R_C}} \quad \text{...(6.36)}$$

Value of stability factor so obtained is less than $(1 + \beta)$, obtained from fixed bias circuit. So this method provides improved stability as compared to that of a fixed bias circuit.

Stability Factor $S_{V_{BE}}$. From Eq. (6.35)

$$I_C\left[\frac{R_B + R_C}{\beta} + R_C\right] = V_{CC} - V_{BE}$$

$$\text{and } S_{V_{BE}} = \frac{\partial I_C}{\partial V_{BE}} = \frac{-\beta}{R_B + R_C (1+\beta)}$$

Though this circuit has a tendency to stabilize the operating point against temperature and β variations but is not much used. The base resistor R_B not only provides a dc feedback for the stabilization of the operating point, but it also causes an ac feedback. As a result of this, voltage gain of the amplifier is reduced—an undesirable feature. This is the reason that this circuit is not much used.

Example 6.18. For the given network shown in the Fig. 6.25 determine I_C, V_{CE}, V_B and V_C.

Solution : Circuit shown in Fig. 6.25 is collector-to-base bias circuit.

Applying KVL in input side, we have

$$V_{CC} = (I_C + I_B) R_C + I_B R_B + V_{BE} = (\beta + 1) I_B R_C + I_B R_B + V_{BE}$$

$$\text{or } I_B = \frac{V_{CC} - V_{BE}}{(\beta+1) R_C + R_B} = \frac{20 - 0.7}{(120+1)\times 4.7\times 10^3 + 680\times 10^3}$$

$$= 15.5\ \mu A$$

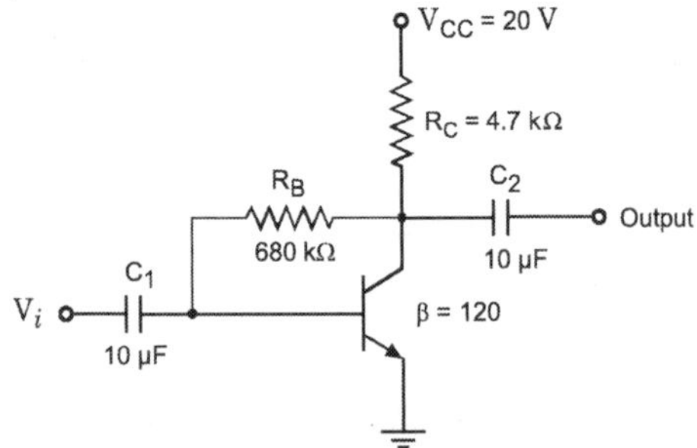

Fig. 6.25

$$\text{or } I_C = \beta I_B = 120 \times 15.5 \times 10^{-6} = \mathbf{1.86\ mA} \quad \textbf{Ans.}$$

$$\text{Base voltage, } V_B = V_{BE} = 0.7\text{ V} \quad (\because \text{ emitter is grounded}) \quad \textbf{Ans.}$$

Applying KVL to output side, we have

$$V_{CE} = V_{CC} - (I_B + I_C)R_C = 20 - (15.5 \times 10^{-6} + 1.86 \times 10^{-3}) \times 4.7 \times 10^3 = \mathbf{11.19\ V} \ \textbf{Ans.}$$

$$\text{Collector voltage } V_C = V_{CE} = \mathbf{11.19\ volts} \quad \textbf{Ans.} \quad (\because \text{ emitter is directly connected to ground})$$

Example 6.19. For the circuit shown in Fig. 6.26 determine Q-point.

[G.G.S.I.P. Univ. Analog Electronics December 2010]

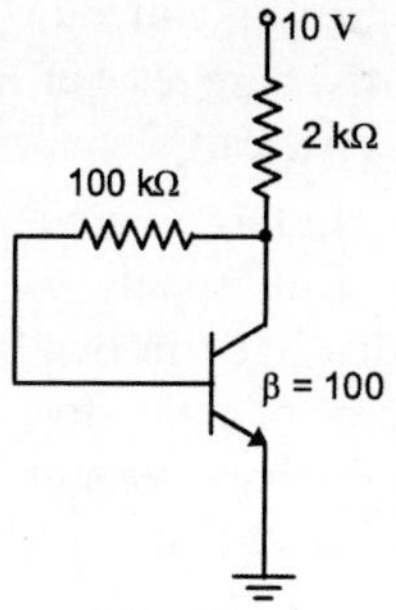

Fig. 6.26

Solution : Applying KVL in input side, we have

$$V_{CC} = (I_C + I_B)R_C + I_B R_B + V_{BE}$$

$$= (\beta + 1)\, I_B R_C + I_B R_B + V_{BE}$$

$$\text{or } I_B = \frac{V_{CC} - V_{BE}}{(\beta + 1) R_C + R_B}$$

$$= \frac{10 - 0.7}{[(100 + 1)2 + 100] \times 10^3}$$

$$= \frac{9.3}{302} \text{ mA} = 0.03 \text{ mA}$$

$$\text{Collector current, } I_C = \beta I_B = 100 \times \frac{9.3}{302} = 3.08 \text{ mA} \quad \textbf{Ans.}$$

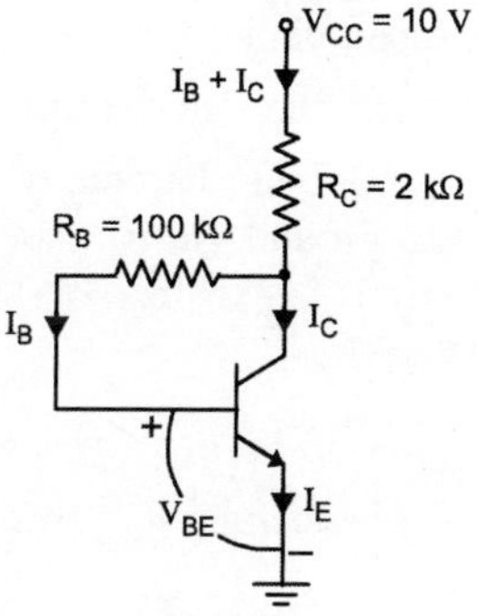

Fig. 6.27

Applying KVL to output side, we have

$$V_{CE} = V_{CC} - (I_B + I_C)\, R_C$$

$$= 10 - 2\ (0.03 + 3.08) \times 10^{-3} \times 2 \times 10^3$$

$$= 3.78 \text{ V} \quad \textbf{Ans.}$$

Example 6.20. In a common emitter circuit an N-P-N transistor having a value of β = 50 is used with V_{CC} = 10 V and R_C = 2 kΩ. If a 100 kΩ resistor is connected between collector and base and V_{BE} = 0, determine (*i*) the position of quiescent point and (*ii*) stability factor S.

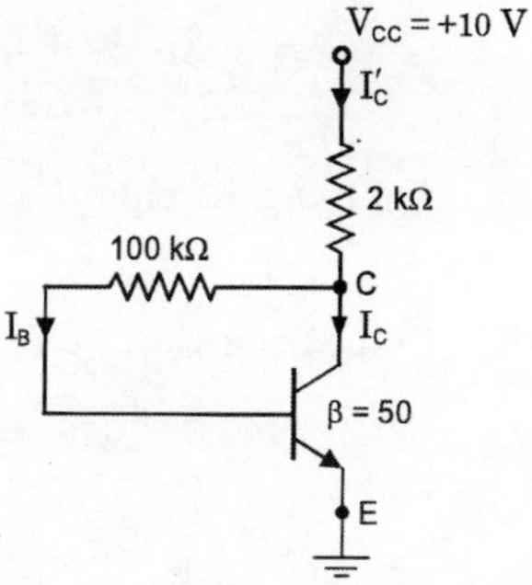

Fig. 6.28

Solution : Applying KVL to base-emitter circuit we have

$$V_{CC} - I'_C R_C - I_B R_B - V_{BE} = 0$$

$$\text{or } V_{CC} - (I_B + I_C)R_C - I_B R_B = 0 \quad \because \ V_{BE} = 0$$

$$\text{or } V_{CC} - (I_B + \beta I_B)R_C - I_B R_B = 0 \quad \because \ I_C = \beta I_B$$

$$\text{or Base current, } I_B = \frac{V_{CC}}{R_B + (1 + \beta)R_C} = \frac{10\text{ V}}{100\text{k}\Omega + (1 + 50) \times 2\text{k}\Omega} = 49.5\ \mu\text{A}$$

$$\text{Collector current, } I_C = \beta I_B = 50 \times 49.5\ \mu\text{A} = 2.475 \text{ mA}$$

Applying KVL to collector-emitter circuit we have

$$V_{CE} = V_{CC} - (I_C + I_B)\, R_C = 10 - (2.475 \text{ mA} + 49.5\ \mu\text{A}) \times 2\text{ k}\Omega = 4.95 \text{ V}$$

Hence Q point lies at

$$I_{CQ} = 2.475 \text{ mA and } V_{CEQ} = 4.95 \text{ V} \quad \textbf{Ans.}$$

$$\text{Stability factor, } S = \frac{1 + \beta}{1 + \beta \dfrac{R_C}{R_B + R_C}} = \frac{50 + 1}{1 + 50 \times \dfrac{2}{2 + 100}} = 25.75 \quad \textbf{Ans.}$$

6.5.6. Base Bias Circuit With Collector and Emitter Feedbacks

In this circuit, as obvious from its name, both collector and emitter feedbacks are used in an attempt to reduce circuit sensitivity to changes in β. Increase in β causes increase in emitter voltage but decreases collector voltage. As a result voltage across base resistor R_B is reduced causing I_B to fall thereby partially compensating the increase in β.

In this circuit a current flows in resistance R_C from the supply V_{CC}. At point A this current divides itself in two parts, one flowing in the collector of transistor, I_C and the second part flows in the base resistor R_B, I_B. Thus current flowing in resistor R_C will be equal to $I_B + I_C$.

No current flows to the output terminals since coupling capacitor C_C offers infinite impedance to it (open circuit for dc).

Let the potential of point A be V_C.

Applying Kirchhoff's second law from point A to ground via R_B, we have

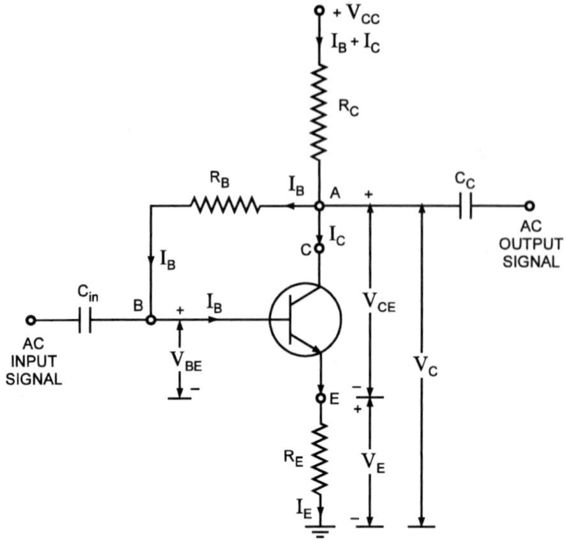

Fig. 6.29

$$V_C - I_B R_B - V_{BE} - I_E R_E = 0$$

$$\text{or } I_B = \frac{V_C - V_{BE} - I_E R_E}{R_B} \qquad ...(6.37)$$

Neglecting V_{BE} in comparison to collector potential V_C which is usually very small.

$$I_B \approx \frac{V_C - I_E R_E}{R_B} \qquad ...(6.38)$$

Now applying Kirchhoff's second law for circuit starting from V_{CC} through collector to emitter terminal, we have

$$V_{CE} = V_{CC} - R_C (I_B + I_C) - I_E R_E \simeq V_{CC} - I_C R_C - I_C R_E \simeq V_{CC} - I_C (R_C + R_E) \qquad ...(6.39)$$

neglecting I_B being too small and I_E being almost equal to I_C.

Applying Kirchhoff's second law to the circuit starting from collector point A to E via R_B

$$V_{CE} = I_B R_B + V_{BE} \qquad ...(6.40)$$

Comparing Eqs. (6.39) and (6.40) we have

$$I_B R_B + V_{BE} = V_{CC} - I_C R_C - I_C R_E$$

$$\text{or } \frac{I_C}{\beta} R_B + V_{BE} = V_{CC} - I_C (R_C + R_E)$$

$$\text{or } I_C \left[\frac{R_B}{\beta} + R_C + R_E\right] = V_{CC} - V_{BE}$$

$$\text{or } I_C = \frac{V_{CC} - V_{BE}}{\frac{R_B}{\beta} + R_C + R_E} \qquad ...(6.41)$$

Neglecting V_{BE}, being very small in comparison to supply voltage V_{CC},

$$I_C = \frac{V_{CC}}{\frac{R_B}{\beta} + R_C + R_E} \qquad ...(6.42)$$

The quiescent point, Q is thus established.

The circuit also stabilizes itself. Let the temperature of transistor junction rise when it is loaded. The increase in temperature causes increase in I_C and β. When I_C increases voltage drop in collector resistance R_C *i.e.* $I_C R_C$ increases and potential of collector drops. This causes reduction in value of base current I_B which in turn reduces I_C being equal to βI_B.

Stability factor S : Applying KVL to the input side, we have

$$V_{CC} = (I_C + I_B)R_C + I_B R_B + V_{BE} + I_E R_E$$
$$= (I_C + I_B)R_C + I_B R_B + V_{BE} + (I_C + I_B)R_E$$
$$= I_C(R_C + R_E) + I_B(R_B + R_C + R_E) + V_{BE}$$

Differentiating above equation w.r.t. I_C we have

$$0 = (R_C + R_E) + \frac{dI_B}{dI_C}(R_B + R_C + R_E) + 0$$

or $$\frac{dI_B}{dI_C} = \frac{-(R_C + R_E)}{R_B + R_C + R_E}$$

Substituting this value in general expression for S, we have

$$S = \frac{1+\beta}{1-\beta\frac{dI_B}{dI_C}} = \frac{1+\beta}{1+\beta\frac{R_C + R_E}{R_B + R_C + R_E}} \qquad ...(6.43)$$

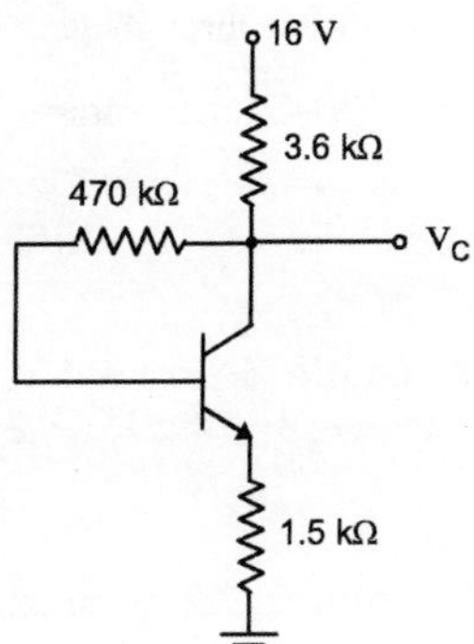

Fig. 6.30

Example 6.21. For the circuit shown in Fig. 6.30, determine (*i*) V_C (*ii*) stability factor $S_{I_{co}}$. Given $\beta = 120$. [*G.G.S.I.P. Univ. Analog Electronics December 2010*]

Solution : Applying KVL to the base-emitter loop in Fig. 6.31, we have

$$V_{CC} = (I_B + I_C)R_C + I_B R_B + V_{BE} + I_E R_E$$

$$= \left(\frac{I_C}{\beta} + I_C\right)R_C + \frac{I_C}{\beta}R_B + V_{BE} + I_C\left(\frac{1+\beta}{\beta}\right)R_E$$

$$= \frac{1+\beta}{\beta}I_C(R_C + R_E) + \frac{I_C}{\beta}R_B + V_{BE}$$

or $$I_C = \frac{V_{CC} - V_{BE}}{\frac{\beta}{1+\beta}(R_C + R_E) + \frac{R_B}{\beta}I_C}$$

$$= \frac{16 - 0.7}{\frac{120}{121}(3.6 + 1.5) \times 10^3 + \frac{470 \times 10^3}{120}} = \frac{15.3}{(5.058 + 3.917) \times 10^3} = 1.7 \text{ mA}$$

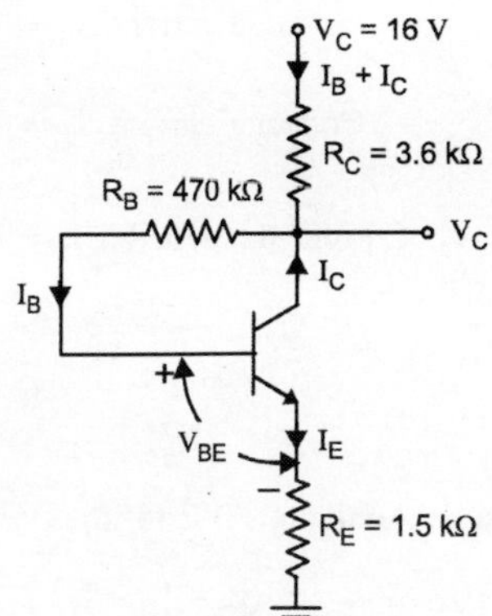

Fig. 6.31

Collector voltage, $V_C = V_{CC} - I_C R_C = 16 - 1.7 \times 10^{-3} \times 3.6 \times 10^3 = 9.88$ V **Ans.**

$$\text{Stability factor, } S = \frac{1+\beta}{1+\beta\frac{R_C + R_E}{R_B + R_C + R_E}} = \frac{1+120}{1+120 \times \frac{3.6+1.5}{470+3.6+1.5}} = \frac{121}{1+120 \times \frac{5.1}{475.1}} = 52.88 \quad \textbf{Ans.}$$

Example 6.22. For the circuit shown in Fig. 6.32, $I_E = 1$ mA, $\beta = 99$ and $V_{BE} = 0.7$ V. Determine (*a*) the current through R_1 and R_C (*b*) the output voltage V_0 and (*c*) the value of R_F.

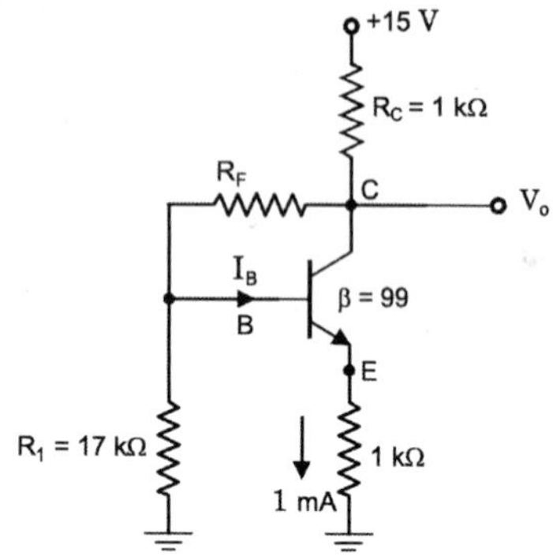

Fig. 6.32

Solution : Emitter current, $I_E = 1$ mA

$$\because \quad I_E = (\beta + 1)I_B$$

$$\therefore \quad \text{Base current, } I_B = \frac{I_E}{\beta+1} = \frac{1\text{mA}}{99+1} = 0.01 \text{ mA}$$

Collector current, $I_C = \beta\ I_B = 99 \times 0.01 = 0.99$ mA

Applying KVL to base-emitter loop we have

$$V_{BE} + I_E R_E = I_1 R_1 \text{ where I is current through } R_1$$

$$\text{or Current through } R_1,\ I_1 = \frac{V_{BE} + I_E R_E}{R_1}$$

$$= \frac{0.7 + 1 \times 10^3 \times 1 \times 10^{-3}}{17} = 0.1 \text{ mA} \quad \textbf{Ans.}$$

(*i*) Current through R_F, I_{R_F} = Current through $R_1 + I_B = 0.1 + 0.01 = 0.11$ mA **Ans.**

Current through R_C, $I'_C = I_C$ + Current through $R_F = 0.99 + 0.11 = 1.1$ mA **Ans.**

(*ii*) Output voltage, $V_0 = V_{CC} - I'_C R_C = 15 - 1.1 \times 10^{-3} \times 1 \times 10^3 = 13.9$ V **Ans.**

$$(iii)\ R_F = \frac{V_0 - V_{BE} - I_E R_E}{I_{R_F}} = \frac{13.9 - 0.7 - 1 \times 10^{-3} \times 1 \times 10^3}{0.11 \times 10^{-3}} = 110.9 \text{ k}\Omega \quad \textbf{Ans.}$$

Example 6.23. Find the value of R_1 in the circuit shown in Fig. 6.33. Assume $\beta = 49$, $I_E = 2$ mA, $V_{BE} = 0.7$ V.

[*G.G.S.I.P. Univ. Analog Electronics December 2010*]

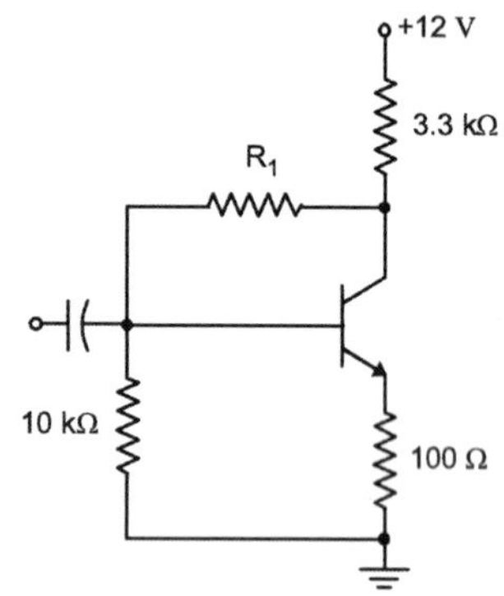

Fig. 6.33

Solution : The given circuit is redrawn as shown in Fig. 6.34.

Emitter current, $I_E = 2$ mA

$$\text{Base current, } I_B = \frac{I_E}{\beta+1} = \frac{2 \text{ mA}}{49+1} = 0.04 \text{ mA}$$

Collector current, $I_C = I_E - I_B = 2 - 0.04 = 1.96$ A

$$\text{Current through } R_2,\ I_1 = \frac{V_{BE} + I_E R_E}{R_2}$$

$$= \frac{0.7 + 2 \times 10^{-3} \times 100}{10 \times 10^3} = 0.09 \text{ mA}$$

Applying KVL to input side, we have

$$V_{CC} = (I_B + I_C + I_1)R_C + (I_B + I_1)R_1 + I_1 R_2$$

$$\text{or} \quad R_1 = \frac{V_{CC} - (I_B + I_C + I_1)R_C - I_1 R_2}{I_B + I_1}$$

$$= \frac{12 - (0.04 + 1.96 + 0.09) \times 10^{-3} \times 3.3 \times 10^3 - 0.09 \times 10^{-3} \times 10 \times 10^3}{(0.04 + 0.09) \times 10^{-3}}$$

$$= \frac{12 - 6.897 - 0.9}{0.13 \times 10^{-3}} = 32.33 \text{ k}\Omega \quad \textbf{Ans.}$$

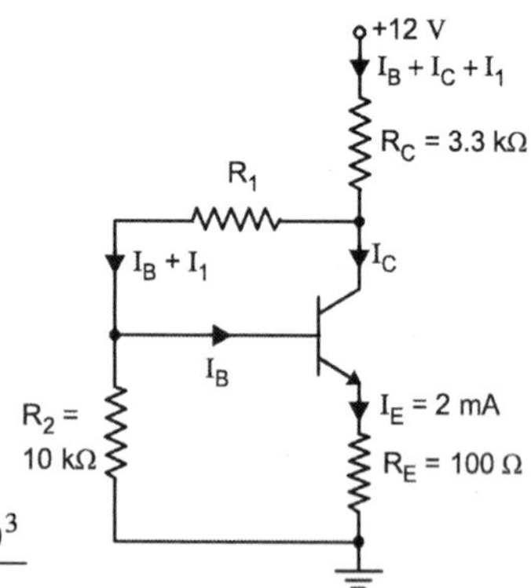

Fig. 6.34

Example 6.24. Find the value of stability factor S with respect to I_{CO} for the circuit shown in Fig. 6.35.

[G.G.S.I.P. Univ. Analog Electronics December 2010]

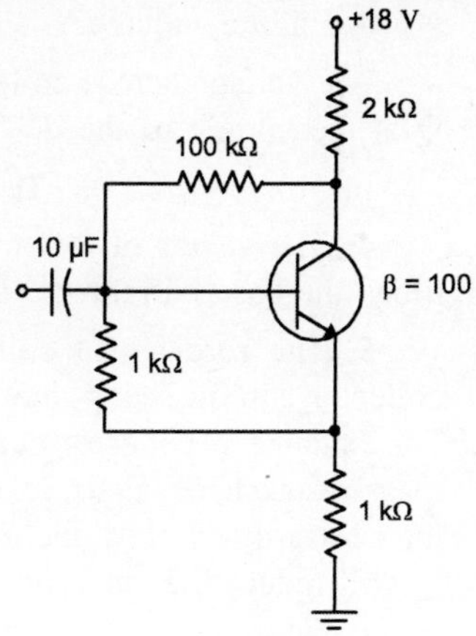

Fig. 6.35

Solution : Drawing dc equivalent circuit, we have circuit as shown in Fig. 6.36.

Applying KVL to the input side of the circuit shown in Fig. 6.36, we have

$$V_{CC} = (I_C + I_B + I_1)R_C + (I_B + I_1)\,R_B + V_{BE} + (I_E + I_1)R_E$$

$$= \left(I_C + I_B + \frac{V_{BE}}{R_1}\right)R_C + \left(I_B + \frac{V_{BE}}{R_1}\right)R_B + V_{BE} + \left(I_C + I_B + \frac{V_{BE}}{R_1}\right)R_E$$

$$\text{or } V_{CC} = V_{BE}\left(1 + \frac{R_C + R_B + R_E}{R_1}\right) + I_C(R_C + R_E) + I_B(R_C + R_B + R_E)$$

Differentiating above equation w.r.t. I_C we have

$$0 = 0 + R_C + R_E + \frac{dI_B}{dI_C}(R_C + R_B + R_E)$$

$$\text{or} \quad \frac{dI_B}{dI_C} = -\frac{R_C + R_E}{R_C + R_B + R_E} = -\frac{2+1}{2+100+1} = \frac{-3}{103}$$

Substituting the value of $\frac{dI_B}{dI_C}$ and β in Eq. (6.5), we have

$$\text{Stability factor, } S = \frac{1+\beta}{1 - \beta\frac{dI_B}{dI_C}}$$

$$= \frac{1+100}{1 - 100 \times \left(\frac{-3}{103}\right)} = \frac{101}{\frac{403}{103}} = 25.8 \text{ Ans.}$$

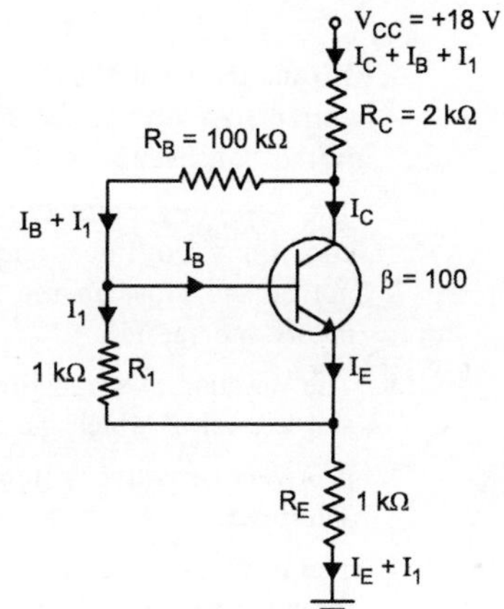

Fig. 6.36

6.6. GUIDELINES FOR DESIGN OF TRANSISTOR BIASING CIRCUITS

In practice, for low power transistors, the following guidelines may be followed for designing the transistor biasing circuits.

1. In common practice R_E is taken to be between 0.5 kΩ to 1.0 kΩ. From the point of view of stabilization larger the value of R_E, better it is but larger value of R_E develops larger voltage drop across it and, therefore, output is reduced. Hence a compromise has to be made in the selection of the value of the emitter resistor R_E.

2. The zero signal collector current I_C is chosen so that the operating point lies at the middle of the dc load line. However, in initial stages of most transistor amplifiers zero signal I_C is taken to be 1 mA. This value of I_C is considered to be most suitable because of the following advantages.

(*i*) The output impedance of a transistor is very high for I_C = 1 mA. So higher voltage gain is obtained.

(*ii*) There is little danger of overheating of collector junction because of small value of I_C.

It is not advisable to operate the transistor below zero signal I_C = 1 mA because of nonlinear characteristics of transistor.

3. Emitter voltage V_E is usually maintained one-tenth of supply voltage V_{CC} *i.e.* $V_E = 0.1\ V_{CC}$.

The voltage across collector-emitter terminals V_{CE} will be 0.5 V_{CC} so as to locate operating point Q at the middle of the dc load line. The remaining voltage 0.4 V_{CC} appears across R_C.

$$\text{Thus } R_C = 4\,R_E$$

4. The values of resistors R_1 and R_2 are selected so as to give current flowing through them 10 times the base current *i.e.* $I_1 \geq 10\ I_B$. This condition is for providing better stabilisation.

5. The zero signal collector current I_C should be slightly more (say 20%) than the maximum collector current swing due to signal otherwise a part of negative half cycle of a signal may be cutoff. For example if collector current swing is expected to be of 2.5 mA due to signal, then zero signal I_C may be taken equal to 3.0 mA. Value of zero signal I_C higher than this value (say 10.0 mA) may unnecessarily overheat the transistor and cause wastage of supply power. Moreover, a higher zero signal I_C will reduce the value of collector resistance R_C for the given V_{CC} and so reduction in voltage gain.

HIGHLIGHTS

1. No matter what type of configuration of a transistor is used in, the basic relationships between the currents are always the same, and the base-emitter voltage V_{BE} is the threshold value if the transistor is in the **"on"** state.
2. The operating point defines where the transistor will operate on its characteristics curves under dc conditions. For linear amplification (amplification with minimum distortion), the operating point should not be too close to the maximum power, voltage, or current rating besides avoiding the regions of cutoff and saturation.
3. The variation in a transistor's dc collector current I_C with temperature in a CE amplifier depends upon β, I_B and I_{CO}, all of which increase with temperature.

   β is very sensitive to temperature, and V_{BE} decreases about 2.5 mV for each 1°C rise in temperature. The reverse saturation current typically doubles for every 10°C rise in temperature.
4. The resulting increase in collector current and shift in operating point may cause clipping and distortion of the input signal, or it may cause thermal runaway in worst condition.
5. The stability factor S is defined as the rate of change of collector current with respect to leakage current I_{CO}, with V_{BE} and β both constant.

   The noteworthy point is that the networks that are the most stable and least sensitive to temperature variations have the smallest stability factors. The larger the value of S, the more likely the circuit to exhibit thermal instability.
6. The fixed bias configuration is the simplest of transistor biasing arrangements, but it is quite unstable due to its sensitivity to β at the operating point.

   The stability factor S for a CE circuit using fixed bias is (β + 1) for either a silicon or germanium transistor.
7. The saturation (maximum) collector current can usually be determined, for any configuration, by superimposing an imaginary short circuit between the collector and emitter terminals of the transistor. The resulting current through the short is then the saturation current.
8. The equation for the load line of a transistor network can be determined by applying Kirchhoff 's voltage law to the output or collector network. The Q-point is then determined by finding the intersection between the base current and the load line drawn on the device characteristics.
9. The stabilization methods involve the reduction of forward bias with temperature rise in order to reduce base current I_B and hence collector current I_C by some means of negative feedback, in different biasing arrangements.
10. The voltage divider bias circuit is the most commonly used biasing arrangement. Its popularity is primarily due to its low sensitivity to variations in β from one transistor to another of the same lot (with the same transistor label). This circuit provides bias stabilization with

$$S = 1+\beta\frac{1+\frac{R_{Th}}{R_E}}{1+\beta+\frac{R_{Th}}{R_E}}$$

11. Collector-to-base bias makes the use of a resistor R_B connected from collector to base and provides bias stabilization with

$$S = \frac{1+\beta}{1+\beta\frac{R_C}{R_B+R_C}}$$

This arrangement may not provide much overall increase in stability in a silicon transistor circuit since the reduced value of R_B will allow a large increase in I_B with temperature than fixed bias.

This arrangement also reduces the voltage gain of the amplifier, an undesirable feature.

EXERCISES

1. What is transistor biasing ? What are the basic conditions which are to be necessarily fulfilled for achieving faithful amplification of input signal in transistor amplifiers.
2. Taking the example of CE amplifier, explain the criterion for selection of a suitable operating point and factors affecting its stability. Hence define stability factor.
3. What is a thermal runaway in transistor amplifier circuit ? Prove that to prevent thermal runaway in CE amplifier, the biasing shall be such that $V_{CE} \le \frac{V_{CC}}{2}$.
4. List and explain the factors responsible for Q-point shift in BJT amplifiers.
[G.G.S.I.P. Univ. Analog Electronics-I December 2004]
5. What is need for stabilization ? List the various factors responsible for shift in Q-point in a transistor. Explain the effect of each parameter on stability of Q-point.
[G.G.S.I.P. Univ. Analog Electronics December 2009]
6. Explain the need for stabilization of any BJT. List the types of stability factors.
[G.G.S.I.P. Univ. Analog Electronics December 2010]
7. What is the significance of 'stability factor' ? Define $S_{(ICO)}$, $S_{(\beta)}$ and $S_{(VBE)}$.
[G.G.S.I.P. Univ. Analog Electronics December 2009]
8. How many types of biasing are done on a BJT to work properly as an amplifier ? Which one is the best and why ?
9. Draw the biasing circuits in cases of (*i*) fixed bias (*ii*) fixed bias with resistor R_E in series with emitter and reference ground (*iii*) potential divider bias.
Derive expressions for stability factor S in terms of the device parameters and circuit components for the above circuits.
10. Prove that the operating point in a potential divider biasing circuit is independent of β. Neglect V_{BE} compared to equivalent voltage levels involved in the circuit.
11. In a common transistor amplifier, how does self bias provide feedback and maintain the operating point stable. Explain.
12. What is self bias ? Draw the circuit showing self bias of an NPN transistor in CE mode. Explain physically how the self bias improves stability. **[G.G.S.I.P. Univ. Analog Electronics December 2010]**
13. Derive the expressions for stability factors $S_{I_{CO}}$ and $S_{V_{BE}}$ for self bias circuit.
[G.G.S.I.P. Univ. Analog Electronics December 2010]

14. Derive the expressions for S, S_β, S_V for a self bias circuit. Compare stability factors for various biasing circuits. **[G.G.S.I.P. Univ. Analog Electronics December 2010]**
15. Draw a practical self-biased CE amplifier circuit. Explain the importance of various capacitors used. **[G.G.S.I.P. Univ. Analog Electronics December 2009]**
16. Explain why self bias circuit is better than a fixed biased circuit. **[G.G.S.I.P. Univ. Analog Electronics December 2009]**
17. Define stability factor. For a collector-to-base biased circuit, find the stability factor S.
18. State the factors to be considered while designing a biasing circuit for a good transistor voltage amplifier.

SHORT ANSWER TYPE QUESTIONS WITH ANSWERS

Q. 1. What is operating point ? Explain its physical significance.

Ans. The zero signal values of I_C and V_{CE} are known as the operating point. It is called operating point because the variations of I_C and V_{CE} take place about this point when signal is applied. It is also known as quiescent point or Q point.

The operating point defines where the transistor will operate on its characteristics curves under dc conditions. For linear amplification (amplification with minimum distortion), the operating point should not be too close to the maximum power, voltage, or current rating besides avoiding the regions of cutoff and saturation.

Q. 2. What is an ac load line ?

Ans. This is the line on the output characteristics of a transistor circuit which gives the values of V_{CE} and I_C when an ac signal is applied. The slope of the ac load line is given as $\frac{-1}{R_C \| R_L}$ where $R_C \| R_L$ represents ac load resistance.

Q. 3. What is faithful amplification ?

Ans. The process of raising the strength of a weak signal without any change in its general shape is referred to as faithful amplification.

Q. 4. What is operating point ? Why is it necessary to stabilize operating point of a transistor amplifier ?

Ans. The zero signal values of I_C and V_{CE} are known as the operating point. It defines where the transistor will operate on its characteristics curves under dc conditions.

The stabilization of operating point is necessary because of (*i*) temperature dependence of collector current I_C, (*ii*) individual variations and (*iii*) thermal runaway.

Q. 5. How does I_{CBO} vary with the temperature ?

Ans. I_{CBO} is exactly like the reverse saturation current of a reverse biased diode, which, theoretically, increases by 8 percent per °C for silicon and 11 per cent per °C for germanium. But from experimental data it is found that the reverse saturation current increases 7 per cent for both silicon and germanium. This is because a surface leakage current component of reverse saturation current is independent of temperature. Since $(1.07)^{10} = 2.0$, the reverse saturation current approximately doubles for every 10°C rise in temperature.

Q. 6. What do you understand by thermal runaway ?

Ans. The collector current I_C, being equal to $\beta I_B + (1 + \beta)\, I_{CO}$, increases with the increase in temperature. This leads to increased power dissipation with further increase in temperature. Being a cumulative process it can lead to *thermal runaway* resulting in burn out of the transistor. The self destruction of an unstabilized transistor is called the *thermal runaway*.

Q. 7. What is bias stabilization ?

Ans. The maintenance of the operating or quiescent point stable (independent of temperature variations or variations in transistor parameters) is known as bias stabilization.

Q. 8. State the condition for thermal stability of operating point of a BJT amplifier.

Ans. The operating point should not shift with temperature variations.

Q. 9. What is meant by midpoint biasing ?

Ans. When an amplifier circuit is so designed that the operating point Q lies at the centre of the dc load line, the amplifier is said to be midpoint biased.

Q. 10. Which two methods make the collector current constant ? Explain them.

Ans. Operating point stability may be achieved by adopting either stabilization techniques or compensation techniques.

Stabilization technique makes use of a resistive biasing circuit that permits such a variation of base current (biasing current) I_B as to maintain collector current I_C almost constant inspite of variations in reverse saturation current I_{CO}, base-emitter voltage V_{BE} and β.

Compensation techniques make use of temperature sensitive devices such as diodes, transistors, thermistors etc. Such devices produce compensating voltages and currents so as to make operating point stable.

Q. 11. What is the requirement of good biasing circuit ?

Ans. *Requirements of Biasing Circuit.* A biasing network associated with a transistor should fulfill the following requirements :

1. Establish the operating point in the middle of the active region of the characteristics, so that on applying the input signal the instantaneous operating point does not move either to the cutoff region or to the saturation region, even at the extreme values of the input signal.
2. Stabilize the collector current I_C against temperature variations.
3. Make the operating point independent of transistor parameters so that replacement of transistor by another transistor of the same type in the circuit does not shift the operating point.

Q. 12. Define stability factor.

Ans. Stability factor is defined as the rate of change of collector current I_C w.r.t. reverse saturation current I_{CO} with V_{BE} and β constant.

Q. 13. What is the significance of stability factor in transistor operation ?

Ans. The stability factor indicates the change in collector current I_C due to change in collector leakage current I_{CO}. Thus a stability factor of high value say 50 means that I_C changes 50 times as much as any change in I_{CO}. In order to achieve greater thermal stability, it is desirable to have as low stability factor as possible. Thus importance of stability factor is with regard to stability of the transistor.

Q. 14. Why fixed bias circuit, inspite of its simplicity, is not much used in practice ?

Ans. Rise in temperature causes increase in leakage current I_{CO}, increase in current amplification factor β and decrease in base-emitter voltage V_{BE}, so operating point is not stabilized. This is the reason that fixed bias circuit is not commonly used.

Q. 15. Explain why collector-to-base bias circuit is seldom used ?

Ans. The base resistor R_B not only provides a dc feedback for the stabilization of the operating point, but it also causes an ac feedback in a collector-to-base bias circuit. This is the reason that this circuit is not much used.

Q. 16. Which value of β is to be used in the analysis of a biasing circuit ?

Ans. Normally, the transistor specification sheet lists minimum and maximum values of β (*i.e.* β_{min} and β_{max}). In this case geometric mean of the two values should be used *i.e.* $\beta = \sqrt{\beta_{min} \times \beta_{max}}$. If the transistor data sheet lists only one value of β, that value should be used.

PROBLEMS

1. In a fixed biasing circuit Fig. 1, R_B = 1 MΩ and R_C = 5 kΩ, V_{CC} = 6 V, β = 100. Determine the Q-point.

[**Ans.** 3 V, 0.6 mA]

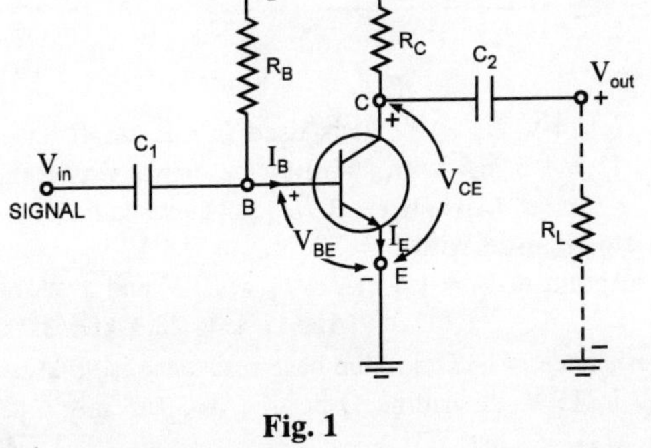

Fig. 1

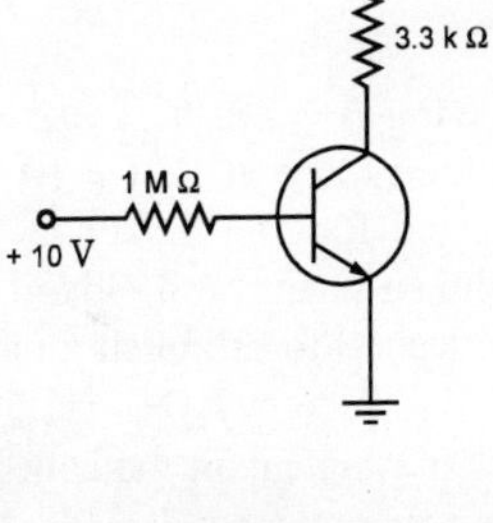

Fig. 2

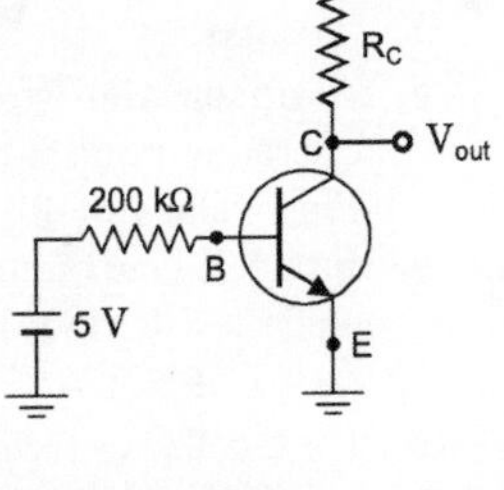

Fig. 3

2. Draw the load line for the Fig. 2. What is I_C at saturation point ? Find V_{CE} at cutoff point.

[**Ans.** 6.06 mA; 20 V]

3. A silicon transistor with $V_{BE\ Sat} = 0.8$ V, $\beta = h_{FE} = 100$, $V_{CE\ Sat} = 0.2$ V is used in circuit shown in Fig. 3. Find the maximum value of R_C for which the transistor remains in saturation.

[**Ans.** 4.667 kΩ]

4. Determine V_C and V_B for the network given below (Fig. 4) with $\beta = 45$ and $V_{BE} = 0.7$ V.

[**Ans.** $V_C = -4.482$ V; $V_B = -8.3$ V]

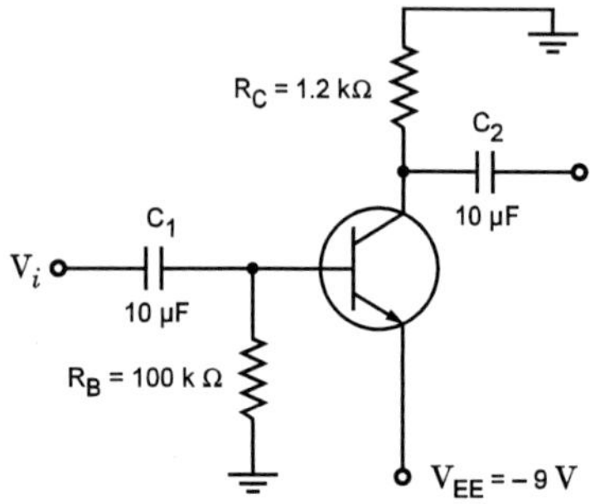

Fig. 4

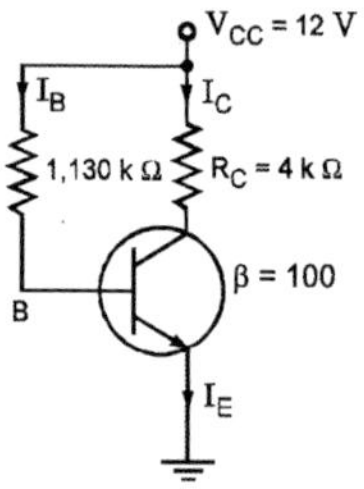

Fig. 5

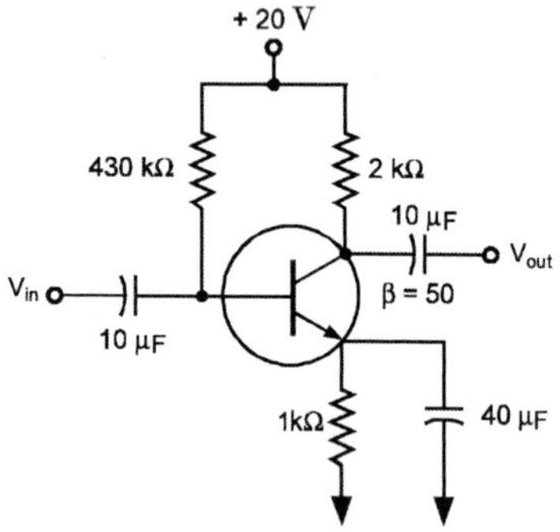

Fig. 6

5. Figure 5 shows the circuit of fixed bias using a silicon transistor with $\beta = 100$. Determine (*i*) base current, (*ii*) collector current (*iii*) V_C, V_B and V_{CB} (*iv*) operating point and (*v*) stability factor S.

[**Ans.** (*i*) 10 μA, (*ii*) 1 mA, (*iii*) 8 V, 0.7 V, 7.3 V, (*iv*) (8 V, 1 mA), (*v*) 101]

6. For the emitter bias network of Fig. 6 determine I_B, I_C, V_{CE}, V_E, V_B and V_C.

[**Ans.** 0.04 mA; 2 mA; 14 V, 2 V, 2.8 V, 16 V]

7. Calculate I_B, I_C, V_E and V_{CE} for the circuit given in Fig. 7 if $\beta = 130$.

[**Ans.** 14 μA, 1.82 mA, –4.24 V, 1.44 V]

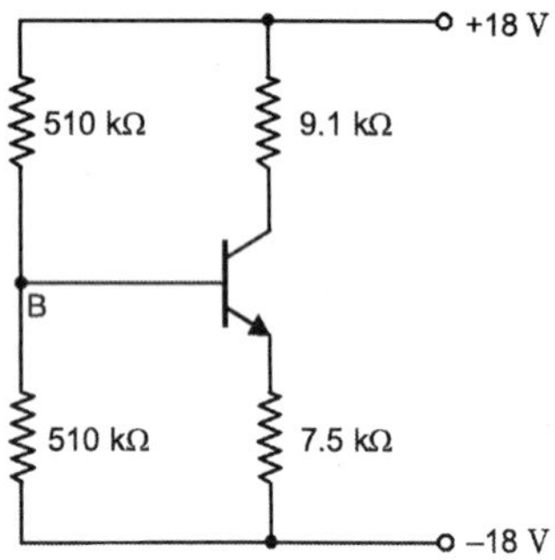

Fig. 7

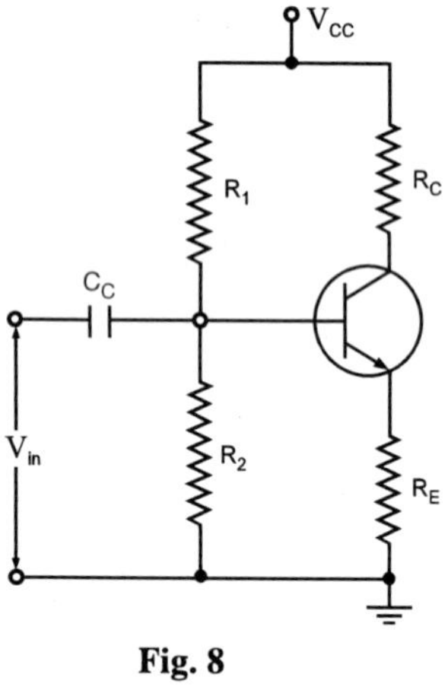

Fig. 8

8. A silicon transistor having $\beta = 40$, $V_{BE} = 2$ V, $V_{CC} = 24$ V, $R_C = 4$ kΩ is used in CE amplifier. The operating point is required to be at $V_{CE} = 10$ V and $I_C = 1.8$ mA with stability factor not exceeding 4. Find values of R_1, R_2 and R_E. [**Ans.** R_E = 3.78 kΩ; 29.94 kΩ; 19.6 kΩ]

9. In a self-bias circuit shown in Fig. 8 a silicon transistor is used with $\beta = 50$, $V_{BE} = 0.6$ V, $V_{CC} = 18$ V and $R_C = 4.3$ kΩ. It is desired to establish a quiescent point at $I_C = 1.5$ mA, $V_{CE} = 10$ V and a stability factor of $S \leq 4$. Find R_E, R_1 and R_2. [**Ans.** 1 kΩ; 26.3 kΩ; 3.7 kΩ]

10. For a collector feedback biasing circuit the collector resistance is 1 kΩ and the base resistance is 90 kΩ. The transistor used is silicon one and has $\beta = 90$. If V_{CC} is 15 V, determine Q-point. [**Ans.** 7.15 mA; 7.85 V]

❑❑❑

CHAPTER 7

Small Signal Amplifiers

INSIDE THIS CHAPTER

7.1. INTRODUCTION

One of the most important characteristics of a signal is its amplitude. Sometimes, a signal may have the desired waveform and frequency, and yet cannot be used because of its small amplitude. This is because the circuits or devices, that are to be operated with such signals as their input, need definite levels of amplitude of signal for proper operation. Thus the amplification or strengthening of the input signal becomes imperative. The device that amplifies the input signal is called the *amplifier*.

Almost no electronic system can work without an amplifier. For instance radio receivers amplify very weak signals until they are strong enough to make the sound audible. The transducers employed in medical and scientific investigations generate signals in microvolt or millivolt range which are incapable to operate indicating instruments. So, these signals need amplification thousands and million times so as to be strong enough to operate indicating instruments.

We have studied that a properly biased transistor increases the amplitude of a weak signal and so acts as an amplifier. So, an amplifier may be defined as a device that increases the current, voltage or power of an input signal with the help of a transistor by furnishing the additional power from a separate source of supply.

When only one transistor with associated circuit is used for increasing the strength of a weak signal, the circuit is known as *single stage transistor amplifier*. It may be emphasised here that a practical amplifier is always a multistage amplifier *i.e.* an amplifier using a number of stages of amplification.

We know that a transistor alone cannot perform the function of amplification and some passive components, such as resistors and capacitors, and a biasing supply is to be connected. When an ac signal is applied between the base and emitter terminals of a properly biased transistor, a small base current starts flowing. Because of transistor action, a much larger ac current (β times the base current) flows through the collector resistor R_C. Since the value of collector resistance is quite high (usually 5–10 kΩ), a large voltage appears across R_C. Thus, a weak signal applied between base and emitter circuit appears in the amplified form between collector and emitter. This explains the *amplifying action of a transistor.* When the input signal is quite weak and produces *small fluctuations* in the collector current in comparison to its quiescent value, the amplifier is called the *small signal amplifier* (or *voltage amplifier*). On the other hand, when the fluctuations produced in collector current are quite large (beyond the linear portion of the output characteristics), the amplifier is called the *large signal* amplifier (or *power amplifier*). Small signal or voltage amplifier is used as the first stage of the amplifier used in radio and TV receivers, tape recorders, stereos and measuring instruments.

A dc voltage source connected to an amplifier is shown in Fig. 7.1. The amplifier consists of transistors that must be biased in the forward active region so that the transistors can act as amplifying devices. It is desirable that the output signal is linearly proportional to the input signal so that the output of the speakers is an exact (as much as possible) reproduction of the signal generated from the CD player. Thus, it is desired that the amplifier be linear one.

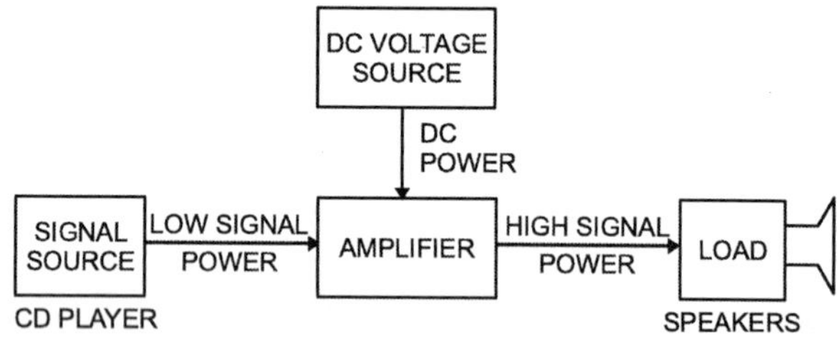

Fig. 7.1. *Block Diagram of a Compact Disc Player System*

Figure 7.1 suggests that there are two types of analysis of the amplifier that must be considered. The first is a dc analysis because of the applied dc voltage source, and the second is a time-varying or ac analysis because of the time-varying signal source. A linear amplifier means that the superposition principle applies. The principle of superposition states : *The response of a linear circuit excited by multiple input signals is the sum of the responses of the circuit to each of the input signals alone.*

For the linear amplifier, then, the dc analysis can be performed with the ac source set to zero. This analysis, called a *large signal* analysis, establishes the operating or Q-point of the transistors in the amplifier. This analysis and design was the primary objective of the previous chapter (Chapter 6). The ac analysis, called a *small-signal analysis,* can be performed with the dc source set to zero. The total response of the amplifier circuit is the sum of the two individual responses.

7.2. TRANSISTOR AMPLIFIER PRACTICAL CIRCUIT

A practical circuit of a single stage transistor amplifier in CE configuration and using self-biasing is shown in Fig. 7.2. The resistors R_1, R_2 and R_E form the *biasing* and stabilization circuit. The biasing circuit must establish a proper operating point otherwise a part of the –ve half cycle of the signal may be cutoff in the output. The resistor R_L connected across the output terminals is called the load. When a number of stages are employed then R_L represents the input resistance for the next stage.

An electrolytic capacitor, called the *input capacitor*, C_{in} of capacity of about 10 μF is used to couple the signal to the transistor base. In the absence of this capacitor, the signal source resistance will come across R_2 and thus change the bias. This capacitor allows only ac signal to flow but isolates the signal source from R_2. Another capacitor, called the emitter bypass capacitor C_E, of capacity of about 100 μF is used in parallel with emitter resistance R_E in order to provide a low reactance path to the amplified ac signal. In the absence of this capacitor, amplified ac signal flowing through R_E will cause a voltage drop across it which in turn will feedback the input side and reduce the output voltage.

For coupling of one stage of amplifier to the next stage another capacitor C_C, called the coupling or blocking capacitor, of capacity of about 10 μF is used. Because of its presence, the output across the load resistance R_L is free from the collector dc voltage. In its absence R_C will come in parallel with the resistor R_1 of the biasing network of next stage and thereby change the biasing conditions of the next stage.

Various currents flowing in different branches of the amplifier circuit are indicated in Fig. 7.2.

With the application of input signal

Base current, i_B = DC base current due to biasing circuit + ac base current due to ac input signal

$$= I_B + i_b \quad ...(7.1)$$

Collector current, i_C = DC collector current due to biasing circuit + ac collector current due to ac input signal

$$= I_C + i_c = \beta (I_B + i_b) \quad ...(7.2)$$

Emitter current i_E = DC emitter current + ac emitter current due to ac input signal

$$= I_E + i_e \quad ...(7.3)$$

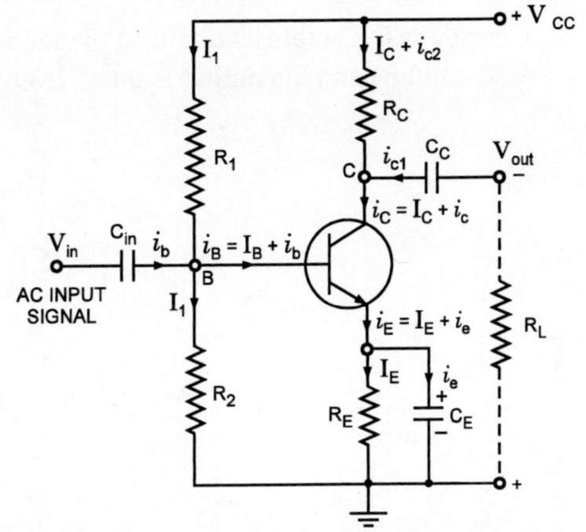

Fig. 7.2. ***Practical Circuit of a Transistor Amplifier in CE Configuration With Self-Biasing***

7.2.1. Emitter Bypass Capacitor

As mentioned above, it is included in the circuit to provide a low reactance path to the amplified ac signal. As shown in Fig. 7.2 it is not connected in series like a coupling capacitor, instead, it is connected across the emitter resistor R_E. The reason for doing so is to bypass or shunt ac current away from the emitter resistor R_E. When the frequency is high enough it looks like a short. Thus emitter bypass capacitor C_E is used to prevent *negative feedback* in the emitter circuit.

Very simply, negative feedback involves feeding back a portion of the amplifier *output* signal to its *input* circuit. Further, the signal is fed back in such a manner that it *opposes* the input signal. As we shall see in chapter 9, the **removal of bypass capacitor** results in an increase in the amplifier's *input resistance,* a reduction in its *voltage gain,* and an *increase* in the *output resistance* "looking into the collector". Obviously, an increase in input resistance is a step in the right direction for voltage amplifier. However, a reduction in the voltage gain and increase in the output resistance are generally undesirable effects. Yet there are some advantages offered by these sacrifices. First, even though the voltage gain is reduced, but *stability* is increased. Specifically, it tends to be less dependent on the transistor's parameters (*e.g.*, its g_m). Second, the increase in the collector-to-ground output resistance makes our approximation of R_{out} equal to R_C even *more realistic.*

7.2.2. AC Ground

You may never have heard of an *ac ground* before, but it really does exist as a separate kind of ground that is different from a *mechanical ground.* A mechanical ground is achieved by connecting a wire between the point to be grounded and ground.

An ac ground is different because it is produced by connecting a *bypass capacitor* between the point to be shorted to ground, as far as ac signals are concerned, and the ground, this kind of ground (*i.e.* ac ground) exists only at high frequencies. In Fig. 7.2 emitter terminal is *ac grounded.* A designer uses an ac ground when any point needs to be frequency sensitive. Such a point remains normal at low frequencies but grounded at high frequencies.

7.3. PHASE REVERSAL

In a common emitter configuration, the output voltage increases in the negative direction when the input signal increases in the positive direction and vice versa. This is called phase reversal* and causes a phase difference of 180° between the input signal voltage V_{in} and output voltage V_{out}.

Consider a CE amplifier circuit shown in Fig. 7.3, the signal being fed at the input terminals (*i.e.* between base and emitter) and the output is taken from the collector and emitter ends.

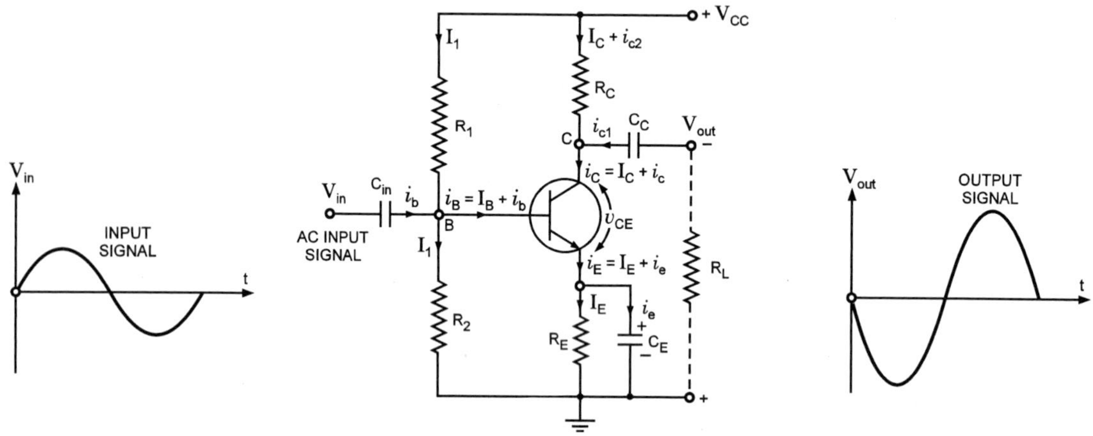

Fig. 7.3. ***Practical Circuit of a Transistor Amplifier in CE Configuration With Self-Biasing***

The instantaneous output voltage v_{CE} is given as

$$v_{CE} = V_{CC} - i_C R_C \quad \text{...(7.4)}$$

Reactance of coupling capacitor C_C is negligible at ordinary signal frequencies and so it may be considered a short for the output signal.

With the increase in the signal voltage in the positive half cycle, the base current increases causing increase in collector current and so the voltage drop $i_C R_C$. As V_{CC} is constant, therefore, output voltage v_{CE} decreases. Thus we can say that as the signal voltage increases in positive direction, the output voltage increases in the negative direction *i.e.* there is a phase difference of 180° between output and input. Conversely on negative half cycle of the input signal voltage, lesser current flows through collector resistance R_C and so voltage drop across it decreases. For this reason output voltage v_{CE} increases.

The above phenomenon can be proved mathematically as below :

Differentiating Eq. (7.4), we have

$$dv_{CE} = 0 - di_C R_C$$

$$\text{or } dv_{CE} = -di_C R_C$$

The negative sign indicates that the output voltage is 180° out of phase with the input signal voltage.

The above fact *i.e.* phase reversal in CE configuration can be illustrated graphically with the aid of output characteristics and ac load line, as shown in Fig. 7.4. The base current varies between I_{B_1} and I_{B_2} with I_B as the zero signal base current. From figure we see that when the base current is maximum in the positive direction, v_{CE} becomes maximum in the negative direction and vice versa. Thus, the input and output voltages are in phase opposition *i.e.* the transistor has produced a phase reversal of output voltage w.r.t. the input signal voltage.

* This will be so when output is taken from collector and emitter ends, as is always done. However, if the output is taken across collector resistance R_C, it will be in phase with the input.

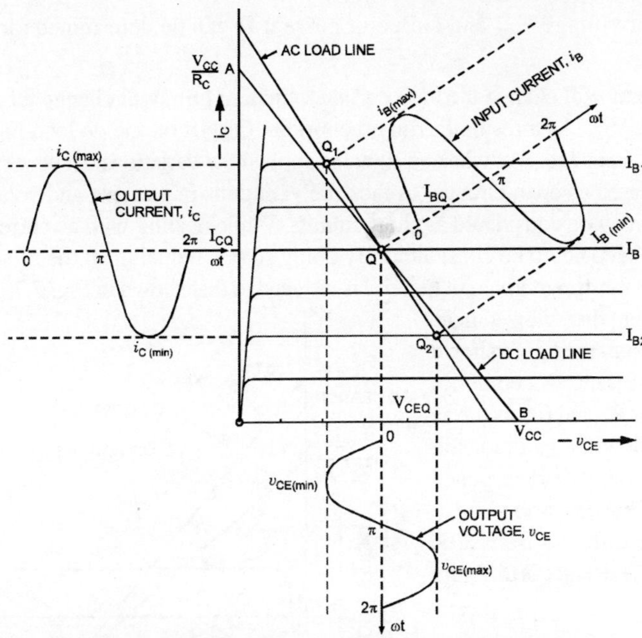

Fig. 7.4. *Graphical Illustration of Phase Reversal*

The noteworthy points are that common base and common collector amplifiers do not produce reversal, and input and output currents are in phase in all the three configurations.

7.4. LOAD LINE ANALYSIS

As already discussed, the relationship between collector-emitter output voltage V_{CE} and collector current I_C is linear and, therefore, it can be represented by a straight line on the output characteristics giving relation between V_{CE} and I_C. This is known as load line. There are two types of load lines, namely dc load line and ac load line. The former determines the values of collector current I_C and collector-emitter voltage V_{CE} corresponding to zero signal conditions whereas the latter gives their (i_C and v_{CE}) values when an ac signal is applied.

Consider the transistor amplifier circuit shown in Fig. 7.2. As regards dc conditions, the amplifier circuit shown in Fig. 7.2 is reduced to that shown in Fig. 7.5. The dc equivalent circuit shown in Fig. 7.5 is so because the capacitors do not allow the flow of the direct current through them and, therefore, act as open circuits. While drawing dc equivalent circuit it is presumed that no signal is applied.

DC equivalent circuit of an amplifier circuit is drawn by reducing all the ac sources to zero and opening all the capacitors.

Applying Kirchhoff 's voltage law to the output section of the basic amplifier circuit shown in Fig. 7.5 we have

$$V_{CC} = I_C R_C + V_{CE} + I_E R_E$$

$$= V_{CE} + I_C (R_C + R_E) \qquad \because I_C = I_E$$

$$\text{or } I_C = \frac{-V_{CE}}{R_C + R_E} + \frac{V_{CC}}{R_C + R_E} \qquad ...(7.5)$$

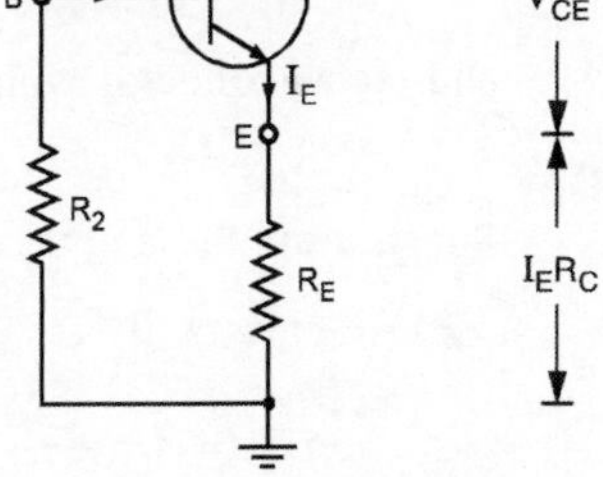

Fig. 7.5. *DC Equivalent Circuit of Circuit Given in Fig. 7.2*

Since V_{CC} and $(R_C + R_E)$ are constant, therefore, it represents a straight line known as dc load line. By plotting this line on the output characteristics, collector-emitter voltage V_{CE} and collector current I_C can be determined for the given value of base current.

Any change in dc bias will change the base current which, in turn, will change the collector current I_C and collector-emitter voltage V_{CE}. As a result, the quiescent point Q shift on the dc load line. This is what roughly happens when ac signal is applied. But when ac signal is applied the variations occur very fast and the capacitors can no longer be considered as open circuits. In fact, the variations in currents and voltages are so fast that the capacitors in the circuit may be considered as short circuits. While dealing with ac currents and voltages (*i.e.* ac conditions), dc supplies need not to be considered. By doing so (*i.e.* reducing all the dc sources to zero and short circuiting all capacitors), the circuit given in Fig. 7.2 is reduced to that shown in Fig. 7.7. This circuit explains the behaviour of the amplifier when viewed in ac conditions. As obvious from the figure shown in Fig. 7.7, the collector resistor R_C comes in parallel with load resistor R_L and forms the ac load for the amplifier. Now the variations in collector current and voltage are determined with the help of ac load line corresponding to ac load. AC load line is different from a dc load line because ac load resistance is different from dc load resistance.

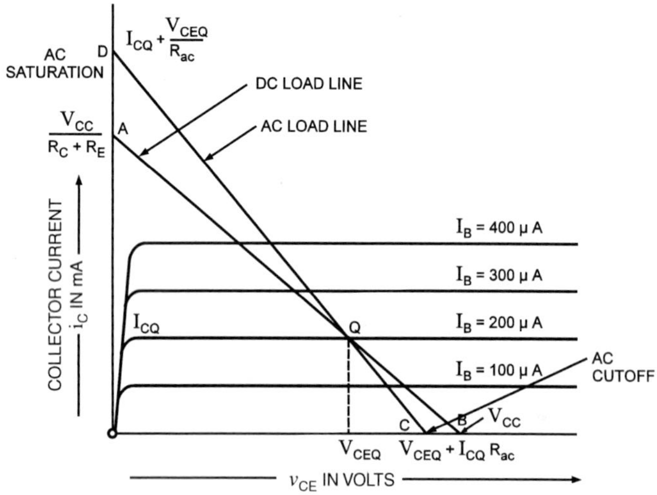

Fig. 7.6

Effective load for ac signal is given as

$$R_{ac} = R_C \,||\, R_L = \frac{R_C R_L}{R_C + R_L} \quad \text{...(7.6)}$$

When no signal is applied, the operating point is Q, as shown in Fig. 7.6. When signal is applied Q point swings along ac load line.

The saturation and cutoff points on the ac load line are different from those of dc load line. Equations for the intercepts of ac load line are derived as below.

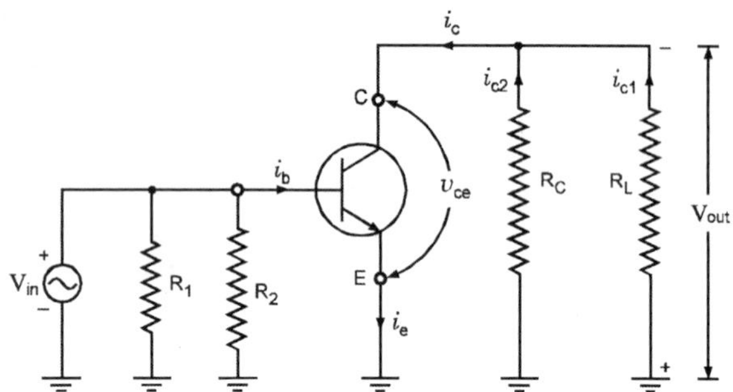

Fig. 7.7. ***AC Equivalent Circuit of Circuit Shown in Fig. 7.2***

Applying Kirchhoff 's voltage law around the collector loop (Fig. 7.7) we have

$$v_{ce} + i_c R_{ac} = 0$$

$$\text{or } i_c = \frac{-v_{ce}}{R_{ac}} \quad \text{...(7.7)}$$

The ac collector current is given by

$$i_c = \Delta I_C = i_C - I_{CQ}$$

and the ac collector voltage is given as

$$v_{ce} = \Delta V_{CE} = v_{CE} - V_{CEQ}$$

Substituting these equations in Eq. (7.7) we have

$$v_{CE} - V_{CEQ} + (i_C - I_{CQ}) R_{ac} = 0$$

$$\text{or } i_C = I_{CQ} + \frac{V_{CEQ}}{R_{ac}} - \frac{v_{CE}}{R_{ac}} \quad \text{...(7.8)}$$

This is the equation of the ac load line. We can find the intercepts in the usual way. When the transistor goes into saturation, v_{CE} is zero and Eq. (7.8) gives

$$i_{C\,(SAT)} = I_{CQ} + \frac{V_{CEQ}}{R_{ac}} \text{ (upper end)} \qquad ...(7.9)$$

where $i_{C\,(SAT)}$ is the ac saturation current, I_{CQ} is dc collector current, V_{CEQ} is dc collector-emitter voltage and R_{ac} is the ac resistance seen by the collector.

When the transistor goes into cutoff, i_C equals zero and we get the ac cutoff voltage of

$$v_{CE\,(OFF)} = V_{CEQ} + I_{CQ} R_{ac} \text{ (lower end)} \qquad ...(7.10)$$

By joining these two points line CQD can be drawn as shown in Fig. 7.6. This is the ac load line because it represents all possible ac operating points. At any instant during the ac cycle, the operating point of the transistor is somewhere along the ac load line, the exact point is determined by the amount of change that takes place from the Q point.

The slope of the ac load line is given as

$$-\frac{OD}{OC} = -\frac{I_{CQ} + \dfrac{V_{CEQ}}{R_{ac}}}{V_{CEQ} + I_{CQ} R_{ac}} = -\frac{I_{CQ} + (V_{CEQ}/R_{ac})}{R_{ac}\left[I_{CQ} + \dfrac{V_{CEQ}}{R_{ac}}\right]} = -\frac{1}{R_{ac}} \qquad ...(7.11)$$

It is seen from Fig. 7.6 that maximum possible positive signal swing is $I_{CQ} R_{ac}$ and maximum possible negative signal swing is V_{CEQ}. It means peak signal handling capacity is limited to $I_{CQ} R_{ac}$ or V_{CEQ} whichever is smaller.

Example 7.1. Draw the dc and ac load lines for the CE circuit shown in Fig. 7.8. What is the maximum peak-to-peak signal that can be obtained ?

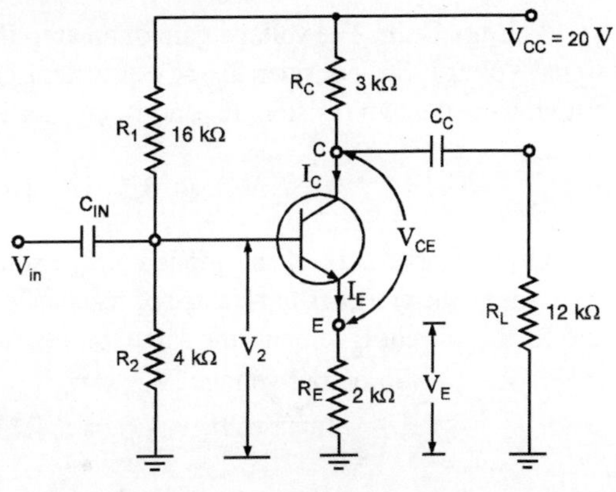

Fig. 7.8

Solution : $V_{CE\,(cutoff)} = V_{CC} = 20$ volts representing point B

$$I_C \text{ (saturation)} = \frac{V_{CC}}{R_C + R_E} = \frac{20}{(3+2)\text{ k}\Omega} = 4 \text{ mA representing point A}$$

Hence line AB gives the dc load line for the given circuit.

Approximate bias conditions can be quickly determined by assuming that base current is too small to effect the base bias.

Voltage across emitter resistance R_E,

V_E = Voltage across resistance R_2, if V_{BE} is neglected

$$\text{So } I_E R_E = V_{CC} \times \frac{R_2}{R_1 + R_2} = 20 \times \frac{4}{16+4} = 4 \text{ V}$$

$$\text{or } I_E = \frac{V_E}{R_E} = \frac{4}{2\text{ k}\Omega} = 2 \text{ mA}$$

Collector current $I_C \simeq I_E = 2$ mA

Hence $I_{CQ} = 2$ mA

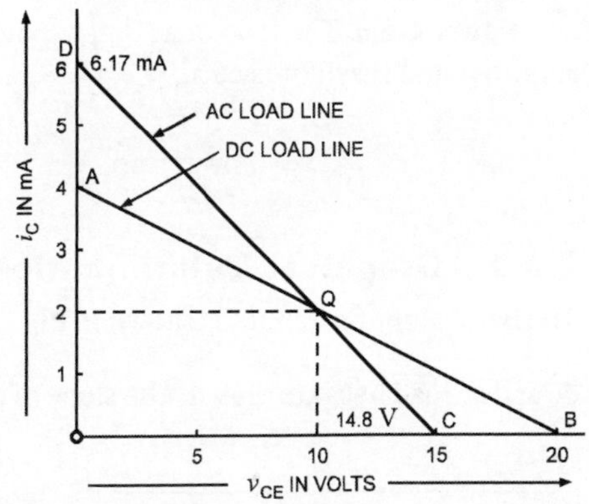

Fig. 7.9

Collector-emitter voltage,

$$V_{CEQ} = V_{CC} - I_{CQ}(R_C + R_E) = 20 - 2 \text{ mA} \times (3+2) \text{ k}\Omega = 10 \text{ V}$$

$$R_{ac} = R_C \,||\, R_L = \frac{R_C R_L}{R_C + R_L} = \frac{3 \times 12}{3+12} = 2.4 \text{ k}\Omega$$

$$i_{C\,(SAT)} = I_{CQ} + \frac{V_{CEQ}}{R_{ac}} = 2 + \frac{10}{2.4} = 6.17 \text{ mA}$$ giving saturation or upper end of ac load line

$$v_{CE\,(OFF)} = V_{CEQ} + I_{CQ} R_{ac} = 10 + 2 \text{ mA} \times 2.4 \text{ k}\Omega = 14.8 \text{ V}$$ giving cutoff point or lower end of ac load line

Hence line joining 14.8 V on V_{CE} – axis and 6.17 mA on I_C – axis gives the ac load line, as illustrated in Fig. 7.9. As expected this line passes through quiescent point Q.

Now, $I_{CQ} R_{ac} = 2 \text{ mA} \times 2.4 \text{ k}\Omega = 4.8 \text{ V}$ and $V_{CEQ} = 10$ V

Taking the small quantity,

Maximum peak output signal = 4.8 V

Hence peak-to-peak value of output signal = 2 × 4.8 = 9.6 V **Ans.**

7.5. AMPLIFIER GAINS

An amplifier is a device which strengthens a weak signal. The strength of a signal may be represented by its current, voltage or power. Accordingly, while studying the *performance* of an amplifier, we determine its current gain, voltage gain and power gain. For this purpose ac equivalent circuit of a transistor amplifier (Fig. 7.7) is considered because only ac values are to be taken into account.

Current Gain. The ratio of output current (collector current) to the input signal current (base current) is known as *current gain* of the amplifier and symbolised as A_i.

$$\text{So current gain, } A_i = \frac{i_c^*}{i_b} \qquad ...(7.12)$$

Voltage Gain. The voltage gain of an amplifier is defined as the ratio of ac output voltage to the ac input signal voltage. So, consider the ac equivalent circuit of the transistor amplifier shown in Fig. 7.7. On the output side, the two resistors R_C and R_L appear in parallel and so the effective load for ac is given as

$$\text{AC load, } R_{ac} = R_C \parallel R_L = \frac{R_C R_L}{R_C + R_L}$$

On the input side, if the input voltage source V_{in} is assumed to be an ideal one (with zero internal resistance), the presence or absence of resistances R_1 and R_2 is immaterial. Whatever may be the values of R_1 and R_2 the current I_b remains the same. So resistors R_1 and R_2 can be ignored altogether.

So output voltage, $V_{out} = i_c R_{ac}$ ...(7.13)

Input voltage, $V_{in} = i_b R_{in}^{**}$ ...(7.14) where R_{in} is the input resistance

$$\text{Voltage gain, } A_v = \frac{V_{out}}{V_{in}} = \frac{i_c R_{ac}}{i_b R_{in}} = \beta \frac{R_{ac}}{R_{in}} \qquad ...(7.15) \because \frac{i_c}{i_b} = \beta$$

Power Gain. The ratio of ac output power (P_{out}) to the ac input power (P_{in}) is called the power gain of the amplifier and is symbolised as A_p

$$\text{Power gain, } A_p = \frac{i_c^2 R_{ac}}{i_b^2 R_{in}} = \beta^2 \frac{R_{ac}}{R_{in}} \qquad ...(7.16)$$

7.5.1. Graphical Determination of Voltage and Current Gains

The basic amplifier circuit is shown in Fig. 7.2 and its output characteristics are shown in Fig. 7.4. First of all dc and ac load lines are drawn. The slope of the dc load line is $-\frac{1}{R_{dc}}$ or $-\frac{1}{R_E + R_C}$ while the slope of ac load

---

* $\frac{i_c}{i_b} = \beta = \frac{\Delta I_C}{\Delta I_B}$ ** $R_{in} = \frac{\Delta V_{BE}}{\Delta I_B}$

line is $-\frac{1}{R_{ac}}$, as determined in Art. 7.4. Let the biasing arrangement give base current I_B. The quiescent point Q is the intersection of dc load line and output characteristic corresponding to base current of I_B (microamperes). Since the operating point Q gives the zero signal condition, the ac load line also passes through this point.

When ac signal is applied, the base current varies between I_{B_1} and I_{B_2}. As a result, the operating point moves on ac load line between points Q_1 and Q_2. The corresponding variations in collector current and collector-emitter voltage are illustrated in Fig. 7.4. Collector current varies between $i_{C\,(max)}$ and $i_{C\,(min)}$ while collector-emitter voltage varies between $v_{CE\,(max)}$ and $v_{CE\,(min)}$, as shown in the figure. If the ac input signal voltage varies between $V_{in\,(max)}$ and $V_{in\,(min)}$, then current gain and voltage gain are given as below:

$$\text{Current gain, } A_i = \frac{i_{C\,(max)} - i_{C\,(min)}}{i_{B\,(max)} - i_{B\,(min)}} \quad \text{...(7.17)}$$

$$\text{Voltage gain, } A_v = \frac{v_{CE\,(max)} - v_{CE\,(min)}}{v_{in\,(max)} - v_{in\,(min)}} \quad \text{...(7.18)}$$

Example 7.2. In a CE transistor amplifier when signal changes by 0.02 V, the base current changes by 10 μA and collector current by 1 mA. If the collector resistance R_C = 5 kΩ and R_L = 10 kΩ. Find (*i*) Voltage gain, (*ii*) Current gain, (*iii*) Input impedance, (*iv*) AC load resistance and (*v*) Power gain.

Solution : (*iii*) Input impedance, $Z_{in} = \frac{\text{Change in } V_{BE}}{\text{Change in } I_B} = \frac{\Delta V_{BE}}{\Delta I_B} = \frac{0.02}{10 \times 10^{-6}} = 2\ \text{k}\Omega$ **Ans.**

(*ii*) Current gain, $\beta = \frac{\text{Change in } I_C}{\text{Change in } I_B} = \frac{\Delta I_C}{\Delta I_B} = \frac{1 \times 10^{-3}}{10 \times 10^{-6}} = 100$ **Ans.**

(*iv*) AC load resistance, $R_{ac} = \frac{R_C R_L}{R_C + R_L} = \frac{5 \times 10}{5 + 10}\ \text{k}\Omega = 3.333\ \text{k}\Omega$ **Ans.**

(*i*) Voltage gain, $A_V = \beta \frac{R_{ac}}{R_{in}} = 100 \times \frac{3.333 \times 10^3}{2 \times 10^3} = 166.7$ **Ans.**

(*v*) Power gain, $A_p = A_V \times A_i = 166.7 \times 100 = 16{,}670$ **Ans.** $\quad \because A_i = \beta = 100$

7.6. AC EMITTER RESISTANCE

The ac or dynamic resistance of the emitter-base junction (or simply emitter) diode is called the *ac emitter resistance*. It is given as change in base-emitter voltage divided by the corresponding change in emitter current *i.e.*

$$r_{ac} = \frac{\Delta V_{BE}}{\Delta I_E} \quad \text{...refer to Fig. 7.10}$$

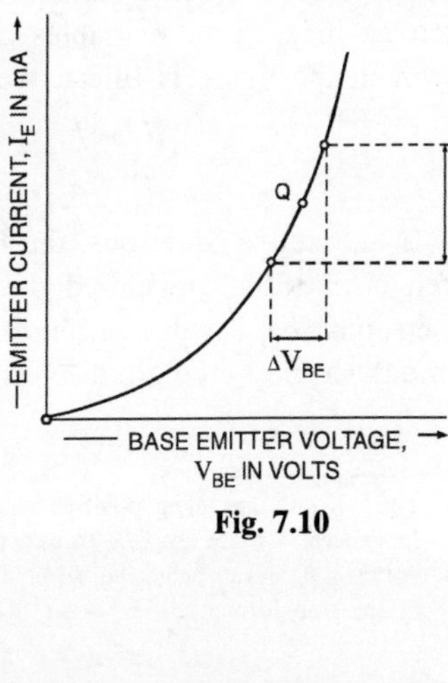

Fig. 7.10

It is possible to derive the formula for ac resistance of the emitter diode by using calculus and ac resistance of the emitter diode is given as $r_{ac} = \frac{25\ \text{mV}}{I_E}$ where I_E is the dc emitter current at Q point.

In derivation of above formula, it has been assumed that the operation is small signal one.

Because the emitter diode is special in transistor analysis, the foregoing formula is usually written as below :

$$r'_e = \frac{25\ \text{mV}}{I_E} \quad \text{...(7.19)}$$

The subscript *e* indicates emitter, the lower case *r* is used to indicate ac or dynamic resistance and the prime, ′, indicates the internal resistance.

7.7. HYBRID PARAMETERS

One of our first concerns in the sinusoidal ac analysis of transistor networks is the magnitude of the input signal (an amplifier may use either small-signal operation or large-signal operation). Study of small-signal operation may be done either graphically or by using small-signal equivalent circuit for the BJT operating in the active region, however, the second method is more convenient. The large-signal operation may be best studied graphically because of involvement of certain nonlinear operation in it. For study of small-signal operation, the transistor may be replaced by its equivalent circuit (or model) and then the usual method of network analysis may be used to obtain expressions for its operating characteristics *e.g.*, input impedance, output impedance, voltage gain, current gain etc.

We here take up the study of hybrid (h) parameter equivalent circuit (or model) for the transistor and analysis of low-frequency small-signal common-emitter (CE) amplifier using this model.

7.8. TWO-PORT DEVICES AND THE HYBRID MODEL

A box representing a two-port network is illustrated in Fig. 7.11. The terminal behaviour of a two-port device may be specified by two voltages and two currents (voltage v_1, and current i_1 at the input port and voltage v_2 and current i_2 at the output port). The conventional positive polarities* of voltages v_1 and v_2 and currents i_1 and i_2 are shown in the figure. Out of four quantities (v_1, v_2, i_1 and i_2), any two may be selected as independent variables** and the remaining two be expressed in terms of the selected independent variables. This leads to various two-port parameters, out of which following three are more important.

1. Open-circuit impedance parameters or Z-parameters Z_{11}, Z_{12}, Z_{21} and Z_{22}
2. Short-circuit admittance parameters or Y-parameters Y_{11}, Y_{12}, Y_{21} and Y_{22}
3. Hybrid parameters or the h parameters.

In transistor amplifier analysis, Z- and Y-parameters were used earlier. But now hybrid or the h parameters alone are used in a transistor circuit analysis and, therefore, only the h parameters will be taken here for discussion.

7.8.1. Hybrid Parameters or h Parameters

For the two-port network illustrated in Fig. 7.11, if input current i_1 and the output voltage v_2 are taken as independent variables and the two-port shown in the figure is linear, we may write

$$v_1 = h_{11}i_1 + h_{12}v_2 \quad ...(7.20)$$

$$i_2 = h_{21}i_1 + h_{22}v_2 \quad ...(7.21)$$

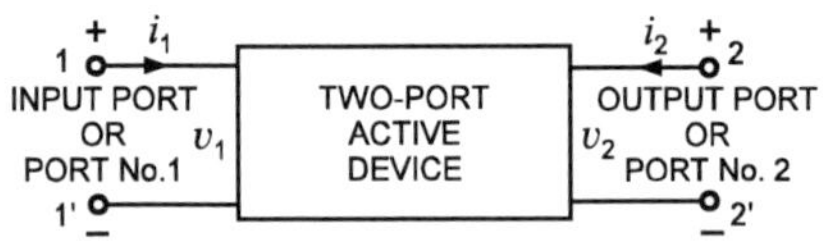

Fig. 7.11. ***Conventional Positive Polarities of Voltages and Currents in a Two-Port Network***

In the above equations, the *hs* are fixed for a given circuit and are called the hybrid or h parameters. Because these four parameters have mixed dimensions (h_{11} has dimension of ohm, h_{12} and h_{21} are dimensionless, and h_{22} has dimension of mho or siemen) so they are called hybrid or h parameters.

* The currents are taken positive when they enter the device and negative when they leave it.

** In general, we are not free to select the independent variables arbitrarily—for example in case of an ideal transformer, two voltages v_1 and v_2 cannot be selected as the independent variables because their ratio is a constant and is equal to the turn-ratio of the transformer.

1. *Meaning of h Parameters.* By assuming that the given two-port network has no reactive element and by applying open circuit ($i_1 = 0$) or short circuit ($v_2 = 0$) conditions to Eqs. (7.20) and (7.21), the h parameters can be defined as below :

If the output terminals are short circuited, (Fig. 7.12) output voltage v_2 becomes zero and Eqs. (7.20) and (7.21) become

$$v_1 = h_{11}i_1 + h_{12} \times 0 = h_{11}i_1$$

and $$i_2 = h_{21}i_1 + h_{22} \times 0 = h_{21}i_1$$

or $$h_{11} = \left.\frac{v_1}{i_1}\right|_{v_2 = 0}$$

and $$h_{21} = \left.\frac{i_2}{i_1}\right|_{v_2 = 0}$$

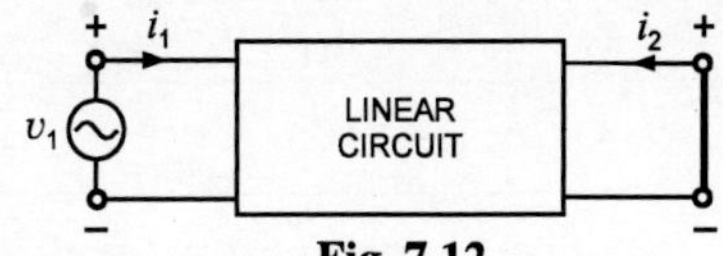

Fig. 7.12

Since h_{11} is the ratio of input voltage and input current with output terminals short circuited, it is called the *input impedance with output short circuited.* The subscript 11 of h_{11} defines the fact that the parameter is determined by the ratio of quantities measured at the input terminals. Its unit is ohm.

Similarly h_{21} is the ratio of output and input currents (*i.e.* i_2/i_1) with output terminals short circuited, so it is called the *forward transfer current gain with output short circuited.* Obviously it is dimensionless quantity.

If the input terminals are open circuited and we drive the output terminals with voltage v_2, as shown in Fig. 7.13, input current i_1 becomes zero and Eqs. (7.20) and (7.21) become

$$v_1 = h_{11} \times 0 + h_{12}\, v_2 = h_{12}\, v_2$$

and $$i_2 = h_{21} \times 0 + h_{22}\, v_2 = h_{22}\, v_2$$

or $$h_{12} = \left.\frac{v_1}{v_2}\right|_{i_1 = 0}$$

and $$h_{22} = \left.\frac{i_2}{v_2}\right|_{i_1 = 0}$$

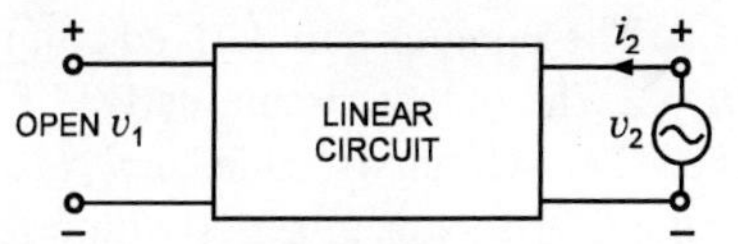

Fig. 7.13

Thus the parameter h_{12} is the ratio of input voltage to the output voltage with zero input current (*i.e.* $i_1 = 0$). It is dimensionless quantity and is called the open-circuit *reverse transfer voltage ratio*, the subscript 12 of h_{12} reveals that the parameter is a transfer quantity determined by the ratio of input to output measurements.

Similarly h_{22} is the ratio of output current to the output voltage with zero input current (*i.e.* $i_1 = 0$). It is called the *open-circuit output admittance* and is measured in siemens. The subscript 22 in h_{22} indicates that it is determined by a ratio of output quantities.

2. *Notations.* The convenient alternative subscript notations recommended by the IEEE Standards are given below :

i = 11 = input $\quad\quad$ o = 22 = output

f = 21 = forward transfer $\quad\quad$ r = 12 = reverse transfer

In case of transistors, another subscript (*b, e* or *c*) is added to designate the type of transistor configuration. For example $h_{ie} = h_{11e}$ = input resistance in common-emitter configuration.

Notations used in transistor amplifier for the three configurations are tabulated in Table 7.1.

TABLE 7.1

S. No.	h parameter	Common Base Configuration	Common Emitter Configuration	Common Collector Configuration
1.	h_{11}	h_{ib}	h_{ie}	h_{ic}
2.	h_{12}	h_{rb}	h_{re}	h_{rc}
3.	h_{21}	h_{fb}	h_{fe}	h_{fc}
4.	h_{22}	h_{ob}	h_{oe}	h_{oc}

Since the two-port network (or the device) described by Eqs. (7.1) and (7.2) is assumed to have no reactive elements, the four parameters h_{11}, h_{12}, h_{21} and h_{22} are real numbers and voltages and currents v_1, v_2, and i_1, i_2 are functions of time. However, if the reactive elements had been included in the device, the excitation would be considered to be sinusoidal, the h parameters would in general be functions of frequency, and the voltages and currents would be represented by phasors V_1, V_2, and I_1, I_2.

7.8.2. Hybrid Model

The hybrid circuit for any two-port network characterized by Eqs. (7.20) and (7.21) is shown in Fig. 7.14. If Kirchhoff's voltage law and Kirchhoff's current law are applied to the input and output ports, Eqs. (7.20) and (7.21) respectively will be obtained. Thus model given in Fig. 7.14 truely satisfies Eqs. (7.20) and (7.21).

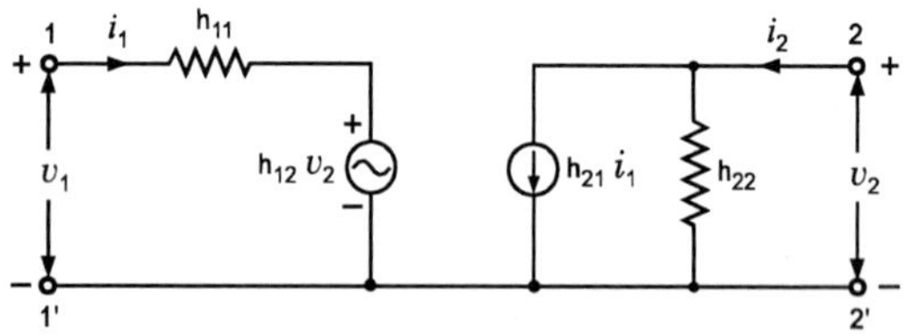

Fig. 7.14. *Hybrid Model For The Two Port Network Shown in Fig. 7.11*

The input circuit derived from Eq. (7.20) appears as a resistance h_{11} in series with a voltage generator $h_{12}v_2$. The output circuit, derived from Eq. (7.21) consists of a current generator $h_{21}i_1$ and shunt resistance h_{22}. This circuit is called *hybrid equivalent* because its input portion is a Thevenin's equivalent (or a voltage generator in series with a resistance) while the output portion is a Norton's equivalent (or a current generator with shunt resistance). Thus it is a mixture or hybrid. The symbol '*h*' is simply the abbreviation of the word hybrid (hybrid means "mixed").

The hybrid equivalent circuit (or model) given in Fig. 7.14 is an extremely important one in the area of electronics today. It will appear over and over again in the analysis to follow. There are two main reasons of popularity of hybrid model. First, it isolates the input and output circuits, their interaction being accounted for by the two controlled voltage and current sources—the effect of output upon input is represented by the equivalent voltage generator $h_{12}v_2$ and the effect of input upon output is represented by the current generator $h_{21}i_1$. The value of the former depends upon the output voltage v_2 while the value for the latter depends upon the input current i_1. Secondly, the two portions of the circuit are in a form which makes it simple to take into account the source and the load circuits.

Example 7.3. Determine the h parameters of the circuit given in Fig. 7.15.

Solution : For determination of forward parameters h_{11} and h_{21}, output terminals are shorted, as illustrated in Fig. 7.16 (*a*).

The input impedance of the network as viewed from input terminals 11′ is given by

$$h_{11} = 4 + 8 \,||\, 8 = 4 + 4 = 8\ \Omega \quad \textbf{Ans.}$$

As seen from the circuit shown in Fig. 7.16 (*a*), the input current is divided into two equal parts at node A. Thus we have

$$i_2 = -\frac{1}{2}i_1$$ negative sign has been placed before right hand side quantity because actually current i_2 is flowing out of the box

The parameter h_{21}, defined as the ratio of output and input currents, is given by

$$h_{21} = \frac{i_2}{i_1} = \frac{-\frac{1}{2}i_1}{i_1} = -0.5 \textbf{ Ans.}$$

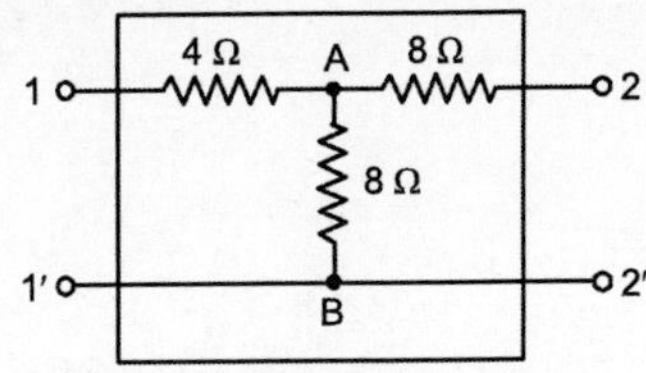

Fig. 7.15

For determination of reverse parameters, input terminals 11′ are kept open and voltage v_2 is applied to the output terminals as shown in Fig. 7.16 (*b*).

The current i_2 produced by voltage v_2 applied across output terminals is given by

$$i_2 = \frac{v_2}{8+8} = \frac{v_2}{16}$$

Voltage appearing across input terminals v_1 is given by

$$v_1 = \text{Voltage across terminals AB (8 Ω resistor)}$$

$$= i_2 R_{AB} = \frac{v_2}{16} \times 8 = \frac{v_2}{2}$$

The parameter h_{12}, defined as the ratio of input voltage to the output voltage is given by

$$h_{12} = \frac{v_1}{v_2} = \frac{v/2}{v_2} = 0.5 \textbf{ Ans.}$$

The input impedance of the network when viewed from the output terminals 22′ with input terminals open [Fig. 7.16 (*b*)] is 8 + 8 = 16 Ω.

So the parameter $h_{22} = \frac{1}{16} = 0.0625 \text{ s}$ **Ans.**

The parameter h_{22}, defined as the ratio of output current to the output voltage with zero input current (*i.e.* $i_1 = 0$), is also given by

$$h_{22} = \left.\frac{i_2}{v_2}\right|_{i_1=0} = \frac{v_2/16}{v_2} = \frac{1}{16} = 0.0625 \text{ s } \textbf{Ans.}$$

(same as determined above)

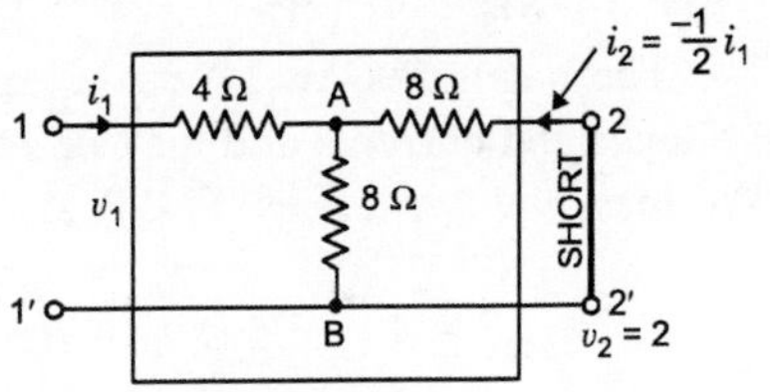

(*a*)

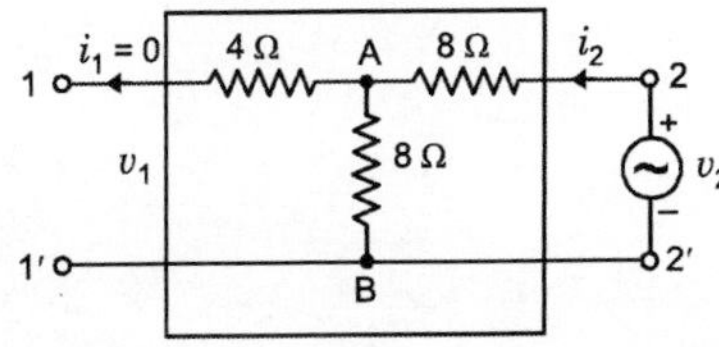

(*b*)

Fig. 7.16

7.9. TRANSISTOR HYBRID MODEL

The basic assumption in arriving at a transistor linear model or equivalent circuit is that the variations about the operating or quiescent point are small and, therefore, the transistor parameters can be considered constant over the small range of operation.

Many transistor models have been proposed, each one having its particular merits and demerits. The transistor model presented here, is given in terms of the h parameters, which are real numbers at audio frequencies, are easy to measure, can also be obtained from the static characteristics of a transistor, and are particularly convenient to use in analysis and design of circuit. Furthermore, a set of h parameters is specified for many transistors by the manufacturers.

To derive a hybrid model for a transistor, let us consider the basic CE amplifier circuit given in Fig. 7.17. The variables i_B, i_C, v_B and v_C represent the total instantaneous values of currents and voltages. We may select the input current i_B and output voltage v_C as independent variables. Since input voltage v_B is some function f_1 of i_B and v_C and output current i_C is another function f_2 of i_B and v_C, we may write

$$v_B = f_1 (i_B, v_C) \quad \text{...(7.22)}$$

$$i_C = f_2 (i_B, v_C) \quad \text{...(7.23)}$$

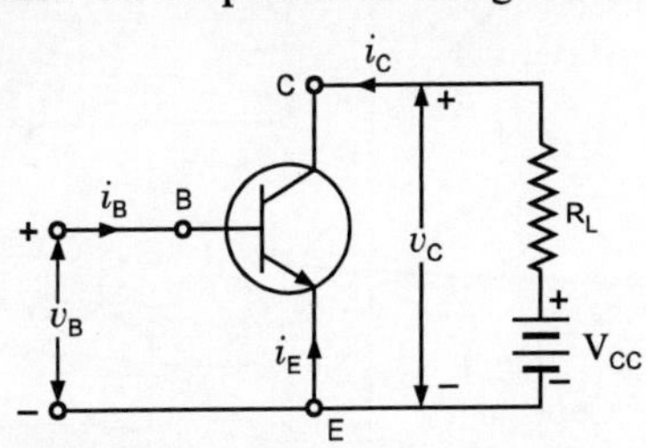

Fig. 7.17. *Basic CE Configuration*

Making a Taylor's series expansion of Eqs. (7.22) and (7.23) about the zero signal operating point (I_B, V_C) and neglecting higher-order terms we have

$$\Delta v_B = \left.\frac{\delta f_1}{\delta i_B}\right|_{V_C} \Delta i_B + \left.\frac{\delta f_1}{\delta v_C}\right|_{I_B} \Delta v_C \qquad ...(7.24)$$

$$\Delta i_C = \left.\frac{\delta f_2}{\delta i_B}\right|_{V_C} \Delta i_B + \left.\frac{\delta f_2}{\delta v_C}\right|_{I_B} \Delta v_C \qquad ...(7.25)$$

where partial derivatives $\frac{\delta f_1}{\delta i_B}$ and $\frac{\delta f_2}{\delta i_B}$ are taken, keeping collector voltage V_C constant while partial derivates $\frac{\delta f_1}{\delta v_C}$ and $\frac{\delta f_2}{\delta v_C}$ are taken, keeping base current I_B constant.

The quantities Δv_B, Δv_C, Δi_B and Δi_C represent the small-signal (incremental) base and collector voltages and currents and may be represented as v_b, v_c. i_b, and i_c respectively as per standard notations. We may now write Eqs. (7.24) and (7.25) as below

$$v_b = h_{ie} i_b + h_{re} v_c \qquad ...(7.26)$$

$$i_c = h_{fe} i_b + h_{oe} v_c \qquad ...(7.27)$$

$$\text{where } h_{ie} \equiv \left.\frac{\delta f_1}{\delta i_B}\right|_{V_C} = \left.\frac{\delta v_B}{\delta i_B}\right|_{V_C} = \left.\frac{v_b}{i_b}\right|_{V_C=0} \qquad ...(7.28a)$$

$$h_{fe} \equiv \left.\frac{\delta f_2}{\delta i_B}\right|_{V_C} = \left.\frac{\delta i_C}{\delta i_B}\right|_{V_C} = \left.\frac{i_c}{i_b}\right|_{V_C=0} \qquad ...(7.28b)$$

$$h_{re} \equiv \left.\frac{\delta f_1}{\delta v_C}\right|_{I_B} = \left.\frac{\delta v_B}{\delta v_C}\right|_{I_B} = \left.\frac{v_b}{v_c}\right|_{I_B=0} \qquad ...(7.28c)$$

$$h_{oe} \equiv \left.\frac{\delta f_2}{\delta v_C}\right|_{I_B} = \left.\frac{\delta i_C}{\delta v_C}\right|_{I_B} = \left.\frac{i_c}{v_c}\right|_{I_B=0} \qquad ...(7.28d)$$

The partial derivatives of Eq. (7.28) define the h parameters for the transistor in common-emitter (CE) configuration.

TABLE 7.2

Configuration	*Circuit Schematic*	*Hybrid Model*	*v-i Equations*
Common Emitter, CE	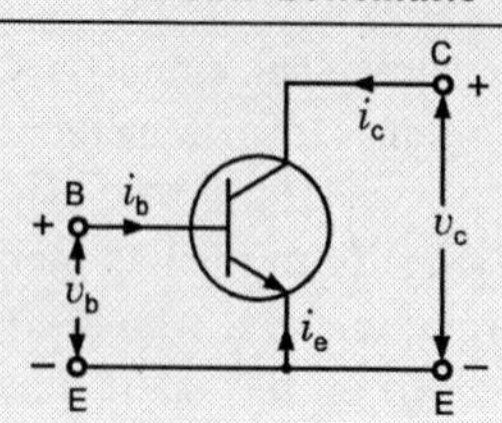	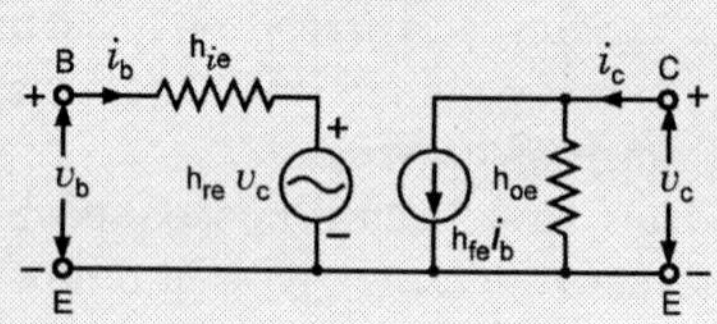	$v_b = h_{ie} i_b + h_{re} v_c$ $i_c = h_{fe} i_b + h_{oe} v_c$
Common Base, CB	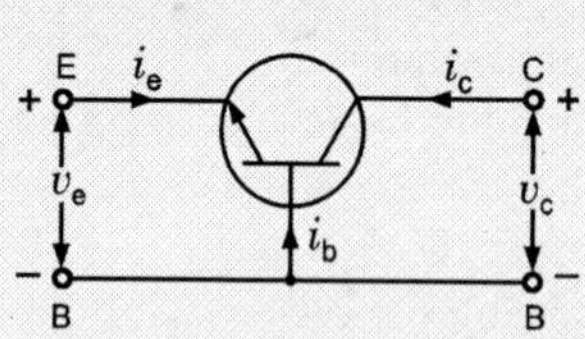	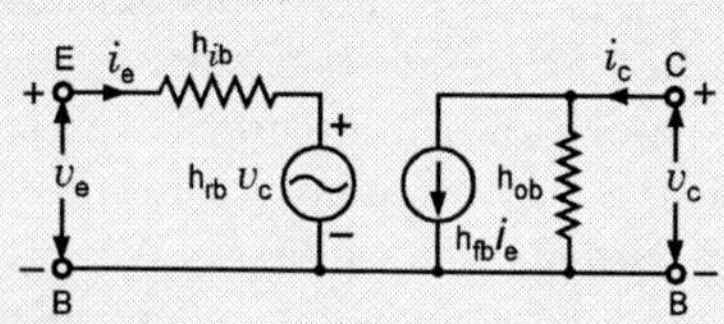	$v_e = h_{ib} i_e + h_{rb} v_c$ $i_c = h_{fb} i_e + h_{ob} v_c$
Common Collector CC	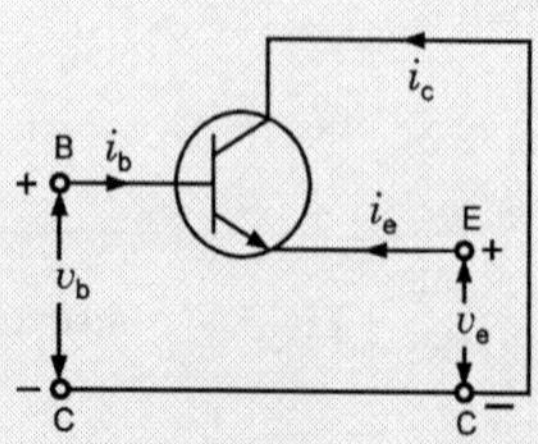	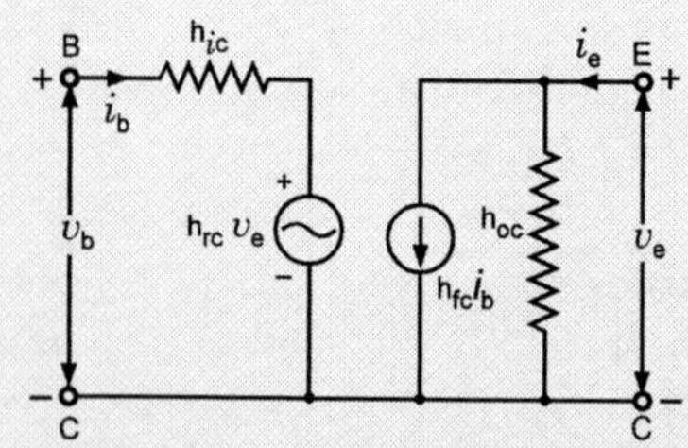	$v_b = h_{ic} i_b + h_{rc} v_e$ $i_e = h_{fc} i_b + h_{oc} v_e$

Equations (7.26) and (7.27) are found to be of exactly the same form as Eqs. (7.20) and (7.21) hence, the model shown in Fig. 7.14 can be used to represent a transistor.

The common-emitter (CE), common-base (CB), and common-collector (CC) configurations, their hybrid models and their terminal volt-ampere equations are summarized in Table 7.2.

The circuits and equations in Table 7.2 are valid for either an N-P-N or P-N-P transistor and are independent of the type of load or biasing method.

7.10. TYPICAL VALUES OF h PARAMETERS FOR A TRANSISTOR

Typical values of h parameters of a transistor for CE, CB and CC configurations at I_E = 1.3 mA are given in Table 7.3.

TABLE 7.3

Configuration / *Parameter*	*Common Emitter CE*	*Common Base CB*	*Common Collector CC*
h_i	1.1 kΩ	21.6 Ω	1.1 kΩ
h_r	2.5×10^{-4}	2.9×10^{-4}	≃ 1
h_f	50	– 0.98	– 51
h_o	25 μS (μA/V)	0.49 μS (μA/V)	25 μS (μA/V)
$1/h_o$	40 kΩ	2.04 MΩ	40 kΩ

7.11. CONVERSION OF HYBRID PARAMETERS IN TRANSISTOR THREE CONFIGURATIONS

Some transistor manufacturers provide only the four CE hybrid parameters while others provide CB hybrid parameters. Sometimes, for a specific purpose, it becomes necessary to convert one set of h parameters in one configuration to another set in another configuration. The conversion formulas can be obtained using the definitions of the parameters involved and Kirchhoff's laws.

For example, let us convert common-emitter h parameters to the common-base h parameters. For this purpose first the CB hybrid model is drawn as shown in Fig. 7.18 (*a*) and then it is redrawn in CE configuration, as shown in Fig. 7.18 (*b*). The latter corresponds in every detail to the former except that the emitter terminal E is made common to the input and output ports.

By definitions

$$h_{re} = \left.\frac{V_{be}}{V_{ce}}\right|_{I_b=0} = \left.\frac{V_{bc}+V_{ce}}{V_{ce}}\right|_{I_b=0} = 1 + \left.\frac{V_{bc}}{V_{ce}}\right|_{I_b=0}$$

If $I_b = 0$, then $I_c = -I_e$, and the current I in h_{ob} in Fig. 7.18 (*b*) is $I = (1 + h_{fb})\, I_e = h_{ob} V_{bc}$

Applying Kirchhoff's voltage law to the output mesh of Fig. 7.18 (*b*) we have

$$h_{ib} I_e + h_{rb} V_{cb} + V_{bc} + V_{ce} = 0$$

Combining the above two equations we have

$$\frac{h_{ib}h_{ob}}{1+h_{fb}} V_{bc} - h_{rb} V_{bc} + V_{bc} + V_{ce} = 0$$

$$\text{or} \quad \frac{V_{bc}}{V_{ce}} = \frac{-(1+h_{fb})}{h_{ib}\, h_{ob} + (1-h_{rb})(1+h_{fb})}$$

$$\text{Thus} \quad h_{re} = 1 + \frac{V_{bc}}{V_{ce}} = \frac{h_{ib}h_{ob} - (1+h_{fb})h_{rb}}{h_{ib}h_{ob} + (1-h_{rb})(1+h_{fb})} \qquad ...(7.29)$$

This is an exact expression. The simpler approximate formula is obtained by noting that, for typical values given in Table 7.3 $h_{rb} << 1$ and $h_{ob}\, h_{ib} << (1 + h_{fb})$

$$\text{So } h_{re} \simeq \frac{h_{ib}\, h_{ob}}{1 + h_{fb}} - h_{rb} \qquad ...(7.30)$$

$$\text{By definition } h_{ie} = \left.\frac{V_{be}}{I_b}\right|_{V_{ce}=0}$$

If terminals C and E are connected together in Fig. 7.18 (*b*), we have circuit shown in Fig. 7.19. From Fig. 7.19, we see that $V_{cb} = -V_{bc} = -V_{be}$

Applying Kirchhoff's voltage law to the left-hand mesh, we have

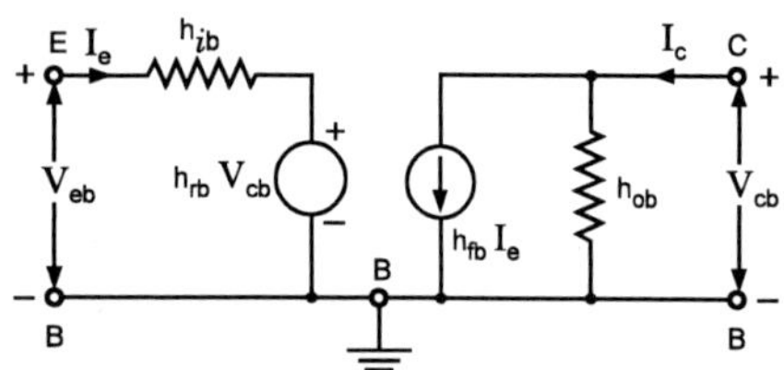

(a) Common Base Hybrid Model

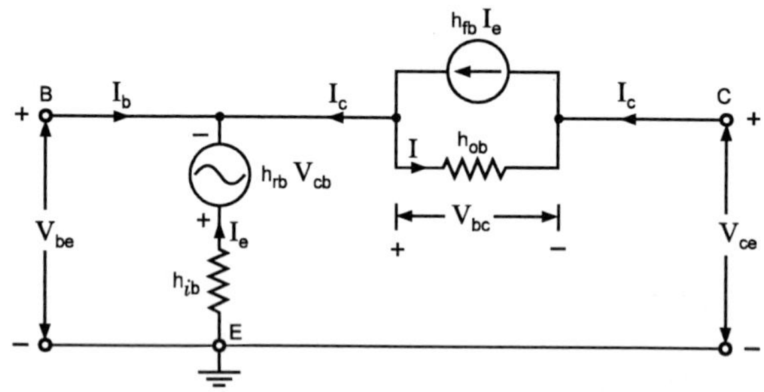

(b) Circuit of Fig (a) Redrawn in CE Configuration

Fig. 7.18

$$V_{be} + h_{ib}\, I_e + h_{rb}\, V_{cb} = 0$$

Combining the above two equations we have

$$I_e = -\frac{1 - h_{rb}}{h_{ib}} V_{be}$$

Applying Kirchhoff's current law to node B in Fig. 7.19, we have

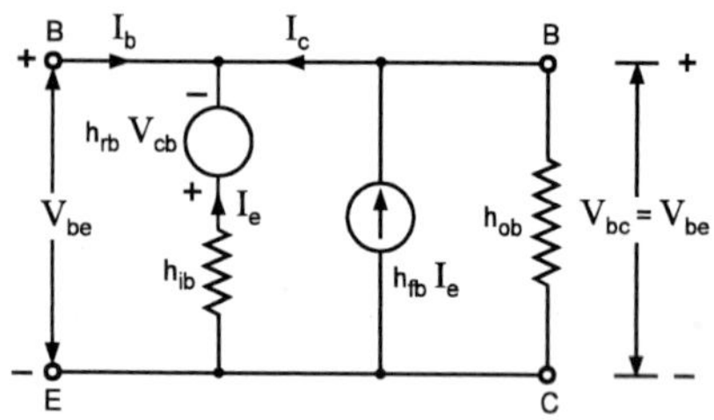

Fig. 7.19

$$I_b + I_e + h_{fb} I_e - h_{ob} V_{be} = 0$$

$$\text{or } I_b = (1 + h_{fb}) \frac{1 - h_{rb}}{h_{ib}} V_{be} + h_{ob} V_{be}$$

$$\text{Thus } h_{ie} = \frac{V_{be}}{I_b} = \frac{h_{ib}}{h_{ib}\, h_{ob} + (1 - h_{rb})(1 + h_{fb})} \qquad ...(7.31)$$

This is an exact expression. Since $h_{rb} << 1$ and $h_{ob}\, h_{ib} << (1 + h_{fb})$, so the above equation is reduced to

$$h_{ie} \simeq \frac{h_{ib}}{1 + h_{fb}} \qquad ...(7.32)$$

Approximate conversion formulas for transistor parameters are given in Table 7.4.

Example 7.4. $h_{fe} = 50$, $h_{ie} = 0.83$ **kΩ, find out the current gain (h_{fb}) and input impedance (h_{ib}) for a transistor in CB configuration.**

Solution :

$$h_{fb} = \frac{-h_{fe}}{1 + h_{fe}} = \frac{-50}{1 + 50} = -0.98 \textbf{ Ans.}$$

$$h_{ib} = \frac{h_{ie}}{1 + h_{fe}} = \frac{0.83 \times 10^3}{1 + 50} = 16.27\ \Omega \textbf{ Ans.}$$

TABLE 7.4

Parameter \ Configuration	Common Emitter, CE	Common Base, CB	Common Collector, CC
h_{ie}	1.1 kΩ	$\dfrac{h_{ib}}{1+h_{fb}}$	h_{ic}*
h_{re}	2.5×10^{-4}	$\dfrac{h_{ib}\,h_{ob}}{1+h_{fb}} - h_{rb}$	$1 - h_{rc}$*
h_{fe}	50	$-\dfrac{h_{fb}}{1+h_{fb}}$	$-(1+h_{fc})$*
h_{oe}	25 μS (μA/V)	$\dfrac{h_{ob}}{1+h_{fb}}$	h_{oc}*
h_{ib}	$\dfrac{h_{ie}}{1+h_{fe}}$	21.6 Ω	$-h_{ic}/h_{fc}$
h_{rb}	$\dfrac{h_{ie}\,h_{oe}}{1+h_{fe}} - h_{re}$	$2.9. \times 10^{-4}$	$h_{rc} - \dfrac{h_{ic}\,h_{oc}}{h_{fc}} - 1$
h_{fb}	$\dfrac{-h_{fe}}{1+h_{fe}}$	− 0.98	$-\dfrac{1+h_{fc}}{h_{fc}}$
h_{ob}	$\dfrac{h_{oe}}{1+h_{fe}}$	0.49 μS (μA/V)	$\dfrac{-h_{oc}}{h_{fc}}$
h_{ic}	h_{ie}*	$\dfrac{h_{ib}}{1+h_{fb}}$	1.1 kΩ
h_{rc}	$1 - h_{re} \simeq 1$*	1	1
h_{fc}	$-(1+h_{fe})$*	$\dfrac{-1}{1+h_{fb}}$	− 51
h_{oc}	h_{oe}*	$\dfrac{h_{ob}}{1+h_{fb}}$	25 μS (μA/V)

* Stands for exact.

7.12. TRANSISTOR AMPLIFIER CIRCUIT PERFORMANCE IN h PARAMETERS

A transistor amplifier can be formed simply by connecting a signal source to the input and an external load to the output terminals of a transistor, as shown in Fig. 7.20 and giving proper bias to it. All transistor amplifiers, connected in any one of the three possible configurations, are basically two-port devices, as shown in Fig. 7.20, that is, there are a pair of input terminals and a pair of output terminals. In Fig. 7.21, the transistor has been replaced with its small-signal hybrid model without

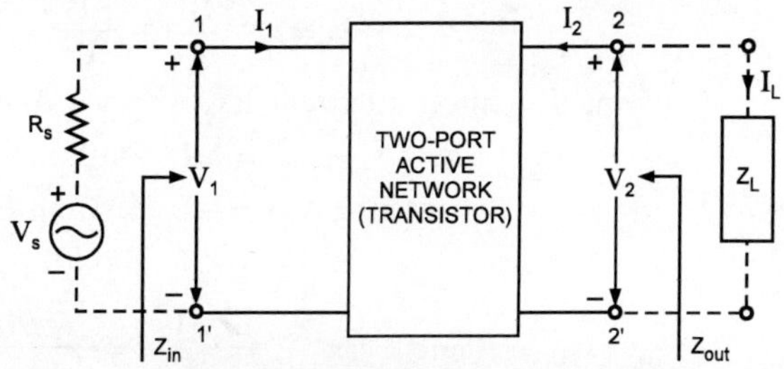

Fig. 7.20. *Basic Amplifier Circuit*

specifying the configuration. The circuit shown in Fig. 7.21 is valid for any type of load whether it be a pure resistance, an impedance or another transistor. This is true because the transistor hybrid model was derived without any regard to the external circuit in which the transistor is incorporated. The only limitation is the requirement that the h parameters remain substantially constant over the working range. For an amplifier there are six quantities of great interest (input impedance, output impedance, current gain, voltage gain, power gain and phase relationship), each of which will be discussed in detail here.

For the analysis of circuit, let us assume sinusoidally varying voltages and currents :

1. *Input Impedance.* The input impedance Z_{in} is defined as the ratio of input voltage to the input current *i.e.*

$$Z_{in} = \frac{V_1}{I_1} \quad ...(7.33)$$

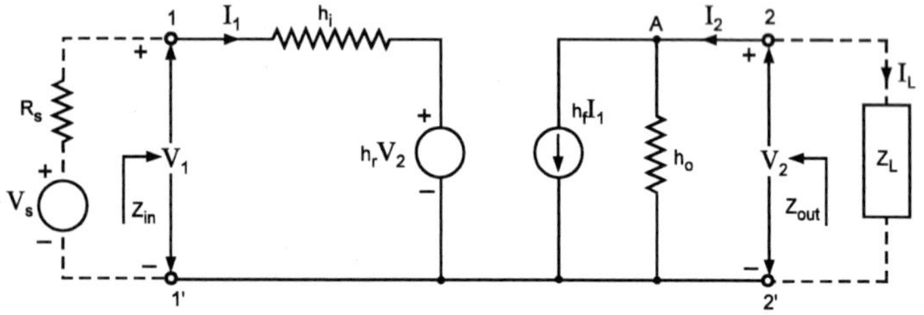

Fig. 7.21. ***Small Signal Hybrid Model of Transistor Amplifier Circuit Shown in Fig. 7.20***

From the input circuit shown in Fig. 7.21 we have

$$V_1 = h_iI_1 + h_rV_2 \quad ...(7.34)$$

Substituting $V_1 = h_iI_1 + h_rV_2$ in Eq. (7.33) we have

$$Z_{in} = \frac{h_iI_1 + h_rV_2}{I_1} = h_i + h_r\frac{V_2}{I_1} \quad ...(7.35)$$

From the output circuit shown in Fig. 7.21 we have

$$I_2 = h_fI_1 + h_oV_2 = \frac{-V_2}{Z_L} \quad ...(7.36)$$

The minus sign is used here because the actual current flows in the direction opposite to that of I_2.

$$\text{So } \frac{V_2}{I_1} = \frac{-h_f}{h_o + \frac{1}{Z_L}} \quad ...(7.37)$$

Substituting value of $\frac{V_2}{I_1}$ from Eq. (7.37) in Eq. (7.35) we have

$$\text{Input impedance } Z_{in} = h_i - \frac{h_rh_f}{h_o + \frac{1}{Z_L}} \quad ...(7.38)$$

$$\simeq h_i \quad \text{...if } h_r \text{ or } Z_L \text{ is very small.}$$

It is seen that Z_{in} depends on Z_L *i.e.* ac resistance of the load across the output terminals of the transistors.

2. *Current Gain.* The current gain is denoted by A_i and is defined as the ratio of output current to input current.

$$i.e.\ A_i = \frac{I_L}{I_1} = \frac{-I_2}{I_1} \quad ...(7.39)$$

Applying Kirchhoff's current law to node A in the output circuit

$$I_2 = h_fI_1 + h_oV_2$$

Substituting $V_2 = -I_2Z_L$ from Eq. (7.36) we have

$$I_2 = h_fI_1 - h_oI_2Z_L$$

$$\text{or Current gain } A_i = \frac{-I_2}{I_1} = \frac{-h_f}{1 + h_oZ_L} \simeq -h_f \text{ if } Z_L \text{ is zero or } h_oZ_L << 1 \quad ...(7.40)$$

Current Gain Taking R_s into Account. The source current is not the transistor input current because it partly flows through R_s and partly through Z_{in}. So the voltage source V_s with series source resistance R_s is replaced by the Norton's equivalent source, shown in Fig. 7.22, consisting of current source I_s with source resistance R_s in shunt. The overall current gain A_{is} is given as

$$A_{is} \equiv \frac{-I_2}{I_s} = \frac{-I_2}{I_1}\cdot\frac{I_1}{I_s} = A_i\frac{I_1}{I_s} \quad \text{...(7.41)}$$

From Fig. 7.22,
$$\frac{I_1}{I_s} = \frac{R_s}{Z_{in}+R_s}$$

So
$$A_{is} = \frac{A_i R_s}{Z_{in}+R_s} \quad \text{...(7.42)}$$

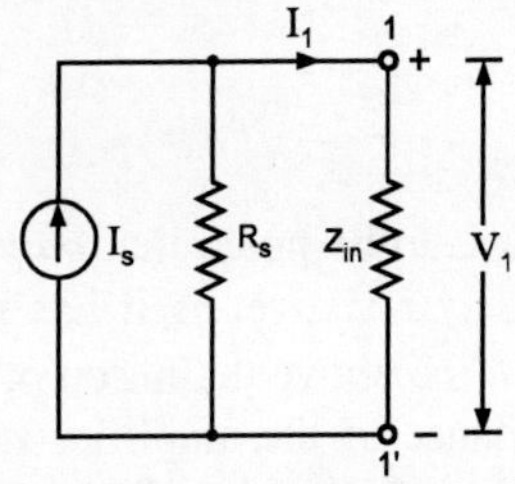

Fig. 7.22. ***Norton's Equivalent For The Source***

Note that if $R_s = \infty$ then $A_{is} = A_i$. Hence A_i is the *current gain for an ideal current source* (one with infinite source resistance).

3. ***Voltage Gain.*** The voltage gain is denoted by A_v and is defined as the ratio of output voltage to the input voltage *i.e.*

$$A_v = \frac{V_2}{V_1} = \frac{V_2}{I_1 Z_{in}} = \frac{-h_f}{\left(h_o + \frac{1}{Z_L}\right)Z_{in}} \quad \text{...(7.43)} \quad \because \text{ from Eq. (7.37) } \frac{V_2}{I_1} = \frac{-h_f}{h_o + \frac{1}{Z_L}}$$

Voltage Gain Taking R_s into Account. Overall voltage gain A_{vs} is given as

$$A_{vs} = \frac{V_2}{V_s} = \frac{V_2}{V_1}\cdot\frac{V_1}{V_s} = A_v\frac{V_1}{V_s} \quad \text{...(7.44)}$$

From the equivalent input circuit of the amplifier given in Fig. 7.23, we have

$$\frac{V_1}{V_s} = \frac{Z_{in}}{Z_{in}+R_s},$$

So
$$A_{vs} = A_v\frac{Z_{in}}{Z_{in}+R_s} \quad \text{...(7.45)}$$

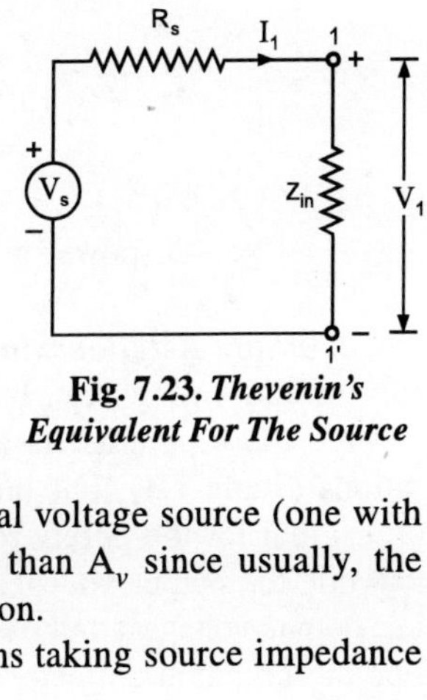

Fig. 7.23. ***Thevenin's Equivalent For The Source***

Note that if $R_s = 0$ the $A_{vs} = A_v$. Hence A_v is the voltage gain for an ideal voltage source (one with zero internal resistance). In practice, the quantity A_{vs} is more meaningful than A_v since usually, the source resistance has significant effect on the overall voltage amplification.

Independent of the transistor characteristics, the voltage and current gains taking source impedance into account, are related as

$$A_{vs} = A_{is}\frac{Z_L}{R_s} \quad \text{...(7.46)}$$

provided that the current and voltage generators have the same source resistance R_s.

4. ***Output Admittance.*** The output admittance Y_{out} is defined as the ratio of the output current to output voltage with $V_s = 0$ *i.e.*

$$Y_{out} \equiv \frac{I_2}{V_2} \text{ with } V_s = 0 \quad \text{...(7.47)}$$

$$= h_f\frac{I_1}{V_2} + h_o \quad \text{...(7.48)} \quad \because \text{ from Eq. (7.36), } I_2 = h_f I_1 + h_o V_2$$

Now from input circuit shown in Fig. 7.21 we have

$$R_s I_1 + h_i I_1 + h_r V_2 = 0 \quad \text{...(7.49) taking } V_s = 0$$

$$\text{or} \quad \frac{I_1}{V_2} = -\frac{h_r}{h_i + R_s} \qquad \text{...(7.50)}$$

Substituting $\frac{I_1}{V_2} = \frac{-h_r}{h_i + R_s}$ from Eq. (7.50) in Eq. (7.48) we have

$$Y_{out} = h_o - \frac{h_f\, h_r}{h_i + R_s} \qquad \text{...(7.51)}$$

It is to be noted that *output admittance is a function of source resistance* R_s. If the source impedance is purely resistive, as it has been assumed, then Y_{out} is real (purely conductance).

In the above definition of Y_{out}, the load Z_L has been considered external to the amplifier. If the output impedance of the amplifier stage with Z_L included is required, the load impedance can be determined as the parallel combination of Z_L and Z_{out} (*i.e.*, $Z_L \parallel Z_{out}$).

5. *Power Gain.* The average power delivered to the load shown in Fig. 7.20, P_2 equals $|V_2|\,|I_L| \cos\phi$ where ϕ is the phase angle between V_2 and I_L. For situation under discussion power delivered, $P_2 = -V_2 I_2 \cos\phi$. The minus sign arises for the same reason discussed in the derivation of input impedance. It indicates that the load is absorbing power and not supplying to the amplifier circuit. For purely resistive load, $\cos\phi = 1$ and $P_2 = -V_2 I_2$. The input power is $V_1 I_1$. So power gain

$$A_P = \frac{P_2}{P_1} = \frac{-V_2}{V_1} \cdot \frac{I_2}{I_1} = -A_v A_i \qquad \text{...(7.52)}$$

$$\text{As } V_2 = -I_2 R_L \text{ and } I_2 = A_i I_1$$

$$\therefore A_v = \frac{V_2}{V_1} = \frac{-A_i I_1 R_L}{V_1} = \frac{-A_i R_L}{V_1/I_1} = \frac{-A_i R_L}{Z_{in}}$$

$$\text{So power gain, } A_p = -A_v A_i = -\left(\frac{-A_i R_L}{Z_{in}}\right) A_i = \frac{A_i^2 R_L}{Z_{in}} \qquad \text{...(7.53)}$$

6. *Phase Relationship.* The phase relationship between the output current or voltage and the input current and voltage can be determined simply examining the Eqs. (7.39), (7.40) and (7.43).

All the h parameters are positive except h_f for the common-base and common-collector configurations (Table 7.3). For the common-emitter configuration it is obvious from Eqs. (7.39), (7.40) and (7.43) that for the purely resistive loads, the transistor output current I_2 is in phase with the input current I_1 while the output voltage, due to the negative sign, has a polarity opposite to that of the input voltage *i.e.* output voltage attains its negative maximum when the input signal attains its positive maximum. For the common-collector and common-base configurations the opposite is true because of negative value of h_f.

The important relations derived above are summarized below :

$$\text{Current gain, } A_i = \frac{-h_f}{1 + h_0 Z_L} \simeq -h_f \quad \text{if } h_0 Z_L << 1$$

$$A_{is} = \frac{A_i R_s}{Z_{in} + R_s}$$

$$\text{Voltage gain, } A_v = \frac{-h_f}{\left(h_0 + \frac{1}{Z_L}\right) Z_{in}} = \frac{A_i Z_L}{Z_{in}}$$

$$A_{vs} = A_V \frac{Z_{in}}{Z_{in} + R_s}$$

$$\text{Input impedance, } Z_{in} = h_i - \frac{h_r h_f}{h_0 + \frac{1}{Z_L}} \simeq h_i \qquad \text{if } h_r \text{ or } Z_L \text{ is very small}$$

$$\text{Output admittance, } Y_{out} = h_o - \frac{h_f h_r}{h_i + R_s} = \frac{1}{Z_{out}}$$

Example 7.5. A CE transistor amplifier circuit has the following characteristics : h_{ie} = 1,000 Ω; h_{fe} = 50; $h_{re} = 2.5 \times 10^{-4}$; $h_{oe} = 25 \times 10^{-6}$ A/V. If the load resistance R_L = 10 kΩ, and source resistance is 100 Ω, find the input resistance, output resistance, voltage, current and power gains.

Solution :

$$\text{Current gain, } A_i = \frac{-h_{fe}}{1 + h_{oe}R_L} = \frac{-50}{1 + 25 \times 10^{-6} \times 10 \times 10^3} = -40 \text{ Ans.}$$

$$\text{Input resistance, } R_{in} = h_{ie} - \frac{h_{re}h_{fe}}{h_{oe} + \frac{1}{R_L}} = 1{,}000 - \frac{2.5 \times 10^{-4} \times 50}{25 \times 10^{-6} + \frac{1}{10 \times 10^3}} = 900\ \Omega \text{ Ans.}$$

$$\text{Voltage gain, } A_v = \frac{A_i R_L}{R_{in}} = -40 \times \frac{10{,}000}{900} = -444.44 \text{ Ans.}$$

$$\text{Overall current gain, } A_{is} = A_i \times \frac{R_s}{R_{in} + R_s} = -40 \times \frac{100}{900 + 100} = -4$$

$$\text{Overall voltage gain, } A_{vs} = A_v \times \frac{R_{in}}{R_{in} + R_s} = -444.44 \times \frac{900}{900 + 100} = -400$$

$$\text{Output conductance, } G_{out} = h_{oe} - \frac{h_{fe}h_{re}}{h_{ie} + R_s} = 25 \times 10^{-6} - \frac{50 \times 2.5 \times 10^{-4}}{1000 + 100} = 13.636 \times 10^{-6} \text{ S}$$

$$\text{Output resistance, } R_{out} = \frac{1}{G_{out}} = \frac{1}{13.636 \times 10^{-6}} = 73.3 \text{ k}\Omega \text{ Ans.}$$

$$\text{Power gain, } A_p = A_{vs} A_{is} = -400 \times -4 = 1{,}600 \text{ Ans.}$$

Example 7.6. A CB transistor amplifier uses a voltage source of internal resistance R_S = 200 Ω and the load resistance is R_L = 1,200 Ω. The h parameters are h_{ib} = 24 Ω, $h_{rb} = 4 \times 10^{-4}$, h_{fb} = –0.98 and h_{ob} = 0.6 μA/V. Calculate the following :

(1) Current gain (2) Input impedance (3) Voltage gain
(4) Overall voltage gain (5) Overall current gain.

Solution :

$$\text{(1) Current gain, } A_i = \frac{-h_{fb}}{1 + h_{ob}.R_L} = \frac{-(-0.98)}{1 + 0.6 \times 10^{-6} \times 1{,}200} = 0.979 \text{ Ans.}$$

$$\text{(2) Input impedance, } R_{in} = h_{ib} - \frac{h_{rb}h_{fb}}{h_{ob} + \frac{1}{Z_L}} = 24 - \frac{4 \times 10^{-4} \times (-0.98)}{0.6 \times 10^{-6} + \frac{1}{1{,}200}} = 24.47\ \Omega \text{ Ans.}$$

$$\text{(3) Voltage gain, } A_v = \frac{A_i R_L}{R_{in}} = \frac{0.979 \times 1{,}200}{24.47} = 48 \text{ Ans.}$$

$$\text{(4) Overall voltage gain, } A_{VS} = \frac{A_V . R_{in}}{R_{in} + R_S} = \frac{48 \times 24.47}{24.47 + 200} = 5.23 \text{ Ans.}$$

$$\text{(5) Overall current gain, } A_{is} = \frac{A_i R_S}{R_{in} + R_S} = \frac{0.979 \times 200}{24.47 + 200} = 0.872 \text{ Ans.}$$

Example 7.7. Given $I_E(dc)$ = 1.2 mA, β = 120 and r_o = 40 kΩ, Sketch the

(*i*) Common-Emitter hybrid equivalent model.

(*ii*) Common-Base hybrid equivalent model.

Solution : Here $\beta = h_{FE} = 120$

$$h_{oe} = \frac{1}{r_{out}} = \frac{1}{40} \times 10^{-3} = 25 \times 10^{-6} \text{ mho}$$

Neglecting h_{re}, being of very small value.

h_{ie} is the resistance between the emitter and the base terminal. This can be approximated as a forward biased diode, *i.e.*

$$\therefore\ h_{ie} = \frac{25\,\text{mV}}{I_E}.\beta = \frac{25\,\text{mV}}{1.2\,\text{mA}} \times 120 = 2.5\,\text{k}\Omega$$

Hence CE hybrid model will be as shown in Fig. 7.24.

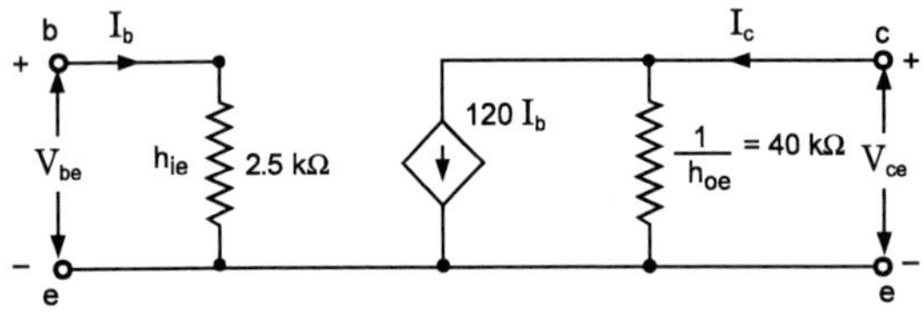

Fig. 7.24

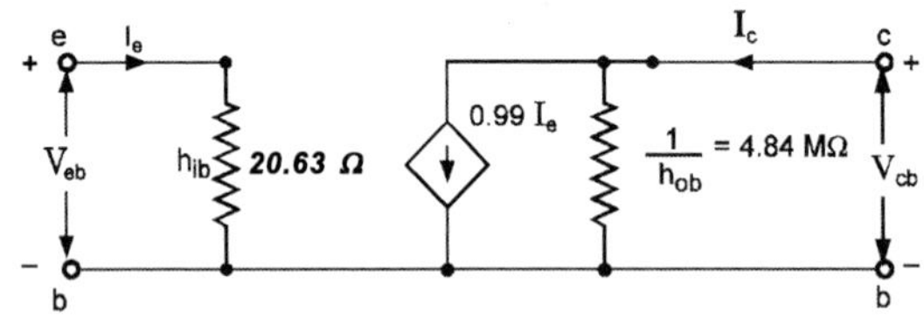

Fig. 7.25

(*ii*) Current amplification factor in CB configuration is α

and $$\alpha = \frac{\beta}{1+\beta} = \frac{120}{120+1} = 0.99 = h_{fb}$$

and $$h_{ob} = \frac{h_{oe}}{1+\beta} = \frac{25 \times 10^{-6}}{1+120} = 0.206 \times 10^{-6} \text{ mho}$$

and, $$h_{ib} = \frac{2.5\,\text{mV}}{I_E}.\alpha = \frac{25\text{ mV}}{1.2\text{ mA}} \times 0.99 = 20.63\,\Omega \quad \textbf{Ans.}$$

Hence CB hybrid model will be as given in Fig. 7.25.

7.13. SIMPLIFIED COMMON-EMITTER HYBRID MODEL

As the h parameters themselves vary widely for the same type of transistor, it is appropriate to make approximations and simplify the expressions for A_i, A_v, A_p, R_{in} and R_{out}. In addition, a better "physical feeling" for the behaviour of transistor circuit may be had from a simple approximate solution than from a more laborious exact calculation. Since the CE configuration is most widely used, it is considered.

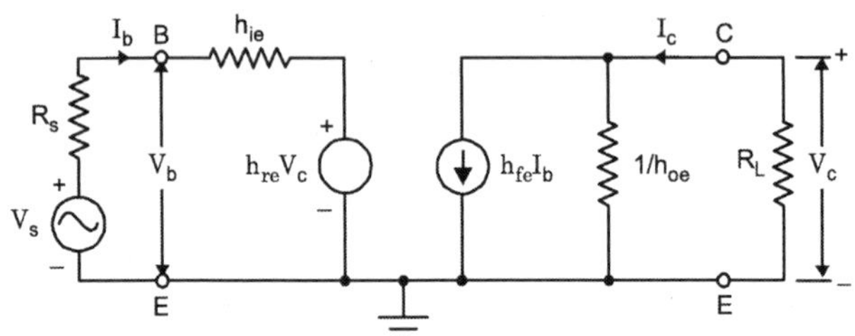

Fig. 7.26. ***Exact CE Hybrid Model***

The exact CE h parameter model is given in Fig. 7.26.

On the output side the resistance $\frac{1}{h_{oe}}$ appears in parallel with the load resistance R_L. The equivalent resistance of the parallel combination of $\frac{1}{h_{oe}}$ and R_L is approximately equal to the lower value *i.e.* R_L provided $\frac{1}{h_{oe}} >> R_L$ *i.e.* the term h_{oe} may be neglected.

Under this condition, the magnitude of the voltage of generator in the emitter circuit also gets modified and is given by

$$h_{re}V_c = h_{re}I_cR_L = h_{re}h_{fe}I_bR_L \qquad \because\ I_c = h_{fe}\,I_b$$

Since $h_{re}h_{fe} = 0.01$, this voltage may be neglected in comparison with the voltage drop across h_{ie} (*i.e.* h_{ie} I_b) provided R_L is not too large.

Thus, we may conclude that, if the load resistance R_L is small, it is possible to neglect the parameters h_{re} and h_{oe} in the circuit of Fig. 7.26 and to have the approximate equivalent circuit as given in Fig. 7.27.

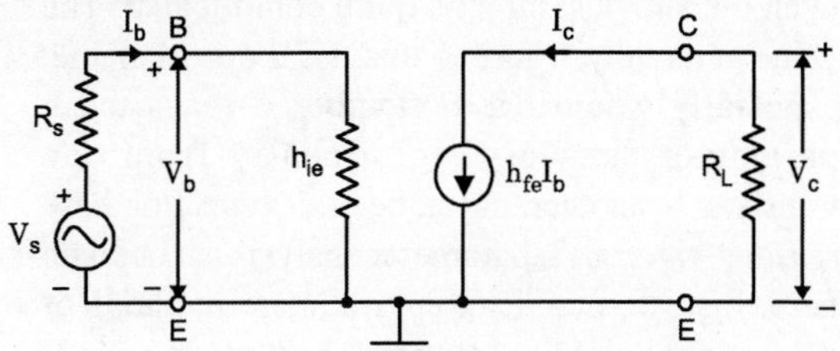

Fig. 7.27. *Approximate CE Hybrid Model*

It can be shown that if $h_{oe}R_L \le 0.1$, the error in computing A_i, A_v, R_{in} and R_{out} for CE configuration is less than 10%.

7.14. ANALYSIS OF A CE TRANSISTOR AMPLIFIER USING h PARAMETERS

A common-emitter transistor amplifier circuit with self biasing is shown in Fig. 7.28. The circuit arrangement has already been explained in Art. 7.2. Various currents flowing in different branches, and various voltages across different components and branches are indicated in the figure.

In the circuit NPN transistor is used. Input signal is applied to the base-emitter circuit while output is taken from the collector-emitter circuit. The emitter-base junction is forward biased and collector-emitter junction is reverse biased by supply V_{CC}. When the input ac signal increases in the positive direction, the voltage between base and emitter V_{be} increases because it is already positive with respect to ground. Thus forward bias is increased and causes increase in base current. Due to transistor action, collector current is increased β times. Due to flow of this current through collector resistance R_C, the voltage drop across it I_cR_C increases considerably. As a result the voltage between collector and emitter V_{ce} decreases. Thus amplified voltage appears across R_L. As explained in Art. 7.3, the positive going input signal appears as a negative going output signal *i.e.*, there is a phase shift of 180° between input and output signals. This is illustrated in Fig. 7.28.

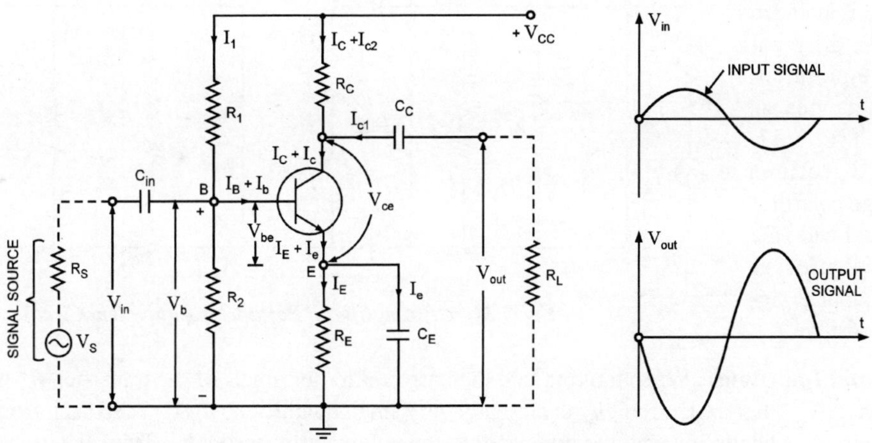

Fig. 7.28. *Common Emitter Transistor Amplifier Circuit With Self Biasing*

Circuit analysis can be done rigorously using h parameters derived from a manufacturer's data sheet for a given transistor. However, the parameters are specified only for a particular set of bias conditions and, therefore, need correction when the circuit bias conditions differ from those mentioned on the data sheet. Correction involves use of graphs showing variation of each parameter with variations in bias conditions

(I_C and V_{CE}). This process makes the analysis of even the simplest circuits quite complicated. The value of lengthy, rigorous analysis is questionable, especially when it is remembered that each h parameter can vary significantly from one transistor to another one of the same type. For these reasons, rigorous h parameter analysis is not taken here. Instead, practical, approximate methods of determining circuit performance are treated.

AC equivalent circuit of a CE transistor amplifier shown in Fig. 7.28 is given in Fig. 7.29. From Fig. 7.29 it is obvious that the circuit input voltage V_{in} is developed across the transistor base and emitter terminals. The output is available across collector and emitter terminals. Since the emitter is common to both input and output, the circuit is designated *common emitter*, or sometimes *ground emitter circuit.*

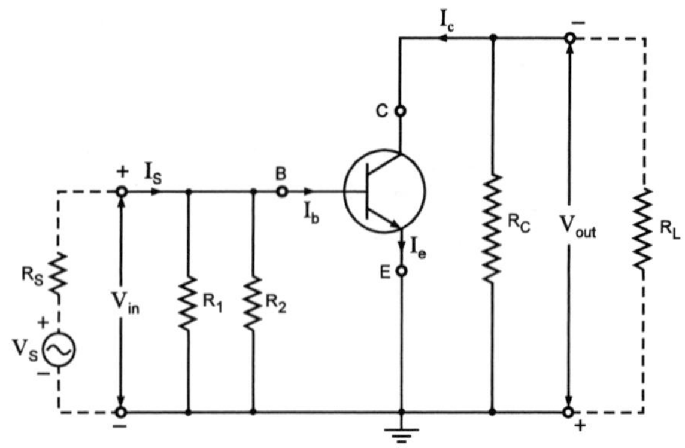

Fig. 7.29. ***CE Amplifier AC Equivalent Circuit***

Some CE circuit may not include bypass capacitor C_E. In such a condition unbypassed resistor R_E exists and it is to be included in the ac equivalent circuit, as shown in Fig. 7.30.

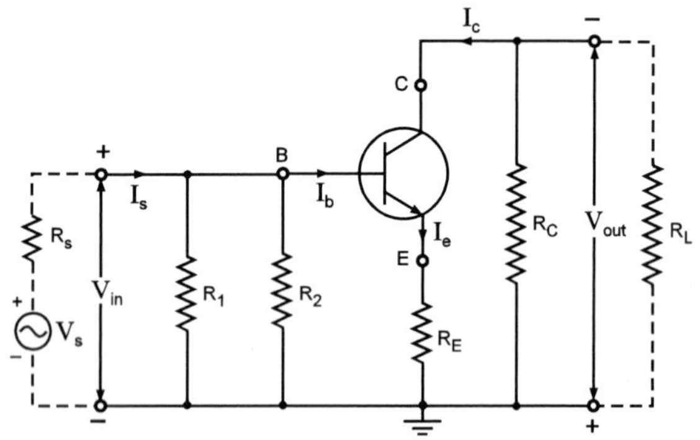

Fig. 7.30. ***CE Amplifier AC Equivalent Circuit With an Unbypassed Emitter Resistor R_E***

The h parameter equivalent of CE circuit given in Fig. 7.28 is given in Fig. 7.31. Figure 7.31 is the same as Fig. 7.29 except that the transistor has been replaced by its h parameter equivalent circuit. When an unbypassed emitter resistor R_E is included (Fig. 7.30), the h parameter equivalent circuit becomes as given in Fig. 7.32. Current directions and voltage polarities in Figs. 7.31 and 7.32 are those that occur when the input ac signal goes positive.

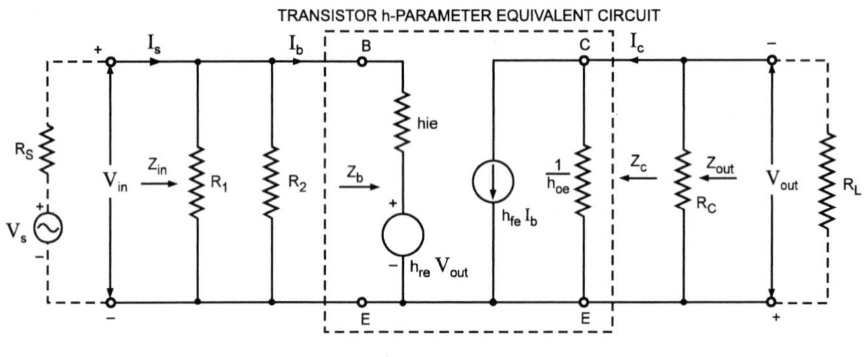

Fig. 7.31. ***CE Amplifier h Parameter Equivalent Circuit***

1. Input Impedance. When looking into the base emitter terminals of the transistor, h_{ie} is seen to be in series with $h_{re} V_{out}$. For a CE circuit h_{re} is normally very small quantity, so $h_{re} V_{out}$ fed back from the output to the input circuit is negligibly small as compared to the voltage drop across h_{ie}. Thus, the input impedance to the transistor base is given as

$$Z_{in\,(base)} \text{ or } Z_b = h_{ie} \qquad ...(7.54)$$

With an unbypassed emitter resistance R_E in the circuit (Fig. 7.32), the computation of Z_b becomes a little more complicated.

From circuit shown in Fig. 7.32, $h_{re} V_{out}$ being neglected, we have

$$V_{in} = h_{ie} I_b + I_e R_E = h_{ie} I_b + R_E (I_b + I_c) \qquad \because I_e = I_b + I_c$$
$$= I_b h_{ie} + I_b R_E + h_{fe} I_b R_E \qquad \because I_c = h_{fe} I_b$$
$$= I_b [h_{ie} + R_E (1 + h_{fe})]$$

and input impedance to the transistor base,

$$Z_b = \frac{V_{in}}{I_b} = h_{ie} + R_E (1 + h_{fe}) \qquad ...(7.55)$$

The actual circuit input impedance is $R_1 \parallel R_2 \parallel Z_b$, so we have

$$Z_{in} \text{ or } Z_{in\,(stage)} = R_1 \parallel R_2 \parallel Z_b \qquad ...(7.56)$$

2. Output Impedance. Since the output voltage variations have little effect upon the input of a CE circuit, only the output of the circuit need be considered in determining the output impedance. Looking back into the collector and emitter terminals of the transistor in Fig. 7.31 a large resistance $\frac{1}{h_{oe}}$ is seen.

Thus, the device output impedance

$$Z_c \simeq \frac{1}{h_{oe}} \qquad ...(7.57)$$

The actual circuit output impedance is Z_c in parallel with R_C *i.e.*

$$Z_{out} = \frac{1}{h_{oe}} \parallel R_C \qquad ...(7.58)$$

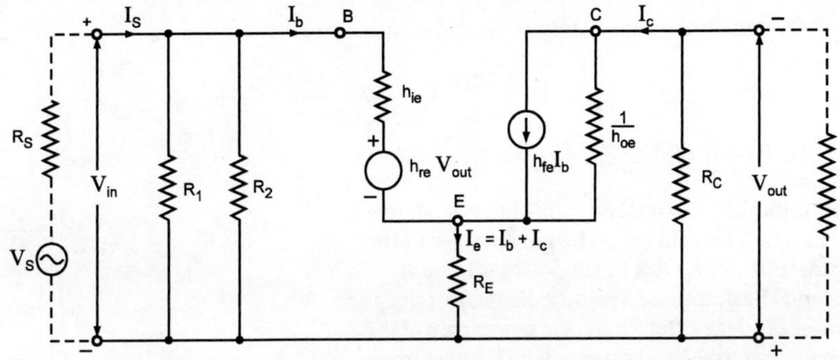

Fig. 7.32. *CE Amplifier h Parameter Equivalent Circuit With R_E Included in The Circuit*

Since $\frac{1}{h_{oe}}$ is typically 1 MΩ or so, and R_C is usually very much less than 1 MΩ, the circuit output impedance Z_{out} is approximately equal to R_C.

3. Voltage Gain. It is given as the ratio of output voltage to the input voltage

$$i.e.\ A_v = \frac{V_{out}}{V_{in}} = \frac{-I_c Z_{ac}}{I_b Z_{in}} = \frac{-I_c (R_C \parallel R_L)}{I_b h_{ie}}$$

$$= \frac{-h_{fe} I_b (R_C \parallel R_L)}{I_b h_{ie}} = \frac{-h_{fe}}{h_{ie}} (R_C \parallel R_L) \qquad ...(7.59)$$

The minus sign indicates that output voltage V_{out} is 180° out of phase with input voltage V_{in}.

With an unbypassed emitter resistance R_E in the circuit (Fig. 7.32).

Ignoring $h_{re} V_{out}$ we have from the input loop of the circuit

$$V_{in} = I_b h_{ie} + I_e R_E = I_b h_{ie} + R_E (I_b + I_c) = I_b h_{ie} + R_E I_b (1 + h_{fe}) = I_b [h_{ie} + R_E (1 + h_{fe})]$$

$$\text{and voltage gain, } A_v = \frac{V_{out}}{V_{in}} = \frac{-I_c (R_C \parallel R_L)}{I_b [h_{ie} + R_E (1 + h_{fe})]} = \frac{-h_{fe} (R_C \parallel R_L)}{h_{ie} + R_E (1 + h_{fe})} \qquad ...(7.60)$$

Usually $R_E (1 + h_{fe}) >> h_{ie}$, therefore,

$$A_v \cong \frac{-R_C \parallel R_L}{R_E} \qquad ...(7.61)$$

4. Current Gain. It is given as the ratio of output current to the input current *i.e.*

$$A_i = \frac{I_L^*}{I_b} = \frac{-I_c}{I_b} = -h_{fe} \quad ...(7.62)$$

The above equation is true for both circuits (with and without R_E). However, it is the device current gain, not the circuit current gain. From Figs. 7.31 and 7.32 it is obvious that the signal current I_S divides between R_B and Z_b where $R_B = R_1 \parallel R_2$. Using current divider equation we have

$$\text{Base current, } I_b = \frac{I_s R_B}{R_B + Z_b} \text{ and } I_c = h_{fe} I_b = \frac{h_{fe} I_s R_B}{R_B + Z_b}$$

Current gain without external load R_L in the circuit,

$$= \frac{-I_c}{I_s} = \frac{-h_{fe} R_B}{R_B + Z_b} \quad ...(7.63)$$

Overall current gain with external load R_L in the circuit,

$$A_i = \frac{I_L}{I_s} = \frac{-h_{fe} R_B R_C}{(R_C + R_L)(R_B + Z_b)} \quad ...(7.64) \because I_L = \frac{-I_c R_C}{R_C + R_L}$$

5. Power Gain. $A_p = A_v A_i$ ...(7.65)

Example 7.8. A transistor amplifier is shown as under [Fig. 7.33 (*a*)]. The parameters of the transistor are h_{ie} = 2 kΩ, h_{oe} = 25 μS and h_{fe} = 55 and the output load resistor dissipates a signal power of 10 mW. Determine the power gain of the amplifier and the input signal emf E. The reactances of the capacitors may be neglected. Draw the equivalent circuit of the amplifier.

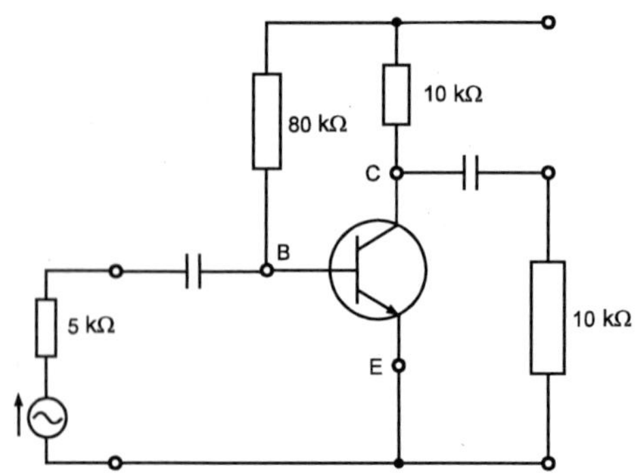

(*a*)

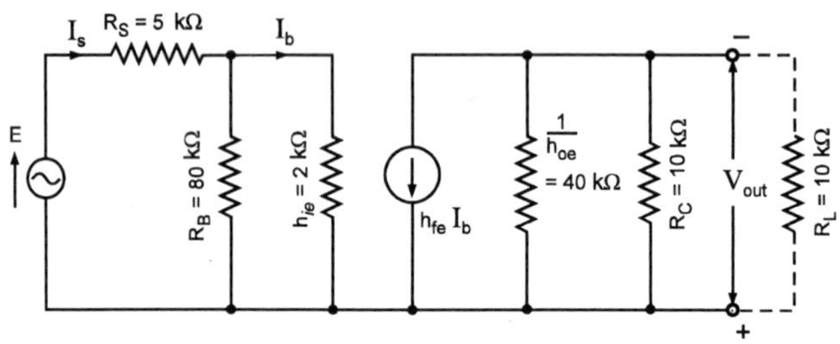

(*b*) *AC Equivalent Circuit of Circuit Shown in Fig. 7.33 (a)*

Fig. 7.33

Solution : $Z_b = h_{ie} = 2\text{ k}\Omega$

$Z_{in} = R_B \parallel h_{ie} = 80 \parallel 2 = 1.95\text{ k}\Omega$

Input impedance to E, $Z_s = R_s + Z_{in}$

$= 5 + 1.95 = 6.95\text{ k}\Omega$

Power drawn from the source,

$$P_{in} = \frac{E^2}{Z_s} = \frac{E^2}{6.95}\text{ mW}$$

$$\text{Base current, } I_b = I_s \times \frac{R_B}{R_B + h_{ie}}$$

$$= \frac{E}{6.95} \times \frac{80}{80+2}$$

$= 0.14\text{ E (mA)}$

$$\text{Output impedance, } Z_{out} = \frac{1}{h_{oe}} \parallel R_C$$

$$= \frac{1}{25 \times 10^{-6}} \parallel 10$$

$= 8\text{ k}\Omega$

AC load resistance, $R_{ac} = Z_{out} \parallel R_L = 8 \parallel 10 = 4.44\text{ k}\Omega$

Output voltage, $V_{out} = -h_{fe} I_b R_{ac} = -55 \times 0.14\text{ E} \times 4.44 = -34.3\text{ E volts}$

* Load current is assumed to flow from top to bottom through R_L as shown in Fig. 7.20.

$$\text{Power dissipated, } P_{out} = \frac{(V_{out})^2}{R_L} = \frac{(-34.3E)^2}{10\times10^3} = 10\text{ mW or } 10\times10^{-3}\text{ W}$$

$$\text{So } E = \frac{\sqrt{10\times10^{-3}\times10\times10^3}}{34.3} = 0.29\text{ V } \textbf{Ans.}$$

$$\text{Power gain, } A_p = \frac{\text{Power dissipated}}{\text{Power input}} = \frac{10\times10^{-3}}{\frac{E^2}{6.95}\times10^{-3}} = \frac{10\times6.95}{(0.29)^2} = 826\textbf{ Ans.}$$

Example 7.9. A BJT amplifier is given in Fig. 7.34. The h parameters are : h_{ie} = 1,500 Ω; $h_{re} = 4\times10^{-4}$; h_{fe} = 100; $h_{oe} = 4\times10^{-4}$ S. Calculate (*i*) Z_{in} (*ii*) Z_{in} with biasing network and (*iii*) voltage gain.

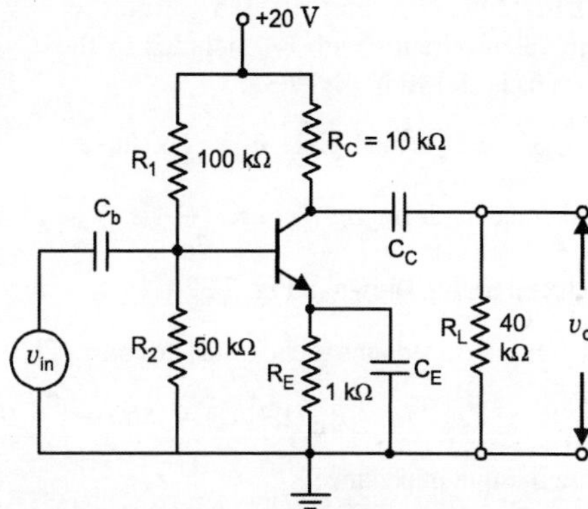

Fig. 7.34

Solution : (*i*) $Z_{in\ (base)}$,

$$Z_b = h_{ie} - \frac{h_{re}h_{fe}}{h_{oe}+\frac{1}{Z_L}} = 1{,}500 - \frac{4\times10^{-4}\times100}{4\times10^{-4}+\frac{1}{R_L\,||\,R_C}}$$

$$= 1{,}500 - \frac{0.04}{4\times10^{-4}+\frac{1}{8\times10^3}}$$

$$\because R_L\,||\,R_C = 40\,||\,10 = \frac{4\times10}{40+10} = 8\text{ k}\Omega$$

$$= 1{,}500 - \frac{0.04}{4\times10^{-4}+1.25\times10^{-4}} = 1{,}500 - 76.2 = 1{,}423.8\ \Omega \text{ or } 1.424\text{ k}\Omega\textbf{ Ans.}$$

(*ii*) Z_{in} with biasing network,

$$Z_{in} \text{ or } Z_{in\ (stage)} = Z_b\,||\,R_1\,||\,R_2 = 1.424\,||\,100\,||\,50 = 1.366\text{ k}\Omega\ \ \textbf{Ans.}$$

$$(iii)\text{ Voltage gain, } A_V = \frac{-h_{fe}}{h_{ie}}(R_C\,||\,R_L) = \frac{-100}{1{,}500}\times8\times10^3 = -533.33\textbf{ Ans.}$$

Example 7.10. Calculate the following using hybrid equivalent model for the following circuit where h_{fe} = 110 and h_{oe} = 20 μA/V.

(*i*) Z_i, (*ii*) Z_0, (*iii*) A_V, (*iv*) A_i

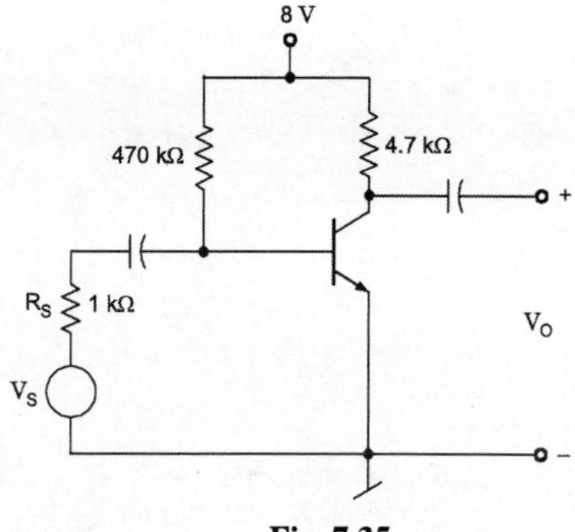

Fig. 7.35

Solution : Given that

R_B = 470 kΩ, R_C = 4.7 kΩ

R_s = 1 kΩ, V_{CC} = 8 V, h_{fe} = 110

h_{oe} = 20 μA/V

and h_{ie} = 1.1 kΩ (standard value)

(*i*) Input impedance $Z_i = R_B\,||\,h_{ie}$ = 470 kΩ || 1.1 kΩ ≃ 1.1 kΩ

$Z_s = R_s + Z_i$ = 1.1 kΩ + 1 kΩ = 2.1 kΩ **Ans.**

$$(ii)\text{ Output impedance } Z_0 = \frac{1}{h_{oe}}\,||\,R_C = \frac{10^6}{20}\,||\,4.7\text{ k}\Omega = 50\text{ k}\Omega\,||\,4.7\text{ k}\Omega = 4.3\text{ k}\Omega\textbf{ Ans.}$$

$$(iii)\text{ Voltage gain } A_V = \frac{-R_C\,.h_{fe}}{h_{ie}} = \frac{-4.7\text{ k}\Omega\times110}{1.1\text{ k}\Omega} = -470\textbf{ Ans.}$$

$$(iv)\text{ Current gain, } A_i = \frac{-h_{fe}\times R_B}{R_B+h_{ie}} = \frac{-110\times470}{470+1.1} = -109.75\textbf{ Ans.}$$

Example 7.11. In a small signal amplifier shown below [Fig. 7.36], h_{fe} = 100, h_{ie} = 560 Ω, R_C = 2 kΩ, R_E = 1 kΩ, R_B = 600 kΩ, h_{re} and h_{oe} are negligible.

(a) Draw the h parameter equivalent for the parameter. Calculate the input and output impedances and the voltage gain.

(b) Give the dc load line of the circuit and find Q-point.

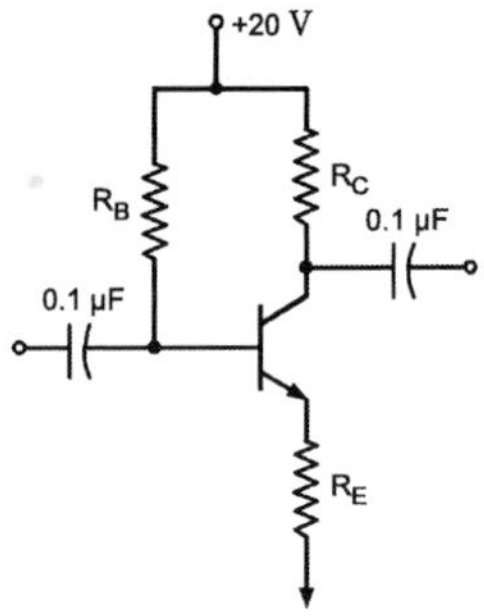

Fig. 7.36

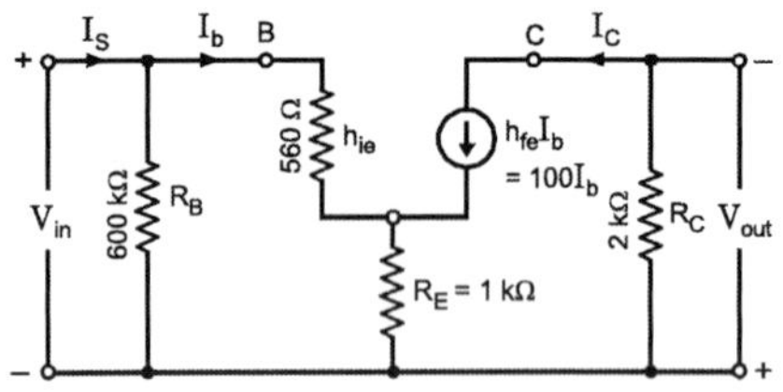

Fig. 7.37

Solution : CE amplifier h parameter equivalent circuit (with R_E included in the circuit) is shown in Fig. 7.32.

Since h_{re} is negligible, $h_{re}V_{out}$ is shorted and since h_{oe} is negligible, *i.e.* $\frac{1}{h_{oe}}$ is close to infinity and, therefore, open. Thus the h parameter equivalent circuit is reduced to that shown in Fig. 7.37.

Input impedance to the transistor base,

$$Z_b = h_{ie} + R_E(1 + h_{fe}) = 560 + 1 \times 10^3 (1 + 100) = 101{,}560\ \Omega = 101.560\ k\Omega \qquad \text{Refer to Eq. (7.55)}$$

Circuit input impedance,

$$Z_{in} = Z_b \,||\, R_B = 101.560 \,||\, 600\ k\Omega = \frac{101.56 \times 600}{101.56 + 600} = 86.86\ k\Omega \textbf{ Ans.} \qquad \text{Refer to Eq. (7.56)}$$

$$\text{Output impedance, } Z_{out} = \frac{1}{h_{oe}} \,||\, R_C \qquad \text{Refer to Eq. (7.58)}$$

$$= R_C = 2\ k\Omega \textbf{ Ans.} \qquad \because \frac{1}{h_{oe}} \simeq \infty,\ h_{oe} \text{ being negligible}$$

$$\text{Voltage gain, } A_V = -\frac{R_C \,||\, R_L}{R_E} = -\frac{2\ k\Omega}{1\ k\Omega} = -2 \textbf{ Ans.} \qquad \text{Refer to Eq. (7.61)}$$

Applying Kirchhoff's voltage law to base-emitter loop (Fig. 7.36), we have

$$V_{CC} = I_B R_B + V_{BE} + I_E R_E = I_B R_B + V_{BE} + (\beta + 1) I_B R_B \qquad \because I_E = (\beta + 1) I_B$$

$$\text{or } I_B = \frac{V_{CC} - V_{BE}}{R_B + (\beta + 1) R_E} = \frac{20 - 0.7}{600 \times 10^3 + (100 + 1) \times 1 \times 10^3}$$

$$\because \beta = h_{fe} = 100 \text{ and assume } V_{BE} = 0.7\ V$$

$$= 0.0275\ mA$$

$$\text{Quiescent collector current, } I_{CQ} = \beta I_B = 100 \times 0.0275\ mA = 2.75\ mA$$

$$\text{Emitter current, } I_E = (\beta + 1) I_B = (100 + 1) \times 0.0275 = 2.78\ mA$$

Applying Kirchhoff's voltage law to the output loop (Fig. 7.37), we have

$$V_{CC} = I_{CQ} R_C + V_{CEQ} + I_E R_E$$

$$\text{So } V_{CEQ} = V_{CC} - I_{CQ} R_C - I_E R_E = 20 - 2.75 \times 10^{-3} \times 2 \times 10^3 - 2.78 \times 10^{-3} \times 1 \times 10^3 = 11.72\ V$$

Thus operating point (V_{CEQ}, I_{CQ}) = (11.72 V, 2.75 mA) **Ans.**

Example 7.12. A transistor amplifier is shown in Fig. 7.38. Compute $A_V = V_0/V_S$, $A_i = I_L/I_S$, R_i and R'_0. Assume $h_{ie} = 1\ k\Omega$, $h_{fe} = 100$, $h_{re} = h_{oe} = 0$.

[*G.G.S.I.P. Univ. Analog Electronics December 2009*]

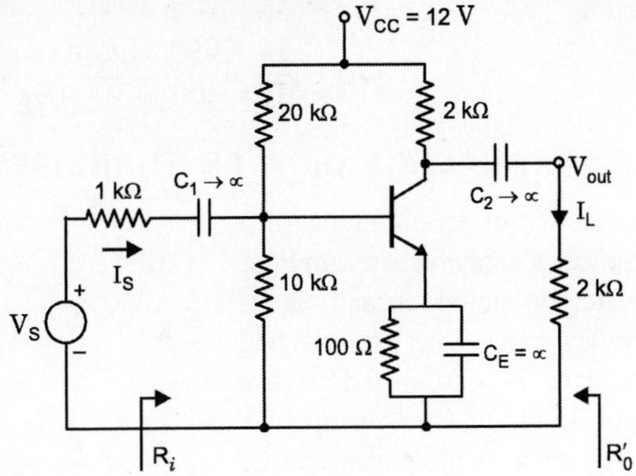

Fig. 7.38

Solution : Input impedance to transistor base,

$$Z_b = h_{ie} + R_E(1 + h_{fe}) = 1{,}000 + 100(1 + 100) = 11{,}100\ \Omega \quad \text{or} \quad 11.1\ k\Omega$$

Actual circuit input impedance,

$$Z_{in} = R_1 \parallel R_2 \parallel Z_b = 20 \parallel 10 \parallel 11.1 = 4.165\ k\Omega \quad \textbf{Ans.}$$

Actual circuit output impedance,

$$R'_{out} = \frac{1}{h_{oe}} \parallel R_C = R_C = 2\ k\Omega \quad \textbf{Ans.}$$

$$\text{Voltage gain, } A_V = \frac{-h_{fe}}{h_{ie}} \times R_C \parallel R_L = \frac{-100}{1\ k\Omega} \times 2\ k\Omega \parallel 2\ k\Omega = -100 \quad \textbf{Ans.}$$

$$\text{Overall voltage gain, } A_{V_S} = A_V \frac{Z_{in}}{Z_{in} + R_S} = -100 \times \frac{4.165}{4.165 + 1} = -80.64 \quad \textbf{Ans.}$$

$$R_B = R_1 \parallel R_2 = 20\ k\Omega \parallel 10\ k\Omega = 6.667\ k\Omega$$

$$\text{Overall current gain, } A_i = \frac{I_L}{I_S} = \frac{-h_{fe} R_B R_C}{(R_C + R_L)(R_B + Z_b)} = \frac{-100 \times 6.667 \times 2}{(2+2)(6.667 + 11.1)} = -18.76 \quad \textbf{Ans.}$$

Example 7.13. The transistor shown in Fig. 7.39 has the following parameters : $h_{fe} = 120$, $h_{ie} = 1.1\ k\Omega$, $h_{re} = 0$, $h_{oe} = 0$. Compute (*i*) R_i (*ii*) $A_{VS} = \frac{V_0}{V_S}$ (*iii*) $A_i = \frac{I_L}{I_i}$ (*iv*) R_0. [*G.G.S.I.P. Univ. Analog Electronics Dec. 2010*]

Solution : Input impedance to transistor base

$$Z_b = h_{ie} + R_E (1 + h_{fe}) = 1.1 \times 1000 + 330(1 + 120) = 1{,}100 + 39{,}930 = 41{,}030\ \Omega \quad \text{or} \quad 41.03\ k\Omega$$

Actual circuit input impedance,

$$Z_{in} = R_1 \parallel R_2 \parallel Z_b = 27 \parallel 15 \parallel 41.03 = 7.8\ k\Omega \quad \textbf{Ans.}$$

Actual circuit output impedance,

$$R'_{out} = \frac{1}{h_{oe}} \parallel R_C = R_C = 2.2\ k\Omega \quad \textbf{Ans.}$$

$$\text{Voltage gain, } A_v = \frac{-h_{fe}}{h_{ie}} \times R_C \parallel R_L = \frac{-120}{1.1\ k\Omega} \times 2.2\ k\Omega \parallel 2\ k\Omega = -114.3$$

Overall voltage gain

$$A_{VS} = A_V \frac{R_{in}}{R_{in} + R_S} = -114.3 \times \frac{7.8}{7.8 + 1} = -101.3 \quad \textbf{Ans.}$$

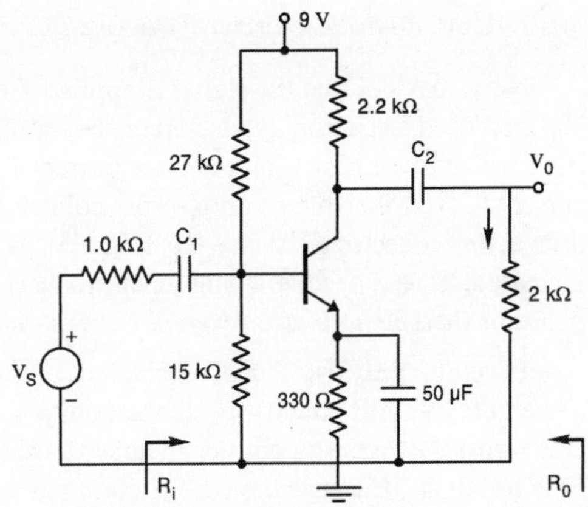

Fig. 7.39

$$R_B = R_1 \parallel R_2 = 27 \text{ k}\Omega \parallel 15 \text{ k}\Omega = 9.643 \text{ k}\Omega$$

$$A_i = \frac{I_L}{I_S} = \frac{-h_{fe} R_B R_C}{(R_C + R_L)(R_B + Z_b)} = \frac{-120 \times 9.643 \times 2.2}{(2.2 + 2)(9.643 + 41.03)} = -11.96 \textbf{ Ans.}$$

7.15. ANALYSIS OF A CB TRANSISTOR AMPLIFIER USING h PARAMETERS

The circuit arrangement of a practical common base amplifier using an emitter current bias is given in Fig. 7.40. The ac signal source is connected to the transistor emitter terminal E through a coupling capacitor C_1 while the external load resistor R_L is connected to the transistor collector terminal C through a coupling capacitor C_2. Another capacitor C_3 constitutes an ac short circuit from the base terminal to ground. The emitter-base junction is forward biased and the collector-base junction is reverse biased as usual. Since the base is ac shorted to ground, all of the input voltage at the emitter terminal E is developed across the base-emitter junction.

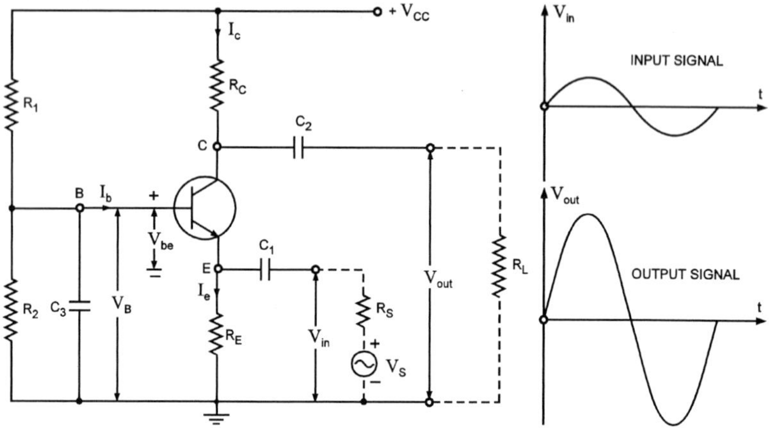

Fig. 7.40. *Common Base Transistor Amplifier Circuit With Emitter Current Bias*

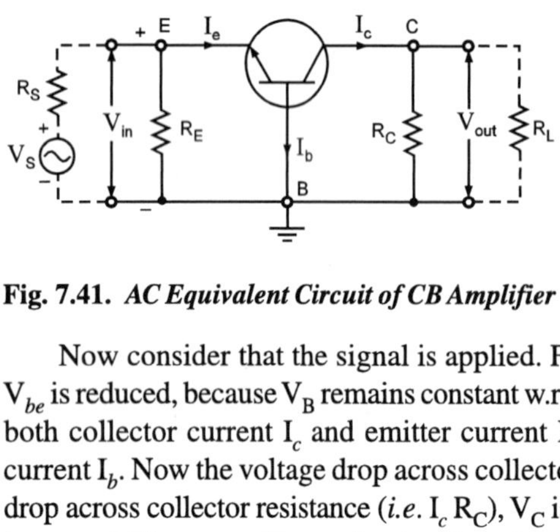

Fig. 7.41. *AC Equivalent Circuit of CB Amplifier*

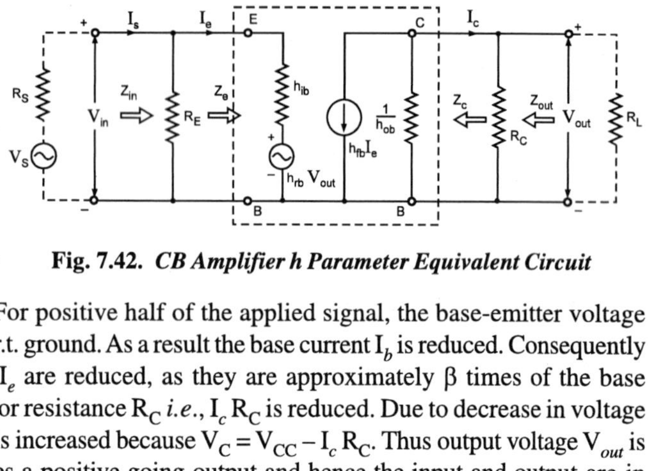

Fig. 7.42. *CB Amplifier h Parameter Equivalent Circuit*

Now consider that the signal is applied. For positive half of the applied signal, the base-emitter voltage V_{be} is reduced, because V_B remains constant w.r.t. ground. As a result the base current I_b is reduced. Consequently both collector current I_c and emitter current I_e are reduced, as they are approximately β times of the base current I_b. Now the voltage drop across collector resistance R_C *i.e.*, $I_c R_C$ is reduced. Due to decrease in voltage drop across collector resistance (*i.e.* $I_c R_C$), V_C is increased because $V_C = V_{CC} - I_c R_C$. Thus output voltage V_{out} is increased. Thus a positive-going input produces a positive going output and hence the input and output are in phase or there is *no phase reversal between input and output.*

Circuit Analysis. The ac equivalent circuit is drawn, as usual, by replacing the dc supply voltage and capacitors by short circuits. From the common base ac equivalent circuit given in Fig. 7.41 it is seen that V_{in} acts across the transistor emitter and base terminals, and output voltage V_{out} is available from the collector and base terminals. Because the base is common to both input and output, the circuit is termed *common base* or sometimes *grounded base circuit.*

The h parameter equivalent of the practical common base circuit given in Fig. 7.40 is shown in Fig. 7.42. Figure 7.42 showing the h parameter equivalent circuit is the same as Fig. 7.41 except that the transistor has been replaced by its h parameter equivalent circuit. Once again, the current directions and voltage polarities in Fig. 7.42 are indicated for a positive-going input signal. The noteworthy point is that it is assumed, while drawing Fig. 7.42, that the base of the transistor is ac shorted to ground through capacitor C_3 (refer to Fig. 7.40). In the absence of capacitor C_3, the parallel combination of R_1 and R_2 appears in series with the transistor base terminal, and the h parameter circuit becomes that given in Fig. 7.43, where $R_B = R_1 \| R_2$. The circuit performance is significantly affected in the absence of capacitor C_3.

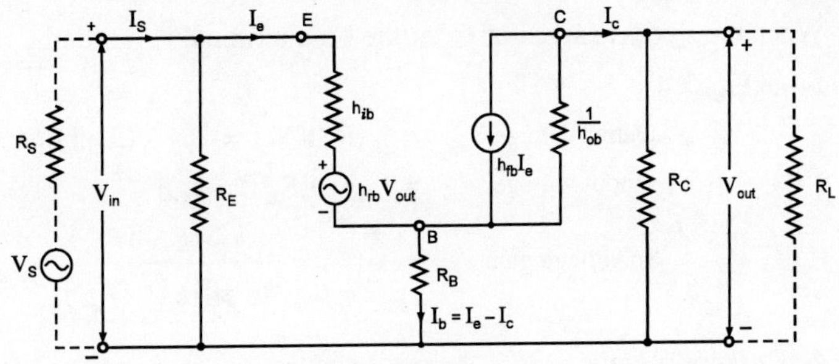

Fig. 7.43. *CB Amplifier h Parameter Equivalent Circuit With Bypass Capacitor At Transistor Base Terminal Omitted*

In the common base circuit, only a fraction of the voltage is fed back to the input *i.e.*, h_{rb} is very small. So $h_{rb}V_{out}$ can be neglected when deriving approximate equations for impedances and gains for a CB circuit.

1. Input Impedance. The input impedance to the transistor emitter is given as

$$Z_e \approx h_{ib} \quad \text{...(7.66)}$$

When the bypass capacitor C_3 at the base terminal is omitted.

From circuit given in Fig. 7.43

$$V_{in} = h_{ib} I_e + I_b \times R_B = h_{ib} I_e + I_e R_B - I_c R_B \quad \because I_b = I_e - I_c$$

$$= h_{ib} I_e + I_e R_B + h_{fb} I_e R_B \quad \because I_c = -h_{fb} I_e$$

$$= I_e [h_{ib} + R_B (1 + h_{fb})] \quad \text{...(7.67)}$$

and input impedance to the transistor emitter,

$$Z_e = \frac{V_{in}}{I_e} = h_{ib} + R_B (1 + h_{fb}) \quad \text{...(7.68)}$$

The actual circuit input impedance is parallel combination of Z_e and R_E, so we have

$$Z_{in} = Z_e \,||\, R_E \quad \text{...(7.69)}$$

2. Output Impedance. Since the output voltage variations have little effect upon the input of a CB circuit, only the output of the circuit need be considered in determining the output impedance. Looking back into the collector and base terminals of the transistor in Fig. 7.42, a large resistance $\frac{1}{h_{ob}}$ is seen. Thus the device output impedance

$$Z_c \approx \frac{1}{h_{ob}} \quad \text{...(7.70)}$$

The actual output impedance is parallel combination of Z_c and R_C so

$$Z_{out} = \frac{1}{h_{ob}} \,||\, R_C \quad \text{...(7.71)}$$

Since $\frac{1}{h_{ob}}$ is of the order of MΩ, and R_C is quite small as compared to $\frac{1}{h_{ob}}$, the circuit output impedance Z_{out} is approximately equal to R_C.

3. Voltage Gain

$$A_v = \frac{V_{out}}{V_{in}} = \frac{I_c Z_{ac}}{I_e Z_{in}} \approx \frac{I_c (R_C \,||\, R_L)}{I_e h_{ib}} \approx \frac{-h_{fb} (R_C \,||\, R_L)}{h_{ib}} \quad \text{...(7.72)} \because I_c = -h_{fb} I_e$$

Because of negative value of h_{fb} the output voltage V_{out} and input voltage V_{in} are in phase.

When the bypass capacitor C_3 at the base terminal is omitted.

From Fig. 7.43

Output voltage, $V_{out} = I_c (R_C \parallel R_L) = -h_{fb} I_e (R_C \parallel R_L)$ $\quad\because I_c = -h_{fb} I_e$

Input voltage, $V_{in} = I_e [h_{ib} + R_B (1 + h_{fb})]$ ...Refer to Eq. (7.67)

$$\text{So voltage gain, } A_v = \frac{V_{out}}{V_{in}} = \frac{-h_{fb}(R_C \parallel R_L)}{h_{ib} + R_B(1 + h_{fb})} \quad ...(7.73)$$

4. Current Gain.

$$A_i = \frac{I_c}{I_e} = -h_{fb} \quad ...(7.74)$$

The above equation is true for both circuits (capacitor C_3 included and without C_3). However, this is the device current gain, not the circuit current gain. From Figs. 7.42 and 7.43 it is obvious that the signal current I_s is divided between R_E and Z_e, and the collector current is divided between R_C and R_L. Thus overall current gain,

$$A_i = \frac{I_L}{I_S} = \frac{-h_{fb} R_E R_C}{(R_E + Z_e)(R_C + R_L)} \quad ...(7.75)$$

5. Power Gain. $A_p = A_v A_i$, as for all other circuits.

Example 7.14. For the common-base circuit given in Fig. 7.40

$R_1 = 20$ kΩ, $R_2 = 5$ kΩ, $R_C = 8$ kΩ, $R_E = 4$ kΩ and $R_L = 6$ kΩ. Determine the input and output impedances and the voltage gain for the circuit if the transistor parameters are as follows :

$h_{ib} = 25\ \Omega$, $h_{ob} = 0.5 \times 10^{-6}$ S and $h_{fb} = -0.98$.

Solution : Input impedance, $Z_{in} = Z_e \parallel R_E = h_{ib} \parallel R_E = 25 \parallel 4{,}000\ \Omega = 24.85\ \Omega$ **Ans.**

Output impedance, $Z_{out} = \dfrac{1}{h_{ob}} \parallel R_C = \dfrac{1}{0.5 \times 10^{-6}} \parallel 8{,}000 = 7.97$ kΩ **Ans.**

Voltage gain, $A_v = \dfrac{-h_{fb}(R_C \parallel R_L)}{h_{ib}} = \dfrac{-(-0.98)(8 \parallel 6)}{25 \times 10^{-3}} = 134.4$ **Ans.**

7.16. ANALYSIS OF A CC TRANSISTOR AMPLIFIER USING h PARAMETERS

The circuit arrangement of a practical common collector amplifier using an emitter current bias is given in Fig. 7.44. The ac signal source is connected to the base terminal through coupling capacitor C_1, while the external

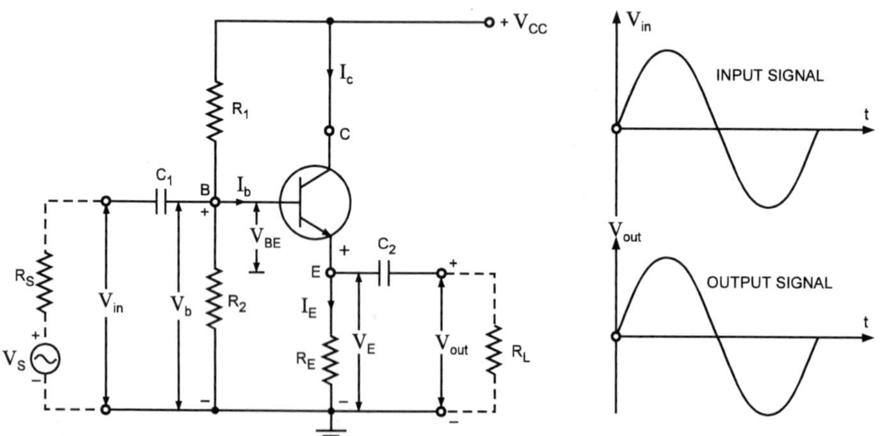

Fig. 7.44. ***Common Collector Transistor Amplifier Circuit With Emitter Current Biasing***

load resistor R_L is connected to the emitter terminal of the transistor through coupling capacitor C_2. Base bias voltage is derived from dc supply V_{CC} by means of potential divider R_1 and R_2 (*i.e.* by emitter current bias). No resistor is connected in series with the transistor collector terminal and no emitter bypass capacitor is used.

When the signal applied to the base terminal of the transistor goes positive, the forward bias V_b is increased causing increase in emitter voltage. This is because emitter voltage V_e is equal to base voltage V_b less base-emitter voltage V_{BE} and V_{BE} remains substantially constant (*i.e.* $V_e = V_b - V_{BE}$). Since emitter is the output terminal, it is seen that the output voltage from a CC circuit is almost* the same as its input voltage.

Because the output voltage at the emitter terminal follows the input signal applied to base terminal, the common collector circuit is also known as an *emitter follower*.

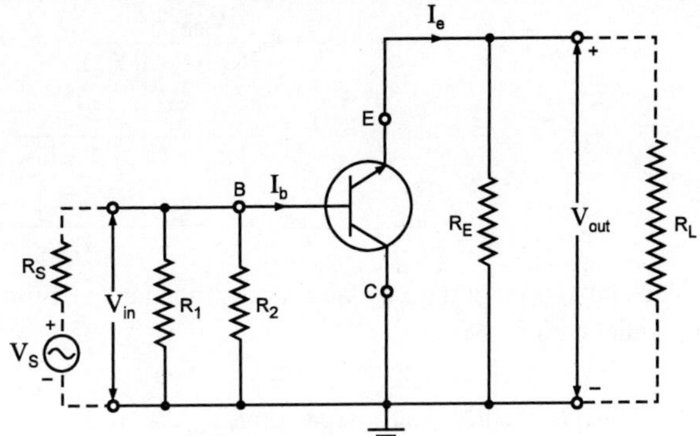

Fig. 7.45. *Common Collector AC Equivalent Circuit*

Circuit Analysis. The common collector ac equivalent circuit is shown in Fig. 7.45, that is obtained by replacing the dc supply voltage and capacitors by short circuits. The input terminals of the ac equivalent circuit are the base and the collector terminals while the output terminals are the emitter and the collector. The collector terminal is common to both input and output, so the circuit is called the *common collector* or *grounded collector circuit*.

The h parameter equivalent of the practical common collector circuit is given in Fig. 7.46, that is drawn by replacing the transistor by its h parameter equivalent circuit. The current directions and voltage polarities in Fig. 7.46 are indicated for a positive-going input signal.

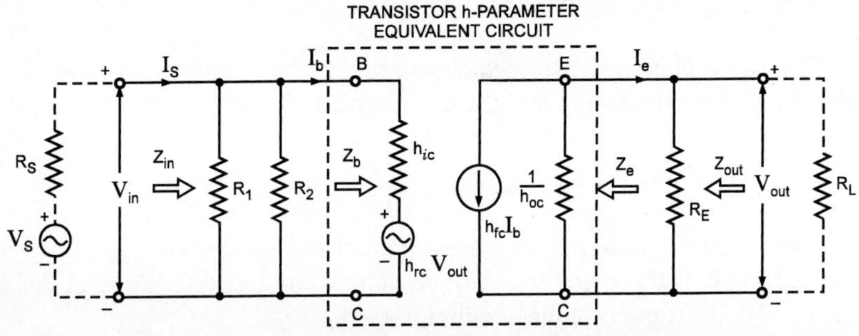

Fig. 7.46. *CC Amplifier h Parameter Equivalent Circuit*

1. Input Impedance. The noteworthy point in CC amplifier is that $h_{rc} = 1$ *i.e.* all of the V_{out} is fed back to the input. Thus

Input voltage, $V_{in} = h_{ic} I_b + V_{out} = h_{ic} I_b + I_e (R_E \parallel R_L) = h_{ic} I_b - h_{fc} I_b (R_E \parallel R_L) = I_b [h_{ic} - h_{fc} (R_E \parallel R_L)]$ ...(7.76)

The input impedance to the transistor base,

$$Z_b = \frac{V_{in}}{I_b} = h_{ic} - h_{fc} (R_E \parallel R_L) \quad ...(7.77)$$

As seen from Fig. 7.46, $Z_{in} = Z_b \parallel R_1 \parallel R_2$ ...(7.78)

2. Output Impedance. In the common collector circuit, any variation in output voltage has a significant effect on the input circuit. So for determination of the output impedance at the transistor emitter (*i.e.*, Z_e), the signal voltage is assumed to be zero and emitter current I_e is determined in terms of output voltage V_{out}

* The output voltage V_{out} can be shown slightly smaller than the input voltage V_{in} because of a very small change in V_{BE}.

$$Z_e = \frac{V_{out}}{I_e} \text{ and } I_e = -h_{fc} I_b$$

With input signal voltage, $V_s = 0$, I_b is produced by V_{out}

$$I_b = \frac{V_{out}}{h_{ic} + (R_1 \| R_2 \| R_s)}$$

$$\text{and } I_e = -h_{fc} I_b = \frac{-h_{fc} V_{out}}{h_{ic} + (R_1 \| R_2 \| R_s)}$$

$$\text{So } Z_e = \frac{V_{out}}{I_e} = -\frac{h_{ic} + (R_1 \| R_2 \| R_s)}{h_{fc}} \quad ...(7.79)$$

The above equation gives the output impedance of the device only. For the circuit output impedance, R_E is in parallel with Z_e so

$$Z_{out} = Z_e \| R_E \quad ...(7.80)$$

Since R_E is usually much larger than Z_e,

$$Z_{out} = Z_e$$

3. Voltage Gain. Output voltage,

$$V_{out} = I_e (R_E \| R_L) = -h_{fc} I_b (R_E \| R_L) \because I_e = -h_{fc} I_b$$

and input voltage $V_{in} = I_b [h_{ic} - h_{fc} (R_E \| R_L)]$ ...refer to Eq. (7.76)

$$\text{So voltage gain } A_v = \frac{V_{out}}{V_{in}} = \frac{-h_{fc} (R_E \| R_L)}{h_{ic} - h_{fc} (R_E \| R_L)} = \frac{-h_{fc} (R_E \| R_L)/h_{ic}}{1 - \dfrac{h_{fc} (R_E \| R_L)}{h_{ic}}} \simeq 1 \quad ...(7.81)$$

The above is in agreement which has already been mentioned about CC amplifier *i.e.,* it has a voltage gain of approximately unity and there is no phase difference between output and input voltages.

4. Current Gain. Current gain, $A_i = \frac{I_e}{I_b} = -h_{fc}$ ...(7.82) $\because I_e = -h_{fc} I_b$

The above Eq. (7.82) gives current gain of the device. The signal current divides between R_B and Z_b, where $R_B = R_1 \| R_2$. Using the same reasoning employed to arrive at the current gain for the CE circuit [Eq. (7.63)], the common collector circuit.

$$\text{Current gain, } A_i = \frac{-h_{fc} R_B R_E}{(R_B + Z_b)(R_L + R_E)} \quad ...(7.83)$$

For the common collector circuit, R_B (*i.e.* $R_1 \| R_2$) is usually very much smaller than Z_b, so that R_B has a significant lesser effect upon the circuit current gain.

5. Power Gain. $A_p = A_v A_i = A_i$ ...(7.84) because $A_v \cong 1$

From the above discussions it is obvious that the CC configuration provides current gain and power gain but no voltage gain. The input impedance is also very high, and the output impedance is quite low. This high input impedance-low output impedance characteristic makes the amplifier suitable for supplying a low impedance load, which is to be supplied with a signal from a high impedance source.

Example 7.15. For an emitter follower, derive expressions for A_i, A_V, R_i and R_0. Compute these for $R_E = 5$ kΩ, $R_S = 1$ kΩ. Assume transistor parameters as $h_{ie} = 1$ kΩ, $h_{fe} = 100$, $h_{re} = 2 \times 10^{-4}$ and $h_{oe} = 20$ μA/V.

[*G.G.S.I.P. Univ. Analog Electronics December 2009*]

Solution :

$$h_{ic} = h_{ie} = 1 \text{ k}\Omega$$

$$h_{fc} = -(1 + h_{fe}) = -(1 + 100) = -101$$

$$Z_{in} = Z_b = h_{ic} - h_{fc}\,(R_E \parallel R_L) = 1 - (-101)(5) = 506\ \text{k}\Omega \quad \textbf{Ans.}$$

$$Z_e = \frac{h_{ic} + R_S}{-h_{fe}} = \frac{1+1}{101} = 0.0198\ \text{k}\Omega \text{ or } 19.8\ \Omega$$

Output impedance, $Z_{out} = Z_e \parallel R_E = 19.8 \parallel 5{,}000 = 19.7\ \Omega$ **Ans.**

Voltage gain, $A_V = 1$ **Ans.**

Current gain, $A_i = -h_{fc} = 101$ **Ans.**

Example 7.16. In a CC amplifier circuit shown in Fig. 7.44, $R_1 = 10$ kΩ, $R_2 = 10$ kΩ, $R_E = 5$ kΩ, $R_L = 5$ kΩ, $R_S = 1$ kΩ.
Determine the circuit input and output impedances and voltage, current and power gains if the transistor parameters are $h_{ie} = 2$ kΩ and $h_{fe} = 100$.

Solution :

$$h_{ic} = h_{ie} = 2\ \text{k}\Omega$$

$$h_{fc} = -(1 + h_{fe}) = -(1 + 100) = -101$$

$$Z_b = h_{ic} - h_{fc}(R_E \parallel R_L) = 2 - (-101)\,(5 \parallel 5) = 254.5\ \text{k}\Omega$$

Input impedance, $Z_{in} = Z_b \parallel R_1 \parallel R_2 = 254.5 \parallel 10 \parallel 10 = 4.9$ kΩ **Ans.**

$$Z_e = \frac{h_{ic} + (R_1 \parallel R_2 \parallel R_s)}{-h_{fc}} = \frac{2 + (1 \parallel 10 \parallel 10)}{-(-101)} = 0.028\ \text{k}\Omega \text{ or } 28\ \Omega$$

Output impedance, $Z_{out} = Z_e \parallel R_E = 28 \parallel 5{,}000 \simeq 28\ \Omega$ **Ans.**

Voltage gain $A_v \simeq 1$ **Ans.**

$$R_B = R_1 \parallel R_2 = 10 \parallel 10 = 5\ \text{k}\Omega$$

$$\text{Current gain, } A_i = \frac{-h_{fc}R_BR_E}{(R_B + Z_B)(R_L + R_E)} = \frac{-(-101)\times 5 \times 5}{(5+254.5)(5+5)} = 0.973$$

Power gain, $A_p = A_i = 0.973$ **Ans.**

Example 7.17. For the circuit shown in Fig. 7.47, find A_V, A_i, Z_i and Z_o

$V_{CC} = 15$ V; $R_B = 100$ kΩ; $R_E = 1$ kΩ; $h_{ie} = 1{,}100$ Ω; $h_{fe} = 50$

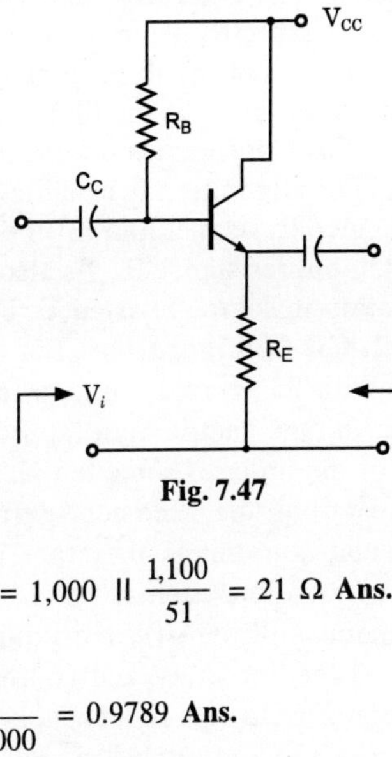

Fig. 7.47

Solution : Since in the given circuit, we get output at the emitter, it is a common collector amplifier circuit and we have

$$h_{ic} = h_{ie} = 1{,}100\ \Omega$$

$$h_{fc} = -(1 + h_{fe}) = -(1 + 50) = -51$$

Input impedance,

$$Z_{in} = [h_{ic} - h_{fc}\,R_E] \parallel R_B$$
$$= [1{,}100 - (-51) \times 1{,}000] \parallel 100{,}000$$
$$= 34.254\ \text{k}\Omega \quad \textbf{Ans.}$$

$$\text{Output impedance, } Z_{out} = R_E \parallel \left(-\frac{h_{ic}}{h_{fe}}\right) = 1{,}000 \parallel \left(\frac{-1{,}100}{-51}\right) = 1{,}000 \parallel \frac{1{,}100}{51} = 21\ \Omega \quad \textbf{Ans.}$$

$$\text{Voltage gain, } A_V = \frac{-h_{fc}R_E}{h_{ic} - h_{fc}R_E} = -\frac{-51 \times 1{,}000}{1{,}100 - (-51)\times 1{,}000} = 0.9789 \quad \textbf{Ans.}$$

$$\text{Current gain, } A_i = A_V \times \frac{Z_{in}}{R_E} = 0.9789 \times \frac{34.254\ \text{k}\Omega}{1\ \text{k}\Omega} = 33.53 \quad \textbf{Ans.}$$

Example 7.18. Input is given to Fig. 7.48 through the base and output is taken from the emitter. Draw the small signal equivalent circuit, using common emitter hybrid parameters. Hence compute the small signal voltage gain and input impedance. Given that h_{ie} = 1 kΩ, h_{fe} = 100, h_{re} = 0 and h_{oe} = 0. What is the configuration of the amplifier ?

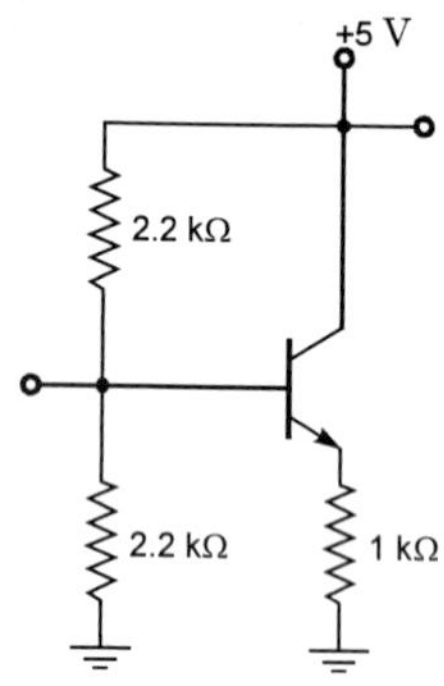

Fig. 7.48

Solution : Since the output is taken from the emitter, it is a common collector amplifier. **Ans.**

$$h_{ic} = h_{ie} = 1 \text{ k}\Omega$$
$$h_{fc} = -(1 + h_{fe}) = -(1 + 100) = -101$$
$$h_{rc} = 1 - h_{re} = 1 - 0 = 1$$
$$h_{oc} = h_{oe} = 0$$

Small signal equivalent circuit using common hybrid parameters is shown in Fig. 7.49.

Input impedance,

$$Z_{in} = [h_{ic} - h_{fc} R_E] \,||\, R_1 \,||\, R_2$$
$$= [1 - (-101) \times 1] \,||\, 2.2 \,||\, 2.2$$
$$= 102 \,||\, 1.1 \text{ k}\Omega = 1.0883 \text{ k}\Omega \text{ **Ans.**}$$

Voltage gain, $A_V = \dfrac{-h_{fc} R_E}{h_{ic} - h_{fc} R_E} = \dfrac{-(-101)\times 1}{1-(-101)\times 1}$

$$= \frac{101}{102} = 0.99 \text{ **Ans.**}$$

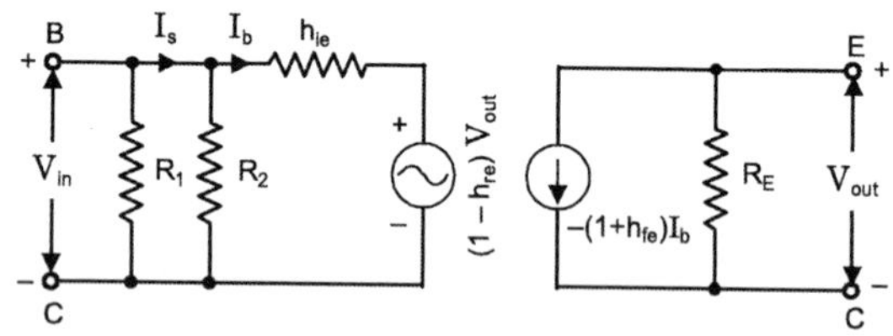

Fig. 7.49. ***Small Signal AC Equivalent Circuit of Emitter Follower Given in Fig. 7.48***

7.17. TRANSISTOR AMPLIFIER CONFIGURATION COMPARISON

It will be quite useful to study the variations of current gain A_i, voltage gain A_v and input impedance Z_{in} as a function of load resistance R_L and variation in output impedance Z_{out} with the change in source resistance R_S. The values of current gain, voltage gain, input impedance are determined as a function of load resistance R_L from the relations discussed in Art. 7.12. These are plotted in Figs. 7.50, 7.51 and 7.52 for each of the three configurations. The output resistance versus source resistance curves of the typical transistor for the three configurations are plotted in Fig. 7.53.

1. CE Configuration. From the curves shown in figures mentioned above it is observed that only the CE configuration is capable of both a voltage gain and a current gain greater than unity. The input resistance R_{in} is medium (of the order of 1 kΩ) and varies nominally with variation in load resistance R_L. Output resistance R_{out} is also moderately high (of the order of 50 kΩ) and varies nominally with the variation in source resistance. This configuration is the most versatile and is widely used.

2. CB Configuration. For this configuration, current gain A_i is less than unity and its magnitude falls with the increase in load resistance R_L beyond 10^4 Ω, reaching a value of about 0.2 for R_L = 10 MΩ; Voltage gain is high (approximately equal to that of CE configuration); input resistance is quite low of the order of some ten ohms (lowest of all the three configurations) and output resistance is the highest of all the three configurations (being greater than 2 MΩ for R_s exceeding 1 kΩ). The common base configuration is, therefore, not suitable as an amplifier for providing current or voltage gain. It has few applications such as (*i*) for matching a very low impedance source to the high input impedance of the main amplifier (*ii*) for driving a high impedance load (*iii*) as a noninverting amplifier with a voltage gain exceeding unity and (*iv*) as a constant-current source for use in a time-base or sweep voltage generator or to charge a capacitor linearly with time.

3. CC Configuration. For common collector configuration, current gain is as high as for CE configuration, voltage gain is less than unity, input resistance is the highest and output resistance is the lowest of all the three configurations. This configuration finds wide application as a buffer amplifier between a high-impedance source and a low-impedance load.

Table 7.5 gives the values of current gain, voltage gain, input resistance and output resistance for a typical transistor (h parameters given in Table 7.3) for all the three configurations using load resistance $R_L = 3\ k\Omega$ and source resistance $R_s = 3\ k\Omega$.

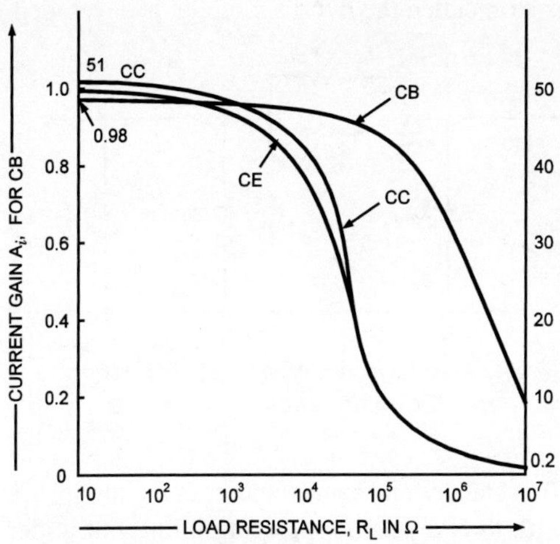

Fig. 7.50. ***Current Gain Vs. Load Resistance Curves***

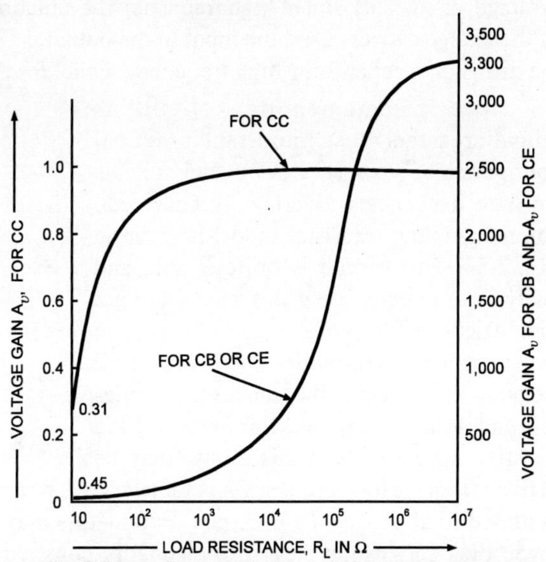

Fig. 7.51. ***Voltage Gain Vs. Load Resistance Curves***

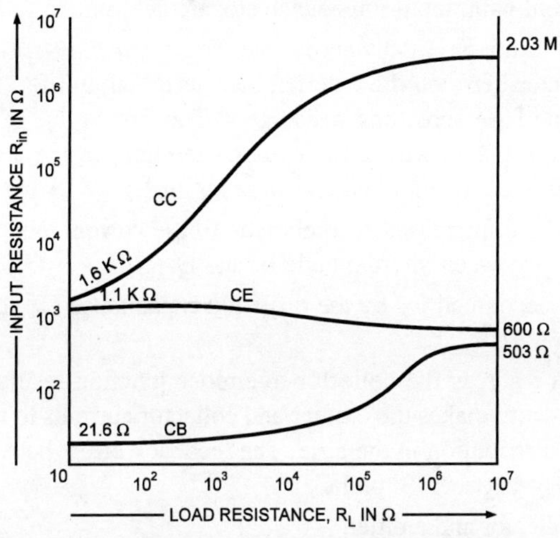

Fig. 7.52. ***Input Resistance Vs. Load Resistance Curves***

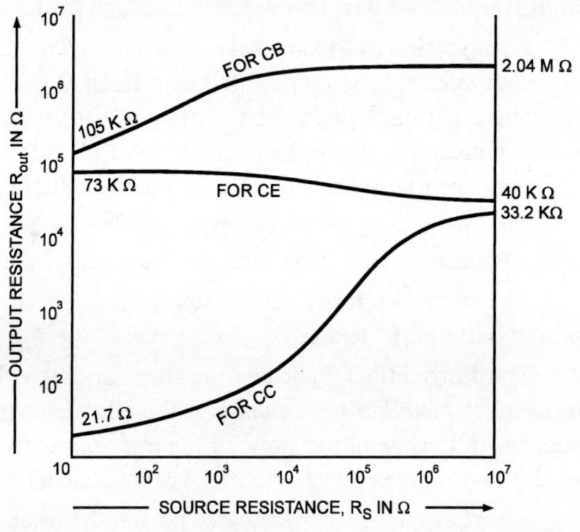

Fig. 7.53. ***Output Resistance Vs. Source Resistance Curves***

TABLE 7.5

Configuration / *Quantity*	*Common Emitter CE*	*Common Base CB*	*Common Collector CC*
Current gain, A_i	High (– 46.5)	Low (0.98)	High (47.5)
Voltage gain, A_v	High (– 131)	High (131)	Low (0.99)
Input resistance, R_{in} for R_L = 3 kΩ	Medium (1.065 kΩ)	Low (22.5 Ω)	High (144 kΩ)
Output resistance, R_{out} for R_s = 3 kΩ	Medium (45.5 kΩ)	High (1.72 MΩ)	Low (80.5 Ω)

7.18. HIGH FREQUENCY HYBRID-π (PI) COMMON EMITTER TRANSISTOR MODEL

At low frequencies it is assumed that the transistor responds instantly* to variations in input signal (voltage or current). But at high frequency the situation is different because the mechanism of the transport of the charge carriers from the input to the output is essentially one of diffusion. Taking into consideration the diffusion mechanism a high frequency model for the transistor, called the hybrid-π model is developed.

Since common-emitter circuit is considered the most important practical configuration, we seek a CE model suitable for high frequencies. Hybrid-π or Giacoletto common emitter transistor model is given in Fig. 7.54. This circuit is quite simple and analysis of circuits using this model are not too difficult and provide results which are in excellent agreement with experiment at all frequencies for which the transistor provides reasonable amplification. Furthermore, the resistive components in this circuit may be derived from the low frequency *h* parameters. All parameters (resistances and capacitances) in the model are assumed to be frequency invariant. Parameters may vary with the quiescent operating point, but under given bias conditions they are reasonably constant for small-signal variations. Typical values of the parameters are indicated in the Fig. 7.54. For high frequency analysis, the transistor is replaced by this high frequency hybrid π-model and voltage gain, current gain, input impedance etc. are determined.

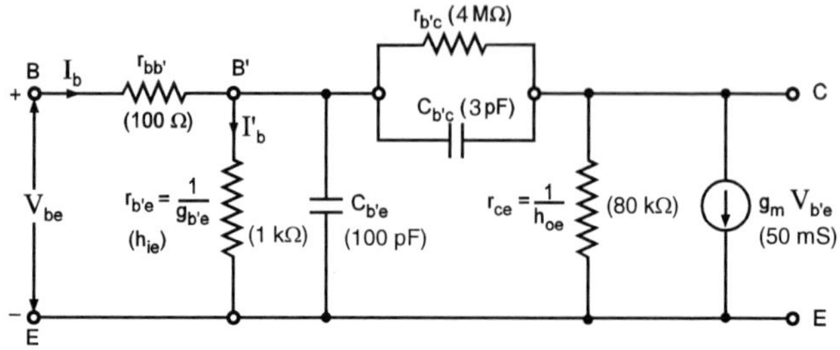

Fig. 7.54. ***Giacoletto or Hybrid-π Model of a Transistor in CE Configuration***

Explanation of Parameters. The internal node B′ is not physically accessible. The *ohmic base-spreading resistance* $r_{bb'}$, is represented as a lumped parameter between the external base terminal and B′. This resistance $r_{bb'}$, includes the base contact, base bulk and base spreading resistances. The first is due to the actual connection to the base, the second includes the resistance from the external terminal to the active region of the transistor, while the last is the actual resistance within the active base region.

The increase in minority carriers in the base results in increased recombination base current, and this effect is taken into account by inserting a resistance $r_{b'e}$ between internal node B′ and E.

The excess-minority carrier storage in the base is accounted for by the diffusion capacitance $C_{b'e}$ connected between B′ and E.

The Early effect indicates that the varying voltage across the collector-to-emitter junction results in *base-width modulation*. A change in the effective base width makes the emitter and collector currents to vary because of change in the slope of the minority-carrier distribution in the base. The feedback effect between output and input is taken into account by connecting $r_{b'c}$ between B′ and C.

Resistance r_{ce} is the resistance present between collector and emitter.

For small variations in the voltage $V_{b'e}$, across the emitter junction, the excess minority-carriers concentration injected into the base is proportional to $V_{b'e}$ and, therefore, the resulting small-signal collector current with the collector shorted to the emitter is proportional to $V_{b'e}$. This effect accounts for the current generator $g_m V_{b'e}$ in the circuit.

Lastly, the collector-junction barrier capacitance is included in the capacitance $C_{b'c}$. It is to be noted that $C_{b'c}$ is a transition capacitance while $C_{b'e}$ is a diffusion capacitance.

* In a physical transistor, the charge carriers (electrons and holes) travel through the process of drift and diffusion. The time of travel of charge carriers from the input to the output and the resulting delay in response is mainly caused due to diffusion process of the charge carriers. At low frequencies, this time delay is so small in comparison to the time period T (= 1/*f*), that it can be neglected.

For the typical h parameters given in Table 7.3, at $I_C = 1.3$ mA and room temperature, component values for a germanium transistor are

$$g_m = 50 \text{ mA/V}\ ;\ r_{b'e} = 1\ k\Omega;\ r_{bb'} = 100\ \Omega,\ r_{b'c} = 4\ M\Omega;\ r_{ce} = 80\ k\Omega;\ C_c = 3\text{ pF},\ C_e = 100\text{ pF}$$

7.19. HYBRID-pI CONDUCTANCES IN TERMS OF LOW-FREQUENCY *h* PARAMETERS

All the resistive components in the hybrid-π model can be had from the *h* parameters in the CE configuration. These *h* parameters are discussed briefly now.

1. Transistor Conductance g_m. Figure 7.55 depicts P-N-P transistor in the CE configuration with the collector shorted to the emitter for time-varying signals. In the active region, the collector current given by expression $I_C = -\alpha I_E + I_{CO}$ is applicable here with $\alpha_N = \alpha_0$.

The transconductance g_m is given by

$$g_m = \left.\frac{\partial I_C}{\partial V_{B'E}}\right|_{V_{CE} = \text{constant}} = -\alpha_0 \frac{\partial I_E}{\partial V_{B'E}} = \alpha_0 \frac{\partial I_E}{\partial V_E} \quad ...(7.85)$$

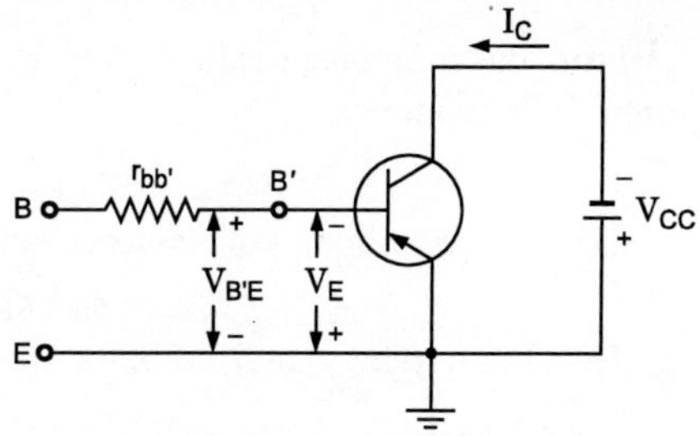

Fig. 7.55. ***P-N-P Transistor in CE Configuration For The Derivation of Transconductance (g_m)***

In the above equation, we have assumed that α_N is independent of V_E. For a P-N-P transistor $V_E = -V_{B'E}$, as indicated in Fig. 7.55

If the emitter diode resistance is r'_e, then

$$r'_e = \frac{\partial V_E}{\partial I_E} \text{ and, therefore,}$$

$$g_m = \frac{\alpha_0}{r'_e} = \frac{\alpha I_E}{V_T} \quad ...(7.86) \quad \because r'_e = \frac{V_T}{I_E}$$

$$= \frac{I_{CO} - I_C}{V_T} \quad ...(7.87) \quad \because \alpha I_E = I_{CO} - I_C$$

For a P-N-P transistor I_C is negative and for an N-P-N transistor I_C is positive. But for the foregoing analysis with $V_E = +V_{B'E}$ we have

$$g_m = \frac{I_C - I_{CO}}{V_T}$$

Hence for either type of transistor g_m is positive.

Since $|I_C| >> |I_{CO}|$, g_m is given by

$$g_m \simeq \frac{|I_C|}{V_T} \quad ...(7.88)$$

We know that $V_T = \dfrac{T}{11{,}600}$

$$\text{So } g_m = \frac{11{,}600\,|I_C|}{T}$$

At room temperature of 27°C (300 K), the value of V_T is 26 mV, so the value of g_m is given by

$$g_m = \frac{|I_C| \text{ in mA}}{26} \quad ...(7.89)$$

For $I_C = 1.3$ mA, $g_m = 0.05$ S (mho) = 50 mA/V and for $I_C = 10$ mA, $g_m = 400$ mS or 400 mA/V

The above values are much larger than the transconductance obtained with tubes and FETs.

2. Input Conductance $g_{b'e}$. Figure 7.56 (*a*) depicts the hybrid-π model valid at low frequencies, where all capacitances are negligible. Figure 7.56 (*b*) depicts the same transistor, using the *h* parameter equivalent circuit.

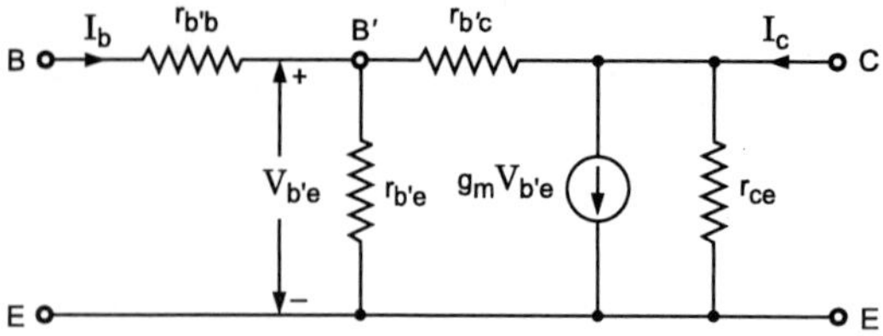

(a) The Hybrid π Model at Low Frequencies

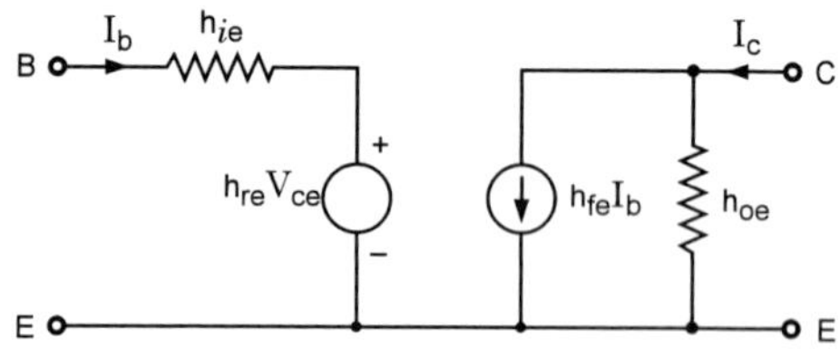

(b) The h Parameter Model at Low Frequencies

Fig. 7.56

From the component values given in Art. 7.18, we note that $r_{b'c} >> r_{b'e}$. Hence almost the complete base current I_b flows into $r_{b'e}$

$$\text{Thus } V_{b'e} \approx I_b r_{b'e} \qquad ...(7.90)$$

and the short-circuit collector current I_c is given by

$$I_c = g_m V_{b'e} \approx g_m I_b r_{b'e}$$

The short-circuit gain h_{fe} is defined as

$$h_{fe} = \left.\frac{I_c}{I_b}\right|_{V_{CE} = \text{Constant}} = g_m r_{b'e}$$

$$\text{or } r_{b'e} = \frac{h_{fe}}{g_m} = \frac{h_{fe} V_T}{|I_c|}$$

$$\text{or } g_{b'e} = \frac{g_m}{h_{fe}} \qquad ...(7.91)$$

The above equation is required expression for the input conductance.

Thus, over the range of currents for which h_{fe} remains fairly constant, $r_{b'e}$ is directly proportional to temperature and inversely proportional to current.

Since $g_m = \dfrac{\alpha_0}{r'_e}$, and $h_{fe} \simeq \dfrac{\alpha_0}{1-\alpha_0}$, $r_{b'e}$ may be expressed in terms of the T-model emitter resistance r'_e as

$$r_{b'e} = \frac{h_{fe}}{g_m} = \frac{r'_e}{1-\alpha_0} \qquad ...(7.92)$$

3. Feedback Conductance $g_{b'c}$. With the input open circuited, h_{re} is defined as the reverse voltage gain, or from Fig. 7.59 (*a*) with $I_b = 0$,

$$h_{re} = \left.\frac{V_{b'e}}{V_{ce}}\right|_{\text{For } I_b = 0} = \frac{r_{b'e}}{r_{b'e} + r_{b'c}} \qquad ...(7.93)$$

$$\text{or } r_{b'e}(1 - h_{re}) = h_{re} \cdot r_{b'c}$$

Since $h_{re} << 1$, then to a good approximation

$$r_{b'e} = h_{re} r_{b'c} \quad \text{or} \quad g_{b'c} = \frac{h_{re}}{r_{b'e}} \qquad ...(7.94)$$

Since $h_{re} \simeq 1 \times 10^{-4}$, Eq. (7.94) verifies that $r_{b'c} >> r_{b'e}$

It is found that h_{re} is quite insensitive to current and temperature. Therefore, $r_{b'c}$ has the same dependence on $|I_C|$ and T as does $r_{b'e}$.

4. Base-Spreading Resistance $r_{bb'}$. Parameter h_{ie} forms the input resistance of the transistor with the output shorted. With reference to the circuit given in Fig. 7.56(*a*), with output shorted, $r_{b'c}$ is in parallel with $r_{b'e}$. Hence

$$h_{ie} = r_{bb'} + r_{b'e} \parallel r_{b'c} = r_{bb'} + r_{b'e} \quad ...(7.95) \quad \because r_{b'c} >> r_{b'e} \text{ and, therefore, } r_{b'e} \parallel r_{b'c} = r_{b'e}$$

or Base-spreading resistance $r_{bb'} = h_{ie} - r_{b'e} = h_{ie} - \dfrac{h_{fe}}{g_m}$ ...(7.96) $\quad \because$ From Eq. (7.92) $r_{b'e} = \dfrac{h_{fe}}{g_m}$

From Eqs. (7.91) and (7.95) it may be noted that the short-circuit input impedance h_{ie} varies with current and temperature in the following manner :

$$h_{ie} = r_{bb'} + \frac{h_{fe} V_T}{|I_C|} \quad ...(7.97)$$

5. Output Conductance g_{ce}. Parameter h_{oe} constitutes the output conductance with open-circuited input

$$i.e., \ h_{oe} = \left. \frac{I_C}{V_{ce}} \right|_{\text{For } I_b = 0}$$

From the circuit shown in Fig. 7.56(*a*) with $I_b = 0$, we have

$$I_c = \frac{V_{ce}}{r_{ce}} + \frac{V_{ce}}{r_{b'c} + r_{b'e}} + g_m V_{b'e} \quad ...(7.98)$$

With $I_b = 0$, we have, from Eq. (7.93), $V_{b'e} = h_{re} V_{ce}$ and from Eq. (7.98), we have

$$h_{oe} = \frac{I_c}{V_{ce}} = \frac{1}{r_{ce}} + \frac{1}{r_{b'c}} + g_m h_{re} \quad ...(7.99)$$

Substituting the values of g_m and h_{re} from Eqs. (7.91) and (7.94) in Eq. (7.99) and assuming $r_{b'c} >> r_{b'e}$ we have

$$h_{oe} = g_{ce} + g_{b'c} + g_{b'e} h_{fe} \frac{g_{b'c}}{g_{b'e}} = g_{ce} + (1 + h_{fe}) g_{b'c} \quad ...(7.100)$$

but if $h_{fe} >> 1$, Eq. (7.100) may be written as

$$g_{ce} \simeq h_{oe} - g_m h_{re} \quad ...(7.101)$$

6. Feedback Conductance $g_{b'c}$:

$$g_{b'c} = h_{re} g_{b'e} \quad ...(7.102)$$

If the CE *h* parameters at low frequencies are known at a given collector current I_c, the conductances or resistances in the hybrid-π model can be calculated from the following equations :

$$g_m = \frac{|I_C|}{V_T} \quad ...(7.103)$$

$$r_{b'e} = \frac{h_{fe}}{g_m} \quad ...(7.104)$$

$$r_{bb'} = h_{ie} - r_{b'e} \quad ...(7.105)$$

$$r_{b'c} = \frac{r_{b'e}}{h_{re}} \quad ...(7.106)$$

$$g_{ce} = h_{oe} - (1 + h_{fe})g_{b'c} = \frac{1}{r_{ce}} \quad ...(7.107)$$

For the typical h-parameters given in Table 7.3, at $I_C = 1.3$ mA and room temperature, the component values obtained are listed on page 255.

Hybrid-π Capacitances. The basic high-frequency hybrid π model, discussed in Art. 7.18, includes two capacitances (*i*) $C_{b'c}$ or C_c and (*ii*) C_e.

1. Collector-Junction Capacitance C_c. The collector-junction capacitance C_c or $C_{b'c}$ is the measured CB output capacitance with the input open ($I_E = 0$), and is usually specified by the manufacturers as C_{ob}. Since in active region the collector junction is reverse biased, then C_c is a transition capacitance, and therefore, varies as V_{CE}^{-n}, when $n = \frac{1}{2}$ or $\frac{1}{3}$ for an abrupt or gradual junction respectively.

Since $C_e = C_{b'e}$ represents, mainly the diffusion capacitance across the emitter junction, it is directly proportional to the current and is almost independent of temperature. Experimentally, C_e is determined from a measurement of frequency f_T at which the CE short-circuit current gain falls to unity. C_e is given approximately by the following relation

$$C_e = \frac{g_m}{2\pi f_T} \quad ...(7.108)$$

Reasonable values for $C_{b'c}$ and $C_{b'e}$ are 3 pF and 100 pF respectively.

Example 7.19. The following low-frequency parameters are known for a given transistor at $I_C = 10$ mA and $V_{CE} = 10$ V and at room temperature $h_{ie} = 500\ \Omega$; $h_{oe} = 4 \times 10^{-5}$ A/V; $h_{fe} = 100$ and $h_{re} = 10^{-4}$. At the same operating point $f_T = 50$ MHz and $C_{ob} = 3$ pF. Calculate the values of all the hybrid parameters.

Solution : From Eq. (7.103);

$$g_m = \frac{I_C}{V_T} = \frac{10 \text{ mA}}{26 \text{ mV}} = 384.6 \text{ mS} \quad \textbf{Ans.}$$

(*ii*) From Eq. (7.104); $$r_{b'e} = \frac{h_{fe}}{g_m} = \frac{100}{384.6 \times 10^{-3}} = 260\ \Omega \quad \textbf{Ans.}$$

(*iii*) From Eq. (7.105); $$r_{b'b} = h_{ie} - r_{b'e} = 500 - 260 = 240\ \Omega \quad \textbf{Ans.}$$

(*iv*) From Eq. (7.106); $$r_{b'c} = \frac{r_{b'e}}{h_{re}} = \frac{260}{10^{-4}} = 2.6 \times 10^6\ \Omega \text{ or } 2.6\text{ M}\Omega \quad \textbf{Ans.}$$

(*v*) From Eq. (7.107); $$g_{ce} = h_{oe} - (1 + h_{fe})g_{b'c}$$

$$= 4 \times 10^{-5} - (1 + 100) \times \frac{1}{2.6 \times 10^6} = 4 \times 10^{-5} - 3.885 \times 10^{-5} = 0.115 \times 10^{-5} \text{ S} \quad \textbf{Ans.}$$

$$C_{b'c} = C_{bo} = 3 \text{ pF} \quad \textbf{Ans.}$$

$$C_{b'e} = \frac{g_m}{2\pi f_T} - C_{b'c} = \frac{384.6 \times 10^{-3}}{2\pi \times 50 \times 10^6} - 3 \times 10^{-12} = 1221 \times 10^{-12} \text{ F or } 1{,}221 \text{ pF} \quad \textbf{Ans.}$$

7.20. VARIATION OF HYBRID π PARAMETERS

The hybrid π parameters vary with collector current magnitude $|I_C|$, collector-to-emitter voltage magnitude $|V_{CE}|$ and temperature T. Increase in $|V_{CE}|$ reduces the effective base width. The decrease in $r_{bb'}$ with $|I_C|$ is because of conductivity modulation of the base with increasing collector current. On the other hand, $r_{bb'}$ increases with increasing temperature because the mobility of majority and minority carrier reduces, and this causes reduction in conductivity. The increase of h_{fe} with temperature has been determined experimentally, whereas the increase with $|V_{CE}|$ is due to the reduction in base width and decrease in recombination which increases the transistor α.

As mentioned in Art. 7.19, the collector junction capacitance C_c or $C_{b'c}$ varies as V_{CE}^{-n} where $n = \frac{1}{2}$ or $\frac{1}{3}$ for an abrupt or gradual junction respectively. The capacitance C_e represents the sum of the emitter diffusion capacitance C_{De} and the transition capacitance C_{Te}. For a forward-biased emitter junction $C_{De} >> C_{Te}$ and, therefore, $C_e \simeq C_{De}$ which is proportional to emitter bias current I_E and is almost independent of temperature.

The dependence of g_m, $r_{bb'}$, $r_{b'e}$, C_e, C_c, h_{fe} and h_{ie} on $|I_C|$, $|V_{CE}|$ and temperature is summarized below in tabular form.

TABLE 7.6. Dependence of Hybrid π Parameters upon $|I_C|$, $|V_{CE}|$ and Temperature

Hybrid π Parameter	*Variation with Increasing*		*Temperature*
	$\lvert I_C \rvert$	$\lvert V_{CE} \rvert$	*T*
g_m	$\lvert I_C \rvert$	Independent	1/T
$r_{bb'}$	Decreases		Increases
$r_{b'e}$	$1/\lvert I_C \rvert$	Increases	Increases
C_e	$\lvert I_C \rvert$	Decreases	
C_c	Independent	Decreases	Independent
h_{fe}	–	Increases	Increases
h_{ie}	$1/\lvert I_C \rvert$	Increases	Increases

Example 7.20. The following transistor measurements made at $I_C = 5$ mA and $V_{CE} = 10$ V and at room temperature. $h_{fe} = 100$; $h_{ie} = 600\ \Omega$; $|A_i| = 10$ at 10 MHz and $C_c = 3$ pF. Find f_B, f_T, C_e, $r_{b'e}$, $r_{bb'}$.

Solution : From Eq. (7.103); $g_m = \frac{I_C}{V_T} = \frac{5\text{ mA}}{26\text{ mV}} = \frac{5}{26}$ S or 192.3 mS

From Eq. (7.104); $r_{b'e} = \frac{h_{fe}}{g_m} = \frac{100}{5/26} = 520\ \Omega$ **Ans.**

From Eq. (7.105); $r_{bb'} = h_{ie} - r_{b'e} = 600 - 520 = 80\ \Omega$ **Ans.**

$$f_T = |A_i| \times f = 10 \times 10\text{ MHz} = 100\text{ MHz} \quad \textbf{Ans.}$$

$$f_\beta = \frac{f_T}{\beta} = \frac{f_T}{h_{fe}} = \frac{100}{100}\text{ MHz} = 1\text{ MHz} \quad \textbf{Ans.}$$

$$C_e = \frac{g_m}{2\pi f_T} - C_{b'c} = \frac{192.3 \times 10^{-3}}{2\pi \times 100 \times 10^6} - 3 \times 10^{-12} \qquad \because C_{b'c} = C_c = 3 \times 10^{-12}\text{ F}$$

$= 303 \times 10^{-12}$ for 303 pF **Ans.**

HIGHLIGHTS

1. The device that amplifies the amplitude of the input signal is called the amplifier. An amplifier may be defined as a device that increases the current, voltage or power of an input signal with the help of a transistor by furnishing the additional power from a separate source of supply.
2. Amplification in the ac domain cannot be had without applying the dc biasing level.
3. For most applications, the BJT amplifier may be considered linear, permitting the use of the superposition theorem for separating the dc and ac analyses and designs.
4. The hybrid equivalent circuit for a transistor consists of a Thevenin's equivalent circuit at the input and a Norton's equivalent circuit at the output.
5. The equations relating the input and output voltages and currents in a transistor are given by

$$v_1 = h_i i_1 + h_r v_2$$

$$i_2 = h_f i_1 + h_0 v_2$$

   where h_i, h_r, h_f and h_0 are the h parameters and have a second subscript of b, e or c according to the configuration under consideration.
6. Typical values of h parameters for CB, CE and CC configurations are given in Table 7.3.
7. If the h parameters are given for one type of amplifier configuration but are required for another type, the approximate conversion formulas in Table 7.4 may be used.
8. Knowing the h parameters for a transistor, the values of current gain A_i and A_{is}, voltage gain A_v and A_{vs}, R_{in} and Y_{out} can be had by using the exact equations given at the end of Art. 7.12 (page 236-237) by using a second subscript appropriate to the amplifier under consideration.
9. The hybrid model for CE, CB and CC configurations is the same. The only difference will be in the magnitudes of the parameters of the equivalent network.
10. When introducing the ac model for a BJT, all dc sources are set to zero and replaced by a short circuit connection to ground, all capacitors are replaced by a short circuit equivalent and all elements in parallel with an introduced short circuit equivalent should be removed from the circuit.

    Elements that were isolated by capacitors for the dc analysis will appear in the ac analysis due to the short circuit equivalent for the capacitive elements.
11. The amplification factor (β or h_{fe}) is the least sensitive to variations in collector current, whereas the output impedance parameter is the most sensitive. The output impedance is also quite sensitive to variations in V_{CE}, whereas the amplification factor is the least sensitive. However, the output impedance is the least sensitive to variations in temperature, whereas the amplification factor is somewhat sensitive.
12. One of the main advantages of a CE amplifier is the ability to operate from a single supply voltage, using a resistor from V_{CC} to the base to provide bias current.

    A CE amplifier enjoys both a voltage and a current gain, with the result that its power gain is usually larger than in either a CB or CC amplifier.

    The input resistance to a CE amplifier is of the order of 1 kΩ and is thus much larger than that in a CB amplifier, but lower than that in a CC amplifier.
13. The CB amplifier has a very low input impedance but it can have significant voltage gain. The current gain is slightly less than unity, and the output impedance is simply R_C.

    The CB amplifier is used primarily to match a very low source impedance (roughly 20 Ω) to a high load impedance (100 kΩ and more).
14. The CC amplifier provides a current gain and power gain but no voltage gain. The input impedance is also very high, and the output impedance is quite low. The high input impedance-low output impedance characteristic makes the amplifier suitable for supplying a load impedance, which is to be supplied with a signal from a high impedance source. Thus, the CC amplifier finds wide application as a buffer amplifier between a high impedance source and a low impedance load.
15. The hybrid-π model of a transistor at high frequencies includes the capacitive effects of the P-N junctions and involves a base-spreading resistance that creates a virtual base.

16. In the hybrid-π model, there are four terminals namely C, B, E and B' out of which C, B and E have their normal meaning and they are user accessible terminals but B' is not a user terminal. Terminal B' is called the 'virtual base' terminal.

17. The resistances and capacitances involved in hybrid-π model are : (*i*) Base spreading resistance $r_{bb'}$ (*ii*) Input resistance $r_{b'e}$ (*iii*) Output resistance r_{ce} (*iv*) Feedback resistance $r_{b'c}$ (*v*) Diffusion capacitance $C_{b'e}$ or C_e (*vi*) Transition capacitance $C_{b'c}$ or C_c.

If the CE *h* parameters at low frequencies are known at a given collector current I_C, the conductance or resistances in the hybrid-π model can be calculated from Eqs. (7.103), (7.104), (7.105), (7.106) and (7.107)

Diffusion capacitance is determined from the relation:

$$C_e = \frac{g_m}{2\pi f_T}$$

EXERCISES

1. Draw the circuit diagram of a single stage transistor amplifier, stating the function of each component used in the circuit.

2. Explain the functions of (*i*) coupling capacitor (*ii*) bypass capacitor.
[G.G.S.I.P. Univ. Analog Electronics December 2010]

3. Explain the phase reversal of output voltage in amplifier.

4. Explain the working of a common emitter amplifier with the help of its circuit diagram. Explain how dc load line is drawn for a common-emitter amplifier.

5. What is a load line and how is it used in the calculation of current and voltage gains for a single stage amplifier.

6. With a circuit diagram, explain the operation of a simple common-emitter amplifier and find its voltage gain expression.

7. Explain how will you determine the voltage gain of the CE amplifier by plotting the dc load line on the output characteristics of the transistor.

8. What is an ac emitter resistance ? Give the formulae for ac resistance of the emitter diode.

9. Draw h parameter model of BJT in (*i*) CE (*ii*) CB (*iii*) CC configurations. Write terminal equations for each model. **[G.G.S.I.P. Univ. Analog Electronics December 2010]**

10. Give the hybrid model of a BJT and derive expressions for current gain, voltage gain, input impedance and output impedance of an amplifier circuit. **[G.G.S.I.P. Univ. Analog Electronics-I December 2004]**

11. Using h-parameters derive expressions for A_I, A_V, Z_i and Z_0 of transistor in CE mode when it is driven by voltage source V_S with source resistance r_s and output voltage V_L is obtained across load resistance R_L.
[G.G.S.I.P. Univ. Analog Electronics December 2009]

12. Draw the hybrid equivalent circuit for common-base configuration and find the expression of current gain, voltage gain and input impedance.

13. With a neat circuit diagram, explain the working of a transistor amplifier in which phase inversion of the input signal does not take place. Obtain the expression for the input resistance of the amplifier.

14. Derive expression for voltage gain, current gain, input resistance, and output resistance of an emitter follower circuit using h parameters. **[G.G.S.I.P. Univ. Analog Electronics December 2010]**

15. Derive expressions of A_v, A_i, Z_{in}, Z_{out} for an emitter follower. List the applications of such circuit.
[G.G.S.I.P. Univ. Analog Electronics December 2010]

16. Compare the important parameters (A_v, A_i, R_i, R_0) of a transistor amplifier in different transistor configurations. **[G.G.S.I.P. Univ. Analog Electronics December 2009]**

17. Compare the performance of CB, CC and CE configurations. Which is most suitable for use as an amplifier ?
[G.G.S.I.P. Univ. Analog Electronics December 2009]

18. Compare the features of transistor amplifier in various configurations.
[G.G.S.I.P. Univ. Analog Electronics December 2010]

19. Draw transistor amplifier in CE, CB and CC configurations and explain the salient features of each type.

[G.G.S.I.P. Univ. Analog Electronics December 2010]

20. Draw the hybrid-π model of a transistor at high frequencies. Explain the significance of each component and give typical values. What is the significance of f_β and f_T. **[G.G.S.I.P. Univ. Analog Electronics December 2010]**

SHORT ANSWER TYPE QUESTIONS WITH ANSWERS

Q. 1. What is an amplifier ?

Ans. The device that amplifies the amplitude of the input signal is called the amplifier. An amplifier may be defined as a device that increases the current, voltage or power of an input signal with the help of a transistor by furnishing the additional power from a separate source of supply.

Q. 2. What is meant by small signal amplifier ?

Ans. When the input signal is quite weak and produces small fluctuations in the output current in comparison to its quiescent value, the amplifier is called the small signal or voltage amplifier.

Q. 3. Write the characteristics of an ideal voltage amplifier.

Ans. Main characteristics of an ideal voltage amplifier is to present a high impedance to the input source so as to minimize the loading effects and to provide a large enough voltage signal to the output device like loudspeaker.

Q. 4. Explain how phase reversal of the signal takes place when it is amplified by a single stage CE voltage amplifier.

[P.T.U. Analog Electronics December 2006]

Ans. In a CE configuration, the output voltage increases in the negative direction when the input signal voltage increases in the positive direction and vice versa. This is called the phase reversal and causes a phase difference of 180° between the input signal voltage and output voltage.

Q. 5. What is an ac emitter resistance ?

Ans. The dynamic resistance of the emitter-base junction diode is called the ac emitter resistance. It is given as $\frac{25 \text{ mV}}{I_E}$ where I_E is the dc emitter current at Q-point.

Q. 6. What is the effect of source resistance on voltage gain of a common base transistor amplifier ?

Ans. The voltage gain of a CB transistor amplifier will decrease if source resistance is considered because in such a case there will be a voltage drop across the source resistance and output voltage will decrease.

Q. 7. Why is coupling capacitor used to connect a signal source to an amplifier ?

Ans. An electrolytic capacitor, called the input capacitor, C_{in} of capacity of about 10 μF is used to couple the signal source to the transistor base. In the absence of this capacitor, the signal source resistance will come across resistor R_2 of the biasing network and thus change the bias. This capacitor allows only ac signal to pass but isolates the signal source from R_2.

Q. 8. What is the function of bypass capacitor ?

Ans. The capacitor C_E, called the emitter bypass capacitor, is connected in parallel with the emitter resistance R_E in order to provide a low reactance path to the amplified ac signal. In the absence of this capacitor, amplified ac signal will cause a voltage drop across it which in turn will feedback the input side and reduce the output voltage.

Q. 9. State what will happen to the voltage gain of an amplifier if the bypass capacitor is open circuited.

[P.T.U. Analog Electronics May 2007]

Ans. Removal of bypass capacitor in a CE amplifier circuit causes excessive degeneration in the amplifier circuit and, therefore, reduction in voltage gain.

Q. 10. What are the alternate notations for 11, 12, 21 and 22 in h-parameters ?

[G.G.S.I.P. Univ. December 2009]

Ans. The alternate notations for 11, 12, 21 and 22 are *i*, *r*, *f* and *o* respectively. In case of transistors another subscript (*b*, *e*, or *c*) is added to designate the type of transistor. For example $h_{ie} = h_{11e}$ = Input resistance in CE configuration.

Q. 11. Why common collector amplifier is called an emitter follower ?

Ans. The CC circuit amplifier is called an emitter follower because in this circuit the output voltage at the emitter terminal follows the input signal applied to the base terminal.

Q. 12. What configuration of bipolar transistor is called as emitter follower ? For what purpose it is used.

Ans. The common collector configuration of bipolar transistor is called as emitter follower because in this circuit the output voltage at the emitter terminal follows the input signal applied to the base terminal. This configuration finds wide application as a buffer amplifier between a high impedance source and a low impedance load.

Q. 13. What are the main purposes for which a common collector amplifier may be used ?

Ans. For a common collector amplifier, current gain is as high as for CE amplifier, voltage gain is less than unity, input resistance is the highest and the output resistance is the lowest of all the three (CE, CC and CB) configurations. This circuit finds wide application as a buffer amplifier between a high impedance source and a low impedance load.

Q. 14. Why is it not possible to use the *h* parameters at high frequencies and large signals.

Ans. *h* parameters cannot be used at high frequencies because at high frequencies the internal capacitances offer very low reactance and they are required to be considered in the analysis.

h parameters cannot be used for large signals because the transistor behaves as a linear device for small signals only.

Q. 15. What are hybrid capacitances ?

Ans. Forward biased P-N junction exhibits a capacitive effect, called the diffusion capacitance. This capacitive effect of the normally forward biased base-emitter junction of transistor is represented as $C_{b'e}$ or C_e. Similarly a reverse biased P-N junction has a transition capacitance and it is shown as $C_{b'c}$ or C_c.

Q. 16. What do you understand by the term 'virtual base' ?

Ans. The internal capacitances of transistor C_e and C_c have a common termination at terminal B', not the input base terminal B (Fig. 7.54). Resistance $R_{bb'}$, referred to as the base-spreading resistance, represents the built resistance of the base and is the portion of base through which the base current flows. Resistance $r_{b'e}$ is that portion of base and emitter which may be thought of as being "in series with" with the collector junction. This establishes a *virtual base* B' for the junction capacitances to be connected to instead of B.

Q. 17. How does h_{ie} vary with I_C if $r_{bb'} << r_{b'e}$.

Ans. If $r_{bb'} << r_{b'e}$, h_{ie} varies as reciprocal of collector current I_C.

PROBLEMS

1. The following test results were obtained in a CE amplifier circuit while measuring h parameters experimentally :

   (*i*) With ac output shorted I_b = 25 μA, I_c = 1.2 mA, V_{be} = 30 mV and V_{ce} = 0

   (*ii*) With ac input open circuited I_b = 0, I_c = 30 μA, V_{be} = 0.3 mV and V_{ce} = 1.2 V

   Determine hybrid parameters of the given transistor.

   [**Ans.** h_{ie} = 1.2 kΩ; h_{fe} = 48; h_{re} = 2.5 × 10^{-4}; h_{oe} = 25 μS]

2. If h_{ie} = 1,500 Ω, h_{fe} = 99, what is h_{ib} ? [**Ans.** 15 Ω]

3. If h_{ie} = 1,500 Ω, h_{fe} = 99, what is h_{ic} ? [**Ans.** 1,500 Ω]

4. A junction transistor has the following h parameters h_{ie} = 2k, h_{re} = 1.6 × 10^{-4}, h_{fe} = 50, h_{oe} = 50 μA/V. Determine the current gain, voltage gain, input resistance and output resistance of CE amplifier if the load resistance is 12 kΩ and source resistance is 500 Ω.

   [**Ans.** A_i = –31.25; R_{in} = 1,940 Ω, A_V = –193.3, A_{is} = –6.4, A_{Vs} = –153.7; R_{out} = 21.37 kΩ]

5. A BJT has the following parameters :

$$h_{ie} = 2000\ \Omega,\ h_{re} = 16 \times 10^{-5},\ h_{fe} = 49 \text{ and } h_{oe} = 50\ \mu A/V.$$

Determine the current gain, voltage gain, input resistance and output resistance of the CE amplifier if the load resistance is 30 kΩ. Neglect source resistance.

[**Ans.** A_i = –19.6; R_{in} = 1.906 kΩ; A_V = –308.5; R_{out} = 21.7 kΩ]

6. A BJT has h_{ie} = 2 kΩ, h_{fe} = 100, $h_{re} = 2.5 \times 10^{-4}$ and h_{oe} = 25 μA/V as parameters in CE configuration. It is used as an emitter follower amplifier with R_s = 1 kΩ and R_L = 500 Ω. Determine for the amplifier the voltage gain $A_{Vs} = V_o/V_s$, the current gain $A_{is} = I_o/I_s$, the input resistance R_i and output resistance R_0.

[**Ans.** 0.9432; 1.886; 51.876 kΩ; 29.68 Ω]

7. A common emitter transistor amplifier circuit has the following characteristics :

h_{ie} = 1,000 Ω ; h_{fe} = 50, $h_{re} = 2.5 \times 10^{-4}$, $h_{oe} = 25 \times 10^{-6}$ A/V. If the load resistance R_L = 10 kΩ and source resistance is 100 Ω, find the input resistance, output resistance and the voltage, current and power gains.

[**Ans.** 10 kΩ ; 73.33 kΩ ; – 39.6, – 40.4, 1,600]

8. A transistor used in CE amplifier configuration has h_{ie} = 2 kΩ, h_{fe} = 100, $h_{re} = 10^{-5}$ and h_{oe} = 25 μA/V. If R_S = 1 kΩ and R_L = 1 kΩ, determine the voltage gain $A_{vs} = \dfrac{V_o}{V_s}$, current gain $A_{is} = \dfrac{I_o}{I_s}$, input resistance R_i and output resistance, R_o.

[**Ans.** A_{vs} = – 32.5; A_{is} = – 32.5; 1.999 kΩ; 40.54 kΩ]

9. A transistor with h_{ie} = 1.5 kΩ and h_{fe} = 75 is used as an emitter follower circuit where resistances R_1 and R_2 are used for normal biasing. Calculate (*i*) A_i, (*ii*) r_{in} (*iii*) r_o and (*iv*) A_v if R_E = 860 Ω, $R_1 \parallel R_2$ = 20 kΩ and R_s = 1 kΩ.

[**Ans.** A_i = 17.5, r_{in} = 66.86 kΩ, r_o = 31.1 Ω, $A_v \simeq 1$]

10. Given the following parameters for a given transistor at I_C = 1.3 mA and at room temperature : h_{oe} = 25 μS; h_{fe} = 50; h_{ie} = 1.1 kΩ and $h_{re} = 2.5 \times 10^{-4}$. Determine g_m; $r_{b'e}$; $r_{bb'}$; $r_{b'c}$ and r_{ce}.

[**Ans.** g_m = 50 mS; $r_{b'e}$ = 1 kΩ; $r_{bb'}$ = 100 Ω; $r_{b'c}$ = 4 MΩ; r_{ce} = 81.6 kΩ]

11. Given the following parameters for a given transistor at I_C = 10 mA, V_{CE} = 10 V and the room temperature : h_{fe} = 100 ; h_{ie} = 500 Ω ; $|A_i|$ = 10 at 10 MHz, and C_c = 3 pF. Find f_β, f_T, C_e, $r_{b'e}$ and $r_{bb'}$

[**Ans.** 1 MHz; 100 MHz; 609 pF; 260 Ω; 240 Ω]

12. A BJT is found to have f_T = 500 MHz, $r_{bb'}$ = 50 Ω; g_m = 100 m mho; $C_{b'c}$ = 5 pF and h_{fe} = 100 as parameters. It is used as an CE amplifier with R_s = 500 Ω and R_L = 1 kΩ.

Determine the amplifier (*i*) midband voltage gain $A_{vs} = \dfrac{V_o}{V_s}$ and (*ii*) the upper 3-dB cutoff frequency.

[**Ans.** (*i*) – 64.5, (*ii*) 5 MHz]

❑❑❑

UNIT III

MULTISTAGE AMPLIFIERS AND FEEDBACK AMPLIFIERS

CONTENTS IN THIS UNIT

CHAPTER 8

Multistage Amplifiers

8.1. INTRODUCTION

INSIDE THIS CHAPTER

The performance of a single stage amplifier has already been discussed in the chapter 7. Although the gain* of an amplifier does depend on the parameters of the device and circuit components, there exists an upper theoretical limit for the gain obtainable from a single stage. The voltage amplification or power gain or frequency response obtained with a single stage of amplification is usually not sufficient to meet the needs of either a composite electronic circuit or load device. For example, a speaker represents a "heavy" load in an audio amplifier system, and several amplifier stages may be required to "boost a signal" originating at a microphone or magnetic tape head to a level sufficient to provide a large amount of power to the speaker. We hear of *preamplifiers, power amplifiers,* and *output amplifiers,* all of which constitute stages of amplification in such a system. Actually, each of these components may itself consist of a number of individual transistor amplifier stages. Amplifiers that produce voltage, current, and/or power gain through the use of two or more stages are called *multistage amplifiers.* It may be emphasized here that a practical amplifier is always a multistage amplifier that may provide a higher voltage or current gain or both.

In a multistage amplifier, the output of first stage is combined to the next stage through a coupling device. The process is known as *cascading*. The coupling device is used to (*i*) transfer the ac output of one stage to the input of the next stage and (*ii*) block the dc to pass from one stage to the next stage *i.e.* to isolate the dc conditions.

For an ideal coupling network the following requirements should be fulfilled.

(*i*) It should not disturb the dc bias conditions of the amplifiers being coupled. This means direct currents should not pass through the coupling network.

(*ii*) The coupling network should transfer ac signal waveform from one amplifier to the next amplifier without any distortion.

(*iii*) Although some voltage loss of signal cannot be avoided in the coupling network but this loss should be minimum, just negligible.

* Gain of an amplifier is the ratio of the output electrical quantity (current, voltage or power) to the input one of the amplifier.

(*iv*) The coupling network should offer equal impedance to the various frequencies of signal wave. In other words the network impedance should not be frequency dependent.

Unfortunately, there is no coupling network which fulfills all the above demands. The four basic methods of coupling are R-C coupling, Transformer coupling, Impedance coupling, and Direct coupling.

The coupling network not only couples two stages; it also forms a part of the load impedance of the preceding stage. Thus, the performance of the amplifier will also depend upon the type of coupling network used. Amplifier is usually named after the type of coupling employed such as R-C coupled amplifier, transformer coupled amplifier, impedance coupled amplifier, and direct coupled amplifier.

In R-C *coupling*, a resistor and a capacitor are used as a coupling device. The capacitor connects the output of one stage to the input of next stage to pass ac signal and to block the dc bias voltages. The amplifier using R-C coupling is called the *R-C coupled amplifier.*

In *transformer coupling*, transformer is used as the coupling device. The amplifier using transformer coupling is called the *transformer coupled amplifier*. In this case there is no need of using a coupling capacitor because the secondary of the coupling transformer conveys the ac component directly to the base of the second stage. Moreover, the secondary winding also provides a base return path and so base resistance is not required.

In *direct coupling* or dc coupling, the individual amplifier stage bias conditions are so designed that the two stages may be directly connected without the necessity of dc isolation. This coupling is used where it is desirable to connect the load directly in series with the output terminal of the active circuit element such as in case of headphones, loudspeakers etc. The amplifier using direct coupling is called the *direct coupled amplifier.*

R-C coupling is the most commonly used coupling between the two stages of a cascaded or multistage amplifier because it is cheaper in cost and very compact circuit and provides excellent frequency response.

The most suitable transistor configuration for cascading is CE configuration because the voltage gain of common emitter amplifier is greater than unity while CC configuration has voltage gain less than unity and the voltage gain of CB configuration using cascading is also less than unity.*

However, for input stage CC or CB configuration may be required for proper impedance matching at the cost of voltage or current gain. It is noteworthy point that for input stage, the consideration is not the maximum voltage gain but the impedance matching of the source with the input impedance of the input stage. Some driving sources may need input circuit to be an almost open circuit while others need an almost short circuit. In certain cases choice of configuration for the input stage is the minimization of noise and maximization of signal/noise power ratio.

8.2. DECIBELS

It is possible to specify amplifiers in terms of their voltage, current and power gains. Although the gain of an amplifier can be expressed as number, yet it is of great practical importance to assign it a unit. The unit assigned is *bel* or *decibel* (dB).

It is often convenient to compare two powers on a logarithmic scale rather than on a linear scale. Accordingly a logarithmic unit, named bel after Alexander Graham Bell was proposed by the telephone industry. The number of bels by which the power output P_{out} exceeds the power input P_{in} is defined as

$$\text{Number of bels} = \log_{10} \frac{P_{out}}{P_{in}} \qquad ...(8.1)$$

For example, a power gain of 100 would be equivalent to $\log_{10} 100$ *i.e.* 2 bels. The bel proved to be too large for most work, so one tenth of a bel the decibel (dB), was adopted. In decibels power gain of 100 is equivalent to

$$= 10 \log_{10} 100 = 10 \times 2 \text{ dB} = 20 \text{ dB}$$

* No doubt, the voltage gain of a single stage CB amplifier is more than unity but the overall voltage gain of multistage amplifier using CB configuration is low, almost equal to the voltage gain of the last stage alone.

In short, to convert a power expressed in bels into decibels, we merely multiply the number of bels by 10. Symbolically

$$\text{Number of dB} = 10 \times \text{number of bels} = 10 \log_{10} \frac{P_{out}}{P_{in}} \quad \text{...(8.2)}$$

8.2.1. Decibel Power Gain

Decibel power gain is defined as

$$A_{p\,(dB)} = 10 \log_{10} \frac{P_{out}}{P_{in}} = 10 \log_{10} A_p \quad \text{...(8.3)} \because \frac{P_{out}}{P_{in}} = \text{Power gain, } A_p$$

To make sure that $A_{p\,(dB)}$ is never confused with A_p, the unit *decibel* (abbreviated dB) is attached to all answers for $A_{p\,(dB)}$.

Decibel power gain has two useful properties :

1. Each time the ordinary power gain increases (or decreases) by a factor of 2, the decibel power gain increases (or decreases) by 3 dB.
2. Each time the ordinary power gain increases (or decreases) by a factor 10, the decibel power gain increases (or decreases) by 10 dB.

These properties, in short, are shown in tabular form in Table 8.1.

TABLE 8.1. ***Properties of Power Gain***

Factor	*Decibel, dB*
× 2	+3
× 0.5	−3
× 10	+10
× 0.1	−10

8.2.2. Decibel Voltage Gain

In practice, it is much easier and convenient to measure voltages than to measure power levels. Therefore, a relationship between the voltage gain and the power gain in decibels was developed. It is derived as below :

$$\text{Voltage gain in dB} = 10 \log_{10} \frac{V_{out}^2/R_L}{V_{in}^2/R_{in}}$$

$$= 10 \log_{10} \left(\frac{V_{out}}{V_{in}}\right)^2 \times \frac{R_{in}}{R_L} = 20 \log_{10} \frac{V_{out}}{V_{in}} + 10 \log_{10} \frac{R_{in}}{R_L} \quad \text{...(8.4)}$$

$$= 20 \log_{10} \frac{V_2}{V_1} \quad \text{...(8.5) provided } R_L = R_{in}$$

$$\text{Similarly current gain in dB} = 20 \log_{10} \frac{I_2}{I_1} + 10 \log_{10} \frac{R_L}{R_{in}} = 20 \log_{10} \frac{I_2}{I_1} \quad \text{...(8.6) provided } R_L = R_{in}$$

It is a common practice in the electronics industry to apply Eqs. (8.5) and (8.6) regardless of the relationship between input and load resistances.

Decibel voltage gain is defined as

$$A_{V\,(dB)} = 20 \log_{10} \frac{V_{out}}{V_{in}} = 20 \log_{10} A_V \quad \text{...(8.7) because } \frac{V_{out}}{V_{in}} = A_V$$

The reason for using 20 instead of 10 in this definition is because power is proportional to the square of voltage.

Decibel voltage gain has two useful properties :

1. Each time the voltage gain increases (or decreases) by a factor 2, the decibel voltage gain increases (or decreases) by 6 dB.
2. Each time the voltage gain increases (or decreases) by a factor of 10, the decibel voltage gain increases (or decreases) by 20 dB.

TABLE 8.2. ***Properties of Voltage Gain***

Factor	*Decibel, dB*
× 2	+6
× 0.5	−6
× 10	+20
× 0.1	−20

Table 8.2 depicts these properties in compact form.

8.2.3. Conversion of Decibel Gain To Ordinary Gain

Decibel gain can be converted into ordinary gain by using the following equations :

$$A_p = \text{antilog}_{10} \frac{A_{p(dB)}}{10} \quad ...(8.8)$$

and
$$A_V = \text{antilog}_{10} \frac{A_{V(dB)}}{20} \quad ...(8.9)$$

8.2.4. Merits of dB Notation

(*i*) Decibel notation permits gains to be directly added in a system involving several stages because use of logarithms changes multiplication into addition.

(*ii*) Decibel notation permits us to denote, both very small as well as very large quantities of linear scale by conveniently small figures. The voltage gain of 10^{-4}, 10^{-2}, 1, 10^2, 10^4, 10^6 may be represented as a voltage gain of –80 dB, –40 dB, 0, 40 dB, 80 dB and 120 dB respectively.

(*iii*) Output of many amplifiers is finally converted into sound and this sound reaches the human ear. Experiments show that our ear response is logarithmic *i.e.* loudness of sound heard by ear is not according to the intensity of sound but according to the log of sound intensity. Thus if the intensity of sound given by the speaker (*i.e.* power) is increased 100 times, our ears hear a doubling effect ($\log_{10} 100 = 2$) *i.e.* as if loudness were doubled instead of made 100 times. Thus the use of dB unit is justified on a psychological basis too.

8.3. MULTISTAGE AMPLIFIER

A multistage amplifier can be represented by a block diagram, as shown in Fig. 8.1. It is to be noted that the output of the first stage makes the input for the second stage, the output of second stage makes the input for third stage and so on. The signal voltage V_s is applied to the input of the first stage and the final output V_{out} is available at the output terminals of the last stage.

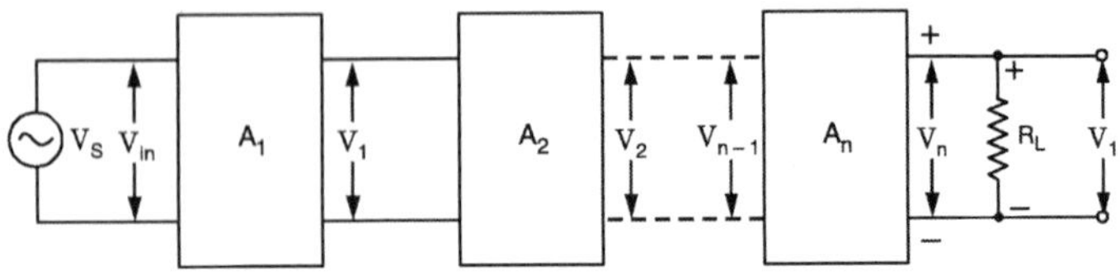

Fig. 8.1. ***n-Stage Amplifier***

Input to the first stage, V_{in} = Signal voltage V_s

Output of first stage or input to the second stage

$$V_1 = A_{V_1} V_{in} \text{ where } A_{V_1} \text{ is the voltage gain of first stage}$$

Output of second stage or input to the third stage

$$V_2 = A_{V_2} V_1 \text{ where } A_{V_2} \text{ is the voltage gain of the second stage}$$

Similarly the output of *n*th stage (or final output),

$$V_{out} = V_n = A_{V_n} V_{n-1} \text{ where } A_{V_n} \text{ is the voltage gain of the last stage}$$

Overall voltage gain of the amplifier is given as

$$A = \frac{V_{out}}{V_s}$$ (visualising the multistage amplifier as a single amplifier with input voltage V_s and output voltage V_{out})

$$= \frac{V_1}{V_s} \times \frac{V_2}{V_1} \times \frac{V_3}{V_2} \times ... \frac{V_{n-1}}{V_{n-2}} \times \frac{V_n}{V_{n-1}}$$

$$= A_{V_1} \times A_{V_2} \times ... \times A_{V_{n-1}} \times A_{V_n} \quad ...(8.10)$$

i.e. the gain of a multistage amplifier is equal to the product of gains of individual stages. It is worthwhile to mention here that in practice total gain A is less than $A_{v_1} \times A_{v_2} \times ... \times A_{v_{n-1}} \times A_{v_n}$ due to the loading effects of the following stages.

When the gains are expressed in dB, the overall gain of a multistage amplifier is given as the sum of gains of individual stages in decibels (dB).

Taking logarithm (to the base 10) of Eq. (8.10) and then multiplying each term by 20 we have

$$20 \log_{10} A_v = 20 \log_{10} A_{v_1} + 20 \log_{10} A_{v_2} + ... + 20 \log_{10} A_{v_{n-1}} + 20 \log_{10} A_{v_n}$$

In the above equation, the term to the left is the overall gain of the multistage amplifier expressed in decibels. The terms on the right denote the gains of the individual stages expressed in decibels. Thus

$$A_{v\,dB} = A_{v\,dB_1} + A_{v\,dB_2} + ... + A_{v\,dB_n} \qquad ...(8.11)$$

Example 8.1. A multistage amplifier consists of three stages. The voltage gains of stages are 60, 100, 160. Calculate the overall voltage gain in dB.

Solution : Voltage gain of first stage in dB = $20 \log_{10} 60 = 35.563$ dB

Voltage gain of second stage in dB = $20 \log_{10} 100 = 40$ dB

Voltage gain of third stage in dB = $20 \log_{10} 160 = 44.082$ dB

Overall voltage gain of the amplifier = $35.563 + 40 + 44.082 = 119.645$ dB **Ans.**

Alternative Method :

Overall voltage gain, $A_v = A_{v_1} \times A_{v_2} \times A_{v_3} = 60 \times 100 \times 160 = 960{,}000$

Overall voltage gain in dB = $20 \log_{10} 960{,}000 = 119.645$ dB **Ans.**

Example 8.2. The overall gain of a 2-stage R-C coupled amplifier is 80 dB. If the voltage gain of the second stage is 150, calculate the voltage gain of the first stage in dB.

[G.G.S.I.P. Univ. Analog Electronics December 2009]

Solution : Overall voltage gain of 2-stage R-C coupled amplifier = 80 dB

Voltage gain of second stage in dB = $20 \log_{10} 150 = 43.52$ dB

Voltage gain of first stage = $80 - 43.52 = 36.48$ dB **Ans.**

8.4. *n*-STAGE CASCADED AMPLIFIER

As already discussed in Art. 8.1, several amplifier stages are usually cascaded to increase the overall voltage gain of the amplifier. However, sometimes cascading is done to obtain the desired output and input impedance for specific applications. Block diagram of an n-stage cascaded amplifier is given in Fig. 8.1. The first stage is driven by a voltage source V_s (or a current source I_s). The output of first stage is fed to the input of the second stage, the output of second stage is supplied to the input of the third stage and so on. The output of the nth or last stage is fed to the load R_L. Actual voltage available at the input of the first stage is V_{in} (V_{in} equals signal voltage V_s if source resistance R_S is negligible) and the voltage available at the output terminals of the last stage is V_{out}. Then the ratio $\frac{V_{out}}{V_{in}}$ gives the voltage gain of the *n*-stage cascaded amplifier. Thus by designing properly a cascaded amplifier, a weak input signal V_{in} of just a few microvolts can be amplified giving output voltage V_{out} of several volts.

Here we take up R-C coupled amplifier, being the most popular amplifier employed for audio-frequency amplification. A typical two-stage R-C coupled amplifier using N-P-N transistors in common emitter configuration (the most suitable transistor configuration for cascading) is given in Fig. 8.2. Detailed discussion of R-C coupled amplifier will be taken up in Art. 8.6. Typical values of circuit components are given in the brackets.

The block diagram of a two stage R-C coupled transistor amplifier shown in Fig. 8.2 indicating the various voltages, currents and resistances involved is given in Fig. 8.3. The biasing arrangements and coupling capacitors have been omitted for sake of simplicity and clarity.

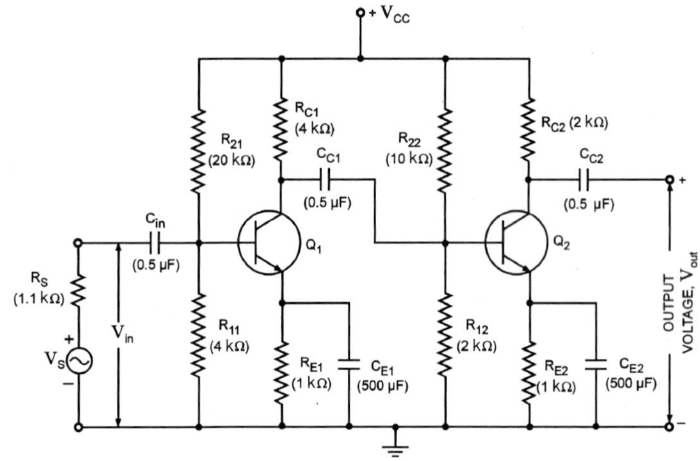

Fig. 8.2. ***Two-Stage R-C Coupled Transistor Amplifier***

1. Voltage Gain. The overall voltage gain of the amplifier is given by the product of the voltage gains of the individual stages. This is proved as below : Voltage gain of first stage,

$$A_{v_1} = \frac{V_1}{V_{in}} = \frac{\text{Output voltage of first stage}}{\text{Input voltage to first stage}}$$

$$= A_{v_1} \angle \theta_1 \qquad ...(8.12)$$

where A_{v1} is the magnitude of the voltage gain of the first stage and θ_1 is the phase angle between output and input voltages of this stage.

Similarly the voltage gain of nth stage, $A_{v_n} = \dfrac{\text{Output voltage of } n\text{th stage}}{\text{Input voltage to } n\text{th stage}} = A_{v_n} \angle \theta_n$ ...(8.13)

Thus the overall voltage gain of the complete n-stage cascaded amplifier is given as

$$A_v = \frac{V_{out}}{V_{in}} = \frac{\text{Output voltage of the } n\text{th stage}}{\text{Voltage input to the first stage}} = A_v \angle \theta \qquad ...(8.14)$$

where A_v is the magnitude of the voltage gain and θ is the phase angle between the output and input voltages of the amplifier.

$$\text{Since } \frac{V_{out}}{V_{in}} = \frac{V_1}{V_{in}} \times \frac{V_2}{V_1} \times \frac{V_3}{V_2} \times ... \times \frac{V_{out}}{V_{n-1}}$$

$$\text{So } \mathbf{A_v} = \mathbf{A_{v_1}} \times \mathbf{A_{v2}} \times \mathbf{A_{v_3}} \times ... \times \mathbf{A_{v_n}} \qquad ...(8.15)$$

$$\text{or } A_v \angle \theta = A_{v_1} \cdot A_{v2} \cdot A_{v_3} ... A_{v_n} \angle \theta_1 + \theta_2 + \theta_3 + ... \theta_n \qquad ...(8.16)$$

$$\text{Hence } A_v = A_{v_1} \cdot A_{v2} \cdot A_{v_3} ... A_{v_n} \qquad ...(8.17)$$

$$\text{and } \theta = \theta_1 + \theta_2 + \theta_3 + ... \theta_n \qquad ...(8.18)$$

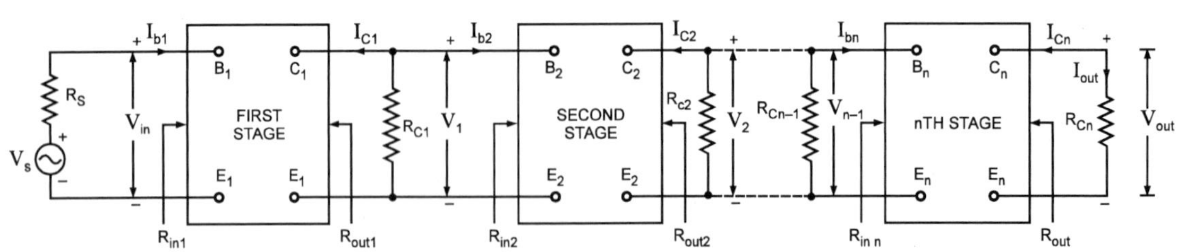

Fig. 8.3. ***n-Stage CE Cascaded Amplifier***

Thus it is concluded that the magnitude of the overall voltage gain equals the product of the magnitudes of the voltage gains of the individual stages and the resultant phase shift of the amplifier equals the sum of the phase shifts introduced by individual stages.

The voltage gain of kth stage (an intermediate stage) of an *n*-stage CE cascaded amplifier shown in Fig. 8.4 is given as

$$A_{vk} = \frac{A_{ik} R_{Lk}}{R_{ink}} \quad ...(8.19)$$

where R_{Lk} is the effective load at the collector of the kth stage.

The quantities in above equation (*i.e.* A_{ik}, R_{Lk} and $R_{in\,k}$) are evaluated by starting with the last stage and proceeding backward. Thus for the *n*th stage, current gain A_{in} and input resistance $R_{in\,n}$ are given as

$$A_{in} = \frac{-h_{fe}}{1 + h_{oe} R_{Ln}} \quad ...(8.20)$$

$$\text{and } R_{in\,n} = h_{ie} + h_{re} A_{in} R_{L\,n} \quad ...(8.21)$$

where $R_{L\,n}$ is the effective load impedance of the final (*i.e.* *n*th) stage and equals R_{cn}.

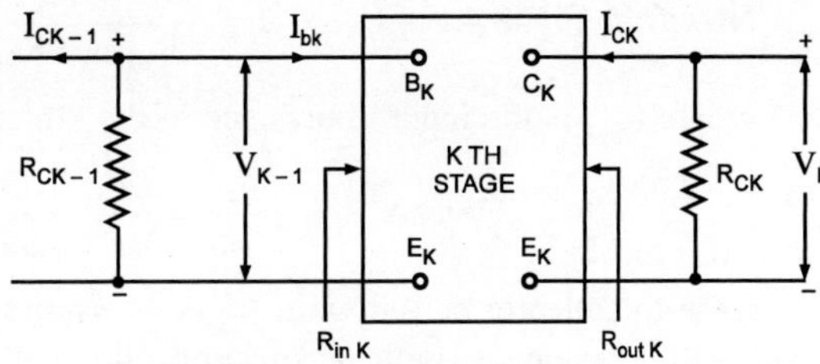

Fig. 8.4. *kth Stage of an n-Stage CE Cascaded Amplifier*

For the (*n* – 1)th stage, effective load impedance $R_{in\,n-1}$ equals $R_{c\,n-1} \parallel R_{in\,n}$. So

$$R_{L\,n-1} = \frac{R_{cn-1} \cdot R_{in\,n}}{R_{cn-1} + R_{inn}} \quad ...(8.22)$$

Now the current gain $A_{i\,(n-1)}$, of the last but one stage [*i.e.* (*n* – 1)th stage] is obtained using Eq. (8.20) by replacing $R_{L\,n}$ by $R_{L\,n-1}$. Similarly the input impedance $R_{in\,n-1}$ of the (*n* – 1)th stage is determined from Eq. (8.21) by replacing A_{in} by $A_{i\,n-1}$ and $R_{L\,n}$ by $R_{L\,n-1}$. Thus proceeding backward, base-to-collector current gains and input impedances of every stage including the first one can be determined. Then the voltage gain of each stage can be determined by using Eq. (8.19).

Overall voltage gain of the *n*-stage cascaded amplifier may then be calculated from Eq. (8.15).

2. Current Gain. Overall voltage gain of an *n*-stage cascaded amplifier can be obtained, without determining the voltage gain of individual stage, from the following equation

$$A_v = A_i \frac{R_{cn}}{R_{in1}} \quad ...(8.23)$$

where A_i is the overall current gain of the *n*-stage cascaded amplifier and

$$A_i = \frac{\text{Output current of the } n\text{th stage}}{\text{Input current (or base current of first stage)}} = \frac{I_{out}}{I_{bi}} = \frac{-I_{cn}}{I_{bi}} = -\frac{I_n}{I_{bi}} \quad ...(8.24)$$

where $I_{cn} \cong I_n$, the collector current of the *n*th stage.

Now let us obtain equations so as to calculate current gain A_i in terms of circuit parameters.

$$\text{Since } \frac{I_n}{I_{bi}} = \frac{I_1}{I_{bi}} \cdot \frac{I_2}{I_1} \cdot ... \frac{I_{n-1}}{I_{n-2}} \cdot \frac{I_n}{I_{n-1}}$$

$$\text{then } A_i = A_{i_1} \cdot A'_{i_2} \cdot A'_{i_3} \; ... \; A'_{i_{n-1}} \cdot A'_{in} \quad ...(8.25)$$

where A_{i_1} is the base-to-collector current gain of the first stage and is equal to $\frac{-I_{ci}}{I_{bi}}$ while A'_{i_2}, A'_{i_3} etc. are the collector-to-collector current gains of second, third etc. stages.

For kth stage, collector-to-collector current gain A'_{ik} is given as

$$A'_{ik} = \frac{I_{ck}}{I_{ck-1}} \qquad \text{...(8.26)}$$

Similarly base-to-collector current gain of kth stage A_{ik} is given as

$$A_{ik} = \frac{-I_{ck}}{I_{bk}} \qquad \text{...(8.27)}$$

Now from Fig. 8.4, $$I_{bk} = -I_{ck-1}\frac{R_{ck-1}}{R_{ck-1}+R_{ink}} \qquad \text{...(8.28)}$$

where $R_{in\,k}$ is the input impedance of the kth stage.

$$\text{Hence } A'_{ik} = \frac{I_{ck}}{I_{ck-1}} = \frac{I_{ck}}{I_{bk}}\cdot\frac{I_{bk}}{I_{ck-1}} = A_{ik}\cdot\frac{R_{ck-1}}{R_{ck-1}+R_{ink}} \qquad \text{...(8.29)}$$

Base-to-collector current gain A_{ik} is determined by starting with the output stage and proceeding backward to the kth stage, as indicated in connection with Eqs. (8.20), (8.21) and (8.22). The collector-to-collector current gains are then determined by using Eq. (8.29) and the overall current gain of the n-stage amplifier is determined by using Eq. (8.25).

3. Input Impedance. The input impedance of the complete cascaded amplifier is determined, as discussed above, by starting with the last stage and proceeding backward as follows.

Determine $$A_{in} = \frac{-h_{fe}}{1+h_{oe}R_{Ln}} \qquad \text{...Refer to Eq. (8.20)}$$

$$R_{in\,n} = h_{ie} + h_{re}A_{in}R_{Ln} \qquad \text{...Refer to Eq. (8.21)}$$

$$R_{L\,n-1} = R_{C\,n-1} \,\|\, R_{in\,n} \qquad \text{...Refer to Eq. (8.22)}$$

$R_{L\,n-1}$ is the effective load impedance of the $(n-1)$th stage.

Calculate $A_{(n-1)}$, $R_{in(n-1)}$ and $R_{L\,n-2}$ from the above equations. Proceed in this manner to find the effective input impedance (R_{in}) of the first stage.

4. Output Impedance. The output impedance of each transistor stage and of the overall amplifier is determined starting with the first stage and using equation

$$Y_{out\,1} = h_{oe} - \frac{h_{fe}\,h_{re}}{h_{ie}+R_S} \qquad \text{...(8.30)}$$

Then $\frac{1}{Y_{out1}}$ gives the corresponding output impedance $R_{out\,1}$.

The output impedance $R'_{out\,k}$ of the kth stage is the parallel combination of the output impedance $R_{out\,k}$ of kth stage and $R_{c\,k}$. The effective source impedance of the (k + 1)th stage is also $R'_{out\,k}$.

5. Power Gain. The overall power gain of the n-stage cascaded amplifier is given as

$$A_p = \frac{\text{Output power}}{\text{Input power}} = \frac{V_{out}\,I_n}{V_{in}\,I_{bi}} = A_v A_i \qquad \text{...(8.31)}$$

$$\text{or } A_p = (A_i)^2\frac{R_{cn}}{R_{in1}} \qquad \text{...(8.32)}$$

8.5. FREQUENCY RESPONSE OF COUPLED AMPLIFIERS

The curve drawn between the voltage gain and signal frequency of an amplifier is known as the *frequency response*. The performance of an amplifier is judged to a considerable extent by its frequency response. In the design of an amplifier, appropriate steps are taken to ensure that gain is essentially uniform over some specified range.

Video amplifiers are almost invariably of the R-C coupled type. For such a stage the frequency characteristics may be divided into three regions: Midband-frequency region, low-frequency region and high-frequency region.

In midband-frequency region, the amplification remains reasonably constant and equal to A_{vm}. For the present discussion midband gain may be assumed to be normalized to unity, *i.e.*, $A_{vm} = 1$.

Low-Frequency Response. In the low-frequency region, below the midband, an amplifier stage behaves like the simple high-pass circuit (Fig. 8.5) of time constant $\tau_1 = R_1C_1$.

The current through the circuit is given by

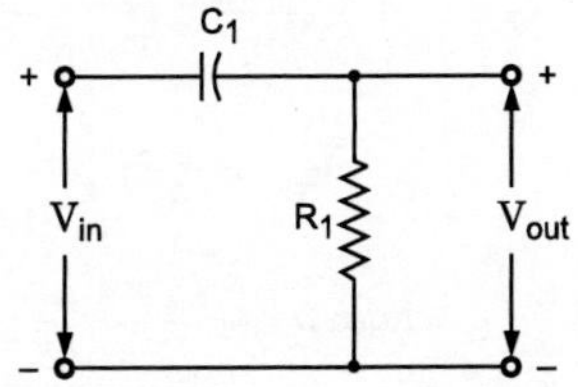

Fig. 8.5. ***A High-Pass R-C Circuit For Determination of Low-Frequency Response of an Amplifier***

$$I = \frac{V_{in}}{R_1 - jX_1}$$

and output voltage,

$$V_{out} = \text{Voltage drop across } R_1 = IR_1$$

$$= \frac{R_1}{R_1 - jX_1} V_{in}$$

$$= \frac{R_1}{R_1 - \dfrac{j}{2\pi f C_1}} V_{in}$$

$$= \frac{V_{in}}{1 - \dfrac{j}{2\pi f C_1 R_1}} \qquad ...(8.33)$$

The voltage gain at low frequencies A_{vl}, is defined as the ratio of the output voltage V_{out}, to the input voltage V_{in}.

$$i.e.\ A_{vl} = \frac{V_{out}}{V_{in}} = \frac{1}{1 - \dfrac{j}{2\pi f C_1 R_1}} = \frac{1}{1 - \dfrac{jf_1}{f}} \qquad ...(8.34)$$

where f_1 is the cutoff frequency and $= \dfrac{1}{2\pi R_1 C_1}$ ...(8.35)

Now magnitude of A_{vl} and phase angle θ are given by

$$|A_{vl}| = \frac{1}{\sqrt{1 - (f_1/f)^2}} \qquad ...(8.36)$$

$$\text{and phase angle } \theta_1 = \text{Tan}^{-1} \frac{f_1}{f} \qquad ...(8.37)$$

At the frequency $f = f_1$, $A_{vl} = \dfrac{1}{\sqrt{2}} = 0.707$ whereas in the midband region ($f >> f_1$), $A_{vl} \rightarrow 1$. Hence f_1 is the frequency at which the gain has fallen to 0.707 times its midband value A_{vm}. From Eq. (8.4) this drop in signal level (assuming equal input and output impedances) corresponds to a decibel reduction of 20 log $\frac{1}{\sqrt{2}}$ or – 3 dB. Accordingly, f_1 is referred to as the *lower 3-dB frequency*. From Eq. (8.35) it is observed that f_1 is that frequency for which the resistance R_1 equals the capacitive reactance *i.e.* $R_1 = \frac{1}{2\pi f_1 C_1}$.

High-Frequency Response. In the high-frequency region, above midband, the stage behaves like a simple low-pass circuit shown in Fig. 8.6.

Proceeding as above, we obtain for the magnitude $|A_{vh}|$ and phase angle θ_2 of the gain

$$|A_{vh}| = \frac{1}{\sqrt{1+(f/f_2)^2}} \quad ...(8.38)$$

$$\theta_2 = \text{Tan}^{-1}\frac{f}{f_2} \quad ...(8.39)$$

$$\text{where } f_2 = \frac{1}{2\pi R_2 C_2} \quad ...(8.40)$$

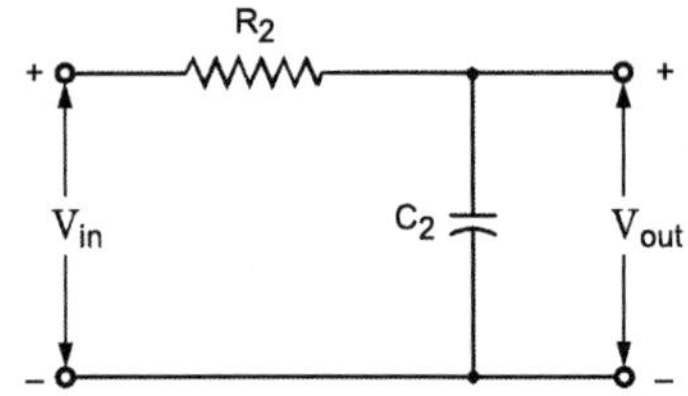

Fig. 8.6. ***A Low-Pass R-C Circuit For Determination of High-Frequency Response of an Amplifier***

Since at $f = f_2$ the gain is reduced to $\frac{1}{\sqrt{2}}$ times its midband value, then f_2 is called the *upper 3-dB frequency*. It also represents that frequency for which the resistance R_2 is equal to the capacitive reactance $\frac{1}{2\pi f_2 C_2}$.

In the above expressions θ_1 and θ_2 represent the angle by which the output lags the input, neglecting the initial 180° phase shift through the amplifier. The frequency dependence of the gains in the high- and low-frequency range is to be seen in Fig. 8.7.

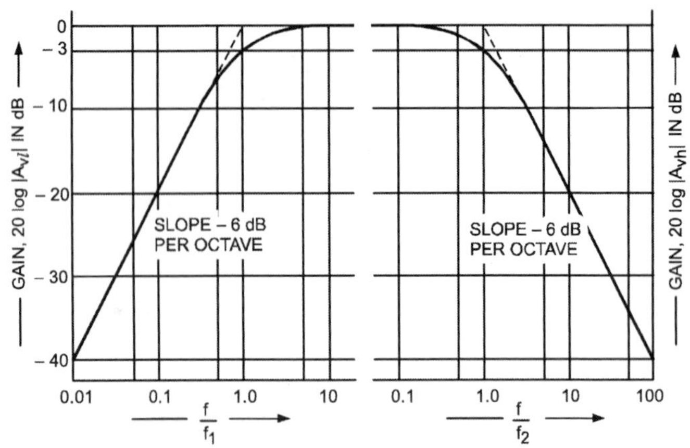

Fig. 8.7. ***A Log-Log Plot of The Amplitude Frequency-Response Characteristic of an R-C Coupled Amplifier***

Bandwidth. The frequency range from f_1 to f_2 is called the *bandwidth* (BW) of the amplifier stage.

8.6. R-C (RESISTANCE-CAPACITANCE) COUPLED TRANSISTOR AMPLIFIER

A two-stage R-C coupled amplifier using N-P-N transistors in CE configuration is shown in Fig. 8.8. The two transistors used are identical and use a common power supply V_{CC}. The resistors R_1, R_2 and R_E form the biasing and stabilization network. In this arrangement, the signal developed across collector resistor R_C of the first stage is coupled to the base of the second stage through the coupling capacitor C_C. As the coupling from one stage to the next is obtained by a coupling capacitor followed by a connection to a shunt resistor, therefore, such amplifiers are called *resistance-capacitance coupled* or *R-C coupled amplifiers*. The input capacitor C_{in} couples ac signal voltage to the base of transistor Q_1. In the absence of C_{in} the signal source will be in parallel with resistor R_2 and the bias voltage of the base will be affected. Thus the function of C_{in} is to allow only the alternating current from signal source to flow into the input circuit.

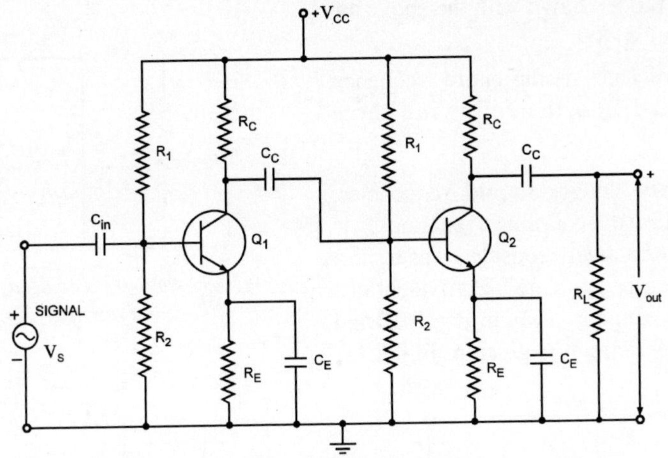

Fig. 8.8. ***Two-Stage R-C Coupled Transistor Amplifier***

The emitter bypass capacitor C_E, offers low reactance path to the signal. If it is not present, then the voltage drop across R_E will reduce the effective voltage available across the base-emitter terminals (the input voltage) and thus reduces the gain.

The coupling capacitor C_C transmits ac signal but blocks the dc voltage of the first stage from reaching the base of the second stage. Thus the dc biasing of the next stage is not interfered with. For this reason, the coupling capacitor C_C is also called the *blocking capacitor.*

8.6.1. Operation

When ac signal is applied to the base of the first amplifier, it appears in the amplified form across collector load R_C. The amplified signal developed across R_C is transmitted to the base of next stage of the amplifier through coupling capacitor C_C. This is further amplified by the next stage and so on. Thus the cascade stages amplify the signal and thus the overall gain is considerably increased. The phase of output is the same as that of input because the phase is reversed twice by two transistors as they are in CE configuration.

It may be mentioned here that the overall gain is less than the product of the gains of individual stages. This is because when a second stage is made to follow the first one, the *effective load resistance* of the first stage is reduced because of the shunting effect of the input resistance of the second stage. This reduces the gain of the stage which is loaded by the next stage. For example, in a 4-stage amplifier the gains of first three stages will be reduced due to loading effect of the subsequent stage. However, the gain of final or last stage (4th stage in this case) which has no loading effect of subsequent stage, remains unaffected. Thus the overall gain shall be less than the product of the gains of four stages.

8.6.2. Analysis

For drawing approximate model of the circuit shown in Fig. 8.8, following assumptions are made.

1. h_{re} is so small that the voltage source $h_{re}V_{out}$ can be neglected.
2. $\frac{1}{h_{oe}}$ is so large that it can be considered as an open circuit.
3. The bias resistors R_1 and R_2 are very large in comparison to h_{ie}.
4. The reactance of emitter bypass capacitor C_E for any given input frequency is so small that parallel combination of R_E and C_E can be considered as a short circuit.

The approximate model drawn with the above assumptions is shown in Fig. 8.9.

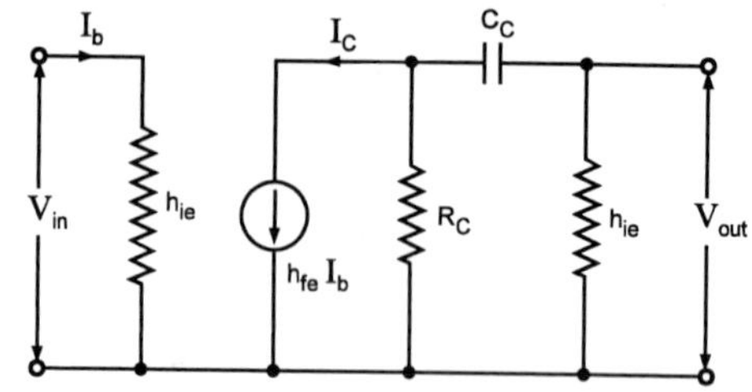

Fig. 8.9. ***Approximate Model of an R-C Coupled Transistor Amplifier***

For the purpose of analysis the entire frequency range may be divided into the following three categories.

(a) **Mid Frequency Range.** At mid frequencies, the impedance offered by the coupling capacitor C_C is so small that it can be effectively considered as a short circuit and its effect can be neglected. Equivalent circuit for mid frequency range is given in Fig. 8.10 and its Thevenin's equivalent circuit is given in Fig. 8.11.

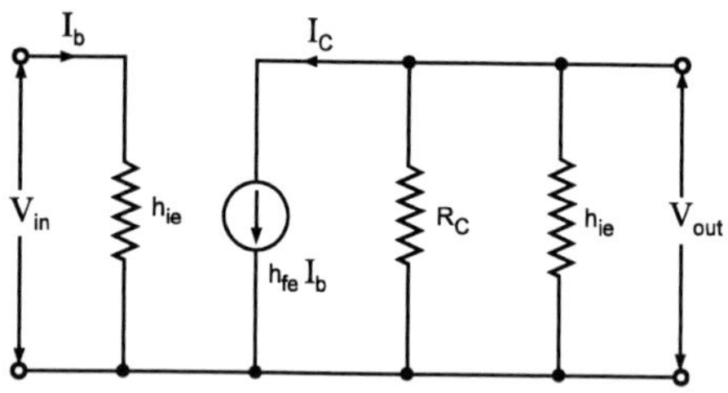

Fig. 8.10. ***Equivalent Circuit For Mid Frequency Range***

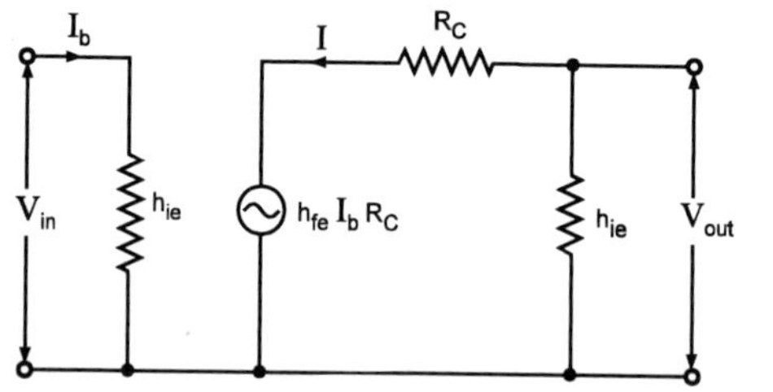

Fig. 8.11. ***Thevenin's Equivalent Circuit***

From circuit shown in Fig. 8.11

$$\text{Current, } I = \frac{h_{fe} I_b R_C}{R_C + h_{ie}} \quad ...(8.41)$$

$$\text{So current gain, } A_{i\,m} = \frac{I}{I_b} = \frac{h_{fe}\, R_C}{R_C + h_{ie}} \quad ...(8.42)$$

$$\text{Output voltage, } V_{out} = h_{ie}\, I = \frac{h_{ie} h_{fe} I_b R_C}{R_C + h_{ie}} \quad ...(8.43)$$

$$\text{Input voltage, } V_{in} = h_{ie}\, I_b$$

$$\text{So voltage gain } A_{v\,m} = \frac{V_{out}}{V_{in}} = \frac{\dfrac{h_{ie} h_{fe} I_b R_C}{(R_C + h_{ie})}}{h_{ie}\, I_b} = \frac{h_{fe}\, R_C}{R_C + h_{ie}} \quad ...(8.44)$$

From Eqs. (8.42) and (8.44) it is obvious that current and voltage gains are equal.

(b) **Low Frequency Range.** In low frequency range, the impedance offered by coupling capacitor C_C is comparable to the collector resistance R_C. It largely affects the current amplification. Therefore, it becomes necessary to include it in the equivalent circuit, as shown in Fig. 8.9.

Thevenin's equivalent circuit is given in Fig. 8.12. From circuit shown in Fig. 8.12

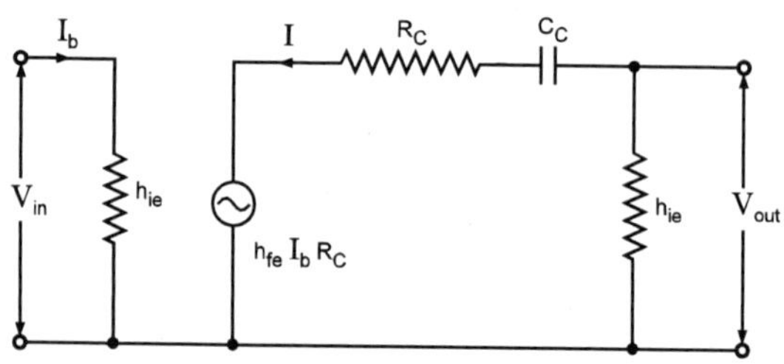

Fig. 8.12. ***Thevenin's Equivalent Circuit For Low Frequency Range***

$$\text{Current, } I = \frac{h_{fe} I_b R_C}{h_{ie} + R_C - j/\omega C_C} \quad ...(8.45)$$

$$\text{So current gain, } A_{il} = \frac{I}{I_b} = \frac{h_{fe}R_C}{h_{ie} + R_C - j/\omega C_C} \quad ...(8.46)$$

$$\text{Output voltage, } V_{out} = h_{ie} I = \frac{h_{fe}\, h_{ie}\, I_b\, R_C}{h_{ie} + R_C - j/\omega C_C} \quad ...(8.47)$$

$$\text{Input voltage } V_{in} = h_{ie} I_b$$

$$\text{So voltage gain, } A_{vl} = \frac{V_{out}}{V_{in}} = \frac{\dfrac{h_{ie}h_{fe}I_bR_C}{h_{ie} + R_C - j/\omega C_C}}{h_{ie}\, I_b} = \frac{h_{fe}\, R_C}{h_{ie} + R_C - j/\omega C_C} \quad ...(8.48)$$

The magnitude of A_{vl} is given by

$$|A_{vl}| = \frac{h_{fe}R_C}{\sqrt{(h_{ie} + R_C)^2 + \left(\dfrac{1}{2\pi f C_C}\right)^2}} \quad ...(8.49)$$

From Eq. (8.49) it is obvious that in the low frequency range voltage gain decreases with the decrease in frequency.

(*c*) High Frequency Range. In high frequency range, the reactance offered by coupling capacitor C_C is very small and it may be considered as a short circuit.

The important factor that comes into picture at high frequencies is the interelectrode capacitances. These capacitances are due to formation of depletion layers at the junctions. The interelectrode capacitances are shown by dotted lines in Fig. 8.13. The capacitance C_{bc} between the base and collector connects the output with the input. Because of this, negative feedback takes place in the circuit and gain is reduced. The feedback effect increases with the increase in frequency because with the increase in frequency reactive impedance of the capacitor decreases. The capacitance C_{be}, between the base and emitter, offers a low impedance path at the input side at high frequencies. This reduces the input impedance of the device, and consequently effective input signal is reduced. Thus gain falls. Similarly, the capacitance C_{ce} produces a shunting effect at high frequencies on the output side. It is noteworthy point that C_{bc} is the most important capacitance because feedback takes place from output circuit to input circuit through this capacitance. This is known as *Miller effect*. Besides these capacitances, there are wiring capacitances C_{W_1} and C_{W_2}, as shown in Fig. 8.13. These are the capacitances between the connecting wires of the circuit and ground.

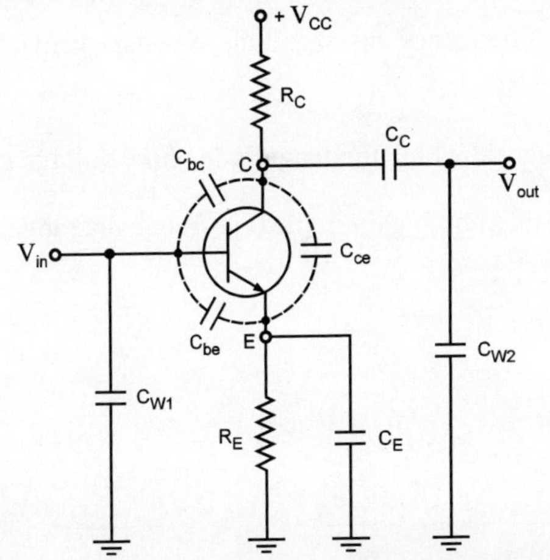

Fig. 8.13. ***R-C Coupled Amplifier at High Frequencies***

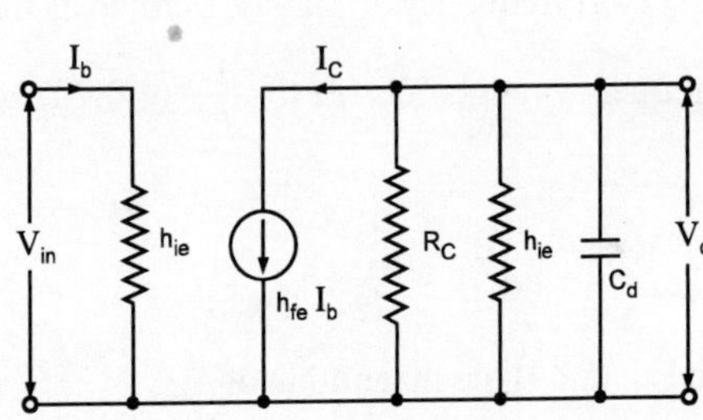

Fig. 8.14. ***Equivalent Circuit For High Frequency Range***

It can be shown that C_{be} and C_{bc} may be replaced with a single capacitor C_d across the input resistance h_{ie} of the transistor. The value of shunt capacitance C_d in the input circuit of the first stage is small because it depends on the output impedance of the first transistor, which is small. But in the output circuit of the first stage C_d is increased by stray capacitance of the wiring. The reactance $\frac{1}{\omega C_d}$ will have appreciable shunting effect on R_2 and h_{ie}. The equivalent circuits are given in Figs. 8.14 and 8.15.

From Thevenin's equivalent circuit shown in Fig. 8.15,

$$\text{Current, } I = \frac{h_{fe} I_b \frac{R_C h_{ie}}{R_C + h_{ie}}}{\frac{R_C h_{ie}}{R_C + h_{ie}} + \frac{1}{j\omega C_d}} = \frac{h_{fe} I_b R_C h_{ie}}{R_C h_{ie} + \frac{1}{j\omega C_d}(R_C + h_{ie})} \qquad ...(8.50)$$

$$\text{So current gain, } A_{ih} = \frac{I}{I_b} = \frac{h_{fe}\, h_{ie}\, R_C}{h_{ie} R_C + \frac{1}{j\omega C_d}(h_{ie} + R_C)} \qquad ...(8.51)$$

$$\text{Output voltage, } V_{out} = I \times \frac{1}{j\omega C_d}$$

$$= \frac{\frac{h_{fe} h_{ie} I_b R_C}{(h_{ie} + R_C)}}{\frac{h_{ie} R_C}{h_{ie} + R_C} + \frac{1}{j\omega C_d}} \times \frac{1}{j\omega C_d}$$

$$= \frac{h_{fe} h_{ie} I_b R_C}{j\omega C_d\, h_{ie} R_C + h_{ie} + R_C}$$

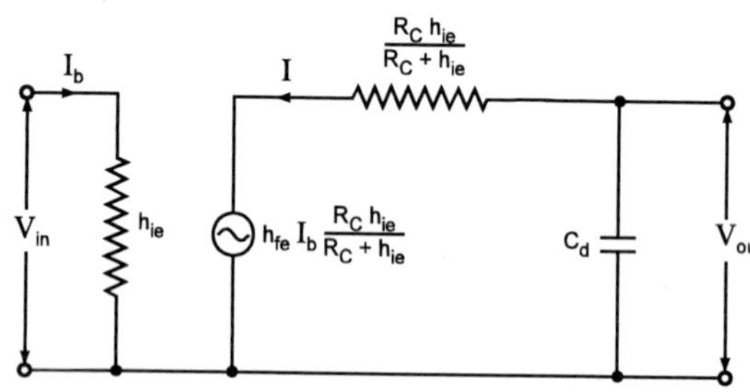

Fig. 8.15. ***Thevenin's Equivalent Circuit***

$$\text{Input voltage, } V_{in} = h_{ie} I_b$$

$$\therefore \text{ Voltage gain, } A_{vh} = \frac{V_{out}}{V_{in}} = \frac{h_{fe} R_C}{j\omega C_d\, h_{ie} R_C + h_{ie} + R_C} \qquad ...(8.52)$$

The magnitude of A_{vh} is given by

$$|A_{vh}| = \frac{h_{fe} R_C}{\sqrt{(h_{ie} + R_C)^2 + (2\pi f C_d h_{ie} R_C)^2}} \qquad ...(8.53)$$

From Eq. (8.53) it is obvious that with the increase in input frequency the magnitude of voltage gain falls off.

8.6.3. Lower Cutoff Frequency

The lower cutoff frequency, f_1 may be defined as the frequency at which the magnitude of the voltage gain in the low frequency range falls off to $\frac{1}{\sqrt{2}}$ or 0.707 times the maximum gain in the mid frequency range. Thus at $f = f_1$

$$|A_{vl}| = \frac{1}{\sqrt{2}} |A_{v_m}|$$

From Eq. (8.49) the magnitude of A_{vl},

$$|A_{vl}| = \frac{h_{fe} R_C}{\sqrt{(h_{ie} + R_C)^2 + \left(\frac{1}{2\pi f C_C}\right)^2}} = \frac{h_{fe} R_C}{(h_{ie} + R_C)\left[1 + \left\{\frac{1}{2\pi f C_C (h_{ie} + R_C)}\right\}^2\right]^{1/2}} \qquad ...(8.54)$$

and from Eq. (8.44), the magnitude of A_{vm},

$$|A_{vm}| = \frac{h_{fe}R_C}{h_{ie} + R_C} \quad ...(8.55)$$

From Eqs. (8.54) and (8.55)

$$\frac{A_{vl}}{A_{vm}} = \frac{1}{\sqrt{1+\left\{\frac{1}{2\pi f C_C (h_{ie} + R_C)}\right\}^2}} \quad ...(8.56)$$

If f_1 is the lower cutoff frequency, then

$$\frac{1}{\sqrt{2}} = \frac{1}{\sqrt{1+\left\{\frac{1}{2\pi f_1 C_C (h_{ie} + R_C)}\right\}^2}}$$

$$\text{or } 1+\left[\frac{1}{2\pi f_1 C_C (h_{ie} + R_C)}\right]^2 = 2$$

$$\text{or } f_1 = \frac{1}{2\pi C_C (h_{ie} + R_C)} \quad ...(8.57)$$

From Eqs. (8.56) and (8.57), we have

$$|A_{vl}| = \frac{|A_{vm}|}{\sqrt{1+\left(\frac{f_1}{f}\right)^2}} = \frac{|A_{vm}|}{\sqrt{1+(\tan\theta_l)^2}} \quad ...(8.58)$$

where $\tan\theta_l = \frac{f_1}{f}$ or $\theta_l = \text{Tan}^{-1}\frac{f_1}{f}$ ....(8.59)

So the total phase shift of the voltage gain A_{vl} in the low frequency range is given by

$$\theta_l = 180° + \theta_l = 180° + \tan^{-1}\frac{f_1}{f} \quad ...(8.60)$$

From above equation it is obvious that with the decrease in frequency θ_l increases and at $f = f_1$

$$\theta_l = 180° + 45° = 225°$$

8.6.4. Upper Cutoff Frequency

The upper cutoff frequency f_2 may be defined as the frequency at which the magnitude of the voltage gain in the high frequency range falls off to $\frac{1}{\sqrt{2}}$ or 0.707 times the magnitude of the gain in the mid frequency range. Thus at $f = f_2$

$$|A_{vh}| = \frac{1}{\sqrt{2}}|A_{vm}|$$

From Eq. (8.53), the magnitude of A_{vh},

$$|A_{vh}| = \frac{h_{fe}R_C}{(h_{ie} + R_C)\sqrt{1+\left\{\frac{2\pi f C_d\, h_{ie}R_C}{h_{ie} + R_C}\right\}^2}} \quad ...(8.61)$$

From Eqs. (8.55) and (8.61) we have

$$\frac{A_{vh}}{A_{vm}} = \frac{1}{\sqrt{1+\left\{\frac{2\pi f C_d h_{ie} R_C}{h_{ie}+R_C}\right\}^2}} \qquad ...(8.62)$$

If f_2 is the upper cutoff frequency then

$$\frac{1}{\sqrt{2}} = \frac{1}{\sqrt{1+\left\{\frac{2\pi f_2 C_d h_{ie} R_C}{h_{ie}+R_C}\right\}^2}}$$

or $$\frac{2\pi f_2 C_d h_{ie} R_C}{h_{ie}+R_C} = 1$$

or Upper cutoff frequency,

$$f_2 = \frac{h_{ie}+R_C}{2\pi C_d h_{ie} R_C} = \frac{1}{2\pi C_d}\left\{\frac{1}{h_{ie}}+\frac{1}{R_C}\right\} \qquad ...(8.63)$$

From Eq. (8.62)

$$\frac{|A_{vh}|}{|A_{vm}|} = \frac{1}{\sqrt{1+\left(\frac{f}{f_2}\right)^2}} = \frac{1}{\sqrt{1+(\tan\theta_h)^2}} \qquad ...(8.64)$$

$$\text{where } \theta_h = \text{Tan}^{-1}\frac{f}{f_2} \qquad ...(8.65)$$

$$\text{So } \phi_h = 180° - \theta_h = 180° - \tan^{-1}\frac{f}{f_2} \qquad ....(8.66)$$

$$\text{At } f = f_2,\ \phi_h = 180° - 45° = 135°$$

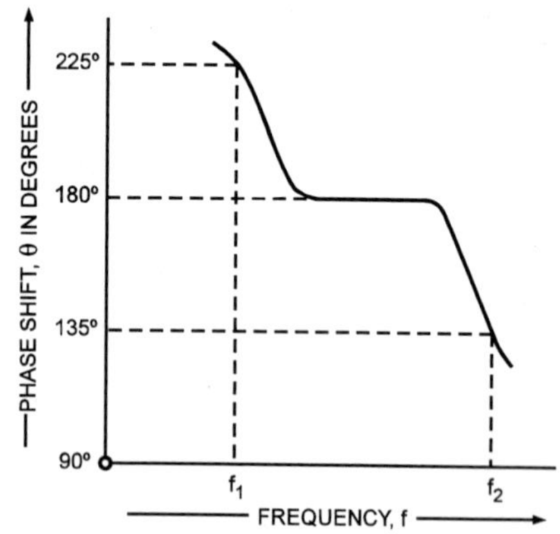

Fig. 8.16. ***Phase Relationship in an R-C Coupled Amplifier***

ϕ_h decreases with the increase in input frequency *f*. The variation of total phase shift between input and output voltages with frequency is shown in Fig. 8.16.

8.6.5. Frequency Response Curve

The frequency response curve of a typical R-C coupled transistor amplifier is shown in Fig. 8.17.

In *mid frequency range* (50 Hz – 20 kHz), the voltage gain of the amplifier is constant, as is obvious from the analysis. With the increase in the frequency in this range, the reactance of the coupling capacitor C_C reduces thereby increasing the gain but at the same time lower capacitive reactance causes higher loading resulting in lower voltage gain. Thus the two effects cancel each other and uniform gain is obtained in mid frequency range.

At low frequencies (below 50 Hz) higher capacitive reactance of coupling capacitor C_C allows very small part of signal to pass from one stage to the next and also because of higher reactance of emitter bypass capacitor C_E, the emitter resistor R_E is not effectively shunted. Thus the voltage gain falls off at low frequencies.

At high frequencies (exceeding 20 kHz), the gain of the amplifier decreases with the increase in frequency. Several factors are responsible for this reduction in gain. At high frequencies, the reactance of coupling capacitor C_C becomes very small and C_C behaves as a short circuit. This increases the loading of next stage and reduces the voltage gain. At high frequencies, capacitive reactance of base-emitter junction is low and so the base current is increased and current gain factor β is reduced. At high frequencies, the interelectrode capacitance C_{bc} connects the output circuit to the input circuit. Thus negative feedback takes place and the gain is reduced. Besides interelectrode capacitances, there are wiring capacitances C_{W_1}, and C_{W_2} and input capacitance C_{in} of the next stage. The effect of the capacitances C_{W_1}, C_{W_2} and input capacitor C_{in} of the next stage can be represented by a single shunt capacitance given as

$$C_s = C_{W_1} + C_{W_2} + C_{in} \quad ...(8.67)$$

The output section of the amplifier is shown in Fig. 8.18 from the ac point of view, for high frequency considerations. The capacitance C_s is the equivalent shunt capacitance as given by Eq. (8.67). In figure coupling and bypass capacitors C_C and C_E are not shown as they effectively represent short circuits at high frequencies.

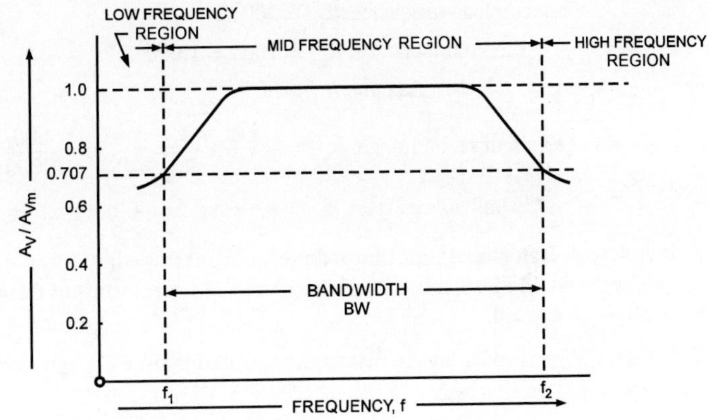

Fig. 8.17. ***Frequency Response Curve of an R-C Coupled Amplifier***

From circuit shown in Fig. 8.18, it is obvious that collector current I_c is made up of three currents I_1, I_2 and I_3. With the increase in frequency of the input signal, the reactive impedance of the shunt capacitance C_s decreases, being inversely proportional to frequency, and as a result the current I_2 through this capacitance C_s increases. This reduces both currents I_1 and I_3 since total current $I_c = (I_1 + I_2 + I_3)$ is almost constant. The output voltage V_{out} being the product of I_3 and R (*i.e.* I_3R) decreases. Thus higher the frequency, the lower is the impedance offered by C_s and lower will be the output voltage V_{out}.

8.6.6. Bandwidth

Frequency response curve of an R-C coupled amplifier is shown in Fig. 8.17. The cutoff frequencies f_1 and f_2 are marked in the figure. The difference of the two frequencies (upper cutoff frequency f_2 and lower cutoff frequency f_1) is called the *bandwidth* (BW)

$$\therefore \text{BW} = f_2 - f_1 \quad ...(8.68)$$

when the bandwidth (BW) is multiplied by the gain at mid frequencies, gain-bandwidth (GBW) is obtained.

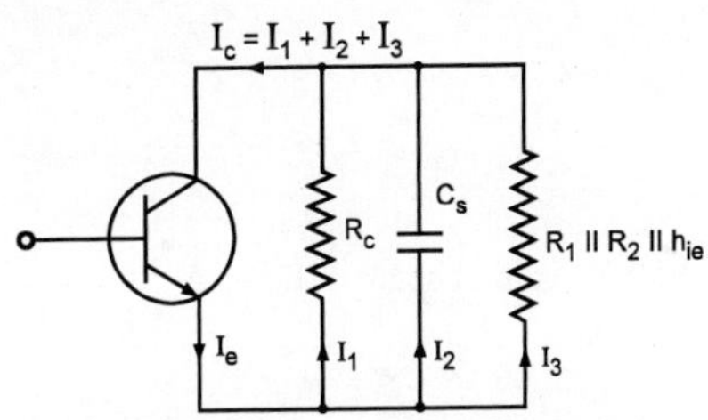

Fig. 8.18. ***Output Section of an R-C Coupled Amplifier at High Frequencies***

8.6.7. Advantages

Excellent frequency response (constant gain over the audio-frequency range, the most important region for speech, music etc.), cheaper in cost (resistors and capacitors used are very cheap) and very compact circuit (resistors and capacitors used being small and extremely light).

8.6.8. Disadvantages

Low voltage and power gain due to low resistance presented by the input of each stage to the preceding stage, tendency of becoming noisy with age, particularly in moist climates, and poor impedance matching due to difference in impedances of the R-C coupled amplifier output and that of the speaker.

8.6.9. Applications

The R-C coupled transistor amplifiers are widely used as voltage amplifiers (*i.e.* in the initial stages of public address systems) because of their excellent audio fidelity over a wide range of frequency. However, because of poor impedance matching, this type of coupling is rarely employed in the final stages.

Example 8.3. A single stage amplifier has a gain of 60. The collector load $R_C = 500\ \Omega$ and the input impedance is 1 kΩ, calculate the overall voltage gain when two stages are cascaded through R-C coupling.

Solution : Voltage gain of first stage, $A_{v_1} = -60$

Collector load resistance, $R_C = 500\ \Omega$

Input impedance, $R_{in} = 1\ k\Omega = 1{,}000\ \Omega$

Assume $h_{fe} = 50$

$$\text{Voltage gain of second stage, } A_{v_2} = -h_{fe} \times \frac{R_C}{R_{in}} = -50 \times \frac{500}{1{,}000} = -25$$

$$\text{Overall voltage gain, } A_v = A_{v_1} \times A_{v_2} = (-60) \times (-25) = 1{,}500 \text{ Ans.}$$

Example 8.4. Determine input impedance, output impedance and overall voltage gain for a 2-stage R-C coupled amplifier shown in Fig. 8.2. Load resistor connected across output terminals of the amplifier is of 10 kΩ. h_{ie} = 1.1 kΩ; h_{fe} = 50, h_{re} = h_{oe} = 0.

Solution : When looking into the base terminals of the transistor Q_1, h_{ie} is seen to be in series with $h_{re}V_{out}$. Since $h_{re} = 0$, so the input impedance to the transistor base,

$$Z_b = h_{ie} \qquad \text{...[Refer to Eq. (7.54)]}$$

(*i*) Input impedance to the circuit, $Z_{in} = R_{11} \,||\, R_{21} \,||\, Z_b$ ...[Refer to Eq. (7.56)]

$= 4 \,||\, 20 \,||\, 1.1\ k\Omega = 0.827\ k\Omega$ **Ans.**

$$\text{The output impedance of transistor } Q_2 = \frac{1}{h_{oe}} = \infty \qquad \text{...[Refer to Eq. (7.57)]}$$

(*ii*) Output impedance of the amplifier circuit,

$$Z_{out} = \frac{1}{h_{oe}} \,||\, R_{C_2} = R_{C_2} \qquad \text{...[Refer to Eq. (7.58)]}$$

$$= 2\ k\Omega \text{ Ans.} \qquad \because h_{oe} = 0$$

$$\text{Voltage gain of second stage, } A_{v_2} = \frac{-h_{fe}}{h_{ie}} (R_{C_2} \,||\, R_L) \qquad \text{...[Refer to Eq. (7.59)]}$$

$$= \frac{-50}{1.1 \times 10^3} \times (2\ k\Omega \,||\, 10\ k\Omega) = \frac{-50}{1.1 \times 10^3} \times 1.667 \times 10^3 = -75.76$$

$$\text{Effective load to the first stage, } R_{ac_1} = R_{C_1} \,||\, R_{22} \,||\, R_{12} \,||\, h_{ie} = 4 \,||\, 10 \,||\, 2 \,||\, 1.1\ k\Omega = 0.5685\ k\Omega$$

$$\text{So voltage gain of first stage, } A_{v_1} = \frac{-h_{fe}}{h_{ie}} R_{ac_1} = \frac{-50}{1.1 \times 10^3} \times 0.5685 \times 10^3 = -25.84$$

$$(iii) \text{ Overall voltage gain, } A_v = A_{v_1} \times A_{v_2} = (-25.84) \times (-75.76) = 1{,}958 \text{ Ans.}$$

Example 8.5. The parameters of the transistor in the circuit shown (Fig. 8.19) are h_{fe} = 50, h_{ie} = 1.1 kΩ, h_{re} = h_{oe} = 0. Find (*i*) midband gain (*ii*) the value of C_b necessary to give a lower 3-dB frequency of 20 Hz.

Solution : (*i*) Midband gain,

$$A_{v_m} = \frac{h_{fe} R_C}{h_{ie} + R_C} = \frac{50 \times 2}{1.1 + 2} = 32.26 \text{ Ans.}$$

...[Refer to Eq. (8.44)]

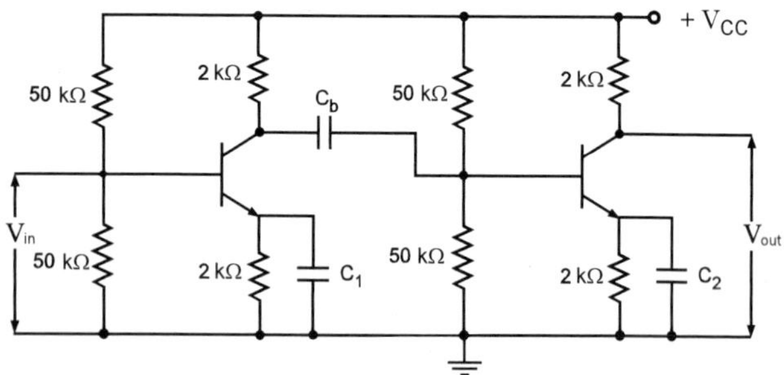

Fig. 8.19

(*ii*) Lower cutoff frequency, $f_1 = \dfrac{1}{2\pi C_b\,(h_{ie} + R_C)}$ ...[Refer to Eq. (8.57)]

$$\text{or } 20 = \frac{1}{2\pi C_b\,(1.1 \times 10^3 + 2 \times 10^3)}$$

$$\text{or } C_b = \frac{1}{2\pi \times 20 \times 3.1 \times 10^3} = 2.567\ \mu\text{F Ans.}$$

Example 8.6. In an amplifier the maximum voltage gain is 1,500 which occurs at 1 kHz. It falls to 1,060.5 at 20 Hz and 20 kHz. Determine (*i*) lower cutoff frequency (*ii*) upper cutoff frequency and (*iii*) bandwidth.

Solution : The maximum gain is 1,500 and 70.7% of this gain is 1,060.5. It is given that gain is 1,060.5 (*i.e.* 70.7% of maximum gain) at frequencies of 20 Hz and 20 kHz, so

(*i*) the lower cutoff frequency, defined as the frequency on lower side at which the voltage gain of the amplifier is exactly 70.7% of the maximum gain, f_1 is 20 Hz **Ans.**

(*ii*) the upper cutoff frequency, defined as the frequency on upper side at which the voltage gain of the amplifier is exactly 70.7% of the maximum gain, f_2 is 20 kHz **Ans.**

(*iii*) Bandwidth is f_1 to f_2 *i.e.* from 20 Hz to 20 kHz **Ans.**

8.7. CASCODE OR CE-CB CONFIGURATION

The CE-CB configuration (usually referred to as *cascode amplifier*) is shown in Fig. 8.20. Figure 8.20 shows a cascode configuration with a common-emitter (CE) stage feeding a common-base (CB) stage. This configuration of course, has basically the input characteristics similar to those of CE amplifier and output characteristics similar to those of CB amplifier. More specifically, it has high output resistance and is inherently more stable. The high output resistance is useful in achieving large voltage gain. This arrangement is designed to provide a high input impedance with low voltage gain to ensure that the input Miller capacitance is at a minimum with the CB stage providing good high-frequency operation.

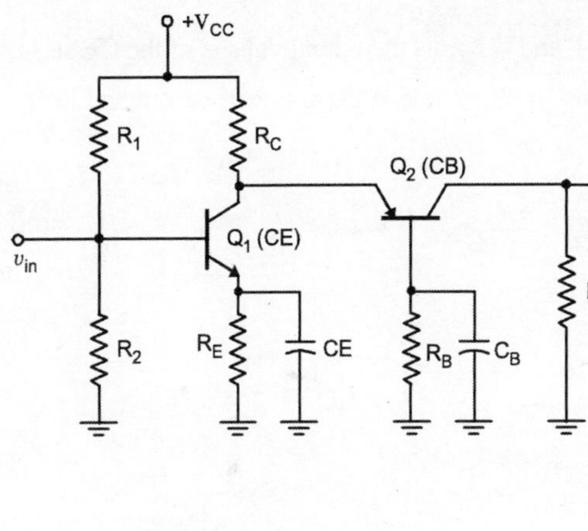

Fig. 8.20. *Cascode Amplifier*

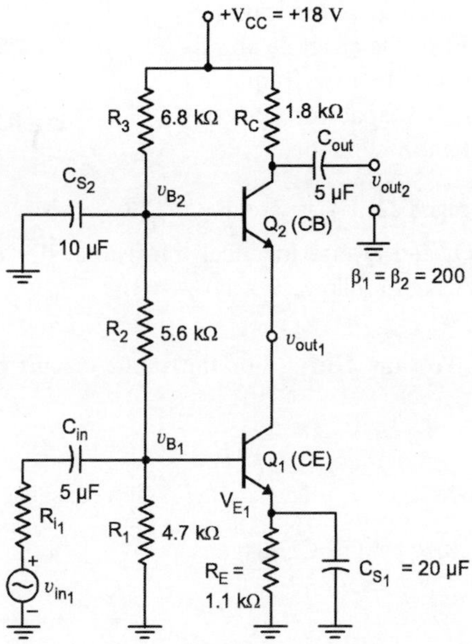

Fig. 8.21. *Practical Cascode Amplifier*

A practical version of a cascode amplifier is given in Fig. 8.21.

8.7.1. DC Analysis

In a cascode amplifier depicted in Fig. 8.21 R_1, R_2 and R_3 form a bias network for transistors Q_1 and Q_2; C_{S_1} and C_{S_2} provide ac signal ground paths from the Q_1 emitter and Q_2 base, respectively; C_{in} is the coupling capacitor. For the proper operation of the circuit it is necessary that transistors Q_1 and Q_2 must be identical.

For dc conditions all capacitors are assumed to be open circuited.

Applying voltage divider rule, we have

$$V_{B_1} = V_{CC} \times \frac{R_1}{(R_1 + R_2 + R_3)} \quad ...(8.69)$$

$$\text{and } V_{E_1} = V_{B_1} - V_{BE_1} = \frac{V_{CC} R_1}{R_1 + R_2 + R_3} - V_{BE_1} \quad ...(8.70)$$

$$\text{Emitter current of transistor } Q_1,\ I_{E_1} = \frac{V_{E_1}}{R_E} = \frac{\dfrac{V_{CC} R_1}{(R_1 + R_2 + R_3)} - V_{BE_1}}{R_E} \quad ...(8.71)$$

Once the emitter current of transistor Q_1 is known, all other currents in and voltages at the three terminals for Q_1 and Q_2 can be readily determined if required.

8.7.2. AC Analysis

Figure 8.22 depicts the small-signal T-equivalent circuit for the cascode amplifier given in Fig. 8.21. Here we will determine low-frequency, small-signal properties (like voltage gain, current gain and input resistance) of the cascode amplifier circuit. In low-frequency analysis, all capacitors are considered short elements.

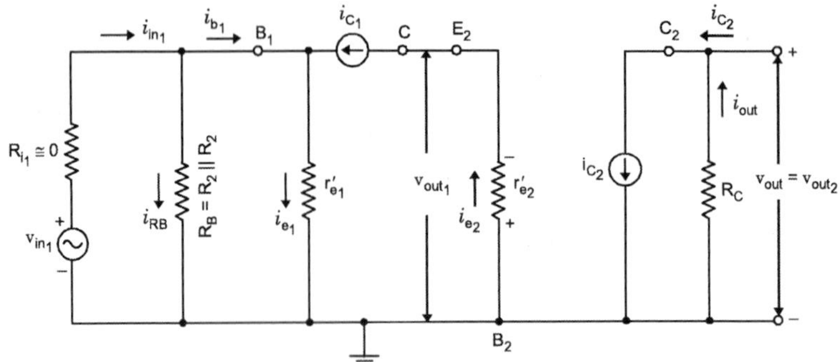

Fig. 8.22. *Small Signal T-Equivalent Circuit of Cascode Amplifier*

In Fig. 8.22, $R_B = R_1 \parallel R_2$; $r'_{e_1} \cong r'_{e_2}$ because $I_{E_1} \cong I_{E_2}$; and v_{out_1} is the output voltage of the CE stage. Since Q_1 and Q_2 are identical transistors, $\beta_{ac_1} = \beta_{ac_2}$ and $\beta_{dc_1} = \beta_{dc_2}$ and subscripts will be omitted in the calculations that follow.

1. Voltage Gain. For the input circuit given in Fig. 8.22, since $R_B >> r'_{e_1}$, $v_{in_1} = r'_{e_1} i_{e_1}$ and $v_{out_1} = -r'_{e_2} i_{e_2}$

Since, $i_{e_2} = i_{C_1}$ and $i_{C_1} \cong i_{e_1}$, it means that $i_{e_1} \cong i_{e_2}$.

Voltage gain of CE stage is

$$A_{v_1} = \frac{v_{out_1}}{v_{in_1}} = \frac{-r'_{e_2} i_{e_2}}{r'_{e_1} i_{e_1}} = -1 \quad ...(8.72) \quad \because r'_{e_1} = r'_{e_2}$$

$$\text{The output voltage, } v_{out} = v_{out_2} = -R_C\, i_{out} \quad ...(8.73)$$

$$i_{out} = i_{C_2} = i_{e_2} \quad \text{and} \quad i_{e_2} = \frac{-v_{out_1}}{r'_{e_2}}$$

Substituting the value of i_{e_2} from above equation into Eq. (8.73), we have

$$v_{out} = -R_C \times \frac{-v_{out_1}}{r'_{e_2}}$$

Then the voltage gain of CB stage is

$$A_{v_2} = \frac{v_{out}}{v_{out_1}} = \frac{R_C}{r'_{e_2}} \quad \text{...(8.74)}$$

Thus overall voltage gain of the cascode (CE-CB) configuration

$$A_v = \frac{v_{out}}{v_{in_1}} = \frac{v_{out_1}}{v_{in_1}} \times \frac{v_{out}}{v_{out_1}}$$

$$= A_{v_1} \times A_{v_2} = -1 \times \frac{R_C}{r'_{e_2}} = \frac{-R_C}{r'_{e_2}} \quad \text{...(8.75)}$$

2. Current Gain. The current gain A_i of the cascode (CE-CB) configuration may be expressed as

$$A_i = \frac{i_{out}}{i_{in_1}} = \frac{i_{b_1}}{i_{in_1}} \times \frac{i_{e_1}}{i_{b_1}} \times \frac{i_{e_2}}{i_{e_1}} \times \frac{i_{C_2}}{i_{e_2}} \times \frac{i_{out}}{i_{C_2}} = \frac{i_{b_1}}{i_{in_1}} \times \frac{i_{e_1}}{i_{b_1}} \quad \text{...(8.76)} \quad \because i_{e_1} = i_{C_1} = i_{e_2} = i_{C_2} = i_{out}$$

Applying the current divider rule to the input circuit, we have

$$i_{b_1} = \frac{R_B\, i_{in_1}}{R_B + \beta_{ac} r'_{e_1}} \quad \text{...(8.77) where } R_B = R_1 \,||\, R_2$$

$$\text{Also,} \quad i_{e_1} = \beta_{ac} i_{b_1}$$

Thus, the current gain Eq. (8.76) becomes

$$A_i = \frac{\beta_{ac} R_B}{R_B + \beta_{ac} r'_{e_1}} \quad \text{...(8.78)}$$

3. Input Resistance. For input circuit,

$$v_{in_1} = r'_{e_1} i_{e_1} = \beta_{ac} r'_{e_1} i_{b_1} \quad \because i_{e_1} = \beta_{ac} i_{b_1}$$

Substituting the value of i_{b_1} from Eq. (8.77) into above equation, we have

$$v_{in_1} = \beta_{ac} r'_{e_1} \times \frac{R_B\, i_{in_1}}{R_B + \beta_{ac} r'_{e_1}}$$

$$\text{and Input resistance,} \quad R_{in} = \frac{v_{in_1}}{i_{in_1}} = \frac{\beta_{ac} r'_{e_1} R_B}{R_B + \beta_{ac} r'_{e_1}} \quad \text{...(8.79)}$$

The cascode amplifier shown in Fig. 8.20 is not encountered very often in discrete circuitry because (*i*) it needs a large number of resistors and capacitors for providing bias and preventing unwanted feedback and (*ii*) it does not offer much flexibility as the IC cascode amplifier does.

8.8. CE-CC CONFIGURATION

In case the output of CE stage is to be used with a relatively low load resistance, then a CC stage should be used as a buffer amplifier because CC stage is used to drive low-resistance loads. The CC stage can be directly coupled to the CE stage as shown in Fig. 8.23. The collector resistance of transistor Q_1 also serves for biasing the base of transistor Q_2.

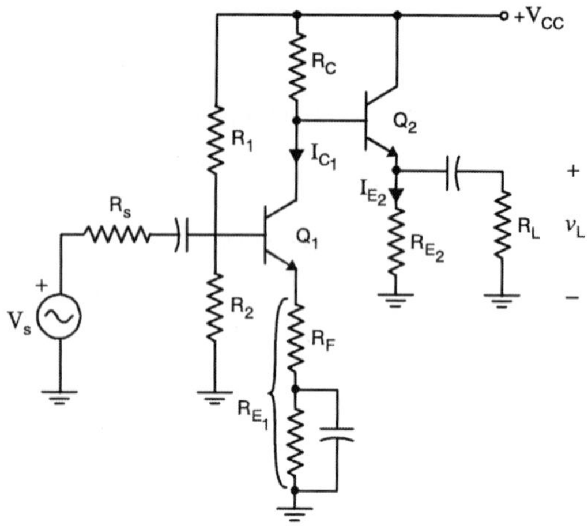

Fig. 8.23. *CE-CC Configuration*

DC Analysis : Collector current of transistor Q_1,

$$I_{C_1} = \frac{V_{Th} - V_{BE}}{R_{E_1} + \dfrac{R_B}{\beta_1}}$$

where $V_{Th} = V_{CC} \times \dfrac{R_2}{R_1 + R_2}$

and $R_B = R_1 \,||\, R_2 = \dfrac{R_1 R_2}{R_1 + R_2}$

The emitter current of transistor Q_2,

$$I_{E_2} = \frac{V_{B_2} - V_{B_E}}{R_{E_2}} = \frac{V_{CC} - (I_{C_1} + I_{B_2})R_C - V_{BE}}{R_{E_2}}$$

Substituting $I_{B_2} = \dfrac{I_{E_2}}{\beta_2 + 1}$ and reducing, we have

$$I_{E_2} = \frac{V_{CC} - V_{BE} - I_{C_1} R_C}{R_{E_2} + \dfrac{R_C}{\beta_2}} \qquad ...(8.80)$$

AC Analysis : AC equivalent circuit is shown in Fig. 8.24.

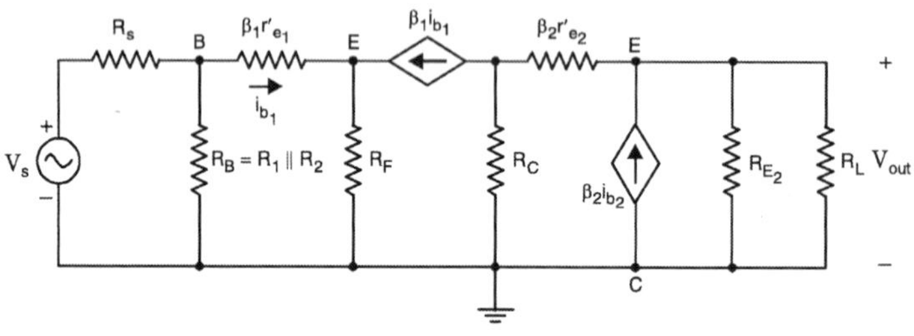

Fig. 8.24. *AC Equivalent Circuit of CE-CC Amplifier*

The overall voltage gain v_{out}/v_s may be expressed as the product of the in-circuit gains of both stages, that is, considering the input resistance of the CC stage as part of the load resistance of the CE stage. Overall voltage gain is obtained by the following expression.

$$\frac{v_{out}}{v_s} = \frac{-r_{L_1}}{\dfrac{R_s}{\beta_1} + (r'_{e_1} + R_F)\left(1 + \dfrac{R_s}{R_B}\right)} \times \frac{r_{L_2}}{r'_{e_2} + r_{L_2}} \qquad ...(8.81)$$

where $r_{L_1} = R_C \,||\, [\beta_2 (r'_{e_2} + r_{L_2})]$ and $r_{L_2} = R_{E_2} \,||\, R_L$.

8.9. BANDPASS OF CASCADED STAGES

Consider an n-stage cascaded amplifier having noninteracting* stages having upper 3-dB frequencies of f'_2, f''_2, f'''_2 and so on. Let $f_2^{(n)}$ be the upper 3-dB frequency for the complete cascaded amplifier. If all stages are considered to be identical, *i.e.* $f'_2 = f''_2 = f'''_2 = ... = f_2$, then $f_2^{(n)}$ can be determined from the following equation

$$\left[1/\sqrt{1+(f_2^{(n)}/f_2)^2}\right]^n = \frac{1}{\sqrt{2}}$$

$$\text{or} \quad \frac{f_2^{(n)}}{f_2} = \sqrt{2^{1/n}-1} \quad ...(8.82)$$

For $n = 2$ and for $n = 3$ we have $f_2^{(2)} = 0.64 f_2$ and $f_2^{(3)} = 0.51 f_2$ respectively. For example if $f_2 = 100$ kHz, then a two-stage amplifier has $f_2^{(2)} = 64$ kHz while three stage amplifier has $f_2^{(3)} = 51$ kHz.

If f_1 be the lower 3-dB frequency of each of the n identical noninteracting cascaded stages, then

$$\frac{f_1^{(n)}}{f_1} = \frac{1}{\sqrt{2^{1/n}-1}} \quad ...(8.83)$$

Thus if lower cutoff frequency $f_1 = 10$ Hz then two stage amplifier will have $f_1^{(2)} = \frac{10}{0.64} = 15.625$ Hz and three stage amplifier will have $f_1^{(3)} = \frac{10}{0.51} = 19.6$ Hz.

Thus it can be concluded that a cascaded amplifier has a lower f_2 and a higher f_1 than a single stage *i.e.* 3-dB bandwidth is reduced.

Example 8.7. Three identical cascaded stages have an overall paper 3-dB frequency of 100 kHz and a lower 3-dB frequency of 25 kHz. Determine f_L and f_H for each stage. Assume noninteracting stages.

Solution : Overall upper 3-dB frequency of 3-stage cascade amplifier.

$$f_2^{(3)} = 100 \text{ kHz}$$

Upper 3-dB frequency of each stage.

$$f_H = \frac{f_2^{(3)}}{\sqrt{(2^{1/3}-1)}} = \frac{f_2^{(3)}}{0.51} = \frac{100}{0.51} \text{ kHz} = 195.15 \text{ kHz} \quad \textbf{Ans.}$$

Overall lower 3-dB frequency of 3-stage cascade amplifier

$$f_1^{(3)} = 25 \text{ kHz}$$

Lower 3-dB frequency of each stage, $f_L = f_1^{(3)} \times \frac{1}{\sqrt{2^{1/3}-1}} = 25 \times 0.51 = 12.75$ kHz. **Ans.**

* In a cascaded amplifier, if the input impedance of, say kth stage, R_{ink} is low, then this R_{ink} being in parallel with the output impedance of the preceding stage [*i.e.* (k – 1)th stage] substantially influences the operation of the (k – 1)th stage. These two stages are thus not isolated and said to be interacting. In such cases, 3-dB frequencies of individual stage cannot be defined.

8.10. DARLINGTON AMPLIFIER

Darlington amplifier with voltage divider bias is shown in Fig. 8.25. In this circuit the output of one amplifier is coupled into the input of the next one by directly joining emitter of one transistor to the base of the other one, as shown in the figure. Obviously, the emitter current of the first transistor becomes the base current for the second transistor.

The biasing analysis is similar to that for one transistor except that two V_{BE} drops are to be considered.

The *Darlington* amplifier circuit in its small signal ac form will appear, as shown in Fig. 8.26.

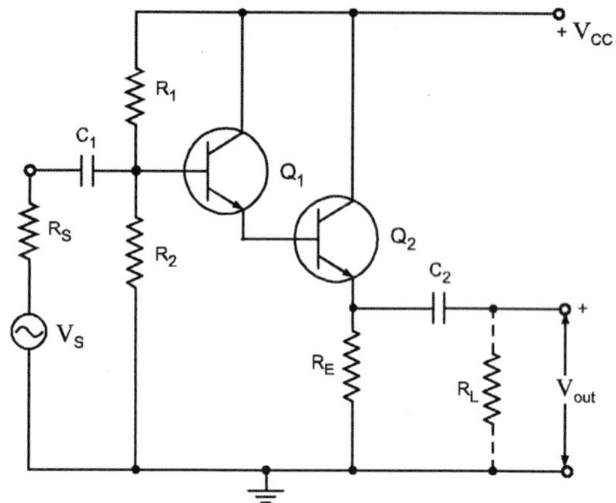

Fig. 8.25. ***Darlington Amplifier***

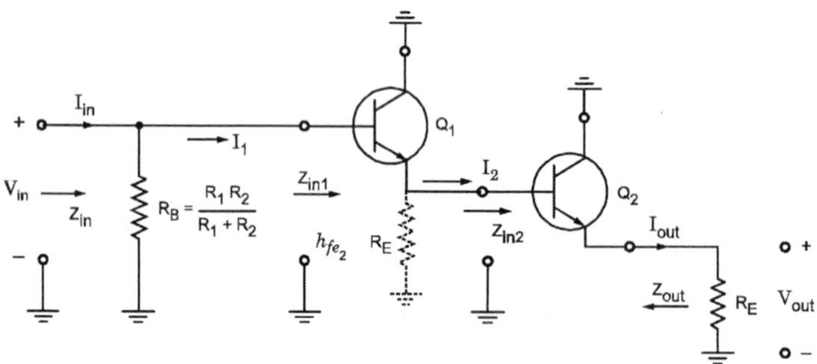

Fig. 8.26. ***Darlington Amplifier Circuit of Fig. 8.25 Redrawn For Determination of Small Signal Response***

8.10.1. Main Characteristics

1. Current Gain. For the second stage :

$$\text{Input impedance, } Z_{in_2} \cong h_{fe_2} R_E$$

$$\text{and current gain, } A_{i_2} = \frac{I_{out}}{I_2} = \frac{I_{e_2}}{I_{b_2}} \simeq h_{fe_2}$$

On a good approximate basis, these equations cannot be applied to the first stage. The "fly in the ointment" is the closeness with which Z_{in_2} compares with $\frac{1}{h_{oe_1}}$. We have seen that $\frac{1}{h_{oe}}$, could be eliminated in the majority of the cases because the load impedance $Z_L << \frac{1}{h_{oe}}$. For the Darlington circuit the input impedance Z_{in_2} is close enough in magnitude to $\frac{1}{h_{oe}}$ to necessitate considering the effects of h_{oe_1}. In chapter 7 it was found that for single stage grounded emitter transistor amplifier, where $\frac{1}{h_{oe}}$ was considered,

$$A_i = \frac{h_{fe}}{1 + h_{oe} Z_L}$$

Applying the above equation to this situation,

$$Z_L = Z_{in2} \cong h_{fe2} R_E \text{ and}$$

$$A_{i_1} = \frac{I_2}{I_1} = \frac{I_{e_1}}{I_{b_1}} \cong \frac{h_{fe_1}}{1 + h_{oe_1}(h_{fe_2} R_E)}$$

$$\text{and current gain, } A_i = \frac{I_{out}}{I_1} = A_{i_1} A_{i_2} = \frac{h_{fe_1} h_{fe_2}}{1 + h_{oe_1}(h_{fe_2} R_E)} \quad ...(8.84)$$

For $h_{fe_1} = h_{fe_2} = h_{fe}$ and $h_{oe_1} = h_{oe}$

$$\text{Current gain, } A_i \cong \frac{h_{fe}^2}{1 + h_{oe} h_{fe} R_E} \quad ...(8.85)$$

For $h_{oe}h_{fe}R_E \leq 0.1$, a fairly good approximation (within 10%) is

$$A_i \cong h_{fe}^2 = \beta^2 \quad ...(8.86)$$

2. Input Impedance. Since $Z_{in_2} \cong h_{fe2} R_E$ is the emitter resistance of the first stage, the input impedance to the first stage is

$Z_{in_1} \cong h_{fe_1}\left(Z_{in_2} \| \frac{1}{h_{oe_1}}\right)$ since $Z_{in_2} = h_{fe_2} R_E$, and $\frac{1}{h_{oe_1}}$ will appear in parallel in the small signal equaivalent circuit. So

$$Z_{in_1} = h_{fe_1}\left(h_{fe_2} R_E \| \frac{1}{h_{oe_1}}\right) = \frac{h_{fe_1} h_{fe_2} R_E / h_{oe_1}}{h_{fe_2} R_E + \frac{1}{h_{oe_1}}} = \frac{h_{fe_1} h_{fe_2} R_E}{h_{oe_1} h_{fe_2} R_E + 1} \quad ...(8.87)$$

If $h_{fe_1} = h_{fe_2} = h_{fe}$ and $h_{oe_1} = h_{oe_2} = h_{oe}$ then

$$\text{Input impedance, } Z_{in_1} = \frac{h_{fe}^2 R_E}{1 + h_{oe} h_{fe} R_E} \quad ...(8.88)$$

$$\text{For } h_{oe} h_{fe} R_E \leq 0.1 \;\; Z_{in_1} \cong h_{fe}^2 R_E = \beta^2 R_E \quad ...(8.89)$$

3. Output Impedance. The output impedance Z_{out} can be determined directly from the emitter equivalent circuits as follows :

$$Z_{out_1} = \frac{R_S + h_{ie_1}}{h_{fe_1}} \quad ...(8.90)$$

$$\text{and } Z_{out_2} = \frac{Z_{out_1} \| \frac{1}{h_{oe_1}} + h_{ie_2}}{h_{fe_2}} \quad ...(8.91)$$

Z_{out_2} is quiet smaller than Z_{out_1} *i.e.*, output impedance is lowered.

Because of this, Darlington amplifier can be used to isolate high impedance source from low impedance load. If high impedance source is directly connected to a low impedance load, most of the signal voltage will be dropped across the high impedance of the source and remaining source signal available may not be able to drive the load.

4. Voltage Gain. Applying Kirchhoff's voltage law to the circuit shown in Fig. 8.25

$$V_{out} = V_{in} - V_{be_1} - V_{be_2}$$

That the output potential is the input potential less the base-to-emitter potential of each transistor. It clearly indicates that $V_{out} < V_{in}$. Voltage gain is closer in magnitude to one than to zero.

On an approximate basis it is given as

$$A_v \cong \frac{1}{1+\dfrac{h_{ie_2}}{h_{fe_2} R_E}} \qquad ...(8.92)$$

8.10.2. Merit

1. It uses very few components and can be readily formed from two adjacent transistors in an IC.
2. It provides excellent characteristics of high input impedance with low output impedance and high current gain, all desirable characteristics for a current gain amplifier.

8.10.3. Applications

It has enormous impedance transformation capability i.e. it can transform a low impedance load to a high impedance load. Hence it is used in a high gain operational amplifier which depends on very high input impedance for its operation as an integrator or summing amplifier in analogue applications.

8.10.4. Darlington Pair

A very popular connection of two BJTs for operation as one "superbeta" transistor is the Darlington pair. In practice, the two transistors are put inside a single package and three terminals E, B and C are brought out, as depicted in Fig. 8.27. The main feature of the Darlington pair (or connection) is that the composite transistor acts as a single unit that has a high current gain (product of the current gains of the individual transistors) and high input impedance. The current gain provided by the Darlington pair is of the order of a few thousand.

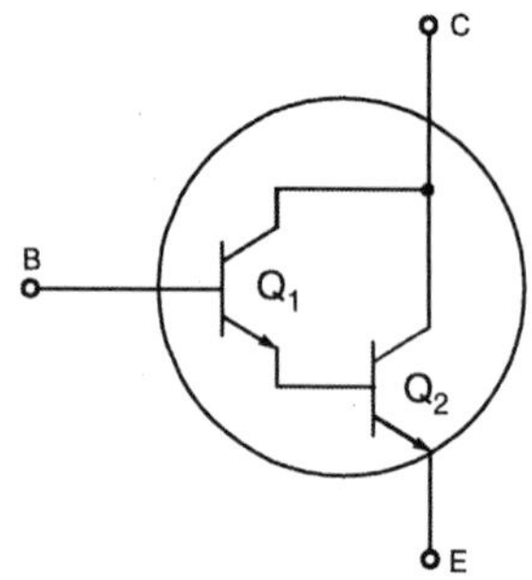

Fig. 8.27. ***Darlington Pair***

HIGHLIGHTS

1. A transistor circuit containing more than one stage of amplification is known as a multistage amplifier. A practical amplifier is always a multistage amplifier.

   In a multistage amplifier, the output of first stage is combined to the input of next stage through a coupling device. The process is known as *cascading*. The coupling device is used to (*i*) transfer the ac output of one stage to the input of the next stage and (*ii*) block the dc to pass from one stage to the next stage *i.e.* to isolate the dc conditions.

   Amplifier is usually named after the type of coupling used such as R-C coupled amplifier, transformer coupled amplifier and direct coupled amplifier.

2. The gain of a multistage amplifier is equal to the product of gains of individual stages. It is worthwhile to mention here that in practice total gain A is less than $A_{v_1} \times A_{v_2} \times ... \times A_{v_{n-1}} \times A_{v_n}$ due to the loading effects of the following stages.
3. For most applications, the BJT amplifier may be considered linear, permitting the use of the superposition theorem for separating the dc and ac analysis and designs.
4. R-C coupled transistor amplifiers have the advantages of excellent frequency response (constant gain over the audio-frequency range, the most important region for speech, music etc.), cheaper in cost, and very compact circuit.
5. R-C coupled transistor amplifiers have the disadvantages of low voltage and power gains, tendency of becoming noisy with age, particularly in moist climates, and poor impedance matching due to difference in impedances of R-C coupled amplifier and that of the speaker.

6. The R-C coupled transistor amplifiers are widely used as voltage amplifiers (*i.e.* in the initial stages of public address systems) because of their excellent audio fidelity over a wide range of frequency. However, because of poor impedance matching, this type of coupling is rarely employed in the final stages.
7. Cascode amplifier is a CE-CB configuration and has basically the input characteristics similar to those of CE amplifier and output characteristics similar to those of CB amplifier. More specifically, it has high output resistance and is inherently more stable. The high output resistance is useful in achieving large voltage gain.
8. CE-CC configuration is used when the amplifier is to be used to drive a low-resistance load.
9. The overall low frequency response $f_1^{(n)}$ for n cascaded stages each having the same value of f_1 is given by

$$f_1^{(n)} = \frac{f_1}{\sqrt{2^{1/n} - 1}}$$

which shows that the low frequency response is poorer than for a single stage.
10. The main feature of Darlington amplifier is that the composite transistor acts as a single unit with a current gain that is the product of the current gains of the individual transistors. Darlington amplifier provides a very large current gain, typically a few thousand.
11. Darlington pair may be considered a device providing externally only three terminals (base, emitter and collector) and having a very high current gain and high input resistance.
12. The feedback pair connection is a two transistor circuit that operates like the Darlington connection and provides a very high current gain (the product of the transistor current gains). Feedback makes use of a PNP transistor driving an NPN transistor.

EXERCISES

1. Explain the need for a multistage amplifier. **[G.G.S.I.P. Univ. Analog Electronics-I, December 2004]**
2. What are multistage amplifiers ? Derive the equations for overall voltage gain of a multistage amplifier in terms of individual voltage gains. **[G.G.S.I.P. Univ. Analog Electronics, December 2009]**
3. Explain in qualitative terms the reasons for fall in the voltage gain of the R-C coupled amplifier at high and low frequencies. Use hybrid π-model for the BJT for your explanation. **[G.G.S.I.P. Univ. Analog Electronics-I, December 2004]**
4. Draw the circuit of an R-C coupled amplifier and explain its frequency response. **[G.G.S.I.P. Univ. Analog Electronics, December 2009]**
5. Find the voltage gain of R-C coupled amplifier in mid-frequency region. Draw neat circuit diagram of such amplifier and give labelled frequency response. **[G.G.S.I.P. Univ. Analog Electronics, December 2010]**
6. Draw the circuit of an R-C coupled amplifier. Explain its frequency response characteristics. **[G.G.S.I.P. Univ. Analog Electronics, December 2010]**
7. Draw frequency response of an RC coupled amplifier and give expression for gain in different frequency regions. Why gain falls at low and high frequencies ? **[G.G.S.I.P. Univ. Analog Electronics, December 2010]**
8. Explain transistor R-C coupled amplifier with special reference to frequency response, advantages, disadvantages and applications.
9. Draw the circuit of a cascode amplifier and explain its operation with its uses.
10. Draw the cascode of a common-emitter stage and common-collector stage and give the expression for input and output resistance of the cascade. **[G.G.S.I.P. Univ. Analog Electronics, December 2010]**
11. Show that cascaded common emitter amplifier stages operating at high frequency have a reduced overall bandwidth given by

$$f_2^{(n)} = f_2\sqrt{2^{1/n} - 1}$$

where n is the number of stages and f_2 is the bandwidth of each stage,
12. Why do you cascade the amplifiers ? Derive the expression for higher 3-dB frequency f_H of an identical noninteracting stages in terms of f_H of one stage.

13. Draw a Darlington emitter follower. Explain why the input impedance is higher than that of a single stage emitter follower.
14. Derive an expression for current gain in Darlington amplifier. **[G.G.S.I.P. Univ. Analog Electronics, December 2010]**

SHORT ANSWER TYPE QUESTIONS WITH ANSWERS

Q. 1. Why do you cascade the amplifiers ?

Ans. The voltage/power gain or frequency response obtainable from a single stage amplifier is usually not sufficient to meet the needs of either a composite electronic circuit or load device, so cascading of amplifiers is required to provide greater voltage or current amplification or both.

Q. 2. What is a multistage amplifier circuit ?

Ans. A transistor circuit containing two or more stages of amplification is known as a multistage amplifier.

Q. 3. List the types of couplings.

Ans. The various coupling schemes used in cascaded amplifiers are
(*i*) R-C coupling (*ii*) transformer coupling (*iii*) impedance coupling and (*iv*) direct coupling.

Q. 4. What are the different types of configurations used in multistage amplifier circuits ?

Ans. The most suitable transistor configuration for cascading is CE configuration because the voltage gain of common emitter amplifier is greater than unity while CC configuration has voltage gain less than unity and the voltage gain of CB configuration using cascading is also less than unity.

However, for input stage CC or CB configuration may be required for proper impedance matching at the cost of voltage or current gain. It is noteworthy point that for input stage, the consideration is not the maximum voltage gain but the impedance matching of the source with the input impedance of the input stage. Some driving sources may need input circuit to be an almost open circuit while others need an almost short circuit. In certain cases choice of configuration for the input stage, is the minimization of noise and maximization of signal/noise power ratio.

Q. 5. Why R-C coupling is the most widely used coupling between the two stages of a cascaded amplifier ?

Ans. R-C coupling is the most commonly used coupling between the two stages of a cascaded or multistage amplifier because it is cheaper in cost, very compact circuit and provides excellent frequency response.

Q. 6. Why the overall gain of a multistage amplifier is less than the product of gains of individual stages ?

Ans. The overall gain of a multistage amplifier is less than the product of gains of individual stages because of the loading effects of the following stages.

Q. 7. Why coupling capacitor provided in a self biased CE R-C coupled amplifier is also called the blocking capacitor ?

Ans. In a self biased R-C coupled CE amplifier, coupling capacitor transmits ac signal but blocks the dc voltage of the first stage from reaching the base of the next stage, so it is also called the blocking capacitor.

Q. 8. Why does R-C coupling give constant gain over mid frequency range ?

Ans. In mid frequency range, the voltage gain of an R-C coupled amplifier remains almost constant. This is because in mid frequency range, with the increase in frequency, the reactance of the coupling capacitor decreases thereby increasing the gain but at the same time lower capacitive reactance causes higher loading resulting in lower voltage gain. Thus the two effects neutralize each other and uniform gain is obtained in mid frequency range.

Q. 9. Why are R-C coupled amplifiers widely used as voltage amplifiers ?

Ans. R-C coupled amplifiers are widely used as voltage amplifiers because of their excellent audio fidelity over a wide range of frequency.

Q. 10. How do you improve the gain of an R-C coupled amplifier at low frequencies ?

Ans. The voltage gain of an R-C coupled amplifier at low frequencies can be improved by (*i*) selecting very large coupling and bypass capacitors so that their reactance is small (*ii*) making $(R_S + R_B) << (R_C + R_L)$ and (*iii*) selecting transistor with small h_{ie}.

Q. 11. What are the main characteristics of cascode amplifier ?

Ans. The main characteristics of cascode amplifier are high input impedance, the same voltage gain as that of a CE amplifier but very low input capacitance so as to provide good high-frequency operation.

Q. 12. What are the main characteristics of a Darlington amplifier ?

Ans. The main characteristics of a Darlington amplifier are high input impedance, low output impedance and high current gain.

Q. 13. What is Darlington pair ?

Ans. Darlington pair may be considered a device providing externally only three terminals (base, emitter and collector) and having a very high current gain and high input resistance.

Q. 14. What is feedback pair ?

Ans. The feedback pair connection is a two transistor circuit that uses a PNP transistor driving an NPN transistor and operates like the Darlington connection and provides a very high current gain.

PROBLEMS

1. A multistage amplifier consists of three stages. The voltage gain of the stages are 30, 50 and 80. Calculate the overall gain in dB. [**Ans.** 101.58 dB]
2. The output power of an amplifier is 100 mW when the signal frequency is 5 kHz. When the frequency is increased to 25 kHz, the output falls to 50 mW. Calculate the dB change in power. [**Ans.** –3 dB]
3. For the two-stage R-C coupled low level audio amplifier shown in Fig. 1, compute the following :

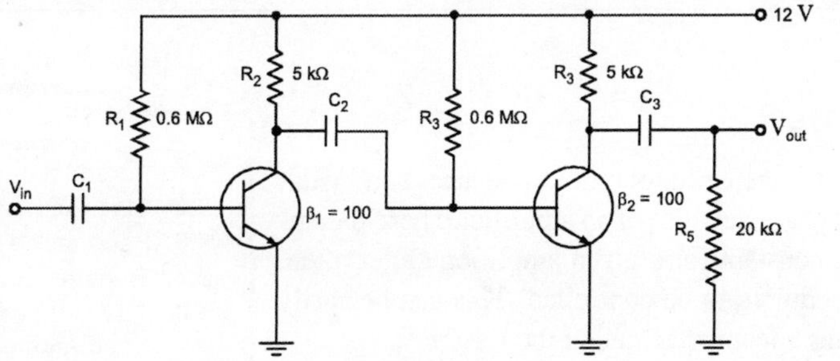

Fig. 1

(*i*) r_i, (*ii*) A_{v1}, (*iii*) A_{v2}, (*iv*) A_v in dB. Neglect V_{BE} and take r_e = 25 mV/I_E

[**Ans.** (*i*) 1.247 kΩ (*ii*) 80 (*iii*) 320 (*iv*) 88 dB]

4. An R-C coupled amplifier has a voltage gain of 200 in the frequency range of 200 Hz and 20 kHz. On either side of these frequencies, the gain falls to 141.5 at 25 Hz and 40 kHz. Determine the bandwidth. [**Ans.** 25 Hz to 40 kHz]

❑❑❑

CHAPTER 9

Feedback Amplifiers

9.1. INTRODUCTION

INSIDE THIS CHAPTER

The voltage gain, input impedance, output impedance, bandwidth etc., are few important characteristics of an amplifier. These parameters are more or less constant for a given amplifier. Quiet often, these parameters are required to be controlled. This can be easily achieved by employing a technique, called the *feedback.*

When a part or fraction of output is combined to the input, feedback is said to exist. Thus the process of combining a fraction of output energy (voltage or current) back to the input is called the *feedback.* When the feedback voltage (or current) is so applied as to increase the input signal (*i.e.* when the feedback is in phase with input signal), it is called the *positive* or *direct* or *regenerative feedback.* Since positive feedback causes excessive distortion and instability, it is seldom used in amplifiers. However, because of its capability of increasing the power of the original signal it is used in oscillator circuits.

When the feedback voltage (or current) is so applied as to weaken the input signal (*i.e.* when the feedback is in phase opposition to the input signal), it is called *negative, inverse* or *degenerative feedback.* Negative feedback reduces the amplifier gain but it has numerous advantages (such as gain stability, reduction in nonlinear distortion, reduction in noise, increase in bandwidth or improvement in frequency response, increase in input impedance and decrease in output impedance) and is frequently used in amplifier circuits. The reduction in gain due to negative feedback, can always be compensated by increasing the number of stages.

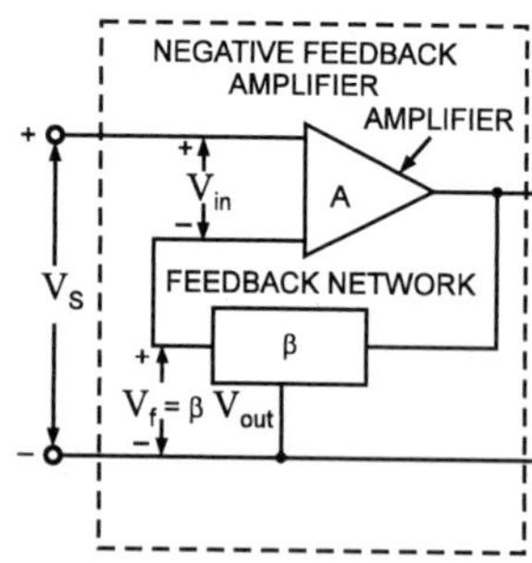

Fig. 9.1

An amplifier with two input terminals and one output terminal, in triangular representation, is shown in Fig. 9.1. The amplifier has a voltage gain A and output voltage V_{out} is applied to a feedback network which reduces V_{out} by a factor β to give a feedback voltage V_f. The feedback network also ensures that feedback voltage V_f is in phase opposition to input signal voltage V_s. The instantaneous voltage at the amplifier input terminals is now $v_{in} = v_s - v_f$. Thus, output voltage will be less than what could be obtained if negative feedback were not employed. This is due to the fact that input voltage has become less than the input signal voltage because of negative feedback. The overall voltage gain $\frac{V_{out}}{V_s}$ is, therefore, reduced. However, as will be demonstrated, the stability of the voltage gain is increased when negative feedback is employed.

A more and more negative feedback is used, the characteristics of a feedback amplifier become more and more independent of the characteristics of the *open-loop* amplifier (amplifier with zero feedback) and ultimately depends on the properties of the feedback network itself.

9.2. CLASSIFICATION OF AMPLIFIERS

The amplifiers without negative feedback are known as *basic* amplifiers. Their gain is unstable [due to (*i*) change in supply voltage, (*ii*) change in parameters, (*iii*) ageing of device and (*iv*) due to rise in leakage current with temperature], distortion is large, noise is more, bandwidth is not sufficiently large and their input and output impedances do not meet the requirements of general applications.

Amplifiers may be usefully classified into four broad categories viz. (*i*) Voltage amplifiers, (*ii*) Current amplifiers, (*iii*) Transconductance amplifiers and (*iv*) Transresistance amplifiers. Their relative magnitudes of input impedance in comparison with the source impedance and that of output impedance in comparison with the load impedance are the determining factors of amplifier category.

9.2.1. Voltage Amplifier

Voltage amplifiers are intended to amplify an input voltage signal and provide an output voltage signal. The voltage amplifier is essentially a voltage-controlled voltage source. The input impedance is required to be high and output impedance is required to be low. Since the signal source is essentially a voltage source, it is convenient to represent it in terms of a Thevenin equivalent circuit (Fig. 9.2). In a voltage amplifier the output quantity of interest is the output voltage.

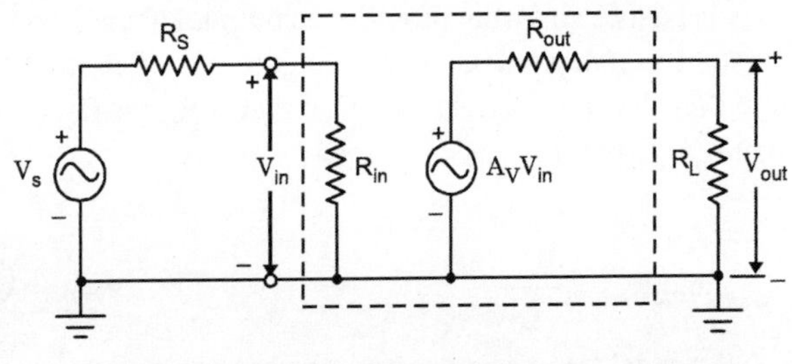

Fig. 9.2. ***Equivalent Circuit of Voltage Amplifier***

From the equivalent circuit of a voltage amplifier depicted in Fig. 9.2, we find that if the input resistance $R_{in} >> R_s$ then $V_{in} \approx V_s$ and if load resistance $R_L >> R_{out}$, then $V_{out} \simeq A_V V_{in}$

$$\text{or Voltage gain, } A_V = \frac{V_{out}}{V_{in}} \simeq \frac{V_{out}}{V_s} \quad ...(9.1)$$

An ideal voltage amplifier should have infinite input resistance and zero output resistance.

The amplifier provides a voltage output proportional to the voltage input and the proportionality constant is independent of the magnitudes of source and load resistances.

9.2.2. Current Amplifier

The input signal in a current amplifier is essentially a current, and thus the signal source is most conveniently represented by its Norton equivalent (Fig. 9.3). The output quantity of interest is current. The current amplifier is essentially a current-controlled current source.

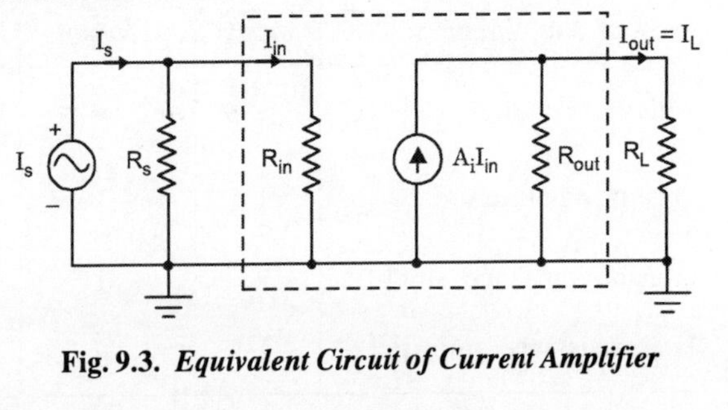

Fig. 9.3. ***Equivalent Circuit of Current Amplifier***

The equivalent circuit of a current amplifier is depicted in Fig. 9.3. In the given circuit if the input resistance $R_{in} << R_s$ then $I_{in} \simeq I_s$ and if $R_L << R_{out}$ then $I_L = I_{out} = A_i I_{in}$

$$\text{Current gain, } A_i = \frac{I_{out}}{I_{in}} = \frac{I_L}{I_s} \quad ...(9.2)$$

An ideal current amplifier should have zero input resistance and infinite output resistance.

An ideal current amplifier is defined as an amplifier which provides an output current proportional to the signal current, and proportionality constant is independent of magnitudes of source and load resistances (R_s and R_L respectively).

9.2.3. Transconductance Amplifier

In transconductance amplifiers the input signal is a voltage and its output signal is a current. The ideal transconductance amplifier provides an output current which is proportional to the signal voltage and the proportionality constant is independent of the magnitudes of R_s and R_L.

From the equivalent circuit of a transconductance amplifier depicted in Fig. 9.4, we find that if $R_s << R_{in}$, then $V_{in} \simeq V_s$ and if load resistance $R_L << R_{out}$,

$$\text{then } I_L = I_{out} = G_m V_{in} = G_m V_s$$

$$\text{and } G_m = \frac{I_L}{V_s} \quad ...(9.3)$$

An ideal transconductance amplifier should have infinite input and output resistances.

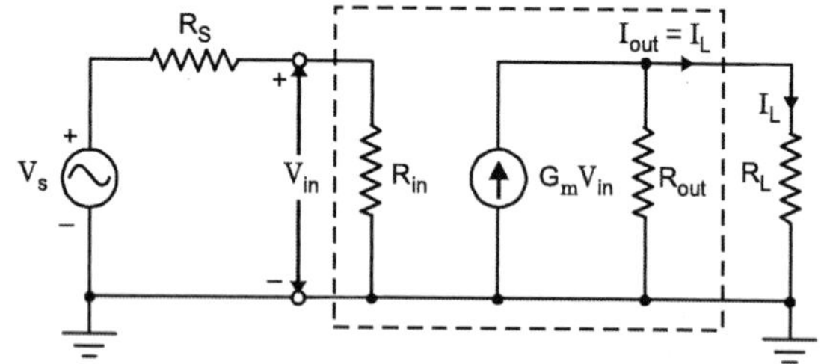

Fig. 9.4. *Equivalent Circuit of Transconductance Amplifier*

9.2.4. Transresistance Amplifier

In transresistance amplifiers the input signal is current and the output signal is voltage. The ideal transresistance amplifier provides an output voltage in proportion to the signal current, and the proportionality constant is independent of the magnitudes of R_s and R_L. From the equivalent circuit of a transresistance amplifier shown in Fig. 9.5, we find that if $R_{in} << R_s$ then $I_{in} \simeq I_s$ and if $R_{out} << R_L$, then

$$V_{out} = R_m I_{in} = R_m I_s$$

$$\text{and } R_m = \frac{V_{out}}{I_s} \quad ...(9.4)$$

An ideal transresistance amplifier should have zero input and output resistances.

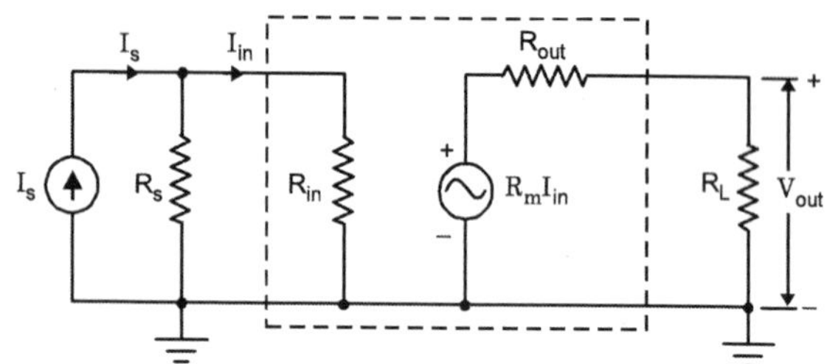

Fig. 9.5. *Equivalent Circuit of Transresistance Amplifier*

The characteristics of ideal amplifiers are summarized below in tabular form :

TABLE 9.1. *Characteristics of Ideal Amplifier*

Type of Amplifier	Input	Output	R_{in}	R_{out}	Transfer Characteristic
Voltage Amplifier	V	V	∝	0	$A_v = \frac{V_{out}}{V_s}$
Current amplifier	I	I	0	∝	$A_i = \frac{I_L}{I_s}$
Transconductance amplifier	V	I	∝	∝	$G_m = \frac{I_L}{V_s}$
Transresistance amplifier	I	V	0	0	$R_m = \frac{V_{out}}{I_s}$

Example 9.1. A current amplifier has an input resistance of 10 Ω, and an output resistance of 10 kΩ and a current gain of 1,000. It is fed by a current source having a source resistance of 10 kΩ and its output is connected to a 10 Ω load resistance.

Find the voltage gain and power gain.

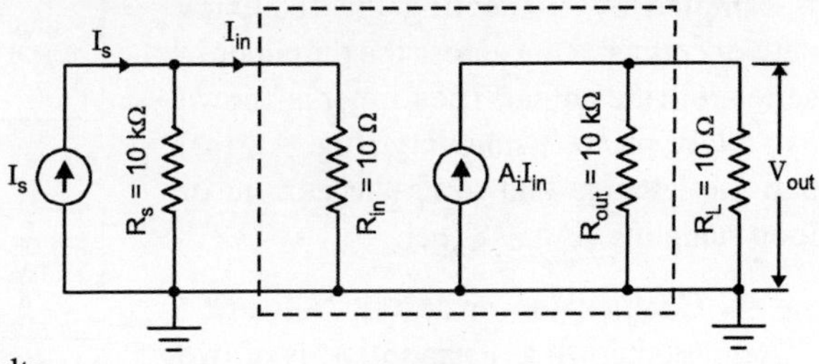

Fig. 9.6

Solution : Source resistance, $R_s = 10\ k\Omega$

Input resistance, $R_{in} = 10\ \Omega$

Output resistance, $R_{out} = 10\ k\Omega$

Load resistance, $R_L = 10\ \Omega$

Current gain, $A_i = 1{,}000$

Input voltage, $V_{in} = I_{in} R_{in} = 10 I_{in}$ volts

Output voltage, $V_{out} = A_i I_{in} (R_{out} \parallel R_L)$

$$= A_i I_{in} \frac{R_{out} \times R_L}{R_{out} + R_L} = 1{,}000 \times I_{in} \times \frac{10{,}000 \times 10}{10{,}000 + 10} = 9{,}990\ I_{in} \text{ volts}$$

$$\text{Voltage gain, } A_V = \frac{V_{out}}{V_{in}} = \frac{9{,}990\ I_{in}}{10 I_{in}} = 999 \textbf{ Ans.}$$

$$\text{Current gain, } A_i = \frac{I_L}{I_s} = \frac{\dfrac{V_{out}}{R_L}}{I_{in} \dfrac{(R_s + R_{in})}{R_s}} \qquad \because I_L = \frac{V_{out}}{R_L} \text{ and } I_{in} = \frac{I_s \times R_s}{R_s + R_{in}}$$

$$= \frac{9{,}990\ I_{in}}{10 \times I_{in} \times \dfrac{10{,}000 + 10}{10{,}000}} = \frac{9{,}990 \times 1{,}000}{10{,}010} = 998$$

$$\text{Power gain, } A_p = A_V \times A_i = 999 \times 998 = 997 \times 10^3 \textbf{ Ans.}$$

9.3. FEEDBACK CONCEPT

Figure 9.7 represents the block diagram of an amplifier with feedback. The output quantity (either voltage or current) is sampled by means of a suitable sampling network (or sampler) and is fed to the *feedback network*. The output of feedback network, which has a fraction of output signal is combined with the external (source) signal through a mixer network and fed into the basic amplifier.

The different blocks of feedback amplifier are explained below :

1. Signal Source : This block is either a signal voltage V_s in series with a resistor R_s or a signal current I_s in parallel with a resistor R_s.

2. Feedback Network : The feedback network is usually in the form of a passive two-port network and may be formed of resistors, inductor and capacitors (most often of resistors). Its function is to return a function of the output energy (voltage or current) to the input of the amplifier.

3. Sampling Network : Sampling networks are of two types, namely, voltage

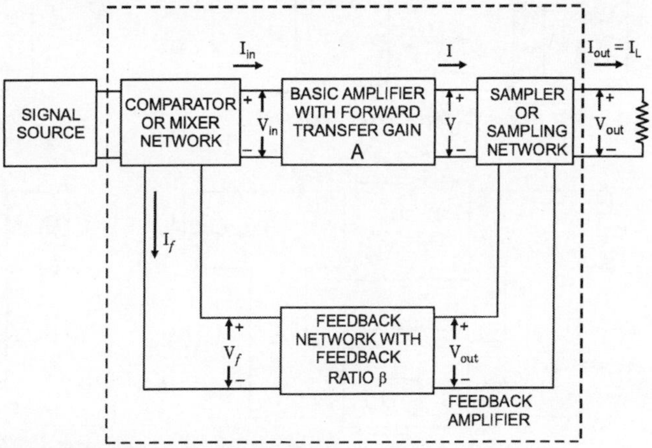

Fig. 9.7. ***Block Diagram of an Amplifier With Feedback***

sampling network and current sampling network, as shown in Fig. 9.8. Output voltage is sampled by connecting the feedback network either shunt across the output or in series with the output. The former is known as *voltage* or *node sampling* [Fig. 9.8 (*a*)] and the latter is known as the current or loop sampling [Fig. 9.8 (*b*)].

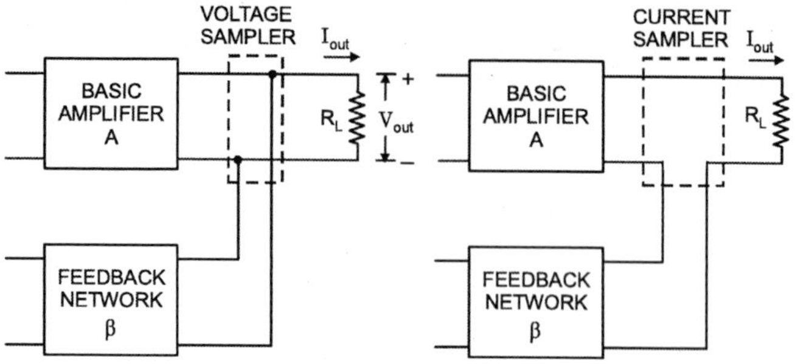

(a) Voltage Sampling Network *(b) Current Sampling Network*

Fig. 9.8

4. Comparator, or Mixer, Network : Mixer, also known as *comparator*, is of two types, namely, series mixer and shunt mixer, as shown in Fig. 9.9. A differential amplifier, which has two inputs and one output proportional to difference between the signals at the two inputs, is usually used as a mixer or comparator.

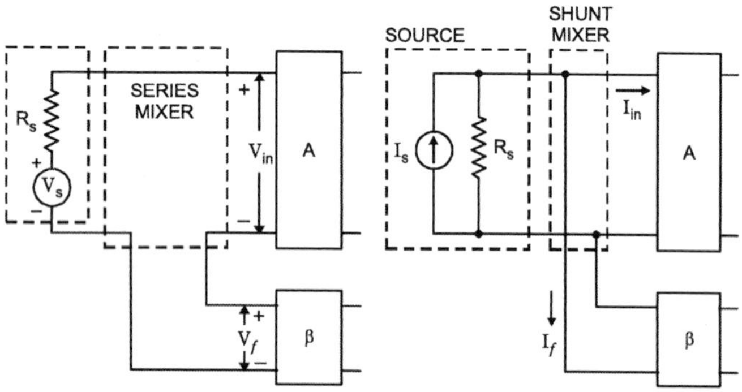

(a) Series Mixing *(b) Shunt Mixing*

Fig. 9.9

Types of Feedback Connections

Any of the output connections shown in Fig. 9.8 may be combined with any of the input connections shown in Fig. 9.9 to form the feedback amplifier shown in Fig. 9.7. The resulting feedback amplifier topologies are given below :

1. Voltage-series feedback, as shown in Fig. 9.10 (*a*)
2. Voltage-shunt feedback, as shown in Fig. 9.10 (*b*).

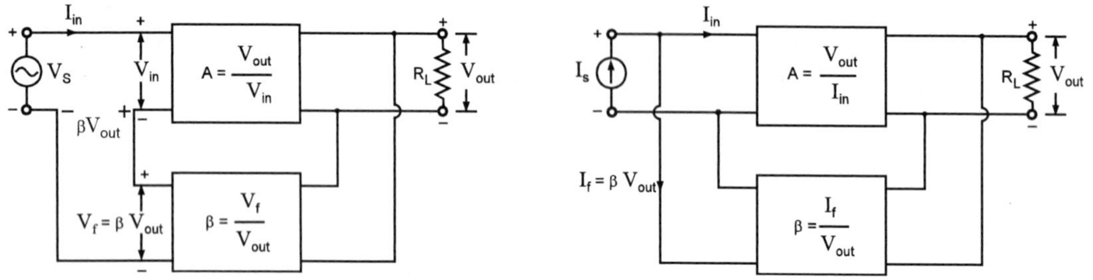

(a) Voltage-Series Feedback *(b) Voltage-Shunt Feedback*

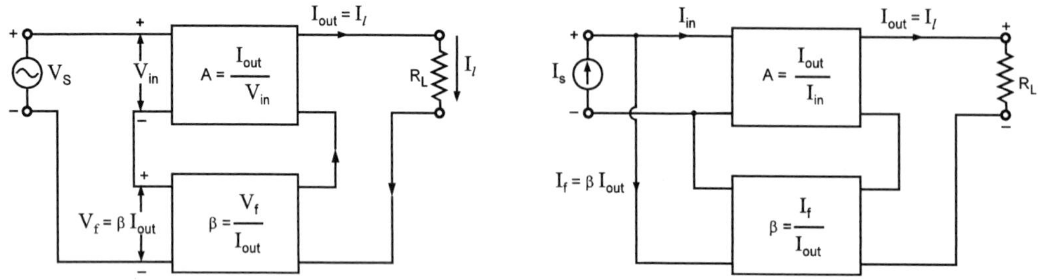

(c) Current-Series Feedback *(d) Current-Shunt Feedback*

Fig. 9.10

3. Current-series feedback, as shown in Fig. 9.10 (*c*).
4. Current shunt feedback, as shown in Fig. 9.10 (*d*).

In the list above, voltage refers to connecting the output voltage as input to the feedback network and current refers to tapping off some output current through the feedback network. Series refers to connecting the feedback signal in series with the input signal voltage; and shunt refers to connecting the feedback signal in parallel (shunt) with an input current source.

Series feedback connections tend to increase the input impedance of the amplifier. This is because in this case feedback signal is returned to the input in series to oppose the applied voltage causing input current to fall and consequently makes the input impedance to increase. In case of shunt or parallel connections, the current drawn from the signal source is increased by an amount equal to feedback current I_f and therefore, input impedance falls. Voltage feedback tends to reduce the output impedance while current feedback tends to increase the output impedance. In cascaded amplifiers, mostly high input impedance and low output impedance are desired. Both of these (*i.e.* high input impedance and low output impedance) are provided by use of voltage-series feedback and so it is most commonly employed.

Transfer Ratio or Gain

The symbol A in Fig. 9.7 represents the ratio of the output signal to the input signal of the basic amplifier. The transfer ratio $\frac{V_{out}}{V_{in}}$, referred to as voltage gain A_v, the transfer ratio $\frac{I_{out}}{I_{in}}$, referred to as current gain A_i, the ratio $\frac{I_{out}}{V_{in}}$, referred to as the transconductance G_m, and the ratio $\frac{V_{out}}{I_{in}}$, referred to as the transresistance R_m, are referred to as a *transfer gain of the basic amplifier without feedback* and represented by symbol A.

The symbol A_f, defined as the ratio of the output signal to the input signal of the amplifier configuration shown in Fig. 9.7, is shown as the *transfer gain of the amplifier with feedback.* Thus, A_f is used to represent any one of the four ratios $\frac{V_{out}}{V_s} = A_{vf}, \frac{I_{out}}{I_s} = A_{if}, \frac{I_{out}}{V_s} = G_{mf}$ and $\frac{V_{out}}{I_s} = R_{mf}$. The relationship between the transfer gain A_f with feedback and the gain A of the ideal amplifier without feedback is derived below (Art. 9.4).

9.4. TRANSFER GAIN WITH FEEDBACK

Figure 9.11 represents a generalized feedback amplifier. The basic amplifier shown in Fig. 9.11 may be an ideal voltage, transresistance, transconductance or current amplifier connected in configurations, as depicted in Fig. 9.11.

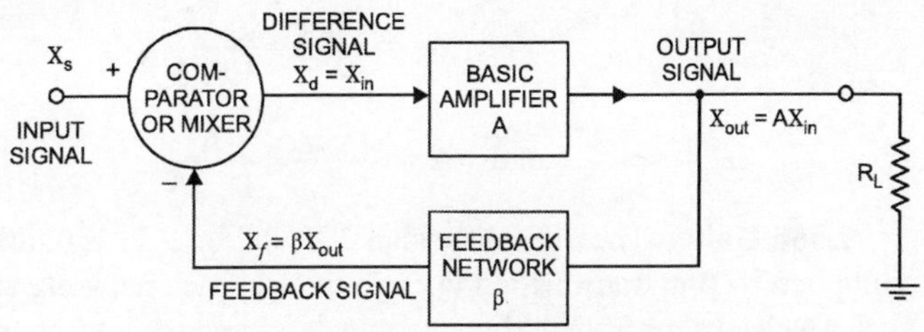

Fig. 9.11. ***Schematic Representation of a Single Loop Feedback Amplifier***

In Fig. 9.11, the source resistance R_s is considered to be the part of the amplifier, and the transfer gain A (A_V, R_M, g_M, A_I) includes the effect of the loading of the β network as well as external load resistance R_L upon the amplifier. The input signal X_s, the output signal X_{out}, the feedback signal X_f and the difference signal X_d, each represents either a voltage or a current. These signals and the ratios A and β are summarized below in tabular form.

The symbol indicated by the circle in Fig. 9.11 represents a mixing or comparison network, whose output is the sum of the inputs, considering the sign indicated at each input. Thus

$$\text{Difference signal, } X_d = X_s - X_f = X_{in} \qquad ...(9.5)$$

X_d is also called the *difference, error,* or *comparison signal* because it represents the difference of applied signal and that fedback to the input.

TABLE 9.2. ***Voltage and Current Signals in Feedback Amplifiers***

Type of Feedback	Signal or Ratio			
	X_{out}	X_s, X_f, X_d	A	β
Voltage-Series Feedback	Voltage	Voltage	A_v	V_f/V_{out}
Voltage-Shunt Feedback	Voltage	Current	R_m	I_f/V_{out}
Current-Series Feedback	Current	Voltage	G_m	V_f/I_{out}
Current-Shunt Feedback	Current	Current	A_i	I_f/I_{out}

The *reverse transmission factor* β, defined by the ratio of feedback signal X_f and output signal, is usually a positive or a negative real number

$$i.e. \quad \beta = \frac{X_f}{X_{out}} \qquad ...(9.6)$$

But in general, β is a complex function of the signal frequency. The symbol β should neither be confused with transistor current gain factor or symbol β used for CE short-circuit gain.

The transfer gain A is defined as the ratio of the output signal X_{out} and the input signal X_{in}, so

$$A = \frac{X_{out}}{X_{in}} \qquad ...(9.7)$$

Substituting Eqs. (9.5) and (9.6) into Eq. (9.7), we have the expression for transfer gain with feedback

$$A_f = \frac{X_{out}}{X_s} = \frac{X_{out}}{X_{in} + X_f} = \frac{\frac{X_{out}}{X_{in}}}{1 + \frac{X_f}{X_{out}} \times \frac{X_{out}}{X_{in}}} = \frac{A}{1 + \beta A} \qquad ...(9.8)$$

The quantity A in Eqs. (9.7) and (9.8) represents the transfer gain of the corresponding amplifier without feedback, but including the loading of β network, R_L and R_s.

A_f will be less than A if the feedback is *negative, inverse* or *degenerative.* A_f will be more than A if the feedback is *positive, direct* or *regenerative.*

Transfer gain with feedback is given by equation

$$A_f = \frac{X_{out}}{X_s} = \frac{A}{1 + \beta A} \qquad \text{for negative feedback} \quad ...(9.9)$$

$$\text{and} \quad A_f = \frac{X_{out}}{X_s} = \frac{A}{1 - \beta A} \qquad \text{for positive feedback} \quad ...(9.10)$$

Loop Gain : The difference signal X_d in Fig. 9.11 is multiplied by A in passing through the amplifier, is multiplied by β in transmission through the feedback network, and is multiplied by –1 in the mixing network. Such a path takes us from the input terminals around the loop consisting of the amplifier and feedback network back to the input; the product (–Aβ) is known as the *loop gain* or *return ratio.* The difference between unity and the loop gain is called the return difference D = 1 – (–Aβ) = 1 + Aβ. Also, the amount of feedback introduced into an amplifier is usually expressed in decibels by the definition

$$N = \text{dB of feedback} = 20 \log_{10} \left| \frac{A_f}{A} \right| = 20 \log_{10} \left| \frac{1}{1 + \beta A} \right| \qquad ...(9.11)$$

N will be a negative number for negative feedback.

In general both A and β are phasor quantities (having magnitude as well as phase). Every amplifier stage introduces a phase shift of 180° (assuming CE configuration, as usual). Thus for a 3-stage amplifier, total phase shift between the output and input will be 3 × 180° = 540° or 180°. β is usually arranged to have a phase value of 0° or 180° (*i.e.* either in phase with input signal voltage or in phase opposition to it).

Since (1 – βA) is a complex quantity, it can be observed that

1. If (1 – βA) is less than unity for Eq. (9.10), then A_f exceeds A. This condition corresponds to positive feedback because voltage feedback adds to input signal voltage and increases input voltage V_{in}. Positive feedback, though, increases the gain but it reduces the stability and increases the distortion and so it is usually avoided.
2. If (1 – βA) is equal to zero then the gain A_f becomes infinite. This is only possible when input is zero. Thus the amplifier is then capable of giving output voltage even with zero input signal. Under such situation the circuit operates as an oscillator.
3. If (1 – βA) is greater than unity then A_f is smaller than A. This means the feedback voltage V_{in} becomes smaller than input signal voltage V_s. This corresponds to negative feedback in an amplifier. Though negative feedback reduces the gain of the amplifier but improves its performance in several aspects, given in the succeeding articles. It may be noted that negative voltage feedback does not affect the current of the circuit.

Note : The assumptions made in investigating the effect of negative feedback on the amplifier characteristics are

1. Basic amplifier is unilateral from input to the output (this is not valid if $h_{re} \neq 0$ for a CE transistor amplifier).
2. Passive feedback network is unilateral and transmits a signal from the output to the input but not in the opposite direction.
3. Feedback network presents no loading on the output of the basic amplifier.
4. There is no forward transmission through the β feedback network.

Example 9.2. The voltage gain of an amplifier without feedback is 3000. Calculate the voltage gain of the amplifier if the negative feedback is introduced in the circuit. Given that feedback fraction *m* = 0.001.

Solution : Open-loop voltage gain (voltage gain without feedback), A = 3000

Feedback fraction, β = 0.01

$$\text{Voltage gain with negative feedback } A_f = \frac{A}{1+\beta A} = \frac{3,000}{1+0.01\times 3,000} = \frac{3,000}{31} = 96.774 \textbf{ Ans.}$$

Example 9.3. Calculate the gain of a negative feedback amplifier with an internal gain A_v = 75 and feedback fraction $m_v = \frac{1}{15}$. What will be the gain if A_v doubles ?

Solution : Internal gain (open-loop voltage gain), A = 75

$$\text{Feedback fraction, } \beta = \frac{1}{15}$$

$$\text{Voltage gain with negative feedback, } A_f = \frac{A}{1+\beta A} = \frac{75}{1+\frac{1}{15}\times 75} = 12.5 \textbf{ Ans.}$$

When A_v doubles *i.e.* when A′ = 2 × 75 = 150

$$A'_f = \frac{A}{1+\beta A'} = \frac{150}{1+\frac{1}{15}\times 150} = 13.64 \textbf{ Ans.}$$

Example 9.4. An amplifier has a voltage gain of 40. The amplifier is now modified to provide a 10% negative feedback in series with the input. Calculate (*i*) voltage gain with feedback, (*ii*) amount of feedback in dB, (*iii*) loop gain.

Solution : Open-loop voltage gain, A = 40

Feedback ratio, β = 10% or 0.1

(*i*) Voltage gain with feedback, $A_f = \dfrac{A}{1+\beta A} = \dfrac{40}{1+0.1\times 40} = 8$ **Ans.**

(*ii*) Amount of feedback in dB $= 20\log_{10}\left|\dfrac{1}{1+\beta A}\right| = 20\log_{10}\left|\dfrac{1}{1+40\times 0.1}\right|$

$= 20\log_{10} 0.2 = -13.98$ **Ans.**

(*iii*) Loop gain $= A\beta = 40\times 0.1 = 4$ **Ans.**

Example 9.5. An amplifier with voltage gain of 60 dB uses $\dfrac{1}{20}$ of its output in negative feedback. Calculate the gain with feedback in dB.

Solution : Open-loop voltage gain, A = 60 dB or antilog $\dfrac{60}{20} = 1{,}000$

Feedback ratio, $\beta = \dfrac{1}{20} = 0.05$

Gain with feedback, $A_f = \dfrac{A}{1+\beta A} = \dfrac{1{,}000}{1+0.05\times 1{,}000} = 19.6$ or $20\log_{10} 19.6$ dB = 25.85 dB **Ans.**

Example 9.6. The voltage gain of an amplifier without feedback is 2,500. If 40 dB of negative feedback is added :

(*i*) what is the gain with feedback ?

(*ii*) for obtaining same output, how much must be the input increased ?

Solution : Open-loop gain, A = 2,500

$1 + \beta A = 40$ dB or antilog $\dfrac{40}{20} = 100$

(*i*) Gain with feedback, $A_f = \dfrac{A}{1+\beta A} = \dfrac{2{,}500}{100} = 25$ **Ans.**

(*ii*) Input for same output will become $\dfrac{A}{A_f}$ times the input without feedback *i.e.* $\dfrac{2500}{25}$ *i.e.*, 100 times the input without feedback. **Ans.**

Example 9.7. Voltage gain of an amplifier without feedback is 60 dB. It decreases to 40 dB with feedback. Calculate the feedback factor.

Solution : Voltage gain of amplifier without feedback, A = 60 dB or 1,000

Voltage gain with feedback, A_f = 40 dB or 100 $\because 20\log A_f = 40$ dB or $A_f = 100$

$\therefore$ Gain with feedback, $A_f = \dfrac{A}{1+\beta A}$

or Feedback factor, $\beta A = \dfrac{A}{A_f} - 1 = \dfrac{1{,}000}{100} - 1 = 9$ **Ans.**

Example 9.8. A single stage transistor amplifier has a voltage gain of 600 without feedback, and 50 with feedback. Calculate the percentage of output which is fedback to the input.

Solution : Voltage gain without feedback, A = 600

Voltage gain with feedback, A_f = 50

$$\because A_f = \frac{A}{1+\beta A}$$

$$\therefore 50 = \frac{600}{1+600\beta}$$

$$\text{or } \beta = 0.01833$$

Percentage of output voltage that is fedback to the input

$$= \frac{V_f}{V_{out}} \times 100 = \beta \times 100 = 0.01833 \times 100 = 1.833\% \text{ Ans.}$$

Example 9.9. An amplifier with negative feedback gives an output of 12.5 volts with an input of 1.5 V. When feedback is removed, it requires 0.25 V input for the same output. Find (*i*) value of voltage gain without feedback, and (*ii*) value of β, if the input and output are in phase and β is real.

Solution : (*i*) Voltage gain without feedback,

$$A_v = \frac{\text{Output voltage}}{\text{Input voltage}} = \frac{12.5}{0.25} = 50$$

∵ Output voltage is same with and without feedback

$$\text{Voltage gain with feedback, } A_{vf} = \frac{\text{Output voltage}}{\text{Input voltage}} = \frac{12.5}{1.5} = 8 \quad \text{with feedback}$$

$$\text{Feedback ratio, } \beta = \frac{\frac{A_v}{A_{fv}} - 1}{A_v} = \frac{\frac{50}{8} - 1}{50} = \frac{21}{200} = 0.105 \text{ or } 10.5\% \text{ Ans.} \qquad \text{Since, } A_{vf} = \frac{A_v}{1+\beta A_v}$$

Example 9.10. An amplifier with negative feedback has a gain of –100. It is found that without feedback an input signal of 50 mV is required to produce a given output whereas with feedback the input signal must be 0.06 V for the same output. Calculate the amount of feedback in dB and the values A_v and β.

Solution : Voltage gain with feedback, $A_f = -100$

Output voltage with negative feedback $= A_f \times$ input voltage $= -100 \times 0.06 = -6$ V

$$\text{Voltage gain without feedback } A = \frac{\text{Output voltage without feedback}}{\text{Input voltage}} = \frac{-6}{50 \times 10^{-3}} = -120$$

∵ Output voltage is same with and without feedback

$$\text{Now since } A_f = \frac{A}{1+\beta A}$$

$$\text{so } 100 = \frac{120}{1+\beta \times 120}$$

$$\text{or } \beta = \frac{120-100}{100 \times 120} = \frac{20}{100 \times 120} = \frac{1}{600} \text{ Ans.}$$

$$\text{Amount of feedback in dB} = 20 \log_{10} \left| \frac{1}{1+A\beta} \right| = 20 \log_{10} \left| \frac{1}{1+100 \times \frac{1}{600}} \right| = -1.34$$

Example 9.11. A negative feedback of β = 0.002 is applied to an amplifier of gain 1,000. Calculate the change in overall gain of the feedback amplifier if the internal amplifier is subjected to a gain reduction of 15%.

Solution : Voltage gain without feedback, $A = 1{,}000$, $\beta = 0.002$

$$\text{Voltage gain with feedback, } A_f = \frac{A}{1+\beta A} = \frac{1{,}000}{1+0.002 \times 1{,}000} = 333.33$$

When open-loop gain is reduced by 15%,

$$A' = (1 - 0.15) \times 1{,}000 = 850$$

$$\text{Voltage gain with feedback, } A'_f = \frac{850}{1 + 0.002 \times 850} = 314.8$$

$$\text{Percentage change in overall gain} = \frac{A_f - A'_f}{A_f} \times 100 = \frac{333.333 - 314.8}{333.333} \times 100 = 5.6\% \textbf{ Ans.}$$

Example 9.12. Find (*a*) feedback ratio (*b*) feedback factor (*c*) voltage gain without feedback (*d*) voltage gain with feedback for a circuit given below. Assume $\beta = 200$ and neglect V_{BE}.

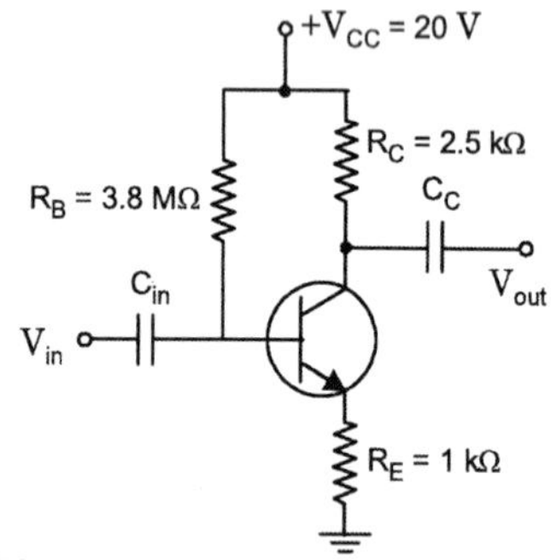

Fig. 9.12

Solution : Collector load resistance,

$$R_C = 2.5 \text{ k}\Omega$$

$$\text{Emitter resistance, } R_E = 1 \text{ k}\Omega$$

$$\text{Assume } h_{ie} = 1.1 \text{ k}\Omega$$

$$\text{and } h_{fe} = \beta = 200$$

$$(c) \text{ Voltage gain without feedback, } A = -\frac{h_{fe}}{h_{ie}} R_C = -\frac{200}{1.1 \text{ k}\Omega} \times 2.5 \text{ k}\Omega = -455 \textbf{ Ans.}$$

$$(d) \text{ Voltage gain with feedback, } A_f = -\frac{R_C}{R_E} = \frac{-2.5}{1} = -2.5 \textbf{ Ans.}$$

$$\text{Since } A_f = \frac{A}{1 + \beta A} \text{ where } \beta \text{ is the feedback ratio}$$

$$(a) \text{ So feedback ratio, } \beta = \frac{A - A_f}{AA_f} = \frac{455 - 2.5}{455 \times 2.5} = 0.398 \textbf{ Ans.}$$

$$(b) \text{ Feedback factor} = A\beta = 455 \times 0.398 = 181 \textbf{ Ans.}$$

9.5. ADVANTAGES OF NEGATIVE FEEDBACK

There are numerous advantages of negative feedback which outweigh its only drawback of reduction in gain. Among the advantages are :

1. **Gain Stability.** The voltage gain of an amplifier with negative feedback is given as $A_f = A/(1 + A\beta)$.

   If $A\beta >> 1$ then the expression becomes $A_f = \frac{1}{\beta}$ *i.e.* overall gain of feedback amplifier A_f is independent of internal gain and depends only on feedback ratio β, and β in turn depends on the passive elements such as resistors. Resistors remain fairly constant and so the gain is stabilised.
2. **Reduced Nonlinear Distortion.** A large signal stage has nonlinear distortion which is reduced by a factor $(1 + A\beta)$ when negative feedback is used.
3. **Reduced Noise.** There is always a noise voltage in the amplifier which is reduced by a factor $(1 + A\beta)$ when negative feedback is used.
4. **Increased Bandwidth (or Improved Frequency Response).** The bandwidth (BW) of an amplifier without feedback is equal to the separation between 3-dB frequencies f_1 and f_2. If A is the gain then gain-bandwidth product is A × BW. With the negative feedback the amplifier gain is reduced and since gain bandwidth product has to remain constant in both cases, so obviously the bandwidth will increase to compensate for the reduction in gain.

5. **Increased Input Impedance.** The input impedance of the amplifier with negative feedback is increased by a factor $(1 + A\beta)$.
6. **Reduced Output Impedance.** The output impedance of the amplifier with negative feedback is reduced by a factor $(1 + A\beta)$.

9.6. GENERAL CHARACTERISTICS OF NEGATIVE FEEDBACK AMPLIFIERS

Inspite of the fact that negative feedback reduces the transfer gain, it is used. Why so ? The simple answer is that it improves many desirable characteristics. We will now examine the effects of negative feedback on different characteristics of an amplifier.

9.6.1. Stabilization of Gain With Negative Feedback

The variations in temperature, supply voltages, ageing of components or variations in transistor parameters with replacement are some of the factors that affect the gain of an amplifier and cause it to change. However, the overall gain of the amplifier can be made independent of these variations if negative feedback is used. This is an important advantage of negative feedback.

From Eq. (9.8), the voltage gain with negative feedback is given as

$$A_f = \frac{A}{1+\beta A} \simeq \frac{A}{\beta A} \simeq \frac{1}{\beta} \quad \text{...(9.12) if } A\beta \text{ is made much larger than unity}$$

The gain is thus independent of internal gain of the amplifier and depends on the passive elements such as resistors. The values of resistors remain fairly constant because they can be chosen very precisely with almost zero temperature coefficient of resistance. Thus the gain is stabilized.

Even if open-loop gain A is not very large, some improvement in gain stability can be achieved.

Differentiating Eq. (9.8) w.r.t. A we have

$$\frac{dA_f}{dA} = \frac{(1+\beta A) - A \times \beta}{(1+\beta A)^2} = \frac{1}{(1+\beta A)^2}$$

$$\text{or} \quad dA_f = \frac{dA}{(1+\beta A)^2}$$

Dividing above equation by Eq. (9.9) we have

$$\frac{dA_f}{A_f} = \frac{dA}{A} \times \frac{1+\beta A}{(1+\beta A)^2} = \frac{1}{(1+\beta A)} \frac{dA}{A} \quad \text{...(9.13)}$$

The term $\frac{dA_f}{A_f}$ represents the fractional change in amplifier transfer gain with feedback and $\frac{dA}{A}$ denotes the fractional change in transfer gain without feedback. The term $\frac{1}{1+A\beta}$ is called the *sensitivity* of the transfer gain.

Thus, the sensitivity of the transfer gain is defined as the ratio of percentage change in transfer gain with feedback to the percentage change in transfer gain without feedback.

$$\text{i.e. Sensitivity} = \frac{\frac{dA_f}{A_f}}{\frac{dA}{A}} = \frac{1}{1+A\beta} \quad \text{...(9.14)}$$

The reciprocal of the term sensitivity is called the *desensitivity,* which is given as

$$\text{Desensitivity} = (1 + A\beta) \quad \text{...(9.15)}$$

In negative feedback $(1 + \beta A) >> 1$, the percentage change in gain with negative feedback is less than the percentage change in gain without feedback. Thus negative feedback improves the gain stability of the amplifier.

The topology determines which transfer ratio (Table 9.2) is stabilized. For instance, for voltage-series feedback Eq. (9.12) signifies that $A_{fv} \simeq \frac{1}{\beta}$ and it is the voltage gain which is stabilized. Similarly, in cases of voltage-shunt feedback, current-series feedback and current-shunt feedback transresistance gain R_m, transconductance gain G_m and current gain A_i respectively are stabilized.

9.6.2. Reduction in Frequency Distortion With Negative Feedback

For a negative feedback amplifier having $\beta A >> 1$ the gain with feedback is given as $A_f \cong 1/\beta$. It follows from this that if the feedback network is purely resistive, the gain with feedback becomes independent of frequency even though the basic amplifier is frequency dependent. Practically, the frequency distortion arising because of varying amplifier gain with frequency is considerably reduced with negative feedback.

9.6.3. Reduction in Nonlinear Distortion With Negative Feedback

Harmonic distortion occurs when a transistor or other device is driven beyond its range of linear operation. Now negative feedback is used and the strength of the input signal is increased by the same factor by which the gain is reduced. Thus the magnitude of output signal remains unaltered and thus there is a reduction in harmonic or nonlinear distortion. It may be proved as below.

Let the amplifier with open-loop gain A produce a distortion D in the output signal without feedback. Now when feedback is applied, output as well distortion is fed back to the input. Let the gain with feedback be A_f and distortion in the output D_f. A part βD_f of this distortion is fed back to the input. It gets amplified by factor A and becomes $\beta A D_f$.

Net distortion, $D_f = D - \beta A D_f$ $\quad\because$ distorted output is in phase opposition

$$\text{or } D_f = \frac{D}{1 + \beta A} \quad \text{...(9.16)}$$

Thus the distortion is reduced by the same factor as the gain.

However, it is to be noted that improvement in distortion is possible only when the distortion is introduced by the amplifier itself, not when it is already present in the input signal.

9.6.4. Reduction in Noise With Negative Feedback

Noise in an amplifier is reduced in the same way as distortion. However, distortion is usually introduced in the last stage of the amplifier and, therefore, is always within the feedback loop whereas noise in an amplifier may be generated or introduced from outside source at any stage that may not be within the feedback loop. The noise can be reduced by negative feedback only if its source lies within the feedback loop after the first stage.

As nonlinear distortion generated within the amplifier is reduced with negative feedback by a factor $(1 + \beta A)$, the noise introduced within amplifier also gets reduced by the same factor $(1 + \beta A)$ on using negative feedback and the performance of the amplifier is thus much improved.

Example 9.13. An amplifier with a closed loop gain of 100 is required and this gain should not vary more than 1% when the inherent gain of the amplifier without feedback varies by 20%. Find the values of A_v and β.

Solution : Variation in open-loop gain, $\frac{dA_v}{A_v} = 20\% = 0.2$

Variation in closed-loop gain, $\frac{dA_{vf}}{A_{vf}} = 1\% = 0.01$

From Eq. (9.14)

$$\text{Feedback factor } \beta A_v = \frac{\dfrac{dA_v}{A_v}}{\dfrac{dA_{vf}}{A_{vf}}} - 1 = \frac{0.2}{0.01} - 1 = 19$$

$$\text{Open-loop voltage gain, } A_v = A_{vf}(1 + \beta A_v) = 100(1 + 19) = 2{,}000 \ \textbf{Ans.}$$

$$\beta = \frac{\dfrac{A_v}{A_{vf}} - 1}{A_v} = \frac{\dfrac{2{,}000}{100} - 1}{2{,}000} = 0.0095 \ \textbf{Ans.}$$

Example 9.14. The open-loop gain of an amplifier changes by 5%, if 10 dB negative feedback is applied. Calculate the percentage change in the closed loop gain. *[G.G.S.I.P. Univ. Analog Electronics December 2010]*

Solution :

$$\beta A_v + 1 = 10 \text{ dB or antilog } \frac{10}{20} = 3.1623$$

$$\text{or } \beta A_v = 3.1623 - 1 = 2.1623$$

$$\text{Since } \beta A_v = \frac{dA_v / A_v}{dA_{vf} / A_{vf}} - 1$$

$$\text{So } 2.1623 = \frac{0.05}{dA_{vf} / A_{vf}} - 1$$

$$\text{or } \frac{dA_{vf}}{A_{vf}} = \frac{0.05}{2.1623 + 1} = 0.0158 \text{ or } 1.58\% \ \textbf{Ans.}$$

Example 9.15. The gain and distortion of an amplifier are 150 and 5% respectively without feedback. If the stage has 10% of its output voltage applied as negative feedback, find the distortion of the amplifier with feedback.

Solution : Open-loop voltage gain, $A = 150$

Feedback ratio, $\beta = 10\% = 0.1$

Distortion without feedback, $D = 5\% = 0.05$

$$\text{Distortion with feedback, } D_f = \frac{D}{1 + \beta A} = \frac{0.05}{1 + 0.1 \times 150} = 0.003125 \text{ or } 0.3125\% \ \textbf{Ans.}$$

Example 9.16. An amplifier has an input of 10 mV and a gain of 200 without feedback. The distortion produced at the output of the amplifier is 10%. It is desired to reduce the distortion to 1% by using negative feedback. Calculate the gain and output voltage with feedback.

Solution : Distortion without feedback, $D = 10\% = 0.1$

Distortion with feedback, $D_f = 1\% = 0.01$

Amplifier gain without feedback, $A = 200$

$$D_f = \frac{D}{1 + \beta A} \qquad \text{...Refer to Eq. (9.16)}$$

$$\text{or } 0.01 = \frac{0.1}{1 + 200\beta}$$

$$\text{or } \beta = \left(\frac{0.1}{0.01} - 1\right) \Big/ 200 = 0.045 \text{ or } 4.5\%$$

$$\text{Gain with feedback, } A_f = \frac{A}{1 + \beta A} = \frac{200}{1 + 0.045 \times 200} = 20 \ \textbf{Ans.}$$

$$\text{New output voltage, } V_{out} = A_f V_s = 20 \times 10 \text{ mV} = 200 \text{ mV or } 0.2 \text{ V} \ \textbf{Ans.}$$

Example 9.17. An amplifier without feedback gives fundamental output 36 V with 7% second harmonic distortion; when the input is 0.028 V.

(*i*) **If 1.2% of the output is fedback into the input in a negative voltage series feedback, what is the output voltage ?**

(*ii*) **If the fundamental output is maintained at 36 V but the second harmonic distortion is reduced to 1%, what is the input voltage ?**

Solution : Amplifier gain without feedback, $A = \frac{\text{Output voltage without feedback}}{\text{Input voltage}} = \frac{36}{0.028} = 1{,}285.7$

(*i*) Feedback ratio, $\beta = 1.2\% = 0.012$

Amplifier gain with feedback, $A_f = \frac{A}{1+\beta A} = \frac{1{,}285.7}{1+0.012 \times 1{,}285.7} = 78.26$

Output voltage, $V_{out} = A_f \times V_{in} = 78.26 \times 0.028 = 2.19$ V **Ans.**

(*ii*) Distortion without feedback, $D = 7\% = 0.07$

Distortion with feedback, $D_f = 1\% = 0.01$

Amplifier gain without feedback, $A = 1{,}285.7$

Since $D_f = \frac{D}{1+\beta A}$

So $\beta = \left(\frac{D}{D_f} - 1\right) \times \frac{1}{A} = \left(\frac{0.07}{0.01} - 1\right) \times \frac{1}{1{,}285.7} = 0.004667$

Gain with feedback, $A_f = \frac{A}{1+\beta A} = \frac{1{,}285.7}{1+0.004467 \times 1{,}285.7} = 183.67$

Input voltage required to provide output voltage of 36 V,

$$V_{in} = \frac{V_{out}}{A_f} = \frac{36}{183.67} = 0.196 \text{ V} \quad \textbf{Ans.}$$

Example 9.18. The output resistance of a voltage series feedback amplifier is 10 Ω. If the gain of the basic amplifier is 100 and feedback fraction is 0.01, what is the output resistance without feedback ?

[*G.G.S.I.P. Univ. Analog Electronics December 2010*]

Solution : Voltage gain without feedback, $A = 100$

Feedback fraction, $\beta = 0.01$

Output resistance with feedback, $R_{out\,f} = 10\ \Omega$

Output resistance without feedback, $R_{out} = R_{out\,f}(1 + \beta A) = 10(1 + 0.01 \times 100) = 20\ \Omega$ **Ans.**

9.7. EFFECT OF NEGATIVE FEEDBACK ON INPUT IMPEDANCE

High input impedance is always desirable in an amplifier as it will not load the preceding stage or the input voltage source. Such a desirable characteristic can be had with the help of negative feedback. The effect of negative feedback on the input impedance of an amplifier is explained below.

9.7.1. Voltage-Series Feedback

A more detailed series feedback connection is shown in Fig. 9.13. The input impedance can be determined as follows :

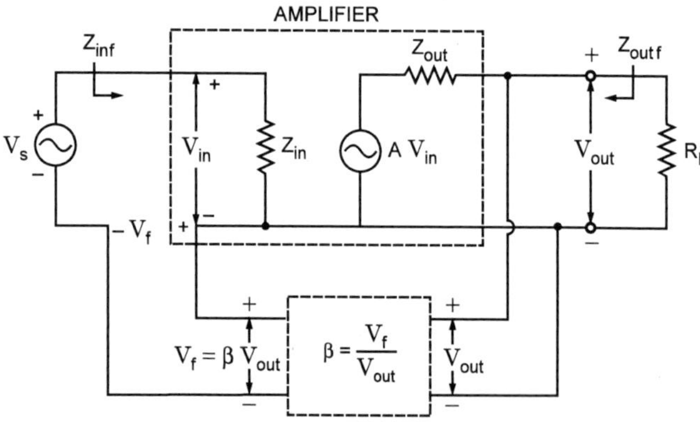

Fig. 9.13

$$I_{in} = \frac{V_{in}}{Z_{in}} = \frac{V_s - V_f}{Z_{in}} = \frac{V_s - \beta V_{out}}{Z_{in}} \qquad \because V_f = \beta V_{out}$$

$$= \frac{V_s - \beta A V_{in}}{Z_{in}} \qquad \because V_{out} = AV_{in}$$

or $I_{in} Z_{in} = V_s - \beta A V_{in}$

or $V_s = I_{in} Z_{in} + \beta A V_{in} = I_{in} Z_{in} + \beta A I_{in} Z_{in} \quad \because V_{in} = I_{in} Z_{in}$

or $\frac{V_s}{I_{in}} = Z_{in} + \beta A Z_{in}$

and $$Z_{inf} = \frac{V_s}{I_{in}} = Z_{in} + (\beta A)\, Z_{in} = Z_{in}(1 + \beta A) \qquad ...(9.17)$$

The input impedance with series feedback is seen to have the value of input impedance without feedback multiplied by the factor $(1 + \beta A)$ and applies to both voltage-series and current series configuration [Figs. 9.10 (*a*) and 9.10 (*c*)].

9.7.2. Voltage-Shunt Feedback

A more detailed shunt feedback connection is depicted in Fig. 9.14.

In this case, $I_s = I_{in} + I_f$ ...(9.18)

and input impedance,

$$Z_{inf} = \frac{V_{in}}{I_s} = \frac{V_{in}}{I_{in} + I_f} = \frac{V_{in}}{I_{in} + \beta V_{out}}$$

$$= \frac{V_{in}/I_{in}}{\frac{I_{in}}{I_{in}} + \frac{\beta V_{out}}{I_{in}}} = \frac{Z_{in}}{1 + \beta A} \qquad ...(9.19)$$

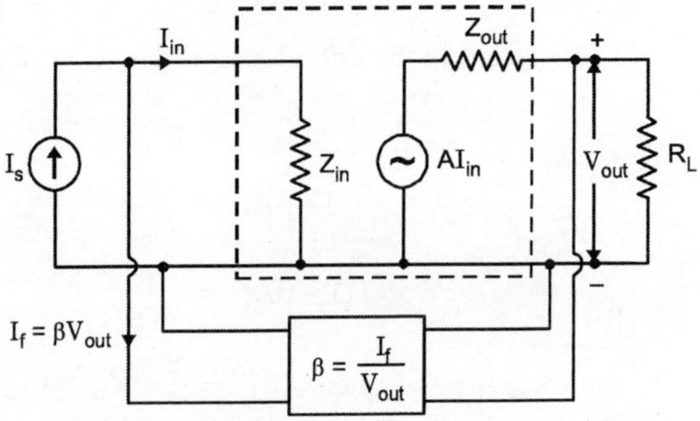

Fig. 9.14. ***Voltage-Shunt Feedback Connection***

So, the input impedance with voltage shunt feedback is reduced by the factor $(1 + \beta A)$.

9.7.3. Current-Shunt Feedback

From Fig. 9.10(*d*) $$I_s = I_{in} + I_f = I_{in} + \beta I_{out} = I_{in} + \beta(AI_{in}) = I_{in}(1 + \beta A) \qquad ...(9.20)$$

and input impedance, $$Z_{inf} = \frac{V_{in}}{I_s} = \frac{V_{in}}{I_{in}(1 + \beta A)} = \frac{V_{in}/I_{in}}{1 + \beta A} = \frac{Z_{in}}{1 + \beta A} \qquad ...(9.21)$$

So, the input impedance with current shunt feedback is reduced by the factor $(1 + \beta A)$.

9.8. EFFECT OF NEGATIVE FEEDBACK ON OUTPUT IMPEDANCE

Just as high input impedance is advantageous to an amplifier, so is low output impedance. With lower output impedance, the amplifier is better suited to drive a low impedance load. Such a desirable characteristic can be had by employing negative feedback. The effect of negative feedback on the output impedance of an amplifier is explained below.

9.8.1. Voltage-Series Feedback

The voltage-series feedback circuit given in Fig. 9.13 provides sufficient circuit detail for determining output impedance with feedback. The input terminals are short circuited so that $-V_f$ is now the only input voltage to the amplifier. Now a voltage source V_{out} is applied at the output terminals so that I_{out} current is drawn from the applied source.

Now $$V_{out} = I_{out} Z_{out} + AV_{in} = I_{out} Z_{out} - AV_f = I_{out} Z_{out} - A(\beta V_{out}) \qquad \because V_{in} = -V_f \text{ and } V_f = \beta V_{out}$$

$$\text{or } V_{out} + A\beta V_{out} = I_{out} Z_{out}$$

$$\text{or } V_{out}(1 + \beta A) = I_{out} Z_{out}$$

$$\text{or } \frac{V_{out}}{I_{out}} = \frac{Z_{out}}{1+\beta A} \quad ...(9.22)$$

$$\text{or } Z_{outf} = \frac{Z_{out}}{1+\beta A} \qquad \because \text{ Output impedance with feedback } Z_{outf} = \frac{V_{out}}{I_{out}}$$

Thus, series voltage negative feedback reduces the output impedance of an amplifier by a factor $(1 + \beta A)$. This is the same factor by which voltage gain is reduced.

9.8.2. Current-Shunt Feedback

The output impedance with current shunt feedback can be determined by disconnecting the load resistance R_L, applying a signal V to the output terminals with current I_s removed. Now output current I is measured. The ratio of V to I gives the output impedance of circuit [Fig. 9.10 (*d*)].

$$\text{With } I_s = 0, I_{in} = -I_f = -\beta I_{out} = -\beta I$$

Since current in the output circuit due to feedback opposes the current I, due to applied voltage, V

$$\text{So, } I = \frac{V}{Z_{out}} - AI_f = \frac{V}{Z_{out}} - A \times \beta I$$

$$\text{or } I + \beta AI = \frac{V}{Z_{out}}$$

$$\text{or } I(1+\beta A) = \frac{V}{Z_{out}} \quad ...(9.23)$$

$$\text{and output impedance, } Z_{outf} = \frac{V}{I} = Z_{out}(1+\beta A) \quad ...(9.24)$$

So, output impedance is increased by the factor $(1 + \beta A)$.

9.8.3. Current-Series Feedback

The output impedance with current-series feedback can be determined by disconnecting load resistance R_L, applying a signal V to the output with V_s shorted out. The ratio of V to I gives the output impedance of the circuit. Now

$$I = \frac{V}{Z_{out}} - AV_{in}$$

$$= \frac{V}{Z_{out}} - AV_f = \frac{V}{Z_{out}} - A\beta I_{out}$$

$$[\because V_{in} = V_f = \beta I_{out}]$$

$$\text{or } Z_{out}(1+\beta A)I = V \qquad \because I_{out} = I$$

or Output impedance,

$$Z_{outf} = \frac{V}{I} = Z_{out}(1+\beta A) \quad ...(9.25)$$

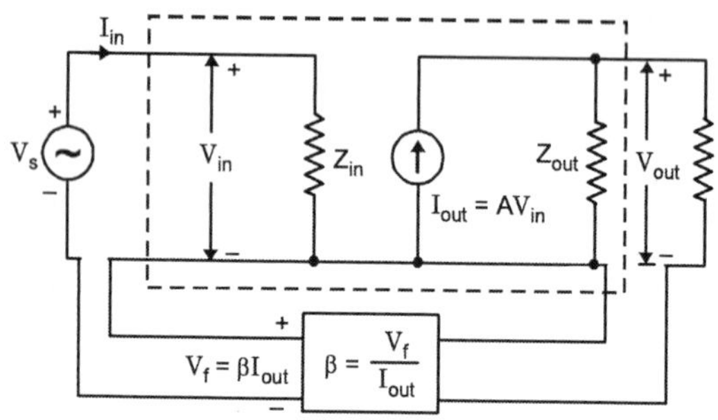

Fig. 9.15. ***Current-Series Feedback Connection***

So, the output impedance is increased by the factor $(1 + \beta A)$.

Example 9.19. Calculate the voltage gain, input and output resistances of a voltage-series feedback amplifier having $A_v = 300$, $R_i = 1.5\ k\Omega$, $R_o = 50\ k\Omega$ and $\beta = \frac{1}{15}$.

Solution : *Without Feedback*

$$\text{Voltage gain, } A = 300$$

$$\text{Input resistance, } R_{in} = 1.5 \text{ k}\Omega$$

$$\text{Output resistance, } R_{out} = 50 \text{ k}\Omega$$

$$\beta = \frac{1}{15}$$

With Feedback

$$\text{Voltage gain, } A_f = \frac{A}{1+\beta A} = \frac{300}{1+\frac{1}{15}\times 300} = 14.3 \textbf{ Ans.}$$

$$\text{Input resistance, } R_{inf} = (1+\beta A)R_{in} = \left(1+\frac{1}{15}\times 300\right)\times 1.5 = 31.5 \text{ k}\Omega$$

$$\text{Output resistance, } R_{out f} = \frac{R_{out}}{1+\beta A} = \frac{50}{1+\frac{1}{15}\times 300} = 2.38 \text{ k}\Omega \textbf{ Ans.}$$

Example 9.20. A negative feedback amplifier in voltage-series configuration feeds 10% of the output back to the input. Voltage gain of the amplifier without feedback is 100. Input and output resistances are 10 kΩ and 1 kΩ respectively. Find percentage reduction in voltage gain, input resistance and output resistance with feedback.

Solution : Given that open-loop gain of amplifier, A = 100; Feedback ratio, β = 10% = 0.1; Input impedance without feedback, Z_{in} = 10 kΩ; Output impedance without feedback, Z_{out} = 1 kΩ

$$\text{Voltage gain with feedback, } A_f = \frac{A}{1+\beta A} = \frac{100}{1+0.1\times 100} = \frac{100}{11} = 9.09$$

$$\text{Percentage reduction in voltage gain} = \frac{100-9.09}{100}\times 100 = 90.91\%. \textbf{ Ans.}$$

$$\text{Input impedance with feedback, } Z_{inf} = (1+\beta A)\, Z_{in} = (1+0.1\times 100)\times 10 = 110 \text{ k}\Omega$$

$$\text{Percentage increase in input impedance} = \frac{Z_{inf}-Z_{in}}{Z_{in}}\times 100 = \frac{110-10}{10}\times 100 = 1{,}000\%$$

i.e., Input impedance with feedback becomes 11 times of that without feedback

$$\text{Output impedance with feedback, } Z_{out f} = \frac{Z_{out}}{1+\beta A} = \frac{1}{1+0.1\times 100} = \frac{1}{11} \text{ k}\Omega$$

$$\text{Percentage reduction in output impedance} = \frac{Z_{out}-Z_{outf}}{Z_{out}}\times 100 = \frac{1-\frac{1}{11}}{1}\times 100 = 90.91\%. \textbf{ Ans.}$$

Example 9.21. An amplifier has a gain of 60 dB (voltage gain). It has an output impedance of Z_0 = 12 kΩ. It is required to modify the output impedance to 600 Ω by applying negative feedback. Calculate the value of feedback factor. Also find the percentage change in overall gain $\frac{dA_f}{A_f}$ for 10% change in gain of internal amplifier $\frac{dA}{A}$.

Solution :

$$\text{Open-loop gain, } A = 60 \text{ dB or antilog}_{10}\frac{60}{20} = 1{,}000$$

$$\text{Output impedance with feedback, } Z_{out f} = \frac{Z_{out}}{1+\beta A}$$

$$\text{Feedback factor, } \beta = \frac{\frac{Z_{out}}{Z_{out f}}-1}{A} = \frac{\frac{12\times 1{,}000}{600}-1}{1{,}000} = \frac{19}{1{,}000} = 0.019 \textbf{ Ans.}$$

For 10% change in the gain of internal amplifier, we have

$$\text{Change in overall gain, } \frac{dA_f}{A_f} = \frac{1}{1+A\beta}\times\frac{dA}{A} = \frac{1}{1+1000\times 0.019}\times 0.1 = 0.005 \text{ or } 0.5\% \textbf{ Ans.}$$

9.9. EFFECT OF NEGATIVE FEEDBACK ON BANDWIDTH

From Eq. (9.8) voltage gain with feedback is given as

$$A_{vf} = \frac{A_v}{1+\beta A_v} \simeq \frac{1}{\beta} \quad \text{if } \beta A_v >> 1 \qquad ...(9.26)$$

and from the above result it may be concluded that the voltage gain may be made to depend entirely on the feedback network. However, it is now important to consider the fact that even if β is constant, the voltage gain A_v is not, since it depends on frequency. This means that at certain low or high frequencies β A_v will not be much larger than unity, and, therefore, Eq. (9.26) will not be valid.

When negative feedback is applied in an amplifier, cutoff frequencies are also affected—lower cutoff frequency is lowered by a factor of (1 + βA) while upper cutoff frequency is raised by the same factor (1 + βA). This is explained below.

(*i*) *Lower Cutoff Frequency* f_1. The voltage gain at a frequency f in low frequency range of R-C coupled amplifier is given as

$$A_{vl} = \frac{A_{vm}}{1 - \frac{jf_1}{f}} \qquad ...(9.27)$$

where f_1 is lower cutoff frequency and $j = \sqrt{-1}$.

When negative feedback is applied

$$A_{vmf} = \frac{A_{vm}}{1+\beta A_{vm}} \qquad ...(9.28)$$

$$\text{and } A_{vlf} = \frac{A_{vl}}{1+\beta A_{vl}} \qquad ...(9.29)$$

Substituting the value of A_{vl} from Eq. (9.27) in Eq. (9.29) we have

$$A_{vlf} = \frac{\dfrac{A_{vm}}{1-jf_1/f}}{1+\beta\dfrac{A_{vm}}{1-jf_1/f}} = \frac{A_{vm}}{1-j\,f_1/f+\beta A_{vm}} = \frac{A_{vm}}{1+\beta A_{vm}-j\,f_1/f}$$

By dividing numerator and denominator by 1 + β A_{vm}, this equation may be rewritten as

$$A_{vlf} = \frac{A_{vmf}}{1-j\,f_1'/f} \qquad ...(9.30)$$

$$\text{where } A_{vmf} = \frac{A_{vm}}{1+\beta A_{vm}} \quad \text{and } f_1' = \frac{f_1}{1+\beta A_{vm}} \qquad ...(9.31)$$

From above equations we see that the midband amplification with feedback A_{vmf} equals the midband amplification without feedback divided by 1 + βA_{vm}. Also the lower cutoff frequency with feedback f_1' equals the lower cutoff frequency without feedback f_1 divided by the same factor (1 + βA_{vm}) *i.e.* lower cutoff frequency is reduced when negative feedback is applied.

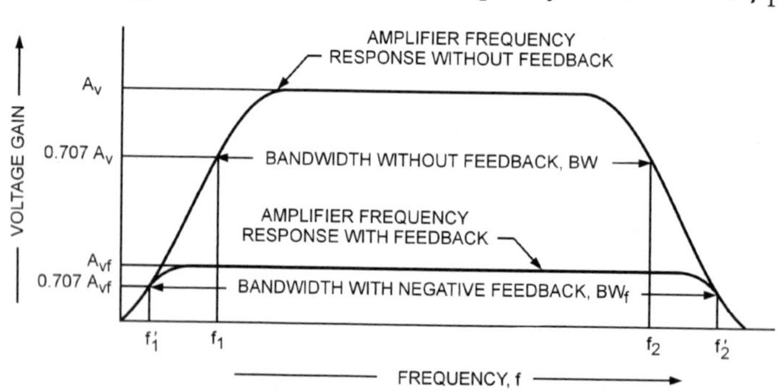

Fig. 9.16. ***Amplifier Frequency Response With and Without Negative Feedback***

(*ii*) *Upper Cutoff Frequency*. The voltage gain at frequency f in the high frequency range of R-C coupled amplifier is given as

$$A_{vh} = \frac{A_{vm}}{1+jf/f_2} \qquad ...(9.32)$$

When negative feedback is applied

$$A_{vhf} = \frac{A_{vh}}{1+\beta A_{vh}} \quad ...(9.33)$$

Substituting the value of A_{vh} from Eq. (9.32) in Eq. (9.33), we have

$$A_{vhf} = \frac{\dfrac{A_{vm}}{1+jf/f_2}}{1+\beta\dfrac{A_{vm}}{1+jf/f_2}} = \frac{A_{vm}}{1+\beta A_{vm}+jf/f_2}$$

By dividing the numerator and denominator by $1 + \beta A_{vm}$, the above equation may be rewritten as

$$A_{vhf} = \frac{A_{vmf}}{1+jf/f_2'} \quad ...(9.34)$$

$$\text{where } A_{vmf} = \frac{A_{vm}}{1+\beta A_{vm}} \quad \text{and} \quad f_2' = f_2(1+\beta A_{vm}) \quad ...(9.35)$$

Thus we see that upper cutoff frequency with feedback equals the corresponding cutoff frequency without feedback f_2 multiplied by the factor $(1 + \beta A_{vm})$ *i.e.* upper cutoff frequency is raised when negative feedback is applied in an amplifier.

The effect of negative feedback on frequency response curve is shown in Fig. 9.16.

Bandwidth. The bandwidth with negative feedback is given as

$$BW_f = f_2' - f_1' = f_2(1+\beta A_{vm}) - \frac{f_1}{1+\beta A_{vm}} \quad ...(9.36)$$

Since $f_2' > f_2$ and $f_1' < f_1$ hence the bandwidth with negative feedback is increased.

Assuming $f_2' >> f_1'$ and $f_2 >> f_1$ bandwidth may be rewritten as

$$BW_f = f_2(1+\beta A_{vm}) = (1+\beta A_{vm}) \text{ times bandwidth without feedback.}$$

Thus we see that bandwidth of an amplifier with negative feedback is increased by the same factor by which voltage gain is reduced and gain bandwidth product remains the same. However, since the amplifier with feedback has lower gain, the net operation is to trade bandwidth for gain.

Example 9.22. An R-C coupled amplifier has a voltage gain of 1,000, f_1 = 50 Hz, f_2 = 200 kHz and a distortion of 5% without feedback. Find the amplifier voltage gain, f_1', f_2' and distortion when negative feedback is applied with feedback ratio of 0.01.

Solution :

Open-loop voltage gain, A = 1,000

Feedback ratio, β = 0.01

Amplifier voltage gain with feedback, $A_f = \dfrac{A}{1+\beta A} = \dfrac{1{,}000}{1+0.01\times 1{,}000} = 90.91$ **Ans.**

Lower cutoff frequency with feedback, $f_1' = \dfrac{f_1}{(1+\beta A)} = \dfrac{50}{1+0.01\times 1{,}000} = 4.55$ Hz

Upper cutoff frequency with feedback, $f_2' = f_2(1+\beta A) = 200\times 10^3(1+0.01\times 1{,}000) = 2.2$ MHz **Ans.**

Distortion with feedback, $D_f = \dfrac{D}{1+\beta A} = \dfrac{5\%}{1+0.01\times 1{,}000} = 0.455\%$ **Ans.**

Example 9.23. An amplifier has the midband gain of 1,500 and a bandwidth of 4 MHz. The midband gain reduces to 150 when a negative feedback is applied. Determine the value of feedback factor and the bandwidth.

Solution : Midband gain $A_{vm} = 1{,}500$

Midband gain with feedback, $A_{vmf} = \dfrac{A_{vm}}{1 + \beta A_{vm}}$

Substituting the value of $A_{vmf} = 150$ and $A_{vm} = 1{,}500$ in above equation, we have

$$150 = \frac{1{,}500}{1 + \beta A_{vm}}$$

or Feedback factor, $\beta A_{vm} = \dfrac{1{,}500}{150} - 1 = 9$ **Ans.**

Bandwidth with feedback, $BW_f = (1 + \beta A_{vm}) \times$ Bandwidth without feedback $= (1 + 9) \times 4 = 40$ MHz **Ans.**

Example 9.24. An amplifier has a mid-frequency gain of 100 and a bandwidth of 200 kHz.

(*i*) What will be the new bandwidth and gain, if 5% negative feedback is introduced ?

(*ii*) What should be the amount of feedback, if the bandwidth is to be restricted to 1 MHz ?

Solution : Mid-frequency gain, $A = 100$

Bandwidth without feedback, $BW = 200$ kHz

Feedback ratio, $\beta = 5\% = 0.05$

(*i*) Bandwidth with feedback, $BW_f = (1 + \beta A) \times BW = (1 + 0.05 \times 100) \times 200 \text{ kHz} = 1{,}200$ kHz or 1.2 MHz **Ans.**

Gain with feedback, $A_f = \dfrac{A}{1 + \beta A} = \dfrac{100}{1 + 0.05 \times 100} = 16.67$ **Ans.**

Feedback ratio to restrict the bandwidth to 1 MHz or 1,000 kHz

$$\beta = \frac{\dfrac{BW_f}{BW} - 1}{A} = \frac{\dfrac{1000}{200} - 1}{100} = 0.04 \text{ or } 4\% \textbf{ Ans.}$$

9.10. EFFECTS OF NEGATIVE FEEDBACK ON AMPLIFIER CHARACTERISTICS

The characteristics of four types of negative feedback amplifiers are summarized below in tabular form.

TABLE 9.3. Effect of Negative Feedback on Amplifier Characteristics.

Type of Feedback/ Characteristics	Voltage Series	Voltage Shunt	Current Series	Current Shunt
Type of amplifier	Voltage	Transresistance	Transconductance	Current
Desensitizes	A_{vf}	R_{mf}	G_{mf}	A_{if}
Transfer gain	Decreases	Decreases	Decreases	Decreases
Nonlinear distortion	Decreases	Decreases	Decreases	Decreases
Noise	Decreases	Decreases	Decreases	Decreases
Bandwidth	Increases	Increases	Increases	Increases
Input impedance	Increases	Decreases	Increases	Decreases
Output impedance	Decreases	Decreases	Increases	Increases

9.11. ANALYSIS OF A FEEDBACK AMPLIFIER

The important characteristics, like A_f, $R_{in\,f}$, $R_{out\,f}$ of the feedback system can be computed with the knowledge of basic amplifier A and feedback network β, so separation of feedback amplifier into two blocks, the basic amplifier and feedback network, will be desirable.

The complete analysis of a feedback amplifier is accomplished by going through the following steps :

Step 1 : Identification of Topology (Type of Feedback)

(*a*) *Identification of Type of Mixing Network*

(*i*) If the feedback signal is subtracted from the externally applied signal as a voltage in the input loop, it is *series mixing.*

(*ii*) If the feedback signal is subtracted from the externally applied signal as a current in the input loop, it is *shunt mixing.*

(*b*) *Identification of Type of Sampling Network :*

(*i*) If the feedback signal becomes zero when the output is shorted, it is *voltage sampling.*

(*ii*) If the feedback signal becomes zero when the output is opened, it is *current sampling.*

Thus, by determination of type of mixing network and sampling network, type of feedback amplifier can be determined. For instance, if an amplifier makes use of a current sampling and series mixing, then it is a current series amplifier.

Step 2 : Determination of Input Circuit

(*i*) For voltage sampling, the output voltage is made zero by shorting the output node.

(*ii*) For current sampling, the output current is made zero by opening the output loop.

Step 3 : Determination of Output Circuit

(*i*) For series mixing, the input current is made zero by opening the input loop.

(*ii*) For shunt mixing, the input voltage is made zero by shorting the input node.

These procedures ensure that the feedback is reduced to zero without altering the loading on the basic amplifier.

Step 4 : Replacement of each active device by its h-parameter model at low frequencies and hybrid π model for a transistor at high frequencies.

Step 5 : Determination of open-loop gain A of the amplifier by applying KVL and KCL to the equivalent circuit.

Step 6 : Indicate X_f (feedback voltage or feedback current) and X_{out} (output voltage or output current) on the circuit and evaluate $\beta = \frac{X_f}{X_{out}}$.

Step 7 : Determination of desensitivity, transfer gain A_f, input resistance $R_{in\,f}$, output resistance $R_{out\,f}$ and $R'_{out\,f}$. Above procedure is summarized below, in tabular form, for reference in analysis of the feedback circuits.

TABLE 9.4. Feedback Amplifiers

Characteristics \ Topology	Voltage-Series (1)	Voltage-Shunt (2)	Current-Series (3)	Current-Shunt (4)
Feedback signal, X_f	Voltage	Current	Voltage	Current
Sampled signal, X_s	Voltage	Voltage	Current	Current
Determination of input loop, set	$V_{out} = 0$	$V_{out} = 0$	$I_{out} = 0$	$I_{out} = 0$
Determination of output loop, set	$I_{in} = 0$	$V_{in} = 0$	$I_{in} = 0$	$V_{in} = 0$
Signal source	Thevenin	Norton	Thevenin	Norton
Reverse transmission factor, $\beta = \frac{X_f}{X_{out}}$.	$\frac{V_f}{V_{out}}$	$\frac{I_f}{V_{out}}$	$\frac{V_f}{I_{out}}$	$\frac{I_f}{I_{out}}$
Transfer ratio, $A = \frac{X_{out}}{X_{in}}$	$A_v = \frac{V_{out}}{V_{in}}$	$R_M = \frac{V_{out}}{I_{in}}$	$G_M = \frac{I_{out}}{V_{in}}$	$A_I = \frac{I_{out}}{I_{in}}$
Desensitivity, $D = 1 + \beta A$	$1 + \beta A_v$	$1 + \beta R_M$	$1 + \beta G_M$	$1 + \beta A_I$
Transfer Gain with feedback, A_f	A_v/D	R_M/D	G_M/D	A_I/D
Input resistance with feedback, $R_{in f}$	DR_{in}	R_{in}/D	DR_{in}	R_{in}/D
Output resistance with feedback $R_{out f}$	$\frac{R_{out}}{1+\beta A_v}$	$\frac{R_{out}}{1+\beta R_m}$	$R_{out}(1 + \beta G_m)$	$R_{out}(1 + \beta A_i)$
$R'_{out f} = R_{out f} \| R_L$	$\frac{R'_{out}}{D}$	$\frac{R'_{out}}{D}$	$R'_{out}\frac{1+\beta G_m}{D}$	$\frac{R'_{out}(1+\beta A_i)}{D}$

9.12. VOLTAGE-SERIES FEEDBACK

This is also called the *shunt-derived series-fed feedback*. Here the amplifier and feedback network are connected in series-parallel. A fraction of the output voltage is applied in series opposition to the input voltage through feedback network, as shown in Fig. 9.10 (*a*).

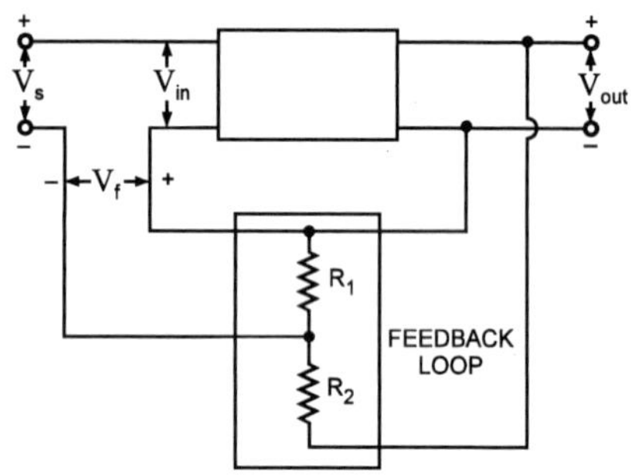

Fig. 9.17. ***Voltage-Series Feedback Circuit***

In a block diagram given in Fig. 9.17, the feedback voltage is derived from the voltage divider circuit formed of resistors R_1 and R_2. Thus feedback voltage is given as

$$V_f = \beta V_{out} = \frac{R_1}{R_1 + R_2} V_{out}$$

Thus $$\beta = \frac{R_1}{R_1 + R_2}$$

As seen, the input impedance of the amplifier and output impedance of the feedback network appear in series to the input and, therefore, input impedance to the amplifier is increased by a factor $(1 + \beta A)$, as shown in Art. 9.7.1.

Similarly input to the feedback network and output of the amplifier appear in parallel to the amplifier output. Thus, so far as V_{out} is concerned, output impedance of the amplifier is reduced, due to the shunting effect of the input to the feedback, by a factor $\frac{1}{1+\beta A}$, as shown in Art. 9.8.1.

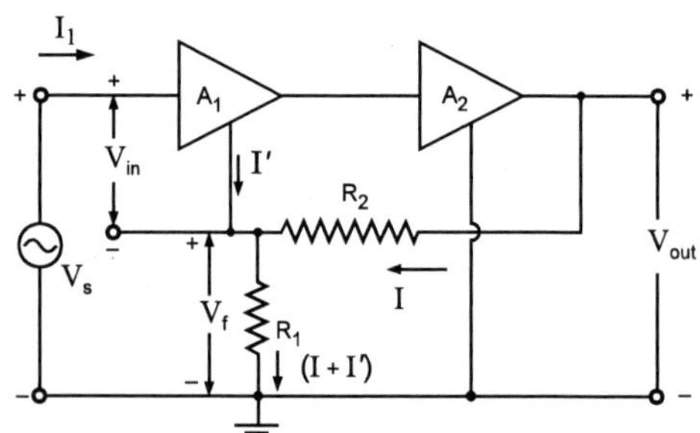

Fig. 9.18

A general two stage amplifier having stage gains of A_1 and A_2 respectively is shown in Fig. 9.18. The output of the second stage is fed back through the feedback network R_1R_2 in opposition to the input signal voltage V_s. Obviously this circuit is a case of *voltage-series negative feedback.* It is expected that input impedance will increase, output impedance will decrease and voltage gain will be stabilised.

An approximate analysis can be made if the open-loop gain is assumed to be very large and $I' << I$. Under such conditions we have

$V_{in} \simeq 0$ for a finite V_{out}, and

$$\text{Supply voltage, } V_s \simeq V_f \simeq I R_1 \simeq R_1 \frac{V_{out}}{R_1 + R_2}$$

$$\text{Thus } \beta = \frac{R_1}{R_1 + R_2}$$

and overall gain with feedback, $$A_f = \frac{V_{out}}{V_s} \simeq \frac{R_1 + R_2}{R_1} = \frac{1}{\beta} \qquad ...(9.37)$$

From the above equation it is obvious that under the above conditions the voltage gain is independent of all parameters except R_1 and R_2. Thus, if these resistors are stable, the circuit amplification is stabilized.

Now let us consider FET common drain amplifier (source follower) and BJT common-collector amplifier (emitter follower) as examples of the voltage-series topology.

FET Source Follower : The circuit is shown in Fig. 9.19. In this circuit, no resistor is connected in series with the drain terminal and no source bycapacitor is used. The input signal is applied to the gate G. The output voltage is developed across source resistor R. Sampled signal is the output voltage V_{out} across source resistor R and feedback signal is the voltage V_f across R. Thus, it is the case of voltage-series feedback and reference should be made to the first topology in Table 9.4.

For drawing basic amplifier without feedback, input circuit is obtained by setting $V_{out} = 0$ and output circuit is obtained by setting $I_{in} = 0$. Thus, V_s appears directly between G and S and R appears only in the output loop, as shown in Fig. 9.19 (*b*). Figure 9.19 (*b*) represents the amplifier circuit without feedback. Replacing FET by its small signal low-frequency model, we have the circuit as shown in Fig. 9.19 (*c*).

From the circuit shown in Fig. 9.19 (*c*).

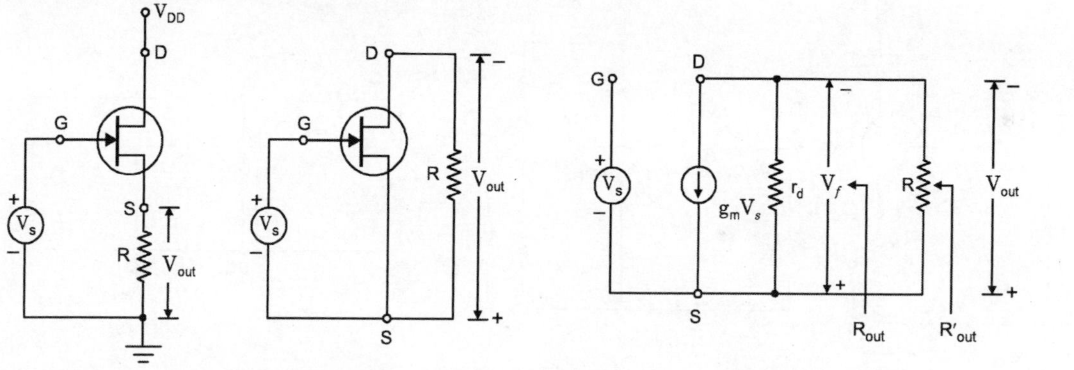

(a) Source Follower (b) Amplifier Without Feedback (c) FET Replaced By Its Small-Signal Low-Frequency Model

Fig. 9.19

Output voltage V_{out} and feedback voltage V_f are equal, so

$$\text{Reverse transmission factor, } \beta = \frac{V_f}{V_{out}} = 1$$

The topology stabilizes the voltage gain.

Input voltage without feedback, $V_{in} = V_s$ Refer to Fig. 9.19 (*b*)

From Fig. 9.19 (*c*)

$$\text{Voltage gain, } A_v = \frac{V_{out}}{V_{in}} = \frac{g_m V_s r_d R}{(r_d + R)V_s} = \frac{\mu R}{r_d + R} \quad \text{...(9.38)} \because \mu = g_m r_d \text{ and } V_{in} = V_s$$

$$\text{Desensitivity, } D = 1 + \beta A_v = 1 + \frac{\mu R}{r_d + R} = \frac{r_d + (1+\mu)R}{r_d + R} \quad \text{...(9.39)}$$

$$\text{and voltage gain with feedback, } A_{vf} = \frac{A_v}{D} = \frac{\mu R}{r_d + (1+\mu R)} \quad \text{...(9.40)}$$

The input impedance of an FET is infinite, $R_{in} = \propto$, and therefore,

$$R_{inf} = R_{in\,D} = \propto$$

We are interested in determination of output resistance looking into the FET source S. Hence R is considered as an external load resistance R_L.

The output resistance, $R_{out} = r_d$

$$\text{and} \quad R_{outf} = \frac{R_{out}}{1 + \beta A_v} = \frac{r_d}{1+\mu} \quad \text{...(9.41) Refer to first column of Table 9.4}$$

$$\because \beta = 1 \text{ and } A_v = \lim_{R \to \infty} A_v = \mu$$

$$R'_{outf} = \frac{R'_{out}}{D} = \frac{R r_d}{R + r_d} \cdot \frac{r_d + R}{r_d + (\mu+1)R} \quad \text{...(9.42)}$$

$$\because R_{outf} = \lim_{R \to \infty} R'_{outf} = \frac{r_d}{\mu + 1} \text{ agreeing with Eq. (9.41)}$$

BJT Emitter Follower : The circuit is depicted in Fig. 9.20. In this circuit no emitter bypass capacitor is used. The feedback signal is the voltage V_f across emitter resistor R_e and the sampled signal V_s is the output voltage across emitter resistor R_e. Thus, it is a case of voltage series feedback and first topology in Table 9.4 should be referred.

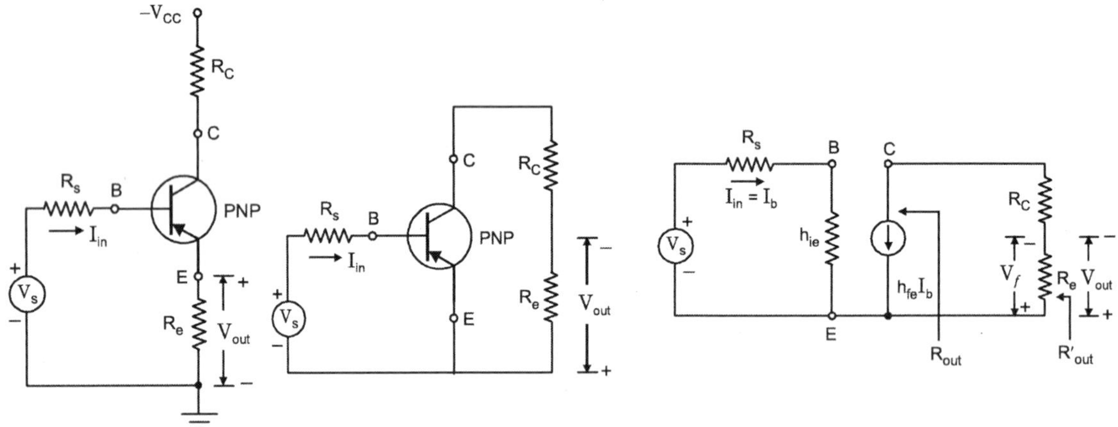

(*a*) *An Emitter Follower* (*b*) *Amplifier Without Feedback* (*c*) *BJT Replaced by Its Small-Signal Low-Frequency Model*

Fig. 9.20. ***Emitter Follower***

For drawing basic amplifier without feedback, input circuit is obtained by setting $V_{out} = 0$ and output circuit is obtained by setting $I_{in} = I_b = 0$. Thus, V_s in series with R_s appears between B and E and R_e appears only in the output loop, as depicted in Fig. 9.20 (*b*). Fig. 9.20 (*b*) represents the amplifier circuit without feedback. Replacing BJT by its small-signal low-frequency model, we have circuit as shown in Fig. 9.20 (*c*).

From the circuit shown in Fig. 9.20 (*c*)

$$V_{out} = V_f \text{ and } \beta = \frac{V_f}{V_{out}} = 1$$

This topology stabilizes the voltage gain.
From inspection of Fig. 9.20 (*c*)

$$\text{Voltage gain, } A_v = \frac{V_{out}}{V_{in}} = \frac{h_{fe} I_b R_e}{V_s} = \frac{h_{fe} R_e}{R_s + h_{ie}} \quad ...(9.43)$$

$\because$ R_s is considered as a part of amplifier and, therefore, $V_{in} = V_s = I_b (R_s + h_{ie})$

$$\text{Desensitivity, } D = 1 + \beta A_v = 1 + \frac{h_{fe} R_e}{R_s + h_{ie}} = \frac{R_s + h_{ie} + h_{fe} R_e}{R_s + h_{ie}} \quad ...(9.44)$$

$$\text{Voltage gain with feedback, } A_{vf} = \frac{A_v}{D} = \frac{h_{fe} R_e}{R_s + h_{ie} + h_{fe} R_e} \quad ...(9.45)$$

For $h_{fe} R_e >> R_s + h_{ie}$, $A_{vf} \approx 1$, as it is expected for an emitter follower.

The input resistance without feedback,

$$R_{in} = R_s + h_{ie}$$

$$R_{inf} = R_{in} D = (R_s + h_{ie}) \times \frac{R_s + h_{ie} + h_{fe} R_e}{R_s + h_{ie}} = R_s + h_{ie} + h_{fe} R_e \quad ...(9.46)$$

We are interested in determination of resistance seen looking into the emitter. So R_e is considered as an external load. From Table 9.4

$$R_{outf} = \frac{R_{out}}{1 + \beta A_v} = \frac{\infty}{\infty} \quad ...(9.47)$$

Because from Fig. 9.20 (*c*), we are looking into a current source, $R_{out} = \infty$ and $A_v = \lim_{Re \to \infty} A_v = \infty$. The indeterminacy in Eq. (9.47) may be resolved by first evaluating R'_{outf} and then going to the limit $R_e \to \infty$. Thus, since $R'_{out} = R_e$,

$$R'_{outf} = \frac{R'_{out}}{D} = \frac{R_e (R_s + h_{ie})}{R_s + h_{ie} + h_{fe} R_e} \quad ...(9.48)$$

and

$$R_{outf} = \lim_{R_e \to \infty} R'_{outf} = \frac{R_s + h_{ie}}{h_{fe}} \quad ...(9.49)$$

Note that the feedback desensitizes voltage gain with respect to variations in h_{fe} and that it increases the input resistance and reduces the output resistance.

The foregoing expressions for A_{vf}, R_{inf} and R_{outf} are based on the assumptions that there is zero forward transmission through the feedback network. But it is not true. On contrary there is such forward transmission due to flow of input current through R_e in Fig. 9.20 (*a*) and so these expressions are only approximately true. Here we have in effect neglected the base current that flows in R_e compared with the collector current.

9.13. VOLTAGE-SHUNT FEEDBACK

It is also known as *shunt-derived shunt-fed feedback* or *voltage shunt inverse feedback*. Here, a small portion of the output voltage is coupled back to the input voltage, as shown in Fig. 9.10 (*b*). Since the feedback network shunt both the input and output of the amplifier, both of the input and output impedances are reduced by a factor $\frac{1}{(1 + \beta A)}$.

Figure 9.21 gives the circuit of a common-emitter amplifier stage with a resistor R_f providing feedback from collector to base. Here the feedback is proportional to the output voltage V_{out} and feedback current I_f gets added in shunt with the input. Thus this circuit forms the case of voltage-shunt inverse feedback amplifier.

From circuit shown in Fig. 9.21.

Feedback current,

$$I_f = \frac{V_{in} - V_{out}}{R_f}$$

$$\simeq \frac{-V_{out}}{R_f} = \beta V_{out} \text{ where } \beta = \frac{-1}{R_f} \quad ...(9.50)$$

$\because V_{in}$ is quite small in comparison with V_{out}

From Eqs. (9.12) and (9.50) and assuming $I_s \simeq I_f$

$$A_f \equiv R_{Mf} \equiv \frac{V_{out}}{I_s} \simeq \frac{1}{\beta} = -R_f \quad ...(9.51)$$

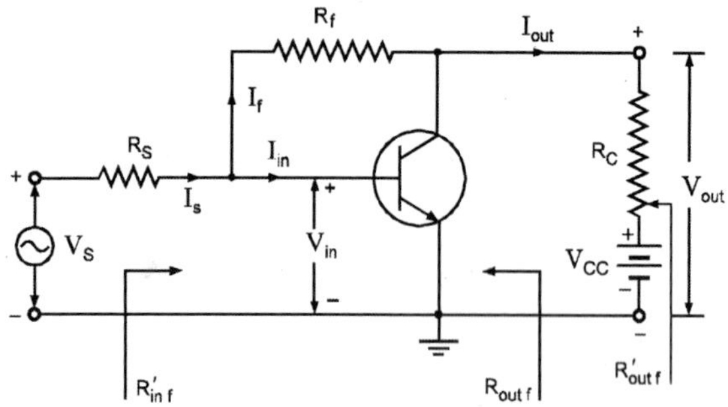

Fig. 9.21. ***Voltage-Shunt Negative Feedback***

Thus the transresistance equals the negative of the feedback resistance from output to input of the transistor and is stable if R_f is a stable resistance. Both the input and output resistances are reduced because the feedback network shunts both the input and output of the amplifier. Assuming input impedance to be zero, voltage gain with feedback.

$$A_f = \frac{V_{out}}{V_s} \simeq \frac{V_{out}}{I_s R_s} = \frac{-R_f}{R_s} \quad ...(9.52)$$

Thus if resistors R_f and R_s are stable elements, then voltage gain is stable (independent of the transistor parameters, the temperature or supply voltage variations).

9.13.1. Amplifier Without Feedback

Referring to second topology in Table 9.4, the input circuit of the amplifier without feedback is obtained by setting $V_{out} = 0$ (shorting the output node). Thus connecting resistor R_f from base to emitter of the transistor. The output circuit is obtained by setting $V_{in} = 0$ (shorting the input node), thus connecting R_f from collector to emitter. Thus, the amplifier circuit without feedback is depicted in Fig. 9.22. Since the feedback signal is a current, the source is represented by a Norton's equivalent with $I_s = \frac{V_s}{R_s}$. The feedback signal is the current I_f in the feedback resistor R_f which is the output circuit.

From Fig. 9.22,
$$\beta = \frac{I_f}{V_{out}} = -\frac{1}{R_f} \quad ...(9.53)$$

which is agreement with Eq. (9.51)

If the amplifier is deactivated by reducing h_{fe} to zero, a current I_f passes through the β network (the resistor R_f) from input to output and this current is

$$I_f = \frac{V_s}{R_s + R_f + R_C} \quad ...(9.54)$$

The output current I_{out} with the amplifier activated is

$$I_{out} = \frac{V_{out}}{R_C} = \frac{A_{vf} V_s}{R_C} \quad ...(9.55)$$

So the condition that the forward transmission through the feedback network can be neglected is $I_{out} >> I_f$

$$\text{or} \quad A_{vf} >> \frac{R_C}{R_s + R_f + R_C} \quad ...(9.56)$$

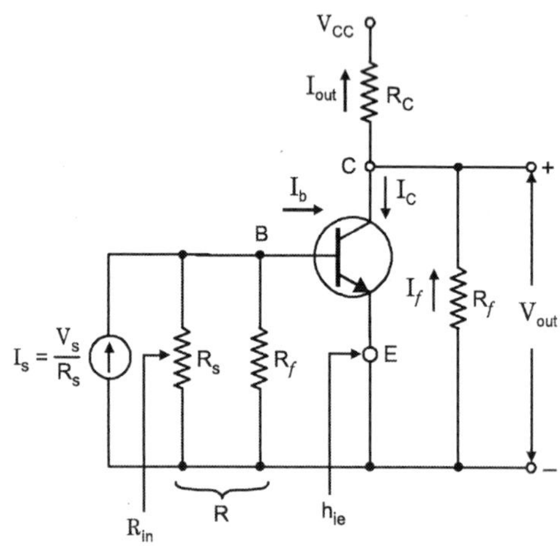

Fig. 9.22. ***Amplifier Without Feedback***

Since this voltage gain is at least unity, this inequality is easily satisfied by selecting $(R_s + R_f) >> R_C$.

Example 9.25. Do the feedback analysis of the following circuit to determine (*i*) feedback topology (*ii*) Feedback factor β (*iii*) A_v, A_{vf}, R_{in}, $R_{in f}$, R_{out}, $R_{out f}$ (*iv*) Desensitivity factor D. Draw the input and output circuit without feedback, but having the effect of feedback. **[G.G.S.I.P. Univ. Analog Electronics, December 2010]**

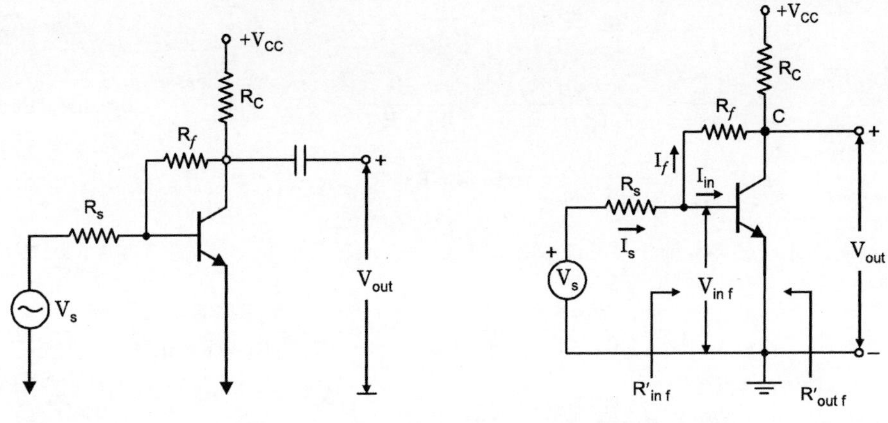

Fig. 9.23

Solution : The circuit is redrawn as shown in Fig. 9.23 (*b*).

(*i*) Figure 9.23 (*b*) indicates that the circuit is a common-emitter amplifier stage with a resistor R_f providing a feedback from collector to base. Here the feedback current is proportional to the output V_{out}, because one end of R_f is connected to the output mode. The feedback current I_f gets added in shunt with the input. Thus, it is a *voltage shunt topology.*

The input circuit of the amplifier without feedback is found by shorting the output node ($V_{out} = 0$). This places R_f between base and ground. The output circuit is obtained by shorting the input node ($V_{in} = 0$), this places R_f between collector and ground.

From circuit shown in Fig. 9.23 (*b*),

$$\text{Feedback current } I_f = \frac{V_{in} - V_{out}}{R_f} = \frac{-V_{out}}{R_f} \quad \because V_{out} > V_{in}.$$

The resultant equivalent circuit is shown in Fig. 9.22.

The input signal source is converted into a Norton's source (current source). Source resistance appears in parallel with the current source I_s where

$$I_s = \frac{V_s}{R_s}$$

If the BJT is replaced by its h-parameter equivalent circuit, we have the circuit, as shown in Fig. 9.24.

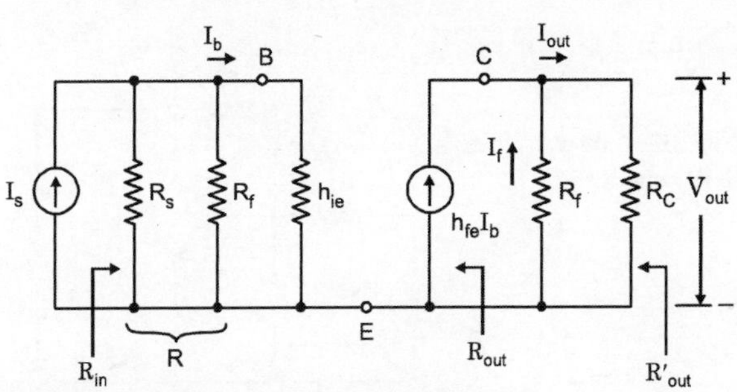

Fig. 9.24

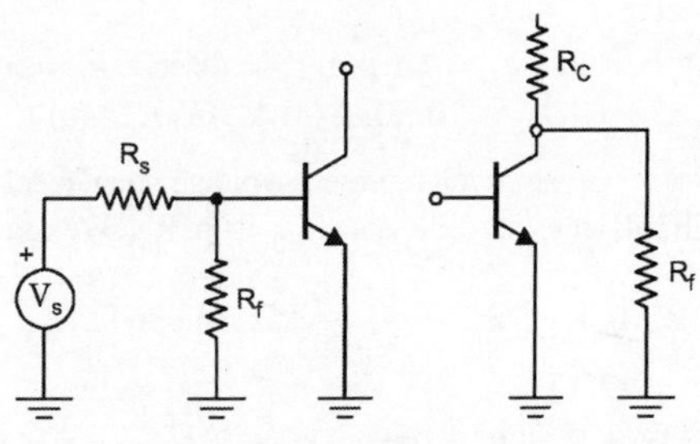

(a) Input Circuit *(b) Output Circuit*

Fig. 9.25

Since the transresistance is stabilised, let us calculate first R_M

$$R_M = \frac{V_{out}}{I_s} = \frac{I_{out} R_C}{I_s} \quad ...(i)$$

where
$$\frac{I_{out}}{I_s} = \frac{I_{out}}{I_b} \times \frac{I_b}{I_s},$$

$$\frac{I_{out}}{I_b} = \frac{-h_{fe} I_b R_f}{I_b (R_f + R_C)} = \frac{-h_{fe} R_f}{R_f + R_C} \quad \text{...by current division rule}$$

and
$$\frac{I_b}{I_s} = \frac{R}{R + h_{ie}} \text{ where } R = R_s \,||\, R_f$$

Substituting these values of $\frac{I_{out}}{I_b}$ and $\frac{I_b}{I_s}$ in Eq. (*i*), we have

$$R_M = \frac{-h_{fe} R_f}{R_f + R_C} \times \frac{R}{R + h_{ie}} \times R_C = \frac{-h_{fe} R_f R R_C}{(R_f + R_C)(R + h_{ie})} \quad ...(ii)$$

From Fig. 9.24 (*ii*) $\beta = \frac{I_f}{V_{out}} = \frac{-1}{R_f}$ **Ans.** ...(*iii*)

Desensitivity factor, $D = 1 + \beta R_M$ **Ans.**

Transresistance with feedback, $R_{Mf} = \frac{R_M}{D}$ **Ans.**

Voltage gain with feedback, $A_{vf} = \frac{V_{out}}{V_s} = \frac{V_{out}}{I_s R_s} = \frac{R_{Mf}}{R_s} = \frac{-R_f}{R_s}$ **Ans.** $\because R_{Mf} = -R_f$

Input resistance $R_{in} = R_s \,||\, R_f \,||\, h_{ie}$ **Ans.**

Input resistance with feedback, $R_{inf} = \frac{R_{in}}{D} = \frac{R_s \,||\, R_f \,||\, h_{ie}}{D}$ **Ans.**

Output resistance, $R_{out} = \infty$

$R'_{out} = R_{out} \,||\, R_f \,||\, R_C$ **Ans.**

Output resistance with feedback, $R_{outf} = \frac{R'_{out}}{D}$ **Ans.**

Input and output circuit without feedback, but having the effect of feedback element are given in Fig. 9.25.

Example 9.26. The circuit of given figure (Fig. 9.26) has the following parameters :

$R_C = 4$ kΩ, $h_{ie} = 1.1$ kΩ, $R' = 40$ kΩ, $h_{fe} = 50$, $R_s = 10$ kΩ, $h_{re} = h_{oe} = 0$. Find (*a*) A_{vf} (*b*) R_{inf} (*c*) R'_{outf}.

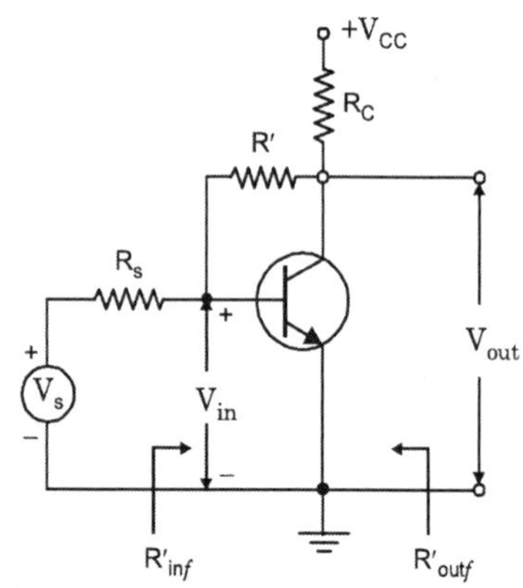

Fig. 9.26

Solution : Given circuit represents voltage-shunt feedback. Since the transresistance is stabilized, let us first calculate R_{MF} from R_M. We can define R'_C and R by

$$R'_C = R_C \,||\, R' = \frac{4 \times 40}{4 + 40} \text{ k}\Omega = 3.636 \text{ k}\Omega$$

$$\text{and } R = R_s \,||\, R' = 10 \,||\, 40 \text{ k}\Omega = \frac{10 \times 40}{10 + 40} = 8 \text{ k}\Omega$$

From circuit shown in Fig. 9.27.

$$R_M = \frac{V_{out}}{I_s} = \frac{-I_C R'_C}{I_s} = \frac{-h_{fe} I_b R'_C}{I_s}$$

or $R_M = \dfrac{-h_{fe}R'_C R}{R + h_{ie}} = \dfrac{-50 \times 3.636 \times 8}{8 + 1.1} = -160\ k\Omega \quad \because I_b = I_s \times \dfrac{R}{R + h_{ie}}$

$$\beta = -\frac{1}{R'} = \frac{-1}{40\ k\Omega} = -0.025\ mA/V$$

Desensitivity factor,

$$D = 1 + \beta R_M = 1 + (-0.025\ mA/V)(-160\ k\Omega) = 5.00$$

$$R_{Mf} = \frac{R_M}{D} = \frac{-160}{5.00} = -32.0\ k\Omega$$

$$A_{vf} = \frac{R_{Mf}}{R_s} = \frac{-32}{10} = -3.2 \ \textbf{Ans.}$$

Input resistance, $R_{in} = \dfrac{Rh_{ie}}{R + h_{ie}} = \dfrac{8 \times 1.1}{8 + 1.1} = 0.967\ k\Omega$

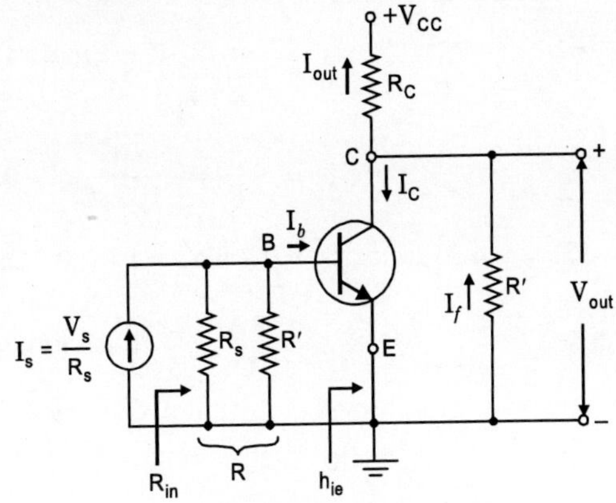

Fig. 9.27. ***Amplifier Without Feedback But Including Loading of R′***

(*b*) $R_{inf} = \dfrac{R_{in}}{D} = \dfrac{0.967}{5} = 0.193\ k\Omega$ or $193\ \Omega$ **Ans.**

Note that the input resistance is quite small as expected. If the input resistance looking to the right of R_s (from base to emitter in Fig. 9.26) is R'_{inf}, then $R_{inf} = R'_{inf} \,||\, R_s$

Solving we have $\quad R'_{inf} = 196\ \Omega \simeq 0.2\ k\Omega$

The impedance seen by voltage source V_s is

$$R_s + R'_{inf} = 10 + 0.2 = 10.2\ k\Omega$$

(*c*) If R_C is considered an external load, the output resistance, neglecting feedback is $R_{out} = R' = 40\ k\Omega$

Since $\quad R_{in} = \lim_{R_C \to \infty} R_M = \dfrac{-h_{fe}R'R}{R + h_{ie}} = \dfrac{-50 \times 40 \times 8}{8 + 1.1} = -1{,}758\ k\Omega$

because $\lim_{R_C \to \infty} R'_C = R'$ and from Table 9.4 (with $R_{out} = R' = 40\ k\Omega$)

$$R_{outf} = \frac{R_{out}}{1 + \beta R_m} = \frac{40}{1 + 0.025 \times 1{,}758}\ k\Omega = 890\ \Omega$$

$$R'_{outf} = R_{outf} \,||\, R_C = \frac{890 \times 4{,}000}{890 + 4{,}000} = 728\ \Omega \ \textbf{Ans.}$$

The output resistance, taking R_C into account but neglecting feedback is, as from Fig. 9.27.

$$R'_{out} = R_C \,||\, R' = R'_C = 3.636\ k\Omega$$

From Table 9.4 $\quad R'_{outf} = \dfrac{R'_{out}}{D} = \dfrac{3.636}{5} = 0.728\ k\Omega = 728\ \Omega$ **Ans.**

This is in agreement with the value computed above.

Example 9.27. For the given feedback amplifier circuit, $h_{fe} = 100$, $h_{ie} = 1\ k\Omega$ and neglect h_{re} and h_{oe}. Find

(*i*) $R_{mf} = \dfrac{V_0}{I_s}$ where $I_s = \dfrac{V_s}{R_s}$ (*ii*) $A_{vf} = \dfrac{V_0}{V_s}$ (*iii*) R_{if} (*iv*) R'_{of}.

Solution : From Fig. 9.28, $R_C = 10\ k\Omega$, $R' = 100\ k\Omega$ and $R_s = 1\ k\Omega$.

We may define R'_C and R by

$$R'_C = R_C \,||\, R' = \frac{R_C \times R'}{R_C + R'} = \frac{10 \times 100}{10 + 100} = 9.09\ k\Omega$$

and $R = R_s || R' = \frac{R_s \times R'}{R_s + R'} = \frac{1 \times 100}{1+100} = 0.99\ k\Omega$

From the analysis of the given network, we have

$$R_M = \frac{-h_{fe} R'_C R}{R + h_{ie}} = \frac{-100 \times 9.09 \times 0.99}{0.99 + 1} = -452\ k\Omega$$

$$\beta = -\frac{1}{R'} = -\frac{1}{100\ k\Omega} = -0.01\ mA/V$$

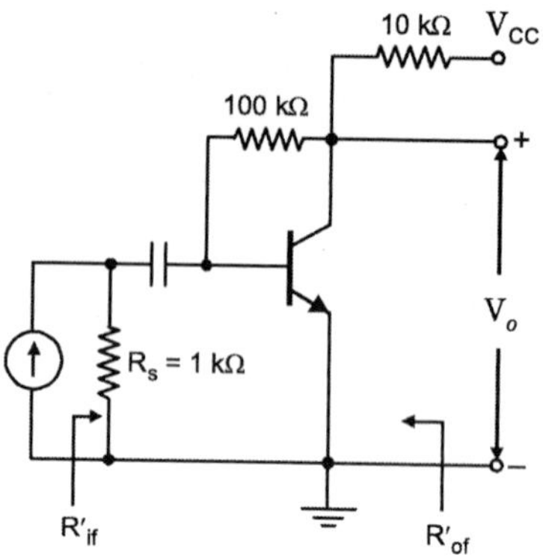

Fig. 9.28

Desensitivity factor, $D = 1 + \beta R_M = 1 + (-0.01\ mA/V)(-452\ k\Omega) = 5.52$

(i) $R_{Mf} = \frac{R_M}{D} = \frac{-452}{5.52} = -81.9\ k\Omega$ **Ans.**

(ii) $A_{vf} = \frac{V_{out}}{V_s} = \frac{R_{Mf}}{R_s} = \frac{-81.9}{1} = -81.9$ **Ans.**

$$R_{in} = \frac{R h_{ie}}{R + h_{ie}} = \frac{0.99 \times 1}{0.99 + 1} = 0.497\ k\Omega = 497\ \Omega$$

(iii) $R_{inf} = \frac{R_{in}}{D} = \frac{497}{5.52} = 90\ \Omega$ **Ans.**

$$R'_{out} = R_C || R' = R'_C = 9.09\ k\Omega$$

(iv) $R'_{outf} = \frac{R'_{out}}{D} = \frac{9.09}{5.52} = 1.65\ k\Omega$ **Ans.**

9.14. CURRENT-SERIES FEEDBACK

It is also known as *series-derived series-fed feedback.* In such a feedback circuit, a part of the output current is made to develop a voltage proportional to the output current and supplied back in series with the input, as shown in Fig. 9.10 (*c*). Since feedback network is in series with the amplifier on the output end as well as on the input end, both input and output impedances are increased with negative feedback. The current feedback can be obtained by removing the bypass capacitor across the emitter resistor R_E, as shown in Fig. 9.29 in case of CE amplifier. The emitter resistance R_E provides the negative current feedback. The current I_e flowing through resistor R_E produces a voltage drop $I_e R_E$ across emitter resistance R_E. This voltage is fedback to the input and opposes the input signal as it is in opposition. This negative voltage feedback is proportional to collector current I_c because $I_e \simeq I_c$. Thus negative current series feedback is provided.

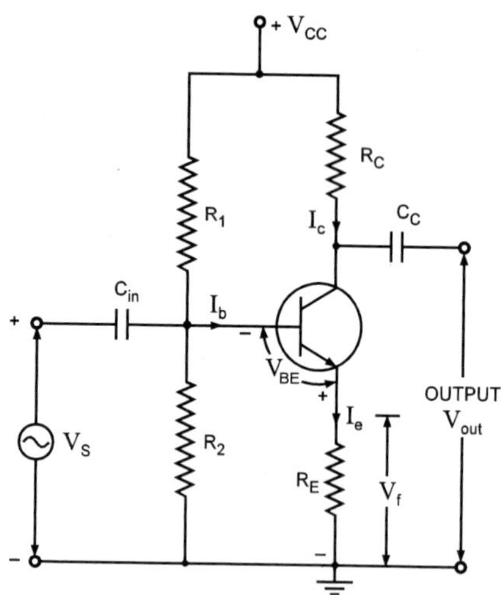

Fig. 9.29

Circuit Analysis*. When an input signal is applied, the resulting collector I_c develops output voltage across the collector load and emitter current I_e develops ac voltage equal to $I_e R_E$ or $I_c R_E$ ($I_e \simeq I_c$) across the emitter.

* For details refer to Art. 7.14.

Thus input voltage to amplifier, which is equal to V_{be}, is given as

$$V_{in} = V_{be} = V_s - I_c R_E = V_s - \frac{V_{out}}{R_C} R_E \quad ...(9.57) \because I_c = \frac{V_{out}}{R_C}$$

$$\text{Thus } V_{in} = V_s - \beta V_{out} \quad ...(9.58) \text{ where } \beta = \frac{R_E}{R_C}$$

Now voltage gain with feedback,

$$A_f = \frac{A}{1+\beta A} \simeq \frac{1}{\beta} \simeq \frac{R_C}{R_E} \quad ...(9.59) \text{ if } \beta A >> 1$$

So effective voltage gain with negative current feedback is given by the expression $A_f = R_C/R_E$. So gain also depends upon the collector load R_C. If R_C depends upon the frequency *i.e.* if R_C is reactive then there will be distortion in the output.

Such a negative feedback is widely employed in practical amplifiers such as in public address systems, record-players, tape recorders, stereo-amplifiers etc. Sometimes, only a part of the emitter resistance is bypassed so that the amplifier gain is not much reduced, and at the same time advantages of negative feedback are achieved.

Example 9.28. Determine A_{vf}, input and output impedances of a current series feedback amplifier.

Solution : Consider the circuit depicted in Fig. 9.30.

Here the feedback signal is a voltage V_f (developed across R_e) which is proportional to the emitter current I_e. Since it can be assumed with reasonable accuracy that $I_e \cong I_c = I_{out}$, so feedback signal is proportional to the output current. Also, the feedback voltage is in series with external input signal V_s, therefore, the topology is a current-series feedback.

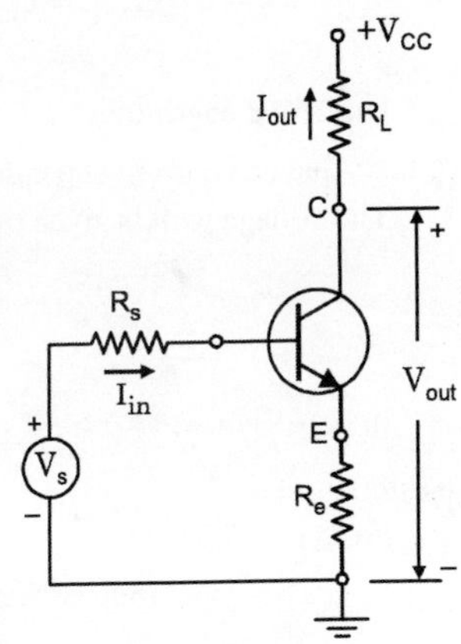

Fig. 9.30

Since this is a current-series feedback we have the input circuit by opening the output loop (setting output current $I_{out} = 0$) and output circuit by opening the input loop [setting input current $I_{in} = 0$ (Refer to Table 9.4)]. Resulting circuit is shown in Fig. 9.31. No ground can be indicated in this circuit because to do so would again couple the input to the output via R_e, that is, it would introduce feedback. Since output current I_{out} is proportional to output voltage V_{out},

$$\beta = \frac{V_f}{I_{out}} = \frac{-I_{out} R_e}{I_{out}} = -R_e \quad ...(i) \because V_f \text{ appears across } R_e \text{ in the output circuit}$$

If BJT is replaced by its h-parameter equivalent circuit, we have circuit, as shown in Fig. 9.32.

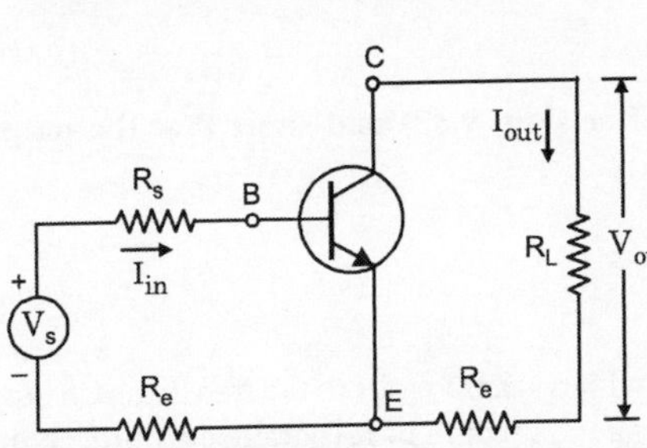

Fig. 9.31. ***Amplifier Without Feedback But Including the Loading of R_e***

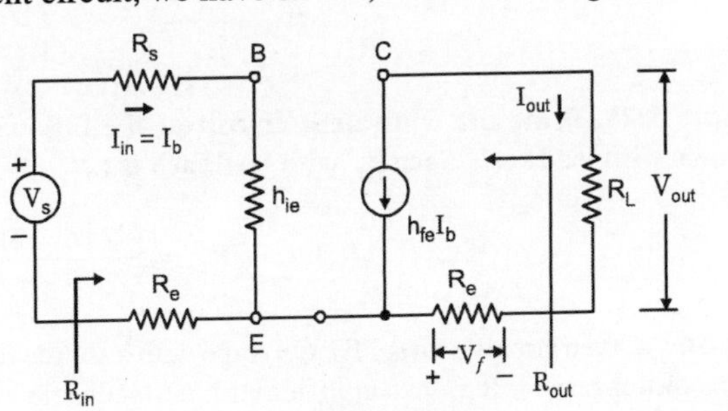

Fig. 9.32

Since input signal V_{in} without feedback is the V_s of Fig. 9.32, then

$$G_M = \frac{I_{out}}{V_{in}} = \frac{-h_{fe}I_b}{V_s} = \frac{-h_{fe}}{R_s + h_{ie} + R_e} \quad \text{...(ii)} \quad \therefore \; V_s = I_b(R_s + h_{ie} + R_e)$$

$$\text{Densensitivity factor, } D = 1 + \beta G_M = 1 + (-R_e) \times \frac{-h_{fe}}{R_s + h_{ie} + R_e} = \frac{R_s + h_{ie} + R_e + h_{fe}R_e}{R_s + h_{ie} + R_e}$$

$$= \frac{R_s + h_{ie} + R_e(1 + h_{fe})}{R_s + h_{ie} + R_e} \quad \text{...(iii)}$$

$$G_{Mf} = \frac{G_M}{D} = \frac{-h_{fe}}{R_s + h_{ie} + R_e(1 + h_{fe})} \quad \text{...(iv)}$$

In case $(1 + h_{fe})\,R_e >> R_s + h_{ie}$ and $h_{fe} >> 1$, then $G_{Mf} \simeq \frac{-1}{R_e}$, in agreement with $G_{Mf} \simeq \frac{1}{\beta}$. If R_e is a stable resistor, the transconductance gain with feedback is stabilized. The load current is given by

$$I_{out} = G_{Mf}V_s = \frac{-h_{fe}V_s}{R_s + h_{ie} + R_e(1 + h_{fe})} \simeq \frac{-V_s}{R_e} \quad \text{...(v)}$$

Under the conditions $(1 + h_{fe})\,R_e >> R_s + h_{ie}$ and $h_{fe} >> 1$, the load current is directly proportional to the input voltage, and this current depends upon R_e and not upon any circuit or transistor parameter.

The voltage gain is given by

$$A_{vf} = \frac{I_{out}\,R_L}{V_s} = \frac{G_{Mf}V_s \times R_L}{V_s} = G_{Mf}\,R_L = \frac{-h_{fe}R_L}{R_s + h_{ie} + R_e(1 + h_{fe})} \quad \text{...(vi)}$$

In case $(1 + h_{fe})\,R_e >> R_s + h_{ie}$ and $h_{fe} >> 1$, $A_{vf} \simeq \frac{-R_L}{R_e}$ and the voltage gain is stable if R_L and R_e are stable resistors.

From Fig. 9.32

$$\text{Input resistance, } R_{in} = R_s + h_{ie} + R_e$$

$$\text{Input resistance with feedback, } R_{inf} = R_{in} \times D = (R_s + h_{ie} + R_e) \times \frac{R_s + h_{ie} + R_e(1 + h_{fe})}{R_s + h_{ie} + R_e}$$

$$= R_s + h_{ie} + R_e\,(1 + h_{fe}) \quad \text{...(vii)}$$

Because R_s is considered to be the part of the amplifier, it appears above as a component of the input resistance.

Since $R_{out} = \infty$, then $\quad R_{outf} = \frac{R_{out}}{(1 + \beta G_m)} = \infty$

So $\quad R'_{outf} = R_L \,||\, R_{outf} = R_L \quad$ ...(viii)

Example 9.29. Draw the equivalent circuit of the following amplifier (Fig. 9.33) and show that the output resistance with load resistance R_L with feedback is :

$$R'_{of} = \frac{R_L\,[r_d + (\mu + 1)R]}{r_d + R_L + (\mu + 1)R}.$$

Solution : Given circuit shows FET CS stage with a source resistor R, which is analogous to the transistor CE stage with an emitter resistor R_e. The amplifier without feedback, but including the loading of R (obtained by proceeding as we did for the BJT) is depicted in Fig. 9.34.

If the FET is replaced by its small-signal low frequency model, we have the circuit shown in Fig. 9.35.

$$\text{Without feedback } V_{in} = V_s$$

$$\text{and } G_M = \frac{I_{out}}{V_{in}} = \frac{I_{out}}{V_s} = \frac{-g_m r_d}{r_d + R_L + R} = \frac{-\mu}{r_d + R_L + R} \qquad (i) \because r_d g_m = \mu$$

$$\beta = \frac{V_f}{I_{out}} = \frac{-I_{out}R}{I_{out}} = -R \qquad ...(ii)$$

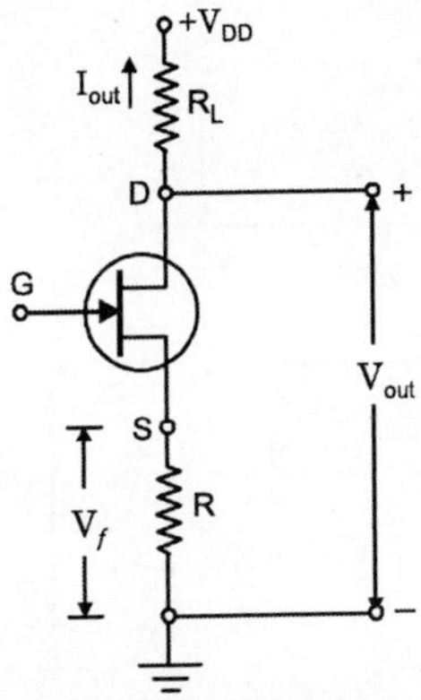

Fig. 9.33. ***An FET Amplifier With a Source Resistor R***

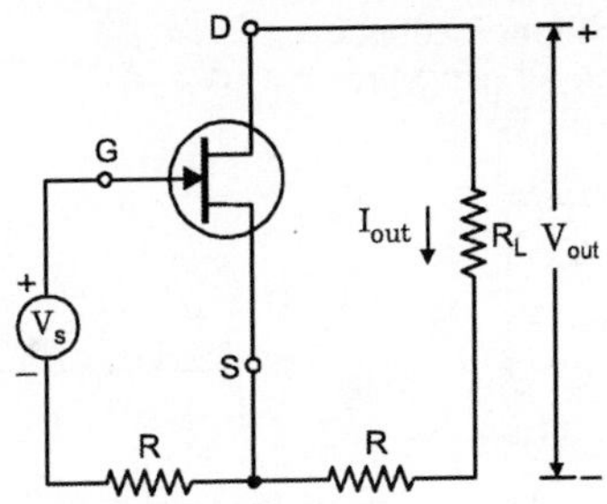

Fig. 9.34. ***Amplifier Without Feedback, But Including the Loading of R***

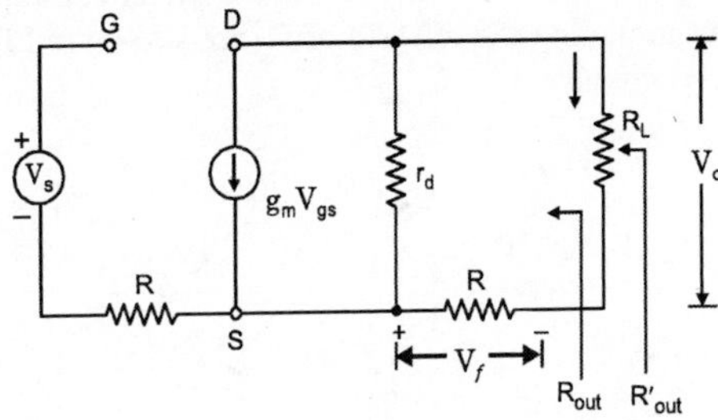

Fig. 9.35

$$\text{Desensitivity factor, } D = 1 + \beta G_M = 1 + (-R) \times \frac{-\mu}{r_d + R_L + R}$$

$$= 1 + \frac{\mu R}{r_d + R_L + R} = \frac{r_d + R_L + R + \mu R}{r_d + R_L + R} = \frac{r_d + R_L + R(1+\mu)}{r_d + R_L + R} \qquad ...(iii)$$

$$G_{Mf} = \frac{G_M}{D} = \frac{-\mu}{r_d + R_L + R} \times \frac{r_d + R_L + R}{r_d + R_L + R(1+\mu R)} = \frac{-\mu}{r_d + R_L + R(1+\mu R)} \qquad ...(iv)$$

Since $R_{in} = \propto$, then $R_{inf} = R_{in} \times D = \propto$ ...(v)

If R_L is considered to be an external load, then from Fig. 9.35

$$R_{out} = r_d + R \qquad ...(vi)$$

For calculation of $R_{out\,f}$ we require G_m and from equation $G_m = \lim_{R_L \to 0} G_M$

Since β is independent of R_L, then using Eq. (*iii*)

$$1 + \beta G_M = \lim_{R_L \to 0} D = \frac{r_d + (\mu+1)R}{r_d + R} \qquad ...(vii)$$

$$R_{out\,f} = R_{out}(1 + \beta\, G_m) = (r_d + R) \times \frac{r_d + (\mu+1)R}{r_d + R} = r_d + (\mu+1)R \qquad ...(viii)$$

$R'_{out\,f}$ is most conveniently calculated as $R_C \,||\, R_{out\,f}$. The same result may be obtained from the expression, with $R'_{out} = R_{out} \,||\, R_C$.

$$\text{Thus, } R'_{out\,f} = R'_{out} \times \frac{1+\beta G_M}{D} = \frac{(r_d + R)\, R_L}{r_d + R_L + R} \times \frac{r_d + (\mu+1)R}{(r_d + R)} \times \frac{r_d + R_L + R}{r_d + R_L + (\mu+1)R}$$

$$= \frac{R_L\,[r_d + (\mu+1)R]}{r_d + R_L + (\mu+1)R} \qquad ...(ix)$$

which is equivalent to R_L in parallel with $R_{out\,f}$

Example 9.30. Find the type of feedback in the following circuit (Fig. 9.36). Also determine the value of feedback factor β.

Solution : The feedback signal is a voltage developed across R_e, which is proportional to emitter current I_e. Since it can be assumed with reasonable accuracy that $I_e \cong I_c = I_{out}$, so feedback signal is proportional to the output current. Also, the feedback voltage is in series with external input voltage V_s, therefore, the topology is a *current-series feedback.*

Since this is a current-series feedback we have the input circuit by opening the output loop and output circuit by opening the input loop. The input circuit, output circuit and circuit without feedback but including the loading of R_e are shown in Figs. 9.37 (*a*) (*b*) (*c*) respectively.

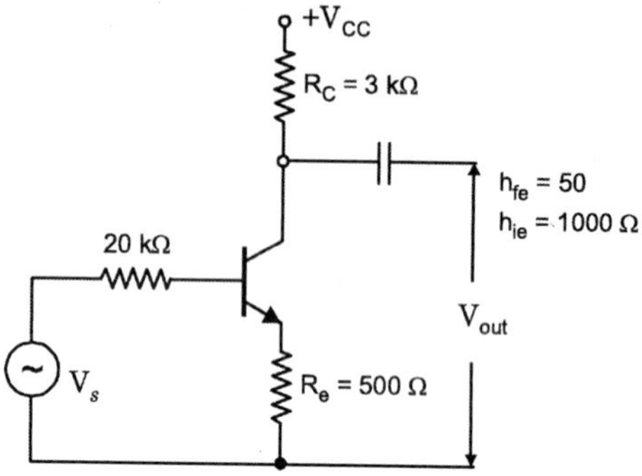

Fig. 9.36

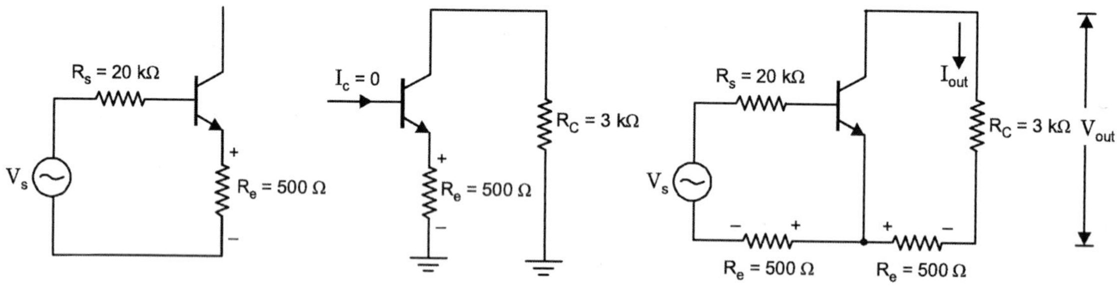

(*a*) *Input Circuit* (*b*) *Output Circuit* (*c*) *Amplifier Without Feedback But Including the Loading of R_e*

Fig. 9.37

$$\text{Feedback factor } \beta = \frac{V_f}{I_{out}} = \frac{-I_{out}R_e}{I_{out}} = -R_e = 500\ \Omega \ \textbf{Ans.}$$

9.15. CURRENT-SHUNT FEEDBACK

It is also known as *series-derived shunt-fed feedback* or *current shunt inverse feedback.* In this circuit, the feedback network picks up a part of the output current and produces a feedback voltage in parallel with the input signal voltage, as shown in Fig. 9.10 (*d*). Since the feedback shunts the input so input impedance is reduced with feedback whereas output impedance is increased because of feedback network being in series with the output.

A two-stage cascaded amplifier using such a feedback is given in Fig. 9.38. Here feedback is from the emitter of second stage to the base of first stage through resistor R_f. In this input voltage (V_{in2}) to second stage is much larger than input voltage (V_{in1}) to first stage, because of amplifying action of transistor Q_1, and is in phase opposition to it (*i.e.* phase difference of 180°). Voltage V_{e2} is only slightly smaller than V_{in2}, because of emitter follower action. Also voltages V_{e2} and V_{in2} are in phase. Thus V_{e2} is larger in magnitude

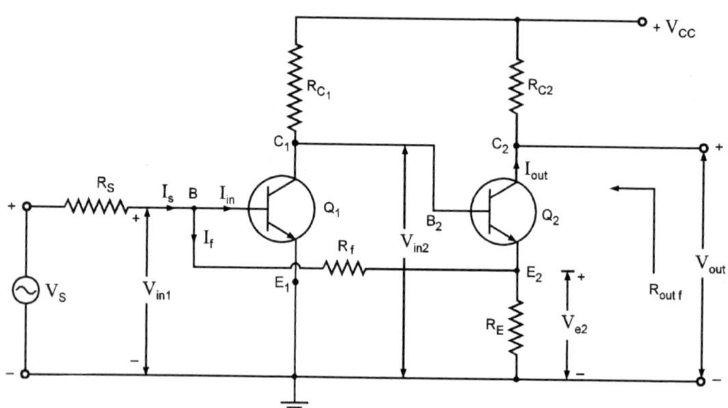

Fig. 9.38. ***Two-Stage CE Amplifier With Current Shunt Feedback***

than V_{in1} and has a phase difference of 180° with V_{in1}. When input signal increases causing I_s to increase, feedback current I_f also increases and so the input current I_{in1} would be smaller than it would be without feedback as it is difference of I_s and I_f. Thus negative feedback is provided.

With R_f much larger than R_E, the circuit shown in Fig. 9.38 approximates a current-shunt feedback pair and feedback current I_f is given as

$$I_f = \frac{V_{in1} - V_{e2}}{R_f} \simeq \frac{-V_{e2}}{R_f} \quad \text{...(9.60)}$$ because V_{in1} is quite small as compared to V_{e2}

Neglecting base current of transistor Q_2 in comparison to its collector current and assuming $R_f >> R_E$

$$V_{e_2} = I_E R_E = (I_f - I_{out}) R_E \quad \text{...(9.61)}$$

$$\text{and so } I_f = \frac{-(I_f - I_{out}) R_E}{R_f} = \frac{R_E\, I_{out}}{R_f + R_E} = \beta I_{out} \quad \text{...(9.62)} \quad \text{where } \beta = \frac{R_E}{R_f + R_E}$$

Since feedback current is proportional to the output current, this circuit is an example of a current-shunt feedback amplifier.

Transfer current gain with feedback,

$$A_{if} = \frac{I_{out}}{I_s} \simeq \frac{1}{\beta} \simeq \frac{R_f + R_E}{R_E} \quad \text{...(9.63)}$$

Thus transfer current gain A_{if} would be stable provided that resistors R_E and R_f are stable.

Assuming input impedance with feedback R_{inf} to be zero, $V_s = I_s R_s$

and the voltage gain with feedback,

$$A_f = \frac{V_{out}}{V_s} = \frac{I_{out} R_{C_2}}{I_s R_s} \simeq \frac{R_f + R_E}{R_E} \times \frac{R_{C_2}}{R_s} = \frac{R_{C_2}}{\beta R_s} \quad \text{...(9.64)}$$

Thus voltage gain with feedback is stable *i.e.* independent of the transistor parameters, the temperature, or supply voltage variations because resistors R_f, R_E, R_{C_2} and R_S are stable elements.

9.15.1. Amplifier Without Feedback

Referring to the fourth topology in Table 9.4, the input circuit of the amplifier without feedback is obtained by opening the output loop at the emitter of Q_2. This places R_f in series with R_E from base to emitter of Q_1. The output circuit is obtained by shorting the input node (base of Q_1). This places R_f in parallel with R_E. The resultant equivalent circuit is shown in Fig. 9.39. Since the feedback signal is a current, the source is represented by a Norton's equivalent circuit with

$$I_s = \frac{V_s}{R_s}$$

The feedback signal is the current I_f in the resistor R_f, which is in the output circuit.

From Fig. 9.39,

With $I_{b_2} < I_{c_2} < I_{out}$

$$\beta = \frac{I_f}{I_{out}} = \frac{R_E}{R_f + R_E} \quad \text{...(9.65)}$$

which is in agreement with Eq. (9.62).

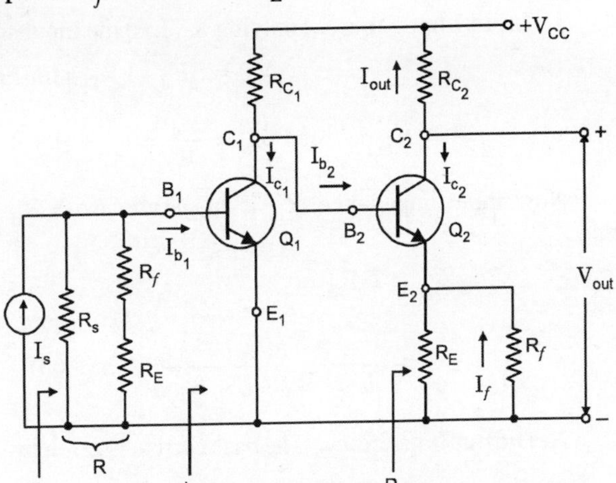

Fig. 9.39. *Amplifier Without Feedback But Including the Loading of R_f*

9.16. EMITTER FOLLOWER

It is also a negative current feedback circuit. This circuit exhibits a large input impedance, a small output impedance, and a voltage gain of approximately unity. Further the output voltage tends to be in phase with the input voltage—hence the term "*follower*".

At first glance, one might be tempted to dismiss such a circuit as useless because of its unity voltage gain. However, its input and output impedances promote its use as input and output *buffer* stages in *amplifier systems.*

The circuit of an emitter follower is shown in Fig. 9.40. The main difference between an emitter follower and conventional amplifier is the absence of collector load and emitter bypass capacitor. The emitter resistance R_E itself acts as the load and the ac output voltage V_{out} is taken across it. The biasing is provided either by base resistor method or potential divider method. The circuit consists of a resistance of the order of 10 kΩ in the emitter itself and the collector is connected directly to the supply. Stabilization of operating point is exceptionally good because of high value of emitter resistance R_E.

Operation. When input signal V_s is applied to the base, the resulting emitter current I_e develops an output voltage V_{out} equal to $I_e R_E$ across the emitter resistance R_E. This voltage opposes the ac signal voltage V_s as it is in phase opposition to V_s. Thus it provides negative current feedback. Moreover, this voltage (V_{out}) fed back to the input is proportional to the emitter current hence this circuit is called a *negative current feedback circuit.*

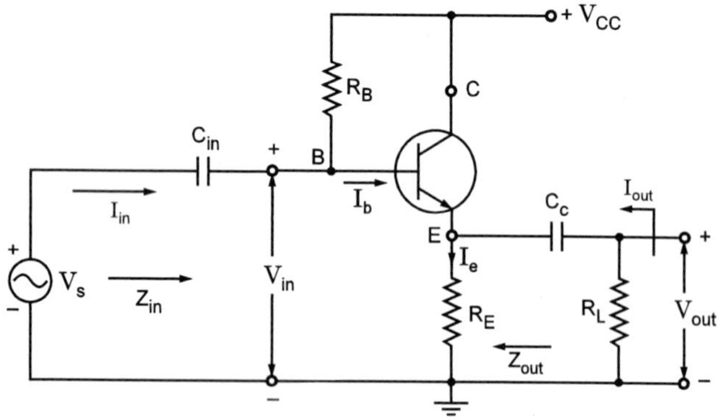

Fig. 9.40. ***Emitter Follower Circuit***

It may be noted that when the input signal voltage v_s goes through its positive half cycle the output voltage v_{out} is also seen to go through its positive half cycle. Hence the output voltage is in phase with the input signal voltage *i.e.* the output voltage (emitter voltage) just follows the input voltage and hence the name *emitter follower.*

Circuit Analysis : For the emitter follower configuration given in Fig. 9.40, the small signal ac model is drawn as shown in Fig. 9.41.

1. **Input Impedance.** Applying KVL to the input side, we have

$$V_{in} = i_b h_{ie} + I_e R_E = I_b h_{ie} + (1 + h_{fe}) I_b R_E$$

$$\text{and } Z_b = \frac{V_{in}}{I_b} = h_{ie} + (1 + h_{fe})R_E \simeq h_{ie} + h_{fe} R_E \qquad \because h_{fe} >> 1$$

Now input impedance, Z_{in} is in parallel with R_B and Z_b

$$\text{So, } Z_{in} = Z_b R_B$$

$$= \frac{(h_{ie} + R_E h_{fe})(R_B)}{h_{ie} + R_E h_{fe} + R_B} \qquad ...(9.66)$$

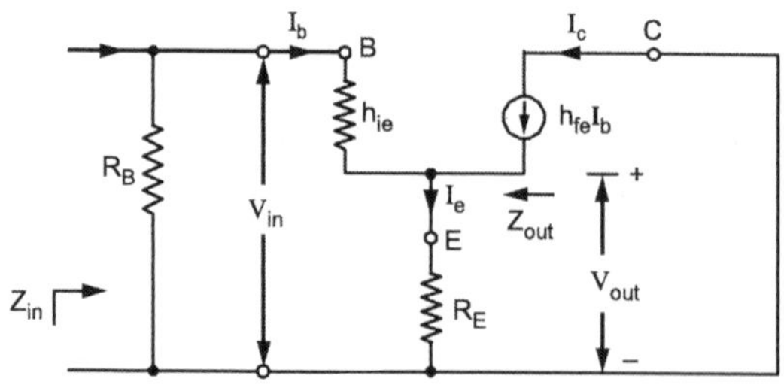

Fig. 9.41. ***Small Signal For Emitter Follower Circuit Shown in Fig. 9.40***

2. **Output Impedance.** The base current is given by

$$I_b = \frac{V_{in}}{Z_b} = \frac{V_{in}}{h_{ie} + (1 + h_{fe})R_E}$$

So emitter current,

$$I_e = (1 + h_{fe})I_b = \frac{(1+h_{fe})V_{in}}{h_{ie} + (1+h_{fe})R_E} = \frac{V_{in}}{\frac{h_{ie}}{1+h_{fe}} + R_E} \quad ...(9.67)$$

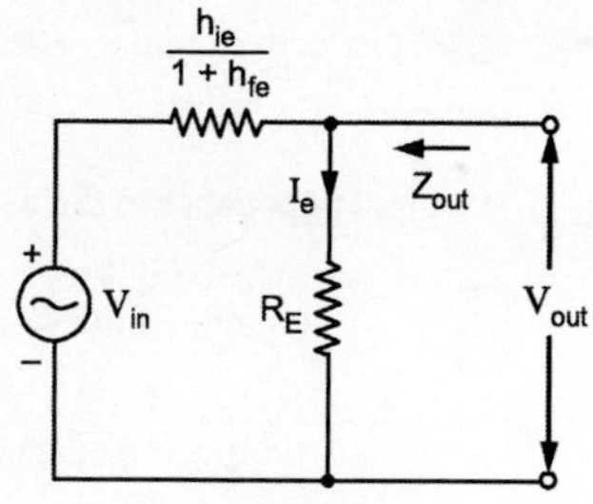

Fig. 9.42

The equivalent circuit defined by above equation is shown in Fig. 9.42.

Thus,
$$Z_{out} = R_E \,||\, \frac{h_{ie}}{1+h_{fe}} = \frac{R_E \times \frac{h_{ie}}{1+h_{fe}}}{R_E + \frac{h_{ie}}{1+h_{fe}}} \quad ...(9.68)$$

3. Voltage Gain.
$$V_{out} = I_e R_E = \frac{V_{in} R_E}{\frac{h_{ie}}{1+h_{fe}} + R_E}$$

$$\text{Voltage gain, } A_v = \frac{V_{out}}{V_{in}} = \frac{R_E}{\frac{h_{ie}}{1+h_{fe}} + R_E} \simeq \frac{R_E}{\frac{h_{ie}}{h_{fe}} + R_E} = \frac{h_{fe} R_E}{h_{ie} + h_{fe} R_E} \quad ...(9.69) \because h_{fe} >> 1$$

4. Current Gain. From Fig. 9.41, we have

$$\text{Output current, } I_{out} = I_e = (1 + h_{fe})\, I_b$$

$$\text{or } \frac{I_{out}}{I_b} = (1 + h_{fe})$$

$$\text{So the current gain, } A_i = \frac{I_{out}}{I_{in}} = \frac{I_{out}}{I_b} \times \frac{I_b}{I_{in}} \quad ...(9.70)$$

From Fig. 9.41,
$$I_b = \frac{R_B I_{in}}{R_B + Z_b}$$

$$\text{or } \frac{I_b}{I_{in}} = \frac{R_B}{R_B + Z_b}$$

$$\text{So } A_i = \frac{I_{out}}{I_b} \times \frac{I_b}{I_{in}} = (1 + h_{fe}) \times \frac{R_B}{R_B + Z_b} = \frac{(1+h_{fe})R_B}{R_B + Z_b} \quad ...(9.71)$$

Advantages. 1. High input impedance and low output impedance, so it can be used for impedance matching. Sometimes it is called *impedance transformer*.

2. Because of 100 per cent feedback, output is distortionless and bandwidth is very large.

3. Relatively high current gain and power gain.

4. Output and input ac voltages are in phase and also approximately equal in magnitude *i.e.* emitter output voltage closely follows the input and so it is called the *emitter follower*.

Applications. Because of high input impedance and low output impedance an emitter follower is capable of transferring maximum power from the high impedance source to low impedance load. When an emitter follower is employed for this purpose, it is called a *buffer amplifier*. Though impedance matching can be achieved by employing a step-down transformer but emitter follower is preferred. It is because emitter follower is not only more convenient than a transformer but it also provides better frequency response.

There are many instances (especially in digital electronics) where an increase in current is required but no increase in voltage is required. For such situations emitter follower, because of its high current gain, can be employed.

Example 9.31. An emitter follower shown in Fig. 9.40, has the following parameters :

V_{CC} = 20 V, R_B = 5 kΩ, R_L = 18.6 kΩ, R_E = 9.3 kΩ, h_{ie} = 1.1 kΩ and h_{fe} = 100.

Determine (*i*) input impedance (*ii*) output impedance (*iii*) voltage gain and (*iv*) current gain.

Solution : Effective emitter resistance

$$R'_E = R_E \,||\, R_L = 9.3 \text{ k}\Omega \,||\, 18.6 \text{ k}\Omega = 6.2 \text{ k}\Omega$$

Input impedance to the base, $Z_b = h_{ie} + (1 + h_{fe})R'_E = 1.1 + (1 + 100) \times 6.2 = 627.3 \text{ k}\Omega$

(*i*) Input impedance, $Z_{in} = R_B \,||\, Z_b = 5 \text{ k}\Omega \,||\, 627.3 \text{ k}\Omega = \dfrac{5 \times 627.3}{5 + 627.3} = 4.96 \text{ k}\Omega$ **Ans.**

(*ii*) Output impedance, $Z_{out} = R'_E \,||\, \dfrac{h_{ie}}{1 + h_{fe}} = 6.2 \text{ k}\Omega \,||\, \dfrac{1.1}{1 + 100} \text{ k}\Omega = 6.2 \text{ k}\Omega \,||\, 10.89 \,\Omega \simeq 10.89 \,\Omega$ **Ans.**

(*iii*) Voltage gain, $A_{vf} = \dfrac{R'_E}{\dfrac{h_{ie}}{1 + h_{fe}} + R'_E} = \dfrac{6{,}200}{\dfrac{1100}{1 + 100} + 6{,}200} = 0.998 \simeq 1$ **Ans.**

(*iv*) Current gain, $A_{if} = \dfrac{(1 + h_{fe})R_B}{R_B + Z_b} = \dfrac{(1 + 100) \times 5 \text{ k}\Omega}{5 \text{ k}\Omega + 627.3 \text{ k}\Omega} = \dfrac{505}{632.3} = 0.798$ **Ans.**

HIGHLIGHTS

1. When a part or fraction of output is combined to the input, feedback is said to exist. Thus the process of combining a fraction of output energy (voltage or current) back to the input is called the *feedback*. When the feedback voltage (or current) is so applied as to increase the input signal (*i.e.* when the feedback is in phase with input signal), it is called the *positive* or *direct* or *regenerative feedback*. Since positive feedback causes excessive distortion and instability, it is seldom used in amplifiers. However, because of its capability of increasing the power of the original signal it is used in oscillator circuits.
2. When the feedback voltage (or current) is so applied as to weaken the input signal (*i.e.* when the feedback is in phase opposition to the input signal), it is called *negative, inverse* or *degenerative feedback*. Negative feedback reduces the amplifier gain but it has numerous advantages (such as gain stability, reduction in nonlinear distortion, reduction in noise, increase in bandwidth or improvement in frequency response, increase in input impedance and decrease in output impedance) and is frequently used in amplifier circuits. The reduction in gain due to negative feedback, can always be compensated by increasing the number of stages.
3. If an amplifier having a voltage gain A has some of the output voltage fed back in series with the input, the voltage gain with feedback is

$$A_f = \frac{A}{1 - \beta A}$$

   where β is the feedback factor, negative for negative feedback and positive for positive feedback. The term βA is called the feedback factor.
4. Though negative feedback reduces the gain of the amplifier but improves its performance in several aspects such as gain stability, reduction in nonlinear distortion, reduction in noise, increase in bandwidth or improvement in frequency response, increase in input impedance and decrease in output impedance.
   (*a*) Gain is stabilized with negative feedback because the voltage gain with negative feedback is given as

$$A_f \simeq \frac{1}{\beta}$$

   (*b*) A large signal stage has nonlinear distortion which is reduced by a factor (1 + Aβ) when negative feedback is used.
   (*c*) With negative feedback the amplifier gain is reduced and since gain bandwidth has to remain constant, obviously the bandwidth will increase to compensate for reduction in gain.
   (*d*) There is always a noise voltage in the amplifier which is reduced by a factor (1 + β) when negative feedback is used.

5. The effect of negative feedback on input and output impedances depend upon the type of negative feedback, as summarized in Tabular form in Art. 9.10. The four types of feedbacks are current and voltage-series, and current and voltage-shunt.

6. An unbypassed resistor in the emitter lead of a CE amplifier is an example of current-series feedback that increases both input and output impedances.

7. The common collector or emitter follower is an example of voltage-series feedback. It inherently uses 100% negative feedback and is the one connection that simultaneously increases input impedance and decreases output impedance. Its voltage gain is slightly less than unity.

8. Voltage-shunt feedback such as a resistor connected from collector-to-base in CE configuration, reduces both input and output impedances.

EXERCISES

1. Classify the basic amplifiers and draw their equivalent circuits and derive expression for transfer characteristic of each of the circuit.

2. What are different types of feedback ?

3. Explain types of negative feedback in trasistor circuits.

4. What do you understand by feedback in amplifiers ? Explain the terms feedback factor and open-loop gain.

5. What do you understand by feedback in amplifiers ? List the four basic negative feedback configurations and indicate the effect of feedback on R_i and R_0 in each of these configurations.

6. Derive an expression for the gain of negative voltage feedback amplifier.

7. State the advantages offered by negative feedback in an amplifier.

[G.G.S.I.P. Univ. Analog Electronics December 2010]

8. Discuss with the help of a circuit example, the purpose of providing (*i*) negative feedback (*ii*) positive feedback in amplifiers.

9. Explain the four types of feedback topologies with the help of schematic diagram.

[G.G.S.I.P. Univ. Analog Electronics December 2010]

10. Prove in a negative feedback amplifier, $\left|\frac{dA_f}{A_f}\right| = \frac{1}{1+\beta A}\left|\frac{dA}{A}\right|$ where A_f = Gain with feedback, A = Transfer gain, β = feedback factor.

11. How the negative feedback changes the impedances ? Explain with mathematical expression.

12. Discuss the effects of negative feedback on gain, stability of gain and noise.

13. Discuss how gain of an amplifier is stabilized with negative feedback.

[G.G.S.I.P. Univ. Analog Electronics December 2010]

14. Mention the advantages and disadvantages of a negative feedback amplifier. Determine the gain bandwidth product of a negative feedback amplifier.

15. Explain the effect of negative feedback on various characteristics of the amplifier.

16. Explain the concept of feedback with the help of block schematic of a signal loop feedback amplifier. Derive expression for the transfer gain with feedback.

17. Define the following :

(*i*) Amount of feedback (in dB) (*ii*) Desensitivity (*iii*) Loop gain.

18. List major advantages of negative feedback. What are the four possible topologies of a feedback amplifier ? Identify the input signal X_0 and feedback signal X_f for each topology (either as current or voltage). Also give dimensions of transfer gain for each topology.

19. Make a distinction between the "voltage" feedback and "current" feedback in amplifier circuits. Discuss the merits in each case and derive the expressions for the net output impedance in each case.

20. Derive expression for output resistance in voltage series feedback amplifier.

21. Prove that the input resistance R_{if} increases for a voltage series feedback topology and current series feedback circuit. **[G.G.S.I.P. Univ. Analog Electronics December 2010]**

22. Does the gain bandwidth product change with feedback ? Using it show that what will be the bandwidth after feedback in (*i*) positive feedback and (*ii*) negative feedback.
23. (*a*) List the steps required to carry out the analysis of a feedback amplifier. Explain in brief.
    (*b*) Draw and explain the circuit of a current-series feedback amplifier.
24. Explain the working of an emitter follower and show how it performs the function of impedance transformation.
25. What type of feedback is the emitter follower an example of ? How much negative feedback is used ?
26. Derive expression for voltage gain, current gain, input resistance, and output resistance of an emitter follower circuit using h parameters. **[G.G.S.I.P. Univ. Analog Electronics December 2010]**

SHORT ANSWER TYPE QUESTIONS WITH ANSWERS

Q. 1. What is feedback in amplifiers ?

Ans. The process of combining a fraction of output energy back to the input is called the feedback.

Q. 2. What is the difference between current and voltage feedbacks ?

Ans. Voltage feedback refers to connecting the output voltage as input to the feedback network while current feedback refers to tapping off some output current through the feedback network.

Q. 3. Why voltage-series feedback is most commonly used in cascaded amplifiers ?

Ans. In cascaded amplifiers, high input impedance and low output impedance are usually the main requirements and both of these requirements can be met by voltage series feedback, so it is most commonly used in cascaded amplifiers.

Q. 4. Define the term stability.

Ans. The variations in temperature, supply voltages, ageing of components or variations in transistor parameters with replacement are some of the factors that affect the gain of an amplifier and cause it to change.

The stability of an amplifier is defined as the ability of the amplifier to maintain the voltage gain constant as possible over as long a time interval as possible.

The overall voltage gain of the amplifier can be stabilized with negative feedback.

Q. 5. Why is negative feedback employed in high gain amplifiers ?

Ans. Negative feedback is employed in high gain amplifiers to stabilise the gain. With the negative feedback voltage gain of an amplifier is given as $A_f = \frac{A}{1 + A\beta}$. If $A\beta >> 1$, then the expression for voltage gain becomes $A_f = \frac{1}{\beta}$ *i.e.* overall voltage gain of feedback amplifier A_f becomes independent of internal gain of the amplifier and depends only on feedback ratio β, and β in turn depends on passive elements such as resistors. Resistors remain fairly constant and so the gain is stabilised. Other advantages of negative feedback are reduced nonlinear distortion, reduced noise, increased bandwidth, increased input impedance and reduced output impedance.

Q. 6. What is the effect of negative feedback on the input and output impedances of a voltage-series feedback amplifier ?

Ans. Series voltage negative feedback increases the input impedance and reduces the output impedance by a factor $(1 + \beta A)$, where βA is the feedback factor.

Q. 7. What is the effect of negative feedback on the bandwidth of an amplifier ?

Ans. Bandwidth of an amplifier is increased, when negative feedback is introduced, by a factor $(1 + \beta A)$, by which its gain is reduced.

Q. 8. How does negative feedback increases bandwidth of an amplifier ?

Ans. The bandwidth (BW) of an amplifier without feedback is equal to the separation between 3-dB frequencies f_1 and f_2. If A is the gain then gain-bandwidth product is $A \times BW$. With the negative feedback the amplifier gain is reduced and since gain bandwidth product has to remain constant in both cases, so obviously the bandwidth will increase to compensate for the reduction in gain.

Q. 9. Why an emitter follower is called so ?

Ans. In an emitter follower, output and input voltages are approximately equal in magnitude and also in phase *i.e.* emitter output voltage closely follows the input and that is why it is called the emitter follower.

Q. 10. What are practical applications of emitter follower ?

Ans. Because of high input impedance and low output impedance an emitter follower is capable of transferring maximum power from the high impedance source to low impedance load. When an emitter follower is employed for this purpose, it is called a *buffer amplifier*. Though impedance matching can be achieved by employing a step-down transformer but emitter follower is preferred. It is because emitter follower is not only more convenient than a transformer but it also provides better frequency response.

There are many instances (especially in digital electronics) where an increase in current is required but no increase in voltage is required. For such situations emitter follower, because of its high current gain, can be employed.

PROBLEMS

1. A feedback amplifier has a voltage gain of 500 without feedback. Determine the voltage gain with feedback if feedback ratio is 0.1. [**Ans.** 9.8]
2. The overall gain of an amplifier is 200. When negative feedback is applied, the gain decreases to 10. Determine the fraction of the output that is fedback to the input. [**Ans.** 0.095]
3. In a negative feedback amplifier $A = 100$; $\beta = 0.02$ and input signal voltage is 40 mV. Determine (*i*) voltage gain with feedback (*ii*) feedback factor, (*iii*) feedback voltage and (*iv*) output voltage.
[**Ans.** (*i*) 33.33 (*ii*) 2 (*iii*) 26.666 m V (*iv*) 1.333 V]
4. To an amplifier of 60 dB gain a feedback (negative) of $\beta = 0.006$ is applied. What would be the change in the overall gain of the feedback amplifier if the gain of the amplifier decreases by 15%. [**Ans**. 2.462%]
5. An amplifier has a gain of 200 and distortion 10% with an input signal voltage of 200 mV. Determine (*i*) output signal voltage (*ii*) distortion voltage and (*iii*) output voltage when an input signal voltage is 200 mV.
[**Ans.** (*i*) 40 V (*ii*) 4 V (*iii*) 44 V]
6. An amplifier has gain of 60 and distortion 10% without feedback. Determine (*i*) gain and (*ii*) distortion when negative feedback is applied, the feedback factor being 6. [**Ans.** (*i*) 8.57 (*ii*) 1.43%]
7. Open-circuit voltage of an amplifier is reduced by 6 when a negative series feedback is applied, by what factor the input and output impedances change ?
[**Ans.** Input impedance is increased to 6 times and output impedance is reduced to 1/6 time]
8. An amplifier with an open loop voltage gain of 1,000 delivers 10 W of output power at 10 per cent harmonic distortion when the input signal is 10 mV. If 40 dB negative voltage-series feedback is applied and the output power is to remain at 10 W, determine required input signal. [**Ans.** 1.01 V]
9. An amplifier has a midband gain of 200 without feedback. The 3-dB frequency width of it is 200 kHz. The amplifier is to be used as a video amplifier that requires 5 MHz bandwidth. What gain can be obtained and what feedback must be used ? What bandwidth could be obtained if the feedback were 100%.
[**Ans.** 0.12; 8; 40.2 MHz]
10. A single-stage amplifier has a voltage gain of 10 and a bandwidth of 1 MHz. Three such stages are cascaded and a negative feed of 10% is applied to the cascade stage. Find the overall voltage gain and bandwidth of the cascaded stage with feedback. [**Ans.** 10; 100 MHz]
11. In a CE amplifier, the emitter resistor is not bypassed the capacitor. Find the voltage gain if $R_C = 20\ k\Omega$ and $R_E = 1.25\ k\Omega$. [**Ans.** 16]
12. An emitter follower is found to have a voltage gain of 0.99. Determine β for the transistor. [**Ans.** 99]

❑❑❑

UNIT

IV

FIELD-EFFECT TRANSISTORS, UJT AND SCR

CONTENTS IN THIS UNIT

FIELD-EFFECT TRANSISTORS : Introduction; Classification; FET Characteristics, Operating Point, Biasing; Enhancement and Depletion Type MOSFETs. Introduction to UJT, SCR, Thyristor-Firing Characteristics.

CHAPTER 10

Field-Effect Transistors (FETs), UJT and SCR

INSIDE THIS CHAPTER

10.1. INTRODUCTION

We know that P-N junctions have two dominant features *viz.* the injection of minority carriers with forward bias and a variation of the depletion width W with reverse bias. These two P-N junction properties are used in two important types of transistors *viz. bipolar junction* transistor abbreviated as BJT and the *junction field-effect transistor* abbreviated as JFET. The BJT discussed in chapter 5 makes use of the injection of minority carriers across a forward biased junction, while JFET discussed in this chapter depends on the control of a junction depletion width under reverse bias. FET is a majority carrier device, and is, therefore, often called a *unipolar* transistor while BJT operates by the injection and collection of minority carriers. Since in this device, both holes and electrons take part in conduction, it is called a *bipolar transistor.* Like its bipolar counterpart, the FET is a three-terminal (namely drain, source and gate) semiconductor device in which current conduction is by one type of majority carriers (electrons in case of an N-channel FET or holes in a P-channel FET). Unlike the BJT, a FET needs virtually no input (bias signal) current. Field-effect devices are controlled by an electric field (or voltage) and give an extremely *high input impedance* – most important advantage over BJT devices.

Field-effect transistors (FETs) are of several types such as junction field-effect transistors (JFETs), metal semiconductor field-effect transistors (MESFETs), and metal-insulator semiconductor field-effect transistors (MISFETs).

In a junction field-effect transistor (JFET) the gate voltage controls the depletion width of a reversed biased P-N junction. A metal semiconductor field-effect transistor (MESFET) is the same basic device except that the P-N junction is replaced by a Schottky barrier rectifying junction. Metal-insulator-semiconductor field-effect transistor (MISFET) is one in which the metal gate electrode may be separated from the semiconductor by an insulator. The metal-oxide semiconductor field-effect transistor (MOSFET) is a special type of MISFET which makes use of an oxide layer as insulator.

The current in a JFET is through a semiconductor region called the *channel,* with ohmic contacts at each end. The basic transistor action is the modulation of the channel conductance by the electric field that is perpendicular to the channel. The modulating electric field is induced in the space charge region of a reverse-biased P-N junction or a reverse-biased Schottky barrier junction and is, therefore, a function of voltage applied to the gate.

Although the bipolar junction transistors (BJTs) reigned supreme in the early days of semiconductor integrated electronics, it has been gradually supplanted in most applications by the silicon MOSFETs. The main reason is, unlike BJTs, the different types of FETs are characterised by high input impedance, as the control voltage is applied to a reverse-biased junction or Schottky barrier, or across an insulator. Such devices are particularly suitable for controlled switching between a conducting state and a nonconducting state and that is why they are quite useful in digital circuits. They are also well suited for integration of many devices on a single chip. The MOSFET, is, without doubt, the core of the integrated circuit (IC) design at present time.

10.2. JUNCTION FIELD-EFFECT TRANSISTORS

Depending upon the majority carriers, JFETs have been classified into two types, namely (*i*) N-channel JFETs with electrons as the majority carriers and (*ii*) P-channel JFETs with holes as the majority charge carriers. N-channel JFETs are preferred because of mobility of electrons higher than that of holes. N-channel and P-channel FETs are shown in Figs. 10.1 (*a*) and 10.1 (*b*) respectively.

10.2.1. Basic Construction

Its structure is quite simple. In an N-channel JFET an N-type silicon bar, referred to as the *channel*, has two smaller pieces of P-type silicon material diffused on the opposite sides of its middle part, forming P-N junctions, as illustrated in Fig. 10.1 (*a*). The two P-N junctions forming diodes or gates are connected internally and a common terminal,* called the *gate terminal*, is brought out. Ohmic contacts (direct electrical connections) are made at the two ends of the channel—one lead is called the *source terminal* S and the other *drain terminal* D.

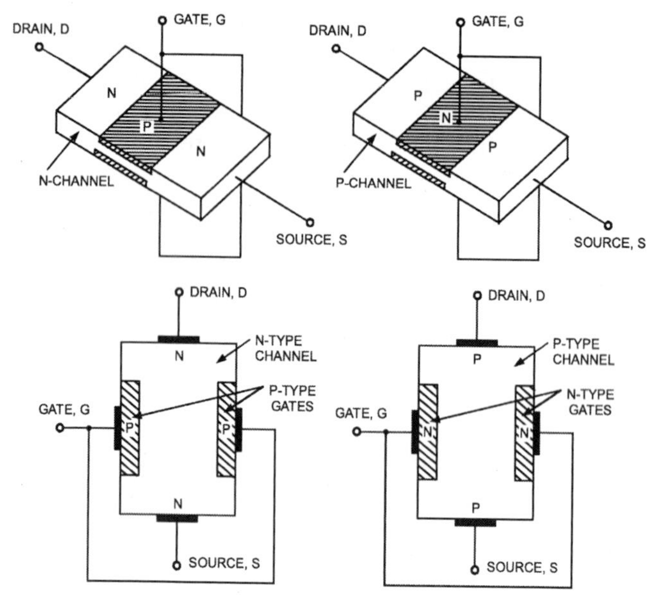

(*a*) *N-Channel JFET* (*b*) *P-Channel JFET*

Fig. 10.1. ***JFETs***

The silicon bar behaves like a resistor between its two terminals D and S. The gate terminal is analogous to the base of an ordinary transistor (BJT). It is used to control the flow of current from source to drain. Thus, source and drain terminals are analogous to emitter and collector terminals respectively of a BJT.

Analogies are seldom perfect and at times can be misleading, but the water analogy of Fig. 10.2 does provide a sense for the JFET control at the gate terminal and the appropriateness of the terminology applied to the terminals of the device. The source of water pressure can be likened to the applied voltage from drain to source, which results in a flow of water (electrons) from the spigot (source). The "gate", through an applied signal (potential) controls the flow of water (charge) to the "drain", the drain and source terminals are at opposite ends of the N-channel as introduced in Fig. 10.1 because the terminology is defined for electron flow.

* When a manufacturer connects a separate lead to each gate, the device is called a dual-gate JFET. The main use of a dual-gate JFET is with a mixer, a special circuit used in communication equipment.

In Fig. 10.1 (*a*), the gate is P-region, while the source and the drain are N-regions. Because of this, a JFET is similar to two diodes. The gate and the source form one of the diodes, and the gate and the drain form the other diode. These two diodes are usually referred to as the *gate-source diode* and the *gate-drain diode*. Since JFET is a silicon device, it takes only 0.7 V for forward bias to get significant current in either diode.

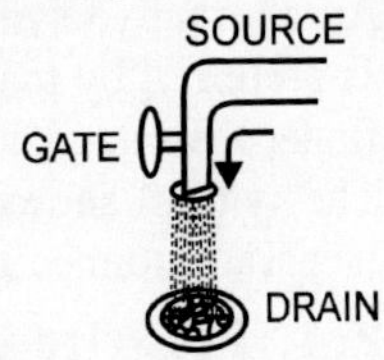

Fig. 10.2. ***Water Analogy For The JFET Control Mechanism***

With the gate terminal not connected, and a potential applied (+ve at the drain and –ve at the source), a current called the *drain current*, I_D flows through the channel located between the two P-regions. This current consists of only majority carriers – electrons in this case.

P-channel JFET is similar in construction to N-channel JFET except that P-type semiconductor material is sandwiched between two N-type junctions, as shown in Fig. 10.1 (*b*). In this case majority carriers are holes.

The basic structure of an N-channel JFET, together with the supply voltages, is shown in Fig. 10.3.

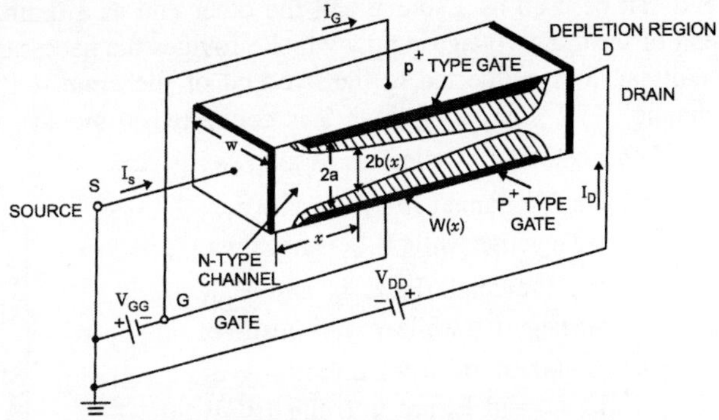

Fig. 10.3. ***Basic Structure of an N-Channel FET***

10.2.2. Standard Notations

The noteworthy FET notations are as follows :

1. ***Source.*** The terminal, through which the majority carriers enter the channel, is called the *source* terminal S and the conventional current entering the channel at S is designated as I_S.
2. ***Drain.*** The terminal, through which the majority carriers leave the channel, is called the *drain* terminal D and the conventional current leaving the channel at D is designated as I_D.
   The drain-to-source voltage is called V_{DS}, and is positive if D is more positive than S.
3. ***Gate.*** There are two internally connected heavily doped impurity regions formed by alloying, by diffusion, or by any other method available to create two P-N junctions. These impurity regions are called the gate G. A voltage V_{GS} is applied between the gate and source in the direction to reverse bias the P-N junction. Conventional current entering the channel at G is designated as I_G.
4. ***Channel.*** The region between the source and drain, sandwiched between the two gates is called the *channel* and the majority carriers move from source to drain through this channel.

10.2.3. Schematic Symbols

The schematic symbols for N-type and P-type JFETs are shown in Figs. 10.4 (*a*) and 10.4 (*b*) respectively. The vertical line in the symbol may be thought as channel and source S and drain D connected to the line.

Note that the direction of the arrow at the gate indicates the direction in which the gate current flows when the gate junction is forward biased. Thus for the N-channel JFET, the arrow at the gate junction points into the device and in P-channel JFET, it is away from the device.

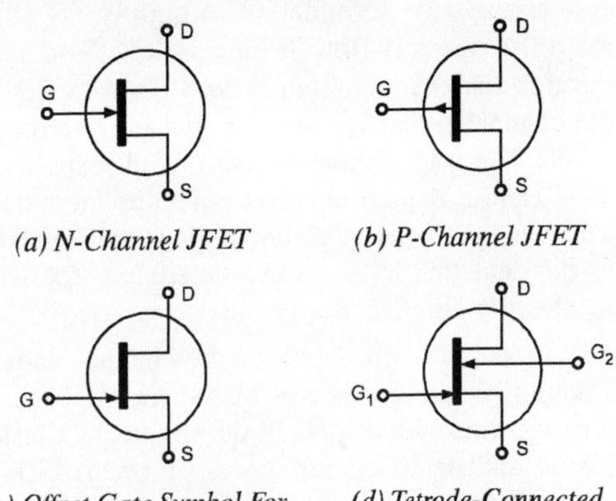

(*a*) *N-Channel JFET* (*b*) *P-Channel JFET*

(*c*) *Offset Gate Symbol For an N-Channel JFET* (*d*) *Tetrode-Connected JFET*

Fig. 10.4. ***Schematic Symbols For JFETs***

An alternative symbol for an N-channel JFET is shown in Fig. 10.4 (*c*). This symbol with its offset gate is preferred by many engineers and technicians. The offset gate points to the source end of the device, a definite advantage in complicated multistage circuit.

The symbol shown in Fig. 10.4 (*d*) is used where the terminals of the two gate regions are provided with separate connecting leads. In this case device is referred to as a **tetrode-connected FET.**

10.2.4. Polarity Conventions

The polarities for N-channel and P-channel JFETs are shown in Figs. 10.5 (*a*) and 10.5 (*b*) respectively. In both of the cases the voltage between gate and source is such that the gate is reverse biased. This is the normal method of connection of JFETs. The drain and source terminals are interchangeable* *i.e.* either end can be used as a source and the other end as a drain. The source terminal is always connected to that end of the drain voltage supply which provides the necessary charge carriers *i.e.* in an N-channel JFET source terminal, S is connected to the –ve end of the drain voltage supply for obtaining electrons whereas in P-channel JFET source terminal, S is connected to the +ve end of the drain voltage supply for getting holes.

Other noteworthy points are as follows :

(*i*) For N-channel JFET, I_D and V_{DD} are positive while V_{GG} is negative.

(*ii*) For P-cannel JFET, I_D and V_{DD} are negative while V_{GG} is positive.

(*iii*) Assumed positive directions of I_S, I_D and I_G are into the JFET.

From the above statements it may be concluded that the action of a P-channel JFET is complementary, that is, all voltages and currents are reversed.

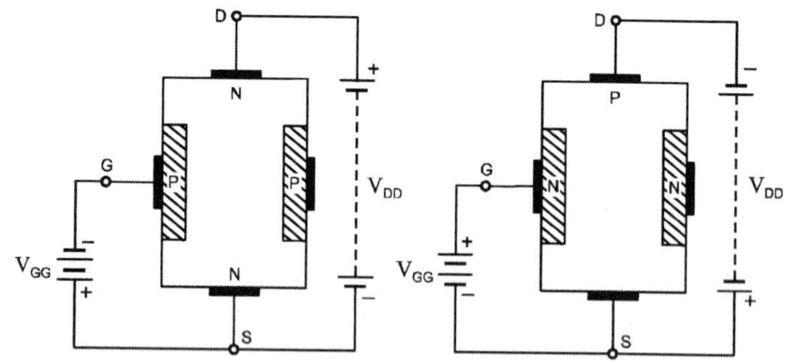

(*a*) *N-Channel JFET* (*b*) *P-Channel JFET*

Fig. 10.5. ***Polarity Conventions For JFETs***

10.2.5. Operation

Let us consider an N-channel JFET for discussing its operation.

(*i*) When neither any bias is applied to the gate (*i.e.* when $V_{GS} = 0$) nor any voltage to the drain w.r.t. source (*i.e.* when $V_{DS} = 0$), the depletion regions around the P-N junctions are of equal thickness and symmetrical.

(*ii*) When positive voltage is applied to the drain terminal D w.r.t. source terminal S without connecting gate terminal G to supply, as illustrated in Fig. 10.6 (*a*), the electrons (which are the majority carriers) flow from terminal S to terminal D whereas conventional drain current I_D flows through the channel from D to S. Due to flow of this current, there is a uniform voltage drop across the channel resistance as we move from terminal D to terminal S. This voltage drop reverse biases the diode. The gate is more negative with respect to those points in the channel which are nearer to D than to S. Hence, depletion layers penetrate more deeply into the channel at points lying closer to D than to S. Thus wedge-shaped depletion regions are formed, as shown in Fig. 10.6 (*a*), when V_{DS} is applied. The size of the depletion layer formed determines the width of the channel and hence the magnitude of current I_D flowing through the channel.

To see how the width of the channel varies with the variation in gate voltage, let us consider the situation when the gate is biased negative with respect to the source while the drain is applied with positive bias with respect to the source, as illustrated in Fig. 10.6 (*b*). Now the P-N junctions are reverse biased and depletion regions are formed. P-regions are heavily doped compared to the N-channel, so the depletion regions penetrate deeply into the channel. Since a depletion region is a region depleted of the charge carriers, it behaves as an insulator. The result is that the channel is narrowed, the resistance

* The drain and source terminals are not interchangeable at high frequencies. Almost always, the manufacturer minimizes the internal capacitance on the drain side of the JFET *i.e.* the capacitance between the gate and the drain is smaller than the capacitance between the gate and the source.

is increased and drain current I_D is reduced. If the negative voltage at the gate is further increased, depletion layers meet at the centre and the drain current I_D is cut off completely. On the other hand, if the negative bias to the gate is reduced, the width of the depletion layers gets reduced causing decrease in resistance and, therefore, increase in drain current I_D. The gate-source voltage V_{GS} at which drain current I_D is cut off completely (pinched off), is called the *pinch-off voltage V_P*. It is also to be noted that

1. The amount of reverse bias is not the same throughout the length of the P-N junction. When the drain current flows through the channel, there is a voltage drop along its length. The result is that the reverse bias at the drain end is more than that at the source end making the width of depletion layer more at the drain end than that at the source end. Thus the channel becomes narrower at the drain end in comparison to that at source end, as shown in Fig. 10.6 (*b*).

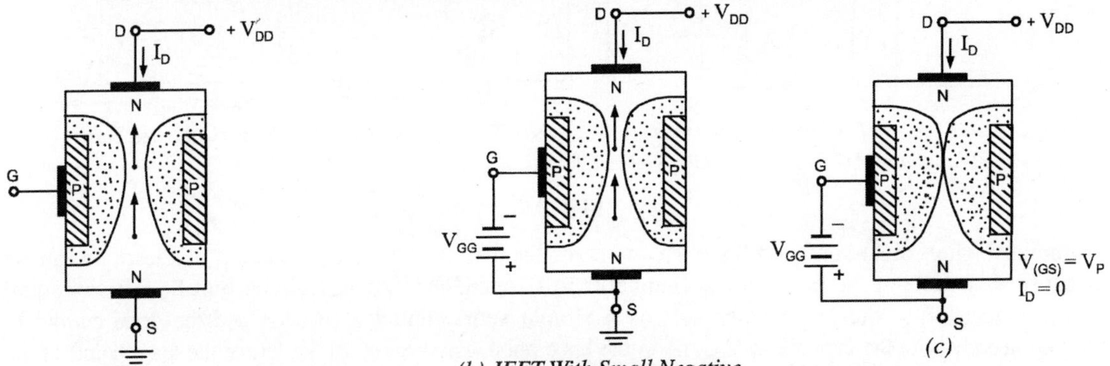

(a) JFET With No Bias Voltage *(b) JFET With Small Negative Gate Source Bias* *(c)*

Fig. 10.6

2. The channel is not completely closed at the drain end. Because in that case there will be no drain current, so there will be no voltage drop along the channel length and amount of reverse bias will become uniform and the wedge shaped depletion region will become rectangular one. The channel will open and the drain current will flow.

   However, at pinch-off voltage, the channel width is reduced to a constant minimum value to allow the flow of drain current.

3. The N-channel JFET behaves as a vacuum tube triode. The drain and source perform the same functions as the plate and cathode, respectively and, like the grid of a triode, the JFET gate controls the drain current. As is also the case with a grid, gate current is to be avoided, so the gate channel junctions are normally never forward biased.

The device is called the *field-effect transistor* (FET) because the drain current (output current) is controlled by the effect of the extension of the field associated with the depletion region developed by the reverse bias at the gate.

P-channel JFET operates in the same manner as an N-channel JFET except that channel current carriers will be the holes in place of electrons and the polarities of V_{GS} and V_{DS} are reversed.

10.3. CHARACTERISTICS OF JFETs

There are two types of static characteristics viz. (1) output or drain characteristic and (2) transfer characteristic.

1. *Output or Drain Characteristic.* The curve drawn between drain current I_D and drain-source voltage V_{DS} with gate-to-source voltage V_{GS} as the parameter is called the *drain* or *output characteristic*. This characteristic is analogous to collector characteristic of a BJT.

***(a)* Drain Characteristic With Shorted Gate.** The circuit diagram for determining the drain characteristic with shorted gate for an N-channel JFET is given in Fig. 10.7 (*a*) and the drain characteristic with shorted gate is shown in Fig. 10.7 (*b*).

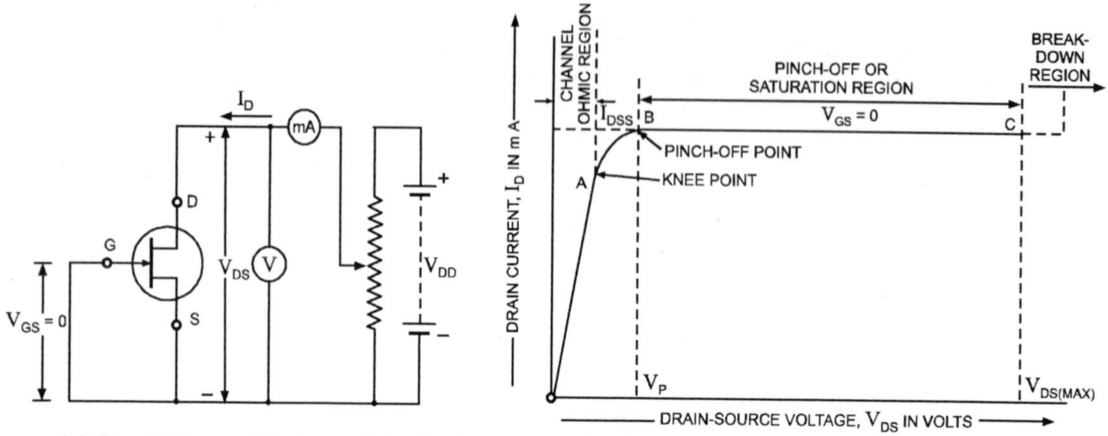

(a) Circuit Diagram For Determining Drain Characteristic With Shorted Gate For an N-Channel JFET

(b) JFET Drain Characteristic With Shorted Gate

Fig. 10.7

Initially when drain-source voltage V_{DS} is zero, there is no attracting potential at the drain, so no current flows inspite of the fact that the channel is fully open. This gives drain current $I_D = 0$. For small applied voltage V_{DS}, the N-type bar acts as a simple semiconductor resistor, and the drain current I_D increases linearly with the increase in V_{DS}, up to the knee point. This region (to the left of the knee point) of the curve is called the *channel ohmic region*, because in this region the FET behaves like an ordinary resistor.

With the increase in drain current I_D, the ohmic voltage drop between the source and channel region reverse biases the gate junction. The reverse biasing of the gate junction is not uniform throughout. The reverse bias is more at the drain end than that at the source end of the channel, so with the increase in V_{DS}, the conducting portion of the channel begins to constric more at the drain end. Eventually a voltage V_{DS} is reached at which the channel is *pinched off*. The drain current I_D no longer increases with the increase in V_{DS}. It approaches a constant saturation value. The value of voltage V_{DS} at which the channel is pinched off (*i.e.* all the free charges from the channel get removed), is called the *pinch-off voltage* V_P. The pinch-off voltage V_P, not too sharply defined on the curve, where the drain current I_D begins to level off and attains a constant value. From point A (knee point) to the point B (pinch-off point) the drain current I_D increases with the increase in voltage V_{DS} following a reverse square law. The region of the characteristic in which drain current I_D remains fairly constant is called the *pinch-off region*. It is also sometimes called the *saturation region* or *amplifier region*. In this region the JFET operates as a *constant current device* since drain current (or output current) remains almost constant. It is the normal operating region of the JFET when used as an amplifier. The drain current in the pinch-off region with $V_{GS} = 0$ is referred to as the *drain-source saturation current*, I_{DSS}.

It is to be noted that in the pinch-off (or saturation) region the channel resistance increases in proportion to increase in V_{DS} and so keeps the drain current almost constant and the reverse bias required by the gate channel junction is supplied entirely by the voltage drop across the channel resistance due to flow of I_{DSS} and not by the external bias because $V_{GS} = 0$.

Drain current in the pinch-off region is given by Shockley's equation

$$I_D = I_{DSS}\left(1 - \frac{V_{GS}}{V_P}\right)^2 = I_{DSS}\left(1 - \frac{V_{GS}}{V_{GS\,(OFF)}}\right)^2 \qquad ...(10.1)$$

where I_D is the drain current at a given gate-source voltage V_{GS}, I_{DSS} is the drain current with gate shorted to source and $V_{GS\,(OFF)}$ is the gate-source cutoff voltage.

If drain-source voltage, V_{DS} is continuously increased, a stage comes when the gate channel junction breaks down. At this point the drain current increases very rapidly, and the JFET may get destroyed. This happens because the charge carriers making up the reverse saturation current at the gate channel junction accelerate to a high velocity and produce an *avalanche effect*.

It will be very interesting to observe that JFET behaves as an ordinary resistor in ohmic region, as a constant current source in pinch-off (or saturation) region and as a constant voltage source in breakdown region [Fig. 10.7 (*b*)].

(*b*) Drain Characteristics With External Bias. The circuit diagram for determining the drain characteristics with different values of external bias is shown in Fig. 10.8 (*a*) and a family of drain characteristics for different values of gate-source voltage V_{GS} is given in Fig. 10.8 (*b*).

It is observed that as the negative gate bias voltage is increased

(*i*) the maximum saturation drain current becomes smaller because the conducting channel now becomes narrower.

(*ii*) Pinch-off voltage is reached at a lower value of drain current I_D than when $V_{GS} = 0$.

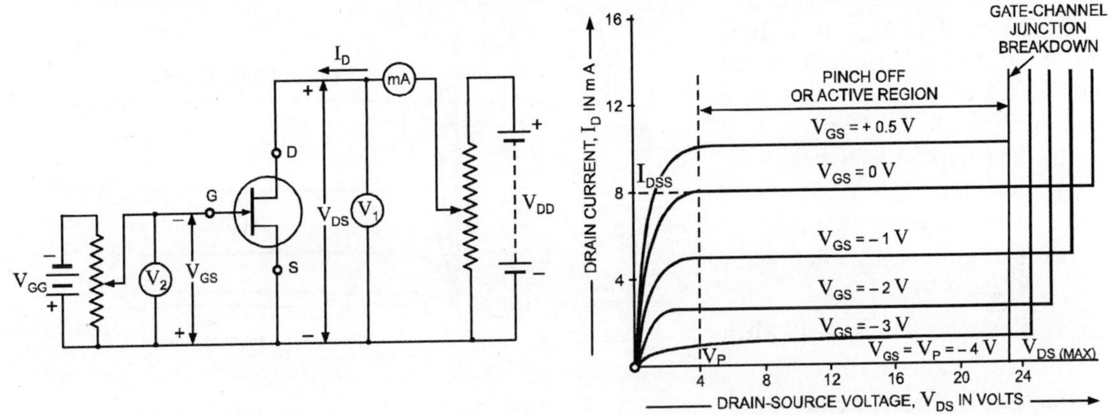

(a) Circuit Diagram For Determining Drain Characteristic With External Bias For an N-Channel JFET *(b) JFET Drain-Characteristics With External Bias*

Fig. 10.8

When an external bias of, say, –1 V is applied between the gate and the source, the gate channel junctions are reverse biased even when drain current, I_D is zero. Hence the depletion regions are already penetrating the channel to a certain extent when drain-source voltage, V_{DS} is zero. Due to this reason, a smaller voltage drop along the channel (*i.e.* smaller than that for $V_{GS} = 0$) will increase the depletion regions to the point where they pinch-off the current. Consequently, the pinch-off voltage V_P is reached at a lower drain current, I_D when $V_{GS} = 0$.

(*iii*) The ohmic region portion decreases.

(*iv*) Value of drain-source voltage V_{DS} for the avalanche breakdown of the gate junction is reduced.

Value of drain-source voltage, V_{DS} for breakdown with the increase in negative bias voltage is reduced simply due to the fact that gate-source voltage, V_{GS} keeps adding to the reverse bias at the junction produced by current flow. Thus the maximum value of V_{DS} that can be applied to a FET is the lowest voltage which causes avalanche breakdown.

It is also observed that with $V_{GS} = 0$, I_D saturates at I_{DSS} and the characteristic shows $V_P = 4$ V. When an external bias of –1 V is applied, the gate channel junctions still require –4 V to achieve pinch-off. It means that a 3 V drop is now required along the channel instead of the previous 4.0 V. Obviously, this drop of 3 V can be achieved with a lower value of drain current. Similarly, when $V_{GS} = -2$ V and – 3 V, pinch-off is achieved with 2 V and 1 V respectively, along the channel. These drops of 2 V and 1 V are, of course, achieved with further reduced values of drain current, I_D. It is further observed that when the gate-source bias is numerically equal to pinch-off voltage, V_P (–4 V in this case), no channel drop is required and, therefore, drain current, I_D is zero. The gate-source bias voltage required to reduce drain current I_D to zero is designated the gate-source cutoff voltage, $V_{GS\,(OFF)}$ and, as explained,

$$V_{GS\,(OFF)} = V_P$$

Hence for working of JFET in the pinch-off or active region it is necessary that the following conditions be fulfilled.

$$V_P < V_{DS} < V_{DS\,(MAX)} \quad ...(10.2)$$

$$V_{GS\,(OFF)} < V_{GS} < 0 \quad ...(10.3)$$

$$0 < I_D < I_{DSS} \quad ...(10.4)$$

2. *Transfer Characteristics.* The transfer characteristics for a JFET can be determined experimentally, keeping drain-source voltage, V_{DS} constant and determining drain current, I_D for various values of gate-source voltage, V_{GS}. The circuit diagram is shown in Fig. 10.8 (*a*). The curve is plotted between gate-source voltage, V_{GS} and drain current, I_D, as illustrated in Fig. 10.9. It is similar to the transconductance characteristic of a vacuum tube or a transistor. It is observed that

(*i*) drain current decreases with the increase in –ve gate-source bias,

(*ii*) drain current, $I_D = I_{DSS}$ when $V_{GS} = 0$

(*iii*) drain current, $I_D = 0$ when $V_{GS} = V_P$

The transfer characteristic follows Eq. (10.1)

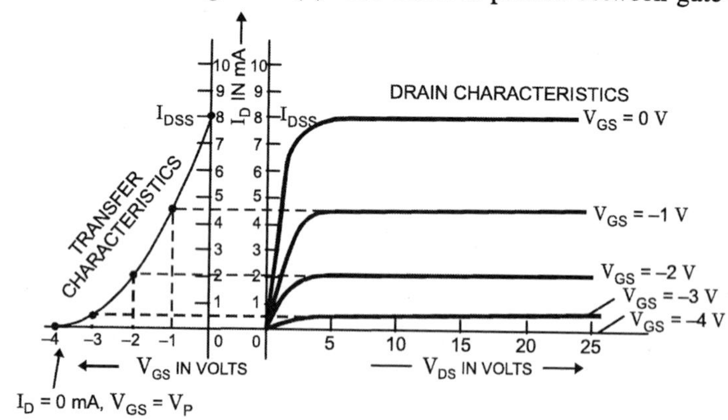

Fig. 10.9. ***Derivation of Transfer Characteristics From Drain Characteristics***

The transfer characteristic can also be derived from the drain characteristic by noting values of drain current, I_D corresponding to various values of gate-source voltage, V_{GS} for a constant drain-source voltage and plotting them.

It may be noted that a P-channel JFET operates in the same way and have the similar characteristics as an N-channel JFET except that channel carriers are holes instead of electrons and the polarities of V_{GS} and V_{DS} are reversed.

10.4. MERITS AND DEMERITS OF JFETs

Junction field-effect transistors combine several merits of both conventional (or bipolar) transistors and vacuum tubes. Some of these are enumerated below :

1. Its operation depends upon the flow of majority carriers only. It is, therefore, a *unipolar* (one type of carrier) device. On the other hand in an ordinary transistor both majority and minority carriers take part in conduction and, therefore, ordinary transistor is sometimes called the *bipolar* transistor. The vacuum tube is another example of a unipolar device.
2. It is simpler to fabricate, smaller in size, rugged in construction and has longer life and higher efficiency. Simpler to fabricate in IC form and space requirement is also lesser.
3. It has high input impedance (of the order of 100 MΩ), because its input circuit (gate to source) is reverse biased, and so permits high degree of isolation between the input and output circuits. However, the input circuit of an ordinary transistor is forward biased and, therefore, ordinary transistor has low input impedance.
4. It carries very small current because of reverse-biased gate and, therefore, it operates just like a vacuum tube where control grid (corresponding to gate in JFET) carries extremely small current and input

voltage controls the output current. This is the reason that JFET is essentially a *voltage driven* device. Ordinary transistor is a current operated device since input current controls the output current.

5. An ordinary transistor uses a current into its base for controlling a large current between collector and emitter whereas in a JFET voltage on the gate (base) terminal is used for controlling the drain current (current between drain and source). Thus an ordinary transistor gain is characterised by current gain whereas the JFET gain is characterised as the *transconductance* (the ratio of drain current and gate-source voltage).
6. JFET has no junction like an ordinary transistor and the conduction is through bulk material current carriers (N-type or P-type semiconductor material) that do not cross junctions. Hence the inherent noise of tubes (owing to high temperature operation) and that of ordinary transistors (owing to junction transitions) is not present in JFET.
7. It is relatively immune to radiation.
8. It has –ve temperature coefficient of resistance and, therefore, has better thermal stability.
9. It has high power gain and, therefore, the necessity of employing driver stages is eliminated.
10. It exhibits no offset voltage at zero drain current and, therefore, makes an excellent signal chopper.
11. It has square-law characteristics and, therefore, it is very useful in the tuners of radio and TV receivers.
12. It has got high frequency response.

The main drawback of JFET is its relative small gain-bandwidth product in comparison with that of a conventional transistor and greater susceptibility to damage in its handling. Other disadvantages of JFET are low voltage gains because of small transconductance and higher cost than that of BJTs.

10.5. FET CONFIGURATIONS

FETs (JFETs and MOSFETs) have three configurations just as BJTs have. The three configurations of FETs—the common gate, common source, and common drain configurations have been given in Fig. 10.10. Input and output terminals of different configurations have been given in Table 10.1.

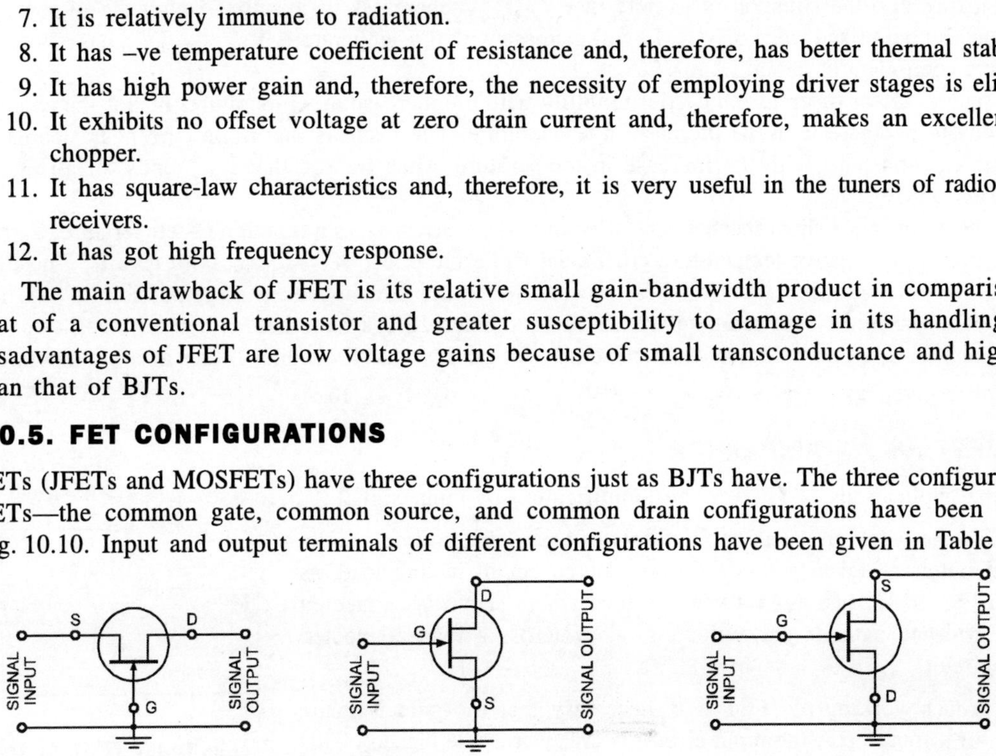

(a) Common Gate Configuration *(b) Common Source Configuration* *(c) Common Drain Configuration*

Fig. 10.10. ***N-Channel FET Configurations***

TABLE 10.1. ***FET Configurations***

Configuration	*Input Terminal*	*Output Terminal*
Common gate	Source	Drain
Common source	Gate	Drain
Common drain	Gate	Source

TABLE 10.2. ***FET-BJT Correspondence***

FET Configuration	*BJT Configuration*
Common gate	Common base
Common source	Common emitter
Common drain	Common collector

The three FET configurations have characteristics that are similar to the three corresponding BJT configurations, as given in Table 10.2.

10.6. JFET TEMPERATURE EFFECTS

The JFET, like other solid state devices, is subject to the effect of temperature. However, it is possible to bias the JFET such that it will exhibit a *zero temperature coefficient i.e.* drain current I_D will be independent of temperature. This is explained below :

There are two primary mechanisms underlying the temperature sensitivity of the conductivity of a JFET channel.

1. Decrease in the depletion region width at the channel-gate P-N junction with the increase in temperature. This results in increase of channel thickness.
2. Decrease in carrier mobility with the increase in temperature.

Increase in channel thickness with the increase in temperature makes drain current I_D to increase. Another way of looking into the situation is to note that $V_{GS\,(OFF)}$ increases in magnitude with the increase in temperature. So far a fixed value of V_{GS}, I_D will increase with the temperature. $V_{GS\,(OFF)}$ [V_P] has a positive temperature coefficient of about 2.2 mV/°C.

The second factor [decrease in carrier mobility with the increase in temperature] makes the channel conductivity to decrease with the increase in temperature. The result is that drain current I_D decreases (assuming V_{DS} constant) with the increase in temperature. Thus we see that I_{DSS} tends to fall as the temperature rises.

Thus we have two distinct mechanisms affecting drain current I_D as a function of temperature. Carrier mobility gives I_D a negative temperature coefficient while the effect of $V_{GS\,(OFF)}$ is to give I_D a positive temperature coefficient. Since both of these mechanisms occur simultaneously, it is possible to bias the JFET so as to exhibit a zero temperature coefficient. Thus JFETs have better thermal stability.

The value of V_{GS} that will give drain current I_D for both N- and P-channel JFETs a zero temperature coefficient is given as $|V_{GS}| \simeq |V_P| - 0.63$ V. ...(10.5)

10.7. JFET AS AN AMPLIFIER

One of the applications of JFET is an amplification of input signal.

In FET the weak signal is applied between gate and source and an amplified output is taken across the drain-source circuit having load, as shown in Fig. 10.11, just as in a vacuum tube where signal is connected between grid and cathode and output is taken across a load connected in plate circuit.

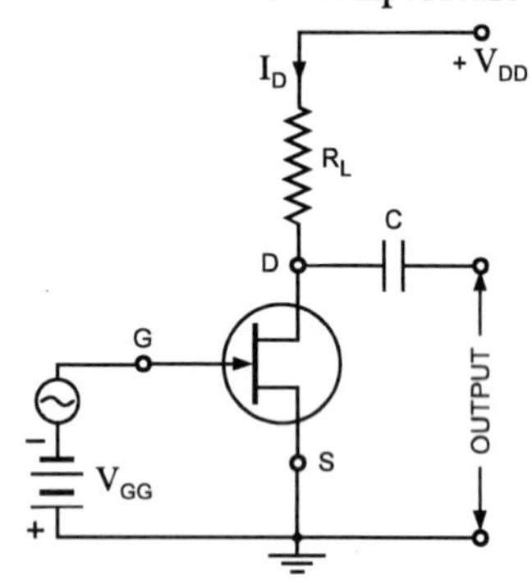

Fig. 10.11. ***JFET As An Amplifier***

For proper working of JFET it is necessary that the gate is made negative w.r.t. source (*i.e.* the input circuit is always reverse biased). This is achieved either by inserting a battery V_{GG} in the gate circuit with the polarity shown in Fig. 10.11 or by providing a biasing circuit.*

On application of an alternating weak signal, the fixed reverse bias of the gate changes and a small change in reverse bias on the gate causes a large change in drain current owing to change in width of the channel. This fact makes the FET capable of amplifying the strength of weak signals. During the positive half cycle of the input signal, the reverse bias on the gate decreases causing increase in channel width and hence the drain current. During the negative half cycle of the input signal, the reverse bias on the gate increases causing drain current to decrease. The result is that the small change in voltage at the gate produces a large change in the drain current. These large variations in drain current give large output across the load R_L and thus, JFET operates as an amplifier.

* Refer to Art. 10.10.

10.8. JFET PARAMETERS

JFET has certain parameters which determine the performance. Such parameters are (*i*) ac drain resistance, (*ii*) transconductance, (*iii*) amplification factor, and (*iv*) dc drain resistance, as explained below :

1. *AC Drain Resistance.* It is defined as the ratio of change in drain-source voltage to change in drain current at constant gate-source voltage and is denoted by r_d.

$$\text{i.e. AC drain resistance, } r_d = \frac{\Delta V_{DS}}{\Delta I_D} \text{ at constant } V_{GS} \qquad ...(10.6)$$

It is also called the *dynamic drain resistance.*

Referring to the output or drain characteristic shown in Fig. 10.8 (*b*), it is clear that in the active region the change in drain current, I_D is very small for change in drain-source voltage, V_{DS} because the characteristic curves are almost flat. Hence ac drain resistance of a JFET is very large ranging from 10 kΩ to 1 MΩ.

2. *Transconductance.* The control that the gate-source voltage has over the drain current, I_D is measured by transconductance. It is denoted by g_{fs}. It may be defined as the ratio of change in drain current to the change in gate-source voltage at constant drain-source voltage.

$$\text{i.e. Transconductance, } g_m = \frac{\Delta I_D}{\Delta V_{GS}} \text{ at constant } V_{DS} \qquad ...(10.7)$$

It is also called the *forward transconductance* (g_{fs}) or forward transadmittance (Y_{fs}). It is measured in mA/volt or millisiemens.

The transconductance measured at I_{DSS} is denoted by g_{mo}.

Mathematical Expression For Transconductance. Differentiating both sides of Shockley equation

$$I_D = I_{DSS}\left(1 - \frac{V_{GS}}{V_P}\right)^2, \text{ we get}$$

$$dI_D = 2I_{DSS}\left(1 - \frac{V_{GS}}{V_P}\right)\left(-\frac{1}{V_P}\right) dV_{GS}$$

$$\text{or} \quad \frac{dI_D}{dV_{GS}} = 2I_{DSS}\left(1 - \frac{V_{GS}}{V_P}\right)\left(-\frac{1}{V_P}\right)$$

$$\text{or} \quad g_m = -\frac{2I_{DSS}}{V_P}\left(1 - \frac{V_{GS}}{V_P}\right) \qquad ...(10.8)$$

Substituting $V_{GS} = 0$ in above equation, we get

$$g_{mo} = -\frac{2I_{DSS}}{V_P} \qquad ...(10.9)$$

From Eqs. (10.8) and (10.9) we have

$$g_m = g_{mo}\left(1 - \frac{V_{GS}}{V_P}\right) \qquad ...(10.10)$$

3. *Amplification Factor.* It is defined as the ratio of change in drain-source voltage to the change in gate-source voltage at constant drain current and is denoted by μ.

$$\text{i.e. Amplification factor, } \mu = \frac{\Delta V_{DS}}{\Delta V_{GS}} \text{ at constant } I_D$$

Amplification factor of a JFET indicates how much more control the gate-source voltage has over drain current in comparison to the drain-source voltage.

$$\text{Amplification factor, } \mu = \frac{\Delta V_{DS}}{\Delta V_{GS}} = \frac{\Delta V_{DS}}{\Delta I_D} \times \frac{\Delta I_D}{\Delta V_{GS}} = r_d \times g_m$$

$$= \text{ac drain resistance} \times \text{transconductance} \qquad ...(10.11)$$

Amplification factor μ of a FET may be as high as 100.

4. *DC Drain Resistance*. It is also called the static or ohmic resistance of the channel and is defined as ratio of drain-source voltage and drain current. It is denoted by R_{DS}

$$i.e., \; R_{DS}^{*} = \frac{V_{DS}}{I_D} \qquad ...(10.12)$$

Example 10.1. When a reverse voltage of 10 V is applied between gate and source of JFET the gate current is 0.1 μA. Determine resistance between gate and source.

Solution : Gate-source voltage, $V_{GS} = 10.0$ V

Gate current, $I_G = 0.1\ \mu A = 1 \times 10^{-7}$ A

$$\text{Gate-to-source resistance, } R_{GS} = \frac{V_{GS}}{I_G} = \frac{10}{1 \times 10^{-7}} = 100\ M\Omega \textbf{ Ans.}$$

Example 10.2. The readings obtained experimentally from a JFET are as follows :

Drain-source voltage, V_{DS} in volts	5	12	12
Gate-source voltage, V_{GS} in volts	**0**	**0**	**– 0.25**
Drain current, I_D in mA	**8**	**8.2**	**7.5**

Determine (*i*) ac drain resistance (*ii*) transconductance and (*iii*) amplification factor.

Solution : (*i*) The drain current increases from 8.0 mA to 8.2 mA when drain-source voltage is increased from 5 V to 12 V, keeping gate-source voltage constant at 0 V.

$$\therefore \text{ AC drain resistance, } r_d = \frac{\Delta V_{DS}}{\Delta I_D} = \frac{(12-5)\text{ V}}{(8.2-8)\text{ mA}} = \frac{7}{0.2 \times 10^{-3}} = 35\ k\Omega \textbf{ Ans.}$$

(*ii*) The drain current changes from 8.2 mA to 7.5 mA when gate-source voltage is changed from 0 to –0.25 V, keeping drain-source voltage constant at 12 V.

$$\therefore \text{ Transconductance, } g_m = \frac{\Delta I_D}{\Delta V_{GS}} = \frac{(7.5-8.2)\text{ mA}}{(-0.25-0)\text{ V}} = 2.8 \text{ mA/V} \textbf{ Ans.}$$

(*iii*) Amplification factor, $\mu = r_d \times g_m = 35 \times 1{,}000 \times 2.8 \times 10^{-3} = 98$ **Ans.**

Example 10.3. Calculate the minimum value of V_{DS} required for an NMOSFET to operate in pinch-off condition when $V_{GS} = 1$ V, $V_P = -2$ V, $I_{DSS} = 10$ mA. Also find the corresponding drain current.

Solution : Drain current in pinch-off region

$$I_D = I_{DSS}\left(1 - \frac{V_{GS}}{V_P}\right)^2 = 10 \times 10^{-3}\left(1 - \frac{1}{(-2)}\right)^2 = 10 \times 10^{-3} \times \left(1\frac{1}{2}\right)^2$$

$$= 22.5 \times 10^{-3} \text{ A or } 22.5 \text{ mA } \textbf{Ans.}$$

The minimum value of V_{DS} required for the MOSFET to operate in pinch-off condition is equal to V_P *i.e.* $V_{DS\,(min)} = V_P = -2$ V **Ans.**

Example 10.4. A JFET has $V_P = -4.5$ V, $I_{DSS} = 10$ mA and $I_{DS} = 2.5$ mA. Determine the transconductance.

Solution : Drain-source saturation current, $I_{DSS} = 10$ mA

Pinch-off voltage, $V_P = -4.5$ V

Drain-source current, $I_{DS} = 2.5$ mA

From Shockley's equation, drain-source current,

$$I_{DS} = I_{DSS}\left[1 - \frac{V_{GS}}{V_P}\right]^2$$

* Drain resistance r_d is not to be confused with the drain-source resistance, R_{DS}. r_d is a dynamic or ac resistance whereas R_{DS} is the dc resistance of the channel when the depletion regions are absent *i.e.* when the device is biased on in the channel ohmic region of the characteristics.

$$\text{or } V_{GS} = V_P\left[1-\sqrt{\frac{I_{DS}}{I_{DSS}}}\right] = -4.5\left[1-\sqrt{\frac{2.5}{10}}\right] = -2.25 \text{ V}$$

From Eq. (10.8), transconductance, $g_m = \dfrac{-2I_{DSS}}{V_P}\left[1-\dfrac{V_{GS}}{V_P}\right] = \dfrac{-2\times10\times10^{-3}}{-4.5}\left[1-\dfrac{-2.25}{-4.5}\right] = 2.22$ mA/V **Ans.**

Example 10.5. A FET has a drain current of 4 mA. If I_{DSS} = 6 mA and $V_{GS\,(off)}$ = –6 V, find the values of V_{GS} and V_P.

Solution : Drain-source current, I_D = 4 mA

Drain-source saturation current, I_{DSS} = 6 mA

$$\text{Gate-to-source voltage, } V_{GS} = V_{GS(off)}\left[1-\sqrt{\frac{I_D}{I_{DSS}}}\right] = -6\left[1-\sqrt{\frac{4 \text{ mA}}{6 \text{ mA}}}\right] = -1.1 \text{ V } \textbf{Ans.}$$

Pinch-off voltage, $V_P = V_{GS(off)} = -6$ V **Ans.**

Example 10.6. A JFET has g_m = 10 mS, I_{DSS} = 10 μA. Calculate $V_{GS\,(OFF)}$.

Solution : Transconductance, g_m = 10 mS = 10 × 10^{-3} S

Drain-source saturation current, I_{DSS} = 10 μA = 10 × 10^{-6} A

$$V_{GS\,(OFF)} \rightarrow V_{GS} = V_P$$

$$g_m = g_{mo} = \frac{-2I_{DSS}}{V_P} = \frac{-2I_{DSS}}{V_{GS(OFF)}}$$

$$\text{or } 10 \times 10^{-3} = \frac{-2\times10\times10^{-6}}{V_{GS(OFF)}}$$

$$\text{or } V_{GS\,(OFF)} = \frac{-2\times10\times10^{-6}}{10\times10^{-3}} = -2 \text{ mV } \textbf{Ans.}$$

Example 10.7. For an N-channel JFET I_{DSS} = 8.7 mA; V_P = –3 V; V_{GS} = – 1 V. Find the values of (*i*) I_D (*ii*) g_{mo} (*iii*) g_m.

Solution : Drain-source saturation current,

I_{DSS} = 8.7 mA = 0.0087 A

Pinch-off voltage, V_P = – 3 V

Gate-source voltage, V_{GS} = – 1V

$$(i) \text{ Drain current, } I_D = I_{DSS}\left(1-\frac{V_{GS}}{V_P}\right)^2 = 0.0087\left(1-\frac{-1}{-3}\right)^2 = 3.8667 \text{ mA } \textbf{Ans.}$$

(*ii*) Transconductance for V_{GS} = 0,

$$g_{mo} = \frac{-2I_{DSS}}{V_P} = \frac{-2\times8.7 \text{ mA}}{-3} = 5.8 \text{ mA/V or } 5.8 \text{ mS } \textbf{Ans.}$$

$$(iii) \text{ Transconductance, } g_m = g_{mo}\left(1-\frac{V_{GS}}{V_P}\right) = 5.8\left(1-\frac{-1}{-3}\right) = 3.867 \text{ mA/V or } 3.867 \text{ mS } \textbf{Ans.}$$

Example 10.8. An N-channel JFET has a pinch-off voltage of – 4.5 V and I_{DSS} = 9 mA. At what value of V_{GS} will I_{DS} be equal to 3 mA? What is its g_m at this I_{DS} ?

Solution : Pinch-off voltage, V_P = – 4.5 V

Drain-source saturation current, I_{DSS} = 9 mA = 9 × 10^{-3} A

Drain-source current, $I_{DS} = 3\text{ mA} = 3 \times 10^{-3}\text{ A}$

From Shockley's equation (Eq. 10.1)

$$V_{GS} = V_P\left[1-\sqrt{\frac{I_{DS}}{I_{DSS}}}\right] = -4.5 \times \left[1-\sqrt{\frac{3\times10^{-3}}{9\times10^{-3}}}\right] = -1.902\text{ V } \textbf{Ans.}$$

Transconductance g_m for $I_{DS} = 3$ mA for which

$$V_{GS} = -1.902\text{ V}$$

$$g_m = \frac{-2I_{DSS}}{V_P}\left[1-\frac{V_{GS}}{V_P}\right] = \frac{-2\times9\times10^{-3}}{-4.5}\left(1-\frac{-1.902}{-4.5}\right)$$

$$= 2.31\text{ mA/V or } 2.31\text{ mS } \textbf{Ans.}$$

10.9. DC LOAD LINE AND BIAS POINT

The dc load line for a FET circuit is drawn upon the output characteristics of the device in exactly the same way as for a BJT circuit.

Consider the FET circuit and characteristics shown in Figs. 10.13 and 10.12 respectively. The drain-source voltage is given as

$$V_{DS} = V_{DD} - I_D R_D \quad ...(10.13)$$

where V_{DD} is the supply voltage and $I_D R_D$ is the voltage drop across R_D.

Consider the following two particular situations :

(*i*) When $I_D = 0$, $V_{DS} = V_{DD}$ ...cutoff point B

(*ii*) When $V_{DS} = 0$, $I_D = \frac{V_{DD}}{R_D}$ ...saturation point A

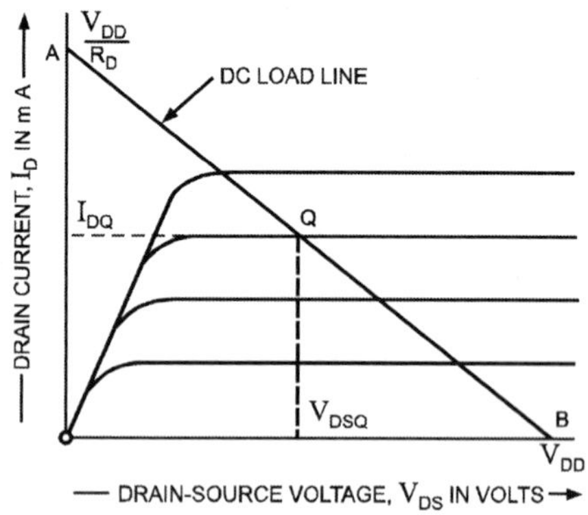

Fig. 10.12

By joining these two points A and B, dc load line is obtained.

A *dc bias point* or *quiescent point* (Q-point) is selected on the FET dc load line. This point defines the dc conditions that exist in a circuit when no input signal is applied. The bias point is selected to give maximum possible variations in output (drain-source) voltage when the drain current is varied by an input signal. In this case the positive-going and negative-going output voltage swings should be as closely equal as possible. Where maximum output voltage swing is not required, the bias point may be at any convenient position on the load line.

For the load line in Fig. 10.12, the bias conditions are $\frac{V_{DD}}{2}$ and $\frac{V_{DD}}{2R_D}$.

For a FET amplifier, V_{DS} must remain in the pinch-off region of the characteristics. This means that it must not be allowed to go below the level of the pinch-off voltage [V_P in Fig. 10.8 (*b*)]. Thus, in the design of a FET bias circuit, the drain source voltage should always be a minimum of $(V_P + 1)$ volt.

$$\textit{i.e.} \quad V_{DS\,(min)} = V_P + 1 \text{ volts} \quad ...(10.14)$$

When an external bias voltage V_{GS} is included, the pinch-off voltage for that bias level on the drain characteristics is $V_P - V_{GS}$, where V_P is the pinch-off voltage at $V_{GS} = 0$ [Fig. 10.8 (*b*)]. Consequently, the minimum drain-source voltage may be reduced to

$$V_{DS\,(min)} = V_P - V_{GS} + 1 \text{ volts} \quad ...(10.15)$$

For the operation of FET as a switch, the device will either be biased *off* or *on* into the channel ohmic region of the drain characteristics.

10.10. FET BIASING

Unlike BJTs, thermal runaway does not occur with FETs, as already discussed in Art. 10.6. However, the wide differences in maximum and minimum transfer characteristics make I_D levels unpredictable with simple fixed-gate bias voltage. To obtain reasonable limits on quiescent drain currents I_D and drain-source voltage V_{DS}, source resistor and potential divider bias techniques must be used. With few exceptions, MOSFET bias circuits are similar to those used for JFETs. Various FET biasing circuits are discussed below :

10.10.1. Fixed Bias

DC bias of a FET device needs setting of gate-source voltage V_{GS} to give desired drain current I_D. For a JFET drain current is limited by the drain-source saturation current I_{DSS}. Since the FET has such a high input impedance that no gate current flows and the dc voltage of the gate set by a voltage divider or a fixed battery voltage is not affected or loaded by the FET.

Fixed dc bias is obtained using a battery V_{GG}. This battery ensures that the gate is always negative w.r.t. source and no current flows through resistor R_G and gate terminal *i.e.* $I_G \cong 0$. The battery provides a voltage V_{GS} to bias the N-channel JFET, but no resulting current is drawn from the battery V_{GG}. Resistor R_G is included to allow any ac signal applied through capacitor C to develop across R_G. While any ac signal will develop across R_G, the dc voltage drop across R_G is equal to $I_G R_G$ *i.e.* 0 volt.

The gate-source voltage V_{GS} is then

$$V_{GS} = -V_G - V_S = -V_{GG} - 0 = -V_{GG} \quad \text{...(10.16)}$$

The drain-source current I_D is then fixed by the gate-source voltage as determined by Eq. (10.1).

This current then causes a voltage drop across the drain resistor R_D and is given as

$$V_{R_D} = I_D R_D$$

$$\text{and output voltage, } V_{out} = V_{DD} - I_D R_D \quad \text{...(10.17)}$$

Since V_{GG} is fixed value of dc supply and the magnitude of gate-to-source voltage V_{GS} is also fixed, hence this circuit is named as *fixed bias circuit.* Since this bias circuit uses two batteries V_{DD} and V_{GG}, it is also known as *two battery bias circuit.*

A FET has a high input impedance. To make advantage of it, R_G should be as large as possible so that the input impedance of the circuit remains high. If R_G is extremely large a charge accumulated on the gate may take a long time to leak off. A reasonable upper limit is 1 MΩ. Normally R_G should not exceed this value.

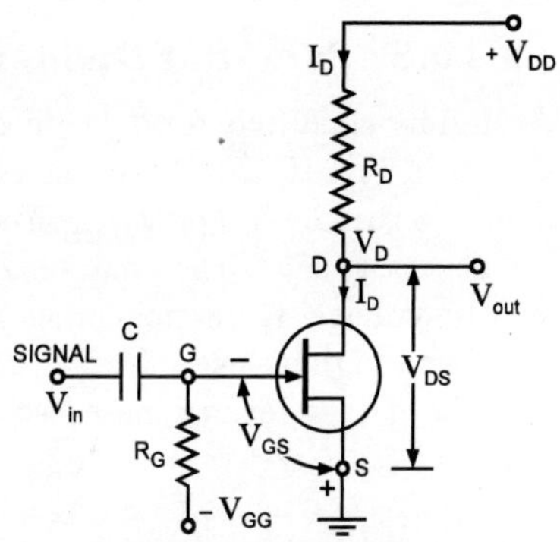

Fig. 10.13. ***Fixed Bias Circuit For JFET***

10.10.2. Self-Bias

This is the most common method for biasing a JFET. Self-bias for an N-channel JFET is shown in Fig. 10.14. This circuit eliminates the requirement of two dc supplies *i.e.*, only drain supply is used and no gate supply is connected. In this circuit, a resistor R_S, known as *bias resistor*, is connected in the source leg. The dc component of drain current I_D flowing through R_S makes a voltage drop across resistor R_S. The voltage drop across R_S reduces the gate-to-source reverse voltage required for FET operation. The capacitor C_S bypasses the ac component of the drain current I_D. The addition of resistor R_G in the circuit does not disturb the dc bias. This is because gate current flowing through it is zero and the gate leakage current is also almost zero. Thus, gate is essentially at dc ground. The resistor R_G is inserted in the circuit to avoid the short circuiting of the ac input voltage. Further, if there is a leakage, the resistor R_G will provide to it an escape route. Otherwise, the leakage current would build up static charge (voltage) at the gate which could change the bias. The resistor R_S, the feedback resistor, also helps in preventing any variation in FET drain current.

Since no gate current flows through the reverse-biased gate, the gate current $I_G = 0$ and, therefore,

$$V_G = I_G R_G = 0\,V$$

With a drain current I_D the voltage at the S is

$$V_S = I_D R_S \qquad ...(10.18)$$

The gate-to-source voltage is then

$$V_{GS} = V_G - V_S = 0 - I_D R_S = - I_D R_S \qquad ...(10.19)$$

So voltage drop across resistance R_S provides the biasing voltage V_{GS} and no external source is required for biasing and this is the reason that it is called *self biasing.*

The operating point (*i.e.* zero signal I_D and V_{DS}) can easily be determined from Eq. (10.1) and equation given below :

$$V_{DS} = V_{DD} - I_D (R_D + R_S) \qquad ...(10.20)$$

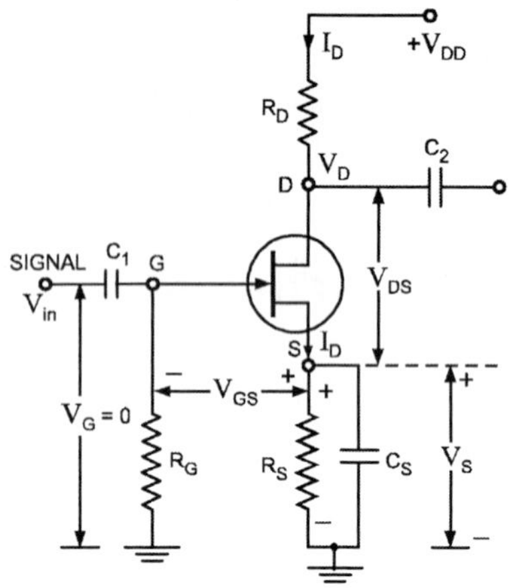

Fig. 10.14. ***Self-Bias Circuit For N-Channel JFET***

Thus dc conditions of JFET amplifier are fully specified.

Self biasing of a JFET stabilises its quiescent operating point against any change in its parameters like transconductance. Let the given JFET be replaced by another JFET having the double conductance then drain current will also try to be double but since any increase in voltage drop across R_S, therefore, gate-source voltage, V_{GS} becomes more negative and thus increase in drain current is reduced.

10.10.3. Potential Divider Bias

A slightly modified form of dc bias is provided by the circuit shown in Fig. 10.15. The resistors R_{G_1} and R_{G_2} form a potential divider across drain supply V_{DD}. The voltage V_2 across R_{G_2} provides the necessary bias. The additional gate resistor R_{G_1} from gate to supply voltage facilitates in larger adjustment of the dc bias point and permits use of larger valued R_S.

The gate is reverse biased so that $I_G = 0$ and gate voltage

$$V_G = V_2 = \frac{V_{DD}}{R_{G_1} + R_{G_2}} \times R_{G_2} \qquad ...(10.21)$$

$$\text{and } V_{GS} = V_G - V_S = V_G - I_D R_S \qquad ...(10.22)$$

The circuit is so designed that $I_D R_S$ is larger than V_G (or V_2) so that V_{GS} is negative. This provides correct bias voltage.

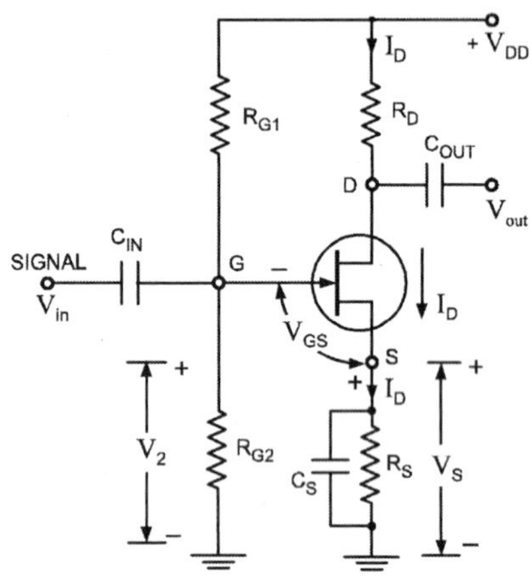

Fig. 10.15. ***Potential Divider Bias Circuit For N-Channel JFET***

The operating point can be determined as

$$I_D = \frac{V_2 - V_{GS}}{R_S}$$

$$\text{and } V_{DS} = V_{DD} - I_D(R_D + R_S) \qquad ...(10.23)$$

Maximum gain is achieved by making resistance R_D as large as possible and for a given level of I_D it needs maximum voltage drop across resistor R_D. However, greatest bias stability is achieved by making R_S as large as possible.

If the gate voltage V_G is very large as compared to gate-to-source voltage V_{GS}, the drain current is approximately constant. In practice, the voltage divider bias is less effective with JFET than BJT. This is because in BJT, $V_{BE} = 0.7$ V (silicon) with only minor variations from one transistor to another transistor. But in a JFET, the V_{GS} can vary several volts from one JFET to another.

Example 10.9. For Fig. 10.16 given that the N-channel JFET has $I_{DSS} = 10$ mA and $V_P = -4$ V. (*i*) Compute the Q-point (I_{DQ}, V_{DSQ}) when $V_{GS} = -1.5$ V. Assume that it is biased in the pinch-off region, (*ii*) Draw the dc load line.

[*G.G.S.I.P. Univ. Analog Electronics December 2009*]

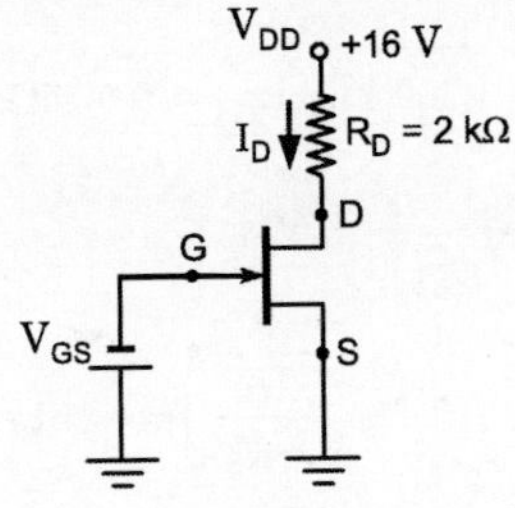

Fig. 10.16

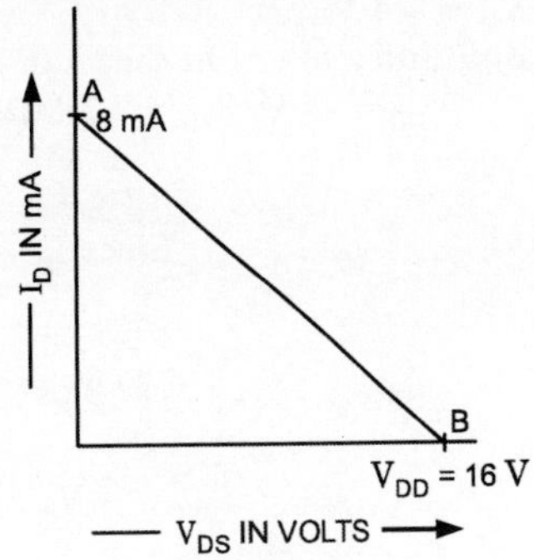

Fig. 10.17

Solution :

$$V_{GSQ} = -1.5 \text{ V}$$

$$I_{DSS} = 10 \text{ mA} = 10 \times 10^{-3} \text{ A}$$

$$V_P = -4 \text{ V}$$

$$\text{Drain current, } I_{DQ} = I_{DSS}\left[1 - \frac{V_{GSQ}}{V_P}\right]^2$$

$$= 10 \times 10^{-3}\left[1 - \frac{-1.5}{-4}\right]^2$$

$$= 3.9 \text{ mA} \quad \textbf{Ans.}$$

$$V_{DSQ} = V_{DD} - I_{DQ}R_D$$

$$= 16 - 3.9 \times 10^{-3} \times 2 \times 10^3 = 8.2 \text{ V} \quad \textbf{Ans.}$$

For DC load line consider the following two particular situations

(*i*) When $I_D = 0$, $V_{DS} = V_{DD} = 16$ V ...cutoff point B

(*ii*) When $V_{DS} = 0$, $I_D = \dfrac{V_{DD}}{R_D} = \dfrac{16}{2 \times 10^3} = 8$ mA ...saturation point A

By joining two points A and B dc load line is obtained (Fig. 10.17)

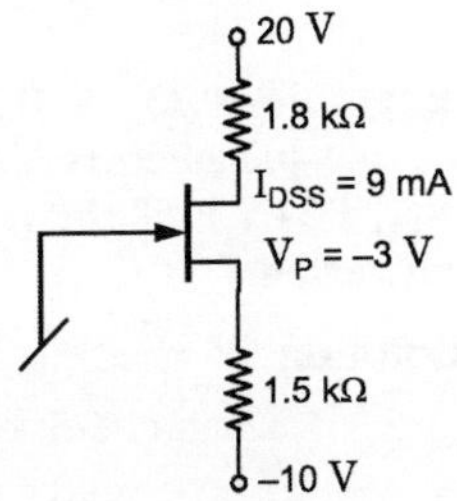

Fig. 10.18

Example 10.10. Determine the following :

(*i*) I_{DQ}, V_{GSQ} (*ii*) V_{DS} (*iii*) V_D (*iv*) V_S

Solution : From Eq. (10.5), we have

(*i*) $|V_{GSQ}| = |V_P| - 0.63 = |3 - 0.63| = |2.37|$ V $= -2.37$ V **Ans.**

$$I_{DQ} = I_{DSS}\left[1 - \frac{V_{GSQ}}{V_P}\right]^2 = 9 \text{ mA}\left[1 - \frac{-2.37}{-3}\right]^2 = 0.4 \text{ mA}$$

(*ii*) $V_{DS} = V_{DD} - I_{DQ}(R_D + R_S) - 10 = 20 - 0.4 \times 10^{-3}(1.8 \times 10^3 + 1.5 \times 10^3) - (-10) = 28.68$ V **Ans.**

(*iii*) $V_D = V_{DD} - I_{DQ}R_D = 20 - 0.4 \times 10^{-3} \times 1.8 \times 10^3 = 19.28$ V **Ans.**

(*iv*) $V_S = V_D - V_{DS} = 19.28 - 28.68 = -9.4$ V **Ans.**

Example 10.11. Determine V_{GS}, I_D, V_{DS}, V_D, V_G and V_S for the following circuit : (Fig. 10.19)

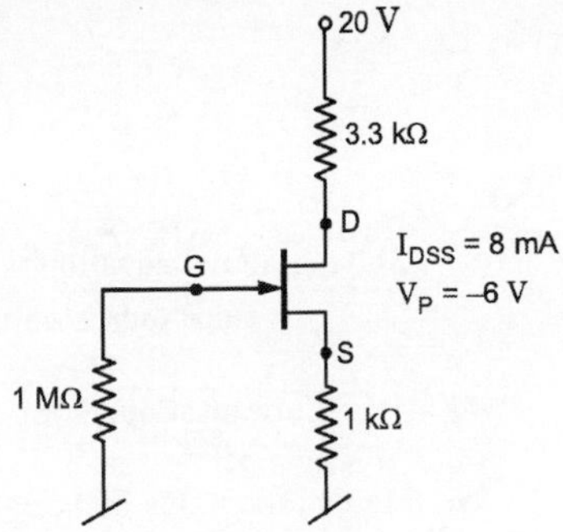

Fig. 10.19

Solution :

$$I_{DQ} = \frac{V_S}{R_S} = \frac{-V_{GSQ}}{R_S} \quad ...(i)$$

$$\text{Also, } I_{DQ} = I_{DSS}\left[1 - \frac{V_{GSQ}}{V_P}\right]^2$$

$$\text{or} \quad \frac{-V_{GSQ}}{1 \times 10^3} = 8 \times 10^{-3}\left[1 - \frac{V_{GSQ}}{-6}\right]^2$$

Solving above equation, we have

$$V_{GSQ} = -2.6 \text{ V} \quad \textbf{Ans.}$$

$$I_{DQ} = \frac{V_S}{R_S} = \frac{-V_{GSQ}}{R_S} = \frac{2.6}{1 \times 10^3} = 2.6 \text{ mA} \quad \textbf{Ans.}$$

$$V_{DS} = V_{DD} - I_{DQ}(R_D + R_S) = 20 - 2.6 \times 10^{-3}(3.3 + 1) \times 10^3 = 8.82 \text{ V} \quad \textbf{Ans.}$$

$$V_G = 0$$

$$V_S = I_{DQ} \times R_S = 2.6 \times 10^{-3} \times 1 \times 10^3 = 2.6 \text{ V} \quad \textbf{Ans.}$$

$$V_D = V_{DD} - I_{DQ} \times R_D = 20 - 2.6 \times 10^{-3} \times 3.3 \times 10^3 = 11.42 \text{ V} \quad \textbf{Ans.}$$

Example 10.12. In an N-channel JFET biased by voltage-divider method, determine the value of R_S to give operating point I_D = 4 mA and V_{DS} = 8 V.

Given that V_{DD} = 25 V, R_{G_1} = 1.2 MΩ, R_{G_2} = 0.6 MΩ. JFET parameters are I_{DSS} = 12 mA and V_P = – 4 V.

Solution : In circuit shown in Fig. 10.15

V_{DD} = 25 V; R_{G_1} = 1.2 MΩ or 1,200 kΩ; R_{G_2} = 0.6 MΩ or 600 kΩ; I_D = 4 mA ; I_{DSS} = 12 mA; V_{DS} = 8 V ; V_P = – 4 V

$$\text{Since } I_D = I_{DSS}\left[1-\frac{V_{GS}}{V_P}\right]^2$$

$$\text{so } V_{GS} = V_P\left(1-\sqrt{\frac{I_D}{I_{DSS}}}\right) = -4\left(1-\sqrt{\frac{4}{12}}\right) = -1.69 \text{ V}$$

$$V_G = \frac{V_{DD}}{R_{G_1}+R_{G_2}} \times R_{G_2} = \frac{25 \times 600}{1{,}200+600} = 8.33 \text{ V}$$

$$R_S = \frac{V_G - V_{GS}}{I_D} = \frac{8.33-(-1.69)}{4 \text{ mA}} = 2.5 \text{ k}\Omega \textbf{ Ans.}$$

Example 10.13. A JFET amplifier with stabilized biasing circuit shown in the Figure (Fig. 10.20) has device parameters V_P = – 2 V, I_{DSS} = 5 mA, R_L = 910 Ω ; R_S = 2.29 kΩ, R_1 = 12 MΩ, R_2 = 8.57 MΩ and V_{DD} = 24 V. Find the value of drain current I_D at the operating point. Also verify that FET will operate in pinch-off region.

Solution : Drain supply voltage, V_{DD} = 24 V

Drain circuit or load resistance, R_L = 910 Ω

Source circuit resistance, R_S = 2.29 kΩ

Drain-source saturation current, I_{DSS} = 5 mA = 5 × 10^{-3} A

Pinch-off voltage, V_P = –2 V

R_1 = 12 MΩ

R_2 = 8.57 MΩ

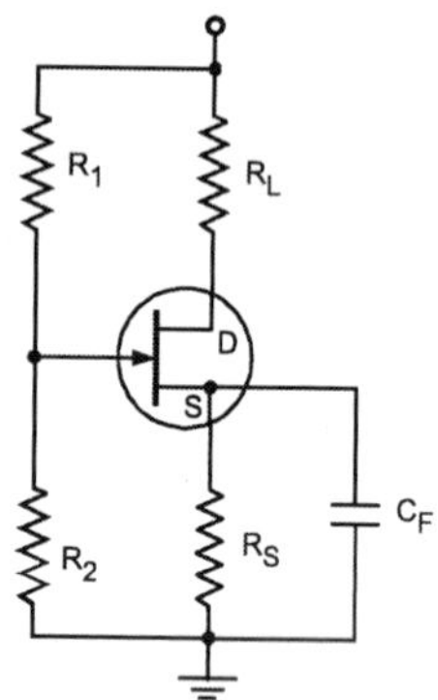

Fig. 10.20

From Eq. (10.21) gate voltage, $V_G = \frac{V_{DD} \times R_2}{R_1 + R_2} = 24 \times \frac{8.57}{12+8.57} = 10$ V

From Eq. (10.22) $V_{GS} = V_G - I_D R_S = 10 - 2.29 \times 10^3 \, I_D$

$$\text{Drain current, } I_D = I_{DSS}\left[1-\frac{V_{GS}}{V_P}\right]^2$$

$$= 5 \times 10^{-3}\left[1-\frac{10-2.29\times10^3 I_D}{-2}\right]^2$$

$$= 5 \times 10^{-3}\,[1 + 5 - 1{,}145\, I_D]^2 = 0.005\,[6 - 1{,}145\, I_D]^2$$

Solving above equation we get, I_D = 4.46 mA

Gate-source voltage, $V_{GS} = V_G - I_D R_S = 10 - 2.29 \times 10^3 \times 4.46 \times 10^{-3} = -0.2134$ V

$$\text{Current at operating point, } I_D = \frac{V_G - V_{GS}}{R_S} = \frac{10-(-0.2134)}{2.29\times10^3} = 4.46 \text{ mA } \textbf{Ans.}$$

We find that the value of I_D at operating point is almost equal to previously computed value of I_D. Hence the FET is operating in pinch-off region.

Example 10.14. An n-channel JFET has V_P = –5 V and I_{DSS} = 12 mA and is used in the circuit as shown in Fig. 10.21. Find the operating point of the circuit. *[G.G.S.I.P. Univ. Analog Electronics, December 2010]*

Solution : Drain supply voltage, V_{DD} = 18 V

Drain circuit or load resistance, R_D = 2 kΩ

Source circuit resistance, R_S = 2 kΩ

Drain-source saturation current, $I_{DSS} = 12 \text{ mA} = 0.012 \text{ A}$

Pinch-off voltage, $V_P = -5 \text{ V}$

$R_1 = 400 \text{ k}\Omega$

$R_2 = 90 \text{ k}\Omega$

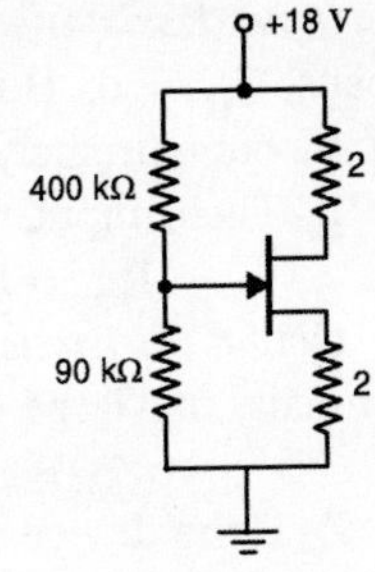

Fig. 10.21

From Eq. (10.21), Gate voltage, $V_G = V_{DD} \times \dfrac{R_2}{R_1 + R_2}$

$$= 18 \times \frac{90}{400 + 90} = 3.306 \text{ V}$$

From Eq. (10.22), $V_{GS} = V_G - I_D R_S = 3.306 - 2{,}000 I_D$

$$\text{Drain current, } I_D = I_{DSS}\left[1 - \frac{V_{GS}}{V_P}\right]^2 = 0.012\left[1 - \frac{3.306 - 2{,}000 I_D}{-5}\right]^2$$

$$= 0.012[1.6612 - 400 I_D]^2$$

Solving above equation, we have $I_{DQ} = 3.085 \text{ mA}$

$$V_{DSQ} = V_{DD} - I_{DQ} R_D = 18 - 3085 \times 10^{-3} \times 2000 = 11.383 \text{ V}$$

So operating point is (11.383 V, 3.085 mA) **Ans.**

10.11. COMMON SOURCE JFET AMPLIFIER

The common source configuration is the most widely employed configuration. It is because this configuration provides high input impedance, good voltage gain and moderate output impedance.

The circuit of a common source N-channel JFET amplifier using self bias is shown in Fig. 10.22. The signal source, V_{in} is connected to JFET gate through coupling capacitor C_1 and external load R_L is connected to the drain terminal D via coupling capacitor C_2. R_G is used to provide leakage path to the gate current, R_S for developing gate bias and C_S for providing ac ground to the input signal.

Note that the polarity of the capacitors in Fig. 10.22—the negative terminal of each (polarized) capacitor must always be at a lower dc potential than the positive terminal. Incorrectly connected capacitors can affect the bias conditions and ac performance of the circuit.

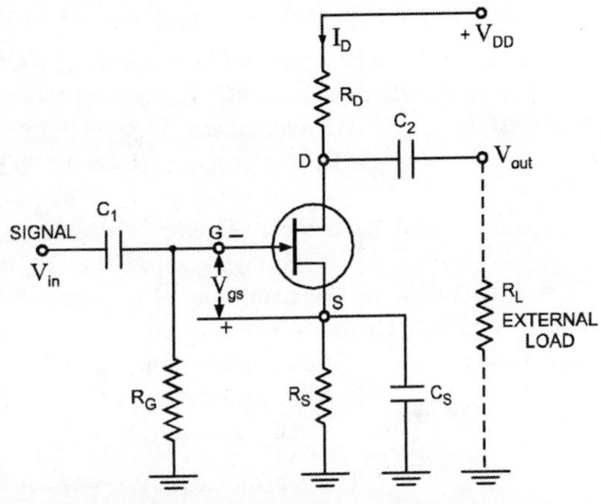

Fig. 10.22. *Common Source Amplifier Circuit*

For ac cosiderations, bypass capacitor C_S effectively ties the FET source terminal to the ground level. Thus, all of the input voltage V_{in} is developed across the gate source terminals. With the increase or decrease in input voltage, gate-source voltage V_{gs} changes and corresponding change in V_{gs} makes the drain current I_D to increase or decrease correspondingly.

The phase relationship between the circuit input and output voltages can be determined by considering the effect of a positive going input signal. When input voltage V_{in} increases in positive direction, it reduces the negative gate-source voltage $-V_{gs}$. The reduction in $-V_{gs}$ raises the level of drain current I_D and consequently increases the voltage drop across drain resistance R_D. Since $V_D = V_{DD} - I_D R_D$, the increase in I_D results in a drop in drain (output) voltage V_D. Thus as V_{in} increases in a positive direction, V_{out} goes in a negative direction. Similarly when V_{in} goes negative, the resultant increase in $-V_{gs}$ reduces the drain current I_D which, in turn, reduces the voltage drop across R_D and develops a positive going output voltage V_{out}. Thus, the amplifier output voltage is 180° out of phase with the input voltage.

Analysis. The first step in ac analysis of the common source circuit is to draw the ac equivalent circuit. This is done by replacing all the capacitors with short circuits and reducing dc supply voltages to zero. AC equivalent circuit for common source amplifier is given in Fig. 10.23.

Input Resistance. In an ideal JFET R_{gs} is infinite because $I_G = 0$. However, in actual device R_{gs} is not infinite but extremely high (100 MΩ or so) in comparison to R_G. Thus, input resistance

$$R_{in} = R_G \parallel R_{gs} \simeq R_G \quad ...(10.24)$$

Output Resistance. Looking into the drain and source terminals, the large drain resistance r_d is seen. Thus,

$$Z_d \simeq r_d \quad ...(10.25)$$

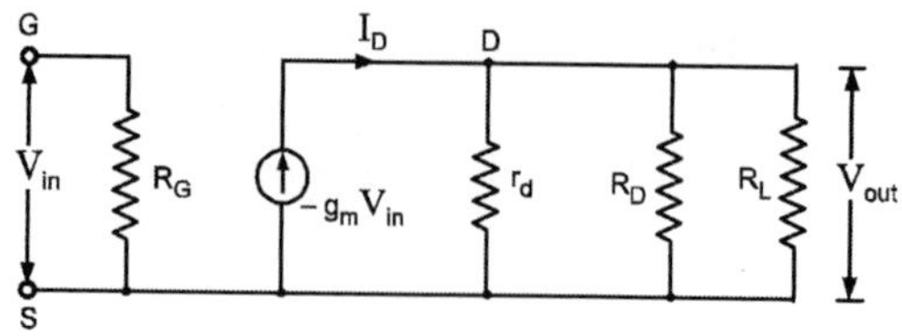

Fig. 10.23. ***AC Equivalent Circuit For Common Source JFET Amplifier***

Z_d is the device output impedance; the circuit output impedance is R_D in parallel with Z_d, so

$$Z_{out} = R_D \parallel Z_d = R_D \parallel r_d \quad ...(10.26)$$

Since usually $r_d >> R_D$, the circuit output impedance is taken to be R_D.

Voltage Gain. Output voltage,

$$V_{out} = I_D (r_d \parallel R_D \parallel R_L) = -g_m V_{in} (r_d \parallel R_D \parallel R_L) \qquad \because I_D = -g_m v_{in}$$

$$\text{and voltage gain, } A_v = \frac{V_{out}}{V_{in}} = -g_m (r_d \parallel R_D \parallel R_L) \quad ...(10.27)$$

usually $r_d >> R_D \parallel R_L$

$$\text{So voltage gain, } A_v = -g_m (R_D \parallel R_L) \quad ...(10.28)$$

The minus sign indicates that output voltage V_{out} is 180° out of phase with input voltage V_{in}.

Example 10.15. The transconductance of a FET used in a voltage amplifier circuit is 2,500 microsiemens and the load resistance is 12 kΩ. Determine the voltage gain of the amplifier circuit. Assume r_d and $R_D >> R_L$.

Solution : Transconductance, $g_m = 2{,}500\ \mu S = 2{,}500 \times 10^{-6}$ S

Load resistance, $R_L = 12$ kΩ $= 12{,}000$ Ω

Voltage gain, $A_v = -g_m \times [r_d \parallel R_D \parallel R_L]$

Since r_d and R_D are very large in comparison to R_L

∴ Voltage gain, $A_v = -g_m R_L = -2{,}500 \times 10^{-6} \times 12{,}000 = -30$ **Ans.**

Example 10.16. In the common source amplifier, drain resistance $R_D = 5$ kΩ, amplification factor μ = 50 and $r_d = 35$ kΩ. Evaluate voltage gain A_V and output resistance R_0.

Solution : Drain resistance, $R_D = 5$ kΩ

AC drain resistance, $r_d = 35$ kΩ

Amplifier factor, μ = 50

$$\text{Transconductance, } g_m = \frac{\mu}{r_d} = \frac{50}{35 \times 10^3} = 1.429 \times 10^{-3}\ \text{S} \qquad \because \text{From Eq. (10.11) } g_m = \frac{\mu}{r_d}$$

$$\text{Voltage gain, } A_V = -g_m \frac{r_d R_D}{r_d + R_D} = -1.429 \times 10^{-3} \times \frac{35 \times 10^3 \times 5 \times 10^3}{35 \times 10^3 + 5 \times 10^3} = -6.25 \quad \textbf{Ans.}$$

$$\text{Output resistance, } R_{out} = R_D \parallel r_d = 5\ \text{k}\Omega \parallel 35\ \text{k}\Omega = \frac{5 \times 10^3 \times 35 \times 10^3}{5 \times 10^3 + 35 \times 10^3} = 4.375\ \text{k}\Omega \quad \textbf{Ans.}$$

Example 10.17. It is desired to operate the JFET shown in Fig. 10.24 at $V_{GS} = -1.0$ V, $V_{DS} = 4.0$ V and $I_{DS} = 1$ mA. Determine the value of R_D and R_S.

Solution : From the dc equivalent circuit given in Fig. 10.25.

$$V_G = I_G R_G = 0 \times 500 \times 10^3 = 0\ \text{V}$$

$$\text{Source terminal voltage, } V_S = V_G - V_{GS} = 0 - (-1.0) = 1\ \text{V}$$

$$\because V_{GS} = V_G - V_S$$

Also $V_S = I_{DS} R_S$

$$\text{So source resistance, } R_S = \frac{V_S}{I_{DS}} = \frac{1\ \text{V}}{1\ \text{mA}} = 1\ \text{k}\Omega \quad \textbf{Ans.}$$

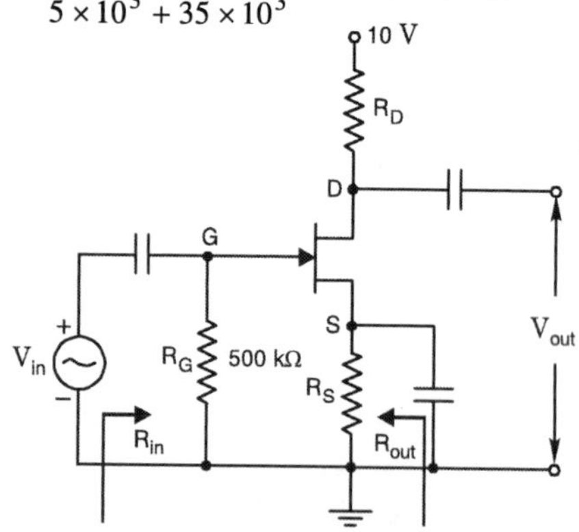

Fig. 10.24

Applying Kirchhoff's voltage law to the output loop, we have

$$V_{DD} = I_{DS} R_D + V_{DS} + I_{DS} R_S = I_{DS}(R_D + R_S) + V_{DS}$$

or $$R_D = \frac{V_{DD} - V_{DS}}{I_{DS}} - R_S$$

$$= \frac{10-4}{1 \times 10^{-3}} - 1 \times 10^3 = 5 \times 10^3\ \Omega \text{ or } 5\ \text{k}\Omega \ \textbf{Ans.}$$

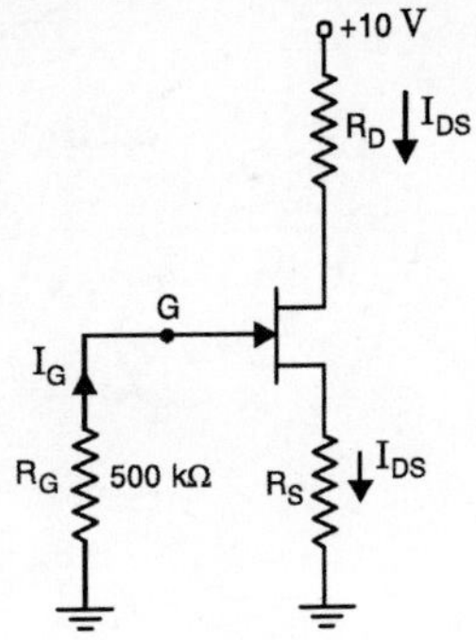

Fig. 10.25

Example 10.18. For the JFET amplifier shown in Fig. 10.26, $\mu = 100$ and $r_d = 50$ kΩ

(i) Draw the ac equivalent circuit.

(ii) Find the voltage gain of the amplifier.

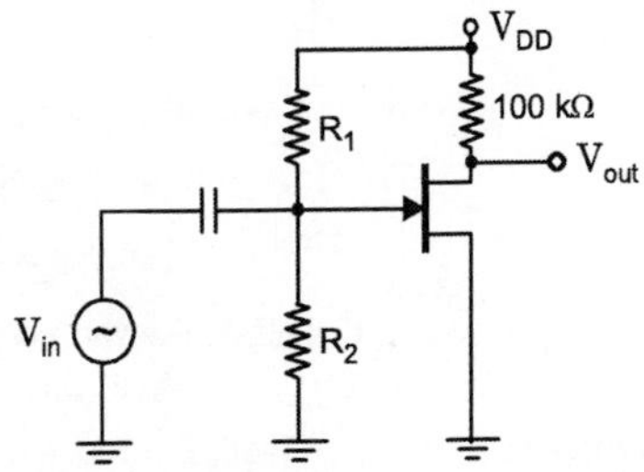

Fig. 10.26

Solution : Amplifier factor, $\mu = 100$

AC drain resistance, $r_d = 50$ kΩ

$$\text{Transconductance, } g_m = \frac{\mu}{r_d} = \frac{100}{50 \times 10^3} = 0.002 \text{ S}$$

(*i*) For drawing ac equivalent circuit all the capacitors are replaced with short circuits and all dc supply voltages are reduced to zero. The ac equivalent circuit is shown in Fig. 10.27.

(*ii*) Voltage gain of the amplifier.

$$A_V = -g_m \times (r_d \| R_D)$$

$$= -g_m \times \frac{r_d \times R_D}{r_d + R_D} = \frac{-\mu R_D}{r_d + R_D} \quad \because g_m r_d = \mu$$

$$= \frac{-100 \times 100 \times 10^3}{50 \times 10^3 + 100 \times 10^3} = -66.67 \ \textbf{Ans.}$$

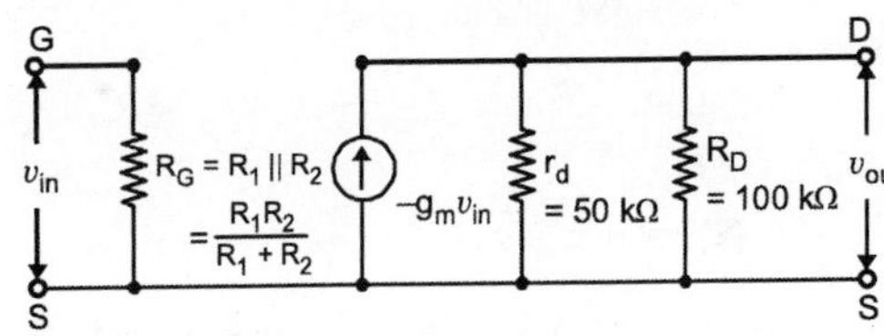

Fig. 10.27

Example 10.19. Determine the voltage gain $A_V = \frac{V_o}{V_i}$ for the amplifier circuit shown in Fig. 10.28. V_i is the input voltage between the gate terminal and ground. Neglect all capacitances.

Solution : Amplification factor, $\mu = 30$

AC drain resistance, $r_d = 5$ kΩ

$$\text{Transconductance, } g_m = \frac{\mu}{r_d} = \frac{30}{5 \times 10^3} = 0.006 \text{ S}$$

Voltage gain of the amplifier,

$$A_V = -g_m \times (r_d \| R_D)$$

$$= -0.006 \times (5\ \text{k}\Omega \| 10\ \text{k}\Omega)$$

$$= -0.006 \times \frac{5 \times 10}{5+10} \times 10^3 = -20 \ \textbf{Ans.}$$

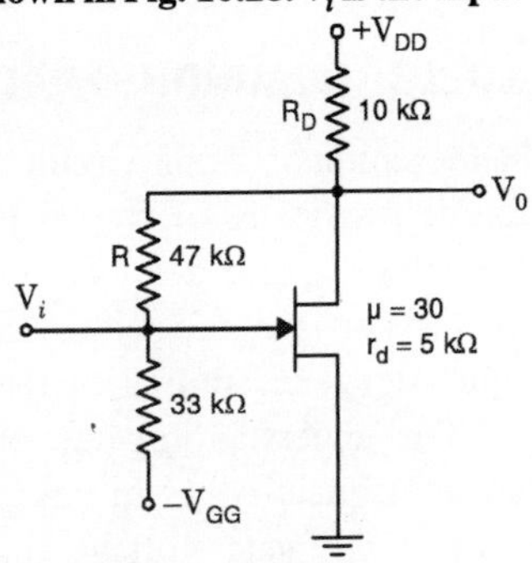

Fig. 10.28

Example 10.20. Consider the following circuit. Determine I_D, V_{GS} and V_{DS} for $|I_{DSS}| = 4$ mA, $V_P = 4$ V.

Solution : For given P-channel FET

Drain supply voltage, $V_{DD} = -60$ V

Load resistance, $R_L = 18$ kΩ

Source circuit resistance, $R_S = 4$ kΩ

Drain source saturation current, $I_{DSS} = 4 \times 10^{-3}$ A

Pinch-off voltage, V_p = 4 V

Voltage applied at gate, $V_G = \dfrac{V_{DD} \times R_2}{R_1 + R_2}$

$$= \frac{-60 \times 200 \times 10^3}{(1300 + 200)\,10^3} = -8\text{V}$$

We know, $V_{GS} = V_G - I_D R_S = -8 + 4 \times 10^3 I_D$

So Drain current, $I_D = I_{DSS}\left[1 - \dfrac{V_{GS}}{V_P}\right]^2$

$$= 4 \times 10^{-3}\left[1 - \frac{(-8 + 4 \times 10^3 I_D)}{4}\right]^2$$

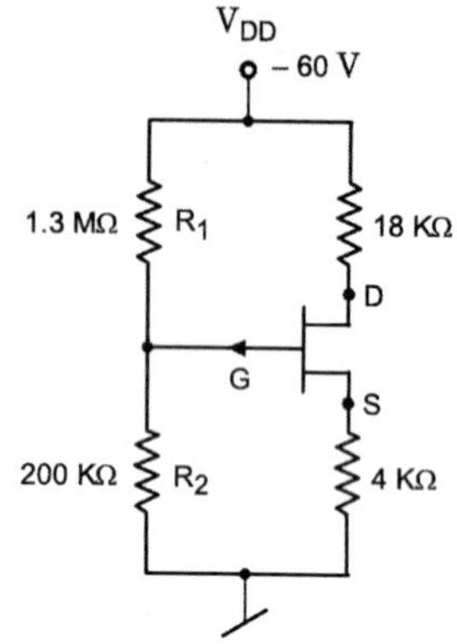

Fig. 10.29

Solving above equation, we have I_D = **2.25 mA Ans.**

Gate-source voltage, $V_{GS} = V_G - I_D R_S = -8 + 2.25 \times 10^{-3} \times 4 \times 10^3 = \mathbf{1\ V}$ **Ans.**

Drain-source voltage, $V_{DS} = V_{DD} - I_D (R_D + R_S) = -60 + 2.25\,(18 + 4) \times 10^3 \times 10^{-3} = \mathbf{-10.5\ V}$ **Ans.**

Example 10.21. Calculate the voltage gain and output resistance of the following circuit. Given that g_m = 2 mA/V and r_d = 10 kΩ.

Solution : Given, R_D = 50 kΩ

Transconductance, g_m = 2 mA/V

drain resistance, r_d = 10 kΩ

(1) Output resistance, $Z_{out} = R_D \| Z_d = R_D \| r_d = \dfrac{50 \times 10}{50 + 10}$ = **8.33 kΩ Ans.**

(2) Voltage gain, $A_V = \dfrac{V_{out}}{V_{in}}$

$= -g_m\,(r_d \| R_D \| R_L)$

$= -g_m\,(r_d \| R_D)\ \because R_L = \infty$ (Refer to Eq. 10.27)

$= -2 \times 10^{-3} \times 8.33 \times 10^3 = \mathbf{-16.66}$ **Ans.**

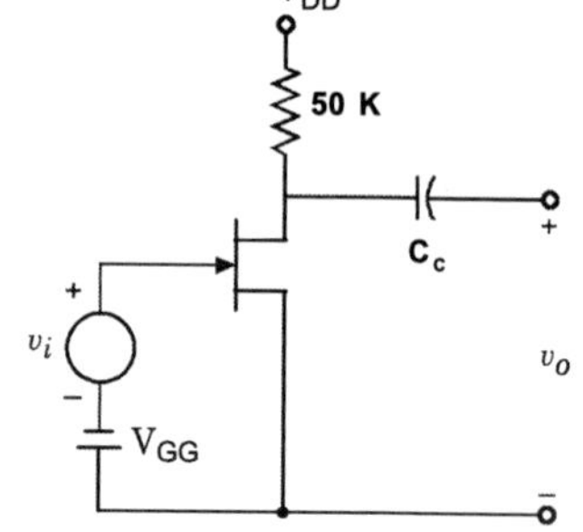

Fig. 10.30

10.12. COMMON DRAIN JFET AMPLIFIER, OR SOURCE FOLLOWER

In the common drain circuit (also called the *source follower*), the output voltage is developed across source resistor R_S. External load resistor R_L is connected to FET source terminal S through coupling capacitor C_2, and gate bias voltage V_G is derived from V_{DD} by means of potential divider R_1 and R_2. No resistor is connected in series with the drain terminal, and no source bypass capacitor is used. The input signal is applied to the gate through coupling capacitor C_1.

For understanding the operation of the common drain circuit, note that gate voltage V_G is a constant quantity, as in any potential divider bias circuit. When a signal is applied to the FET gate through C_1, gate voltage increases and decreases as the input signal goes positive and negative respectively. Also V_{gs} remains substantially constant, therefore, the source voltage V_s increases or decreases with the increase or decrease in gate voltage. Since the FET source is the output terminal, it is seen that the output voltage from a common drain circuit is approximately the same as the input voltage. Thus, the common drain circuit can be said to have approximately unity gain. Because the output voltage at the FET source terminal follows variations in the signal voltage applied to the gate, the common drain circuit is also known as a *source* follower.

The common drain ac equivalent circuit is shown in Fig. 10.32. The current generator is $g_m V_{gs}$ where $V_{gs} = V_{in} - V_{out}$. Moreover $R_G = R_1 \parallel R_2$.

$$V_{out} = I_D (r_d \parallel R_S \parallel R_L) \qquad \text{where } I_D = g_m V_{gs}$$

or $$V_{out} = g_m V_{gs} (r_d \parallel R_S \parallel R_L)$$

and $$V_{in} = V_{gs} + V_{out} = V_{gs} + V_{gs} g_m (r_d \parallel R_S \parallel R_L)$$

$$\because V_{gs} = V_{in} - V_{out}$$

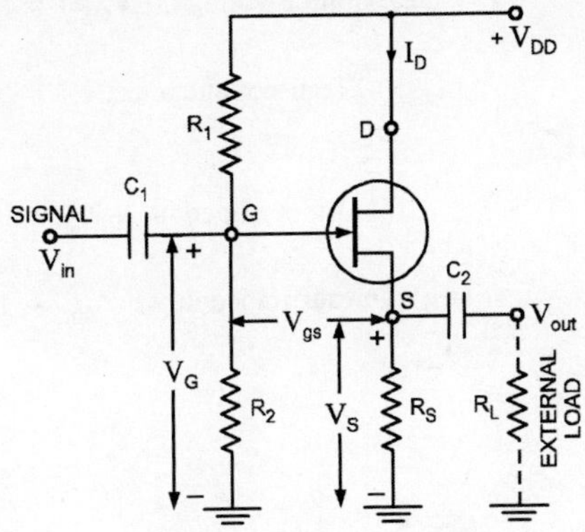

Fig. 10.31. ***Common Drain Amplifier Circuit***

$$\text{Voltage gain } A_v = \frac{V_{out}}{V_{in}} = \frac{g_m (r_d \parallel R_S \parallel R_L)}{1 + g_m (r_d \parallel R_S \parallel R_L)}$$

Normally $r_d >> R_S \parallel R_L$

$$\therefore A_v = \frac{g_m (R_S \parallel R_L)}{1 + g_m (R_S \parallel R_L)} \qquad \text{...(10.29)}$$

If $g_m (R_S \parallel R_L) >> 1$ then $A_v \approx 1$

So value of voltage gain becomes approximately equal to unity (slightly less than unity).

Input Impedance. The circuit input impedance is $R_G = R_1 \parallel R_2$.

Output Impedance. In a common drain circuit, any variation in output voltage V_{out} affects upon the gate-source voltage V_{gs} and V_{gs} is given as

$$V_{gs} = V_{out} \times \frac{R_{gs}}{(R_s \parallel R_G) + R_{gs}}$$

Normally $R_{gs} >> R_s \parallel R_G$, so $V_{gs} = V_{out}$

and $I_D = g_m V_{gs}$

$$\text{Also } Z_s = \frac{V_{out}}{I_D} = \frac{V_{gs}}{g_m V_{gs}} = \frac{1}{g_m} \qquad \text{...(10.30)}$$

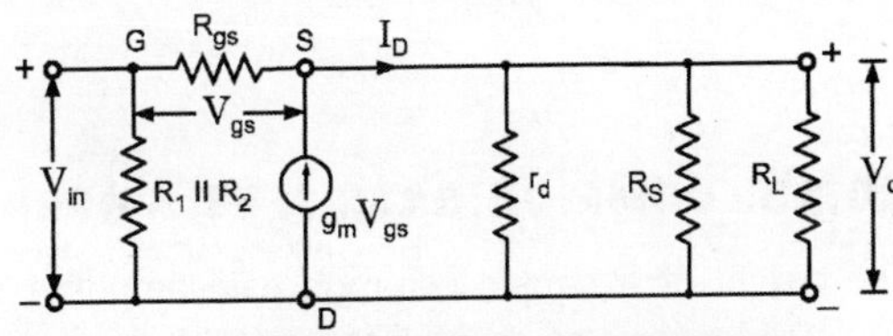

Fig. 10.32. ***AC Equivalent Circuit For Common Drain JFET Amplifier***

Actually, r_d is in parallel with $\frac{1}{g_m}$, but since $r_d >> \frac{1}{g_m}$, it is neglected. R_s is the output impedance of the device. The circuit output impedance

$$Z_{out} = R_S \parallel \frac{1}{g_m} \qquad \text{...(10.31)}$$

Example 10.22. The common drain circuit given in Fig. 10.31 has R_1 = 4 MΩ, R_2 = 2 MΩ, R_S = 2.5 kΩ, R_L = 25 kΩ and g_m is 2,500 μS. Determine *(i)* input impedance, *(ii)* output impedance and *(iii)* voltage gain.

Solution :

(i) Input impedance, $Z_{in} = R_G = R_1 \parallel R_2$ = 4 MΩ || 2 MΩ = 1.33 MΩ **Ans.**

(ii) Output impedance, $Z_{out} = R_S \parallel \frac{1}{g_m} = 2.5\text{ k}\Omega \parallel \frac{1}{2{,}500\ \mu\text{S}} = 2.5\text{ k}\Omega \parallel 0.4\text{ k}\Omega = 345\ \Omega$ **Ans.**

(iii) Voltage gain, $A_v = \frac{g_m (R_S \parallel R_L)}{1 + g_m (R_S \parallel R_L)} = 2{,}500 \times 10^{-6} \times \frac{2.5\text{ k}\Omega \parallel 25\text{ k}\Omega}{1 + 2{,}500 \times 10^{-6} \times (2.5\text{ k}\Omega \parallel 25\text{ k}\Omega)} = 0.85$ **Ans.**

Example 10.23. Calculate Z_{in}, Z_{out} and A_V for the following circuit (Fig. 10.33)

Solution : Source resistance, $R_S = 2.2\text{ k}\Omega = 2.2 \times 10^3\ \Omega$

Gate resistance, $R_G = 10\text{ M}\Omega = 10 \times 10^6\ \Omega$

AC drain resistance, $r_d = 40\text{ k}\Omega = 40 \times 10^3\ \Omega$

Drain-source saturation current, I_{DSS} = 9 mA

Pinch-off voltage, V_P = – 4.5 V

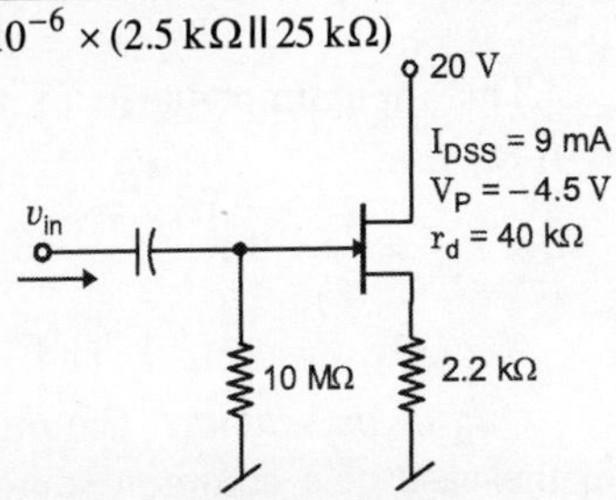

Fig. 10.33

Gate-source voltage, $| V_{GS} | = | V_P | - 0.63 \text{ V} = | 4.5 - 0.63 | \text{ V} = | 3.87 | \text{ V} = -3.87 \text{ V}$

$$\text{Transconductance, } g_m = g_{mo}\left[1-\frac{V_{GS}}{V_P}\right] = \frac{-2I_{DSS}}{V_P}\left[1-\frac{V_{GS}}{V_P}\right] = -2\times\frac{9\times10^{-3}}{-4.5}\left(1-\frac{-3.87}{-4.5}\right)$$
$$= 0.56 \times 10^{-3} \text{ S}$$

(*i*) Input impedance, $Z_{in} = R_G = 10 \text{ M}\Omega$ **Ans.**

$$(ii) \text{ Output impedance, } Z_{out} = R_S \,||\, \frac{1}{g_m} \,||\, r_d = 2.2\times10^3 \,||\, \frac{10^3}{0.56} \,||\, 40\times10^3$$
$$= (2.2 \,||\, 1.786 \,||\, 40)\times10^3 \ \Omega$$
$$= \frac{2.2\times1.786\times40}{1.786\times40+2.2\times40+2.2\times1.786}\times10^3 \ \Omega$$
$$= \frac{2.2\times1.786\times40}{163.37} \text{ k}\Omega = 0.962 \text{ k}\Omega \quad \textbf{Ans.}$$

$$(iii) \text{ Voltage gain, } A_V = \frac{g_m \times R_S \,||\, r_d}{1+g_m(R_S \,||\, r_d)} = \frac{0.56\times10^{-3}\times\dfrac{2.2\times10^3\times40\times10^3}{2.2\times10^3+40\times10^3}}{1+0.56\times10^{-3}\times\dfrac{2.2\times10^3\times40\times10^3}{2.2\times10^3+40\times10^3}}$$
$$= \frac{1.16}{1+1.16} = 0.54 \quad \textbf{Ans.}$$

10.13. COMMON GATE JFET AMPLIFIER

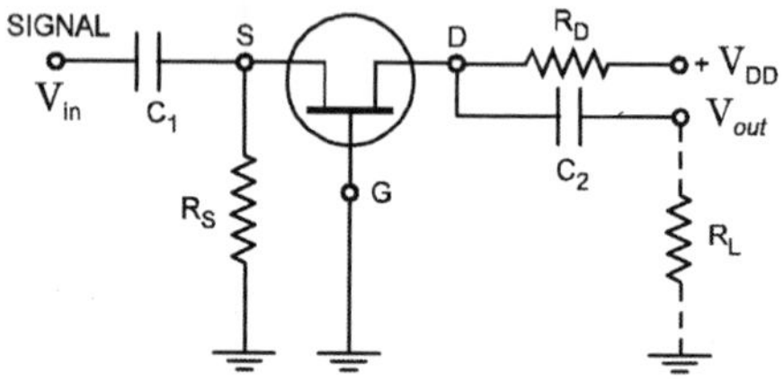

Fig. 10.34. *Common Gate Amplifier Circuit*

The circuit of a simple common gate amplifier is shown in Fig. 10.34 and its ac equivalent circuit is shown in Fig. 10.35. In this amplifier circuit, input signal is applied to the source terminal S and output is taken from the drain terminal D, as illustrated in Fig. 10.34. The gate is grounded and external load resistor R_L is connected to the drain terminal D through coupling capacitor C_2.

In ac equivalent circuit, current source is connected between the drain and source terminals, as usual. However, since source and drain are the input and output terminals respectively, $g_m V_{gs}$ appears between the input and output.

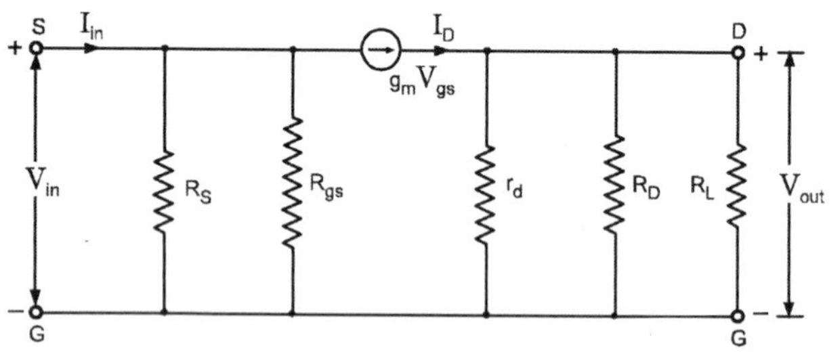

Fig. 10.35. *AC Equivalent Circuit For Common Gate JFET Amplifier*

Input Impedance. Ignoring current through R_S and R_{gs}, the input current

$$I_{in} = I_D = g_m V_{gs} = g_m V_{in}$$

So input impedance of FET,

$$Z_s = \frac{V_{in}}{I_{in}} = \frac{V_{in}}{g_m V_{in}} = \frac{1}{g_m} \qquad ...(10.32)$$

The input impedance of the circuit is $Z_s \,||\, R_S$. So,

$$Z_{in} = \frac{1}{g_m} \,||\, R_S \qquad ...(10.33)$$

Actually speaking r_d and R_D are also involved in Z_{in} but their effect is negligible.

Output Impedance. The output of a common gate circuit is taken from the drain terminal, just as in the case of a common source circuit. So the output impedance at the drain terminal is

$$Z_d = r_d \qquad ...(10.34)$$

Again this equation gives the impedance looking into the device drain terminal. The circuit output impedance

$$Z_{out} = R_D \parallel Z_d = R_D \parallel r_d \qquad ...(10.35)$$

$$V_{out} = I_D (r_d \parallel R_D \parallel R_L) \quad \text{where } I_D = g_m V_{gs}$$

$$= g_m V_{gs} (r_d \parallel R_D \parallel R_L) = g_m V_{in} (r_d \parallel R_D \parallel R_L) \quad ...(10.36) \because V_{gs} = V_{in}$$

$$\text{Voltage gain, } A_v = \frac{V_{out}}{V_{in}} = g_m (r_d \parallel R_D \parallel R_L) \qquad ...(10.37)$$

Usually $r_d >> R_D \parallel R_L$

$$\text{So } A_v = g_m (R_D \parallel R_L) \qquad ...(10.38)$$

The above equation for voltage gain is the same as that in case of a common source circuit (Eq. 10.28), except that there is no minus sign. It means that output voltage and input voltage are in phase.

10.14. APPLICATIONS OF FETs

FET has a very high input impedance (100 MΩ in case of JFETs and 10^4 to 10^9 MΩ in case of MOSFETs), the major shortcomings of an ordinary transistor *i.e.* low input impedance with consequent of loading of signal source is eliminated in FET. Hence FET is an ideal device for use in almost every application in which transistors can be used.

1. FETs are widely used as input amplifiers in oscilloscopes, electronic voltmeters and other measuring and testing equipment, because of their high input impedance.
2. FETs are used in RF amplifiers in FM tuners and communication equipments for the low noise level.
3. FETs are used as voltage variable resistors (VVRs) in operational amplifiers and tone controllers etc. because it is a voltage controlled device.
4. FETs are used in mixer circuits in FM and TV receivers, and communication equipments because of their low intermodulation distortion.
5. FETs are used in low frequency amplifiers in hearing aids and inductive transducers because of the small coupling capacitors.
6. FETs are used in digital circuits in computers, LSI and memory circuits because of very small size.

10.15. METAL-INSULATOR-SEMICONDUCTOR FIELD-EFFECT TRANSISTORS (MISFETs)

The metal-insulator-semiconductor field-effect transistor (MISFET) is one of the most widely used electronic devices, particularly in digital circuits, because of relatively small size—thousands of devices can be fabricated in a single integrated circuit (IC). Since it is constructed with the gate terminal insulated from the channel, it may be referred to generally as an *insulated-gate field-effect transistor* (IGFET). However, since most of such devices are made using silicon for the semiconductor, SiO_2 for the insulator, and metal or heavily doped polysilicon for the gate electrode, the term *metal-oxide-semiconductor field-effect* transistor (MOSFET) or simply MOS is commonly used. MISFET is the more general term, where the insulator is not necessarily silicon dioxide (SiO_2) and the semiconductor is not necessarily silicon (Si). Here discussion will be about MOSFETs although the same physics applies to the MISFETs.

Like, a JFET, a MOSFET is also a three terminal (source, gate and drain) device and drain current in it is also controlled by gate bias. The operation of MOSFET is similar to that of JFET. It can be employed in any of the circuits covered for the JFET and, therefore, all the equations apply equally well to the MOSFET and JFET in amplifier connections. However, MOSFET has lower capacitance and input impedance much more than that of a JFET owing to small leakage current. In case of a MOSFET the positive voltage may be applied to the gate and still the gate current remains zero.

MOSFETs are of two types namely (*i*) *enhancement type* MOSFET or E-MOSFET and (*ii*) depletion enhancement MOSFET or DE-MOSFET. In the depletion-mode construction a channel is physically constructed and a current between drain and source is due to voltage applied across the drain-source terminals. The enhancement MOSFET structure has no channel formed during its construction. Voltage is applied to the gate, in this case, to develop a channel of charge carriers so that a current results when a voltage is applied across the drain-source terminals.

The enhancement type MOSFET (E-MOSFET) is widely used in both discrete and integrated circuits. In discrete circuits, the main use is in power switching, which means turning largest currents *on* and *off*. In ICs, the main use is in digital switching, the basic process behind modern computers. Although their use is declined, depletion-mode MOSFETs (DE-MOSFETs) are still found in high-frequency front-end communication circuits as RF amplifiers.

10.16. DEPLETION-ENHANCEMENT MOSFETs (DE-MOSFETs)

In Art. 10.2.5 we have seen that when the gate is biased negative w.r.t. the source in an N-channel JFET, the depletion-region widths are increased. The increase in the depletion regions reduces the channel thickness, which increases its resistance. The net result is that drain current I_D is reduced.

If the polarity of V_{GG} were reversed so as to apply a positive bias to the gate w.r.t. source, the P-N junctions between the gate and the channel would then be forward biased. Since a forward bias reduces the width of a depletion region, the thickness of channel would increase with a corresponding decrease in channel resistance. As a result, drain current I_D would increase beyond the JFET's I_{DSS} value.

The normal operation of a JFET is in its depletion mode of operation. However, as discussed above, it is also possible to *enhance* the conductivity of the JFET channel. However, the forward bias of the silicon P-N junction is usually restricted to a maximum of 0.5 V (more conservative limit is 0.2 V) so as to limit the gate current.

As we have seen in Art. 10.8, the greater the I_D is compared to I_{DSS}, the greater the transconductance g_m will be. In Arts. 10.11 and 10.13 we have seen that the voltage gain is directly proportional to g_m. So, in general, the higher the g_m, the better it is. This is one of the advantages of being able to enhance the channel.

As its name suggests, the *depletion-enhancement* MOSFET (DE-MOSFET) was developed to be used in either or both the depletion and enhancement modes.

10.16.1. Construction

Figure 10.36 shows the construction of an N-channel DE-MOSFET. It consists of a lightly doped P-type substrate into which two blocks of heavily doped N-type material are diffused forming the source and drain. An N-channel is formed by diffusion between the source and drain. The type of impurity for the channel is the same as for the source and drain. Now a thin layer of SiO_2 dielectric is grown over the entire surface and holes are cut through the SiO_2 (silicon dioxide) layer to make contact with the N-type blocks (Source and Drain). Metal is deposited through the holes to provide drain and source terminals, and on the surface area between drain and source, a metal plate is deposited. This layer constitutes the gate. SiO_2 layer results in an extremely high input impedance of the order of 10^{10} to 10^{15} Ω for this device. The chip area of a MOSFET is typically 0.003 μm^2 or less which is about only 5% of the area required by a BJT.

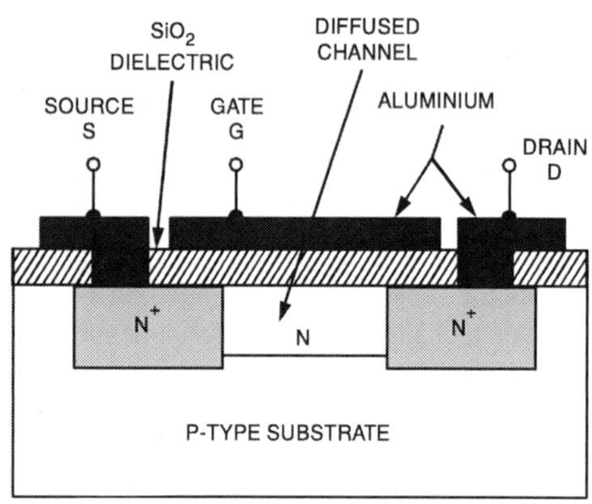

Fig. 10.36. ***N-Channel DE-MOSFET Structure***

A P-channel DE-MOSFET is constructed like an N-channel DE-MOSFET, starting with an N-type substrate and diffusing P-type drain and source blocks and connecting them internally by a P-dopped channel region.

In some cases the substrate is internally connected to the source terminal S and in others an additional terminal labelled SS is provided resulting in a four terminal device.

Silicon dioxide (SiO_2) is a type of insulator referred to as a *dielectric,* which develops opposing (as indicated by the prefix di-) electric fields within the dielectric when exposed to an externally applied field. The fact that the SiO_2 layer is an insulating layer means that there is no direct electrical connection between the gate terminal G and the channel of a MOSFET. It is the insulating layer of SiO_2 in the MOSFET construction that accounts for the very desirable high input impedance of the device.

In fact, the input resistance of a MOSFET is usually much higher than that of a typically JFET, even though the input impedance of most JFETs is sufficiently high for applications. Because of very high input impedance, the gate current I_G is essentially zero for dc-biased configurations.

10.16.2. Operation

It is obvious from Fig. 10.36 that there is no P-N junction between gate and channel. Here the diffused channel N, the insulating dielectric SiO_2 layer and the metal layer of the gate forms a parallel plate capacitor. DE-MOSFET can be operated with either a positive or a negative gate. When gate is positive with respect to the source it operates in the enhancement—or E-mode and when the gate is negative w.r.t. the source, as illustrated in Fig. 10.37, it operates in depletion-mode.

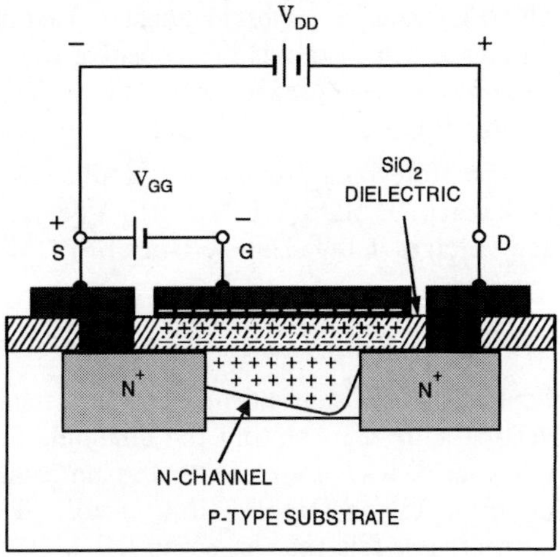

Fig. 10.37. ***Depletion Mode Operation***

When the drain is made + ve with respect to source, a drain current will flow, even with zero gate potential and the MOSFET is said to be operating in E-mode. In this mode of operation gate attracts the – ve charge carriers from the P-substrate to the N-channel and thus reduces the channel resistance and increases the drain current. The more positive the gate is made, the more drain current flows.

On the other hand when the gate is made – ve with respect to the substrate, the gate repells some of the –ve charge carriers out of the N-channel. This creates a depletion region in the channel, as illustrated in Fig. 10.37, and, therefore, increases the channel resistance and reduces the drain current. The more – ve the gate, the less the drain current. In this mode of operation the device is referred to as a *depletion-mode MOSFET.* Here too much negative gate voltage can pinch-off the channel. Thus operation is similar to that of JFET.

10.16.3. Characteristics

Typical *drain characteristics*, for various levels of gate-source voltage, of an N-channel MOSFET are shown in Fig. 10.38. The upper curves are for positive V_{GS} and the lower curves are for negative V_{GS}. The bottom drain curve is for $V_{GS} = V_{GS\,(OFF)}$. For a specified drain-source voltage V_{DS}, $V_{GS\,(OFF)}$ is the gate-source voltage at which drain current reduces to a certain specified negligibly small value, as shown in Fig. 10.38. This voltage corresponds to the pinch-off voltage V_P of JFET. For V_{GS} between $V_{GS\,(OFF)}$ and zero, the device operates in *depletion-mode* while for V_{GS} exceeding zero the device operates in *enhancement mode*. These drain curves again display an ohmic region, a constant current source region

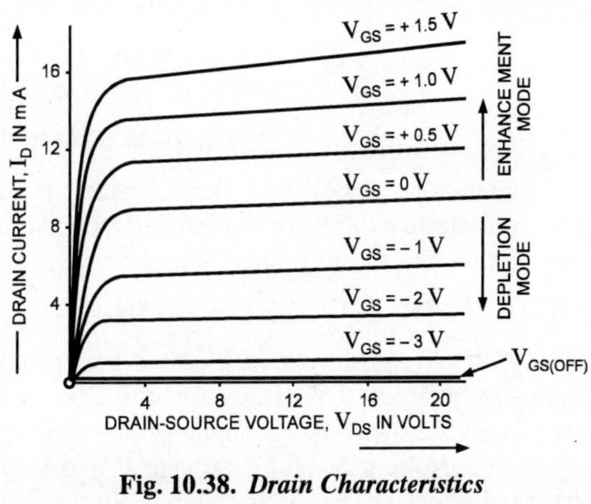

Fig. 10.38. ***Drain Characteristics***

and a cutoff region. MOSFET has two major applications: a constant current source and a voltage variable resistor.

The *transfer* (or *transconductance*) characteristic for an N-channel DE-MOSFET is shown in Fig. 10.39. I_{DSS} is the drain current with a shorted gate. Since the curve extends to the right of the origin, I_{DSS} is no longer the maximum possible drain current. Mathematically, the curve is still part of a parabola and the same square-law relation exists as with a JFET. In fact, the depletion-mode MOSFET has a drain current given by the same transconductance equation as before, Eq. (10.1). Furthermore, it has the same equivalent circuits as a JFET. Because of this, the analysis of a depletion-mode MOSFET circuit is almost identical to that of a JFET circuit. The only difference is the analysis for a positive gate, but even here the same basic formulas are used to determine the drain current I_D, gate-source voltage V_{GS} etc.

The foregoing discussion is applicable in principle also to the P-channel DE-MOSFET. For such a device the sign of all currents and voltages in the characteristics [Figs. 10.38 and 10.39] must be reversed.

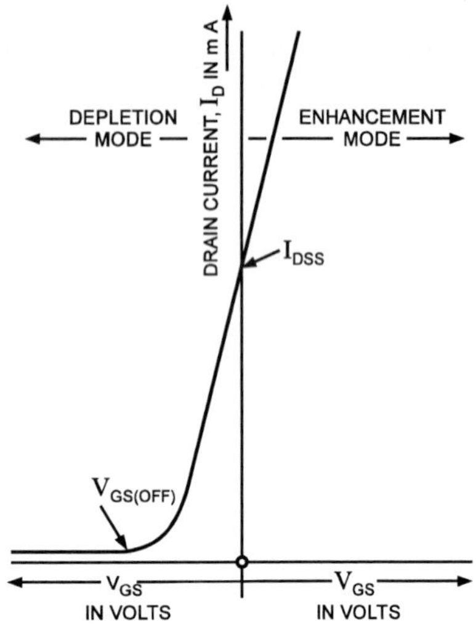

Fig. 10.39. ***Transfer Characteristic***

10.16.4. Schematic Symbols

Figure 10.40 (*a*) shows the schematic symbol for a DE-MOSFET. Just to the right of the gate is the thin vertical line representing the channel. The drain lead comes out from the top of the channel and the source lead connects to the bottom. The arrow is on the P-substrate and points to the N-material. In some applications, a voltage can be applied to the substrate for added control of drain current. For this reason, some DE-MOSFETs have four terminal leads. But in most applications, the substrate is connected to the source. Usually the substrate is connected to the source internally by the manufacturer. This results in a three terminal device whose schematic symbol is shown in Fig. 10.40 (*b*).

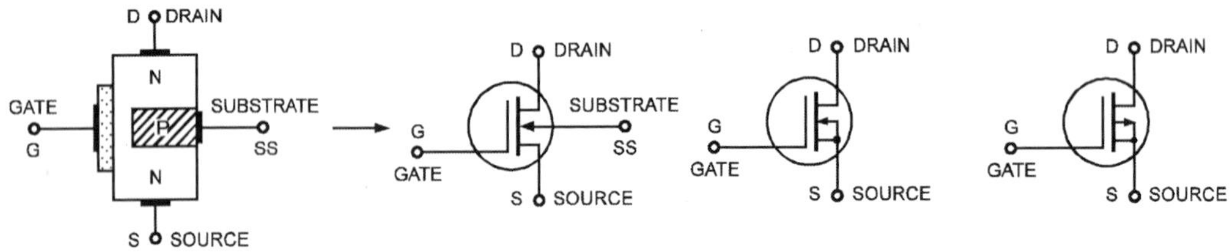

(*a*) (*b*) *N-Channel DE-MOSFET* (*c*) *P-Channel DE-MOSFET*

Fig. 10.40. ***Schematic Symbols For DE-MOSFETs***

Schematic symbol for a three terminal P-channel DE-MOSFET device is shown in Fig. 10.40 (*c*). The schematic symbol of a P-channel DE-MOSFET is similar to that of an N-channel DE-MOSFET, except that the arrow points outward.

10.16.5. DE-MOSFET Amplifiers

DE-MOSFET can be operated either with positive or negative gate, so its operating or quiescent point can be set at $V_{GS} = 0$ V, as illustrated in Fig. 10.41 (*b*). In Fig. 10.41 (*a*) the gate G is at ground potential for dc since ac signal source appears to be short circuited for dc and sets Q-point at $V_{GS} = 0$ V. The ac input signal causes the gate-source voltage V_{gs} to vary above and below zero developing drain current I_D and drain voltage V_{out}. The biasing circuit given in Fig. 10.41 (*a*) has $V_{gs} = 0$ V, $I_D = I_{DSS}$ and dc drain voltage

$$V_{DS} = V_{DD} - I_{DSS} R_D \qquad ...(10.39)$$

The zero bias of Fig. 10.41 (*a*) is unique with depletion-mode MOSFETs. Although any of the JFET biasing method will work with a depletion-mode MOSFET, the zero-bias method shown in Fig. 10.41 (*a*) is preferred as it does the job simply and adequately.

After biasing the depletion-mode MOSFET to a Q-point, it can amplify small signals. The JFET formulas for voltage gain can be applied directly to a MOSFET amplifier. But like the JFET, the depletion-mode MOSFET has a relatively low voltage gain. MOSFETs have excellent low-noise properties, a definite advantage for any stage near the front end of a system where the signal is weak, such as a TV receiver. As with a JFET, the transconductance g_m in case of MOSFET can be controlled by varying the gate-source voltage V_{gs}. Due to the fact that a DE-MOSFET is normally an *on* device, it is also possible to use self-bias by adding a source resistor. The operation becomes the same as a self-biased JFET circuit.

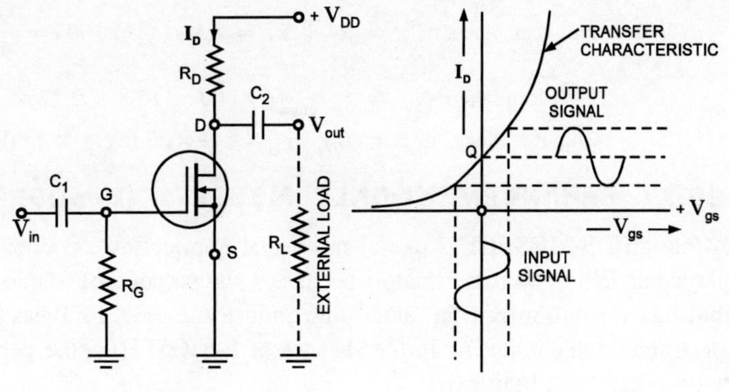

Fig. 10.41

Example 10.24. Sketch the transfer characteristic for an N-channel depletion type MOSFET with I_{DSS} = 10 mA and V_P = – 4 V.

Solution : From Eq. (10.1)

At $V_{GS} = 0$, $I_D = I_{DSS} = 10$ mA

At $V_{GS} = 1$, $I_D = 10 \text{ mA} \left(1 - \dfrac{+1\text{ V}}{-4\text{ V}}\right)^2 = 15.625$ mA

At $V_{GS} = -1$ V, $I_D = 10 \text{ mA} \left(1 - \dfrac{1}{4}\right)^2 = 5.625$ mA

At $V_{GS} = -2$ V, $I_D = 10 \text{ mA} \left(1 - \dfrac{2}{4}\right)^2 = 2.5$ mA

At $V_{GS} = -4$ V $I_D = 10 \text{ mA} \left(1 - \dfrac{4}{4}\right)^2 = 0$ mA

The transfer characteristic curve is drawn as shown in Fig. 10.42.

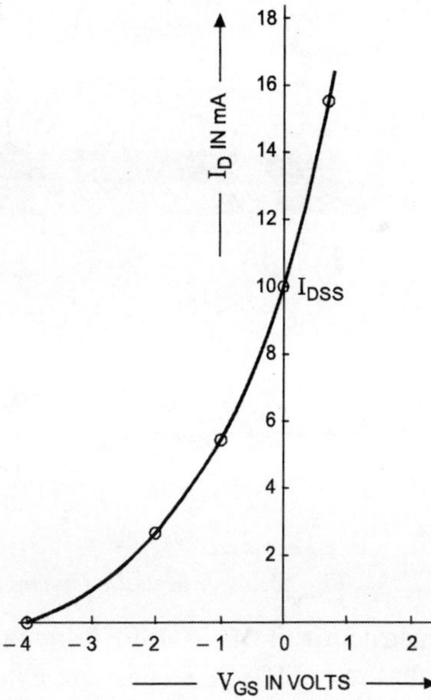

Fig. 10.42. Transfer Characteristic For An N-Channel Depletion Type MOSFET With I_{DSS} = 10 mA and V_P = –4 V

Example 10.25. The DE-MOSFET amplifier given in Fig. 10.43 has $V_{GS(off)}$ = –2 V, I_{DSS} = 4 mA, and g_{mo} = 2,000 µS. Determine the output voltage of the circuit.

Solution : With the source grounded, $V_{GS} = 0$ and $I_D = 4$ mA

Drain-to-source voltage,

$$V_{DS} = V_{DD} - I_D R_D = 20 - 4 \times 10^{-3} \times 2.5 \times 10^3 = 10 \text{ V}$$

Since $V_{GS} = 0$ V, $g_m = g_{mo} = 2{,}000 \times 10^{-6}$ S

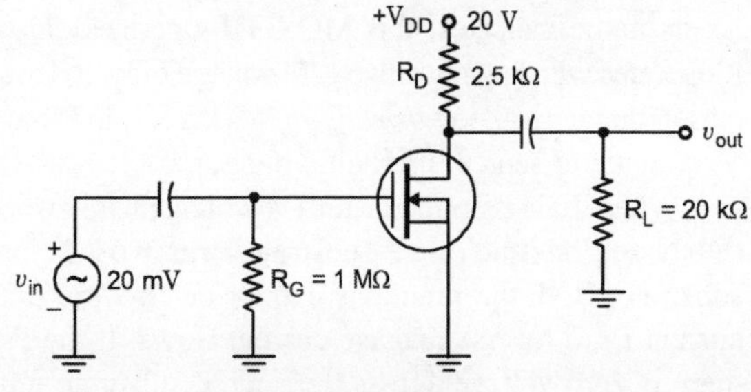

Fig. 10.43

$$\text{The output resistance, } r_{out} = R_D \,||\, R_L = 2.5 \text{ k}\Omega \,||\, 20 \text{ k}\Omega = \frac{2.5 \times 20}{2.5 + 20} \text{k}\Omega = 2.22 \text{ k}\Omega$$

$$\text{Voltage gain, } A_V = g_m r_{out} = 2{,}000 \times 10^{-6} \times 2.22 \times 10^3 = 4.44$$

$$\text{Output voltage, } v_{out} = A_V \times v_{in} = 4.44 \times 20 \text{ mV} = 88.8 \text{ mV}$$ **Ans.**

10.17. ENHANCEMENT-ONLY MOSFETs (E-MOSFETs)

Although DE-MOSFET is useful in special applications, it does not enjoy widespread use. However, it played an important role in history because it was part of the evolution towards the E-mode MOSFET, a device that has revolutionized the electronic industry. E-MOSFET has become enormously important in digital electronics and computers. In the absence of E-MOSFETs, the personal computers (PCs) that are now so widespread would not exist.

10.17.1. Construction

Figure 10.44 shows the construction of an N-channel E-MOSFET. The main difference between the

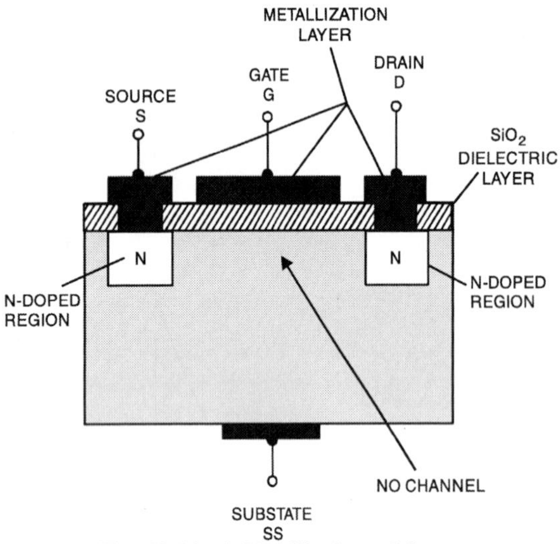

Fig. 10.44. ***A Metallic Crystal Structure***

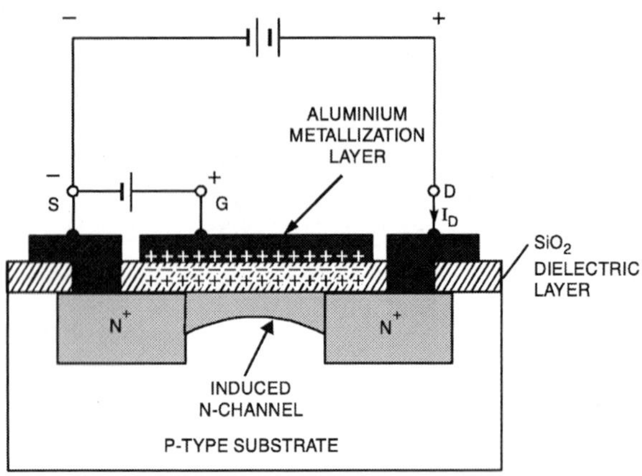

Fig. 10.45. ***Operation of N-Channel E-MOSFET***

construction of DE-MOSFET and that of E-MOSFET, as we see from Figs. 10.36 and 10.44 respectively, is that in E-MOSFET substrate extends all the way to the silicon dioxide (SiO_2) and no channels are doped between the source and the drain. Channels are electrically induced in these MOSFETs, when a positive gate-source voltage V_{GS} is applied to it.

10.17.2. Operation

As its name indicates, this MOSFET operates only in the *enhancement mode* and has no depletion mode. It operates with large positive gate voltage only. It does not conduct when gate-source voltage $V_{GS} = 0$. This is the reason that it is called *normally-off* MOSFET. In these MOSFETs drain current I_D flows only when V_{GS} exceeds V_{GST} [gate-to-source threshold voltage].

When drain is applied with +ve voltage with respect to source and no potential is applied to the gate, the two N-regions and one P-substrate form two P-N junctions connected back to back with a resistance of the P-substrate. Both the junctions cannot be forward biased at the same time, so only an extremly small drain current *i.e.* a reverse leakage current flows. If the P-type substrate is now connected to the source terminal, there is zero voltage across the source-substrate junction, and the drain-substrate junction remains reverse biased. When the gate is made positive with respect to the source and the substrate, negative (*i.e.* minority) charge carriers within the substrate are attracted to the +ve gate and accumulate close to the surface of the substrate. As the gate voltage is increased, more and more electrons accumulate under the gate. Since these electrons cannot flow across the insulated layer of silicon dioxide to the gate, they accumulate at the surface of the substrate just below the gate. These accumulated minority charge carriers make an N-type channel stretching from drain to source. When this occurs, a channel is induced by forming what is termed an *inversion layer* (N-type). Now a drain current starts flowing. The strength of the drain current depends upon the channel resistance, which in turn depends on the number of charge carriers attracted to the positive gate. Thus drain current is controlled by the gate potential.

Since the conductivity of the channel is enhanced by the +ve bias on the gate, this device is called the *enhancement MOSFET* or E-MOSFET.

The minimum value of gate-to-source voltage V_{GS} that is required to form the inversion layer (N-type) is termed the *gate-to-source threshold voltage* V_{GST}. For V_{GS} below V_{GST}, the drain current $I_D = 0$. But for V_{GS} exceeding V_{GST} an N-type inversion layer connects the source to drain and the drain current I_D is large. Depending upon the device being used, V_{GST} may vary from less than 1 V to more than 5 V.

JFETs and DE-MOSFETs are classified as the depletion-mode devices because their conductivity depends on the action of depletion layers. E-MOSFET is classified as an enhancement-mode device because its conductivity depends on the action of the inversion layer. Depletion-mode devices are normally ON when the gate-source voltage $V_{GS} = 0$, whereas the enhancement-mode devices are normally OFF when $V_{GS} = 0$.

10.17.3. Characteristics

Drain characteristics of an N-channel E-MOSFET are shown in Fig. 10.46. The lowest curve is the V_{GST} curve. When V_{GS} is lesser than V_{GST}, I_D is approximately zero. When V_{GS} is greater than V_{GST}, the device *turns on* and the drain current I_D is controlled by the gate voltage. The characteristic curves have almost vertical and almost horizontal parts. The almost vertical components of the curves correspond to the ohmic region, and the horizontal components correspond to the constant current region. Thus E-MOSFET can be operated in either of these regions *i.e.* it can be used as a variable voltage resistor (VVR) or as a constant current source.

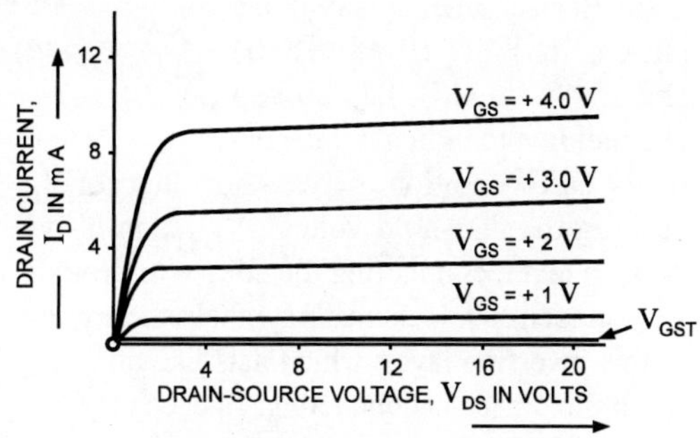

Fig. 10.46. ***Drain Characteristics***

Figure 10.47 shows a typical transconductance curve. The current I_{DSS} at $V_{GS} \le 0$ is very small, being of the order of a few nanoamperes. When the V_{GS} is made positive, the drain current I_D increases slowly at first, and then much more rapidly with an increase in V_{GS}. The manufacturer sometimes indicates the *gate-source threshold voltage* V_{GST} at which the drain current I_D attains some defined small value, say 10 μA. The current $I_{D\,(ON)}$, corresponding approximately to the maximum permissible value of drain current given on the drain characteristics and the values of V_{GS} required to give this current $V_{GS\,(ON)}$ are also usually given on the manufacturers data sheet.

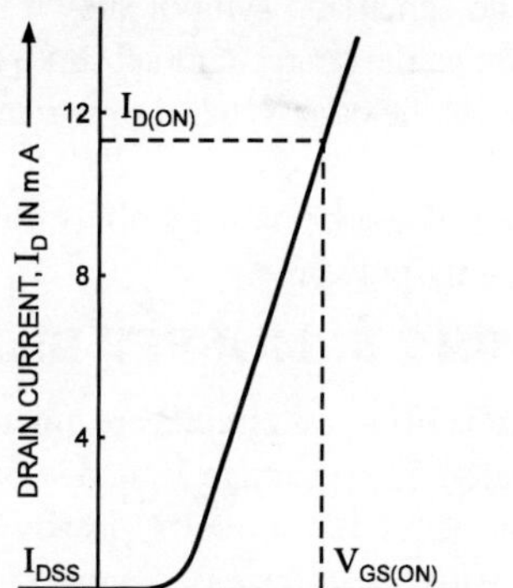

Fig. 10.47. ***Transfer Characteristic***

The equation for the transfer characteristic does not obey Eq. (10.1). However, it does follow a similar "square-law type" of relationship. The equation for the transfer characteristic of E-MOSFETs is given as

$$I_D = K(V_{GS} - V_{GST})^2 \qquad ...(10.40)$$

where K, typically 0.3 mA/V^2, is a property of the device construction. Note that I_{DSS} cannot be associated with an E-MOSFET, because no drain current flows with $V_{GS} = 0$ volt.

With JFETs and DE-MOSFETs, the value of I_{DSS} and $V_{GS\,(OFF)}$ are the key quantities required for analysis. With E-MOSFETs, the key quantities are $I_{D\,(ON)}$, V_{GST}, and $V_{GS\,(ON)}$ indicated in Fig. 10.47. These three quantities are the first items to look for on a data sheet. By substituting these quantities in Eq. (10.40), the equation can be arranged in a more useful form:

$$I_D = KI_{D\,(ON)} \qquad ...(10.41)$$

$$\text{where } K = \left(\frac{V_{GS} - V_{GST}}{V_{GS\,(ON)} - V_{GST}}\right)^2 \qquad ...(10.42)$$

The input impedance of the E-MOSFET is very high (of the order of $10^{15}\,\Omega$) as the gate of the E-MOSFET is insulated from the channel. Transconductance for this device typically ranges from 1,000 μS to 2,000 μS (or 1 to 2 mA per volt).

10.17.4. Schematic Symbols

Schematic symbols for an N-channel E-MOSFET are shown in Fig. 10.48 (*a*). For zero value of V_{GS}, the E-MOSFET is OFF because there is no conducting channel between source and drain. Each of schematic symbols shown in Figs. 10.48 (*a*) (*ii*) and 10.48 (*a*) (*iii*), has broken channel line to indicate this normally OFF condition. As we know that for V_{GS} exceeding the threshold voltage V_{GST}, an N-type inversion layer, connecting the source to drain, is created. In each of the schematic symbols, the arrow points to this inversion layer, which acts like an N-channel when the device is conducting. In each case, the fact that the device has an insulated gate is indicated by the gate not making direct contact with the channel. The schematic symbol shown in Fig. 10.48 (*a*) (*iii*) shows the source and substrate internally connected, while the other symbol shown in Fig. 10.48 (*a*) (*ii*) shows the substrate connection brought out separately from the source.

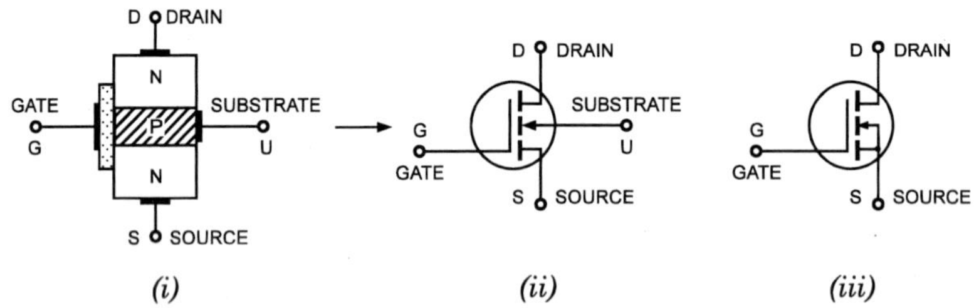

(*a*) *N-Channel E-MOSFET*

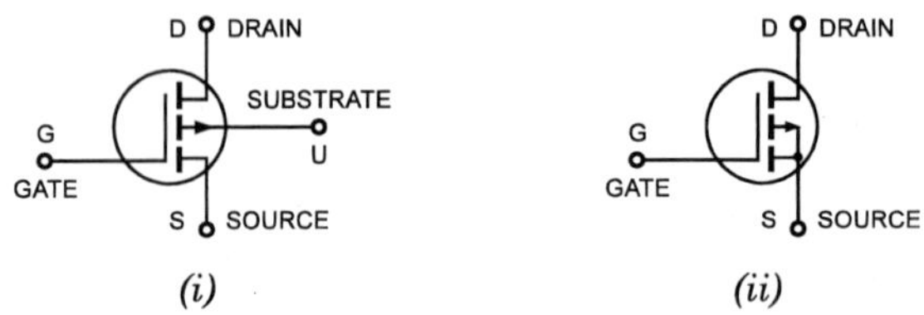

(*b*) *P-Channel E-MOSFET*

Fig. 10.48. ***Schematic Symbols For E-MOSFETs***

The schematic symbols for a P-channel E-MOSFET are shown in Fig. 10.48 (*b*). In these cases the arrow points outwards.

10.17.5. MOSFET Biasing

MOSFET bias circuits are quite similar to the circuits used for JFET biasing. In the case of DE-MOSFETs, the gate-source voltage V_{GS} may be either positive or negative in both N-channel and P-channel devices. Any of the JFET bias circuits already discussed can be used to give a negative V_{GS} for an N-channel device, or a positive V_{GS} for a P-channel device. To operate an N-channel MOSFET in enhancement mode (either a DE-MOSFET or E-MOSFET), the gate is required to be biased positively with respect to the source. The self-bias circuit is unsuitable in such a case, and if fixed voltage bias is to be used, V_{GS} must be a positive quantity. The same remarks apply for a P-channel MOSFET operating in E-mode, except that the gate must be made negative w.r.t. the source.

A popular circuit arrangement for biasing the N-channel E-MOSFET is shown in Fig. 10.49. Resistor R_f brings a suitably large voltage ($V_{GS} > V_{GST}$) to the gate to drive the E-MOSFET *on*. The drain current I_D increases until an equilibrium condition of V_{DS} (or V_{GS}) and drain current I_D is established. The MOSFET drain current is fixed by V_{GS} as given by Eq. (10.40). The drain current I_D causes a voltage drop across R_D, so that

$$V_D = V_{DD} - I_D R_D \quad ...(10.43)$$

Since the gate current $I_G = 0$, the voltage $V_{GS} = V_{DS}$, so Eq. (10.40) may be written as

$$I_D = K(V_{DS} - V_{GST})^2 \quad ...(10.44)$$

With $V_S = 0$ volt, the value of V_{DS} is

$$V_{DS} = V_D - V_S = V_{DD} - I_D R_D \quad ...(10.45)$$

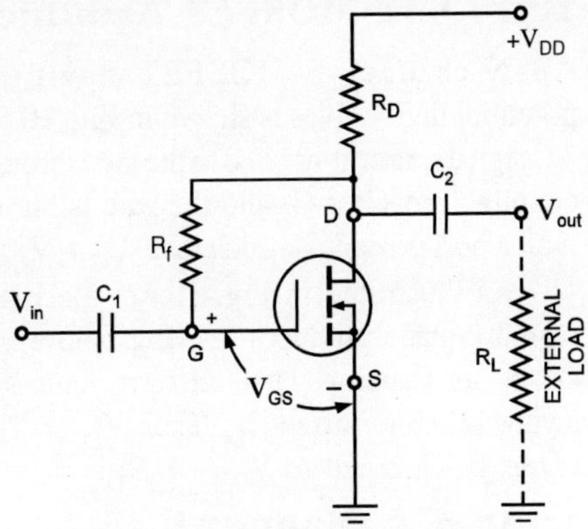

Fig. 10.49. ***Drain Feedback Biasing Circuit For E-MOSFET***

If I_D tries to increase due to some reason (say due to increase in temperature) V_{DS} will decrease due to larger voltage drop in R_D and cause V_{GS} to decrease. So I_D will decrease to overcome the original increase in I_D. Thus operating or quiescent point remains fixed.

If for reasons of linearity in device operation or maximum output voltage, equality of V_{GS} and V_{DS} is not desirable, then the circuit shown in Fig. 10.50, known as drain-to-gate bias circuit may be used. Here, the gate bias voltage is $V_{GS} = \frac{R_1}{R_1 + R_f} V_{DS}$. This circuit also, like that shown in Fig. 10.49, offers the dc stabilization through the feedback resistor R_f. However, the input resistance is reduced because of the Miller effect, R_f corresponds to an equivalent resistance, $R_{in} = R_f(1 - A_V)$ shunting the amplifier input.

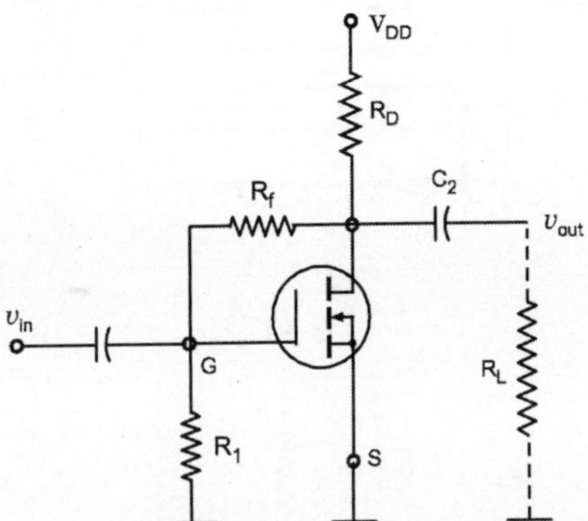

Fig. 10.50. ***Improved Version of Circuit Shown in Fig. 10.49***

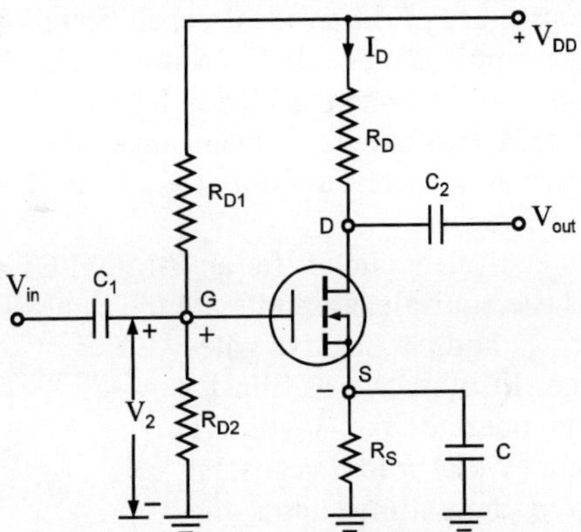

Fig. 10.51. ***Potential Divider Biasing Circuit For E-MOSFET***

Also, the voltage divider biasing technique (circuit shown in Fig. 10.51) can be used for E-MOSFET. Here the dc stability is accomplished by the feedback through source resistance R_S.

$$\text{Here } V_{GS} = V_2 = \frac{V_{DD}}{R_{D_1} + R_{D_2}} \times R_{D_2} \quad ...(10.46)$$

and $V_{DS} = V_{DD} - I_D R_D$ ...(10.47) where $I_D = K(V_{GS} - V_{GST})^2$

The value of resistors used in this divider circuit is quite high as E-MOSFET gate draws very little current due to the insulation layer.

10.17.6. E-MOSFET Amplifier

An N-channel E-MOSFET having a potential divider bias is shown in Fig. 10.51. AC signal source is applied to the gate through coupling capacitor C_1 and the gate is biased with a positive voltage such that $V_{GS} > V_{GST}$.

As illustrated in Fig. 10.52, the input signal voltage produces a swing below and above its Q-point. This, in turn, causes a swing in drain current I_D. Thus V_D or V_{out} varies being equal to $V_{DD} - I_D R_D$.

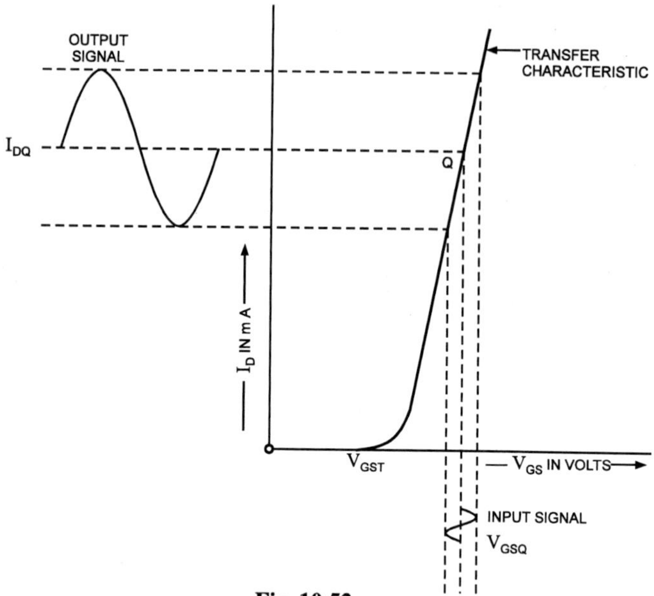

Fig. 10.52

10.17.7. E-MOSFET Equivalent Circuit

Ideal drain curves for an E-MOSFET are shown in Fig. 10.53. All drain curves superimpose in the ohmic region to give a single, almost vertical line and are horizontal in the current-source region. There is no breakdown region. These ideal drain curves are similar to depletion-mode curves, except for the proportional knee voltage V'_K, which is given as

$$V'_K = I_D R_{DS} \qquad ...(10.48)$$

This voltage V'_K is the border between the ohmic region and the current source region in an ideal enhancement-mode device. This border concept is identical to V_P. Since E-MOSFET does not have a pinch-off voltage where depletion layers meet together, V'_P is not used. Instead, they have an inversion layer. Because a different physical mechanism is involved, symbol V'_K is used for the border between the two regions.

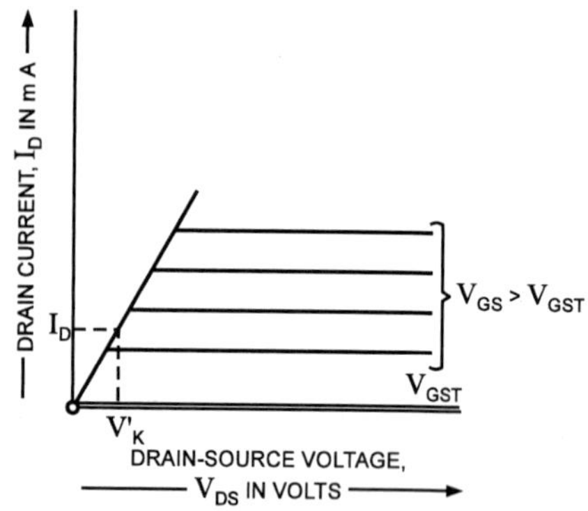

Fig. 10.53

Two ideal equivalent circuits for an E-MOSFET are shown in Fig. 10.54. These equivalent circuits are the same as for a JFET, except for $I_{D(ON)}$ and the positive gate-source voltage. In other words, the E-MOSFET can operate like a voltage variable resistor (VVR) or like a constant current source depending upon the location of the operating point. The proportional knee voltage is the guide. When V_{DS} exceeds V'_K, the device is a constant current device and V_D below V'_K, the device is a resistor.

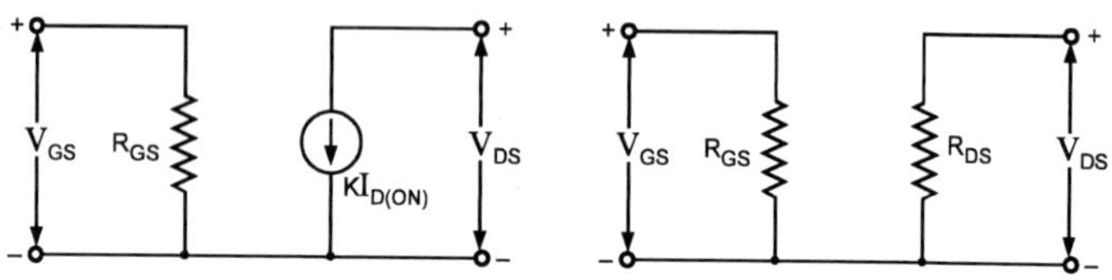

(a) *(b)*

Fig. 10.54. *Ideal Equivalent Circuits For E-MOSFETs*

10.18. COMPARISON OF N-CHANNEL FETs WITH P-CHANNEL FETs

S. No.	*Particulars*	*N-Channel FETs*	*P-Channel FETs*
1.	Current carriers	Electrons	Holes
2.	Mobility	Large	Poor
3.	Noise	Less	More
4.	Transconductance	Larger	Lesser

10.19. COMPARISON OF JFETs AND MOSFETs

JFETs and MOSFETs are quite similar in their operating principles and in their electrical characteristics. However, they differ in some aspects, as detailed below :

1. In MOSFETs (DE-MOSFETs as well as E-MOSFETs), the transverse electric field induced across an insulating layer deposited on the semiconductor material controls the conductivity of the channel whereas in the JFETs the transverse electric field across the reverse-biased PN junction controls the conductivity of the channel.
2. JFETs can only be operated in the depletion mode whereas MOSFETs can be operated in either depletion or in enhancement mode. In a JFET, if the gate is forward biased, excess-carrier injunction occurs and the gate current is substantial. Thus channel conductance is enhanced to some degree due to excess carriers but the device is never operated with gate forward biased because gate current is undesirable.
3. MOSFETs have input impedance much higher than that of JFETs because of small leakage currents. The input impedance of the MOSFET is of the order of 10^{10} to 10^{15} Ω whereas the input impedance of JFET is of the order of 10^8 Ω.
4. JFETs have characteristic curves more flatter than those of MOSFETs indicating a higher drain resistance.

   The drain resistance of a JFET is of the order of 10^5 to 10^6 Ω whereas the drain resistance of a MOSFET is of the order of 1 to 50 kΩ.
5. MOSFET is very susceptible to overload voltage and needs special handling during installation. It may get damaged easily if not properly handled.
6. When JFET is operated with a reverse bias on the junction, the gate current I_G is larger than it would be in a comparable MOSFET. The current caused by minority carrier extraction across a reverse biased junction is greater, per unit area, than the leakage current that is supported by the oxide layer in a MOSFET. Thus MOSFET devices are more useful in electrometer applications than are the JFETs.

For the above reasons, and also because MOSFETs are somewhat easier to manufacture, they are more widely used than are the JFETs.

10.20. COMPARISON BETWEEN NMOS AND PMOS

1. P-channel is much easier and cheaper to produce than the N-channel device.
2. The drain resistance of P-channel MOSFET is three times higher than that of an identical N-channel device.
3. The N-channel MOSFET is smaller for the same complexity than the P-channel MOSFET.
4. The N-channel MOSFET has the higher packing density which makes it faster in switching applications due to the smaller junction areas and low inherent capacitances.
5. A P-channel MOSFET occupies a larger area than the N-channel MOSFET for the given drain current rating. This is because the electron mobility is 2.5 times of that of an hole.
6. The N-channel MOSFET has higher false turn-on possibility than the P-channel device because of the positively charged contaminants.

10.21. UNIJUNCTION TRANSISTOR (UJT)

Unijunction transistor (abbreviated as UJT), also called the *double-base diode,* is a 2 layer, 3 terminal solid-state (silicon) switching device. The device has a unique characteristic that when it is triggered, its emitter current increases regeneratively (due to negative resistance characteristic) until it is restricted by emitter power supply. The low cost per unit, combined with its unique characteristics, have warranted its use in a wide variety of applications. A few include oscillators, pulse generators, sawtooth generators, trigger circuits, phase control, timing circuits, and voltage- or current-regulated supplies. The device is, in general, a low-power absorbing device under normal operating conditions and provides tremendous aid in the continual effort to design relatively efficient systems.

10.21.1. Construction

The basic structure of a unijunction transistor is shown in Fig. 10.55 (*a*). It essentially consists of a lightly-doped N-type silicon bar with a small piece of heavily doped P-type material alloyed to its one side to produce single P-N junction.

The single P-N junction accounts for the terminology *unijunction*.

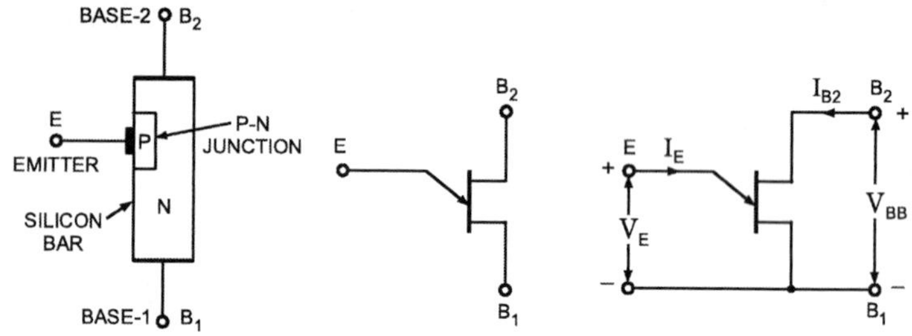

(*a*) *Basic Structure* (*b*) *Schematic Symbol* (*c*) *Basic Arrangement*

Fig. 10.55. ***Unijunction Transistor***

The silicon bar, at its ends, has two ohmic contacts designated as base-1 (B_1) and base-2 (B_2), as shown and the P-type region is termed the emitter (E). The emitter junction is usually located closer to base-2 (B_2) than base-1 (B_1) so that the device is not symmetrical, because symmetrical unit does not provide optimum electrical characteristics for most of the applications.

The symbol for unijunction transistor is shown in Fig. 10.55 (*b*). The emitter leg is drawn at an angle to the vertical line representing the N-type material slab and the arrowhead points in the direction of *conventional current* when the device is forward biased, active or in the conducting state. The basic arrangement for the UJT is shown in Fig. 10.55 (*c*).

A complementary UJT is formed by diffusing an N-type emitter terminal on a P-type base. Except for the polarities of voltage and current, the characteristics of a complementary UJT are exactly the same as those of a conventional UJT.

The noteworthy points about UJT are given below :

1. The device has only one junction, so it is called the *unijunction* device.
2. The device, because of one P-N junction, is quite similar to a diode but it differs from an ordinary diode that it has three terminals.
3. The structure of a UJT is quite similar to that of an N-channel JFET. The main difference is that P-type (gate) material surrounds the N-type (channel) material in case of JFET and the gate surface of the JFET is much larger than emitter junction of UJT.
4. In a unijunction transistor the emitter is heavily doped while the N-region is lightly doped, so the resistance between the base terminals is relatively high, typically 4 to 10 kΩ when the emitter is open.
5. The N-type silicon bar has a high resistance and the resistance between emitter and base-1 is larger than that between emitter and base-2. It is because emitter is closer to base-2 than base-1.
6. UJT is operated with emitter junction forward biased while the JFET is normally operated with the gate junction reverse biased.
7. UJT does not have ability to amplify but it has the ability to control a large ac power with a small signal.
8. It exhibits a negative resistance characteristic and so it can be employed as an oscillator.

10.20.2. Equivalent Circuit

The equivalent circuit of a UJT is shown in Fig. 10.56. The P-N junction is represented in the emitter by a diode.

Silicon bar being lightly doped has a high resistance and can be represented as two resistors connected in series–R_{B_1} from base-1 to point A and R_{B_2} from base-2 to point A, as shown in Fig. 10.56. The resistance R_{B_1}

is shown as a variable resistor since its magnitude varies with the emitter current I_E. In fact, for a typical UJT, R_{B_1} may vary from 5 kΩ down to 50 Ω for a corresponding change of emitter current I_E from 0 to 50 μA. The total resistance of silicon bar from one end to the other end *i.e.*, the resistance between B_2 and B_1 with emitter terminal E open is called the *interbase resistance* R_{BB}. From the equivalent circuit shown in Fig. 10.56 it is obvious that

$$R_{BB} = R_{B_1} + R_{B_2} \text{ with emitter terminal E open} \quad ...(10.49)$$

R_{BB} is typically within the range of 4 kΩ and 10 kΩ. Usually R_{B_1} is greater than R_{B_2} and R_{B_1} = 60 per cent of R_{BB}.

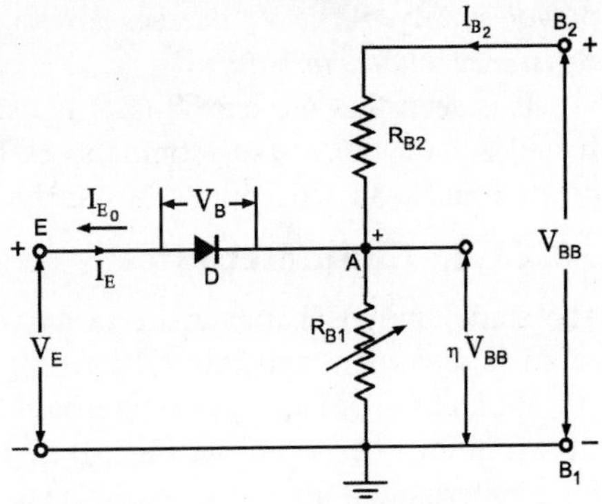

Fig. 10.56. ***Equivalent Circuit of a UJT***

With a voltage V_{BB} applied between B_2 and B_1, as shown in the figure, the voltage at the junction of R_{B_1} and R_{B_2} (point A) is

$$V_A = V_{BB} \times \frac{R_{B_1}}{R_{B_1} + R_{B_2}} = \frac{R_{B_1}}{R_{BB}} V_{BB} \quad ...(10.50)$$

$$= \eta V_{BB} \quad ...(10.51)$$

The Greek letter η (eta) is called the *intrinsic stand-off* ratio of the device and is defined as

$$\eta = \frac{R_{B_1}}{R_{B_1} + R_{B_2}} \text{ with terminal E open } = \frac{R_{B_1}}{R_{BB}} \quad ...(10.52)$$

The intrinsic stand-off ratio is the property of a UJT and is always less than unity. The typical range of η is from 0.5 to 0.8.

Voltage V_A is also the voltage at the cathode of the diode representing the P-N junction. With the emitter terminal open circuited, the only current flowing is

$$I_{B_2} = \frac{V_{BB}}{R_{BB}} \quad ...(10.53)$$

If the emitter terminal is grounded, the P-N junction is reverse biased and a small *emitter reverse current* I_{E_0} flows.

10.21.3. Operation

Imagine that the emitter supply voltage is turned down to zero. Then the intrinsic stand-off voltage reverse biases the emitter diode, as mentioned above. If V_B is the barrier voltage of the emitter diode, then the total reverse bias voltage is $V_A + V_B = \eta V_{BB} + V_B$. For silicon V_B = 0.7 V.

Now let the emitter supply voltage V_E be slowly increased. When V_E becomes equal to ηV_{BB}, I_{E_0} will be reduced to zero. With equal voltage levels on each side of the diode, neither reverse nor forward current will flow. When emitter supply voltage is further increased, the diode becomes forward biased as soon as it exceeds the total reverse bias voltage ($\eta V_{BB} + V_B$). This value of emitter voltage V_E is called the *peak point voltage* and is denoted by V_P. When $V_E = V_P$, emitter current I_E starts to flow through R_{B_1} to ground (*i.e.* B_1). This is the minimum current that is required to trigger the UJT. This is called the *peak point emitter current* and denoted by I_P. I_P is inversely proportional to the interbase voltage, V_{BB}. Now when the emitter diode starts conducting, charge carriers are injected into the R_{B_1} region of the bar. Since the resistance of a semiconductor material depends upon doping, the resistance of region R_{B_1} decreases rapidly due to additional charge carriers (holes). With this decrease in resistance, the voltage drop across R_{B_1} also decreases, causes the emitter diode to be more heavily forward biased. This, in turn, results in larger forward current, and consequently more charge carriers are injected causing still further reduction in the resistance of the R_{B_1} region. Thus the emitter current goes on increasing until it is limited by the emitter power supply. Since V_A decreases with the increase in emitter current, the UJT is said to have *negative resistance characteristic*.

It is seen that the base-2 (B_2) is used only for applying external voltage V_{BB} across it. Terminals E and B_1 are the active terminals. UJT is usually triggered into conduction by applying a suitable positive pulse to the emitter. It can be turned off by applying a negative trigger pulse.

10.21.4. Characteristics

The static emitter characteristic (a curve showing the relation between emitter voltage V_E and emitter current I_E) of a UJT at a given interbase voltage V_{BB} is shown in Fig. 10.57.* From Fig. 10.57 it is noted that for emitter potentials to the left of peak point, emitter current I_E never exceeds I_{EO} (measured in microamperes). The current I_{EO} corresponds very closely to the reverse leakage current I_{CO} of the conventional BJT. This region, as shown in the figure, is called the *cutoff* region. Once conduction is established at $V_E = V_P$ the emitter potential V_E starts decreasing with the increase in emitter current I_E. This corresponds exactly to the decrease in resistance R_{B_1} for increasing current I_E. This device, therefore, has a *negative resistance region* which is stable enough to be used with a great deal of reliability in the areas of applications listed earlier. Eventually, the valley point reaches, and any further increase in emitter current I_E places the device in the saturation region, as shown in the figure. In this region characteristic approaches that of a semiconductor diode in the equivalent circuit shown in Fig. 10.56. Three other important parameters for the UJT are I_P, V_V and I_V (shown in the Fig. 10.57) and are defined below :

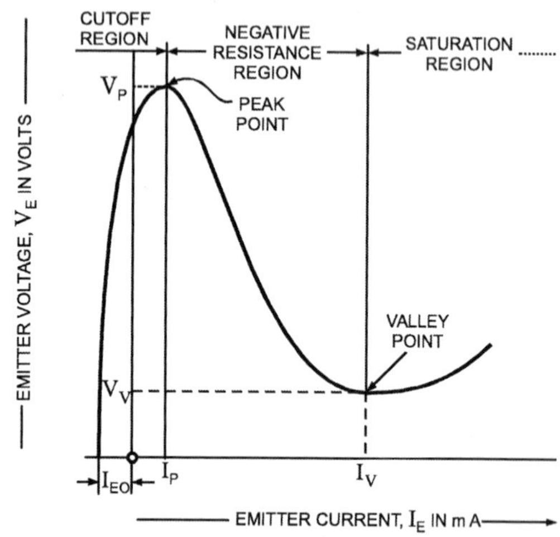

Fig. 10.57. *Static Emitter-Characteristic For a UJT*

Peak Point Emitter Current I_P. It is the emitter current at the peak point. It represents the minimum current that is required to trigger the device (UJT). It is inversely proportional to the interbase voltage V_{BB}.

Valley Point Voltage V_V. The valley point voltage is the emitter voltage at the valley point. The valley point voltage increases with the increase in interbase voltage V_{BB}.

Valley Point Current I_V. The valley point current is the emitter current at the valley point. It increases with the increase in interbase voltage V_{BB}.

10.21.5. Special Features of UJT

The special features of a UJT are :

1. A stable triggering voltage (V_P)—a fixed fraction of applied interbase voltage V_{BB}.
2. A very low value of triggering current.
3. A high pulse current capability.
4. A negative resistance characteristic.
5. Low cost.

Example 10.26. A silicon UJT has an interbase resistance of 8 kΩ and R_{B_1} of 5 kΩ with $I_E = 0$. Determine (*i*) R_{B_2} and (*ii*) intrinsic stand-off ratio.

Determine the UJT current, stand-off voltage and peak-point voltage when a voltage of 15 V is applied between the bases. Take $V_B = 0.7$ V.

* V-I characteristic of UJT is one of the very few curves, in which current I is taken on X-axis and voltage V is taken on Y-axis, rather say, it is a I-V characteristic curve.

Solution : Interbase resistance, $R_{BB} = 8\ k\Omega$

$R_{B_1} = 5\ k\Omega$

(*i*) $R_{B_2} = R_{BB} - R_{B_1} = 8 - 5 = 3\ k\Omega$ **Ans.**

(*ii*) Intrinsic stand-off ratio, $\eta = \dfrac{R_{B_1}}{R_{BB}} = \dfrac{5}{8} = 0.625$ **Ans.**

Assuming V_E less than peak point voltage

UJT current, $I_{B_1} = I_{B_2} = \dfrac{V_{BB}}{R_{BB}} = \dfrac{15}{8\ k\Omega} = 1.875$ mA **Ans.**

Stand-off voltage $= \eta V_{BB} = 0.625 \times 15 = 9.375$ V **Ans.**

Peak point voltage, $V_P = \eta\ V_{BB} + V_B = 9.375 + 0.7 = 10.075$ V **Ans.**

10.22. THYRISTOR

Thyristor is the general name given to a family of semiconductor devices having four layers with a control mechanism, although this term is most commonly applied to the SCR (silicon controlled rectifier). This term is derived from *thyratron* and *transistor* because the device combines the rectification action of thyratron and control action of transistor. Thyristor stands for *THYRatron transISTOR*.

The device was first developed at the Bell Laboratories in the USA and was commercially introduced by the General Electrical Company in the USA in 1957. The device appeared in the market under different names such as SCR, thyristor, thyrode etc. Since its inception the thyristor has come to stay as a basic building block in many industrial and power system applications. Its ability to be controlled, compactness, fast response, high reliability, better efficiency, large power handling capacity, high voltage and current ratings, good trigger sensitivity, static operation, large power gain, sturdy construction, long life, very little maintenance and low cost of fabrication—due to advancement in the field of fabrication—have given the thyristor a colourful reception in every field. Today thyristors are finding applications in the control of dc/ac motors; for the improvement of power factor; and as switching devices. They have become an integral part of HVDC transmission. They are so advantageous that our major steel plants had decided to modernise their plants by replacing MG sets and mercury-arc rectifiers by thyristor converters. Thyristors have helped in further cost reduction and in the development of drive system by changing the emphasis from dc motors to ac motors. With cycle converters and inverters, the speed of an ac motor can also be controlled with ease and reliability. Apart from these main applications it finds use as a switching device, particularly in the improvement of power factor of transmission lines and mains. Thyristor can be used as a power switching device with a power handling capacity ranging from a few watts to as high as 4 MW (2,500 A at 1,600 V). Some thyristors have a rating as high as 400 A, 10,000 V for use in hv dc transmission lines.

Thyristors, with its large number of advantages and tremendous control capabilities, have numerous applications and have completely replaced the electromagnetic control systems. Thyristor basically serves two functions viz. electronic switching and electronic control. Some of the applications of thyristors are listed below:

1. Speed control of dc and ac motors.
2. As rectifier for conversion of ac into dc.
3. As inverter for conversion of dc into ac.
4. As dc chopper or dc to dc converter for converting dc at one level to dc at another level.
5. As cycloconverter for converting ac of one frequency into ac of another frequency.
6. Control of temperature, level, position and illumination.
7. Power switches (dc and ac circuit breakers).
8. HVDC transmission lines.

9. Improvement of power factor in transmission lines.
10. As static switches.
11. Control of induction heating.
12. Relay control.
13. Phase control.
14. As special power supplies for aircraft and computers etc.

SCRs and triacs having high voltage and current ratings are widely employed for power control applications whereas other members of thyristor family are employed for small power applications and for switching in control and digital circuits.

10.22.1. Thyristor Family

The most important member of the thyristor family is the silicon controlled rectifier (SCR).* SCR is a four layer (P-N-P-N), three junction semiconductor device with three terminals, namely, the anode, the cathode and the gate. It is a unilateral device and conduction takes place from anode to cathode under proper bias conditions (forward bias).

10.23. SILICON CONTROLLED RECTIFIER (SCR)

As the terminology indicates, the SCR is a controlled rectifier constructed of a silicon semiconductor material with a third terminal for control purposes. Silicon was chosen because of its high temperature and power capabilities. The basic operation of the SCR is different from that of an ordinary two layer semiconductor diode in that a third terminal, called a *gate*, determines when the rectifier switches from the open-circuit to short-circuit state. It is not enough simply to forward bias the anode-to-cathode region of the device. In the conduction state the dynamic resistance of the SCR is typically 0.01 to 0.1 Ω and reverse resistance is typically 100 kΩ or more. It is widely used as a switching device in power control applications. It can control loads by switching *on* and *off* up to many thousand times a second. It can switch *on* for a variable lengths of time duration, thereby delivering desired amount of power to the load. Thus, it possesses the advantage of a rheostat as well as a switch with none of their drawbacks. A schematic diagram and symbolic representation of an SCR are shown in Figs. 10.58 (*a*) and (*b*) respectively. As illustrated in Fig. 10.58 (*a*), SCR is a three terminal four layer semiconductor device, the layers being alternately of P-type and N-type. The junctions are marked J_1, J_2 and J_3 (junctions J_1 and J_3 operate in forward direction while middle junction J_2 operates in the reverse direction) whereas the three terminals are anode (A), cathode (C) and gate (G) which is connected to the inner P-type layer. The function of the gate is to control the firing of SCR. In normal operating conditions, anode is positive with respect to cathode.

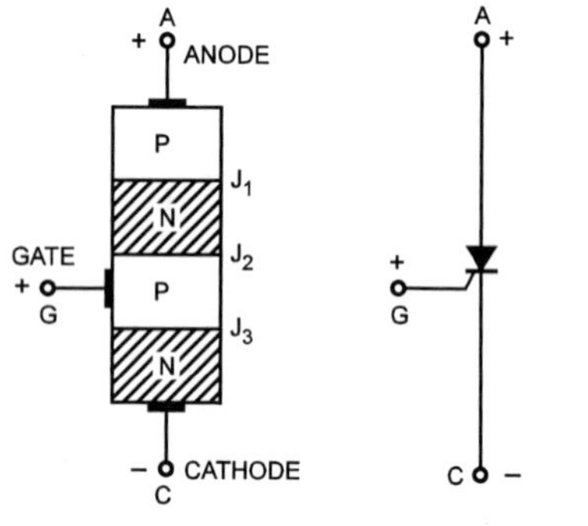

(*a*) Schematic Diagram (*b*) Symbolic Representation

Fig. 10.58. ***Silicon Controlled Rectifier***

10.23.1. Construction

From Fig. 10.59 (*a*) it is clear that SCR is essentially an ordinary rectifier (PN) and a junction transistor (N-P-N) combined in one unit to form PNPN device. Three terminals are taken: one from the outer P-type material, known as *anode*, second from the outer N-type material, known as *cathode* and the third from the base of transistor section known as the *gate*.

* It is called SCR because it is a silicon (not germanium) device and is used as a rectifier and that rectification can be controlled.

This device is made of silicon only because leakage current in silicon is much smaller than that in germanium and it is necessary that leakage current should be minimum possible as this device is used as a switch.

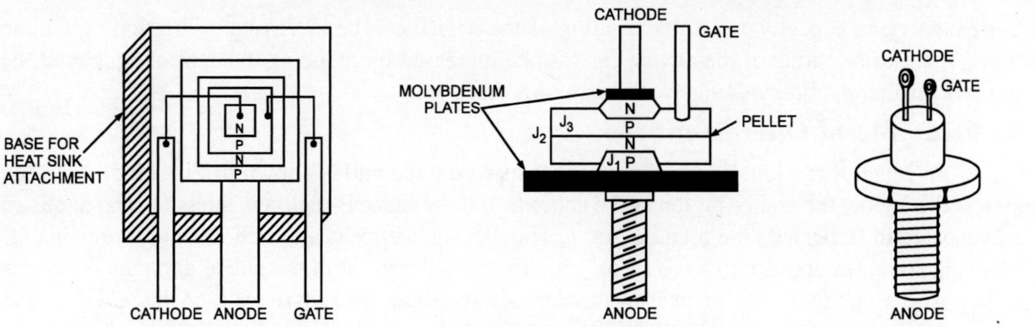

(a) *Planar Type Construction* (b) *Mesa Type Construction* (c) *External Construction*

Fig. 10.59. ***Silicon Controlled Rectifier***

The basic material used for fabrication of an SCR is N-type silicon. It has a specific resistance of about 6 Ω-mm. Silicon is the natural choice as base material because of the following advantages.

(*i*) ability to withstand high junction temperature of the order of 150ºC;
(*ii*) high thermal conductivity;
(*iii*) less variations in characteristics with temperature; and
(*iv*) less leakage current in P-N junction.

It consists, essentially, of a four layer pellet of P and N type silicon semiconductor materials. The junctions are diffused or alloyed. The material which may be used for P diffusion is aluminium and for N diffusion is phosphorus. The contact with anode can be made with an aluminium foil and through cathode and gate by metal sheet. Diffusion must be carried out at a proper temperature and for necessary duration to provide correct concentration because this decides the properties of the device. The main advantage of planar construction is that junction formation always takes place underneath a thin layer of silicon dioxide grown over the silicon wafer, before diffusion commences, which prevents contamination of silicon surfaces. Thus, planar, pellets are protected to a large extent from outside environments. Drawbacks of planar construction are that more silicon is required per ampere of current carrying capability, and that more wafer processing steps are required. Planar structures are best suited, therefore, to low current devices where many pellets can be obtained from a single wafer, and to complex structures where photoresist techniques are needed for geometry control.

The other technique is the mesa construction shown in Fig. 10.59 (*b*). This technique is used for high power SCRs. In this technique, the inner junction J_2 is obtained by diffusion, and then the outer two layers are alloyed to it. The PNPN pellet is properly braced with tungsten or molybdenum plates to provide greater mechanical strength and make it capable of handling large currents. One of these plates is hard soldered to a copper or an aluminium stud, which is threaded for attachment to a heat sink. This provides an efficient thermal path for conducting the internal losses to the surrounding medium. The uses of hard solder between the pellet and backup plates minimises thermal fatigue when the SCRs are subjected to temperature induced stresses. For medium and low power SCRs, the pellet is mounted directly on the copper stud or casing, using a soft solder which absorbs the thermal stresses set up by differential expansion and provides a good thermal path for heat transfer. For a larger cooling arrangement, which is required for high power SCRs, the press pack or hockey puck construction is employed, which provides for double-sided air or water cooling.

The salient features to be considered, while designing an SCR, are the diameter and thickness of wafer, composition of the base material, type and amount of the material to be diffused into the wafer, shape, position and contact area of the gate, shape and size of the SCR, type of heat sink etc.

Fabrication technology determines various properties of the device. The voltage rating of a device can be increased by lightly doping the inner two layers and increasing their thickness. But due to this increased resistance, forward voltage drop increases and large triggering currents are required causing greater power dissipation accompanied by smaller current ratings. The heat dissipation of silicon falls from 1.5 W/cm^2 at 25 ºC to 1.25 W/cm^2 at 125 ºC. A high voltage power device can seldom be used beyond 125 ºC.

The current carrying capacity and voltage rating of the device can be increased by irradiating silicon with neutrons. The current rating of the device can also be increased by reducing the current density at the junction but this results in a bulky device with large turn-on time.

10.23.2. Principle of Operation

The SCR is a four layer, three junction and a three terminal device and is shown in Fig. 10.58 (*a*). The end P-region is the anode, the end N-region is the cathode and the inner P-region is the gate. The anode to cathode is connected in series with the load circuit. Essentially the device is a switch. Ideally it remains off (voltage blocking state), or appears to have an infinite impedance until both the anode and gate terminals have suitable positive voltages with respect to the cathode terminal. The thyristor then *switches on* and current flows and continues to conduct without further gate signals. Ideally the thyristor has zero impedance in conduction state. For switching off or reverting to the blocking state, there must be no gate signal and the anode current must be reduced to zero. Current can flow only in one direction.

In absence of external bias voltages, the majority carriers in each layer diffuse until there is a built-in voltage that retards further diffusion. Some majority carriers have enough energy to cross the barrier caused by the retarding electric field at each junction. These carriers then become minority carriers and can recombine with majority carriers. Minority carriers in each layer can be accelerated across each junction by the fixed field, but because of absence of external circuit in this case the sum of majority and minority carrier currents must be zero.

A voltage bias, as shown in Fig. 10.60, and an external circuit to carry current allow internal currents which include the following terms :

The current I_1 is due to (*i*) majority carriers (holes) crossing junction J_1; (*ii*) minority carriers crossing junction J_1; (*iii*) holes injected at junction J_2 diffusing through the N-region and crossing junction J_1; and (*iv*) minority carriers from junction J_2 diffusing through the N-region and crossing junction J_1.

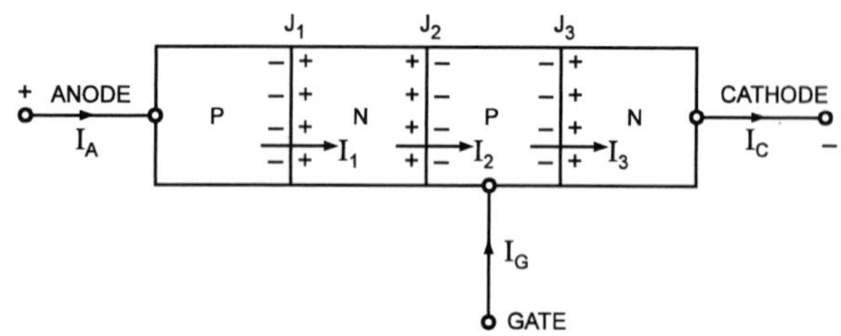

Fig. 10.60. ***Diagrammatic Representation Showing Current Flow and Voltage Bias in An SCR***

Similarly I_2 is due to six terms and I_3 is due to four terms.

The two simple analogues to explain the basic action for the thyristor are those of the diode and the two transistor models.

1. *Diode Model.* The thyristor is similar to three diodes in series as there are three P-N junctions. Without gate bias, there is always at least one reverse biased junction to prevent conduction irrespective of the polarity of an applied voltage between anode and cathode. If the anode is made positive and the gate is also biased positively with respect to cathode, the P-layer at the gate is flooded by the electrons from the cathode and loses its identity as a P-layer. Accordingly the thyristor becomes equivalent to a conducting diode.

2. *Two Transistor Model.* Imagine the SCR cut along the dotted line, as shown in Fig. 10.61 (*a*). Then we can have two devices, as shown in Fig. 10.61 (*b*). These two devices can be recognised as two transistors. The upper left one is P-N-P transistor and the lower right N-P-N type. Further it can be recognized that the base of the P-N-P transistor is joined to the collector of the N-P-N transistor while the collector of P-N-P is joined to the base of N-P-N transistor, as illustrated in Fig. 10.61 (*c*). This construction has been conceived merely to explain the working of SCR, otherwise in physical shape the SCR has four solid layers of P-N-P-N type only.

Now we can see that the two transistors are connected in such a manner that the collector of Q_1 is connected to the base of Q_2 *i.e.* the output collector current of Q_1 becomes the base current for Q_2. In the similar way the collector of Q_2 is joined to the base of Q_1 which shows that the output collector current of Q_2 is fed to Q_1, as input base current. These are back to back connections of transistors in such a way that the output of one goes into as input of other transistor and vice-versa. This gives net gain of loop circuit as $\beta_1 \times \beta_2$ where β_1 and β_2 are current gains of two transistors respectively.

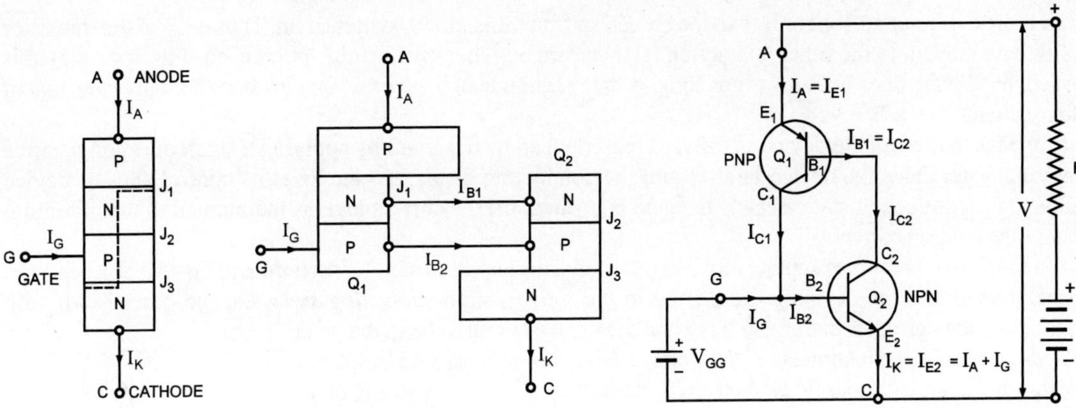

(a) *(b) SCR Split-up into Two Transistors* *(c) Two Transistor Equivalent Circuit of an SCR*

Fig. 10.61. ***Two Transistor Model of an SCR***

When the gate current is zero or the gate terminal is open, the only current in circulation is the leakage current, which is very small in case of silicon devices specially and the total current is a little higher than sum of individual leakage currents. Under these conditions P-N-P-N device is said to be in its forward blocking or high impedance 'off' state. As soon as a small amount of gate current is given to the base of transistor Q_2 by applying forward bias to its base-emitter junction, it generates the collector current as β_2 times the base current. This collector current of Q_2 is fed as input base current to Q_1 which is further multiplied by β_1 times as I_{C_1} which forms input base current of Q_2 and undergoes further amplification. In this way both transistors feedback each other and the collector current of each goes on multiplying. This process is very quick and soon both the transistors drive each other to saturation. Now the device is said to be in *on-state*. The current through the on-state SCR is controlled by external impedance only.

Analysis of Two Transistor Model. If I_{E_1} and I_{B_1} are the emitter current and base current respectively of the P-N-P transistor; I_A and I_K are the respective anode current and cathode current of the SCR; I_{C_1} and I_{C_2} are the collector currents of P-N-P and N-P-N transistors respectively; I_G is the gate current of the SCR; and α_1 and α_2 are the fraction of holes and electrons injected from anode and cathode on junction J_2 respectively, then neglecting leakage currents of transistors we have

$$I_{B_1} = I_A - I_{C_1} = I_A - \alpha_1 I_A = (1 - \alpha_1)\, I_A \qquad \text{...(10.54)}$$

$$\text{Also } I_{C_2} = \alpha_2 I_K$$

$$\text{and } I_{B_1} = I_{C_2}$$

$$\text{or } (1 - \alpha_1)\, I_A = \alpha_2 I_K \qquad \text{...(10.55)}$$

$$\text{or } \frac{(1 - \alpha_1) I_A}{\alpha_2} = I_A + I_G \qquad \text{Since } I_K = I_A + I_G,$$

$$\text{or } (1 - \alpha_1) I_A = \alpha_2\, (I_A + I_G)$$

$$\text{or } I_A = \frac{\alpha_2 I_G}{1 - (\alpha_1 + \alpha_2)} \qquad \text{...(10.56)}$$

From above Eq. (10.56) it is obvious that if $(\alpha_1 + \alpha_2)$ becomes equal to unity, then anode current I_A attains infinite value. In other words, the device suddenly latches into conduction state from the original non-conducting state. The reason for such a behaviour is the regenerative manner in which the two constituent transistors Q_1 and Q_2 are interconnected. Thus if the gate current I_G is of such a magnitude that $\alpha_1 + \alpha_2 = 1$, the SCR will be triggered.

The value of $(\alpha_1 + \alpha_2)$ can be made almost unity by a positive gate current of proper magnitude and for short duration at the gate. Signal I_G when applied to the gate causes a flow of current in transistor Q_2. This gives rise to the collector current I_{C_2} in transistor Q_2. Now the transistor Q_1 base is energized because its base current is the collector current of transistor Q_2 and so first transistor is switched on. Thus each of the transistor supplies base current to the other and action is, therefore, regenerative. At this point even if the gate signal is removed, the device does not turn off as long as the magnitude of the current flowing does not fall below that of holding current.

It implies that when the device is forward biased it can be triggered by applying a small duration positive pulse at the gate. Once the device latches into the conducting state, the gate loses its control and the device remains in the conducting state, even if the gate is opened, till the current level is maintained to the minimum of that of the holding current.

When $(\alpha_1 + \alpha_2)$ becomes appreciably less than unity, the anode current I_A, according to Eq. (10.56), becomes extremely small and the device is said to be in the *'off'* or *non-conducting state*. On the other hand, with $(\alpha_1 + \alpha_2) \simeq 1$, the current is extremely large and the device is said to be in the *on* or *conducting state*. In conducting state, the voltage drop across the device drops to a low value and a large current flows through it, limited only by the resistor R in the external circuit. Thus the existence of the device either in the *'on' state* or in the *'off' state* depends on the applied voltage and the gate current.

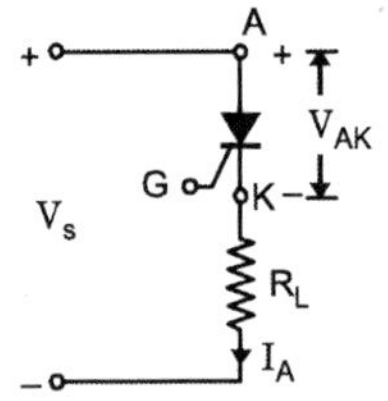

(a) Circuit Diagram

10.23.3. Volt-Ampere Characteristics

As already mentioned, the SCR is a four layer device with three terminals, namely, the anode, the cathode and the gate. When the anode is made positive with respect to the cathode, junctions J_1 and J_3 are forward biased and junction J_2 is reverse biased and only the leakage current will flow through the device. The SCR is then said to be in the *forward blocking state* or in the *forward mode* or *off state*. But when the cathode is made positive with respect to the anode, junctions J_1 and J_3 are reverse biased, a small reverse leakage current will flow through the SCR and the SCR is said to be in the *reverse blocking state* or *in reverse mode*.

When the anode is positive with respect to cathode *i.e.* when the SCR is in forward mode, the SCR does not conduct unless the forward voltage exceeds certain value, called the *forward breakover voltage*, V_{FBO}. In non-conducting state, the current through the SCR is the leakage current which is very small and is negligible. If a positive gate current is supplied, the SCR can become conducting at a voltage much lesser than forward breakover voltage. The larger the gate current, lower the breakover voltage [Fig. 10.62 (*c*)]. With sufficiently large gate current, the SCR behaves identical to PN rectifier. Once the SCR is switched *on*, the forward voltage drop across it is suddenly reduced to very small value, say about 1 volt. In the *conducting* or *on-state*, the current through the SCR is limited by the external impedance.

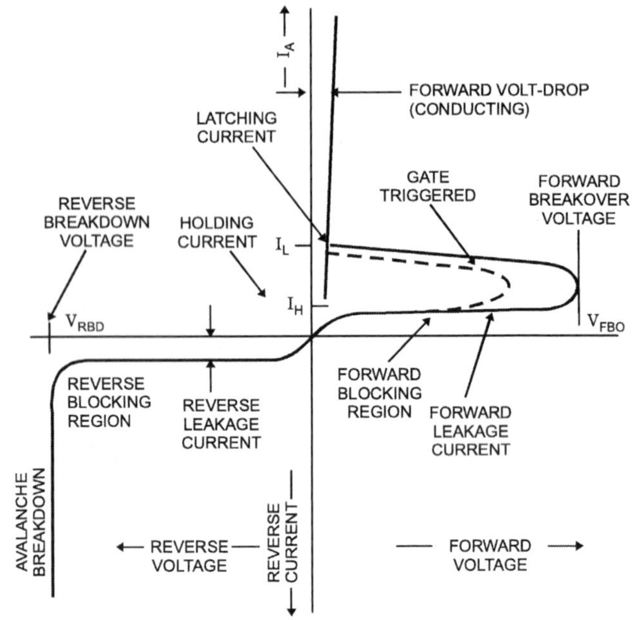

(b) V-I Characteristics

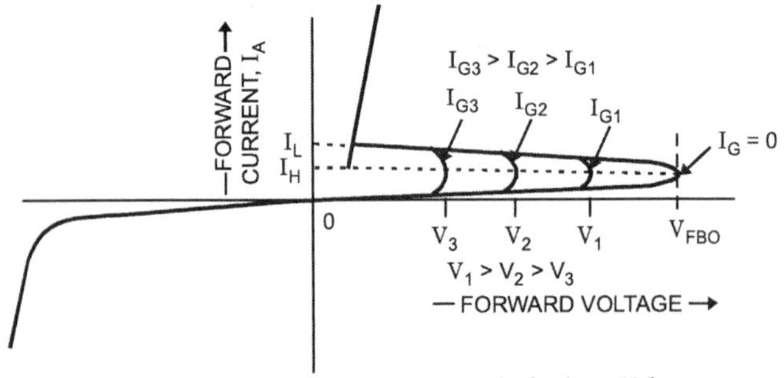

(c) Effects of Gate Current On Forward Blocking Voltage

Fig. 10.62

When the anode is negative with respect to cathode *i.e.* when the SCR is in reverse mode or in blocking state no current flows through the SCR except very small leakage current of the order of few microamperes. But if the reverse voltage is increased beyond a certain value, called the *reverse breakdown voltage*, V_{RBD}, *avalanche breakdown* takes place. Forward breakover voltage V_{FBO} is usually higher than reverse breakdown voltage, V_{RBD}.

From the foregoing discussion, it can be seen that the SCR has two stable and reversible operating states. The change over from *off-state* to *on-state*, called *turn-on*, can be achieved by increasing the forward voltage beyond V_{FBO}. A more convenient and useful method of turning *on* the device employs the gate drive. If the forward voltage is less than the forward breakover voltage, V_{FBO}, it can be turned *on* by applying a positive voltage between the gate and the cathode. This method is called the *gate control*. Another very important feature of the gate is that once the SCR is triggered to on-state the gate loses its control.

The switching action of gate takes place only when

(*i*) SCR is forward biased *i.e.* anode is positive with respect to cathode, and

(*ii*) suitable positive voltage is applied between the gate and the cathode.

Once the SCR has been switched *on*, it has no control on the amount of current flowing through it. The current through the SCR is entirely controlled by the external impedance connected in the circuit and the applied voltage. There is, however, a very small, about 1 V, potential drop across the SCR. The forward current through the SCR can be reduced by reducing the applied voltage or by increasing the circuit impedance. There is, however, a minimum forward current that must be maintained to keep the SCR in conducting state. This is called the *holding current* rating of SCR. If the current through the SCR is reduced below the level of holding current, the device returns to off-state or blocking state.

The SCR can be switched *off* by reducing the forward current below the level of holding current which may be done either by reducing the applied voltage or by increasing the circuit impedance.

Note : *The gate can only trigger or switch on the SCR, it cannot switch off.*

Alternatively the SCR can be *switched off* by applying negative voltage to the anode (reverse mode), the SCR naturally will be *switched off*.

Here one point is noteworthy, the SCR takes certain time to switch *off*. The time, called the *turn-off time*, must be allowed before forward voltage may be applied again otherwise the device will switch *on* with forward voltage without any gate pulse. The turn-off time is about 15 microseconds, which is immaterial when dealing with power frequency, but this becomes important in the inverter circuits, which are to operate at high frequency.

10.23.4. SCR Turn-on Methods

As mentioned earlier, the SCR can be switched *on* either by increasing the forward voltage beyond forward breakover voltage V_{FBO} or by applying a positive gate signal when the device is forward biased. Of these two methods, the latter, called the *gate-control method*, is used as it is more efficient and easy to implement for power control. The following points have to be noted when designing the gate control circuit.

1. Appropriate gate-to-cathode voltage must be applied for turn-on when the device is forward biased.
2. The gate signal must be removed after the device is turned *on*.
3. No gate signal should be applied when the device is reverse biased.
4. When the device is in the off-state, a negative voltage applied between the gate and the cathode will improve the characteristics of the device. In such an instance, a large positive voltage will be required to overcome this negative bias for turn-on.

There are three methods of triggering the device by gate control.

(a) Triggering By a DC Gate Signal. In this method, a dc voltage of proper polarity and magnitude is applied between the gate and the cathode when the device is to be turned *on*. It must, however, be noted that the SCR is a current operated device and it is the gate

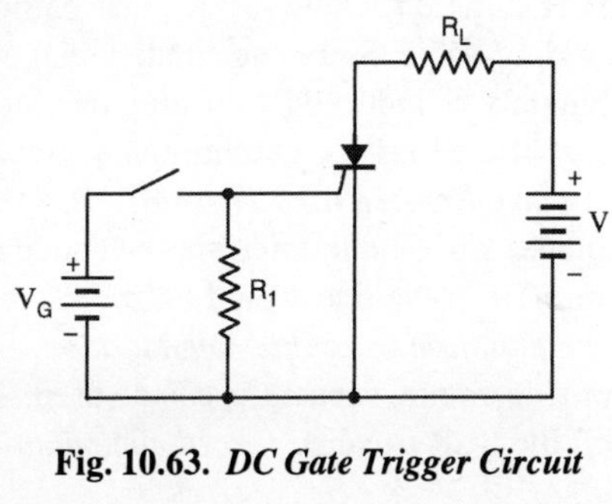

Fig. 10.63. ***DC Gate Trigger Circuit***

current that turns *on* the device. The drawback of this method is that gate signal has to be continuously applied resulting in increase in internal power dissipation and that there is no isolation of the gate control circuit from the main power circuit. A typical circuit used for triggering an SCR with a dc gate bias is shown in Fig. 10.63. The resistance R_1 is connected in the circuit to provide noise suppression and improve turn-on time. The turn-on time primarily depends upon the magnitude of gate current—the higher the gate current, the shorter the turn-on time.

(b) Triggering By an AC Gate Signal. In many power control circuits that use ac input, the gate-to-cathode voltage is obtained from a phase-shifted ac voltage derived from the main supply. The main advantage of this method is that proper isolation of power and control circuits can be provided. The firing angle control is obtained very conveniently by changing the phase angle of the control signal. However, the gate drive is maintained for one half-cycle after the device is turned *on* and a reverse voltage is applied between the gate and the cathode during the negative half cycle.

(c) Triggering By a Pulsed Gate Signal. In this method, the gate drive consists of a single pulse appearing periodically, or a sequence of high frequency pluses. This is called the carrier *frequency gating*. A pulse transformer is used for isolation. The gate losses are very much reduced since the drive is discontinuous.

Trigger Current and Trigger Voltage. An SCR, as shown in Fig. 10.64, has a gate voltage of V_G. When this gate voltage is in the vicinity of 0.7 V, the SCR will turn *on* and the output voltage will drop from $+V_{CC}$ to a low value. When a gate resistor R_G is used as shown, the input voltage required to trigger an SCR can be determined from the following equation :

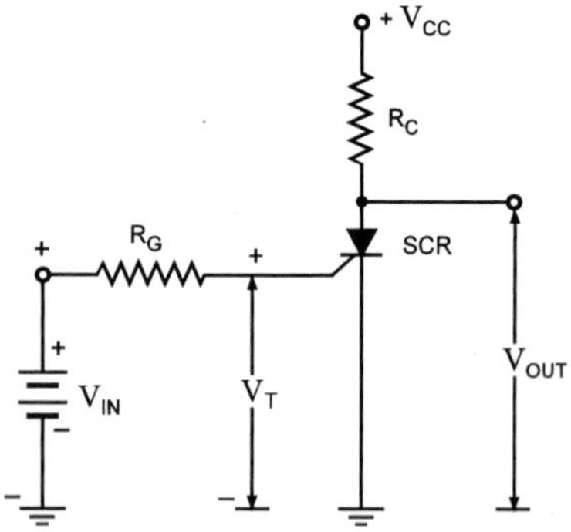

Fig. 10.64. ***An SCR Circuit***

$$V_{IN} = V_T + I_T R_G \qquad ...(10.57)$$

where V_T and I_T are the trigger voltage and trigger current respectively, required for the device gate. This information is available on data sheets. For instance, the data sheet of a 2N4441 gives $V_T = 0.75$ V and $I_T = 10$ mA. Sometimes a gate resistor is not used. In this case, R_G is the Thevenin's resistance of the circuit driving the gate. Unless Eq. (10.57) is satisfied, the SCR cannot turn *on*. After the SCR has turned *on*, it stays on even if V_{IN} is reduced to zero. In this case, the output voltage remains low indefinitely.

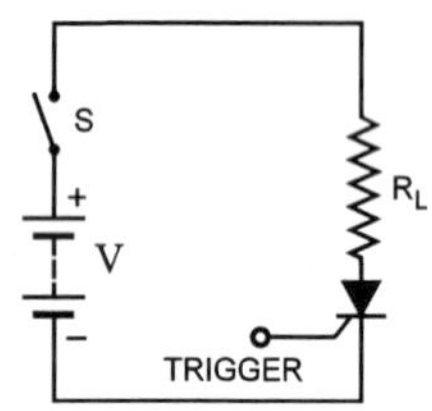

Fig. 10.65

10.23.5. SCR Turn-off Methods

As already mentioned, once the SCR is fired, it remains *on* even when triggering pulse is removed. This ability of the SCR to remain *on* even when gate current is removed is referred to as *latching*. So SCR cannot be *turned off* by simply removing the gate pulse.

There are three methods of switching *off* the SCR, namely *natural commutation*, *reverse bias turn-off*, and *gate turn-off*.

(a) Natural Commutation. When the anode current is reduced below the level of the holding current, the SCR turns *off*. However, it must be noted that rated anode current is usually larger than 1,000 times the holding value. Since the anode voltage remains positive w.r.t. the cathode in a dc circuit, the anode current can only be reduced by opening the line switch S (Fig. 10.65), increasing the load impedance R_L or shunting part of the load current through a circuit parallel to the SCR, *i.e.*, short circuiting the device.

(b) Reverse Bias Turn-off. A reverse anode to cathode voltage (the cathode positive with respect to the anode) will tend to interrupt the anode current. The voltage reverses every half cycle in an ac circuit, so that an SCR in the line would be reverse biased every negative cycle and would turn *off*. This is called *phase commutation* or ac *line commutation*. To create a reverse biased voltage across the SCR, which is in the line of a dc circuit, capacitors can be used. The method of discharging a capacitor in parallel with an SCR to turn *off* the SCR is called *forced commutation*.

In power electronic applications one advantage of using SCRs is that they are compact. The control equipment is also compact if integrated circuits are used. There has also been an attempt to miniaturize capacitors used for forced commutation and for filtering. The former use is important because the currents can be high and thermal dissipation takes high priority in design considerations. Small sizes of capacitors are at present being achieved by the use of metallized plastic film or a plastic film and aluminium foil.

(c) ***Gate Turn-Off***. In some specially designed SCRs the characteristics are such that a negative gate current increases the holding current so that it exceeds the load current and the device turns *off*. The current ratings are presently below 10 A and this type will not be considered further.

10.23.6. Normal Operating Conditions

For normal operation of an SCR, the points to be kept in view are :

1. The supply voltage is to be kept much less than breakover voltage, otherwise it will get damaged.
2. The SCR is turned *on* by passing an appropriate gate current and not by applying supply voltage more than breakover voltage.
3. In case of operation with ac supply, the peak reverse voltage should not exceed the reverse breakdown voltage.
4. While turning *off* the SCR from the on-state, the anode current should be reduced to holding current.
5. For high value of gate current, the SCR will close at much reduced supply voltage.

Example 10.27. An SCR has a breakover voltage of 250 V, a trigger current of 12 mA and holding current of 12 mA. Explain the inferences you make from these datas. What will happen if the gate current is 16 mA ?

Solution : As defined above breakover voltage is the minimum forward voltage, with gate open, at which SCR starts conducting heavily. So, breakover voltage of 250 V means that if the gate is open the SCR will not start conducting until and unless the voltage applied becomes equal to or exceeds 250 V.

Trigger current of 12 mA means that the SCR will start conducting heavily only when a minimum gate current of 12 mA is passed in cases the applied voltage is less than breakover voltage of 250 V. In other words the SCR will not start conducting if trigger current is less than 12 mA and voltage applied is less than 250 V.

Holding current of 12 mA means the SCR will not stop conducting, even if triggering current is removed, until and unless the anode current is reduced to a value below 12 mA by reducing the supply voltage.

If the gate current is increased to 16 mA, the SCR will start conducting with lower supply voltage.

Example 10.28. An ac voltage $v = 240 \sin 314\,t$ is applied to an SCR. If the SCR has a forward breakdown voltage of 180 V, find the time during which the SCR remains off.

Solution : The applied voltage is given as

$$v = 240 \sin 314t$$

Let the instantaneous voltage be 180 V at time t_1 then

$$180 = 240 \sin 314t$$

$$\text{or} \quad t_1 = \frac{1}{314} \sin^{-1} \frac{180}{240} = 2.7 \text{ ms}$$

$$\text{Turn-off period, } T_{off} = \frac{T}{2} + t_1 = \frac{2\pi}{\omega} - t_1 = \frac{2\pi}{314} - 2.7 \times 10^{-3} = 17.3 \text{ ms } \textbf{Ans.}$$

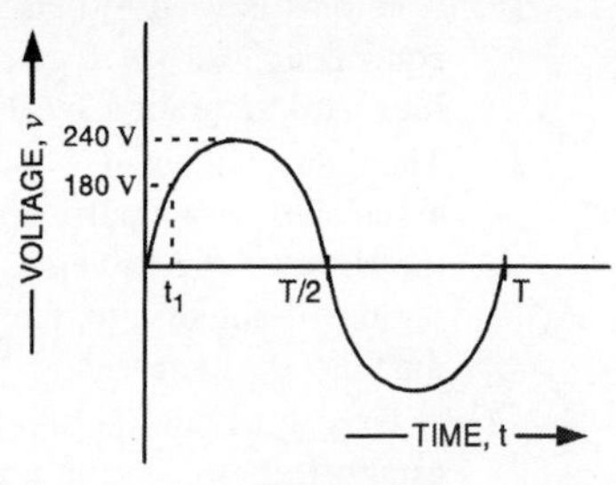

Fig. 10.66

∵ SCR will remain *off* during negative half cycle also.

HIGHLIGHTS

1. The FET is a majority carrier device, and is, therefore, often called a unipolar transistor. Like its bipolar counterpart, the FET is a three-terminal (namely drain, source and gate) semiconductor device in which current conduction is by one type of majority carriers (either electrons or holes).

2. Field effect devices are controlled by an electric field (or voltage) and give an extremely high input impedance—most important advantage over BJT devices. FETs are of several types such as JFETs, MESFETs and MISFETs.
3. The JFET is a voltage controlled device and has two diodes; the gate-source diode and the gate-drain diode. For normal operation, the gate-source diode is reverse biased. Then the gate voltage controls the drain current.
4. The JFET has a high input impedance (of the order of 100 MΩ), a high degree of isolation between input and output, and a low noise figure. The main disadvantage of a JFET is its relatively low gain-bandwidth product, due to the junction capacitive effects.
5. The JFET can actually be used as a voltage-controlled resistor because of unique sensitivity of the drain-to-source resistance R_{DS} to the gate-to-source voltage V_{GS}.
6. The curve drawn between drain current I_D and drain-to-source voltage V_{DS} with gate-to-source voltage V_{GS} as the parameter is called the drain characteristic.
   Maximum drain current for any JFET labelled as I_{DSS} occurs when the gate-source voltage V_{GS} is zero. The minimum drain current for a JFET occurs at pinch-off defined by $V_{GS} = V_P$. The pinch-off voltage separates the ohmic and active regions for $V_{GS} = 0$. The gate-source cutoff voltage has the same magnitude as the pinch-off voltage. $V_{GS(off)}$ turns the JFET off.
7. The relationship between the drain current I_D and the gate-to-source voltage V_{GS} of a JFET is a nonlinear defined by Shockley's equation

$$I_D = I_{DSS}\left[1-\frac{V_{GS}}{V_P}\right]^2 = I_{DSS}\left[1-\frac{V_{GS}}{V_{GS(off)}}\right]^2$$

   As the current level approaches I_{DSS}, the sensitivity of I_D to variations in V_{GS} increases significantly.
8. The curve drawn between drain current I_D and gate-to-source voltage V_{GS} for a given value of drain-to-source voltage V_{DS} is called the transfer characteristic. The transfer characteristics are the characteristics of the device itself and are not sensitive to the circuit in which the JFET is used. The drain current increases more rapidly as V_{GS} approaches zero. Because the drain current equation contains a squared quantity, JFETs are referred to as square-law devices. The normalized transconductance curve indicates that I_D equals one-quarter of the maximum when V_{GS} equals half of the cutoff $\left(i.e.\, I_D = \frac{I_{DSS}}{4} \text{ when } V_{GS} = \frac{V_P}{2}\right)$ and at a point where $I_D = \frac{I_{DSS}}{2}$, $V_{GS} \cong 0.3$ V.
9. Maximum operating conditions are determined by the product of the V_{DS} and I_D.
10. Gate bias is used to bias a JFET in the ohmic region. When it operates in the ohmic region, a JFET is equivalent to a small resistance of R_{DS}. To ensure operation in the ohmic region, the JFET is driven into hard saturation by using $V_{GS} = 0$ and $I_{D(sat)} << I_{DSS}$.
11. The transconductance parameter g_m is determined by the ratio of the change in drain current ΔI_D associated with a particular change in gate-to-source voltage ΔV_{GS} in the region of interest. The steeper the slope of the I_D versus V_{GS} curve, the greater is the value of g_m. In addition, the closer the point or region of interest to the saturation current I_{DSS}, the greater is the transconductance parameter. The quantity g_m increases as V_{GS} approaches zero. Data sheets may list g_m or g_{fs} in siemens.
12. A fixed bias configuration has, as the label implies, a fixed dc voltage applied from gate to source to establish the operating point.
    When the gate voltage is much larger than V_{GS} (*i.e.* $V_G >> V_{GS}$), voltage-divider bias can set up a stable Q-point in the active region. When positive and negative supply voltages are available, two-supply source bias can be employed to swamp out the variations in V_{GS} and set up a stable Q-point. When supply voltages are not large, current-source bias can be used to have a stable Q-point. Self-bias is used only with small-signal amplifiers because the Q-point is less stable than with the other biasing methods.
13. A common-source amplifier has a voltage gain of $g_m r_d$ and produces an inverted output signal.
14. The common-drain amplifier (or source follower) has a voltage gain less than unity, but a higher input resistance with a bootstrapped bias circuit. One of the most applications of a JFET amplifier is the source follower, which is often used at the front end of the systems because of high input resistance.

15. The common-gate amplifier has a characteristically low input resistance and a high output resistance, although its voltage gain may be quite high. It has few applications.
16. JFETs are employed in multiplexers, chopper amplifiers, buffer amplifiers, voltage-controlled resistors, cascade amplifiers, current limiters, current sources and AGC circuits.
17. The operation of a metal-semiconductor field-effect transistor (MOSFET) is identical to that of a JFET. In the MOSFET, however, a metal-semiconductor rectifying contact, instead of a P-N junction, is used for the gate electrode.
18. The MOSFET is a field-effect transistor in which the metal gate is insulated by a very thin oxide layer from the semiconductor channel. It is acually a four-terminal device, with the substrate or body being the fourth terminal.

    There are two types of MOSFETs—the N-channel in which current is due to the flow of electrons in the inversion layer, and the P-channel in which current is due to the flow of holes in the inversion layer. Each of these devices can be either enhancement mode (E-mode) in which device is normally "off" and is turned on by applying a gate voltage, or depletion mode (DE-mode) in which the device is normally "on" and is turned off by applying a gate voltage.
19. The current in a MOSFET is due to the flow of carriers in the inversion layer between the source and drain terminals. The inversion layer charge density and channel conductance are controlled by the gate voltage, which means that the channel current is also controlled by the gate voltage.
20. The depletion-mode MOSFET, has a source, gate, and drain. The gate is insulated from the channel. Because of this, the input resistance is very high. The DE-MOSFET has limited use, mainly in RF circuits.
21. The drain characteristic curves for a DE-MOSFET are similar to those of a JFET when operating in depletion mode. Unlike JFETs, DE-MOSFETs can also operate in the enhancement mode. When operating in enhancement mode, the drain current is greater than I_{DSS} (*i.e.* $I_D > I_{DSS}$).

    DE-MOSFETs are mainly employed as RF amplifiers. They have good high frequency response, generate low levels of electrical noise, and maintain high input impedance values when V_{GS} is negative or positive.
22. The enhancement type MOSFET is normally off. When the gate voltage is equal to the threshold voltage (V_{GST}), an N-type inversion layer connects the source to the drain. When the gate voltage exceeds the V_{GST}, the device conducts heavily. Because of the thin insulating layer, MOSFETs are easily destroyed unless precautions are taken in their handling.
23. Since the E-MOSFET is primarily a switching device, it usually operates between cutoff and saturation. When it is biased in the ohmic region, it acts like a small resistance.
24. Unijunction transistor (UJT) is a 2 layer, 3 terminal solid state (silicon) switching device. One common application of the UJT is the triggering of other devices like SCR, triac etc. Other applications include oscillators, pulse generators, sawtooth generators, phase control, timing circuits, and voltage- or current-regulated supplies.
25. Thyristor is the general name given to a family of semiconductor devices having four layers with control mechanism, although the term is most commonly applied to the SCR.
26. The silicon controlled rectifier (SCR) is the most widely used thyristor. The SCR may be thought of as a PNP and an NPN transistor interconnected, with regenerative feedback that is initiated by a gate current pulse. It can switch very large currents *on* and *off*. On application of forward voltage to SCR, a small forward leakage current flows. When the applied voltage exceeds the breakover voltage, the device turns *on* and a large current flow. The SCR may be triggered into a conducting state by a pulse gate current, but the gate loses control of anode current after firing. An SCR may be turned *off* by reducing the current below the holding value or by reducing the anode-to-cathode voltage to zero. Gate triggering of an SCR requires the application of gate voltage and current that must lie in the preferred area of the graphical gate characteristics is to ensure reliable switching. SCR has turn-on time of 1–4 μs and turn-off time of 10–300 μs. An SCR has a number of voltage, current and power ratings. These ratings should not be exceeded.

    The applications of SCRs include power control, switching, zero voltage switching, overvoltage protection, pulse circuits, battery charging regulator, heater control circuit, emergency lighting system etc.

EXERCISES

1. With the help of neat diagram, describe the construction and operation of a JFET.
[G.G.S.I.P. Univ. Analog Electronics December 2009]
2. Draw the basic structure of an N-channel JFET. Draw and explain the drain characteristics qualitatively.
[G.G.S.I.P. Univ. Analog Electronics December 2009]
3. Explain the construction, working and characteristics of a JFET.
[G.G.S.I.P. Univ. Analog Electronics December 2010]
4. Explain the principle of operation of an N-channel JFET. Draw a family of drain characteristics and mark the various regions of operation. **[G.G.S.I.P. Univ. Analog Electronics December 2010]**
5. Sketch the basic structure of an N-channel JFET and draw the volt-ampere (V/I) characteristics and explain about each region of the volt-ampere (V/I) characteristics qualitatively.
[U.P. Technical Univ. Solid State Devices & Circuits 2009-10]
6. State the advantages of FET over BJT.
[Pb. Technical Univ. Electronic Devices and Circuits May 2008]
7. Prove that the transconductance g_m of a JFET is given by

$$g_m = \frac{2}{|V_P|}\sqrt{I_{DS}\, I_{DSS}}$$

where V_P = pinch-off voltage, I_{DS} = drain current, I_{DSS} = maximum I_{DS} corresponding to V_{GS} = 0 volt.
8. Explain the meaning of pinch-off voltage in a JFET. How does the current flow in a JFET after pinch-off ? Explain the meaning of the terminologies (*i*) transconductance, g_m (*ii*) drain resistance, r_d and (*iii*) amplification factor, μ of a JFET.
9. Discuss (*i*) Fixed bias (*ii*) Self bias (*iii*) Voltage divider biasing arrangements used for JFET amplifier circuits. **[G.G.S.I.P. Univ. Analog Electronics December 2010]**
10. Discuss the three configurations of FET biasing. Explain any one in detail.
11. Discuss methods used to bias FET against device and temperature variations.
12. Draw the circuit diagram for a common source FET amplifier and the corresponding equivalent circuit with appropriate labelling. Find expressions for voltage gain, output resistance in terms of FET parameters and circuit elements. What is the order of magnitude for voltage gain for a common drain FET amplifier ?
13. Sketch the circuit of a source follower. At low frequencies what is
    (*i*) The maximum value of the voltage gain ?
    (*ii*) The order of magnitude of the output impedance.
14. Explain the characteristics of N-channel depletion type MOSFET.
[G.G.S.I.P. Univ. Analog Electronics December 2010]
15. Draw the structure of an N-channel depletion type MOSFET. Explain its working with the help of output drain characteristics and transfer characteristics. **[G.G.S.I.P. Univ. Analog Electronics December 2010]**
16. Define channel length modulation. **[G.G.S.I.P. Univ. Analog Electronics December 2010]**
17. Draw the structure and transfer curve for enhancement MOSFET and depletion MOSFET. Choose P or N type. **[G.G.S.I.P. Univ. Analog Electronics December 2009]**
18. Draw the cross-section of a P-channel enhancement MOSFET and explain its characteristics.
[G.G.S.I.P. Univ. Analog Electronics-I December 2004]
19. Explain the difference between enhancement and depletion type MOSFETs.
[G.G.S.I.P. Univ. Analog Electronics-I December 2004]
20. Describe briefly, the construction of a MOSFET in enhancement mode. Draw its characteristics. Draw the equivalent circuit of a MOSFET opeating in enhancement mode.
21. Draw and explain the biasing circuit for a depletion type MOSFET.
22. Draw the biasing circuit of an enhancement mode MOSFET and explain it.

23. Explain the transfer characteristics for N-channel and P-channel MOSFETs alongwith their output characteristics of an N-channel enhancement MOSFET.
24. Describe the construction and operation of a MOSFET in enhancement mode. Draw its characteristics and equivalent circuit of the device.
25. Give a comparison between JFET and MOSFET. **[G.G.S.I.P. Univ. Analog Electronics-I, December 2004]**
26. Write short notes on the following :
    (*i*) JFET. (*ii*) Biasing of FET.
27. Describe the construction and working of UJT ? Also draw its characteristics.
28. Explain the working principle of UJT. Also explain why UJT has a negative resistance region.
29. What are the differences between a transistor and a thyristor ? Explain the operation of a thyristor with the help of its characteristics.
30. Draw two transistor model of an SCR and explain its turn-on mechanism. Why does SCR not turn *on* in reverse biased condition ?
31. Explain the two transistor analogy of thyristor.
32. By drawing suitable diagrams, explain the construction details and working of a silicon controlled rectifier (SCR).
33. Explain the working of an SCR giving its construction and VI characteristics.
    [G.G.S.I.P. Univ. Analog Electronics December 2010]
34. Explain various triggering methods of SCR.
35. Explain the turn-off mechanism of SCR.

SHORT ANSWER TYPE QUESTIONS WITH ANSWERS

Q. 1. What is the major difference between a bipolar and a unipolar device ? Explain with example.

Ans. In a bipolar device, current conduction is by both majority carriers and minority carriers (both by holes as well as electrons), such as in a case of BJT. On the other hand, in a unipolar device, the current conduction is by only one type of majority carriers (either by electrons or by holes), such as in case of FET.

Q. 2. Explain why BJTs are called bipolar devices while FETs are called unipolar devices.

Ans. In ordinary transistors, current conduction is by both majority carriers and minority carriers (both by holes as well as electrons) and, therefore, these are called bipolar transistors. In field-effect transistors current conduction is by only one type of majority carriers (either by electrons or by holes) and, therefore, these are called unipolar transistors.

Q. 3. Define "current-controlled device" and "voltage-controlled device".

Ans. A current-controlled device is one in which a current defines the operating conditions of the device, whereas a voltage-controlled device is one in which a particular voltage defines the operating conditions.

Q. 4. Why is terminology "field-effect" appropriate for this terminal device ?

Ans. The device is called field-effect transistor (FET) because the drain current (output current) is controlled by the effect of extension of the field associated with the depletion region developed by the reverse bias at the gate.

Q. 5. Why I_G is effectively zero ampere for a JFET transistor ?

Ans. The gate channel junction in a JFET is a P-N junction and since it is normally reverse biased, a minority charge carrier current flows. This is gate-source cutoff current, known as gate reverse current. So the gate current, I_G can be said to be effectively zero ampere.

Q. 6. What do you understand by terms channel and drain in JFET ?

Ans. ***Channel.*** The region between the source and drain, sandwiched between the two gates is called the *channel* and the majority carriers move from source to drain through this channel.

Drain. The terminal, through which the majority carriers leave the channel, is called the *drain* terminal D and the conventional current leaving the channel at D is designated as I_D.

Q. 7. Why the channel of a JFET is never completely closed at the drain end ?

Ans. If the channel is completely closed in JFET, then there will be no drain current, so there will be no voltage drop along the channel length and amount of reverse bias will become uniform and wedge-shaped depletion region will become rectangular one.

Q. 8. How is drain current controlled in a JFET ?

Ans. In a JFET drain current is controlled by controlling the reverse bias given to its gate (*i.e.* V_{GS}).

Q. 9. What is meant by drain characteristic of FETs ?

Ans. The curve drawn between drain current and drain-source voltage with gate-to-source voltage as the parameter is called the drain characteristic.

Q. 10. What happens to the drain current of a P-channel JFET when a positive voltage is applied on its gate ?

Ans. Whenever the positive voltage is applied to the gate of P-channel JFET, the depletion region width increases because of which channel is narrowed, the resistance is increased and drain current is reduced.

Q. 11. What is meant by transfer characteristic of FETs ?

Ans. The curve drawn between drain current and gate-source voltage for a given value of drain-source voltage is called the transfer characteristic.

Q. 12. No current flows through the channel of a JFET when $V_{GS} = 0$ and $V_{DS} = 0$, though channel is fully-open. Why ?

Ans. When drain-source voltage is zero, there is no attracting potential at the drain, so no current flows inspite of the fact that the channel is fully open.

Q. 13. Explain "pinch-off" voltage in an N-channel JFET. **[G.G.S.I.P. Univ. December 2004]**

Ans. The value of drain-source voltage at which channel is pinched-off (*i.e.* all the free charges from the channel get removed) is called the pinch-off voltage.

Q. 14. What are the parameters that control the pinch-off voltage of JFET ?

Ans. Parameters that control the pinch-off voltage of JFET are electron charge, donor/acceptor concentration density, permittivity of channel material and half width of channel bar.

Q. 15. Why FET is called a voltage controlled device ?

Ans. In a FET, drain current is controlled by the effect of the extension of the field associated with the depletion region developed by the reverse bias on the gate, so it is called a voltage controlled device.

Q. 16. How does the FET behave (*i*) for small values of $|V_{DS}|$ and (*ii*) for large values of $|V_{DS}|$?

Ans. (*i*)FET behaves as an ordinary resistor for small value of $|V_{DS}|$ *i.e.* in ohmic region.

(*ii*)FET behaves as a constant current source for large values of $|V_{DS}|$ till breakdown occurs.

Q. 17. How does the current vary with the gate voltage in the saturation region ?

Ans. Drain current decreases with the increase in gate-source bias $|V_{GS}|$. When $V_{GS} = 0$; drain current $I_D = I_{DSS}$, drain-source saturation current and when $V_{GS} = V_P$; drain current $I_D = 0$.

Q. 18. How does the transconductance vary with drain current ?

Ans. Transconductance g_m of a FET varies with the drain current I_D as given by the following equation

$$g_m = \frac{2}{|V_P|}\sqrt{I_{DS} \cdot I_{DSS}}$$

Q. 19. What is Shockley equation ?

Ans. Drain current in the pinch-off region of JFET is given by the equation, known as Shockley equation,

$$I_D = I_{DSS}\left[1 - \frac{V_{GS}}{V_P}\right]^2 = I_{DSS}\left[1 - \frac{V_{GS}}{V_{GS(OFF)}}\right]^2$$

where I_D is the drain current at a given gate-source voltage V_{GS}, I_{DSS} is the drain current with gate shorted to source and V_{GS} is the gate-source cutoff voltage.

Q. 20. Explain how the mutual conductance of a FET varies with I_D and V_G.

Ans. Mutual conductance g_m of a FET is determined by the ratio of the change in drain current ΔI_D associated with a particular change in gate-to-source voltage ΔV_{GS} in the region of interest. The steeper the slope of the I_D versus V_{GS} curve, the greater is the value of g_m. The quantity g_m increases as V_{GS} approaches zero.

Q. 21. What is meant by gate-source cutoff voltage ?

Ans. The gate-source bias voltage required to reduce the drain current to zero is designated the gate-source cutoff voltage $V_{GS\ (OFF)}$. It is equal to pinch-off voltage V_P.

Q. 22. What is meant by saturation region ?

Ans. The region of drain characteristic of a FET in which drain current remains fairly constant is called the saturation or pinch-off region.

Q. 23. What is meant by drain-source saturation current I_{DSS} ?

Ans. The drain current in pinch-off or saturation region with zero gate-source voltage ($V_{GS} = 0$) is referred to the drain-source saturation current I_{DSS}.

Q. 24. What is physical significance of high input impedance of an FET ?

Ans. Input circuit of an FET (gate to source) is reverse biased and the input gate current is very small (of the order of few nA), so the input impedance is very high. It permits high degree of isolation between the input and output circuits.

Q. 25. What is dynamic resistance of a JFET ?

Ans. The ratio of change in drain-source voltage to change in drain current at a given gate-source voltage is known as ac drain resistance or dynamic resistance r_d

$$i.e.\ r_d = \frac{\Delta V_{DS}}{\Delta I_D} \text{ at constant } V_{GS}.$$

Q. 26. What is meant by transconductance with reference to JFET ?

Ans. The control that gate-source voltage has over the drain current is measured by the transconductance of a JFET. It may be defined as the ratio of change in drain current to the change in gate-source voltage for a given value of drain-source voltage *i.e.*

$$g_m = \frac{\Delta I_D}{\Delta V_{GS}} \text{ at constant } V_{DS}.$$

Q. 27. Why is MOSFET called sometimes IGFET ?

Ans. MOSFET is constructed with the gate terminal insulated from the channel so it is sometimes called insulated gate FET (or IGFET).

Q. 28. What is meant by gate-to-source threshold voltage in E-MOSFET ?

Ans. The minimum value of gate-to-source voltage V_{GS} that is required to form the inversion layer is termed the gate-to-source threshold voltage V_{GST}.

Q. 29. Why MOSFETs are never connected or disconnected in the circuit when power is ON ?

Ans. If a MOSFET is connected or disconnected in circuit while the power is ON, transient voltages caused by inductive kickback and other effects may exceed $V_{GS\ (max)}$ and thus wipe out the MOSFET.

Q. 30. Name the factors which make the JFET superior to BJT ?

Ans. The high input impedance, low output impedance and low noise level make JFET far superior to the BJT.

Q. 31. Why enhancement mode MOSFETs are preferred over the depletion mode ones in the fabrication of integrated circuits ?

Ans. Since the conductivity of the channel is enhanced by the positive bias given to the gate in case of enhancement mode MOSFETs, the E-MOSFETs are preferred over the DE-MOSFETs in the fabrication of ICs.

Q. 32. What does UJT stand for ? Justify the name UJT.

Ans. UJT stands for unijunction transistor. It essentially consists of a lightly doped N-type silicon bar with a small piece of heavily doped P-type material alloyed to its one side to produce single P-N junction. The single P-N junction justifies the name UJT.

Q. 33. What is negative resistance region in UJT ? Explain its physical significance.

Ans. Once conduction is established at emitter voltage $V_E = V_P$, peak value of emitter voltage, the emitter potential V_E starts decreasing with the increase in emitter current I_E. This corresponds to a negative resistance region which is stable enough to be used with a great deal of reliability in the area of applications—oscillator.

Q. 34. How is current limited in conducting state of an SCR ?

Ans. The current in conducting state of an SCR is controlled by external impedance.

Q. 35. What is the effect of negative gate current on a normal SCR ?

Ans. Negative gate current will increase the holding current of the SCR.

Q. 36. What is false triggering ?

Ans. False triggering is unintended turn-on of an SCR either through gate due to noise pick-up or excessive anode voltage.

Q. 37. What are the factors on which 'turn-off time' of an SCR depends ?

Ans. Turn-off time of an SCR depends upon doping densities, recombination time, junction temperature, on-state current, rate of decay of on-state current.

Q. 38. How the forced turn-off of an SCR is different from natural turn-off ?

Ans. In forced turn-off of an SCR current is brought below the holding current by an additional circuit employing energy storage elements while in natural turn-off, it is the natural voltage/current variation.

Q. 39. Why 'pulse triggering' is preferred and when does it fail ?

Ans. Pulse triggering is preferred as it is effective to turn *on* an SCR keeping the device gate dissipation low and ensures a fast turn-on keeping the *di/dt* stress on the device low.

Pulse triggering shall fail if the pulse is of short duration and the load circuit contains large inductance. With highly inductive load, the device current builds up slowly. If the pulse duration is small and the anode current is not able to build up to latching current, I_L, the device turns *off* on removal of gate pulse.

Q. 40. What is holding current in an SCR ?

Ans. Holding current is the minimum on-state current required to keep the SCR in conducting state without any gate drive.

Q. 41. Define latch current of an SCR.

Ans. Latching current is the minimum device current, which must be attained by the device, before the gate drive is removed.

Q. 42. Give the typical values of current ratings of SCR.

Ans. Forward breakover current—less than a few hundred micro-amperes.

Peak forward current 30 A to over 100 A.

Holding current—a few mA to few hundred milliamperes.

PROBLEMS

1. When gate-source voltage is –12 V and gate current is 0.1 μA, determine the resistance between the gate and source of a given JFET. [**Ans.** 120 MΩ]
2. In case of a JFET the variation in drain current is of 100 μ A when drain-source voltage is varied by 1.25 V, keeping gate-source voltage V_{GS} constant, determine the dynamic resistance of the JFET. [**Ans.** 12.5 kΩ]
3. In a JFET the drain current is changed by 0.25 mA when the gate-source voltage is changed by 0.125 V, keeping drain-source voltage constant. Calculate the transconductance of the given JFET. [**Ans.** 2 mS]
4. A JFET has drain current of 10 mA. If I_{DSS} = 20 mA and V_P = 4 V, find V_{GS}. [**Ans.** 1.172 V]

5. For the given measurement $V_S = 1.7$ V for the network as shown in the Fig. 1. Determine
(*i*) I_{DQ} (*ii*) V_{GS}, (*iii*) I_{DSS}, (*iv*) V_D, (*v*) V_{DS}.
[**Ans.** (*i*) 3.33 mA, (*ii*) –1.7 V, (*iii*) 10 mA, (*iv*) 11.33 V, (*v*) 9.63 V]

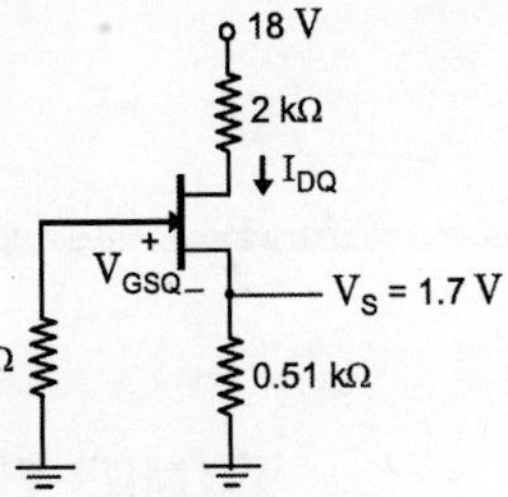

Fig. 1

6. In the circuit shown in Fig. 10.14, $V_{DD} = 30$ V, $R_D = 2.5$ kΩ, $R_S = 500$ Ω and $I_D = 2.5$ mA. Determine V_{DS} and V_{GS}. [**Ans.** (*i*) 22.5 V (*ii*) –1.25 V]
7. In the circuit shown in Fig. 10.13, $V_{DD} = 25$ V, $V_{GG} = 1.6$ V, $R_D = 2.5$ kΩ, $R_G = 2.5$ MΩ, $I_{DSS} = 10$ m A and $V_P = -3$ V. Determine V_{GS}, I_D and V_{DS}. [**Ans.** –1.6 V; 2.18 mA, 19.55 V]
8. In an N-channel JFET biased by voltage divider method, determine the value of R_S to give operating point $I_D = 3$ mA and $V_{DS} = 10$ V. Given that $V_{DD} = 30$ V, $R_{G_1} = 1.5$ MΩ, $R_{G_2} = 0.5$ MΩ. JFET parameters are $I_{DSS} = 12$ mA and $V_P = -4$ V. [**Ans.** 3.167 kΩ]
9. The transconductance of a FET used in a voltage circuit is 4,000 microsiemens. The load resistance is 15 kΩ and drain circuit resistance is 10 MΩ. Calculate the voltage gain of the amplifier circuit. [**Ans.** –59.9]
10. The common drain circuit given in Fig. 10.31 has $R_1 = 3.5$ MΩ, $R_2 = 1.5$ MΩ, $R_S = 2$ kΩ, $R_L = 20$ kΩ, g_m is 2.5 mS. Determine input impedance, output impedance and voltage gain. [**Ans.** 1.05 MΩ, 333.3 Ω; 0.82]
11. For an N-channel zero-biased DE-MOSFET shown in Fig. 10.41 (*a*), determine drain-source voltage if $I_{DSS} = 10.5$ mA, $V_{DD} = 30$ V, $R_D = 1.5$ kΩ, $R_G = 12$ MΩ and $V_{GS\,(OFF)} = -4$ V. [**Ans.** 14.25 V]
12. The N-channel E-MOSFET used in the common source amplifier, shown in Fig. 2, has the following parameters :

$$I_{D\,(ON)} = 4 \text{ mA at } V_{GS(ON)} = 10 \text{ V}, V_{GST} = 4 \text{ V and } g_m = 5{,}000\ \mu\text{S}.$$

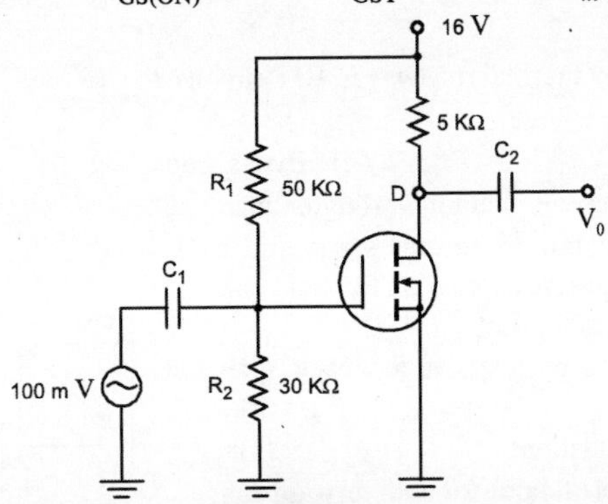

Fig. 2

Calculate V_{GS}, I_D, V_{DS} and v_o. [**Ans.** $V_{GS} = 6$ V; $I_D = 0.44$ mA; $V_{DS} = 13.78$ V and $v_o = 2.5$ V]
13. The intrinsic stand-off ratio for a UJT is determined to be 0.65. If the interbase resistance is 7.5 kΩ, determine R_{B_1} and R_{B_2}. [**Ans.** 4.875 kΩ; 2.625 kΩ]
14. A silicon UJT has an interbase resistance of 10 kΩ and R_{B_1} of 6 kΩ with $I_E = 0$. Voltage of 20 V is applied between the bases. Determine (*i*) R_{B_2} (*ii*) intrinsic stand-off ratio (*ii*) UJT current (*iv*) stand-off voltage and peak-point voltage. Take $V_B = 0.7$ V. [**Ans.** (*i*) 4 kΩ (*ii*) 0.6 (*iii*) 2 mA (*iv*) 12 V (*v*) 12.7 V]
15. An SCR has a breakover voltage of 400 V, a trigger current of 10.5 mA and holding current of 10.5 mA. What do you infer from it ? What will happen if gate current is made 12.5 mA ?

EXAMINATION PAPERS

B.TECH.

THIRD SEMESTER EXAMINATION, DECEMBER 2010

ANALOG ELECTRONICS-I (ETEE-203)

Time : 3 Hours *Total Marks : 75*

Note : *Attempt **five** questions including Q. 1 which is compulsory. Internal choice is indicated.*

1. Answer the following :
   (*a*) A 1 kΩ resistor is to be fabricated as a narrow strip of p-type Si, 4 mm thick. If the strip is 400 μm long and 20 μm wide, what concentration of acceptor atoms is required ? Given hole mobility at room temperature as 480 $cm^2 v^{-1} s^{-1}$. (2)
   (*b*) Why reverse saturated current in Ge diode is more as compared to Si diode ? (2)
   (*c*) Explain the significance of diffusion capacitance and derive expression for it. (3)
   (*d*) The current flowing through a Si diode at room temperature is 9×10^{-7}A, when a large reverse voltage is applied. Calculate the current through the diode when 0.4V forward bias is applied. Consider $\eta = 2$, where is ideality factor. (2)
   (*e*) What do you understand by thermal runaway in BJT circuits ? (2)
   (*f*) Explain the function of (3)
   (*i*) Coupling capacitor (*ii*) Bypass capacitor.
   (*g*) Derive an expression for current gain for Darlington pair. (3)
   (*h*) Draw the cascade of a common-emitter stage and a common-collector stage and give expression for input and output resistance of the cascade. (3)
   (*i*) State the advantages offered by negative feedback in an amplifier. (2)
   (*j*) Define channel length modulation. (3)

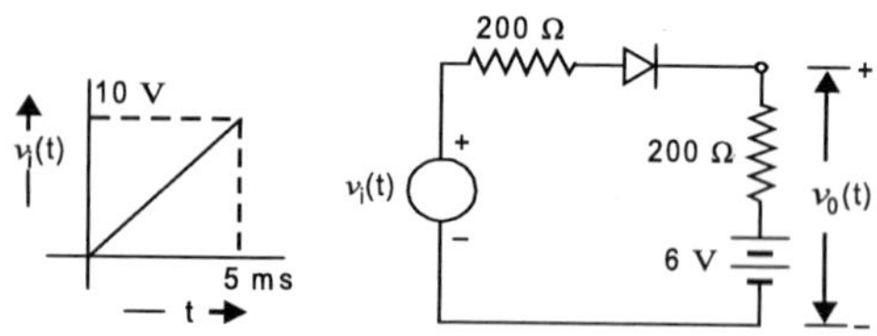

Fig. 1

2. (*a*) Draw the large signal model of diode for both forward and reverse bias and explain the model. (3)
   (*b*) For the circuit shown in Fig. 1 find and sketch the output voltage $v_0(t)$ for $0 \le t \le 5$ ms. Assume that the diode is ideal. (3.5)
   (*c*) For full wave rectifier derive expression for average current, rms current, ripple factor and rectification efficiency. (6)

OR

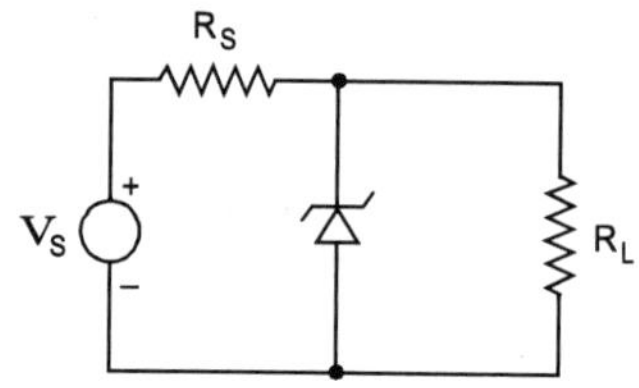

Fig. 2

3. (*a*) In the circuit shown in Fig. 2 a 5V Zener diode is used which provides regulation for $50mA \le I_Z \le 1A$. Determine the range of load currents for which regulation is achieved if the unregulated voltage V_s varies between 7.5V and 10V. The resistance $R_S = 4.75\Omega$. (4)
   (*b*) A diode, whose internal resistance is 20Ω, has to supply power to a 200Ω load from a 110V (rms) source of supply. Calculate the (5)
   (*i*) Peak load current (*ii*) DC load current (*iii*) AC load current (*iv*) Rectification efficiency
   (*v*) Peak inverse voltage.
   (*c*) What do you understand by drift current and diffusion current ? Write an equation for the net electron current in a semiconductor. (3.5)

4. (*a*) Write Eber's Moll equations for an NPN transistor. Based on these equations draw and explain the output characteristics of NPN transistor in CE configuration. (6)
   (*b*) What is self bias ? Draw the circuit showing self bias of an NPN tansistor in CE mode. Explain physically how the self bias improves stability. (4)
   (*c*) Consider the circuit shown in Fig. 3. Find the value of R_C required to obtain $V_C = 5V$. (2.5)

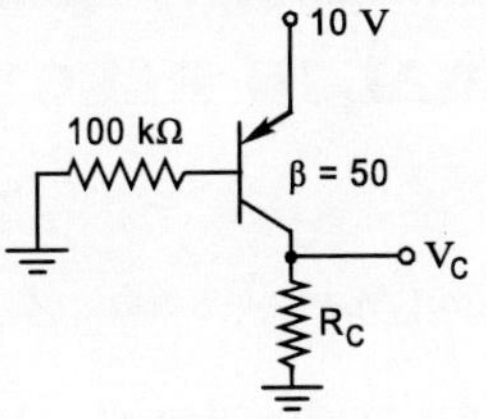

Fig. 3

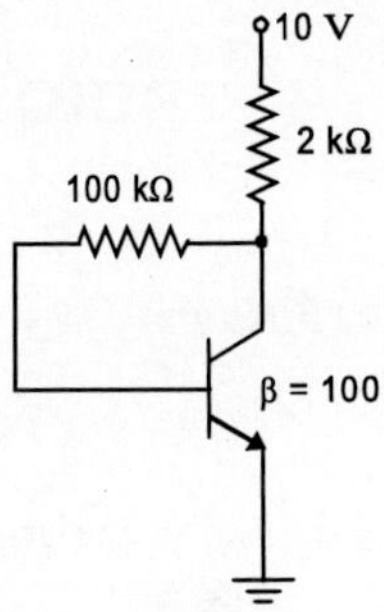

Fig. 4

Or

5. (*a*) For the circuit shown in Fig. 4 determine Q point. (4)
   (*b*) Derive expressions for stability factors S_{ICO} and S_{VBE} for self bias circuit. (5)
   (*c*) Explain Early effect. (3.5)
6. (*a*) Draw Hybrid π model of BJT and explain significance of each element. Also relate transconductance (g_m) and $*r_\pi$ with β_0, where terms have their usual meaning. (6)
   (*b*) Derive expression for voltage gain, current gain, input resistance and output resistance of an emitter follower circuit using h parameters. (6.5)

OR

7. (*a*) For the circuit shown in Fig. 5(*a*), determine current gain (I_0/I_s), voltage gain (V_0/V_s), input resistance (R_i) and output resistance (R_0). The transistor has following parameters : $h_{ie} = 2k\Omega$, $h_{fe} = 50$, $h_{re} = 2 \times 10^{-4}$ and $h_{oe} = 20 \times 10^{-6}$ A/V. (6.5)
   (*b*) Draw frequency response of an RC coupled amplifier and give expression for gain in different frequency region. Why gain falls at low and high frequencies ? (6)
8. (*a*) For the circuit shown in Fig. 6, identify the feedback topology employed and hence determine A_F, R_{IF} and R_{OF}, where terms have their usual meaning. (6)
   (*b*) Explain the principle of operation of an n-channel JFET. Draw a family of drain characteristics and mark various regions of operation. (6.5)

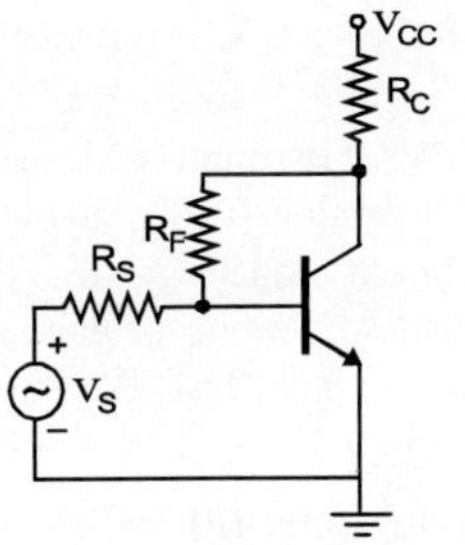

Fig. 6

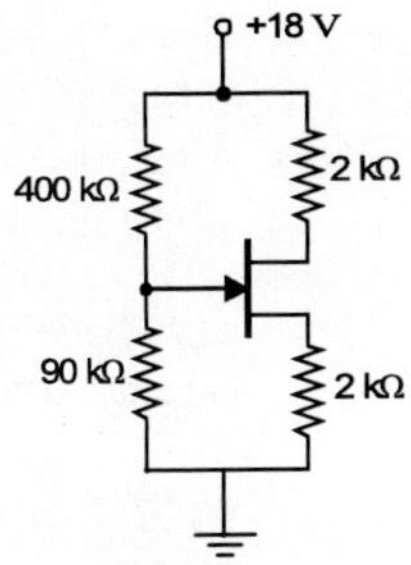

Fig. 7

OR

9. (*a*) (*i*) Discuss how gain of an amplifier is stabilized with negative feedback. (3)
   (*ii*) The open-loop gain of an amplifier changes by 5% if 10 dB negative feedback is applied. Calculate the percentage change in the closed-loop gain. (2)
   (*iii*) The output resistance of a voltage series feedback amplifier is 10Ω. If the gain of the basic amplifier is 100 and feedback fraction is 0.01, what is the output resistance without feedback ? (2)
   (*b*) An n-channel JFET has $V_P = -5V$ and $I_{DSS} = 12$ mA and is used in circuit shown in Fig. 7. Find the operating point for the circuit. (5.5)

B.TECH.

THIRD SEMESTER EXAMINATION, DECEMBER 2010

ANALOG ELECTRONICS-I (ETEC-207)

Time : 3 Hours *Total Marks : 75*

Note : *Attempt one question from each unit. Q. 1 is compulsory*

1. (*a*) Explain the large signal model and small signal model of a pn junction diode in both forward bias and reverse bias conditions.
   (*b*) Draw transistor amplifier in CE, CB and CC configuration and explain the salient feature of each type.
   (*c*) Draw the hybrid-π model of a transistor at high frequencies. Explain the significance of each component and give typical value. What is the significance of f_β and f_T?
   (*d*) Draw the circuit of an RC coupled amplifier. Explain its frequency response characteristics.
   (*e*) Explain the four types of feedback topologies with the help of schematic diagrams. (5 × 5 = 25)

UNIT–I

2. (*a*) Prove that the transition capacitance C_T for a step graded pn junction diode is given by $C_T = C_0\left(1 - \frac{V_D}{V_0}\right)^{-1/2}$.
   where C_0 = transition capacitance of an open circuited diode, V_D = diode voltage, V_0 = potential barrier voltage. (6.5)
   (*b*) A full wave rectifier shown in Fig. 1, uses ideal silicon diodes. Determine (*i*) V_{dc} at the load ; (*ii*) PIV rating of each diode ; (*iii*) I_{max} through each diode ; (*iv*) power rating of each diode ; (*v*) efficiency. Given : n = 10 : 1, R_L = 5Ω. (6)

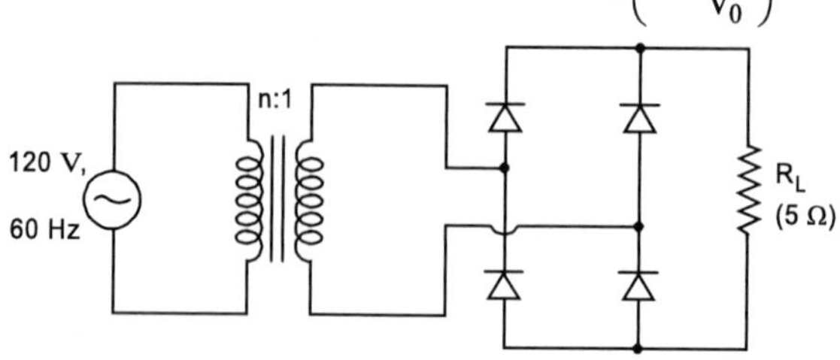

Fig. 1

OR

3. (*a*) Explain (*i*) Storage time (*ii*) transition time (*iii*) reverse recovery time in a pn junction diode with the help of waveforms. How the storage time effects can be reduced ? (6.5)
   (*b*) A zener voltage regulator has variable load R_L requiring load current to vary from 10 mA to 85 mA. It is given that V_Z = 10V, $I_{Z, min}$ = 15 mA, $I_{Z, max}$ = 100 mA and the series resistance R_S = 40Ω. Calculate the range of V_{dc} variation permissible and zener power dissipation $P_{Z, max}$. (6)

UNIT–II

4. (*a*) Draw the output characteristics of a transistor in CE configuration and show the total region where transistor can be used as an amplifier. Explain why the slope of characteristic is more in CE mode compared to CB mode ? (6.5)
   (*b*) Find the value of R_1 in the circuit shown in Fig. 2. Assume β = 49, I_E = 2 mA, V_{BE} = 0.7V. (6)

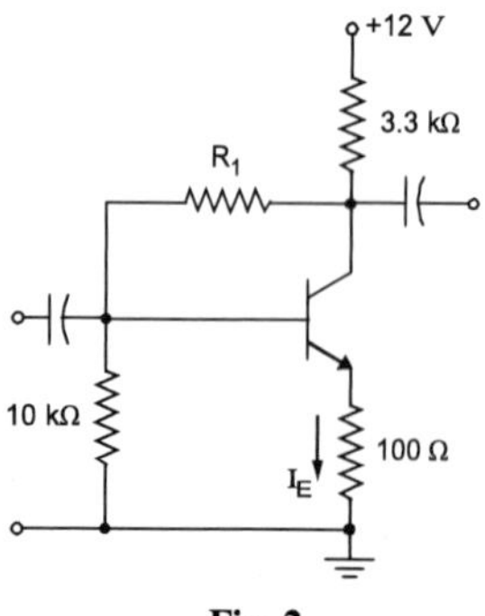

Fig. 2

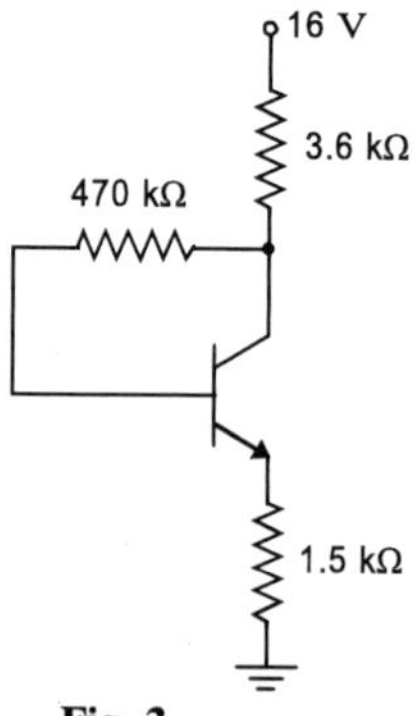

Fig. 3

OR

5. (*a*) For the circuit shown in Fig. 3, determine (*i*) V_C; (*ii*) Stability factor $S(I_{CO})$. Given $\beta = 120$. (5)

(*b*) The transistor shown in Fig. 4 has the following parameters : $h_{fe} = 120, h_{ie} = 1.1\ k\Omega, h_{re} = h_{oe} = 0$. Compute (*i*) R_i; (*ii*) $A_{vs} = V_0/V_s$; (*iii*) $A_I = I_L/I_i$; (*iv*) R_0. (7.5)

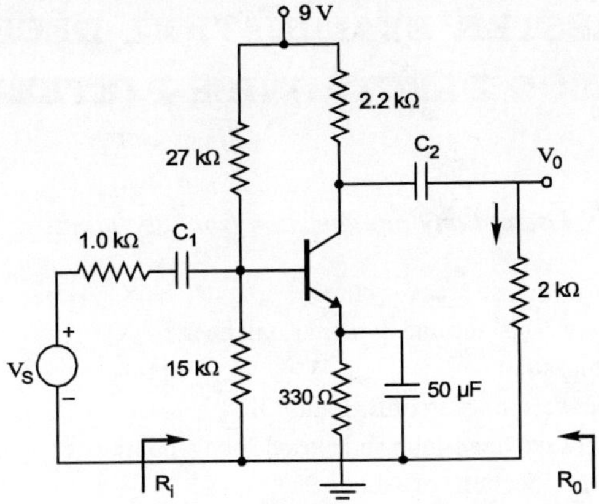

Fig. 4

UNIT–III

6. (*a*) Prove that the input resistance, R_{if} increases for a voltage series feedback topology and current series feedback circuit. (6.5)

(*b*) Calculate the overall input impedance for the BJT cascaded circuit shown in Fig. 5. Assume identical transistors with $\lambda_{ie} = 0$. (6)

OR

7. (*a*) A cascaded amplifier stage is shown in Fig. 6. Calculate the overall voltage gain and current gain. Transistors are identical with $h_{ie} = 2\ k\Omega$, $h_{fe} = 50$, $h_{re} = h_{oe} = 0$. (6.5)

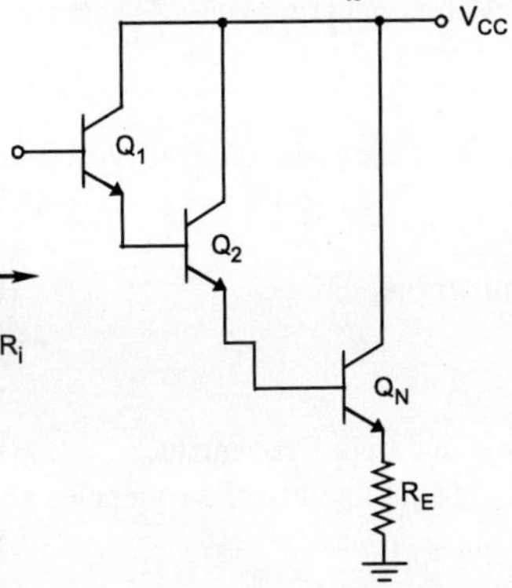

Fig. 5.

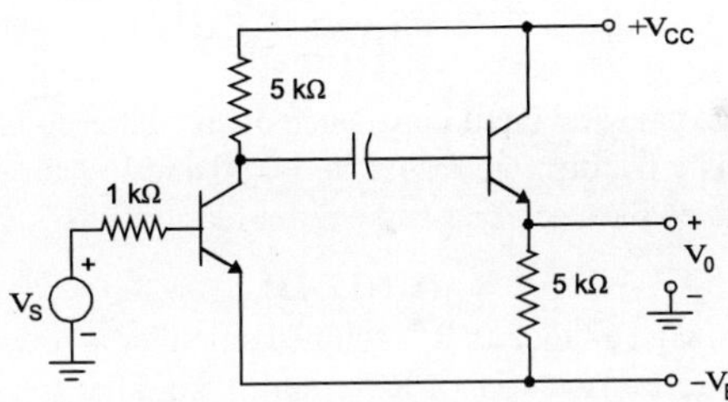

Fig. 6

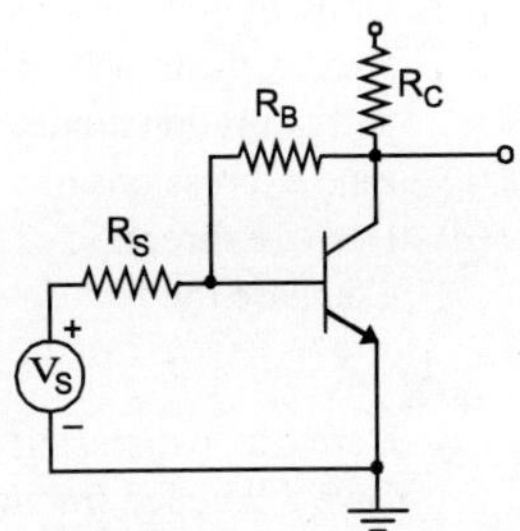

Fig. 7

(*b*) For a feedback circuit shown in Fig. 7. Identify the type of feedback used. Using feedback principles, compute (*i*) β ; (*ii*) A_{vf}. (6)

UNIT–IV

8. (*a*) Draw the structure of an n-channel depletion type MOSFET. Explain its working with the help of output drain characteristics and transfer characteristics. (6.5)

(*b*) Explain the working of an SCR giving its construction and VI characteristics. (6)

OR

9. (*a*) Explain the construction, working and characteristics of a JFET. (6.5)

(*b*) Discuss the (*i*) Fixed-bias ; (*ii*) Self-bias ; (*iii*) Voltage divider biasing arrangements used for JFET amplifier circuits. (6)

❑❑❑

B.TECH.

THIRD SEMESTER EXAMINATION, DECEMBER 2011

ANALOG ELECTRONICS-I (ETEE-203)

Time : 3 Hours *Maximum Marks : 75*

Note : *Q. No. 1 is compulsory, Attempt any one question from each unit.*

1. (*a*) Explain the significance of reverse recovery time. (3)

(*b*) What is the difference between static and dynamic load lines ? (3)

(*c*) Explain the need for stabilization. (3)

(*d*) What is the difference between biasing of BJT and FET ? (3)

(*e*) Explain the difference between open-loop and closed-loop amplifier configurations. (3)

(*f*) What are the important specifications of a diode ? (3)

(*g*) Give the hybrid models for transistor in CE, CB, CC configurations at low frequencies. (4)

(*h*) List the characteristics of an ideal Amplifier. (3)

UNIT-I

2. (*a*) Derive expressions for ripple factor and rectification efficiency of half-wave and full-wave rectifiers. (6)

(*b*) Derive an expression for the transition capacitance of an abrupt PN junction. Explain the significance of each symbol in the expression. (6.5)

3. (*a*) Explain the V-I characteristics of PN junction and Zener diode. Give expressions of dc and ac resistances in each case. (6.5)

(*b*) Explain the difference between drift and diffusion phenomenon. (2)

(*c*) Explain the performance characteristics of LED. Give its important features and typical applications. (4)

UNIT-II

4. (*a*) Draw Eber-Moll's model of an NPN transistor. Give necessary equations. (4)

(*b*) Explain the transistor characteristics in CE configuration. Explain the behaviour of the transistor in active, cut-off and saturation mode. (4.5)

(*c*) Find expressions of voltage gain and input impedance of an emitter follower. (4)

5. (*a*) Derive expression of stability factors with respect to I_{CO}, β and V_{BE} for emitter bias circuit. (8)

(*b*) Compare the features of fixed bias, self bias and emitter bias circuits. (4.5)

UNIT-III

6. (*a*) Derive an expression of voltage gain of an RC-coupled amplifier at low, mid and high frequencies. (8)

(*b*) Give a labelled frequency response curve of RC-coupled amplifier clearly indicating cut-off frequencies and bandwidth. (4.5)

7. (*a*) Explain the functioning of Darlington Pair. (6)

(*b*) Drive and write the formula to calculate the impedance gain of an amplifier. (6.5)

UNIT-IV

8. (*a*) Explain the output and transfer characteristics of enhancement and depletion type MOSFET. (8)

(*b*) Compare the features of BJT and FET. (4.5)

9. (*a*) What are feedback amplifiers ? State and explain properties of negative feedback. (6.5)

(*b*) Explain different types of feedback amplifiers. (6)

B.TECH.

THIRD SEMESTER EXAMINATION, DECEMBER 2011

ANALOG ELECTRONICS-I (ETEC-207)

Time : 3 Hours *Maximum Marks : 75*

Note : *Attempt one question from each unit. Q. 1 is compulsory. Graph paper is required.*

1. (*a*) Explain the transient response of a diode when driven from ON to OFF state or vice versa.
   (*b*) Discuss "Base Width Modulation" and explain its effects.
   (*c*) Define the various stability factors and explain how to compute the total change in the collector current due to these.
   (*d*) Derive the approximate conversion formulas for 'h' parameters from CE to CC configuration.
   (*e*) Draw the equivalent circuit for four types of controlled amplifiers and explain the conditions under which these can be made to achieve ideal conditions. $(5 \times 5 = 25)$

UNIT–I

2. (*a*) The circuit shown in Fig. 1 uses a silicon diode at room temperature. Given : $V_v = 0.6$ V, $R_f = 10\ \Omega$. Compute (*i*) the ac voltage across load R_L, (*ii*) the total voltage across R_L, (*iii*) the total current in the circuit. (6.5)

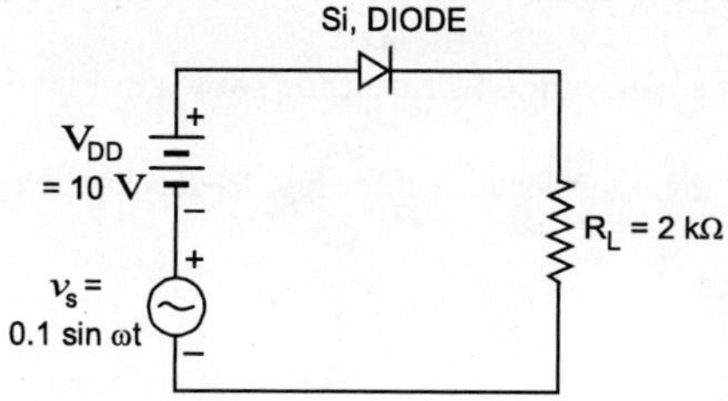

Fig. 1

   (*b*) Explain the reason for the 'Diffusion Capacitance, C_D' in a pn junction diode. Derive an expression for it. If a silicon diode has forward biased current, $I_{DQ} = 1$ mA and given that $D_P = 13$, $L_P = 2.6$ cms. Compute the value of C_D. (6)

OR

3. (*a*) For a centre-tapped full-wave rectifier, show the current waveform for non-ideal diodes and derive the expression for I_{dc}. Also, find ripple factor, rectifier efficiency and PIV. (6.5)
   (*b*) Explain the working of (*i*) Zener diode, (*ii*) LED. Discuss one application for each. (6)

UNIT–II

4. (*a*) Show the various current components for a pnp transistor in active mode and explain its operation. (3)
   (*b*) Explain the terms : (*i*) Early voltage, V_A, (*ii*) I_{CEO}, (*iii*) I_{CBO}. (3)
   (*c*) Find the region of operation of the circuit shown in Fig. 2. Assume $\beta_F = 100$, $V_{BE,\ active} = 0.7$ V, $V_{BE,\ sat} = 0.8$ V, $V_{CE,\ sat} = 0.2$ V, $V_{BE,\ cut\ off} = 0$ V. (6.5)

OR

5. (*a*) Derive an expression for the stability factor $S(I_{CO})$ for the circuit shown in Fig. 3. (6)
   (*b*) For the amplifier shown in Fig. 4, the transistor parameters are: $h_{ie} = 1\ k\Omega$, $h_{fe} = 50$, $h_{oe} = 25\ \mu A/S$, and $h_{re} = 2.5 \times 10^{-4}$. Compute $A_{rs} = V_0/V_s$.

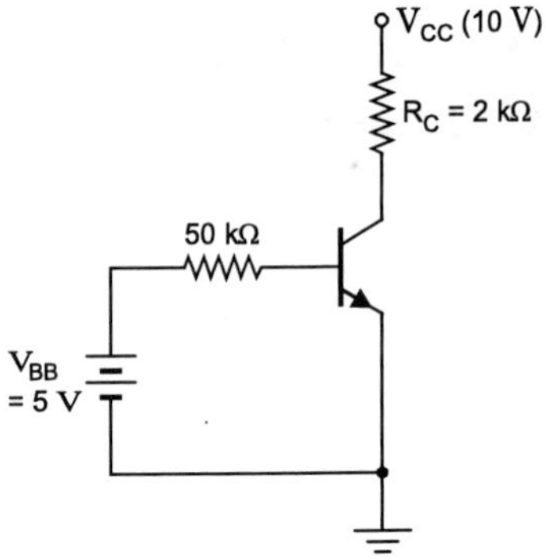

Fig. 2

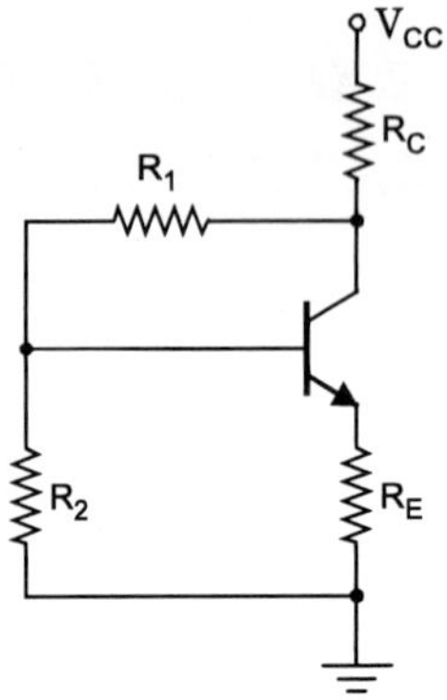

Fig. 3

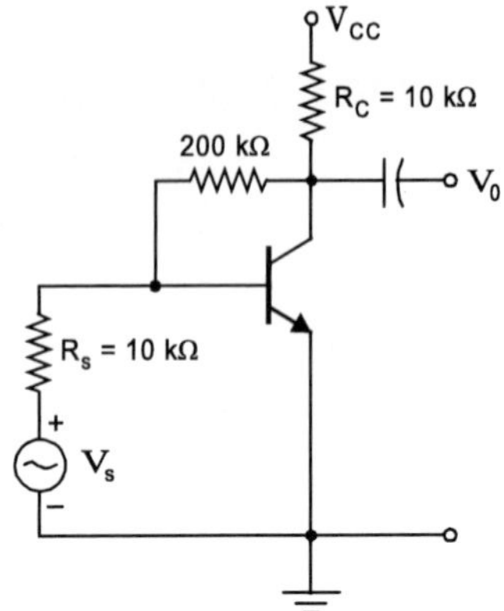

Fig. 4

UNIT–III

6. (*a*) Draw the circuit diagram of a two stage RC-coupled amplifiers. Explain the importance of each component. Draw and explain its frequency response. (6)
   (*b*) Discuss the various feedback topologies. Draw a general feedback amplifier diagram and derive the expression for the closed loop gain. Discuss the advantages of negative feedback. (6.5)

OR

7. (*a*) Calculate the overall voltage gain of the BJT cascade shown in Fig. 5. Assume h_{ie} = 1 kΩ, h_{fe} = 100, h_{oe} = h_{re} = 0. (6.5)
   (*b*) Draw the circuit of feedback amplifiers for each of the four feedback topologies and justify the type of feedback. (6)

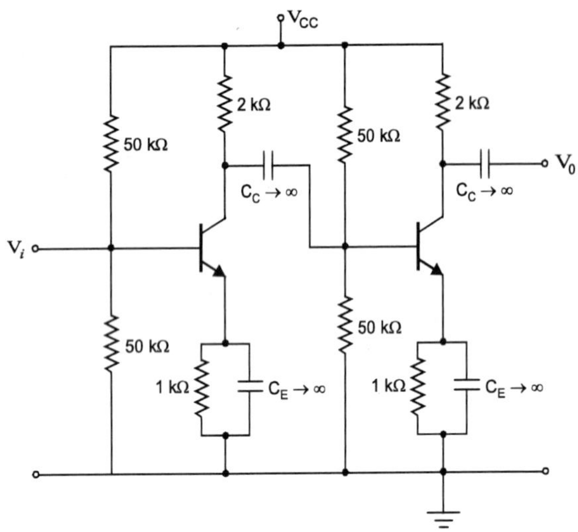

Fig. 5

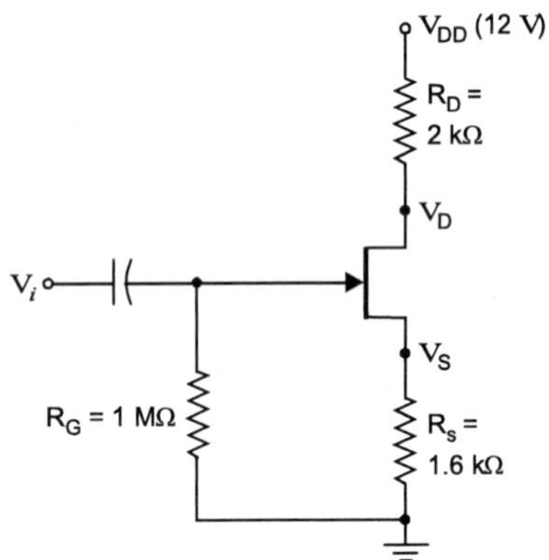

Fig. 6

UNIT–IV

8. (*a*) Explain the construction, operation and volt-ampere characteristics of a JFET. (6.5)
   (*b*) Explain the working and V-I characteristics of UJT. Discuss any one application. (6)

OR

9. (*a*) Draw the structure of an enhancement type NMOS transistor. Draw and explain the drain and transfer characteristics. (6)
   (*b*) For the *n*-channel JFET amplifier shown in Fig. 6, determine V_{DSQ}, I_{DSQ}, V_D and V_S. Given I_{DSS} = 6 mA and V_P = –6 V. (6.5)

□□□

INDEX

A

B

C

D

E

F

G

H

I

J

L

M

N

O

P

U

V

Z

❑❑❑

Win Prizes

ATTENTION READERS

Win Prizes

We request for your frank assessment regarding various aspects of the book as given below :

ANALOG ELECTRONICS-I

By : J.B. GUPTA **(GGSIPU)**

Please Fill up the Sheet in Neat and Clean Writing

(i) What topics of your syllabus, important for your examination point of view, are not covered in this book ?

....

....

....

....

(ii) What are the chapters wherein the treatment of the subject matter is not systematic or organised or updated ?

....

....

....

....

(iii) Have you come across misprints/mistakes/factual inaccuracies in the book ? Please specify the chapters and the page numbers.

....

....

....

....

(iv) What other books on the same subject have you found/heard better than the present book ? Please specify in terms of price and quality.

....

....

....

....

(v) Further suggestions and comments for the improvement of this book.

....

....

....

....

(PS : If need be please attach a separate sheet to write your views)

OTHER DETAILS

(i) Who recommended you this book ? (Please tick (✓) in the box near the option relevant to you.

☐ Teacher ☐ Friends ☐ Bookseller

....

....

....

....

(ii) Name of the recommending teacher, his designation and address:

....

....

....

....

(iii) Name and address of the bookseller you purchased the book from :

....

....

....

....

(iv) Name and address of your institution (Please mention the University or Board as the case may be)

....

....

....

....

(v) Your name and complete address :

....

....

....

....

(vi) Course you are enrolled in :

....

....

....

....

The best assessment will be awarded each month : The award will be in the form of our publications as decided by the Editorial Board.

Please mail the filled up coupon at the vearliest to :

S.K. KATARIA & SONS®

Publisher of Engineering Books

4885/109, Prakash Mahal, Dr. Subhash Bhargav Lane,

Opposite Delhi Medical Association, Daryaganj, New Delhi–110002 (INDIA)

Phone: +91-11-43551243; *Mobile:* +91-9871775858, 7678425209

e-mail: katariabook@gmail.com, katariabooks@yahoo.com

Website: www.skkatariaandsons.com